Global Geomorphology

Global Geomorphology

An introduction
to the study of landforms

MICHAEL A SUMMERFIELD

Department of Geography, University of Edinburgh

PEARSON

Prentice
Hall

Harlow, England • London • New York • Boston • San Francisco • Toronto
Sydney • Tokyo • Singapore • Hong Kong • Seoul • Taipei • New Delhi
Cape Town • Madrid • Mexico City • Amsterdam • Munich • Paris • Milan

Pearson Education Limited
Edinburgh Gate
Harlow
Essex CM20 2JE
England

and Associated Companies throughout the world

Visit us on the World Wide Web at:
www.pearsoned.co.uk

First published 1991

British Library Cataloguing-in-Publication Data
Summerfield, Michael A.
  Global geomorphology
  1. Landforms
  I. Title
  551.41

ISBN-10: 0-582-30156-4
ISBN-13: 978-0-582-30156-6

Library of Congress Cataloging-in-Publication Data
A catalog entry for this title is available from the Library of Congress.

20 19
08 07 06

Set in 10/11.5 Times
Printed in Malaysia, VVP

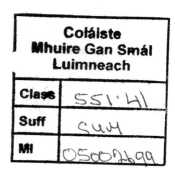

To my parents

Brief table of contents

Detailed table of contents

Preface

If it is to present its subject matter coherently a textbook must have a point of view. The perspective of this book is that an adequate appreciation of landform genesis must encompass a knowledge of the large-scale framework of landscapes as well as an understanding of the smaller-scale processes which create individual landforms. An emphasis on small-scale, surface processes and their associated landforms has been pervasive in geomorphology since the 1960s, to the point where the larger-scale aspects of landform genesis, and in particular the role of internal mechanisms in influencing the development of major morphological features, have come to be regarded as almost incidental to the main thrust of research in the subject. To borrow an analogy from Richard Lewontin, commenting on the relationship between population genetics and evolutionary biology, 'this has led to a kind of auto mechanics concerned with carburettor settings and tyre pressures, but not with how the car was manufactured or where it is is going'.

There are growing signs that this situation is changing and this book attempts to redress the balance by giving due weight to problems of long-term, large-scale landscape development. In particular I have attempted to integrate ideas on global tectonics fully into landscape analysis and to incorporate the results of newly applied dating techniques and of research on the offshore sedimentary record. The examination of surface processes and the landforms they create still accounts for the bulk of the text, but the title of the book is intended to convey the global perspective that I wish to present.

Geomorphology has grown in scope and depth over recent years to the extent that even in a fairly lengthy text I have been forced to be selective in the topics discussed. Perhaps the major omission is applied geomorphology, but to have done this topic justice would have extended the length of the book by a further 20 or 30 per cent. With its emphasis on the way the large-scale components of landscapes develop over the long term this text is concerned essentially with the development of naturally-created landforms, an understanding of which is, of course, vital if we are correctly to assess the impact of human activity. A second significant omission is submarine geomorphology; this again is largely for reasons of space since the history of the ocean basins and the operation of submarine processes raise some quite distinct issues which I felt could not be adequately tackled without a fairly lengthy treatment.

Guide to the reader

My aim is that this book should be used as a resource which provides both basic background information and guidance towards the more advanced study of particular topics. In order to make the book more readable I have omitted references from the text, but each chapter concludes with a detailed guide to further reading and a list of references. These literature guides will be useful to those pursuing topics for essay or project preparation. The references have been selected on the basis of their importance, readability and accessibility, although the nature of the coverage in the literature of some subjects means that I have not always been successful in meeting these criteria.

I have assumed that readers have a basic knowledge of the physical environment including the main types of climate and vegetation, and the major rock types and rock-forming minerals. An introductory course in physical geography or physical geology would provide the appropriate background. Nevertheless, I have attempted to define all but the most elementary technical terms and these are printed in bold type in the text. Where possible, terms have been defined where they first occur in the text, but in all cases a page number in bold type in the index indicates where the definition of a term can be found. The mathematical competence of students taking courses in geomorphology varies enormously and this provides a major dilemma for the textbook author. The view I have gained from my own experience and discussions with numerous colleagues both in the UK and the USA is that while a fairly rigorous mathematical treatment might be desirable, such an approach would not be appropriate for the majority of students. Moreover, a strong mathematical emphasis can in some cases create the impression that certain geomorphic processes are more fully understood than they actually are. I

have decided, therefore, to rely on verbal description and discussion, but I have included a mathematical representation of selected concepts and processes in boxes throughout the text. Furthermore, the text *An Introduction to Quantitative Geomorphology: An Exercise Manual* by Larry Mayer (Prentice-Hall, Englewood Cliffs and London, 1990) provides a mathematical treatment of a number of the topics covered in this book along with a range of relevant field and laboratory exercises.

Part I of the book (Chapter 1) provides an introduction to some of the major concepts applied in the analysis of landscapes; other important concepts are introduced at appropriate points in subsequent chapters. Part II, comprising Chapters 2–5, examines the effects of internal processes on the form of the landscape, while Part III (Chapters 6–14) looks at the wide range of surface processes and their associated landforms. The chapters in Part IV (15–18) consider the ways in which internal and external geomorphic processes interact and the book concludes in Part V with a survey of planetary geomorphology. Extensive cross-referencing means that to some extent the chapters do not have to be read in the order they are presented. In particular, Chapters 6–14 can just as easily precede Chapters 2–5 as follow them. Chapters 15–19, however, are to a large extent founded on material in earlier chapters. Readers with a background in physical geology might want to skip much of Chapter 2 as it provides an introduction to global tectonics. For those without a geological training some useful background material is provided in the appendices. These also contain information on the units of measurement used in the book and a brief discussion of dating techniques relevant to geomorphology.

Acknowledgements

Preparing a textbook is a major undertaking which can only be accomplished through the efforts of many individuals besides the author. The diagrams are a crucial component of the book and I have been fortunate in having the services of Ray Harris and Anona Lyons of the cartographic staff at the Department of Geography of the University of Edinburgh. Initial assistance with the typing of some of the draft chapters was provided by Mrs B. L. Summerfield and Mrs W. Rust and most of the tables were typed by Miss S. Smith. A number of individuals and organizations have been kind enough to provide me with photographs and other imagery in areas where my own resources have proved inadequate; these are acknowledged in the relevant figure captions. I would, however, particularly like to thank Nick Short who made available to me several Landsat images originally published in his co-edited book *Geomorphology from Space*. I would also like to thank those individuals and organizations who granted permission for copyright material to be reproduced or used as the basis for diagrams or tables. Specific acknowledgements are listed below.

Some of the material presented in this book relates directly, or indirectly, to my own research experience which has included fieldwork in many parts of the world, and I am grateful to those organizations that have provided financial support for this research. These include the Natural Environment Research Council, the Royal Society, Texaco Inc., British Aerospace p.l.c., the Carnegie Trust for the Universities of Scotland, the University of Oxford and the University of Edinburgh. I also owe a considerable debt to those numerous colleagues who have influenced my approach to geomorphology, both through their published work and informal discussions. Naturally this book draws extensively on the work of a great number of earth scientists, only a small proportion of whom are specifically mentioned in the main body of the text. The guides to further reading will, however, enable readers to follow-up work by individual researchers.

I am especially grateful to those colleagues who put considerable time and effort into providing comments on the draft manuscript, or parts thereof. These individuals are listed separately. The responsibility for the final text is, of course, solely my own.

Finally, I would like to thank Sue Smith for her support and encouragement.

The author and publishers are grateful to the following for permission to reproduce copyright figures and tables:

Academic Press Inc. (London) Ltd. and the respective authors for figs. 4.13 (Burke & Whiteman, 1973), 6.28 (McFarlane, 1983), 6.29 (Goudie, 1983), 13.10 (Hayes, 1976), 15.6 (Gansser, 1983); American Association of Petroleum Geologists for figs. 17.1 & 17.2 (Vail *et al.*, 1977), 17.8 & 18.17 (Kominz, 1984) and table 13.4 (Wright & Coleman, 1973); American Association for the Advancement of Science and the respective authors for figs. 16.6 (Burnett & Schumm, 1983), 17.4 (Haq *et al.*, 1987) © 1983 & 1987 by the AAAS; American Journal of Science and the respective authors for figs. 3.22 (Le Fort, 1975), 15.15 (Ruxton & McDougall, 1967), 15.18 (Ahnert, 1970), 18.17 (Berner *et al.*, 1983); the editor, *American Scientist* for fig. 19.21 (Head *et al.*, 1981); American Society of Civil Engineers for fig. 8.10 (Simons & Richardson, 1963) and table 9.3 (Benedict *et al.*, 1971); Annual Reviews Inc. for fig. 9.8 (Schumm, 1985) © 1985 by Annual Reviews Inc.; Edward Arnold (Publishers) Ltd. for figs. 5.7 (Clapperton, 1977), 10.4, 10.23, 10.24, 10.25, 10.27 & 10.28 (Warren, 1979), 10.32 (Pye, 1984), 15.24 (Goudie, 1983), 19.5 (Baker, 1981); Edward Arnold (Publishers) Ltd. and the respective authors and figs. 11.18, 11.19, 11.20, 11.22 & 14.3 (Sugden & John, 1976), 12.3, 12.5, 12.7, 12.11 & 14.6 (Washburn, 1979) and tables 11.4 (Sugden & John, 1976), 12.3, 12.6 & 14.3 (Washburn, 1979), 12.2 (Embleton & King, 1975); A. A. Balkema Publishers (Netherlands) for fig. 17.13 (Dingle *et al.*,

1983); the author, P. Bishop for fig. 15.17 (Bishop, 1985); Basil Blackwell Ltd, for fig. 6.23 (Jennings, 1985); Blackwell Scientific Publications Ltd. for figs. 8.14 (Miller *et al.*, 1977), 13.24 (Elliott, 1978); California Division of Mines & Geology for figs. 3.27 (Crowell, 1975), 16.7 (Jenkins, 1964), 16.11 (Wallace, 1975); Cambridge University Press for fig. B7.3 (Carson & Kirkby, 1972) and figs. 2.16 & 3.23 (Smith, Hurley & Briden, 1981), 7.4, 7.5, 7.27, 7.28 & 7.29 (Carson & Kirkby, 1972); Cambridge University Press and the respective authors for figs. 13.25 (Stoddart, 1969), 14.5 (Harland *et al.*, 1982); Canadian Society of Petroleum Geologists for table 14.5 (Baker, 1978); Carnegie Institution of Washington for fig. 5.16 (Williams, 1942); Chapman & Hall Ltd. for fig. 16.21 (Holmes, 1978); the authors, D. Chapman & H. Pollock for fig. 2.8 (Chapman & Pollock, 1977); the author, J. Chappell for fig. 15.5 (Chappell, 1974); the editor, *Clay Minerals* for fig. 6.15 (Bayliss & Loughnan, 1964); Columbia University Press (New York) for fig. 18.13 (Johnson, 1931); Croom Helm Ltd. for figs. 6.10 & 7.1 (Crozier, 1986), 15.25 (Yoshikawa, 1985), 17.3 (Pillans, 1987); the author, J. Demek for fig. 12.17 (Demek, 1969); the author, D. Drewry for fig. 11.8 (Drewry, 1983); DSIR Publishing (New Zealand Depart. of Scientific & Industrial Research) for fig. 3.26 (Kingma, 1958) the editor, *Ecologae Geologicae Helvetiae* for fig. 3.19 (Collet, 1935); Elsevier Science Publishers (Physical Sciences & Engineering Div.) for fig. 10.13 (Wilson, 1972); Elsevier Science Publishers (Physical Sciences & Engineering Div.) and the respective authors for figs. 9.20 (Allen, 1964), 13.16 (Ingle, 1966), 14.21 (Lancaster, 1981), 18.17 (Barron, 1983) and table 6.4 (Loughnan, 1969); Environment Canada and the author, M. Church for fig. 12.6 (Church, 1974); W. H. Freeman & Co. for fig. 1.12 (Leopold, Wolman & Miller, 1964) Copyright © 1964 W. H. Freeman & Co; Gebrüder Borntraeger for figs. 7.19 (Dalrymple *et al.*, 1968), 10.29 (Tsoar, 1984), 18.8 (Brunsden, 1980) and table 7.2 (Selby, 1980); the editor, *Geographica Helvetica* for fig. 16.6 (Scheidegger, 1979); the editor, *Geological Journal* for fig. 9.20 (Allen, 1970); Geological Society and the respective authors for figs. 3.11 (Gansser, 1973), 16.12 & 16.13 (Frostick & Reid, 1987); the editor and The Geologists Association for fig. 3.20 (Oxburgh, 1974); Gulf Coast Association of Geological Societies for fig. 13.24 (Frazier, 1967); the author, W. Hamilton for fig. 3.14 (Hamilton, 1977); Institute of British Geographers for fig. B17.1 (Chappell, 1987) and fig. 7.24 (Kirkby, 1971); International Glaciological Society and the author, J. Nye for fig. 11.6 (Nye, 1952); the author, D. James for fig. 3.13 (James, 1971); Longman Group UK Ltd. for figs. 1.15 (Tricart & Cailleux, 1972), 5.17 (Sparks, 1971), 6.13 (Ollier, 1984), 9.25 (Sparks, 1972), 12.6 & 12.7 (French, 1976), 13.6, 13.9, 13.14, 13.19 & 13.25 (Davies, 1980), 14.14 (Lowe & Walker, 1984), 16.1 (Morisawa, 1985), 16.9 (Ollier, 1981) and tables 15.3 (Morgan, 1986), 16.1 & 16.2 (Morisawa, 1985); Longman Group UK Ltd. for table 11.8 (Price, 1973) published by Oliver & Boyd; Macmillan Magazines Ltd. and the respective authors for figs. 3.20 (Ox-

burgh, 1972), 3.31 (Coney *et al.*, 1980), 4.14 (Watts, 1982), 10.22 (Wasson & Hyde, 1983), 14.6 (Hovan *et al.*, 1989), 19.22 (McGill, 1982) Copyright © 1972, 1980, 1982, 1983 & 1989 Macmillan Magazines Ltd.; Macmillan Publishers Ltd. for fig. 18.4 (Thomas, 1974); Mineralogical Society of America for fig. 6.15 (Loughnan & Bayliss, 1961) Copyright © by Mineralogical Society of America; Methuen & Co. for figs. 6.1 (Carson, 1969), 9.16 (Richards, 1982 & Dury, 1969), 9.17 (Richards, 1982), 10.2 (Bagnold, 1954), 14.9 (Chorley *et al.*, 1984), 14.18 (Schumm, 1969), 16.25 (Heinzelin, 1964), 18.1 (Thornes & Brunsden, 1977); National Research Council of Canada for fig. 12.3 (Mackay, 1971); the author, A. Nur for fig. 3.32 (Nur & Ben-Avraham, 1983); the author, H. Ohmori for figs. 15.19 & 15.20 (Ohmori, 1983); The Open University for fig. 2.15 (Oxburgh, 1971) Copyright © 1971 The Open University Press; the author, S. Ouchi for table 16.3 (Ouchi, 1985); Oxford University Press for figs. E2 (Statham, 1977), 15.26 (Selby, 1982) and tables 7.1 & 7.3 (Selby, 1982), 14.1 (Douglas, 1978) Copyright © 1977, 1978 & 1982 Oxford University Press; Pergamon Press plc for figs. 2.12 (Heirtzler *et al.*, 1966), 11.23 (Baker & Bunker, 1985) Copyright © 1966 & 1985 Pergamon Press plc; Pergamon Press Australia for figs. 10.2 & 10.10 (Mabbutt, 1977); the author, W. Pitman for figs. 17.6 & 17.7 (Pitman, 1978); Prentice-Hall (Englewood Cliffs) for figs. 5.16 (MacDonald, 1972), 16.2 & 16.4 (Bloom, 1978) and table 5.2 (MacDonald, 1972); the editor, *Quaternary Research* for fig. 17.9 (Clark *et al.*, 1978); Routledge Ltd. for figs. 7.18 & 17.20 (Parsons, 1988); The Royal Society and the author, C. Beaumont for fig. 4.17 (Beaumont *et al.*, 1982); the author, S. A. Schumm for fig. 9.13 (Schumm & Khan, 1972); Scientific American Inc. for fig. 19.28 (Soderblom & Johnson, 1982) Copyright © 1982 by Scientific American Inc. All rights reserved; Society of Economic Paleontologists & Mineralogists for figs. 3.26 (Crowell, 1974), 8.11 (Middleton & Southard, 1984), 9.9 (Schumm, 1981) and table 13.5 (Morgan, 1970); Seismological Society of America for fig. 2.10 (Barazangi & Dorman, 1969); Springer-Verlag and the author, G. Scott for fig. 17.20 (Scott & Rotondo, 1983); Minister of Supply and Services Canada for table 11.7 (Prest, 1968); Syracuse University (Dept. of Geography) for fig. 16.19 (Oberlander, 1965); Taylor & Francis Ltd. (Wykeham Publications) and the author, J. Guest for table 19.2 (Guest & Greeley, 1977); Transportation Research Board for figs. 7.6, 7.9, 7.10, 7.11 and table 7.5 (Varnes, 1978); The Unesco Press for fig. 8.3 (Unesco, 1978); United States Geological Survey for figs. 3.34 (Wesson *et al.*, 1975), 5.19 (Hunt, 1953), 9.22 (Lustig, 1965), 9.26 (Leopold & Miller, 1954), 10.17 (McKee, 1979), 19.5 (Scott & Carr, 1978); The University of Chicago Press for figs. 1.11 (Brunsden & Kesel, 1973), 2.1 (Mörner, 1976), 13.1 & 13.2 (Inman & Nordstrom, 1971), 14.19 (Baker & Penteado-Orellana, 1977), 18.15 (Bishop *et al.*, 1985) and tables 13.1, 13.2 & 13.3 (Inman & Nordstrom, 1971) © 1971, 1973, 1976, 1977 & 1985 by the University of Chicago. All rights reserved; University of Natal Press (Pietermaritzburg) for fig. 17.16 (King, 1982); University of Texas (Texas Bureau of Economic Geology) for fig. 13.21 (Fisher *et al.*, 1969); University of Uppsala (Geological Institute) for fig. 8.13 (Hjulström, 1935); Unwin Hyman Ltd. for fig. B11.3 (Boulton, 1974) and figs. 7.17 (Young & Saunders, 1986), 8.9 & 9.11 (Allen, 1970), 15.23 (Middleton *et al.*, 1986), 16.19 & 16.20 (Oberlander, 1985), 18.6 (Schumm, 1975), 18.14 (Summerfield, 1985), 19.13 (Greeley, 1985) Copyright © 1970, 1974, 1975, 1976, 1985 & 1986 Unwin Hyman Ltd.; Van Nostrand Reinhold for fig. 12.12 (Muller, 1968) Copyright © 1968 Van Nostrand Reinhold; the author, P. Vogt for fig. 4.4 (Vogt, 1981); John Wiley & Sons Ltd. for figs. 6.2, 6.4 & 6.16 (Curtis, 1976), 7.14 (Atkinson, 1978), 7.21 (Abrahams & Parsons, 1987), 7.23 (Carson, 1976), 10.13 (Wasson & Hyde, 1983), 15.17 (Brown *et al.*, 1980), 15.9 (Walling & Webb, 1983), 15.13 (Clayton & Megahan, 1986), 15.14 (Foster *et al.*, 1985), 15.21 (Walling & Webb, 1986), 15.22 (Schmidt, 1985) Copyright © 1976, 1978, 1980, 1983, 1985, 1986 & 1987 John Wiley & Sons Ltd.; John Wiley & Sons Inc. for figs. 2.2, 2.3 & 2.5 (Wyllie, 1976), 5.9 (Rittman, 1962), 7.13 (Elter & Trevisan, 1973), 9.19, 9.26 & 16.7 (Schumm, 1977), 9.25 (Thornbury, 1969), 11.16 & 11.26 (Flint, 1971), 18.2 (Strahler, 1969) and tables 5.3 (Rittman, 1962), 9.2 & 14.4 (Schumm, 1977) Copyright © 1962, 1969, 1971, 1973, 1976 & 1977 John Wiley & Sons Inc.; Yale University Press for fig. 19.18 (Carr, 1981) Copyright © 1981 Yale University Press. Whilst every effort has been made to trace the owners of copyright material, in a few cases this has proved impossible and we take this opportunity to offer our apologies to any copyright holders whose rights we may have unwittingly infringed.

We are grateful to the following for permission to reproduce copyright photographs:

Aerofilms Ltd. for figs. 3.18 & chapter heading photograph for Chapter 16; J. Chappell for chapter heading photograph for Chapter 17; K. G. Cox for cover photograph. Geodetic Institute, Copenhagen for chapter heading photograph for Chapter 11 & fig. 11.1; Geological Survey of Canada for fig. 12.13; Goddard Space Flight Center for figs. 1.10A & 3.28; K. Mulligan for fig. 10.18; D. Munro for figs. 3.12 & 6.25; G. M. Robinson for fig. 11.15; © Royal Geographical Society/M. J. Day for fig. 6.24; Satellite Remote Sensing Centre for chapter heading photographs for Chapters 1, 14 & fig. 18.12; S. J. Smith for fig. 9.6; D. E. Sugden for figs. 11.2 & 11.12; J. T. Teller for fig. 10.19; USGS for fig. 1.10C; C. Warren for cover photograph; A. L. Washburn for figs. 12.8, 12.10 & 12.16; M. I. Whitney for fig. 10.8; World Data Center A for Rockets and Satellites for figs. 1.1, 1.2, chapter heading photograph for Chapter 3, Figs. 19.2, 19.3, 19.6, 19.7, 19.8, 19.10, 19.11, 19.12, 19.14, 19.15, 19.16, 19.17, 19.19, 19.23, 19.24 & cover photograph, 19.25 & 19.26.

List of reviewers

Athol D. Abrahams, Department of Geography, State University of New York at Buffalo

Victor R. Baker, Department of Geosciences, University of Arizona

Paul Bishop, Department of Geography and Environmental Science, Monash University

Peter A. Bull, School of Geography, University of Oxford

Chalmers M. Clapperton, Department of Geography, University of Aberdeen

Ian Douglas, School of Geography, University of Manchester

Thomas Dunne, Department of Geological Sciences, University of Washington

Andrew S. Goudie, School of Geography, University of Oxford

Barbara A. Kennedy, School of Geography, University of Oxford

Patrick D. Nunn, Department of Geography, University of the South Pacific

Kenneth Pye, Postgraduate Research Institute for Sedimentology, University of Reading

Stanley A. Schumm, Department of Earth Resources, Colorado State University

Alan G. Smith, Department of Earth Sciences, University of Cambridge

David E. Sugden, Department of Geography, University of Edinburgh

Stanley W. Trimble, Department of Geography, University of California, Los Angeles

Tjeerd H. Van Andel, Department of Earth Sciences, University of Cambridge

Andrew Warren, Department of Geography, University College, London

Chapter heading
plate captions

1. View of the Earth showing landforms at a global scale. The image shows a nearly cloud-free Africa with southern Europe towards the top and the north-east of South America on the left. Landscape features evident at this scale are primarily related to the major horizontal motions of the continents over the past 200 Ma. Such movements have created the Atlantic Ocean and led to the more recent opening of the Red Sea (top right corner). (Meteostat image courtesy Satellite Remote Sensing Centre, RSA.)

2. Apollo 7 image showing a view north-east across the Sinai Peninsula and the northern end of the Red Sea with the Gulf of the Suez on the left and the Gulf of Aqaba on the right. This region is seeing the early stages of the development of a divergent plate boundary separating the African and the Arabian Plates with a spreading ridge extending northward up the Red Sea. The linear feature in the top right corner of the image is the Dead Sea Rift, a tranform plate boundary which extends northward from the Gulf of Aqaba. The eastern side of this rift is moving northward relative to the western side. (Image courtesy of NASA, Lyndon B. Johnson Space Center, Houston, Texas.)

3. Looking west towards the Himalayas and Tibetan Plateau. The dark area to the left is the lower forested slopes of the Nepal Himalaya. Mount Everest is located a little below the middle–centre of the view. (Apollo 7 image courtesy R. J. Allenby Jr and the World Data Center A for Rockets and Satellites.)

4. The summit of the Drakensburg Escarpment in Natal, South Africa, capped at this location by basalt lavas more than 1 km thick. This part of the Great Escarpment of southern Africa lies about 160 km inland from the coast and has formed along a sheared margin.

5. The snow-covered summit of Mount Taranaki (Egmont), an almost perfectly symmetric strato-volcano, North Island, New Zealand.

6. Rectangular joint system on Checkerboard Mesa, Zion National Park, Utah, USA. The fractures may have resulted from cyclic near-surface volume changes resulting from temperature fluctuations, wetting and drying or freeze–thaw.

7. Compound slope developed in flat-lying, alternating beds of shale and resistant, massive sandstone, Utah, USA.

8. Highly turbulent flow in a boulder-bed channel, River Inn, Austria.

9. The finely dissected terrain of the Loess Plateau, Shanxi Province, China. The high drainage density in combination with the ease with which the wind-borne silt (loess) covering the region can be eroded by running water causes extraordinarily high rates of erosion. The Huang He River visible flowing north to south on the right of the image is estimated to carry an annual sediment load of between 1.5 and 1.9×10^9 t where it leaves the loess plateau. The area covered is about 130 km across. (Landsat image courtesy N. M. Short.)

10. Part of the Rub'al Khali, or Empty Quarter, of southern Saudi Arabia, the largest sand sea on Earth. The large-scale regularity that can be achieved by wind-shaped depositional forms is clearly evident in this image which spans an area 140 km across. Three categories of dune form are visible; complex barchanoid ridges which interconnect to form a dune network or aklé pattern cover the bottom and right half of the image, while complex linear dunes occupying the top left quarter pass into star dunes in the extreme top left corner. Note that north is at the bottom of the image. (Landsat image courtesy A.S. Walker.)

11. Crevasse patterns and medial and lateral moraine visible on a glacier in northern Milne Land, eastern Greenland. (Photo courtesy of the Geodetic Institute, Copenhagen, Denmark.)

12. Ice-wedge polygons near Barrow on the northern

coastal plain of Alaska. The polygons range from 7 to 15 m across. (Photo courtesy United States Geological Survey Photo Library, Denver, Colorado, T. Péwé 845.)

13. Oblique, northward-looking view of the coastline of North Carolina and Virginia, USA taken with a hand-held camera on the Apollo 9 Earth-orbital mission. Cape Hatteras is the eastern extremity of the line of barrier islands cleary visible running down the centre of the image. These barrier islands, which have a total length of over 300 km in the area covered by the image, form part of a nearly continous chain of barrier islands extending from Long Island, New York, to Florida. Cape Hatteras is just one of a number of cuspate fore-lands along the eastern coast of the USA. Cape Look-out is the promontory at the bottom of the image and Pamlico Sound is the large body of water to the west of Cape Hatteras. (Image courtesy NASA, Lyndon B. Johnson Space Center, Houston, Texas.)

14. Landsat image of the north-central Kalahari showing a range of landforms created under alternating humid and arid conditions. East-west orientated relict dune systems evident from vegetation contrasts on the image and indicative of past aridity have been cut by river channels which thus post-date them and demonstrate later humidity. Flooded pans apparent from their dark tone are visible in the top right corner of the image. The dark channel in the extreme bottom left corner of the image is the Okavango River which has a perennial discharge fed from the Angolan Highlands far to the north-west. The broad light-toned channel running across the centre of the image is of uncertain origin but may represent a periodically active overflow from the Okavango. The area covered by the image which is about 180 km across is located within the area of Group A dunes in Figure 14.21. (Landsat image processed by the Satellite Remote Sensing Centre of the CSIR, RSA.)

15. View from the Tasman Sea eastwards to the Southern Alps, South Island, New Zealand. The Southern Alps have one of the highest sustained rates of crustal uplift known, reaching a maximum of around 10 000 m Ma^{-1}. These rapid rates are matched by equally high rates of erosion.

16. Anticline in the Zagros Mountains, Iran, at an early stage of fluvial dissection. (Photo © Aerofilms Ltd.)

17. A flight of uplifted coral reef terraces recording sea-level highstands during the past 120 ka, Huon Peninsula, Papua New Guinea. (Photo courtesy J. Chappell.)

18. Structural control of the Great Escarpment in western Cape Province, South Africa, provided by a massive sandstone unit (see Figure 18.12.)

19. The south polar region (left part of image) of Triton, one of Neptune's moons. The morphology of Triton seems to be a collage of landscapes observed on other bodies in the Solar System — certainly Triton has the most enigmatic surface of any planetary body yet explored. On this exceedingly cold world the surface appears to be formed of frozen methane and nitrogen. The dark streaks visible at the bototm left of the image may be formed by eruptions of nitrogen from volcanoes, some of which may still be active. The long features on the right side of the image are probably down-faulted blocks analogous to rift valleys on Earth. (Voyager 2 mosaic courtesy JPL and NASA.)

Part I

Introduction

1

Approaches to geomorphology

1.1 The science of landforms

Geomorphology is the science concerned with the form of the landsurface and the processes which create it. It is extended by some to include the study of submarine features, and with the advent of planetary exploration must now incorporate the landscapes of the major solid bodies of the Solar System. One focus for geomorphic research is the relationship between landforms and the processes currently acting on them. But many landforms cannot be fully explained by the nature and intensity of geomophic processes now operating so it is also necessary to consider past events that may have helped shape the landscape. To a significant extent, then, geomorphology is a historical science.

Since the landsurface is located at the interface of the Earth's lithosphere, atmosphere, hydrosphere and biosphere, geomorphology is closely related to a wide range of other disciplines (Table 1.1). While having a central interest in landforms, geomorphologists must, none the less, be aware of those aspects of allied disciplines that bear on their subject. Equally, geomorphology has a potential, as yet only partially realized, of making significant contributions to these other areas of knowledge.

1.2 The development of ideas

The way geomorphologists approach the study of landforms at the present time can only be seen in a proper context if we appreciate how the central concepts of geomorphology have been developed. Long before the term geomorphology itself was introduced in the 1880s people had speculated on the forces and mechanisms that had created the natural landscape around them. Aristotle, Herodotus, Seneca and Strabo, among other Greek and Roman philosophers, wrote on phenomena such as the origin of river valleys and deltas, and the relationship between earthquakes and deformation

Table 1.1 Examples of relationships between geomorphology and allied disciplines

DISCIPLINE	EXAMPLE OF CONTRIBUTION TO GEOMORPHOLOGY	EXAMPLE OF CONTRIBUTION FROM GEOMORPHOLOGY
Geophysics	Mechanisms and rates of uplift	Erosional response of landsurface to uplift
Sedimentology	Reconstruction of past erosional events from a sedimentary sequence	Form of alluvial channels in interpretation of fluvial sediments
Geochemistry	Rate and nature of chemical reactions in rock weathering	Mobilization of elements in earth surface environments
Hydrology	Frequency and intensity of flooding	Sediment concentration in streams
Climatology	Effect of climatic elements on rate and nature of geomorphic processes	Effect of surface deposits and morphology on climatic variables
Pedology	Effect of soil properties on slope stability	Topographic control over soil-forming processes
Biology	Role of vegetation cover in affecting rates of erosion	Topographic control over micro-environments of plant growth
Engineering	Techniques for analysis of slope instability	Identification of morphological features indicative of slope instability
Space science	Context for understanding special characteristics of landform-creating environment on the Earth	Interpretation of planetary landscapes by analogy with terrestrial landforms

of the landsurface. The idea that streams have sufficient power to erode their valleys was appreciated to some extent by Seneca, and certainly by Leonardo da Vinci in the fifteenth century, but it was not until the late eighteenth century that the implications of this fundamental concept began to be fully explored.

1.2.1   The age of Hutton and Lyell

In 1785 James Hutton presented a paper to the Royal Society of Edinburgh in which he argued that the landsurface had been shaped by the slow, unremitting erosive action of water rather than by the catastrophic events advocated by biblical scholars; to the history of the Earth Hutton saw 'no vestige of a beginning, no prospect of an end'. His ideas disseminated only slowly until in 1802, five years after his death, John Playfair, his friend and Professor of Mathematics at the University of Edinburgh, restated and elaborated his views with an elegance and clarity that has rarely been matched in scientific writing. In his *Illustrations of the Huttonian Theory of the Earth* Playfair provided the first detailed and closely reasoned account of several important aspects of landform genesis, most notably the relationship between rivers and their valleys:

Every river appears to consist of a main trunk, fed from a variety of branches, each running in a valley proportioned to its size, and all of them together forming a system of valleys, communicating with one another, and having such a nice adjustment of their declivities, that none of them join the principal valley, either on too high or too low a level, a circumstance which would be infinitely improbable if each of these valleys were not the work of the stream which flows in it.

Hutton's methodology, founded on the belief that the slow but continuous operation of processes observable at the present day provided a sufficient basis for explaining the present configuration of the Earth's surface, was taken up and developed by Charles Lyell in his idea of **uniformity**. Through his highly influential work, *Principles of Geology* (1830–33), Lyell became the 'great high priest' of what became known as the principle of **uniformitarianism**, a concept frequently (but inadequately) summarized by the phrase 'the present is the key to the past'. Lyell's notion of uniformity was far more complex than is often appreciated by many earth scientists and this has led to much confusion as writers have failed to distinguish between its various meanings. In fact four distinct meanings can be identified in Lyell's *Principles*.

1. **Uniformity of law:** this is the assumption that natural laws are constant in time and space.
2. **Uniformity of process:** this is the proposition that if past events can be explained as the consequence of processes now known to be operating then additional unknown causes should not be invoked. In essence, this is the principle of simplicity adopted in all scientific explanation; if known processes are capable of explaining natural phenomena additional 'exotic' mechanisms should not be introduced. For Lyell, and Hutton before him, this principle was in fundamental opposition to notions of divine intervention as an explanation for the Earth's surface form.
3. **Uniformity of rate (gradualism):** this is the proposition that changes on the Earth's surface are usually slow, steady and gradual. Although Lyell did acknowledge that major events, such as floods and earthquakes, do take place, he maintained that such phenomena are local in extent and that they occurred in the past with the same average frequency as they do today.
4. **Uniformity of state:** this is the idea that, although change occurs, it is directionless; that is the Earth always looked and behaved much as it does at the present time. This concept was a central pillar in Lyell's grand vision of earth history as an endless succession of cycles.

These multiple meanings of uniformity led to much confusion in the vigorous debate which Lyell provoked after 1830 because it was possible to accept some of the propositions embodied in uniformitarianism while at the same time rejecting others. Opposition to uniformitarianism came from geologists who subscribed to **catastrophism** – the idea that many of the features of the landscape were to be explained by rapidly occurring events, rather than by gradual change. The great majority of catastrophists were not, as is often portrayed, believers in a landscape created by acts of divine intervention, since, by the 1830s, few serious geologists accepted a 6000 year biblical time scale for Earth history. In fact they had no argument with the uniformity of law and uniformity of process, but, on the basis of their interpretation of the available field evidence, they firmly rejected Lyell's ideas on the uniformity of rate and uniformity of state. One important type of evidence the catastrophists pointed to was the so-called 'drift' deposits formed of a mixture of boulders, gravel and sand which were known to blanket large areas of northern Europe. These materials, which were to be found on hilltops as well as in lowlands, had earlier been cited as evidence of the biblical Flood but, scriptural arguments aside, they provided a powerful argument against Lyell's extreme notions of gradualism and uniformity of state. The argument in effect centred around the extent to which the intensity of particular landscape-forming processes might change over time, and this debate has continued in various guises to the present day where the primary concern is the relative significance of rare, large magnitude (catastrophic) events in landform genesis.

The idea that these drift deposits had been laid down by glaciers gradually emerged in the early nineteenth century; but a glacial theory did not become a widely accepted explanation for drift deposits, and other landforms which were apparently inexplicable in terms of normal fluvial erosion, until after the publication in 1840 of Louis Agassiz's *Etudes sur les Glaciers*. His notion of a Great Ice Age was soon being applied by other workers to the landscapes of northern Britain, while Agassiz himself, who eventually moved to the USA from his native Switzerland, extended his glacial theory to North America. Concurrent with the acceptance of the idea of continental glaciation was a continuing debate over the relative importance of marine and fluvial erosion. By the 1870s this had been resolved firmly in favour of the predominance of rivers in shaping the landscape.

1.2.2 Developments in North America

If the foundations of the scientific study of landforms were laid in Europe in the first half of the nineteenth century, much of the conceptual structure of the modern discipline was erected in the second half of that century by a remarkably gifted group of American geologists led by John Wesley Powell and Grove Karl Gilbert. Exploring the mostly semi-arid terrain of the western USA where the detailed relationships between rock structures and landforms were largely unobscured by soil and vegetation, Powell was able to develop a structural and genetic classification of mountains, as well as classifications of valleys and drainage systems. His greatest conceptual contribution was the recognition of the importance in landform development of **base level** – the lower limit in the landscape, ultimately represented by sea level, below which rivers cannot erode.

Of even greater significance to the later development of the subject was the pioneering research of Gilbert. He created the first systematic analysis of the mutual interaction between the driving forces of erosion and the resisting forces represented by the rocks and superficial deposits of the Earth's surface. The resulting series of laws of landscape development, founded upon Gilbert's concept of dynamic equilibrium (see Section 1.3.4), were brilliantly presented in his classic monograph, *Report on the Geology of the Henry Mountains,* and developed further in a novel quantitative treatment of fluvial processes published in 1914.

Whereas Gilbert emphasized the adjustment between present forms and present processes, his compatriot and contemporary William Morris Davis founded a school of geomorphology based on the concept of a systematic progression of landform change through time initiated by rapid uplift of the landsurface. This evolutionary sequence, termed the **cycle of erosion** (see Section 1.3.4), was enthusiastically extended and applied by Davis's students and other researchers in the USA and the UK. But it was never accepted by the majority of European geomorphologists who reacted against what they saw as the overly theoretical and idealized nature of the model, as well as the way it underplayed the importance of climate in influencing landform development. A further European challenge came from Walther Penck, who rejected the crucial assumption inherent in the cycle of erosion that the Earth's surface can be stable for a sufficient period of time after an episode of rapid uplift for an evolutionary sequence of landforms to be developed. He argued instead that the overall form of the landscape would depend primarily on whether the rate of uplift was increasing, decreasing or constant through time.

1.2.3 The modern era

A lack of empirical evidence as to the nature and rate of landscape change through time, coupled with a poor level of understanding of the processes responsible for landform genesis, led to increasing doubts among many geomorphologists as to the viability of historical explanation in geomorphology. Foreshadowed by R. E. Horton's remarkable synthesis of drainage basin hydrology published in 1945, the following decades witnessed a growing emphasis, especially in the UK and North America, on both the quantitative analysis of landform morphology (landform **morphometry** or **geomorphometry**) and on the field measurement of geomorphic processes. These developments were not so evident in Continental Europe where the earlier tradition of geomorphology founded on the relationship between landform characteristics and climatic zones was strengthened after the Second World War.

The 1960s and 1970s saw a major reorientation of geomorphology in the UK and USA towards the development of predictive models of short-term landform change. These were based on a much improved knowledge of geomorphic processes founded on a greater understanding of the basic physical principles involved. Indeed, these models often reflected a significant input from research by engineers, particularly with respect to slope stability, the flow of water in river channels and the entrainment and transport of sediment. During this period there was also a rapid growth in applied geomorphology with predictive models being used to assess the likely response of the landscape to changing conditions brought about by human activities, such as land use changes and dam construction.

1.2.4 Future directions

Although greatly advancing our knowledge of surface processes, much current research in geomorphology contributes little to our understanding of how extensive areas of a landscape change over long periods of time. But revolutionary changes since the late 1960s in disciplines allied to geomorphology, together with the application of new techniques, now provide the opportunity to look at this problem anew and to develop a global perspective for geomorphology to accompany the existing disciplinary focus on small-scale, surface process studies. What new directions, then, can we see for geomorphology within such a global perspective?

The late 1960s and early 1970s saw a dramatic change in our understanding of the Earth as the notion of continental mobility embodied in the concept of plate tectonics was developed and refined (see Ch. 2). Davis's scheme of long-term landform development had no comprehensive tectonic theory with which to work, and consequently his assumptions about uplift and landsurface stability were inadequately grounded. We now have a model capable of making predictions of general patterns of uplift and stability for the

Earth's surface through time and space and know that a mode of landscape development applicable to one tectonic setting is not necessarily relevant to another. Moreover, significant advances in dating techniques mean that in many cases long-term rates of uplift and denudation can be estimated with some accuracy. One of the results of the application of these dating techniques is the realization that in relatively stable regions landsurfaces may survive without significant modification by erosional processes for several tens of millions of years.

A related advance in the earth sciences of great potential value to geomorphologists has come from the exploration of the oceans. Since the 1960s an active programme of drilling in the deep ocean basins and in the shallower waters around the margins of the continents has produced a wealth of data on the quantities of sediment that have accumulated in these environments over periods of 100 Ma or more (see Ch. 15). These sedimentary sequences provide a unique repository of information recording the erosion of adjacent land masses. There have been numerous studies by geomorphologists which have used the volume of sediment accumulating in lakes and reservoirs to quantify rates of denudation over periods of up to a few thousand years, so the approach is not novel; what is novel is the temporal and spatial scale which can now be addressed.

A third development of importance to problems of long-term landform evolution has been the revolution in our understanding of climatic change (see Ch. 14). We now know that climatic fluctuations, especially during the past 2 Ma or so, have been both more frequent and more extensive in their effects than was appreciated up to the 1960s. The implications of these new ideas of global climatic change for models of long-term landscape development have been especially appreciated by geomorphologists investigating humid tropical landscapes where earlier ideas of climatic stability have had to be dramatically revised. Given the increasing importance now being attached to the interactions between the atmosphere, the oceans and the landsurface in climatic modelling, we can anticipate that geomorphology will come to play a more central role in the understanding of climatic change itself. We now know, for instance, that atmospheric carbon dioxide is a crucial factor in determining global temperatures and that its concentration in the atmosphere is partly controlled by the rate of weathering reactions on the landsurface; this, in turn, is influenced by a range of geomorphic factors.

One further opportunity for developing a broader, global perspective for geomorphology lies in the comparative analysis of the planets and moons of the Solar System (see Ch. 19). The period since the 1960s has seen a new research frontier open for geomorphologists with the beginning of the exploration of the Solar System. As with the exploration of the then unfamiliar arid landscapes of the south-west USA which led to the pioneering work of Gilbert and Powell, the alien landscapes of the Moon (Fig. 1.1), Mars (Fig. 1.2) and the other planetary bodies have presented a major challenge to geomorphologists.

The study of planetary surfaces is a 'two-way street'; on the one hand terrestrial analogues have been invaluable in the interpretation of the surface forms of other planetary bodies (Fig. 1.3), but on the other hand planetary landscapes provide an invaluable perspective with which to consider long-term landform development on the Earth. Many planetary surfaces have experienced relatively little change for billions of years and so retain the effects of very rare, but cata-

Fig. 1.1 *Hadley Rille, a sinuous channel more than 115 km in length cutting across a lava plain adjacent to the Montes Apennines (mountains on the right of the image) on the edge of the Moon's Imbrium Basin. This rille, like other similar channels on the Moon, is thought to be a lava channel on the basis of analogous landforms occurring in volcanic regions on the Earth (although the lunar versions are many times larger than their terrestrial equivalents). (Apollo 15 image, World Data Center A for Rockets and Satellites.)*

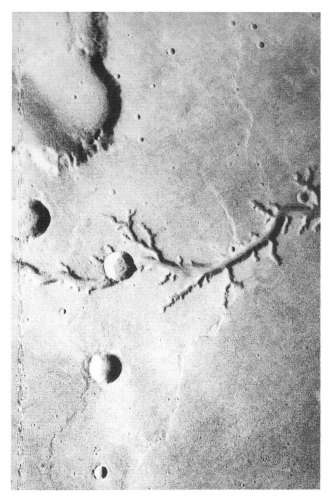

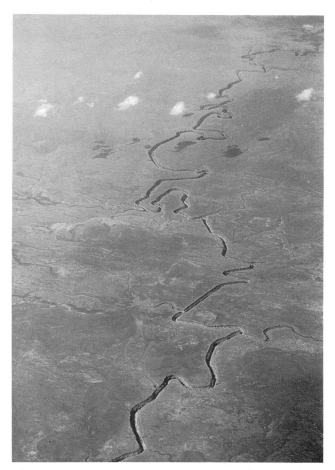

Fig. 1.2 *Part of Nirgal Vallis, an 800 km long channel system on Mars incised into old cratered terrain (see Section 19.3.8). The tributary heads have an amphitheatre-like morphology and through analogies with equivalent terrestrial forms this supports an origin involving runoff emerging from springs (a process known as spring sapping or ground water sapping). Note the crater located on the channel near the left-hand edge of the image which must have been formed after the channel. The area shown is about 55 km across. (Viking 1 image, World Data Center A for Rockets and Satellites.)*

Fig. 1.3 *Oblique view from a height of about 10 km of the canyon of the Little Colorado River in Arizona, USA, a tributary of the Colorado River. This canyon, like many others in the Colorado Plateau in Utah and Arizona, has a 'box' form very similar to some of the channels on Mars. Such landforms provide valuable analogies for interpreting equivalent Martian features (see Figure 1.2). It is thought that such **box canyons** on Earth form as a result of water percolating through a thick permeable sandstone unit and emerging where it encounters a less permeable underlying lithology. The resulting spring sapping causes the canyon head to recede.*

clysmic, landscape-forming events. On the Earth the effects of these events, such as the impact of large objects and catastrophic floods, are usually rapidly obliterated by erosional, tectonic and volcanic processes, but in some cases we can use the landscape history of other planets to assess their role in Earth history. For instance, although only about 120 craters formed by impacts rather than volcanic activity have been identified on the Earth (Fig. 1.4), we can estimate the approximate frequency of major impact events in the past caused by comets, meteorites and asteroids by using the largely preserved history of bombardment experienced by the Moon and some other planetary bodies.

1.3 Some key concepts

1.3.1 Endogenic and exogenic processes

The Earth's detailed form at any instant in time represents the net effect of surface, or **exogenic processes**, and internal, or **endogenic processes**. Exogenic (also termed exogenetic) processes, including the action of water, ice and wind, predominantly involve **denudation**, that is, the removal of material, and thus generally lead to a reduction in elevation and **relief** (Fig. 1.5). (Note that the term relief refers to a difference in height and must be distinguished from elevation and altitude which refer to height above some datum

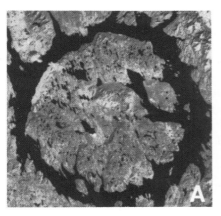

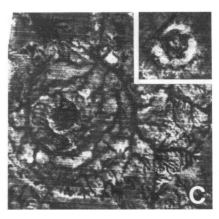

Fig. 1.4 *Examples of impact structures identified on Earth. (A) Manicougan (Quebec, Canada), formed around 210 Ma BP, is about 70 km across and has a broad central upwarp and a surrounding depression now filled with water after the construction of a dam. (B) Clearwater Lakes (Quebec, Canada), a pair of depressions 20 and 30 km across which have been much eroded. (C) Serra da Cangalha (Brazil), a 12 km diameter ring structure, and the smaller 4 km diameter Riachao Ring (inset). It has been estimated that a 10 km-diameter object hits the Earth on average every 100 Ma at a velocity of about 20 km s⁻¹ forming a 150 km wide crater. (Landsat images courtesy of J. McHone and N. M. Short.)*

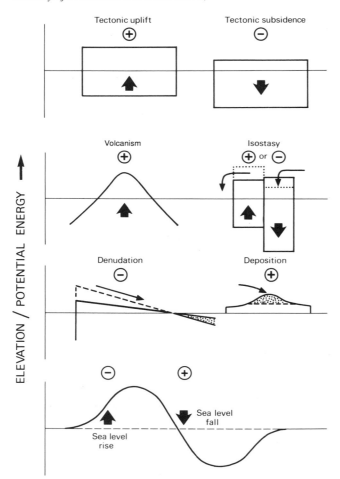

Fig. 1.5 *Schematic illustration of the change in elevation (height relative to a datum) associated with various exogenic and endogenic processes. The positive and negative signs indicate, respectively, increases and decreases in elevation and potential energy. The mechanism of isostasy is discussed in Section 2.2.4.*

– normally mean sea level.) An exception is the localized deposition of material, to form sand dunes for instance, which causes an increase in relief. Denudation can involve the removal of both solid particles and dissolved material. In this book we will generally refer to the former as **erosion** or **mechanical denudation**, and to the latter as **chemical denudation**.

The two sources of energy which power the various exogenic processes are solar radiation and the potential energy arising from the gravitational attraction of the Earth which, in the absence of sufficient resisting forces, causes the downslope movement of water, ice and particles of rock and soil. Solar radiation acts in diverse ways, providing the energy for biological activity, the evaporation of water and the functioning of the Earth's atmospheric circulation.

Endogenic (alternatively endogenetic) processes are generally constructional in that they usually lead to an increase in elevation and relief. Three major types of processes are involved. **Igneous activity** consists of the movement of molten rock, or magma, on to, or towards, the Earth's surface. **Orogenesis (orogeny)** is the formation of mountain belts which are typically arcuate or linear in plan form. **Epeirogenesis (epeirogeny)** is the uplift of usually large areas of the Earth's surface without significant folding or fracture. The broad structures of the Earth's crust and the processes of deformation and faulting which give rise to them are described by the term **tectonics**, while the term **morphotectonics** is applied to the interaction between tectonics and landform genesis. **Neotectonics** refers to the processes and effects of recent tectonic activity and is usually applied to Late Cenozoic events.

1.3.2   Geomorphic systems

While the application of **systems analysis** to geomorphic phenomena has, arguably, not in itself led to any great advances in understanding, there has been a widespread use of systems concepts in geomorphology and an extensive adoption of its terminology. A **system** can be defined as a set of objects or characteristics which are related to one another and operate together as a complex entity. Systems analysis focuses on the relationships between these objects or characteristics.

In geomorphology three kinds of system can be identified. Statistical relationships between the morphological properties of landform elements are represented by **morphological systems**, while movements of mass and flows of energy through the landscape are described by **cascading systems**. Interactions between these two types of system resulting from adjustments between process and form are represented by **process–response systems** (Fig. 1.6).

Before a system can be analyzed its boundaries must be defined. In an **open system** there is a movement of both energy and matter across the system boundary, whereas in a **closed system** only energy is transferred. An **input** of mass or energy into a system is transmitted through it (**throughput**) and leaves as an **output**. Changes in inputs of energy or mass usually produce changes in outputs, but may also give rise to adjustments in the structure of a part of the system (**subsystem**). Changes in the flow of energy and mass as well as adjustments to the structure of the system are controlled by the relationships between variables within the system. These variables, representing the form of the landscape, the rate of geomorphic processes acting upon it

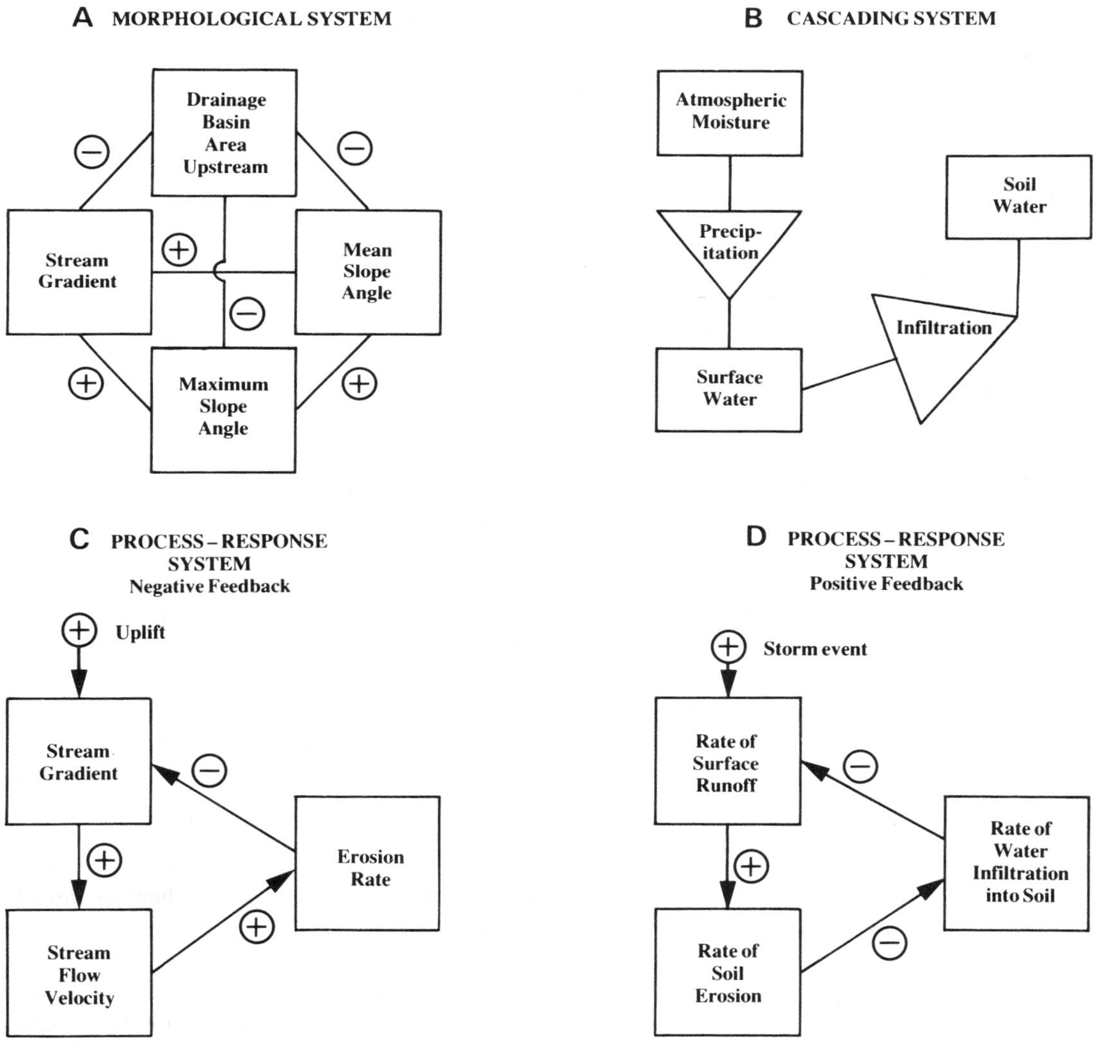

Fig. 1.6 *Examples of simple morphological (A), cascading (B), and process–response systems (C) and (D). Negative and positive signs indicate the nature of the relationship between variables. In (B) rectangles represent storages and triangles transfers between storages. Note that negative feedback systems (C) have an odd number of negative relationships, whereas positive feedback systems (D) have either an even number or none.*

and the environmental factors influencing these geomorphic processes, may be either independent (causal) or dependent (responding to causal variables).

A very common characteristic of geomorphic systems is **negative feedback**, a condition whereby the structure of the system is capable of adjusting in a way that minimizes the effect of externally generated changes. Such an ability of self-regulation or **homeostasis**, means that a system can maintain a state of balance or equilibrium. If faulting across a river bed causes an instantaneous increase in channel gradient, for instance, the resulting increase in river flow velocity will tend to promote a local increase in the rate of channel downcutting and, as a consequence, a reduction in channel gradient (Fig. 1.6 (C)).

In other cases an input change may engender a system response which produces an output which reinforces the original input and eventually causes a shift in the system to a new equilibrium state. This 'snowball effect', as it is colloquially called, occurs in systems exhibiting **positive feedback** and is precipitated by the breaching of a **threshold** in the system. When a severe storm leads to the erosion of the uppermost, permeable layer of the soil the less permeable subsoil is exposed (Fig. 1.6(D)). As this horizon cannot absorb the water running off the surface so effectively the depth of runoff increases, the rate of erosion accelerates and subsoil material with an even lower capacity for water absorption is exposed. In this way it is possible for an entire soil profile to be removed in a single severe storm. The new equilibrium is reached when all the loose, readily erodible soil has been removed and the much more resistant underlying weathered bedrock is exposed.

A final important characteristic of systems is their hierarchical property. A specific system may be composed of numerous smaller systems, but itself form a part of a larger system. Perhaps the clearest example of a system hierarchy in geomorphology is provided by drainage basins; a medium-sized catchment may contain numerous individual streams, each with its own small basin, but at the same time form part of a much larger drainage system.

1.3.3   Magnitude and frequency

There are great variations in the rates at which different geomorphic processes operate. The rate of some processes, such as the flow of ice within an ice sheet, may be fairly constant over millennia, whereas other mechanisms, such as landsliding, are inactive for long periods, although when they do operate they act with great rapidity. Moreover, the same process may exhibit a markedly different intensity and degree of variability under contrasting climatic conditions. In a constantly humid environment lacking high-intensity storms, river discharges may not vary by more than an order of magnitude (by a factor of 10) over several years. By contrast, under a semi-arid climatic regime where predomi-

nantly dry conditions are occasionally punctuated by violent storms, river discharges may vary by two or three orders of magnitude (by a factor of 100 or 1000).

Understanding the frequency with which geomorphic events of different magnitudes occur is clearly a crucial component of any explanation of landform genesis. Measurements of the operation of various geomorphic processes over a range of time scales show that extreme, high-intensity events are rare and that low- to medium-intensity events prevail for the great majority of the time. Accordingly we find that plots of the frequency distribution of many variables which influence the operation and intensity of geomorphic processes, such as wind speeds and river discharges, have a positive (right)-skewed form (Fig. 1.7). The rarity of many types of extreme event, such as major river floods, arises from the fact that these generally require a specific combination of conditions which, while individually not uncommon, are very unlikely to coincide. A simple analogy is provided by the throwing of dice. The probability of throwing a six with one die is 1 in 6 (16.67 per cent), but the probability of obtaining three sixes from three dice thrown simultaneously is only 1 in 216 (0.46 per cent).

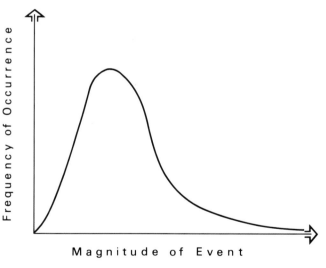

Fig. 1.7 *Characteristic right-skewed frequency distribution of variables (such as wind speed, river discharge and wave height) affecting geomorphic processes. The highest frequency of events occurs in the lower to middle magnitude range, whereas there is only a small proportion of high magnitude events.*

The frequency of an event of a specific magnitude is expressed as the average length of time between events of that magnitude and is known as its **return period** or **recurrence interval** (Fig. 1.8). If a particular maximum annual discharge of a river has a recurrence interval of 20 a this means that there is a 1 in 20 (5 per cent) chance that a flood of this magnitude or greater will occur in any one year

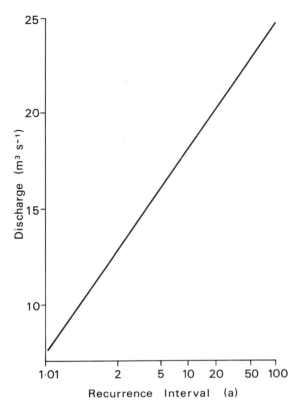

Fig. 1.8 *A typical relationship between stream discharge and recurrence interval.*

and that, *on average*, there will be one such flood every 20 a. The accurate calculation of such probabilities requires a record of measurements which is at least as long as the recurrence interval being estimated. Consequently the more extreme and rarer an event, in general the more difficult it is to estimate accurately its recurrence interval. A further problem is that over the long term the average state of a variable may change significantly. Climatic changes, for instance, can involve marked shifts in average temperatures and precipitation and these may in turn modify the **magnitude – frequency** relationship of geomorphic processes.

It is important to emphasize that for virtually all geomorphic events such recurrence intervals are averages only and do not indicate *when* we should expect an event of a particular magnitude to occur. We could, for instance, observe five separate floods with a recurrence interval of 20 a in a single year, although this would be highly unlikely. In the case of some processes, however, a degree of regularity is evident in the occurrence of events of a similar magnitude. Earthquakes, for example, represent the release of stresses which gradually build up within the Earth's crust. A large earthquake, with its related aftershocks, is therefore unlikely to be followed within a period of several years by another large earthquake in the same locality since much of the stress would have been released in the earlier event.

1.3.4 Equilibrium and evolution

Earlier in this chapter (Section 1.2) reference was made to two contrasting approaches to the understanding of landform genesis. In the first of these, pioneered by Gilbert, the emphasis is placed on the mutual adjustment between present forms and processes. This is sometimes described as the **functional approach** to geomorphology, and the kinds of geomorphic systems with which it is usually concerned commonly exhibit negative feedback and the tendency for an equilibrium form to be restored after a disturbance by a high magnitude – low frequency event. The second mode of explanation, which is indelibly linked with the work of Davis, focuses on progressive changes in the landscape through time and is usually described as the **evolutionary**, or **historical approach** to geomorphology. We clearly need to examine further this apparent paradox that landforms can be considered both to retain an equilibrium form and undergo a progressive change in form through time.

If we imagine observing the boulder-strewn bed of a mountain stream for a period of a few hours we would be very unlikely to witness any measurable change in its form (assuming, of course, that the river is not in flood). Water flows in the channel transporting fine sediment and dissolved material, but its gradient, form and elevation above sea level remain essentially the same. We can describe this situation as one of **static equilibrium** (Fig. 1.9(A)).

If we were to remain at our observation point for several months or years we might well see a major flood which, by causing a significant amount of erosion, lowers the bed of the channel by a small, but measurable, amount. Although this temporarily causes a change in the form of the channel, over the succeeding years the slow deposition of sediment will lead to a restoration of the original channel bed elevation; in these circumstances the channel can be said to be in **steady-state equilibrium** since there is no change in its mean elevation over the period in question (Fig. 1.9(B)). Note that this is essentially the kind of adjustment envisaged by Gilbert in his concept of dynamic equilibrium (see Section 1.2.2).

Stretching our powers of imagination further we could envisage continuing our observations over thousands or hundreds of thousands of years. Over such a period of time there would be hundreds of floods including many events of very high magnitude. Episodes of severe erosion would lead to a progressive lowering of the channel floor and possibly a gradual decrease in channel gradient. This situation is one which is now generally termed **dynamic equilibrium** (Fig. 1.9(C)).

Finally, we can imagine what might happen over a period of millions of years. As the altitude of the mountain carrying our stream decreases we would see a progressive reduction in the elevation of the channel bed. However, we would expect the rate of decrease in elevation to decline

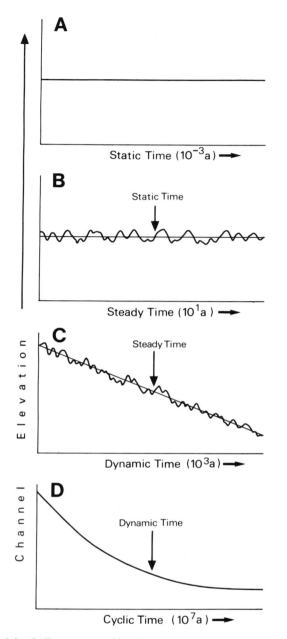

Fig. 1.9 *Different types of landform equilibrium illustrated with reference to schematic changes in channel elevation over different time spans: (A) static equilibrium; (B) steady-state equilibrium; (C) dynamic equilibrium; (D) decay equilibrium. The time scales shown are merely suggestive and will vary by orders of magnitude depending on the type of landform being considered and the nature and intensity of the prevailing geomorphic processes.*

through time as the channel bed approaches base level and the channel gradient becomes so low that rates of erosion reach a minimum. Over this longest time scale the landscape evolves towards a **decay equilibrium** and a landsurface of subdued relief, termed an **erosion surface** or

planation surface is created. (Fig. 1.9(D)). This is essentially the notion of landscape change which forms the basis of Davis's cycle of erosion.

1.3.5   Scale in geomorphology

1.3.5.1   Temporal scale

From our discussion of equilibrium and evolution it will be apparent that the time scale over which the development of a landform is being considered is fundamental in determining what kind of conception of equilibrium is relevant. The time scales indicated in Figure 1.9 serve as only a very general guide since the precise period of time required to attain any particular type of equilibrium depends on a number of factors. In environments in which rates of erosion and deposition are high, landforms can be modified rapidly and the effects of occasional high-magnitude – low-frequency events can generally be obliterated fairly quickly. In such environments steady-state equilibrium can be attained in a short period of time. A second factor is the degree of resistance of the material experiencing erosion. The form of a channel in unconsolidated alluvium may change in a few hours in response to a marked change in river discharge, whereas a channel of similar size cut in densely cemented, resistant bedrock may only adjust after hundreds or thousands of years (assuming the change in discharge persists). A third factor is the size of the landform being considered. The shape of a ripple on a sand dune will adjust much more rapidly to a change in wind direction than the form of the sand dune on which it lies simply because a far smaller amount of sediment has to be moved by the wind to accomplish the adjustment. Similarly, we would expect a small mountain mass a few hundred metres high to reach decay equilibrium in a few million years, whereas tens of millions of years would probably be required for a major mountain range several thousands of metres high.

In more general terms we can talk of the speed with which a change in an input to a geomorphic system, such as an increase in rate of uplift or decrease in river discharge, is fully reflected in a change in form. This is referred to as the **relaxation time** of the system and can range from a few minutes for changes in a small section of an alluvial channel to tens of millions of years for the uplift of a major mountain range.

1.3.5.2   Spatial scale

From our discussion of temporal scale it is clear that there is a close relationship between the temporal and spatial scales at which landform change occurs. Returning to our illustration of the different types of equilibrium in Section 1.3.4, we can see that while it is appropriate to consider a small section of a stream channel in terms of static, steady-state, and perhaps even dynamic equilibrium, it is inappropriate to discuss long-term landform evolution and the attainment of decay equilibrium in terms of such a restricted spatial scale. Not all parts of a landscape can be simultaneously in a steady-state equilibrium since sediment is continuously being removed and relief progressively reduced.

Geomorphology is concerned with phenomena over an enormous range of scales from the form of an individual boulder to the morphology of the Earth's major relief features (Fig. 1.10). It is often useful to be able to categorize this range of scale and to talk, for instance, of microscale or macroscale forms. Such a classification is presented in Table 1.2 with some suggested ranges of linear and areal scale, although it must be emphasized that the divisions between each scale are somewhat arbitrary. Examples of landforms at each scale are indicated together with the main endogenic and exogenic factors influencing landform genesis at the different scales. A very approximate equivalence between these spatial scales and the temporal scales associated with the different types of equilibrium is also suggested; the exact nature of this relationship will, however, depend on the characteristics of the particular landform being considered.

1.3.5.3 Scale and causality

As the temporal scale of analysis changes so the status and interrelationships of variables in geomorphic systems alter accordingly. If we examine landforms over dynamic time then variables which only exhibit measurable change over a longer period can be regarded as fixed; that is, they are independent variables of the external environment. Over dynamic time such variables include lithology and structure and the initial relief of the region. Those factors which change over dynamic time, such as hillslope morphology and channel gradient, are elements of the geomorphic system and are therefore dependent variables at this scale. Variables such as water discharge and rate of sediment transport in a river channel which change very rapidly over dynamic time can be regarded as having mean values which are part of the system, but around which occur random fluctuations which are irrelevant to the system at this scale.

A further consideration in understanding the factors controlling landform development at different temporal and spatial scales is the relative importance of endogenic and exogenic processes. If we are examining a drainage basin of a few square kilometres in a recently uplifted mountain range the tectonic history of the area will be essentially

Table 1.2 Hierarchy of spatial and temporal scales in geomorphology

SPATIAL SCALE	DIMENSIONS		EXAMPLES OF LANDFORMS				MAJOR CONTROLLING FACTORS		TEMPORAL DURATION SCALE	
	Linear (km)	Areal (km²)	Endogenic	Fluvial	Exogenic Glacial	Aeolian	Endogenic	Exogenic		
Micro	<0.5	<0.25	Minor fault scarps	Pools and riffles in a small river channel	Small moraine ridges	Sand ripples	Individual earthquakes and volcanic eruptions	Microclimates; meteorological events	Steady time	10^1a
Meso	0.5–10	0.25–10^2	Small volcanoes	Meanders	Small glacial valleys	Dunes	Local and regional isostatic uplift; localized volcanism and seismicity	Local climates; short-term climatic change	Dynamic time	10^3a
Macro	10–10^3	10^2–10^6	Block–faulted terrain	Floodplains of major rivers	Highland ice caps	Sand seas	Regional uplift and subsidence	Regional climates; long-term climatic change (glacial-interglacial cycles)	Cyclic time	10^7a
Mega	>10^3	>10^6	Major mountain ranges	Major drainage basins	Continental ice sheets	Large sand seas	Long-term patterns of uplift, subsidence and continental motion	Major climatic zones; very long-term climatic change (ice ages)		

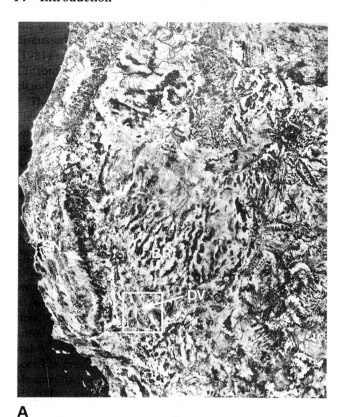

A

Fig. 1.10 *Images illustrating the range of spatial scales considered in geomorphology. (A) The megascale landscape assemblages of the western part of the USA covering an area of about 2.25×10^6 km² (linear dimension around 1200 km). At this scale the overall tectonic framework of the region is clearly visible. Among the most prominent features are the Sierra Nevada (SN) and the Basin and Range Province of Nevada and southern California (BR), a region characterized by a succession of mountain ranges and intervening basins which includes Death Valley (DV). (Part of mosaic of ERTS-1 (Landsat) imagery prepared for NASA Goddard Space Flight Center by the USDA Soil Conservation Service.) (B) The macroscale landscape of Death Valley (DV) and adjoining NW–SE-orientated basins containing extensive lake deposits (light tones) laid down during more humid periods in the past (boxed area on (A)). The western side of Death Valley is flanked by a coalescing series of alluvial fans (A) formed by sediment brought down from the adjacent mountain range. At this scale both detailed tectonic controls and the broad effects of exogenic processes are evident. The area covered is about 34×10^3 km² (linear dimension 185 km). (Landsat image courtesy A.S. Walker.) (C) Alluvial fans at the northern end of Death Valley (boxed area on (B)). The area covered is about 70 km² (linear dimension about 8 km). At this mesoscale the effects of exogenic processes (in this case erosion and the deposition of sediment) can be clearly seen. (Air photo by United States Geological Survey.) (D) A view across the surface of one of the fans shown in (C) (arrowed). At this scale the detailed effects of exogenic processes are apparent.*

B

C

D

irrelevant to understanding the present-day morphology of landform elements in the basin. The prevalence of steep slopes and rapid rates of erosion related to its history of uplift would be regarded as elements of the external environment outside the geomorphic system being studied. If the scale were to be expanded to a few thousands of square kilometres a full understanding of the landscape would require some consideration of its history over the previous million years or so and an assessment of the interaction between endogenic and exogenic processes over that period. At the largest scale we might be dealing with an entire mountain range covering tens of thousands of square kilometres; in this case it would be the endogenic processes that would have exerted the major control over the gross morphology of the landscape with the operation of exogenic processes playing a role in shaping the details.

1.3.6   Explanation in geomorphology

The vast range of temporal and spatial scales means that no one methodological approach to explanation is appropriate for all research in geomorphology. At very short time scales we may be concerned solely with the operation of processes and their relationships with presently existing landforms; at the other extreme we may be aiming to establish a historical sequence of landform development over a period of millions of years and relating this to long-term changes in endogenic processes. This distinction involves two aspects of reality termed by the palaeontologist G.G. Simpson the **immanent** and the **configurational**. By immanent he meant those aspects of reality to do with the inherent properties of the Universe, that is, the physical laws that govern the behaviour of matter. By configurational he meant those forms (or configurations) which arise from the operation of the physical laws of the Universe at a particular point in time.

The importance of this distinction lies in the contrasting approaches to explanation that it implies. When looking at a landscape we can either try to discover what processes are currently active and attempt to explain its present form with reference to these processes, or we can endeavour to unravel the history of the landscape and understand its present form in terms of a sequence of landscapes through time (see Section 1.3.4). The first of these (the functional approach) emphasizes the immanent aspects of reality, the second (the evolutionary or historical approach) emphasizes the configurational aspects. Whereas relating present forms to currently active processes may be a successful strategy if we are working at the small scale, or where landforms are adjusting very rapidly to the operation of geomorphic processes, this is not an adequate approach where we are considering landscapes at the large scale or which have long relaxation times (Table 1.3).

An important question raised by the historical approach to explanation in geomorphology concerns the assumptions we make about the rate at which processes have operated in the past. As we have already seen (see Section 1.2.1) Lyell emphasized the uniformity of both the rates of geomorphic processes through time and the average relief of the Earth's surface (he considered uplift at one place would be roughly balanced by subsidence at another). Such a view is, of course, untenable in the light of our knowledge of major climatic changes during the Earth's recent history. Indeed, some geomorphologists have argued that the great magnitude of these changes casts doubt on the whole idea of uniformitarianism and have instead advocated **neocatastrophism** in its place. This rejection of uniformitarianism arises from a confusion over the diverse concepts encompassed by the term – a confusion which, as we have seen (Section 1.2.1), began immediately on the introduction of the concept in the 1830s.

A valuable attempt to clarify the term has been made by S. J. Gould who distinguishes between two fundamentally different types of uniformitarianism. The first, which he calls **methodological uniformitarianism** and which encompasses uniformity of law and of process, is the proposition that natural laws are invariant – that is, they are constant in

Table 1.3 Relationship between spatial and temporal scale and approaches to explanation in geomorphology

| | SPATIAL SCALE | | | |
	Micro	Meso	Macro	Mega
Predominant genetic mechanisms	Exogenic	Primarily exogenic	Exogenic/ endogenic	Primarily endogenic
Time required for adjustment of form to process	Short	Moderate	Long	Very long
Temporal scale	Steady	Dynamic	Cyclic	Cyclic
Appropriate explanatory basis	Immanent	Immanent/ configurational	Configurational	Configurational

space and time – and that those observable at the present time are sufficient to explain past events. The second, which he terms **substantive uniformitarianism** and which incorporates uniformity of rate and of state, postulates rates of natural processes and material conditions that are essentially constant through time. This proposition involves a claim about the world which can be tested and which we now know to be false (in any strict sense), whereas methodological uniformitarianism, or **actualism** as it is also called, is a statement of scientific method and is a fundamental element of any attempt to provide scientific explanations of how landscapes have changed through time.

1.4 Methods of analysis

A battery of techniques and instruments are now available to monitor the day-to-day operation of a wide range of geomorphic processes, such as the gradual downslope movement of debris on a slope, the transport of sediment in a river or the movement of ice at the bed of a glacier. These methods of data acquisition, which include laboratory experiments as well as field measurements, have generated a wealth of data on short-term landform change, but we can rarely apply these results directly to the problem of long-term landscape change. Primarily this is because there are almost invariably changes in the magnitude–frequency relationships of geomorphic processes in the long term. Although there are now techniques which can be applied to estimate average denudation rates over millions of years, indirect strategies have to be adopted if we are to determine how the form of the landscape has changed over long periods of time. We can do this in two ways; by **space–time substitution**, where variations in form over space are interpreted in terms of changes through time, or by simulating landform changes either mathematically or through the use of hardware models.

1.4.1 Direct observations

With a few rare exceptions direct observations of changes in form are confined to features of limited dimensions over periods of months or years. Significant changes in form can occur where readily mobilized unconsolidated sediments, such as beach sand or alluvium, are subject to frequent and intense geomorphic activity. Such changes can be instrumented and monitored over periods of weeks, months or years. Occasionally it is possible to observe landform changes which, although normally occurring very slowly, under certain conditions take place sufficiently rapidly to produce measurable changes in a short period of time. In a classic study examining slope evolution in the clay badlands at Perth Amboy, New Jersey, USA, S. A. Schumm measured the depth of erosion on slope profiles over a period of ten weeks. The rate of erosion on this impermeable and largely

unvegetated terrain was sufficiently high for a lowering of slopes of over 20 mm to be observed in this short period of time.

For periods longer than a few years other methods have to be employed to document landform change. Aerial photography can be particularly valuable in areas where repeated surveys are available. In the U K aerial surveys extend back to the 1940s and a similar period of coverage is available for parts of Europe and North America. This source of information can be especially valuable for tracking landscape features such as those associated with landsliding where significant landform changes may take several decades but in which the rate of modification varies enormously through time. Satellite remote sensing is now beginning to provide an important additional means of monitoring landform change, especially where extensive or remote areas are being studied.

Topographic maps provide another valuable source for documenting landscape change. In some regions accurate topographic surveying extends well back into the nineteenth century and sequences of maps have been used to reconstruct various landform changes including those affecting alluvial channels and coastal features. It is necessary to use such sources with care, however, as some early surveys may not be sufficiently accurate. In rare instances a particular landscape-forming event may be anticipated and valuable measurements recorded before a significant change in form occurs. A notable example is provided by the fall of Threatening Rock in New Mexico, U S A in 1941. The progressive movement of this vertical column away from a cliff face had been monitored over several years prior to it eventually toppling over.

1.4.2 Space–time substitution

Space–time substitution in landform analysis was pioneered by Charles Darwin in the testing of his hypothesis of coral reef formation (see Section 17.6.3). Barrier reefs, fringing reefs and atolls occurring at various locations in the world's oceans were considered by Darwin to represent different evolutionary stages of island development applicable to any particular subsiding volcanic peak in tropical waters where coral growth could occur. Influenced by Darwin's methodology, Davis found support for his evolutionary scheme of landform change through time in the form of the landscape in different localities which he considered to represent particular temporal stages of development.

Clearly this approach has its dangers. A researcher might, for instance, endeavour to fit landforms in different places into an assumed temporal sequence simply to satisfy a preconceived notion of how such landforms change through time, even though other sequences of change might be equally justified by the evidence. The essential guard against this kind of erroneous reasoning is a sufficiently specific model of landform change linked to a causal mechanism which indicates particular changes in form. In Darwin's case his causal mechanism was the growth of coral as the volcanic substrate subsided; no other explanation could so adequately account for the different forms of oceanic islands he observed. A second problem is the danger of assuming that a temporal sequence exists simply because spatial variablity in form is evident. In reality spatial variations in landforms may arise simply from random fluctuations around an equilibrium form and this possibility must be eliminated before temporal sequences are proposed. A third difficulty is that factors other than time may be responsible for systematic variations in form over space. The form of a slope, for instance, may be significantly related to lithological controls which may vary spatially in a consistent fashion.

Space–time substitution is most safely applied where there is unambiguous evidence for the sequential development of adjacent landform features. This might occur where, for example, the gradual downstream migration of a meander leads to the progressive elimination of active channel erosion at the foot of a meander bluff (Fig. 1.11). In this case it is possible to observe in the spatial sequence the progressive change in slope form after the cessation of active removal of basal debris by the river.

Some geomorphologists have attempted to set the procedure of space–time substitution in an apparently more rigorous framework by invoking the **ergodic hypothesis**. This notion was originally applied in the field of statistical mechanics and proposes that sampling in space is the equivalent of sampling through time. In order for this assumption to be valid the statistical distribution of objects or events over space and through time must be the same. If this is the case then the probabilities of sampling a property

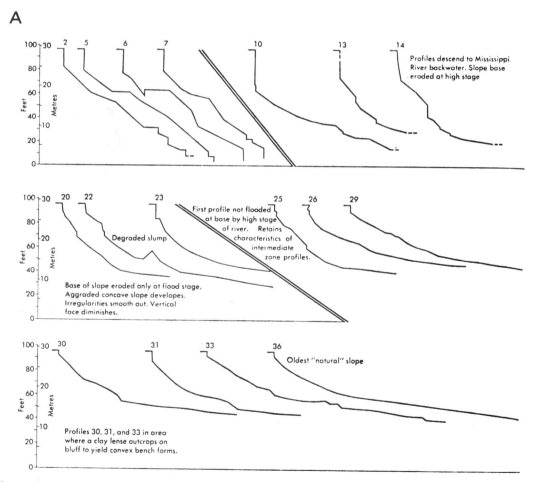

Fig. 1.11 *Sequence of slope profiles along Port Hudson bluff on the Mississippi River in Louisiana, USA (A), and map showing location of profiles (B). The entire bluff segment was being undercut by the Mississippi River in 1722 since when the channel has shifted downstream about 3 km, being at the location of profile 36 in 1849, 34 in 1883, 25 in 1909 and 23 in 1941. Profiles 2–7 in (A) are being actively undercut by the river; profiles 10–14 are undercut during high flows only; profiles 20–36 are subject to basal aggradation. The profiles, which show a decline in mean slope angle from 44 to 20°, represent a temporal sequence reflecting changes in basal conditions. (From D. Brunsden and R. H. Kesel (1973)* Journal of Geology **81**, *Figs. 4 and 6, pp. 581 and 584.)*

B

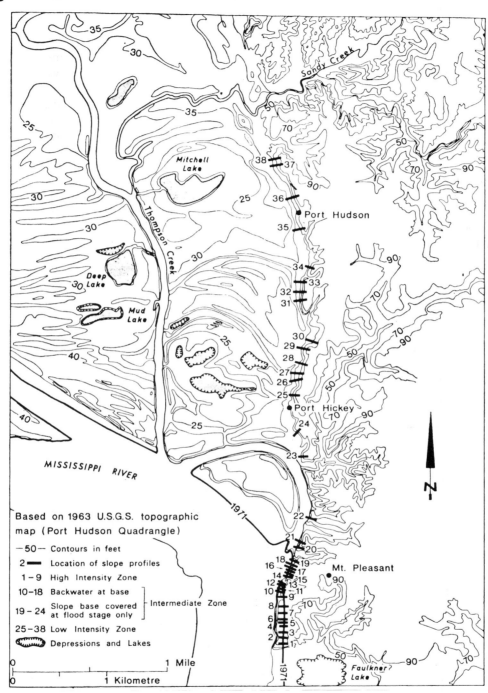

Based on 1963 U.S.G.S. topographic
map (Port Hudson Quadrangle)

—50— Contours in feet

2 ▬ Location of slope profiles

1 – 9 High Intensity Zone

10–18 Backwater at base

19 – 24 Slope base covered
at flood stage only ⎫ Intermediate Zone

25–38 Low Intensity Zone

〰️ Depressions and Lakes

0 ——————— 1 Mile

0 ——————— 1 Kilometre

of the landscape through time and over space are inter-changeable. It is clear that there are few instances in geomorphology where such an assumption can be shown to be justified. Space–time substitution as an approach to understanding the way landforms change over periods of time beyond those accessible to direct observation must, therefore, be based in most circumstances on specific pre-dictions of the morphological changes arising from the operation of well-defined processes. This, of course, is a familiar notion to earth scientists in general who are con-cerned with the interpretation of geological structures or sedimentary bodies which they are often forced to interpret with reference to examples in different stages of develop-ment in different localities.

1.4.3   Simulation

An alternative, but also complementary, approach to determining the nature of landscape change through time is through the use of **simulation models** of which there are essentially three types: hardware models, analogue models and mathematical models.

Hardware models, of which laboratory flumes are perhaps the best known example, are frequently used to study the operation of specific geomorphic processes, such as the transport of sediment by wind or water; but they can also be used to explore the sequence of landform adjustments that arise as a result of changes in external variables influencing these processes. An excellent example of this kind of application of hardware models is the assessment of the effects of a lowering of base level on an artificial channel created in a flume.

The major problem with hardware models is that of scale. Such models are almost invariably constructed at a size much smaller than reality and usually the aim is to accelerate specific geomorphic effects under controlled conditions so that substantial changes in form can be observed over a relatively short period of time. Unfortunately, though, changes in scale affect the relationships between the various properties of hardware models and the real world in different ways. Scale ratios for length are different for those involving velocities and acceleration, while properties such as mass and inertia which are critical in processes such as sediment transport are also not simply a linear function of length. A partial solution to this problem is to maintain the correct scaling for a particular variable of interest by distorting the scale relationships of other variables which are not of specific interest.

A second type of simulation is carried out using **analogue models**. These involve the replacement of materials which occur in the real world with other materials which enable the effects of processes to be more readily observed. Examples include the use of clay to synthesize the effects of glacier flow, or the development of fault patterns. Such models have now been largely replaced by mathematical models which represent geomorphic phenomena and the relationships between them in terms of mathematical expressions.

These mathematical models constitute the third kind of simulation model. **Deterministic mathematical models** are based on exact relationships between independent (causal) variables and dependent (response) variables, such as those that predict the way in which a slope profile is transformed over time from some initial form. By contrast **stochastic mathematical models** incorporate a random component which allows different possible outcomes to arise from a given set of initial conditions. One of the most widely applied types of stochastic model is that used to simulate the development of stream networks where the growth of individual channel links in nature is affected by numerous minor

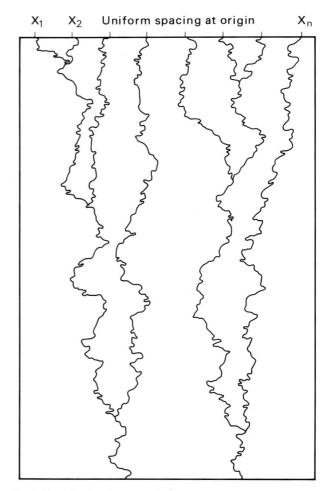

X$_1$ X$_2$ Uniform spacing at origin   X$_n$

Fig. 1.12 *The development of rills on a hillslope simulated using a stochastic (random) model. The model specified that in each time increment the 'random walk' could proceed forward at any angle, but could not go backward. Note the equal spacing between the points of origin (X$_1$–X$_n$) for each rill. (After L. B. Leopold, M. G. Wolman and J. P. Miller (1964) Fluvial Processes in Geomorphology. W. H. Freeman, San Francisco, Fig. 10–4, p. 416.)*

factors, the net effects of which can be adequately treated as random (Fig. 1.12).

Given the complexity of most geomorphic processes, the construction of mathematical models necessitates a good deal of abstraction and selection, and the geomorphic phenomena considered are almost invariably represented in a highly idealized fashion. Landform properties and processes must be strictly defined if they are to be represented mathematically. Mathematical models simplify the complexity of form existing in the real world and in so doing enable us to predict how landforms will change through time. How successful such predictions are depends on how good a representation of the real world our model is. The most accurate simulations of landform change are generally those which treat relatively simple geomorphic systems, and which are based on a detailed knowledge of the geomorphic processes involved.

There are two major problems in using mathematical models to simulate landform change over time. One is **equifinality** – different models may generate similar results and unless the geomorphic processes and landforms involved are very well understood it may not be possible to decide which version most adequately describes the processes that are operating. The second problem is that we need to have a detailed knowledge of the actual changes in form that occur in the landscape in order that mathematical models predicting particular changes may be tested. This is an often neglected but crucially important issue.

1.5 Endogenic and exogenic factors

1.5.1 Sources of energy

The processes that shape the world's landscapes are powered by two major sources of energy (Fig. 1.13). The energy for endogenic mechanisms comes primarily from geothermal heat, although small contributions are also made by tidal energy generated by the gravitational attraction of the Sun and the Moon and by rotational energy derived from the momentum of the Earth's rotation. The ultimate sources of energy for exogenic processes are the potential energy arising from the height of material above base level, and that proportion of solar radiation received by the Earth.

1.5.1.1 *Internal energy*

Evidence of the Earth's internal energy is provided by the **geothermal heat flow** which can be measured, and in some cases observed, at the surface. Volcanoes, of course, represent local areas of higher than average heat flow, but the Earth's internal heat is also the ultimate source of energy for virtually all tectonic processes and the associated horizontal and vertical movements of the crust. Although the energy involved in the various endogenic mechanisms is not known, the present rate of heat flow to the surface is fairly

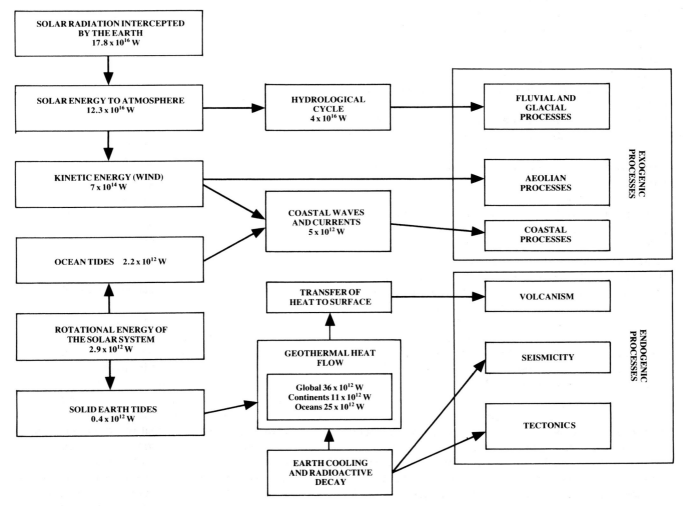

Fig. 1.13 *Estimated energy flows relevant to various geomorphic processes (data from various sources).*

well established. It can be measured from the increase in temperature with depth (averaging around 20–30 °C km⁻¹), although allowances must be made for the thermal conductivity of the rock. Heat flow rates within the continents average around $56\,mW\,m^{-2}$ and range up to $200\,mW\,m^{-2}$ in areas of active volcanic activity. In the ocean basins the average is about $78\,mW\,m^{-2}$, but rates as high as $250\,mW\,m^{-2}$ have been recorded. These variations in heat flow provide important clues as to the nature of the Earth's interior and the operation of endogenic processes.

The major source of geothermal heat is the radioactive decay of the long-lived isotopes of uranium, thorium and potassium. About 83 per cent of the heat flow to the Earth's surface is attributable to this process, the remainder being provided by the continued cooling of the Earth which has been proceeding since its formation some 4.6 Ga ago. Because the half-lives of the major heat-producing isotopes are in the range 10^9–10^{10} a the supply of heat from radioactive decay has been more or less constant over the past several hundred million years. Eventually, though, this heat supply will gradually diminish and the endogenic processes arising from it will become less active and ultimately cease. Here, then, we have a system which, although in a steady state for a geologically significant period of time, is in fact, in the very long term, moving towards a decay equilibrium.

1.5.1.2   Solar radiation

Solar radiation provides an enormous source of energy, but only a very small proportion of this is utilized in the operation of geomorphic processes. The upper atmosphere intercepts about $17.8 \times 10^{16}\,W$ of largely short-wave radiant energy from the Sun of which about 30 per cent is immediately reflected back into space (Fig. 1.13). The remainder heats the atmosphere and the surface and generates a mean global surface temperature of about 15 °C. Of the $12.3 \times 10^{16}\,W$ of solar energy received by the atmosphere a significant amount (about 33 per cent) drives the **hydrological cycle** (see Section 1.5.2), the continuous movement of water in its gaseous, liquid and solid states between and within the atmosphere, oceans and landsurface. Latent heat is absorbed by the vaporization (evaporation) of water from the surface of the oceans and continents and through transpiration (the loss of water vapour from plant cells) by vegetation. A small proportion (about 1 per cent) is converted into kinetic energy and powers the circulation of the air in the atmosphere (winds) and water in the oceans (ocean currents). A further minute, but highly significant, proportion (about 0.1 per cent) is consumed in photosynthesis (the fixation of radiant energy by plants).

The receipt of solar radiation varies over the Earth's surface both temporally and spatially. Taking a global view there is an excess of incoming over outgoing radiation between about latitude 40° N and S and a deficit polewards of these latitudes. The resulting latitudinal temperature gradient gives rise to the general circulation of the atmosphere which contributes, along with ocean currents, to the redistribution of heat from the equator towards the poles and is the major factor in the climatic zonation of the Earth. At the regional scale marked temperature differences are generated by the contrasting thermal properties of the continents and oceans, with the higher heat capacity of the oceans helping to moderate in coastal areas the extremes of temperature characteristic of continental interiors. At a local scale altitude becomes a primary factor affecting the heat budget. With increasing elevation there is a progressively greater heat loss through long-wave radiation from the Earth's surface. This leads to an overall decrease in mean temperature but an increase in diurnal range.

1.5.2   The hydrological cycle

The hydrological cycle can be conceived as a system of storages between which water is transferred. The oceans represent by a considerable margin the largest storage, but ice sheets and glaciers account for a significant proportion of fresh water (Fig. 1.14). Of more geomorphic significance are the magnitudes of the transfers between storages. Approximately $517 \times 10^3\,km^3$ of water is annually evaporated and reprecipitated over the globe as a whole; this is roughly equivalent to a layer of water 1 m thick over the entire surface of the Earth. Water is transferred to the atmosphere as a vapour by evaporation from the oceans and a combination of evaporation and transpiration (**evapotranspiration**) from the continents, and is returned to the surface in a liquid or solid state as various forms of precipitation (Fig. 1.14). The significant excess of precipitation over evapotranspiration on the continents (around 40 per cent) means that large amounts of water are returned to the oceans as surface runoff. Although only 0.0001 per cent of all water (and 0.004 per cent of fresh water) is to be found in rivers at any one time, runoff is by far the most important element of the hydrological cycle in terms of landform genesis.

In addition to transfers between major storages there are movements within them (that is, transfers between substorages). Most precipitation, for instance enters the soil water or ground water storage before returning to the oceans as surface runoff. The rate at which water moves through the hydrological cycle also varies significantly. Water which becomes part of the shallow ground water storage may remain there for several years while water entering deep ground water storage may be removed from the active circulation of the hydrological cycle for hundreds of thousands of years. Similarly, water frozen into large ice sheets may travel hundreds of kilometres over tens of thousands of years before eventually returning to the ocean.

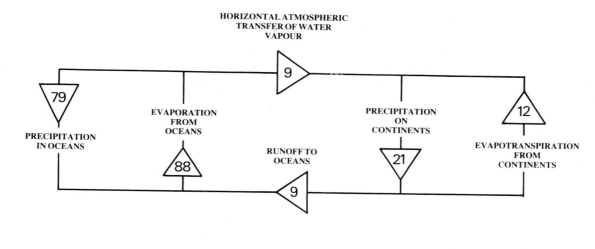

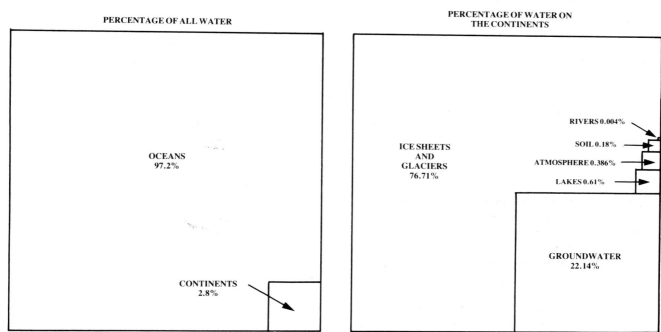

Fig. 1.14 *The hydrological cycle. The figures representing transfers are percentages of the mean annual global precipitation of approximately 1000 mm. The proportion of water in the various storages is also illustrated. (Data from various sources.)*

1.5.3 Climatic controls

The wide range of climatic conditions over the Earth's surface, together with the role of climatic factors in influencing the nature and rate of operation of geomorphic processes, has led some geomorphologists to suggest that different climates are associated with characteristic landform assemblages. More specifically, it has been argued that particular climatic conditions can have an effect on geomorphic processes sufficient to outweigh the influence on landform development of differences in tectonic setting, rock type and relief, and over a period of time can generate a distinctive association of landforms of regional extent. This is the essential assumption of climatic geomorphology which seeks to establish the nature of landforms associated with distinctive climatic regimes and to identify the specific combinations of geomorphic processes which give rise to them. Areas characterized by landforms associated with a particular climatic environment are called **morphoclimatic zones** (or regions). The term morphogenetic region is also used, but this is arguably less appropriate as it implies that climatic factors alone control landform genesis.

The validity of the assumptions behind climatic geomorphology have been severely challenged by many geomorphologists who see little evidence of a close relationship between climate and landform morphology except under

Table 1.4 The Earth's major morphoclimatic zones (based on various sources)

MORPHOCLIMATIC ZONE	MEAN ANNUAL TEMPERATURE (°C)	MEAN ANNUAL PRECIPITATION (mm)	RELATIVE IMPORTANCE OF GEOMORPHIC PROCESSESS
Humid tropical	20–30	>1500	High potential rates of chemical weathering; mechanical weathering limited; active, highly episodic mass movement; moderate to low rates of stream corrasion but locally high rates of dissolved and suspended load transport
Tropical wet–dry	20–30	600–1500	Chemical weathering active during wet season; rates of mechanical weathering low to moderate; mass movement fairly active; fluvial action high during wet season with overland and channel flow; wind action generally minimal but locally moderate in dry season.
Tropical semi–arid	10–30	300–600	Chemical weathering rates moderate to low; mechanical weathering locally active especially on drier and cooler margins; mass movement locally active but sporadic; fluvial action rates high but episodic; wind action moderate to high
Tropical arid	10–30	0–300	Mechanical weathering rates high (especially salt weathering); chemical weathering minimal; mass movement minimal; rates of fluvial activity generally very low but sporadically high; wind action at a maximum
Humid mid–latitude	0–20	400–1800	Chemical weathering rates moderate, increasing to high at lower latitudes; mechanical weathering activity moderate with frost action important at higher latitudes; mass movement activity moderate to high; moderate rates of fluvial processes: wind action confined to coasts
Dry continental	0–10	100–400	Chemical weathering rates low to moderate; mechanical weathering, especially frost action, seasonally active; mass movement moderate and episodic; fluvial processes active in wet season; wind action locally moderate
Periglacial	<0	100–1000	Mechanical weathering very active with frost action at a maximum; chemical weathering rates low to moderate; mass movement very active; fluvial processes seasonally active; wind action rates locally high
Glacial	<0	0–1000	Mechanical weathering rates (especially frost action) high; chemical weathering rates low; mass movement rates low except locally; fluvial action confined to seasonal melt; glacial action at a maximum; wind action significant
Azonal mountain zone	Highly variable	Highly variable	Rates of all processes vary significantly with altitude; mechanical and glacial action become significant at high elevations

the most extreme climatic contrasts. Suffice it to say here that with the exception of landforms which have very short relaxation times, most elements of the landscape are likely to be, to a greater or lesser extent, out of equilibrium with prevailing climatic conditions because of the rapidity and magnitude of global climatic changes that have occurred, especially over the past 2–3 Ma. Indeed in some regions, such as central Australia and central southern Africa, there is generally a very low rate of geomorphic activity and the landscape is dominated by **relict landforms** developed under climates quite different from those prevailing now.

Notwithstanding uncertainties about the precise relationship between climate and landform genesis, it is clear that there are major contrasts in the kind of geomorphic processes active under certain climatic regimes and that all climatically related processes vary in the intensity with which they operate from one climatic region to another (Table 1.4). Indeed a number of attempts have been made to produce global maps of morphoclimatic zones, and Fig. 1.15 shows a modified version of the widely cited map by the French geomorphologists J. Tricart and A. Cailleux.

The boundaries between morphoclimatic zones are somewhat arbitrary, but some climatic parameters are rather specifically related to the operation of particular geomorphic processes. Frost action can obviously occur only where

ground temperatures fall below freezing, while permanently frozen ground will only develop where the mean annual temperature is at least below 0 °C (Fig. 1.16(A)). Wind action is most active and widespread in arid regions within a zone fairly well defined by the 200 mm mean annual isohyet (Fig. 1.16(B)), although wind can be an important process on coasts in all climatic zones. Global variations in precipitation undoubtedly have some influence on rates of weathering and fluvial activity, but the frequency of high-intensity rainfall events is probably far more significant in affecting rates of erosion than overall rainfall amounts (Fig. 1.16(B)).

1.5.4   Human agency

Landforms, and the processes that create them, impinge on human activity in many ways. High magnitude geomorphic events, such as large landslides or major floods, become natural hazards if they affect people. Whether this happens depends on the distribution of people and geomorphic events in time and space. By examining the landforms themselves it is often possible to tell which areas are likely to be at risk from high-magnitude–low-frequency geomorphic events, while a knowledge of the conditions that occur immediately before a major event, such as high-intensity

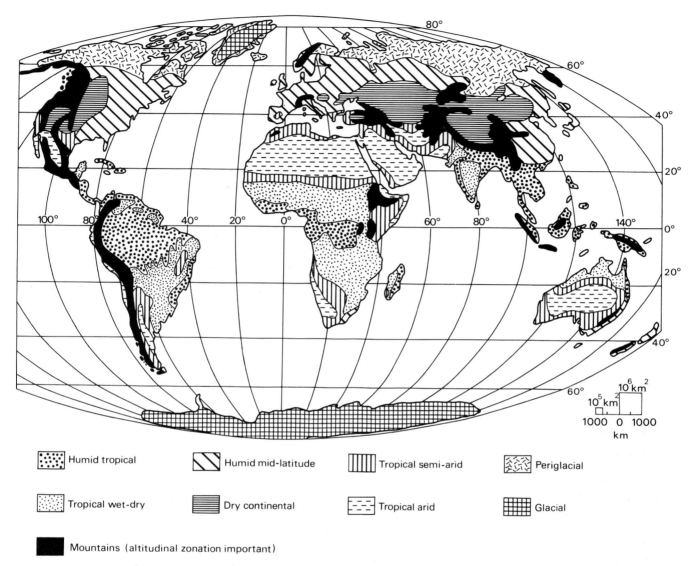

Fig. 1.15 *Global distribution of morphoclimatic zones (see Table 1.4 for details of climatic parameters and geomorphic processes.) (Modified from J. Tricart and A. Cailleux (1972)* Introduction to Climatic Geomorphology. *Longman, London, Map III.)*

rainfall preceding a flood, can provide the opportunity to issue warnings for evacuation. A knowledge of geomorphology can also contribute to environmental management in regions where special account has to be taken of particular geomorphic phenomena; examples include the impact of pipeline and road construction on permanently frozen ground in high-latitude environments, and the encroachment of sand dunes on human settlements in arid regions. In some cases an understanding of the factors controlling geomorphic processes can be directly relevant to engineering problems, an example being the application of geomorphic research on weathering to the problem of the deterioration of buildings in arid regions due to the presence of salts. Finally, a knowledge of landform genesis can assist in prospecting for certain kinds of mineral resources.

This book does not specifically consider these topics

which together constitute the subdiscipline of applied geomorphology. None the less, it is vital to take account of the way human activity itself has affected the natural landscape. Human agency as a factor in landform genesis began with the first fires lit by our ancestors which burnt the vegetation and temporarily left the soil more vulnerable to erosion. The impact of human activity on the landscape has been increasing at an accelerating rate, especially since the Industrial Revolution and the introduction of modern farming techniques.

Human activity can affect the landscape in two ways; by the direct creation of new landforms, and by an alteration in the rate at which geomorphic processes operate. Examples of the creation of new features in the landscape include reservoirs formed by the construction of dams (Fig. 1.17), and the effects of large-scale strip mining. In the

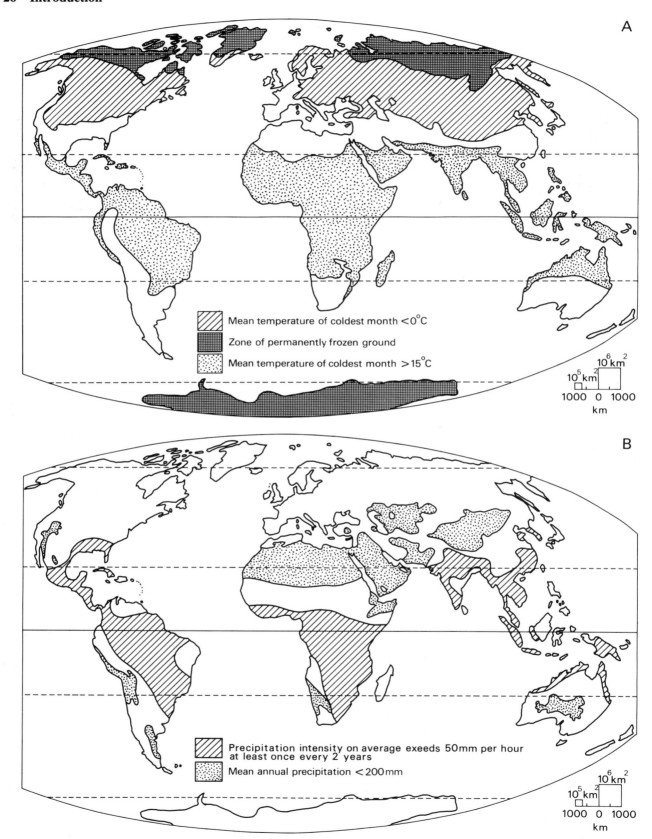

Fig. 1.16 *Global distribution of some key climatic variables: (A) important temperature limits; (B) areas experiencing high precipitation intensities and regions with low mean annual precipitation. (Based on data in R. Common (1966) in: G. H. Dury (ed.)* Essays in Geomorphology. *Heinemann, London 53–81; and B. M. Reich, 1963,* Journal of Hydrology *1, 3–28).*

Fig. 1.17 Oblique aerial view of the western end of Lake Mead on the Arizona–Nevada border, USA, a large reservoir formed through the impoundment of the Colorado River by the Hoover Dam (visible in the centre of the photograph).

Fig. 1.18 Terraced hillsides in southern Cyprus dating back to at least the period of Roman occupation.

are in danger of interpreting the past in terms of an atypical present.

Further reading

A wide range of material is available for those interested in the history of geomorphology. An excellent recent survey is that by Tinkler (1985) while Chorley *et al.* (1964) provide a detailed assessment of developments up to the turn of the century. The second volume of this work (Chorley *et al.*, 1973) is devoted to W. M. Davis but two further planned volumes have yet to appear. A valuable study of the early years of geomorphology in Britain is that by Davies (1969) who has also recently presented a sober re-evaluation of Hutton's original contribution (Davies, 1985). Gould (1987) contains brilliant and highly readable essays exploring the conceptions of time embodied in the models of Earth history presented by Hutton and Lyell; this is required reading for those who really wish to grasp the philosophical framework of landform analysis established in the late eighteenth and early nineteenth centuries. Rudwick (1985) is (in spite of its title) also worth consulting for its discussion of the various meanings of uniformitarianism.

G. K. Gilbert is the subject of an excellent biographical study by Pyne (1980) emphasizing his methodological approach; briefer assessments of Gilbert's work are provided by Pyne (1975) and Yochelson (1980) while his key monographs are Gilbert (1877, 1914). Other original works of particular importance include: Hutton (1788) (contained in Eyles, 1970); Playfair (1802), which is available as a facsimile edition (White, 1956); Lyell (1830–33); Agassiz (1840), available in an English translation (Carozzi, 1967); Davis (1909); and Penck (1924), which is also available in translation (Penck, 1953). More recent developments, with a particular emphasis on fluvial geomorphology are reviewed by Gregory (1985), while Horton's seminal paper is well worth consulting (Horton, 1945).

Appalachian region of the U S A strip mining has produced benches on hillsides formed of rock waste which, it has been estimated, extend for a total distance of around 30 000 km. The modification of hillslopes is not, however, only a recent phenomenon, as a visit to many areas in the Mediterranean region would demonstrate (Fig. 1.18).

Of greater global significance than the direct creation of landforms is the effect of human activities on the rates of geomorphic processes. Agriculture, lumbering and the construction of settlements can all cause dramatic changes in land use which in turn may greatly change the susceptibility of the landsurface to erosion. Not surprisingly, the extension of activities such as agriculture into new regions over the past 100 years or so has led to some dramatic changes in the quantity of sediment being transported by rivers in the areas affected. Fluvial systems have, in many instances, undergone dramatic and complex adjustments as a result. If we are to understand how the natural landscape behaves it is vital that these effects are taken into account, otherwise we

Recommended reading on the various topics considered in the discussion of future directions in geomorphology is indicated in the various chapters which develop these themes, but it is worth mentioning here the thought-provoking essays by Baker (1986) and Hayden *et al.* (1986), the assessment of impact cratering on the Earth by Grieve (1987), the review of comparative planetary geomorphology by Sharp (1980) and the excellent case study of interpreting planetary landforms in terms of terrestrial analogues contained in Howard *et al.* (1988)

Thorn (1988) provides a useful introduction to concepts in geomorphology (although I feel he severely underplays the importance of endogenic factors in landscape development) while Baker (1988), Chorley (1978), Ritter (1988) and Scheidegger (1987) discuss from various perspectives key principles and methodological approaches. On specific concepts in geomorphology, Scheidegger (1979) considers the 'antagonism' between endogenic and exogenic processes and Summerfield (1981) briefly discusses their relative significance as the spatial scale is altered. Magnitude–frequency relationships are treated in a classic paper by Wolman and Miller (1960) and reconsidered by Wolman and Gerson (1978). The application of systems analysis to landform studies is discussed by Chorley (1962), Chorley and Kennedy (1971) and Huggett (1985). The problem of temporal scale in geomorphology is treated from various perspectives by Cullingford *et al.* (1980), Montgomery (1989), Schumm and Lichty (1965), Thorn (1982) and Thornes and Brunsden (1977). The related issue of appropriate conceptions of equilibrium in geomorphic research is considered by Chorley and Kennedy (1971) and Schumm (1977) (especially Chapter 1), although it is important to be aware of the different uses of the terms dynamic equilibrium and steady state in the latter in comparison with some earlier work. The modern usage of the term uniformitarianism is critically evaluated by Gould (1965) and Simpson (1970) provides a detailed discussion which introduces the notion of the immanent and configurational aspects of reality. There are relatively few general treatments of the problem of landform change through time, but Thornes and Brunsden (1977) cover a broad range of issues.

On methods of analysis Gardiner and Dackombe (1981) and Goudie (1981) between them comprehensively cover field and laboratory techniques in geomorphology. Schumm's detailed investigation of the evolution of microscale badland topography (Schumm, 1956) and the study by Schumm and Chorley (1964) of the fall of Threatening Rock provide classic examples of the value of direct observations. The general principles of space–time substitution are thoroughly reviewed by Paine (1985) while Brunsden and Kesel (1973) provide an example of the application of this approach. A useful introduction to simulation modelling is provided by Chorley and Kennedy (1971) and examples of the application of simulation modelling to fluvial and coastal land-

forms are presented by Howard (1971) and King and McCullagh (1971). Mosley and Zimpfer (1978) and Schumm *et al.* (1987) provide examples of the use of hardware models and Anderson (1988) contains detailed presentations of mathematical models.

The Earth's internal energy and the present global heat flow are treated at an introductory level by Pollack and Chapman (1977) while Williams (1982) outlines the surface energy budget (see especially Chapters 1 and 3). Barry (1969) provides a useful summary of the global hydrological cycle, although some of the data quoted have subsequently been revised. The principles of climatic geomorphology are critically summarized by Stoddart (1969) and presented in detail by Büdel (1982) and Tricart and Cailleux (1972). There is a large literature on applied geomorphology and the impact of human activities on the landscape with useful starting points being provided by Coates (1981), Cooke and Doornkamp (1974), Craig and Craft (1980), Gregory and Walling (1987) and Hails (1977).

Finally, it is appropriate here to mention some of the more important general sources of research and reference information in geomorphology. The four major journals covering geomorphic research are *Catena, Earth Surface Processes and Landforms, Geomorphology* and *Zeitschrift für Geomorphologie*, although articles on geomorphology are to be found in a wide range of geography and earth science journals. The reviews of recent developments in various fields within geomorphology contained in *Progress in Physical Geography* are particularly useful. Fairbridge (1968) is still a valuable reference work, but this has recently been supplemented by the less detailed but broader Goudie (1985). Snead (1981) is a useful geomorphic atlas while Short and Blair (1986) is an invaluable source of images of landforms taken from space with detailed accompanying commentaries.

References

Agassiz, L. J. (1840) *Etudes sur les Glaciers*. Neuchâtel.
Anderson, M. G. (ed.) (1988) *Modelling Geomorphological Systems*. Wiley, Chichester and New York.
Baker, V. R. (1986) Introduction: Regional landform analysis. In: N.M. Short and R. W. Blair Jr (eds) *Geomorphology From Space: A Global Overview of Regional Landforms*. NASA, Washington D C, 1–26.
Baker, V. R. (1988) Geological fluvial geomorphology. *Geological Society of America Bulletin* **100**, 1157–1167.
Barry, R. G. (1969) The world hydrological cycle. In: R. J. Chorley (ed.) *Water, Earth and Man*. Methuen, London; Barnes and Noble, New York, 8–26.
Brunsden, D. and Kesel, R.H. (1973) The evolution of a Mississippi river bluff in historic time. *Journal of Geology* **81**, 576–97.
Büdel, J. (1982) *Climatic Geomorphology* (translated by L. Fischer and D. Busche). Princeton University Press, Princeton and Guildford.
Carozzi, A. V. (1967) *Studies on Glaciers by Louis Agassiz*. Hafner, New York.

Chorley, R. J. (1962) Geomorphology and general systems theory. *United States Geological Survey Professional Paper* 500B.

Chorley, R. J. (1978) Bases for theory in geomorphology. In: C. Embleton, D. Brunsden and D. K. C. Jones (eds) *Geomorphology: Present Problems and Future Prospects*. Oxford University Press, Oxford and New York, 1–24.

Chorley, R. J., Beckinsale, R. P. and Dunn, A. J. (1973) *The History of the Study of Landforms or the Development of Geomorphology*. Volume Two: *The Life and Work of William Morris Davis*. Methuen, London; Harper and Row, New York.

Chorley, R. J., Dunn, A. J. and Beckinsale, R. P. (1964) *The History of the Study of Landforms or the Development of Geomorphology*. Volume One: *Geomorphology Before Davis*. Methuen, London; Wiley, New York.

Chorley, R. J. and Kennedy, B. A. (1971) *Physical Geography: A Systems Approach*. Prentice-Hall, London and Englewood Cliffs.

Coates, D. R. (1981) *Environmental Geology*. Wiley, New York and Chichester.

Cooke, R. U. and Doornkamp, J. C. (1974) *Geomorphology in Environmental Management: An Introduction*. Clarendon Press, Oxford and New York.

Craig, R. G. and Craft, J. L. (eds) (1980) *Applied Geomorphology*. Allen and Unwin, London and Boston.

Cullingford, R. A., Davidson, D. A. and Lewin, J. (eds) (1980) *Timescales in Geomorphology*. Wiley, Chichester and New York.

Davies, G. L. (1969) *The Earth in Decay: A History of British Geomorphology 1578–1878*. MacDonald, London.

Davies, G. L. H. (1985) James Hutton and the study of landforms. *Progress in Physical Geography* 9, 382–9.

Davis, W. M. (1909) *Geographical Essays* (edited by D. W. Johnson). Ginn, Boston.

Eyles, V. A. (1970) *James Hutton's 'System of the Earth, 1785; Theory of the Earth, 1788; Observations on Granite, 1794'*. Hafner, Darien.

Fairbridge, R. W. (ed.) (1968) *Encyclopedia of Geomorphology*. Reinhold, New York.

Gardiner, V. and Dackombe, R. V. (1981) *Geomorphological Field Manual*. Allen and Unwin, London and Boston.

Gilbert, G. K. (1877) *Report on the Geology of the Henry Mountains*. United States Department of the Interior, Washington D C.

Gilbert, G. K. (1914) The transportation of debris by running water. *United States Geological Survey Professional Paper* 86.

Goudie, A. (ed.) (1981) *Geomorphological Techniques*. Allen and Unwin, London and Boston,

Goudie, A. (ed.) (1985) *The Encyclopaedic Dictionary of Physical Geography*. Blackwell, Oxford and New York.

Gould, S. J. (1965) Is uniformitarianism necessary? *American Journal of Science* 263, 223–8.

Gould, S. J. (1987) *Time's Arrow, Time's Cycle: Myth and Metaphor in the Discovery of Geological Time*. Harvard University Press, Cambridge and London.

Gregory, K. J. (1985) *The Nature of Physical Geography*. Arnold, London.

Gregory, K. J. and Walling, D. E. (eds) (1987) *Human Activity and Environmental Processes*. Wiley, Chichester and New York.

Grieve, R. A. F. (1987) Terrestrial impact structures. *Annual Review of Earth and Planetary Sciences* 15, 245–70.

Hails, J. R. (ed.) (1977) *Applied Geomorphology*. Elsevier, Amsterdam and Oxford.

Hayden, R. S., Blair, R. W. Jr, Garvin, J. and Short, N. M. (1986) Global geomorphology: outlook for the future. In: N. M. Short and R. W. Blair Jr (eds) *Geomorphology From Space: A Global Overview of Regional Landforms*. NASA, Washington D C., 657–72.

Horton, R. E. (1945) Erosional development of streams and their drainage basins: hydrophysical approach to quantitative morphology. *Bulletin of the Geological Society of America* 56, 275–370.

Howard, A. D. (1971) Simulation model of stream capture. *Geological Society of America Bulletin* 82, 1355–76.

Howard, A. D. and Kochel, R. C. (1988) Introduction to cuesta landforms and sapping processes on the Colorado Plateau. In: A. D. Howard, R. C. Kochel and H. R. Holt (eds) *Sapping Features of the Colorado Plateau* NASA SP–491, NASA, Washington, 6–56.

Huggett, R. J. (1985) *Earth Surface Systems*. Springer-Verlag, Berlin and New York.

Hutton, J. (1788) Theory of the earth; or an investigation of the laws observable in the composition, dissolution, and restoration of land upon the globe. *Transactions of the Royal Society of Edinburgh* 1, 209–304.

King, C. A. M. and McCullagh, M. J. (1971) A simulation model of a complex recurved spit. *Journal of Geology* 79, 22–37.

Lyell, C. (1830–33) *Principles of Geology* (3 vols) Murray, London.

Montgomery, K. (1989) Concepts of equilibrium and evolution in geomorphology: the model of branch systems. *Progress in Physical Geography* 13, 47–66.

Mosley, M. P. and Zimpfer, G. L. (1978) Hardware models in geomorphology. *Progress in Physical Geography* 2, 461–83.

Paine, A. D. M. (1985) 'Ergodic' reasoning in geomorphology: time for a review of the term? *Progress in Physical Geography* 9, 1–15.

Penck, W. (1924) *Die Morphologische Analyse: Ein Kapital der Physikalischen Geologie*. Engelhorn, Stuttgart.

Penck, W. (1953) *Morphological Analysis of Landforms* (translated by H. Czech and K. C. Boswell) Macmillan, London.

Playfair, J. (1802) *Illustrations of the Huttonian Theory of the Earth*. William Creech, Edinburgh.

Pollack, H. N. and Chapman, D. S. (1977) The flow of heat from the earth's interior. *Scientific American* 237(2), 60–76.

Pyne, S. (1975) The mind of Grove Karl Gilbert. In: W. N. Melhorn and R. C. Flemal (eds) *Theories of Landform Development*. Allen and Unwin, London and Boston, 277–98.

Pyne, S. J. (1980) *Grove Karl Gilbert: A Great Engine of Research*. University of Texas Press, Austin.

Ritter, D. F. (1988) Landscape analysis and the search for geomorphic unity. *Geological Society of America Bulletin* 100, 160–71.

Rudwick, M. J. S. (1985) *The Meaning of Fossils: Episodes in the History of Palaeontology* (2nd edn) University of Chicago Press, Chicago and London.

Scheidegger, A. E. (1979) The principle of antagonism in the Earth's evolution. *Tectonophysics* 55, T7–T10.

Scheidegger, A. E. (1987) The fundamental principles of landscape evolution. *Catena Supplement* 10, 199–210.

Schumm, S. A. (1956) Evolution of drainage systems and slopes in badlands at Perth Amboy, New Jersey. *Bulletin of the Geological Society of America* 67, 597–646.

Schumm, S. A. (1977) *The Fluvial System*. Wiley, New York and London.

Schumm, S. A. and Chorley, R. J. (1964) The fall of Threatening Rock. *American Journal of Science* 262, 1041–54.

Schumm, S. A. and Lichty, R. W. (1965) Time, space and causality in geomorphology. *American Journal of Science* 263, 110-119.

Schumm, S. A., Mosley, M. P. and Weaver, W. E. (1987)

Experimental Fluvial Geomorphology. Wiley, New York and Chichester.

Sharp, R. P. (1980) Geomorphological processes on terrestrial planetary surfaces. *Annual Review of Earth and Planetary Sciences* **8**, 231–61.

Short, N. M. and Blair, R. W. Jr (eds) (1986) *Geomorphology From Space: A Global Overview of Regional Landforms.* NASA, Washington D C.

Simpson, G. G. (1970) Uniformitarianism. An inquiry into principle, theory, and method in geohistory and biohistory. In: M. K. Hecht and W. C. Steere (eds) *Essays in Evolution and Genetics in Honor of Theodosius Dobzhansky.* Appleton-Century-Crofts, New York, 43–96.

Snead, R. E. (1981) *World Atlas of Geomorphic Features.* Reinhold, New York.

Stoddart, D. (1969) Climatic geomorphology: review and re-assessment. *Progress in Geography* **1**, 159–222.

Summerfield, M. A. (1981) Macroscale geomorphology. *Area* **13**, 3–8.

Thorn, C. E. (ed.) (1982) *Space and Time in Geomorphology.* Allen and Unwin, Boston and London.

Thorn, C. E. (1988) *An Introduction to Theoretical Geomorphology.* Unwin Hyman, Boston and London.

Thornes, J. B. and Brunsden, D. (1977) *Geomorphology and Time.* Methuen, London.

Tinkler, K. J. (1985) *A Short History of Geomorphology.* Croom Helm, London.

Tricart, J. and Cailleux, A. (1972) *Introduction to Climatic Geomorphology* (translated by C. J. Kiewiet de Jonge). Longman, London.

White, G. W. (1956) *'Illustrations of the Huttonian Theory of the Earth', by John Playfair.* University of Illinois Press, Urbana.

Williams, P. J. (1982) *The Surface of the Earth: An Introduction to Geotechnical Science.* Longman, London and New York.

Wolman, M. G. and Gerson, R. (1978) Relative scales of time and effectiveness of climate in watershed geomorphology. *Earth Surface Processes* **3**, 189–208.

Wolman, M. G. and Miller, J. P. (1960) Magnitude and frequency of forces in geomorphic processes. *Journal of Geology* **68**, 54–74.

Yochelson, E. L. (ed.) (1980) The scientific ideas of G. K. Gilbert. *Geological Society of America Special Paper* 183.

Part II

Endogenic processes and landforms

2

Global morphology and tectonics

2.1 Global morphology

2.1.1 The geoid

The Earth is only approximately a sphere. As a result of the centrifugal force of rotation it bulges at the equator and its polar radius (6378 km) is 21 km shorter than its equatorial radius (6397 km); thus the Earth is more accurately described as an oblate spheroid. Even this description, how-ever, is not completely accurate as inhomogeneities in the distribution of mass in the Earth's interior produce further small but measurable irregularities on its surface. These irregularities have been determined with great precision over recent years through the very accurate measurement of deviations in the orbits of artificial satellites.

The shape of the Earth determined in this way is known as the **geoid** and is represented by the surface defined by mean sea level over the oceans and the extension of sea level along imaginary canals across the continents. As is evident in Figure 2.1 this surface shows many irregularities compared with a simple oblate spheroid; there is, for instance, a bulge of 76 m near New Guinea and a depression of some 104 m to the south of India. The sig-nificance of the geoid for the operation of geomorphic processes is that it defines the ultimate base level for denudation; any change in the geoid would therefore cause consequential changes in base levels. This is an important issue in studies of sea-level change and is examined further in Chapter 17.

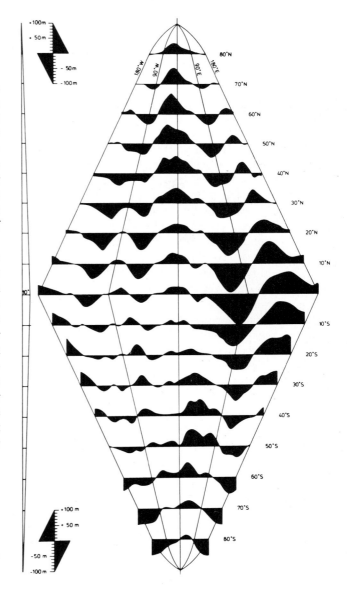

Fig. 2.1 *Geodetic sea level profiles at intervals of 10° latitude. The horizontal latitudinal lines separate positive geoid values (above) from negative geoid values (below). (From N. –A. Mörner (1976),* Journal of Geology **84**, *Fig. 1, p. 124.)*

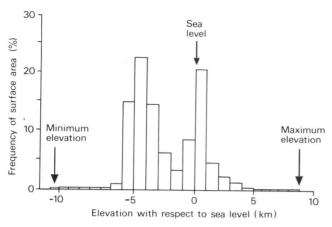

Fig. 2.2 *Proportionate areal distribution of the solid surface of the Earth between successive elevations. Note the two peaks representing the ocean basins and the continental platforms. (Modified from P. J. Wyllie 1976, The Way the Earth Works. Wiley, New York, Fig. 3–11(a), p. 38.*

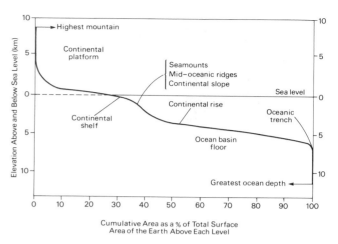

Fig. 2.3 *Global hypsometric curve indicating the main morphological features characterizing the Earth's surface at different elevations. (Modified from P. J. Wullie (1976) The Way the Earth Works. Wiley, New York, Fig. 3–11(b), p. 38.).*

2.1.2 Global hypsometry

If the overall distribution of height of the Earth's solid surface above and below sea level is plotted as a frequency distribution its most obvious feature is the bimodal form of the curve (Fig. 2.2). The two peaks in the distribution represent the **continental platforms** and the **ocean basins**. Between these two maxima there is a lower frequency of depths representing the more shallow regions of the oceans. At either end of the histogram are thin tails indicating that very little of the Earth's solid surface is found more than 5 km above, or 6 km below, mean sea level. The extremes of the distribution mark the greatest elevation of 8.8 km above sea level (Mount Everest) and the greatest depression at more than 11 km below sea level (the Marianas Trench in the western Pacific Ocean).

Other features are more clearly revealed if the data are plotted as a cumulative frequency distribution indicating the proportion of the surface above any given height (Fig. 2.3). This produces a **hypsometric curve** which shows the elevations of those regions lying between the continental platforms and the ocean basins and suggests that the true break between the continents and oceans lies not at present sea level but some 200 m or so lower. As defined by present mean sea level the continents account for 29 per cent of the Earth's surface area and the oceans 71 per cent, but if the break of slope represented on the global hypsometric curve is used the figure for the continents is nearer 37 per cent.

2.1.3 Major morphological features

Although the global hypsometric curve provides a valuable statistical summary of the Earth's relief as a whole it is also necessary to identify the major components of global morphology that give rise to the curve (Fig. 2.4). Beginning with the more familiar form of the continents we can distinguish between the continental platforms formed by plateaus and lowlands, and the major linear mountain systems known as **orogenic mountain belts**, or simply **orogens**. One major mountain system cuts across central and southern Eurasia extending from the Alps in the west, through the Himalayas to the mountains of western China. Another, formed by the Andes and the North American Cordillera, runs along the entire western margin of the Americas. Other older and more subdued ranges include the Appalachians in eastern North America and the Urals in western Eurasia. Extensive areas within continental platforms are formed of **basement**, a complex of metamorphic and igneous rocks of Palaeozic or Precambrian age. In some localities, notably East Africa, continental platforms are traversed by **rift valleys** consisting of linear troughs formed by the subsidence of crust between parallel systems of faults.

The submarine extension of a continent is called the **continental shelf**. This is bordered by a **continental slope** which inclines at an angle of around 3–6° towards the ocean basin and which is separated from it by a **continental rise** (Fig. 2.5). The ocean basins themselves are traversed by a vast system of **mid-oceanic ridges** up to 1000 km in width and tens of thousands of kilometres in length. On average they rise some 2 km above the surrounding ocean floor but in a few localities, notably at Iceland, they reach above sea level. Although the surface of the ocean basins is relatively uniform it is punctuated in places by volcanoes. Some of these rise above sea level, but the vast majority are submarine features known as **seamounts**, or if they are flat-topped, **guyots**. The greatest ocean depths are

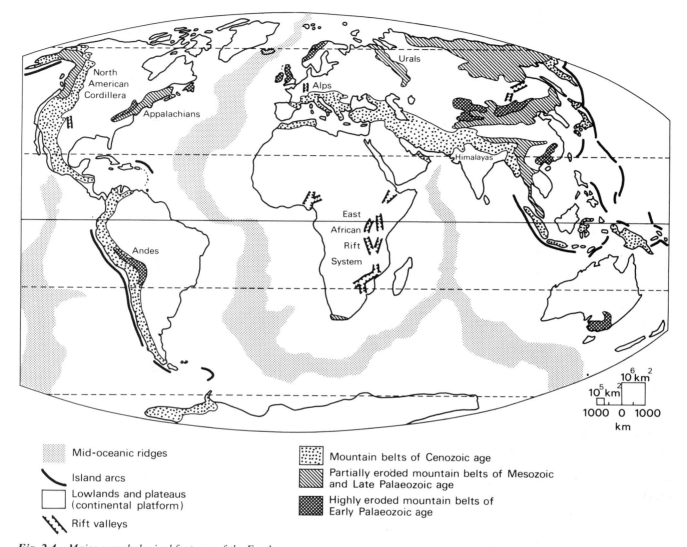

Fig. 2.4 Major morphological features of the Earth.

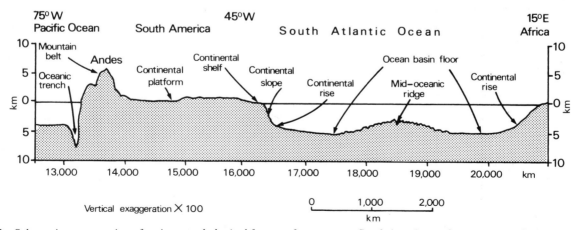

Fig. 2.5 Schematic cross-section of major morphological features from western South America to the west coast of Africa at the latitude of the Tropic of Cancer. Note that the curvature of the Earth is not shown. (Modified from P. J. Wyllie (1976) The Way the Earth Works. Wiley, New York, Fig. 3–10, pp. 36–7.)

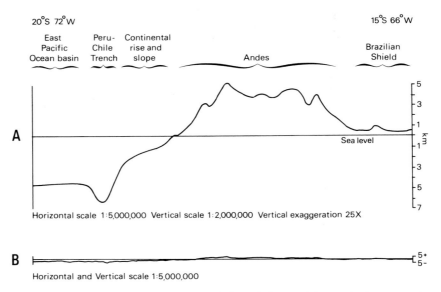

Fig. 2.6 *Cross-section of the Andes Mountains and the Peru–Chile Trench drawn with a vertical exaggeration of 25 times (A), and with the same horizontal and vertical scale (B).*

achieved in **oceanic trenches** which are to be found both along some continental margins, notably western South America (Fig. 2.6), and in association with **island arcs** which are usually located close to the edges of continents.

The relief exhibited by the highest mountains and the deepest oceanic trenches, although apparently spectacular, is in fact trivial when compared with the dimensions of the Earth as a whole. The maximum difference in altitude associated with the Earth's solid relief is about 20 km, or only 0.31 per cent of the planet's mean diameter. This is why substantial vertical exaggeration must nearly always be employed in the depiction of extensive areas of relief in cross-section (Fig. 2.6). Only with the most dramatic topography can large-scale relief forms be clearly drawn with identical horizontal and vertical scales.

2.2 Earth structure

2.2.1 Seismic evidence

Studies of the way in which earthquake waves are transmitted through the Earth have shown that its interior has a concentric structure. When an earthquake occurs the energy released is transmitted in wave form in all directions. The velocities of earthquake (seismic) waves, together with the paths they follow, are determined by the properties of the material through which they pass. Two major types of wave are produced and these travel at different speeds. Primary, or **P-waves**, are the fastest. They are compressional waves, the energy being transmitted by an initial compression of particles which is then passed on to adjacent particles. This produces a sequence of zones of compression and expansion which travel away from the source. Secondary, or

S-waves, are shear waves and they transmit energy by an 'up and down' motion of particles normal to the direction of propagation. The density and elastic properties of rock are both important in affecting the passage of seismic waves. P-waves can be transmitted through any material (sound waves are of this type), but S-waves can only be transmitted through solids. From the response of these waves as they travel through the Earth it is possible to build up a general picture of the properties of its interior.

Two major discontinuities in the variation of the velocity of earthquake waves with depth show that the Earth's interior can be divided into three zones; the **crust**, the **mantle** and the **core**. The boundary between the core and the mantle is very sharp and is located at a depth of about 2900 km. The mantle-crust boundary is marked by the **Mohorovičić discontinuity** (or the **Moho** as it is usually abbreviated). Its depth below the continents averages around 35 km, but below the ocean basins it is only 5–10 km. The crust, therefore, is an extremely thin surface layer representing only about 0.5 per cent of the Earth's radius.

2.2.2 Mantle, asthenosphere, crust and lithosphere

The mantle is mainly solid and appears to comprise minerals of high density, rich in magnesium and iron. The boundary between the mantle and crust represents a significant change in rock density but there may also be differences in chemical composition. Detailed studies of the velocity of seismic waves in the upper mantle show that after the marked increase corresponding to the Moho at the crust–mantle boundary their velocity gradually becomes greater to a depth of around 100 km, where a small, but

Table 2.1 Summary of the properties of the crust, mantle, lithosphere and asthenosphere

	THICKNESS	VELOCITIES OF P- AND S-WAVES (km s^{-1})	MEAN DENSITY (kg m^{-3})	BEHAVIOUR	
Crust	Continental crust Mean: 35 km Min: <30 km Max: 70 km Oceanic crust: 5–10 km	Increase with depth to 6.6 P-waves 3.8 S-waves but considerable spatial variation	Continental crust: 2700 Oceanic crust: 3000	Solid to top of low velocity zone. Elastic deformation under vertical crustal loading.	Lithosphere
	— Moho —				
Mantle	Base of crust to depth of 2900 km	Immediately below Moho: 8.5 P-waves 4.8 S-waves	3320 immediately below Moho to 5600 at base of mantle		
		Decrease to minimum of 7.8 P-waves 4.2 S-waves in low velocity zone between depth of 100 and 300 km		Very weak in low velocity zone (depth range 100–300 km). Inelastic deformation or 'flow'.	Asthenosphere
		Below 300 km increase to 14.0 P-waves 7.5 S-waves at base of mantle		Below low velocity zone progressively less weak, but convection probably possible deep into the mantle.	

significant, decrease is found which continues to a depth of about 300 km (Table 2.1).

This layer of attenuated velocity is known as the **low velocity zone** and the reduction in seismic wave velocities is considered to be due to partial melting in this region of the mantle. It partly corresponds to the **asthenosphere**, the plastic-like properties of which permits slow 'flow' of material in response to forces applied over long periods of time (Fig. 2.7). Lubrication may be provided by melting between mineral boundaries, but molten material must only form a small proportion of the asthenosphere since it is capable of transmitting S-waves.

Compared with the mantle, the crust forms a very thin layer. Beneath the oceans it is thought to extend to a depth of around 10–15 km, that is to only 5–10 km below the ocean floor. Below the continents the crust is much thicker, averaging about 35 km but reaching up to 70 km beneath some mountain ranges.

Of greatest significance to the creation of macroscale landforms is the boundary of the asthenosphere with the solid mantle above. The crust and the immediately underlying mantle located above the asthenosphere in fact appear to behave as a coherent semi-rigid layer called the **lithosphere**. Two types can be distinguished: **oceanic lithosphere** is capped by thin oceanic crust, while **continental lithosphere** is capped by much thicker continental crust. The thickness of the lithosphere itself shows considerable variability and the boundary with the underlying asthenosphere is gradational and difficult to define precisely. In addition to the analysis of seismic data, heat

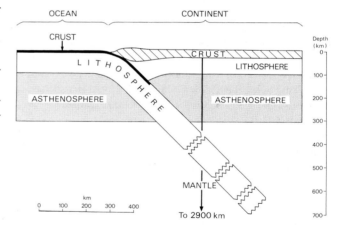

Fig. 2.7 *Schematic representation of the relationship between crust, mantle, lithosphere and asthenosphere. Reincorporation of the lithosphere into the underlying mantle is also shown (see Section 2.4.)*

flow measurements have been employed to map variations in lithospheric thickness on a global basis (Fig. 2.8). These indicate generally thin lithosphere beneath the oceans, especially in the vicinity of mid-oceanic ridges, and a marked thickening in certain continental areas, including Antarctica, North Africa, Brazil and north-eastern North America.

2.2.3 Gravity anomalies

Measurement of variations in the gravitational attraction at

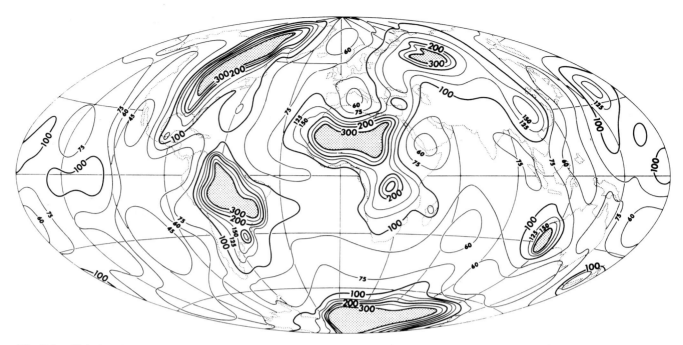

Fig. 2.8 *Global variation in the thickness of the lithosphere estimated from heat flow measurements. (From D. S. Chapman and H. N. Pollack (1977) Geology 5, Fig. 4, p. 268.)*

the Earth's surface has provided much useful information about the characteristics of the crust and the lithosphere. The force of gravity measured at the surface depends on the altitude of the location, the gravitational pull of the underlying rock and the attraction exerted by any areas of highland near by (this last effect being negligible except close to large mountain masses).

The reference point for the calculation of gravity anomalies (which are measured in milligals) is an international standard for the gravitational attraction at the Earth's surface which takes into account the effects of the rotation of the Earth. Correction for altitude alone gives the **free-air gravity anomaly**. To allow for the mass present in a mountain range above the reference surface the theoretical pull of the rock, based on an assumed average rock density, is subtracted from the free-air anomaly (or a correction is made for the presence of sea water if the point is over the ocean). This gives the **Bouguer anomaly** which is the most widely used measure of gravitational deviations over the Earth's surface.

2.2.4 Isostasy

As all but the uppermost part of the mantle appears capable of viscous flow when subject to prolonged stress, the semi-rigid lithosphere, capped by continental or oceanic crust, can be viewed as 'floating' on the underlying asthenosphere (the most easily deformable viscous part of the mantle). To attain hydrostatic equilibrium the position of the lithosphere adjusts vertically in accordance with its

density and thickness. The term **isostasy** (meaning 'equal standing') was introduced by the American geologist Dutton in 1889 to describe this state of equilibrium. Until quite recently it was considered that it was the crust that attained equilibrium with respect to the mantle, but it is now known that isostatic adjustments also involve the rigid mantle forming the lower part of the lithosphere. Nevertheless it is differences in the density and thickness of the crust that are largely responsible for variations in the isostatic adjustment of the lithosphere as a whole.

If it is in isostatic equilibrium one section of lithosphere will stand higher than another because it is of lower density (the crustal density or **Pratt model**), of the same density but thicker (the crustal thickness or **Airy model**), or through a combination of both a lower density and greater thickness (Fig. 2.9). Continental lithosphere stands higher than oceanic lithosphere because continental crust is both of greater thickness and lower density than oceanic crust. The great differences in elevation within the continents are, in most cases, related to variations in crustal thickness with areas at high elevations generally being underlain by deep roots of buoyant crustal rock (Box 2.1).

Bouguer anomalies are generally negative on land and positive over the oceans. This would be expected from differences in the density and thickness of continental and oceanic crust. Bouguer anomalies include a correction for altitude but only on the basis of an average crustal density. The continents, particularly where elevations are high, exhibit lower than predicted (negative) gravity values because of the greater thickness of underlying crust of

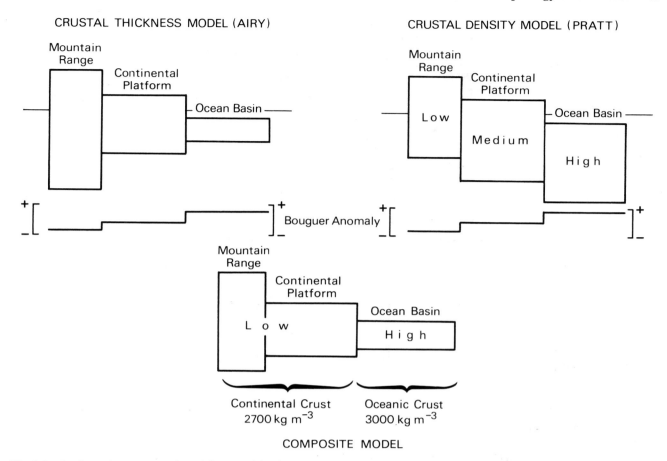

CRUSTAL THICKNESS MODEL (AIRY)

CRUSTAL DENSITY MODEL (PRATT)

COMPOSITE MODEL

Fig. 2.9 *A schematic representation of three models of isostatic equilibrium. Although differences in the density and thickness of the crust are illustrated, the lithosphere as a whole is involved in isostatic compensation.*

relatively low density. Higher than predicted (positive) gravity values occur over the oceans because they are underlain by thin and relatively dense crust.

When corrections are made to take account of these differences in crustal density the magnitude of gravity anomalies is significantly reduced. Remaining discrepancies are termed **isostatic anomalies**, but why should such anomalies exist when the lithosphere is apparently free to attain isostatic equilibrium with respect to the asthenosphere?

One reason is that the lithosphere is not divided into small discrete blocks able to move freely up and down with respect to each other. The lithosphere possesses a certain degree of rigidity so that the mass of any load (such as an ice sheet) placed on it is supported over a greater area than that covered by the load itself. In other words the lithosphere experiences flexure, just as a springboard does when a diver walks along it. This kind of behaviour, known as **flexural isostasy**, is especially important for oceanic lithosphere and for thick continental lithosphere with a high rigidity (see Section 4.2.3).

A second reason for isostatic anomalies is that the lithosphere is not capable of adjusting instantaneously to a change in load. Although the great ice sheets which

covered much of North America and northern Europe in the recent past had largely disappeared by 10 000 a BP, the landsurface in these regions is still rising rapidly in response to the removal of this load.

A third reason for isostatic anomalies is that there are dynamic forces present in the sub-lithospheric mantle which are capable of actively dragging down or pushing up the lithosphere. These forces also play a key role in the large horizontal movements experienced by the lithosphere which we will be examining in detail later in this chapter.

2.2.5   Crustal structure

Seismic and gravity data, together with direct evidence from the rocks themselves, allow us to identify the structural and compositional differences between oceanic and continental crust. Oceanic crust has a mean density of about 3000 kg m-3 and is composed of layers of basic rocks, broadly basaltic and gabbroic in composition, with a thin veneer of sediments. Over most of the ocean floor this sedimentary cover is only 1–2 km thick and in the vicinity of mid-oceanic ridges it becomes very thin indeed. One of the most remarkable discoveries arising from oceanographic research

Box 2.1   Surface elevation and crustal thickness

Regions of high elevation on the continents that are in approximate isostatic equilibrium are known to be 'supported' by crust of greater than average thickness (although compensation actually occurs at the base of the lithosphere). The increase in crustal thickness required to support topography of a given elevation is determined by the relative densities of the continental crust and the sub-lithospheric mantle displaced at depth. Assuming isostatic equilibrium has been attained the relationship between crustal thickness (c) and elevation (h) is given by

$$c = \bar{c} + \frac{h\rho_m}{\rho_m - \rho_c}$$

where $\bar{c}$ is the average thickness of continental crust, and ρ_c and ρ_m are the densities of the crust and the sub-lithospheric mantle respectively. (The average elevation of continental crust, including that part below sea level, is assumed to be 0 km; the actual figure is around 120 m above sea level.) Taking the average thickness of continental crust to be 35 km, and crustal and mantle densities to be 2700 kg m^{-3} and 3300 kg m^{-3} respectively, we can calculate values to show the dependence of elevation on crustal thickness (Fig. B2.1).

These results broadly accord with what we know about the thickness of the crust under regions of high elevation from other lines of evidence. For instance, the Tibetan Plateau, which has extensive areas over 6 km in elevation, is thought to be underlain by crust around 70 km thick. In some elevated regions, such as the Basin and Range Province of the south-west USA, this relationship does not seem to apply because the lithosphere is apparently much thinner than average and is underlain by unusually hot asthenosphere of lower than normal density.

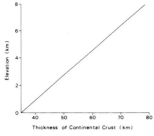

Fig. B2.1   Elevation of continental areas above sea level as a function of crustal thickness.

depth of between 10 and 30 km and this divides the continental crust into upper and lower layers. The upper crust is composed of rocks of a broadly granitic composition, with a highly variable covering ranging from unmetamorphosed sedimentary and volcanic rocks to medium- and high-grade metamorphosed sedimentary strata. The composition of the lower continental crust is less well known but probably consists mainly of granulite, an intermediate to basic rock formed in very high pressure and temperature environments and containing mostly calcium-rich feldspars and pyroxene. Another possibility is the presence of amphibolite rocks with an intermediate to basic composition since this would also give the seismic wave velocities recorded at this depth. The earlier notion that the lower continental crust is largely basaltic in composition has received little support from more recent studies.

In addition to differences in the thickness and density of continental and oceanic crust identified from seismic evidence and the nature of gravity anomalies, it is possible, through a more detailed analysis of such data together with heat flow measurements and geological information indicative of the history of crustal stability, to identify several major structural regions within the continents and oceans (Table 2.2). Most of these regions are also topographically distinct, but some, such as the continental shield, or craton, and mid-continent types, are primarily differentiated on the basis of their detailed structure and the thickness of their sedimentary cover. The validity of any model of global tectonics must be judged by its ability to account for the distribution and characteristics of these major structural regions.

2.3   Development of ideas on global tectonics

Serious scientific attempts to explain the major structural and relief features of the Earth began only in the nineteenth century. In 1829 Elie de Beaumont put forward the idea that the Earth is contracting and argued that compressional stresses set up in the crust as a result of the cooling of the Earth's interior would give rise to faulting, folding and thickening of the crust, and eventually to the formation of mountain ranges. This proposal found support in the work of Lord Kelvin, the pioneer Victorian geophysicist, who attempted to calculate the age of the Earth from its probable rate of cooling on the assumption that it had formed as a molten offshoot of the Sun. However, as the role of radioactive decay in generating heat in the Earth's interior (and thus drastically reducing the previously hypothezised rate of cooling) became appreciated the notion of a contracting Earth was rejected. Various global tectonic models were proposed during the nineteenth and the first part of the twentieth century, with most attention being focused on the origin of the major mountain systems of the Earth's surface. The majority of the models assumed that the

over the past two decades or so has been the youth of oceanic sediments and the underlying basaltic crust. The oldest known rocks from the ocean floor come from the western Pacific, but these are only of Jurassic age. This can be contrasted with the antiquity of rocks exposed over large areas of the continents which in some cases are more than 3000 Ma old.

Like the oceanic crust, continental crust has a layered structure, but one which is much more complex and less clearly defined. A seismic boundary, known as the **Conrad discontinuity**, underlies some continental regions at a

Table 2.2 Structural classification of crustal types

CRUSTAL TYPE	TYPICAL CRUSTAL THICKNESS (km)	HEAT FLOW (mW m^{-2})	BOUGUER ANOMALY (mgal)	DEGREE OF STABILITY	CHARACTERISTICS
Continental shield, or craton	35	29–38	–10 to –30	Very stable	Low to moderate elevation. Composed of highly deformed Precambrian metamorphic and plutonic rocks, unaffected by post-Precambrian tectonism. No covering of post-Precambrian sediments
Mid-continent	38	33–50	–10 to –40	Stable	Generally similar to continental shields, the major difference being the development since the Precambrian of broad undulations which have led to the accumulation of thick sedimentary sequences in extensive basins, particularly near continental margins. The mid-continent structural type commonly occurs adjacent to continental shields and together they account for the majority of the area of continental platforms
Basin and range	30	71–105	–200 to –250	Very unstable	Named after the type area of the Basin and Range Province of the western U S A, this structural type is characterized by great instability associated with significant extension of the crust giving rise to a series of basins and intervening, usually significantly eroded, upland areas. Notable are the very thin crust (for continental regions), the high mean elevation, high heat flow, strongly negative gravity anomalies and marked volcanic and seismic activity
Young mountain belt or active orogen	55	29–84	–200 to –300	Very unstable	Comparatively narrow, elongated regions attaining elevations in excess of 3 km which have experienced relatively recent and often rapid uplift, in many cases preceded by intense folding and thrusting of thick sedimentary sequences as a result of crustal compression. Crustal thicknessess are highly variable, ranging up to a maximum of about 70 km. There is also a wide range of heat flow rates with higher rates being characteristic of younger mountain belts
High plateau	35?	84?	–150 to –250	Very unstable	This crustal type lacks a distinct geophysical character, but is typified by high elevations resulting from uplift lacking associated folding or thrust faulting. The Colorado Plateau provides an example
Island arc	30	29–167	–50 to +100	Very unstable	This type includes a wide range of structural forms, including significant 'continental' fragments such as Japan and New Zealand, as well as arcs formed of numerous individual volcanic peaks of predominantly andesitic composition. Crustal thickness, heat flow rates and Bouguer anomalies are all highly variable both between and within island arc systems. Volcanic and seismic activity is intense
Oceanic trench	?	Low	Strongly negative	Very unstable	Closely associated with island arcs but are best considered separately as they may be found adjacent to young mountain belts. Earthquake activity is marked, particularly towards the adjacent island arc or mountain belt
Ocean basin	11	54	+250 to +350	Very stable	Covers extensive areas of the ocean. Ocean basins are broken by long linear fractures or faults and are punctuated in places by volcanoes of basaltic composition
Mid-oceanic	10	42–335	+200 to +250	Unstable	Crustal type typified by shallow earthquakes and abundant volcanic activity and composed of basaltic lava. Some ridges have a central rift valley and at a limited number of points they break the surface of the ocean to form islands

Source: Based mainly on J. N. Brune (1969) *American Geophysical Union Monograph* **13**, 230–42; and E. W. Spencer (1977) *Introduction to the Structure of the Earth* (2nd edn) McGraw-Hill, New York.

positions of the continents were fixed and that the ocean basins were ancient features.

2.3.1 Continental drift

The notion of **continental drift** is not new. It was can-vassed on a number of occasions in the nineteenth century, but its most influential advocate was Alfred Wegener who intially presented his ideas in 1912. He outlined three main lines of supporting evidence. First, he pointed out the 'fit' of coastlines now separated by thousands of kilometres of ocean, in particular those of South America and Africa. He reconstructed the present land masses into a single super-continent called Pangaea and suggested that this had initially split into two continents, Gondwanaland (Gondwana) to the south and Laurasia to the north, before further rupture and drift resulted in the familiar shape and location of the continents today.

Secondly, Wegener, a climatologist by training, referred to the global distribution of rocks characteristic of particular climatic environments, particularly ancient glacial deposits (**tillites**). He maintained that the distribution of tillites and patterns of glacial striations produced by ice sheets during the Late Palaeozoic glaciation (now termed the Gondwana Ice Age) found in the now widely dispersed continental areas of southern Africa, Australia, South America, India and Antarctica indicated that these land masses were contiguous at that time and probably located fairly close to the South Pole. He also pointed out, though somewhat less convincingly, that some geological structures could be traced from one continent to another across what are now wide stretches of ocean.

Wegener's third main category of evidence was palaeontological. He argued that there were marked similarities between fossil terrestrial fauna and flora from Palaeozoic strata in the various southern continents and between North America and Europe, suggesting that during this era there was free movement over a single large land mass. Fossils from more recent strata showed progressively less similarities with time in the different continents; this he attributed to the contrasting evolutionary paths of groups of animals and plants separated by continental drift.

The greatest obstacle to the acceptance of the drift hypothesis was not so much the nature of this supporting evidence (although alternative explanations were readily proffered) but rather the failure to find a convincing mechanism by which it could occur. From an examination of global hypsometry Wegener had noted the concentration of large areas of the Earth's surface around two levels representing the continental platforms and the ocean basins. He saw this as being compatible with the crust being made up of two layers, the upper one of relatively low density, the lower layer of higher density. He suggested that tidal forces resulted in the lighter continental crust 'ploughing' through the substratum of denser crust underlaying the oceans. This proposal was readily dismissed by geophysicists who were easily able to demonstrate that the Earth is far too strong to be deformed by such tidal forces. Suggestions made by Arthur Holmes and others that continents could be moved by convection currents within the mantle were ignored rather than countered by opponents of drift. While information about the Earth's interior and the ocean floors was so sparse, most geologists preferred the safety of the established doctrine of stationary continents.

2.3.2   Palaeomagnetic evidence

During the mid-1950s, at a time when continental drift was not seriously considered by most earth scientists, new evidence in the form of palaeomagnetic data from rocks again began to bring into question the notion of stationary continents. S.K. Runcorn and his associates, working in Britain, conducted an intensive programme of data collection involving the measurement of **remanent magnetism** in rocks of various ages from around the world. Earlier studies of such palaeomagnetism in France and Japan had shown that iron-rich volcanic rocks, such as basalt, record the magnetic field prevailing at the time they are formed. As basaltic lavas cool through the temperature interval 500 to 450 °C (the **Curie point**) the atomic groups within the iron minerals they contain become aligned parallel to the magnetic lines of force acting upon them. Once the temperature of such rocks falls below 450 °C this magnetic orientation becomes 'frozen' into the individual minerals, only being subsequently disturbed by marked heating. It was found that the palaeomagnetism of young rocks tended to be close to that expected from the present magnetic field, but older rocks showed marked deviations.

As a rough approximation the Earth's present magnetic field can be represented by a regular dipolar pattern (similar to that produced when a bar magnet is held beneath iron filings scattered on a sheet of paper). If it is assumed that the Earth's magnetic field was of this form in the past, then it is possible to estimate the position of the magnetic pole for rocks of known ages (dated by radiometric methods) by measuring their palaeomagnetism. On the basis of such measurements on rocks from Europe, Runcorn and his colleagues demonstrated an apparent movement of the magnetic pole over the past 500 Ma. This phenomenon was termed **polar wandering** and is now thought to arise from the movements of the continents themselves rather than from any significant shift in the location of the magnetic pole itself.

After extending palaeomagnetic investigations to rocks from North America it was found that while these too showed an apparent movement of the magnetic pole the path of polar wandering appeared to differ systematically from that determined for Europe. This discrepancy could be explained if North America and Europe had been moving with respect to each other as well as with respect to the magnetic pole. Further work began to suggest that palaeomagnetic data supported the model of continental drift proposed by Wegener although there were initial uncertainties over the accuracy of the technique. Palaeomagnetic data could indicate north–south movements (palaeolatitudes) but they could not give past longitudinal positions. Moreover, the weakness of the remanent magnetism contained within volcanic rocks together with

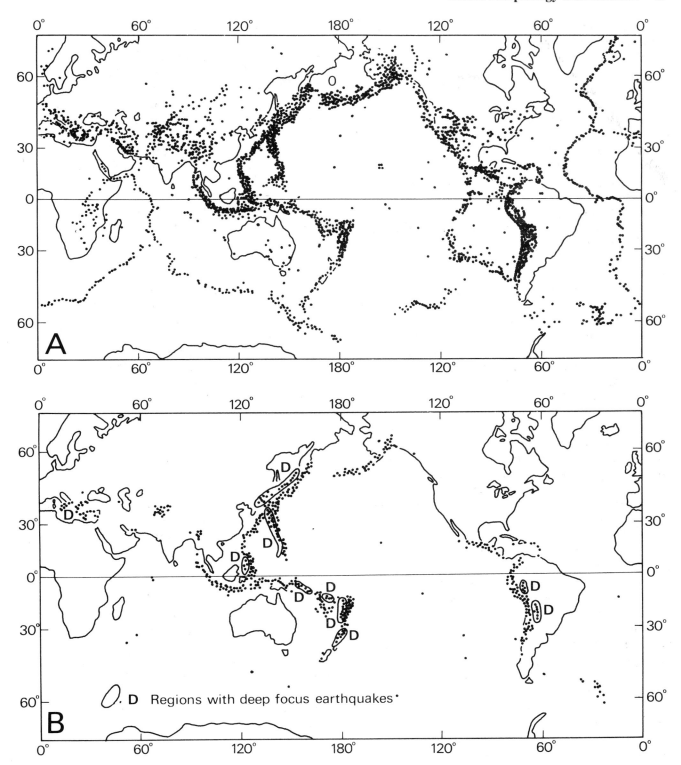

Fig. 2.10 *Location of approximately 30 000 earthquakes recorded by the US Coast and Geodetic Survey between 1961 and 1967. In (A) all the registered earthquakes are shown, whereas (B) records only those earthquakes generated at depths in excess of 100 km, with a separate indication of those regions experiencing deep focus seismic events (300–700 km). (Modified from M. Barazangi and J. Dorman (1969)* Bulletin of the Seismological Society of America **59**, 370–6.)

various sources of error meant that large numbers of individual measurements had to be averaged in order to provide a reasonably precise estimate of past continental positions. Consequently the palaeomagnetic evidence did not lead to an immediate resurrection of the continental drift hypothesis.

2.3.3   Global seismicity

Although the general distribution of earthquakes around the world had been known for many years, it was only with the establishment of a comprehensive network of seismographic stations to monitor underground nuclear weapons testing in the early 1960s that a detailed global picture emerged (Fig. 2.10). Far from being randomly distributed, nearly all seismic activity is concentrated in relatively narrow zones. This pattern coincides with a number of structural features, including mid-oceanic ridges, mountain belts, oceanic trenches and volcanic island arcs.

Looking in more detail, it is possible to classify seismic activity into four zones. The first is located along the mid-oceanic ridges. Earthquakes here are generated at shallow depths (less than 70 km), and they coincide with the high heat flows and volcanic activity of the ridges. The second category is closely related to oceanic trenches and their associated island arcs or mountain belts. Such a zone extends more or less continuously around the western, northern and eastern margins of the Pacific Ocean. The most notable feature of this seismic zone is the relationship between the depths of earthquake generation and their location with respect to the adjacent island arc or mountain belt (Fig. 2.11). On their oceanic margin earthquake foci

are shallow, but they become progressively deeper further away from the trench up to a maximum depth of about 700 km. This relationship gives rise to a dipping planar region of seismic activity, known as a **Wadati-Benioff zone**, inclined at an angle of between about 20 and 55°.

A third area of earthquake activity extends as a rather diffuse belt from the Mediterranean region, through the Himalayas and into Burma. The earthquakes here, which are associated with a long and complex mountain belt, are generally shallow and generated primarily by compressive stresses. The final zone of seismic activity is confined to relatively small areas within the continents and is characterized by shallow earthquakes which are not associated with marked volcanic activity. The most notable example is the famous San Andreas Fault Zone of California, where crustal movement involves large lateral displacements.

2.3.4   Sea-floor spreading

The idea that the polarity of the Earth's magnetic field has reversed through geological time was proposed early this century by Brunhes in France and Matuyama in Japan. Measurements of remanent magnetism in a variety of volcanic rocks found on the continents had shown that some were magnetized in the opposite direction to the Earth's present magnetic field. Further data on these **geomagnetic reversals** were collected and by the mid-1960s a chronology of reversals had been established through the radiometric dating of a large number of samples from all over the world. The resulting detailed palaeomagnetic record showed that during the past 4 Ma the average duration of 'normal' fields (like the present one) had been about 420 ka and that of 'reversed' fields a little longer (about 480 ka). Periods of a particular polarity are termed **epochs**, with shorter phases of opposite magnetization within these being referred to as **events**.

Geophysical surveys during the late 1950s involving the measurement of the magnetic characteristics of the crust of the ocean floor indicated the presence of anomalous linear patterns (Fig. 2.12). These **magnetic anomalies** were found to be hundreds of kilometres long and typically 20–30 km across, and it initially seemed that they were caused by alternating bands of weakly and strongly magnetized rocks. No convincing explanation was initially produced to explain these features.

In 1962 H. H. Hess, of Princeton University in the USA, published a radical explanation for much of the geological data that had been accumulating over the previous decade or so. This proposed that mid-oceanic ridges represent regions where new oceanic crust is being generated by the upwelling of hot mantle material. He suggested that this new crust spreads laterally away from these ridges until it reaches an island arc or mountain belt where it descends into the mantle along the adjacent oceanic trench. This

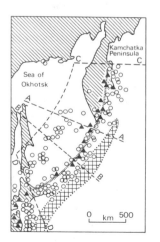
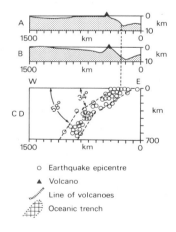

Fig. 2.11   *Map of earthquake epicentres in the northern Japan–Kurile–Kamchatka region of the north-western Pacific. Earthquakes with a magnitude of 6 or more on the Richter scale are shown. The composite profile plots the depth of all the earthquakes on the map in relation to their distance from the oceanic trench. (After H. Benioff, 1954,* Bulletin of the Geological Society of America **65**, *Fig. 2, p. 388.)*

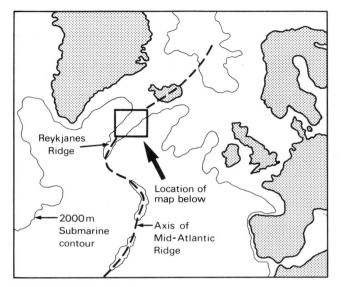

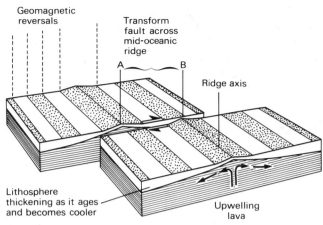

Fig. 2.13 *Schematic illustration of the sea-floor spreading hypothesis of Hess as related to the record of palaeomagnetic reversals by Vine and Matthews. A displacement across the ridge representing a transform fault is also shown.*

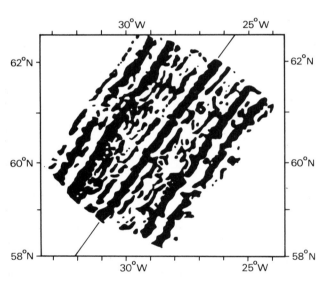

Fig. 2.12 *Magnetic anomalies across the Reykjanes Ridge in the North Atlantic. This is a summary diagram of several traverses across the ridge, the axis of which is indicated by the diagonal line. Areas of positive anomalies are shown in black. (Magnetic anomaly map after J. R. Heirtzler et al. (1966) Deep Sea Research 13, Fig. 7, p. 435.)*

'conveyor belt' view of the oceanic crust, which owed something to the earlier ideas of Holmes, was termed **sea-floor spreading** by Dietz, another American geologist.

In 1963 Vine and Matthews, two Cambridge geophysicists, took up the idea of sea-floor spreading and linked it with data on the palaeomagnetic anomalies observed along mid-oceanic ridges (Fig. 2.12). They proposed that these anomalies, which seemed to be arranged symmetrically on either side of, and roughly parallel with, the ridge crest, were not a result of variations in the *intensity* of magnetization, as earlier suggested, but rather a consequence of the

direction of magnetization. Positive anomalies represented underlying rocks with normal polarity, negative anomalies those with reversed polarity. As new crust emerged at mid-oceanic ridges, they suggested, it became magnetized with the polarity of the prevailing magnetic field (Fig. 2.13).

Subsequently the widths of the linear magnetic zones were related to the duration of each successive epoch of normal and reversed polarity on the assumption that the ocean floor spreads at a constant rate away from mid-oceanic ridges. The close relationship observed between the width of each magnetic anomaly and the corresponding time scale of magnetic reversals, which by now had been accurately established by the radiometric dating of terrestrial lavas, convinced many earth scientists of the reality of sea-floor spreading. Further corroborative evidence was provided by the palaeomagnetic record of ocean sediments containing particles of iron-rich minerals which lay on the basaltic oceanic crust.

When the patterns of magnetic anomalies across mid-oceanic ridges were mapped out it was noticed that in several cases the parallel bands of similarly magnetized crust were broken by numerous offsets or displacements which seemed to be aligned approximately at right angles to the axis of the ridge. Some of these displacements were very large, amounting to 1000 km or more in places. It appeared that these offsets marked some kind of lateral movement between adjacent sections of oceanic crust. In 1965 J. T. Wilson of the University of Toronto proposed that these were not simple strike-slip faults, as had been previously suggested, but were an entirely new variety, which he termed a **transform fault** (Fig. 2.13).

During the late 1960s considerable effort was invested in establishing the history of the ocean floor, much of this involving the activities of the Deep Sea Drilling Project.

The chronology of geomagnetic reversals was extended back into the Mesozoic and this together with the dating of ocean floor sediments enabled the age of large areas of oceanic crust to be determined. Little was found to be more than 150 Ma old.

2.4   Plate tectonics

By the late 1960s a number of lines of evidence were being brought together into a new, radical model of global tectonics. Palaeomagnetic studies on the continents had indicated that continental drift had, in fact, occurred, while more detailed information on the Earth's crustal and subcrustal structure had led many to accept the existence of a relatively mobile zone within the mantle (the asthenosphere). Detailed mapping of the mid-oceanic ridges, together with magnetic and seismic surveys, had indicated that these were zones of crustal tension where new crust was being extruded from the mantle below. The history of polarity reversals of the Earth's magnetic field recorded in this newly generated crust as it cooled indicated that, once formed, it moved away from ridges towards oceanic trenches. On the basis of the much more complete seismic data available by the late 1960s it was being suggested that

the inclined zone of seismicity (the Wadati-Benioff zone) associated with island arcs and active continental margin mountain belts could be explained by the existence of a slab of lithosphere plunging down into the mantle. Analysis of motions associated with earthquakes generated in the Wadati–Benioff zone suggested that lithosphere was indeed moving downwards into the mantle along what became termed **subduction zones**. This supported Hess's idea of crustal formation along mid-oceanic ridges and its reabsorption or **subduction**, into the mantle below oceanic trenches.

All these ideas were incorporated into the theory of **plate tectonics** which was more or less separately developed by a number of workers at about the same time in the late 1960s. The outer layer of the Earth was seen as consisting of a rigid lithosphere, composed of several 'plates', which moved over the underlying more mobile asthenosphere. Sea-floor spreading and continental drift were seen as involving these lithospheric **plates** rather than just oceanic or continental crust.

2.4.1   Outline of the plate tectonics model

The plate tectonics model proposes that the Earth's surface

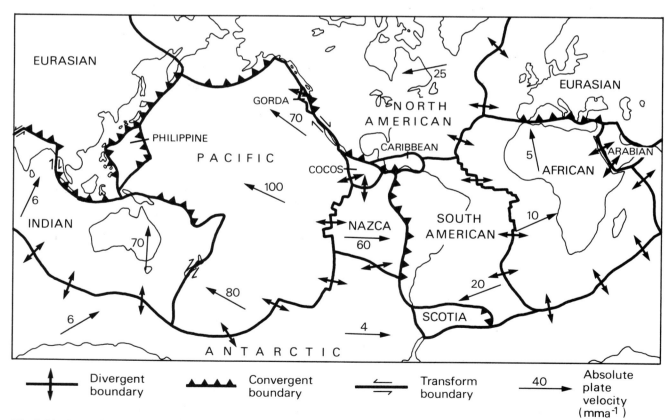

Fig. 2.14   *Map of the major lithospheric plates. The various types of plate boundary are shown and the estimated current rates and directions of plate movement are indicated by arrows (rates in mm a⁻¹.)*

comprises seven major, and at least a dozen minor, lithospheric plates composed of the crust and the upper, more rigid, part of the mantle (Fig. 2.14). These plates are constantly in motion with respect to one another and to the Earth's axis of rotation, and the motion of one plate influences the movement of others. Rates of movement range up to 100 mm a^{-1} (100 km Ma^{-1}) and average around 70 mm a^{-1} (70 km Ma^{-1}). Much, though not all, of the Earth's tectonic, volcanic and seismic activity is directly or indirectly associated with movement between neighbouring plates. The narrow zone marking the relative movement between two plates, which in most cases is fairly clearly demarcated by seismic activity, is termed a **plate boundary**, with the peripheral region adjacent to this boundary being referred to as a **plate margin**.

There are three types of plate boundary. At a **divergent boundary**, such as the Mid-Atlantic Ridge, new crust is formed and attached to the upper part of adjoining lithospheric plates while new upper mantle is accreted to the lower part. Plate movement is laterally away from the **spreading ridge**. At a **convergent boundary**, such as the western coast of South America, two plates are in motion towards each other, with one plate slipping down below the other along a subduction zone. The surface area of a plate is reduced at a subduction zone whereas it is increased along a spreading ridge. Along a **transform boundary**, two plates simply move laterally past each other along a transform fault without any major element of divergence or convergence. The most famous example of this kind of boundary is the San Andreas Fault System in California, USA. At some localities three plates may come into contact. Such a boundary is known as a **triple junction**, an example being the junction of the Pacific, Nazca and Cocos Plates (Fig. 2.14).

The upper layer of a plate is composed of either oceanic or continental crust or both. There is a fundamental difference between the behaviour of lithosphere capped with oceanic crust (oceanic lithosphere) and that covered with continental crust (continental lithosphere) since only oceanic lithosphere can be generated at mid-oceanic ridges and subducted at oceanic trenches.

Oceanic lithosphere has a mean density rather close to that of the immediately underlying asthenosphere. When newly formed along mid-oceanic ridges it is hot and thin and it probably has a slightly lower density. However, as oceanic lithosphere ages, cools and thickens it becomes more dense than the asthenosphere and rests upon it in an unstable state. Along zones of plate convergence, especially where these occur along the margins of continents (such as along the west coast of South America), buckling of the oceanic lithosphere causes it to founder and sink into the underlying asthenosphere, thereby forming a subduction zone. Alternatively, it seems possible that old, cold and thick oceanic lithosphere can attain a density

sufficient for it to subside spontaneously into the asthenosphere. This may have occurred in the western Pacific Ocean where the age of the lithosphere presently being subducted indicates that cooling over a period of about 180–200 Ma is required for this process of spontaneous subduction to occur. By contrast continental lithosphere has a significantly lower mean density than oceanic lithosphere and its buoyancy with respect to the asthenosphere prevents all but limited subduction.

The two fundamental assumptions underlying the plate tectonics model are that the surface area of the Earth has not changed significantly with respect to the rate of generation of new oceanic crust, and that there is a lack of internal deformation within plates compared to the relative motion between them. If the Earth's surface area has increased significantly, sea-floor spreading could occur without associated plate subduction, the rate of crustal formation at mid-oceanic ridges matching the increase in surface area. There are, however, good reasons to believe that the Earth's radius has not increased significantly during the past 500 Ma or so.

Other evidence supports the idea of a lack of major deformation within plates, at least over plates composed of oceanic lithosphere. Oceanic crust has a layered structure which shows remarkably little disturbance and this would not be the case if significant deformation had occurred. In order for plates to behave in the manner proposed the lithosphere of which they are composed must be sufficiently rigid compared with the underlying asthenosphere for stress to be transmitted from one side to the other. If this were not the case significant deformation would occur within plates rather than being concentrated along plate margins. The movements of plates would be greatly complicated if they were to undergo significant internal deformation rather than acting as thin but rigid caps in motion around a sphere. It is quite clear, however, that marked deformation does in fact occur within some continental regions which are thousands of kilometres away from the nearest plate boundary.

In the remainder of this book we will be using plate tectonics as a framework for the interpretation of the Earth's large-scale topographic features. As will become apparent, however, there are several instances where it will be necessary to modify the basic model outlined in this chapter.

2.4.2 Classification of plate boundaries

In Table 2.3 each form of plate boundary is classified according to the type of interacting lithosphere involved. The most well-defined divergent boundaries are associated with mid-oceanic ridges, but they may also occur in an incipient form within continents as rift valleys comprising linear systems of faults associated with tension in the crust. Some rift valleys are analogous to mid-oceanic ridges in

Table 2.3 Classification and primary characteristics of plate boundary types

Boundary type	Stress	MORPHOLOGICAL AND STRUCTURAL FEATURES		
		Oceanic-oceanic lithosphere	Oceanic-continental lithosphere	Continental-continental lithosphere
Divergent	Tensional	Mid-oceanic ridge Volcanic activity	—	Rift valley Volcanoes
Convergent	Compressional	Oceanic trench and volcanic island arc	Oceanic trench and continental-margin mountain belt and volcanicity	—
		Complex island arc collision zone	Modified continental-margin mountain belt	Mountain belt, limited volcanic activity
Transform	Shear	Ridges and valleys normal to ridge axis	—	Fault zone, no volcanicity

experiencing shallow earthquakes and volcanicity.

Convergent boundaries between oceanic lithosphere are marked by an oceanic trench, a volcanic island arc and a Wadati–Benioff zone. Subduction of oceanic lithosphere beneath continental lithosphere is also associated with a trench and an inclined zone of seismicity, but in this case volcanic activity occurs within the active mountain belt located on the continental lithosphere fringing the trench. Where two continents collide neither experiences significant subduction but some crustal thickening occurs and a mountain belt is formed.

Transform boundaries, marked by transform faults, occur both as fracture zones along mid-oceanic ridges and as strike-slip fault zones within continental lithosphere. Plate movements can be oblique as well as simply divergent, convergent or transform. Oblique divergence is most commonly accommodated by transform offsets along a mid-oceanic ridge crest, while oblique convergence is resolved by the complex adjustment of lithospheric fragments along the plate boundary.

2.4.3 Plate motion

Plates vary greatly in area (Fig.2.14). The Pacific Plate, for instance, underlies almost the whole Pacific Ocean whereas a number of minor plates within the continents have an area of less than $1 \times 10^6 \, km^2$. Except where subduction zones lie adjacent to mountain belts on continental margins, plate boundaries do not coincide with continental coastlines. Many continental margins are not separated by subduction zones from the divergent boundaries marked by mid-oceanic ridges. For example, apart from its northern margin along the Mediterranean Sea, the African Plate is surrounded by divergent or transform boundaries. This means that it must be growing in area and that, consequently, the total area of all the other plates must be declining if the 'constant area' assumption of plate tectonics is to be satisfied. Moreover, since sea-floor spreading is taking place from both the Mid-Atlantic Ridge and the Carlsberg Ridge (in the Indian Ocean), one or both of these ridges must be experiencing movement with respect to the Earth's rotational axis.

If plates are regarded as rigid caps involved in relative motion around a sphere their movement can be analysed in terms of specific geometric controls. In Figure 2.15A we can imagine a segment of the Earth's surface (ABE) which has split into two, the two halves (ABC and ADE) remaining in contact only at point A. Point A is the **pole of rotation** about which movement occurs, and the separation of originally adjacent points on each segment occurs along small circles about this point. If we consider irregularly shaped areas on the sphere (X and Y), rather than the regular segments, these can be seen to move apart in such a way that their motion can similarly be described with respect to the same pole of rotation at A. The effect is clearly illustrated in the alignment of transform faults which mark the off-sets along either side of divergent plate boundaries (Fig. 2.15B). These represent segments of small circles concentric about the pole of rotation with respect to which the diverging plates are moving. A consequence of this is that the rate of spreading along a ridge will be a function of the sine of the angular distance from the pole of rotation; the greater the distance from the pole of rotation: the greater the rate of spreading (up to an angle of $90°$). At convergent boundaries the rate of plate subduction is determined in a similar way.

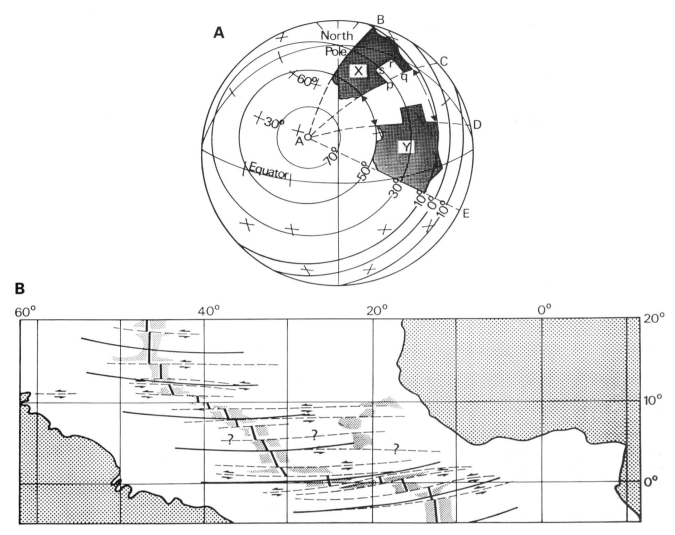

Fig. 2.15 *The geometry of plate motion: (A) the movement of two segments of the Earth's surface with respect to a common pole of rotation; (B) the alignment of transform faults (dashed lines) in the equatorial Atlantic compared with small circles (solid lines) concentric about a pole located at 58°N, 36°W. (After E. R. Oxburgh (1971) in: I. G. Gass, P. J. Smith and R. C. L. Wilson (eds)* Understanding the Earth. *Open University Press, Milton Keynes, Figs. 19.2 and 19.3, p. 266.)*

By comparing the alignments of transform faults, the poles of rotation of the major plates have now been established. Moreover, by relating this information to data on spreading rates, the directions of relative plate motions derived from the analysis of earthquakes, and palaeo-magnetic evidence of continental palaeolatitudes, it has been possible to establish the direction and speed of present-day plate motions and what is probably a fairly accurate history of movements over the past 200 Ma or so (Fig. 2.16).

Since the mid-1980s a direct means of measuring present rates and directions of plate motion has become available. This technique, known as **very-long-baseline interferometry (VLBI)**, can determine with great accuracy the distance between two points on the Earth's surface by using radio telescopes. The measurement is derived from the very small difference in the time of arrival of signals received at different telescopes from a distant radio source (usually a quasar). The accuracy of this technique has been steadily improving and measurements can now be made over thousands of kilometres to an accuracy of a few milli-metres. Preliminary results have confirmed that plates are moving in the directions, and at the rates, indicated by already acquired geophysical data. These results are important for plate tectonic theory because they show that the Earth is not currently expanding as has been suggested by a small minority of earth scientists, at least not at a rate which is significant compared with rates of plate motion. The 1990s will see further refinements to these measure-ments as the **Global Positioning System** becomes fully

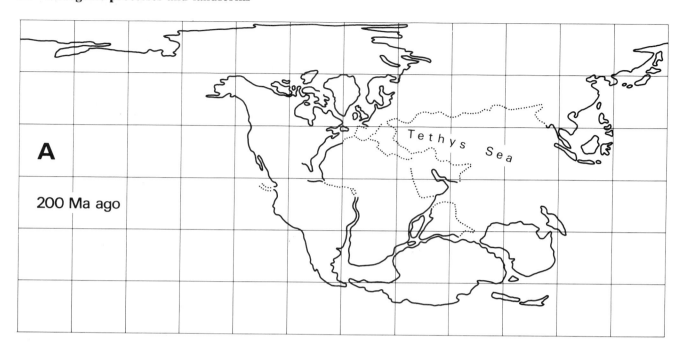

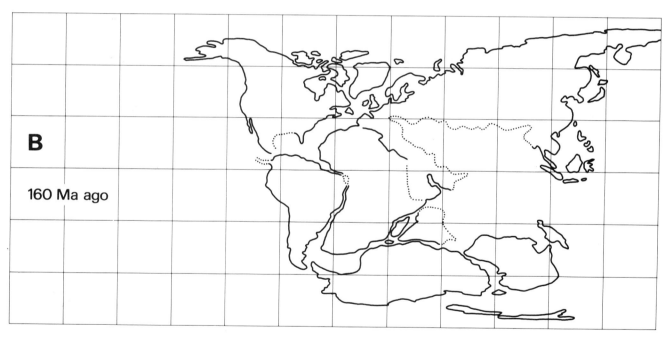

Fig. 2.16 *The pattern of continental drift over the past 200 Ma. The six maps show the estimated past positions of the continents at 40 Ma intervals. Note the assemblage of the continents into one major landmass (Pangaea) 200 Ma ago with the Tethys Sea as a large embayment on its eastern flank (A). By 160 Ma* BP *this landmass had begun to rupture between North Africa and North America (B) and 40 Ma later the separation of Gondwana (the southern continent) and Laurasia (the northern continent) was complete (C). Over the past 120 Ma the Atlantic Ocean has opened and the various continents making up Gondwana have drifted apart (D–F). Particularly noticeable are the rapid northward movement of India and Australia away from Antarctica. For purposes of comparison all the maps have been produced on a cylindrical equidistant projection. Such a projection, however, grossly distorts areas and shapes in high latitudes. (Modified from A. G. Smith, A. M. Hurley and J. C. Briden (1981)* Phanerozoic Paleocontinental World Maps. *Cambridge University Press, Cambridge, Maps 1 and 2, pp. 8–9, Maps 13 and 14, pp. 20–1, Maps 21 and 22, pp. 28–9, Maps 29 and 30, pp. 36–7, Maps 37 and 38, pp. 44–5 and Maps 45 and 46, pp. 52–3.)*

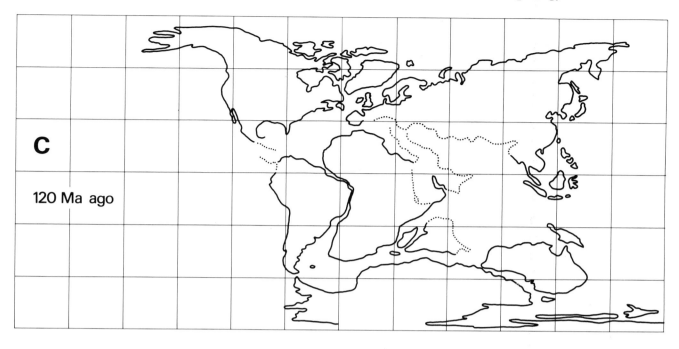

C

120 Ma ago

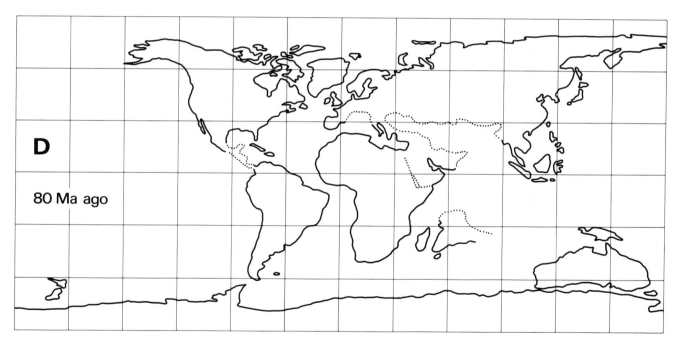

D

80 Ma ago

operational. This method of measuring distances between points on the Earth's surface is similar to VLBI but uses signals from Earth-orbiting satellites rather than extra-terrestrial radio sources.

2.4.4   Mechanisms of plate movement

In spite of the wide range of supporting evidence for the existence of plate tectonics the question of what causes plates to move has yet to be fully resolved. One idea is that lateral flow in the asthenosphere, arising from **convection** in the mantle, drags along the overlying lithosphere (Fig. 2.17(A–C). Convection currents rise and diverge below mid-oceanic ridges and converge and descend along subduction zones. Convection occurs when too much heat is present at depth to be conveyed upwards solely by

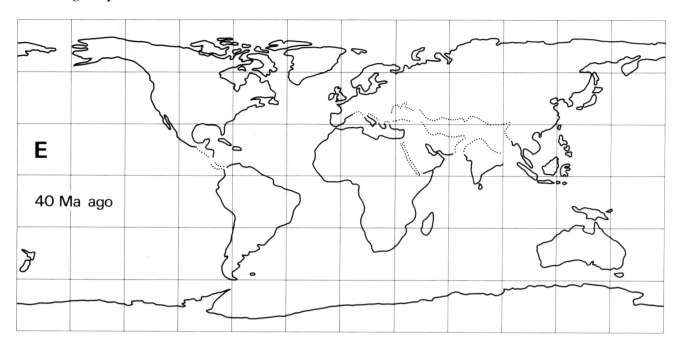

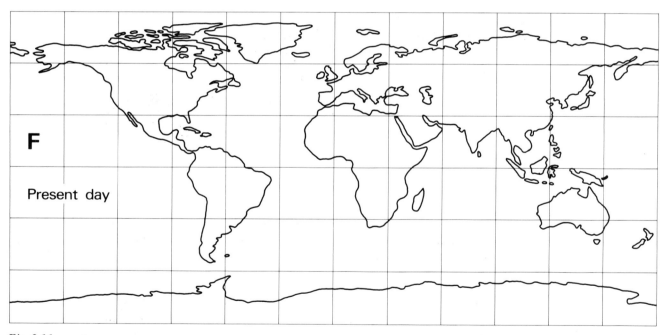

Fig. 2.16   *cont.*

thermal conduction. Temperatures increase and material begins to rise since it is less dense than the surrounding medium; cooler, denser material moves laterally at depth to take its place. If, however, the medium is too viscous convection will be inhibited, or even prevented altogether. Although our knowledge of the physical properties of the Earth's interior is at present far from complete, most geophysicists now support the idea that convection does in fact occur in the mantle; whether this convection is solely, or even primarily, responsible for plate movement is another matter.

Most of the dispute about the nature of convection within the Earth revolves around whether the circulation pattern involves all, or only part, of the mantle. On the basis of certain assumptions about the chemical heterogeneity of the mantle and a postulated marked increase in viscosity with

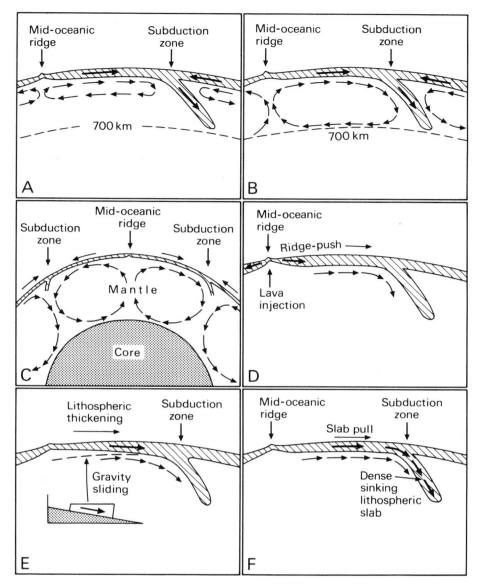

Fig. 2.17 *Possible mechanisms of plate movement: (A) shallow convection confined to the asthenosphere (maximum depth about 300 km); (B) convection in the upper mantle extending to a depth of about 700 km; (C) whole mantle convection; (D) ridge push; (E) gravity sliding; (F) slab pull.*

depth, it has been argued that convection would be confined to the asthenosphere, that is to a maximum depth of about 300 km (Fig. 2.17(A)). Some researchers, however, have suggested that 700 km may represent the lower limit of convection (Fig. 2.17(B)).

Recently the idea of **whole-mantle convection** has gained support, largely as a result of new estimates of the viscosity of the mantle below the asthenosphere. It has been suggested that a hot layer at the core–mantle boundary, generated by convection in the mobile outer core, produces a circulating system extending up through the entire mantle (Fig. 2.17(C)). An important point is that these large-scale convection cells fit in with the dimensions of plates. Most

convection models predict that convection cells have similar horizontal and vertical dimensions, although arguments have been presented against this idea. As some plates have horizontal dimensions of thousands of kilometres it has been argued that convection cells of equivalent horizontal and vertical dimensions would be required to move them. An alternative view is that convection involves a two-scale pattern consisting of large-scale convection in the upper mantle responsible for plate movement superimposed on smaller-scale convection cells confined to the topmost 650 km of the mantle.

Although convection in the mantle has attracted considerable support as a likely cause of plate motion, it has also been argued that flow in the asthenosphere may simply be a consequence of the drag exerted by the movement of overlying plates. Such movement may be caused by at least three different mechanisms. The injection of lava along mid-oceanic ridges, for instance, could push the adjacent plates apart and thereby contribute to plate motion (Fig. 2.17(D)). Perhaps more significant is the thickening of oceanic lithosphere as it moves away from mid-oceanic ridges. Such a configuration would tend to make a plate slide under the force of gravity downwards from a mid-oceanic ridge towards a subduction zone (Fig. 2.17(E)). This would drag the underlying asthenosphere along and promote a compensatory upward movement within the mantle under mid-oceanic ridges.

Cool, thick, old oceanic lithosphere is gravitationally unstable as it is generally denser than the asthenosphere over which it lies. Consequently a cold, dense lithospheric slab descending up to 700 km into the mantle at a subduction zone will tend to pull the remainder of the plate with it (Fig. 2.17(F)). The low conductivity of lithosphere contributes to this effect, it being calculated that the centre of a descending lithospheric slab may be 1000°C cooler than the surrounding mantle at a depth of 400 km. Warming occurs so slowly that the descending lithosphere retains its high density characteristics for a very long time.

Although there is still disagreement as to the relative significance of the various possible mechanisms of plate motion, recent assessments have indicated that the pull of descending lithospheric slabs is the predominant driving force. This interpretation is supported by the relationship observed between the rate of movement of individual plates and the length of subduction zones along their margins. Those attached to long subduction zones, such as the Pacific Plate, have been moving at comparatively high velocities (60–90 mm a^{-1}), whereas plates lacking extensive subducting boundaries have experienced slower rates of movement (below 40 mm a^{-1}).

Further reading

Since the intention of this chapter has been to provide an introduction to global tectonics and the major elements of the Earth's morphology and structure, the suggested reading is mainly limited to texts which will be useful in providing more details on these topics. For those readers with little or no background in geology there are a number of excellent introductory texts. Two of the best are those by Press and Siever (1986) and Skinner and Porter (1987). A more advanced coverage of the Earth's internal structure is provided by Brown and Mussett (1981) while Turcotte and Schubert (1982) give a thorough quantitative grounding in tectonic processes.

A continent-by-continent survey of the Earth's major morphological features is to be found in the book by King (1967) (although its treatment of tectonic processes and landform development has been superseded by more recent research). Cogley (1984, 1985) and Harrison *et al.* (1983) contain detailed analyses of continental hypsometry. An accessible summary of the properties of the mantle is provided by McKenzie (1983), while Pollack and Chapman (1977) discuss the use of heat flow data in the estimation of global variations in the thickness of the lithosphere. A much more advanced treatment of the Earth's interior with particular reference to isostasy is to be found in the conference proceedings edited by Mörner (1980).

The developments which led to the formulation of the plate tectonics model are charted in the collection of original articles edited by Cox (1973). This incorporates all the classic papers including those by Hess, Vine and Matthews, and Wilson. Clear introductions to each topic make this an ideal starting point for those interested in following the birth of plate tectonics. A briefer historical survey is provided by Hallam (1973) who has also written an assessment of Wegener's pioneering contribution (Hallam, 1975). For those with a deeper interest in the history of the plate tectonics revolution Le Grand (1988) is an important source. An excellent guide to the study of plate motions, which has a strong practical bias, is that by Cox and Hart (1986). There are a number of reconstructions of past continental movements available but the series of maps produced by Smith *et al.* (1981) is particularly useful.

The application of the plate tectonics model to problems of Earth structure and topography are considered in the following two chapters and guidance to further reading will be found there. However, for those wanting a fuller account of the mechanisms of plate movement, McKenzie and Richter (1976) discuss the possibility of two scales of convection while Boss (1983) reviews research on the convection question. The book edited by Davies and Runcorn (1980) contains a number of papers on the various suggested mechanisms. Finally, Van Andel (1984) and Molnar (1988) present brief but thought-provoking assessments of the current status of plate tectonics theory and highlight the remaining problems of the model particularly when applied to the continents.

References

Boss, A. P. (1983) Convection. *Reviews of Geophysics and Space Physics* **21**, 1511–20.

Brown, G. C. and Mussett, A. E. (1981) *The Inaccessible Earth.* Allen and Unwin, London and Boston.

Cogley, J. G. (1984) Continental margins and extent and number of the continents. *Reviews of Geophysics and Space Physics* **22**, 101–22.

Cogley, J.G. (1985) Hypsometry of the continents. *Zeitschrift für Geomorphologie Supplementband* **53**, 1–48.

Cox, A. (ed.) (1973) *Plate Tectonics and Geomagnetic Reversals.* W. H. Freeman, San Francisco.

Cox, A. and Hart, R. B. H. (1986) *Plate Tectonics: How It Works.* Blackwell, Palo Alto and Oxford.

Davies, P. A. and Runcorn, S. K. (eds) (1980) *Mechanisms of Continental Drift and Plate Tectonics.* Academic Press, London and New York.

Hallam, A. (1973) *A Revolution in the Earth Sciences: From Continental Drift to Plate Tectonics.* Oxford University Press, Oxford and New York.

Hallam, A. (1975) Alfred Wegener and the hypothesis of continental drift. *Scientific American* **232**(2), 88–97.

Harrison, C. G. A., Miskell, K. J., Brass, G. W., Saltzman, E. S. and Sloan, J. L. II. (1983) Continental hypsography. *Tectonics* **2**, 357–77.

King, L. C. 1967. *The Morphology of the Earth: A Study and Synthesis of World Scenery* (2nd edn). Oliver and Boyd, Edinburgh.

Le Grand, H. E. 1988. *Drifting Continents and Shifting Theories: The Modern Revolution in Geology and Scientific Change.* Cambridge University Press, Cambridge and New York.

McKenzie, D. P. (1983) The Earth's mantle. *Scientific American* **249**(3), 50–62.

McKenzie, D. P. and Richter, F. (1976) Convection currents in the Earth's mantle. *Scientific American* **235**(5), 60–76.

Molnar, P. (1988) Continental tectonics in the aftermath of plate tectonics. *Nature* **335**, 131–7.

Mörner, N–A. (ed.) 1980. *Earth Rheology, Isostasy and Eustasy.* Wiley, Chichester and New York.

Pollack, H. N. and Chapman, D. S. (1977). The flow of heat from the earth's interior. *Scientific American* **237**(2), 60–76.

Press, F. and Siever, R. (1986) *Earth* (4th edn) W.H. Freeman, San Francisco.

Skinner, B. J. and Porter, S. C. (1987). *Physical Geology.* Wiley, New York and Chichester.

Smith, A. G., Hurley, A. M. and Briden, J. C. (1981) *Phanerozoic Paleocontinental World Maps.* Cambridge University Press, Cambridge and New York.

Turcotte, D. L. and Schubert, G. (1982). *Geodynamics: Applications of Continuum Physics to Geological Problems.* Wiley, New York and Chichester.

Van Andel, T. H. (1984) Plate tectonics at the threshold of middle age. *Geologie en Mijnbouw* **63**, 337–41.

3

Landforms and tectonics of plate margins

3.1 Convergent plate margins: general characteristics

The fundamental types of landform created at convergent plate margins are orogens and island arcs. Two distinct mechanisms are responsible for their development into significant morphological features. One involves collision and occurs where the motion of two plates towards each other brings into contact two continents, a continent and an island arc or two island arcs. The other mechanism involves the effects of heating associated with plate subduction. Such consumption of lithosphere can take the form of either subduction of oceanic lithosphere beneath a plate capped by continental crust or the subduction of oceanic lithosphere beneath a second oceanic plate.

The subduction of oceanic lithosphere is a steady-state process in that it can continue until a continent or island arc arrives at the subduction zone. The west coast of South America provides an example of oceanic lithosphere subduction (Nazca Plate) beneath continental lithosphere. The distinction between steady-state subduction and plate collision at convergent boundaries is due to the resistance of continental crust and island arcs to subduction, although it is possible that the latter can be subducted. The collision of two continents leads to the eventual cessation of subduction, whereas other types of collision may cause a change in the location of the subduction zone. The main types of convergent plate margin and their possible modes of development are summarized in Table 3.1. They are classified into steady-state and collision margins and then further subdivided according to the nature of the crust on the interacting plates, the mode of plate interaction and the type of orogen or island arc formed. Also listed are possible courses of subsequent development for each margin subtype.

3.1.1 Steady-state margins

Figure 3.1 illustrates the two types of **steady-state margin**. Where oceanic lithosphere is subducted beneath another oceanic part of a plate the associated volcanic activity and other thermal effects produce an **intra-oceanic island arc.**

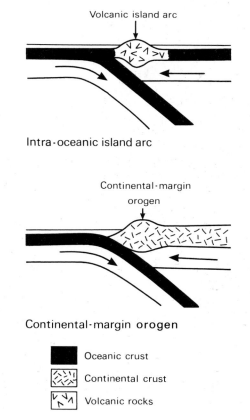

Fig. 3.1 *Major elements of the two types of steady-state margin.*

Table 3.1 Classification of convergent plate margins

MARGIN TYPE	STEADY STATE		COLLISION			
TYPE OF CRUST ON INTERACTING PLATES	Oceanic/oceanic (though may contain sliver of continental type crust)	Oceanic/continental	Continental/continental	Oceanic (island arc type)/continental		Oceanic (island arc type)/oceanic (island arc type)
NATURE OF INTERACTION	Oceanic → oceanic	Oceanic → continent	Continent → continent	Island arc → continent	Continent → island arc	Island arc → island arc
TYPE OF OROGEN FORMED	Intra-oceanic island arc	Continental-margin orogen	Inter-continental collision orogen	Modified continental-margin orogen	Modified 'passive' continental-margin orogen (subduction halted)	Compound intra-oceanic island arc orogen
POSSIBLE EVOLUTION	Island arc → continent collision OR continent → island arc collision	Continent → continent collision OR island arc → continent collision	No further development as convergent boundary since subduction halted	Back-arc subduction halted but forearc subduction may continue	Subduction halted OR polarity reversal with new subduction zone in back arc	Subduction of 'underthrusting arc' continues OR polarity reversal of subduction behind 'overriding arc'

These are particularly extensively developed in the western Pacific Ocean. Their typical arcuate plan form is probably due to the relationship between the angle of subduction and the Earth's curvature. Some idea of the effect can be gained by depressing the surface of a table tennis ball with the thumb. Subduction of oceanic lithosphere beneath a plate carrying continental crust gives rise to a **continental-margin orogen.** The Andes provide an excellent example and this type of orogen is sometimes referred to as an Andean type or Cordilleran type. However, such an orogen may also form a **continental-margin island arc** if the continental crust behind the arc is below sea level. This is the case with the Sumatra–Java part of the Sunda Arc in the East Indies.

3.1.2   Collision margins

A **collision margin** can take a number of forms, depending on the characteristics of the interacting plate boundaries. The first type involves the collision of two continents which gives rise to an **intercontinental collision orogen** (Fig. 3.2), the most notable example of which are the Himalayas. This type must begin with a continental-margin orogen on the overriding plate and be transformed into an intercontinental collision orogen when continental crust on the underthrusting plate reaches the subduction zone. The changes experienced by the leading edge of the continent on the subducted plate are dramatic since there is a conversion from a 'passive' continental margin (see Chapter 4) to an active plate boundary. Eventually the subduction zone is converted into a **suture zone** as the two continental blocks are welded together.

A second type of collision involves the migration of an intra-oceanic island arc towards a subduction zone bounded by continental crust (Fig. 3.3). Unless the island arc is subducted it will be accreted to the continental-margin orogen previously formed along the edge of the continent. Subduction along the continental margin would eventually cease, but it would probably continue along the outer edge of the island arc as oceanic lithosphere would still be available for subduction there. The resulting orogen would be a modified continental margin type.

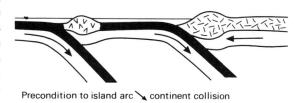

Precondition to island arc ⬊ continent collision

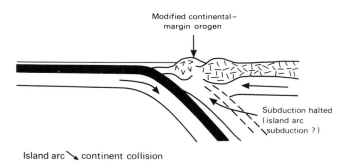

Island arc ⬊ continent collision

Figure 3.3   *Possible consequence of an intra-oceanic island arc moving towards a subduction zone adjacent to a continent (key as for Figure 3.1).*

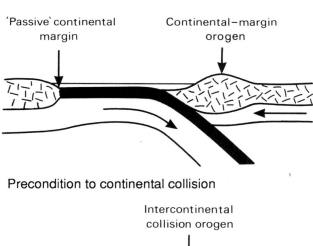

Precondition to continental collision

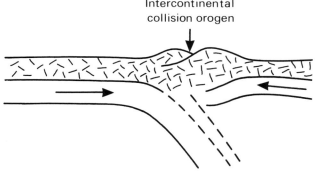

Continental collision

Fig. 3.2   *Development of an intercontinental collision orogen (key as for Figure 3.1.)*

Where a continent moves towards a subduction zone associated with an intra-oceanic island arc the consequences are rather different because of the resistance of continental crust to significant subduction. The interaction can be accommodated in two ways (Fig. 3.4). A possible example of such a continent – island arc collision is provided by northern New Guinea. The collision of intra-oceanic arcs can also take two forms (Fig. 3.5). Some of these plate boundary configurations and stages of development are largely conjectural since clear examples do not exist at the present day. There is, for example, uncertainty as to whether island arcs can be subducted and whether the direction of subduction can be reversed, as illustrated in Figure 3.4.

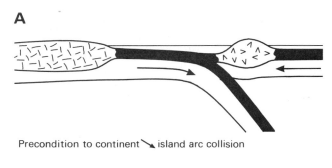

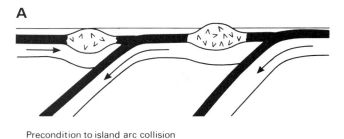

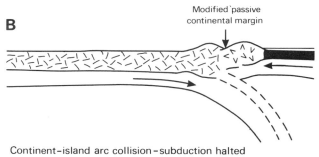

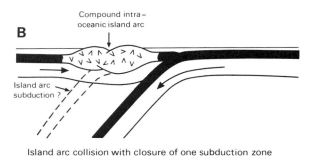

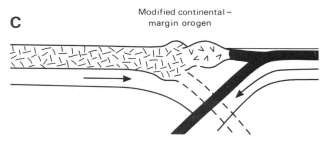

Fig. 3.4 *Possible development of a convergent margin where continental crust reaches a subduction zone adjacent to an intra-oceanic island arc. One possibility is that the collision between the island arc (on the overriding plate) and the continental margin (on he subducting plate) will eventually stop subduction (B). An alternative mode of development which maintains the relative opposing motion between the two plates is a reversal of the polarity of subduction, a process sometimes called 'flipping' (C). The original subduction zone is closed off but a new one forms behind the advancing island arc. Both processes would give rise to a modified continental-margin orogen. In (B) a 'passive' continental margin would gradually be created since subduction would eventually stop. In (C) an active modified continental-margin orogen would be formed with an accreted island arc along its edge and a new subduction zone beyond the island arc (key as for Figure 3.1).*

Fig. 3.5 *Possible evolutionary sequences following the collision of two intra-oceanic island arcs. When the island arcs collide, assuming that one is not subducted below the other, the subduction zone at the collision site will be closed off (B). Part of the original relative motion between the two island arcs could be maintained by continued subduction adjacent to one of them. However, another possibility is a partial polarity reversal of the system by the establishment of a new subduction zone behind the island arc on the overriding plate (C) (key as for Figure 3.1).*

such as the Hawaiian Islands, which are located far from plate boundaries and have a different origin (see Section 4.2.1).

3.2.1 General characteristics and formation

At a subduction zone the downgoing oceanic lithosphere finds its topographic expression in the trench formed on the frontal side of the island arc (Fig. 3.7). Much of the veneer of sediments resting on the oceanic crust of the downgoing plate will not be subducted since it is not firmly attached to the underlying crust and because it is not sufficiently dense to sink of its own accord into the asthenosphere. Consequently it either accumulates in the trench or is scraped off the downgoing plate and forms wedges of sediment on the inner (arc side) of the trench.

3.2 Intra - oceanic island arcs

There are about 20 intra-oceanic island arcs located along subduction zones at the present day, the great majority being found in the western Pacific Ocean (Fig. 3.6). Such volcanic island arcs must be distinguished from other volcanic chains,

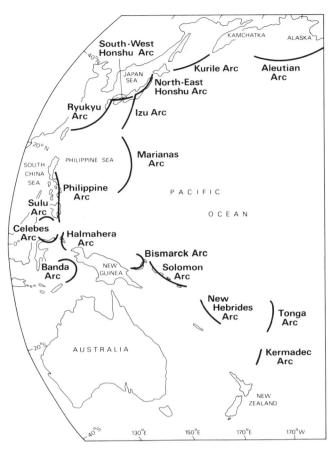

Fig. 3.6 *Location of intra-oceanic island arcs in the western Pacific Ocean. Other intra-oceanic arcs are the Lesser Antilles in the Caribbean, the South Sandwich Islands (Scotia Arc) in the South Atlantic and the Andaman and Nicobar Islands between Burma and Sumatra (see Figure 3.9).*

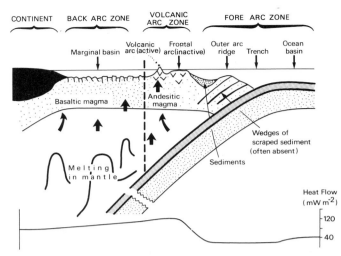

Fig. 3.7 *Schematic representation of the processes operating during the formation of intra-oceanic island arcs and the major morphological elements common to most arcs.*

As the plate descends frictional heat is generated between its upper surface and the surrounding mantle. Heating probably begins at quite shallow depths but because the downgoing plate is cold virtually all of this heat is initially absorbed. This is why low heat flows are recorded at the surface immediately behind the trench (Fig. 3.7). At a depth of about 100 km melting begins to occur along the edge of the subducted lithosphere and heat begins to be transferred upwards by rising pockets of magma. The dramatic surface expression of this marked increase in heat flow is the formation of volcanoes along the arc although relief is also generated by the emplacement of large igneous intrusions (see Section 5.1) in the underlying crust. The most common form of volcanic activity in this zone is the eruption of andesitic lavas, although there is a broad compositional range from basalt to rhyolite.

Fairly consistent patterns are evident in the location of the volcanic activity with respect to the fringing trench. The volcanoes tend to be located between 80 and 150 km vertically above the Wadati–Benioff zone marking the position of the downgoing lithosphere. This is related to the angle of subduction and the depth at which melting of the subducted lithosphere becomes significant. With typical angles of subduction of 30–50° this means that there is a gap of some 75–175 km between the trench and the arc of volcanoes. The active volcanic belt may consist of a single chain of volcanoes or form a broad belt up to 150 km in width. At any one time, however, volcanic activity is generally confined to a zone no more than 50 km across.

3.2.2 Morphological and structural elements

If we look at intra-oceanic island arcs in more detail it is possible to identify several major morphological and structural elements that are common to many (Fig. 3.7). The fundamental division is into the **forearc**, **volcanic-arc** and **back-arc** zones. Beginning in the forearc zone and moving towards the volcanic arc the first feature encountered is the trench which rises to the **outer-arc ridge**. This is located over the subduction zone and consists largely of thrust sheets of sediment and slivers of oceanic crust, known as **ophiolite**, emplaced by the subducted plate. The outer-arc ridge is usually a submarine feature but in a few cases, such as Barbados in the West Indies and Middleton Island in the Aleutian Arc in Alaska, parts of it are exposed above sea level. Between the outer-arc ridge and the volcanic arc a **forearc basin** may be formed. The volcanic-arc zone itself may consist of two elements: the active volcanic arc and an inactive **frontal arc** located on its forearc side and composed of older volcanic rocks. The whole entity is sometimes referred to as a **magmatic arc.** In most island arcs only a relatively small proportion of the individual volcanoes actually rise above sea level.

A complicated pattern of basins and ridges is located behind the volcanic arc. This back-arc zone may simply consist of a large marginal basin separating the intra-oceanic island arc from an adjacent continent, or it may be a more complex feature containing one or more **remnant arcs,** separated from the volcanic arc and each other by **interarc basins.** All these structures are submarine.

A number of processes have been proposed to explain the formation of back-arc basins but no one mechanism seems to be applicable to all cases. One favoured process is known as **back-arc spreading** and this has been applied to several such basins. Japan, for example, is composed of relatively thick continental-type crust, some of it quite old (Early Palaeozoic), rather than the more usual modified oceanic crust of most other intra-oceanic island arcs. This is difficult to explain unless Japan has been detached from the eastern margin of the Eurasian continent and has migrated eastwards. The process thought to be responsible is broadly analogous to sea-floor spreading along mid-oceanic ridges. Indeed, linear magnetic anomalies have been identified behind a number of island arcs; those behind the Scotia Arc in the southern Atlantic Ocean, for instance, indicate that spreading has occurred here over the past 7–8 Ma. Back-arc spreading may be initiated within the volcanic arc because the high heat flux from below weakens the lithosphere. If the lithosphere of the back-arc zone has a component of motion away from the volcanic arc relative to the underlying asthenosphere the arc may be split apart. This eventually leads to the creation of a remnant arc separated by an interarc basin from an active volcanic arc.

3.3   Continental-margin orogens and continental-margin island arcs

Continental-margin orogens and continental-margin island arcs are broadly similar to intra-oceanic island arcs in their tectonic setting. The differences between them arise from the presence of continental crust on the margin of the overriding plate. This influences not only the nature of volcanic activity and landforms produced but also the processes tending to initiate back-arc spreading behind the volcanic arc. Continental-margin orogens are found along the western edge of South America (the Andes) and along part of the west coast of North America (the Cascade Mountains of Oregon and Washington, USA.). The western part of the Sunda Arc in the East Indies is an example of a continental-margin island arc since its back-arc zone comprises drowned continental crust (the Sunda Shelf).

As with an intra-oceanic island arc the oceanward side of a continental-margin orogen is marked by an oceanic trench associated with the downgoing oceanic lithosphere (Fig. 3.8). An outer-arc ridge, formed of wedges of deformed oceanic sediment and slivers of oceanic crust in addition to sediment originating from the adjacent continent, may undergo exten-

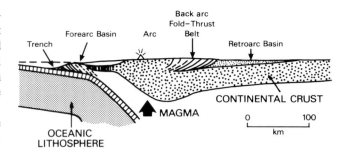

Fig. 3.8 *Main structural and morphological features of a continental-margin orogen. The type illustrated is the 'contracted' form (see text). (After W. R. Dickinson (1977) in: M. Talwani and W. C. Pitman III (eds) Island Arcs, Deep Sea Trenches and Back-Arc Basins. Copyright by the American Geophysical Union, Washington, DC, Fig. 5, p. 38.)*

sive vertical development and in some cases rise above sea level. The Mentawai Islands off the west coast of Sumatra represent the highest parts of the outer-arc ridge of the Sunda Arc and this structure can be traced northwards as far as the Indoburman Ranges in Burma (Fig. 3.9). Parts of the coastal ranges of Mexico, Peru and Chile similarly represent outer-arc ridges. Volcanicity and magma emplacement, largely in

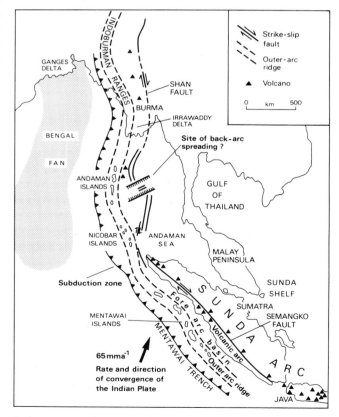

Fig. 3.9 *Map of the major structural and morphological elements of the eastern margin of the Indian Ocean (based on various sources).*

the form of large granite intrusions, characterize the volcanic arc. This is usually located between about 175 and 275 km from the trench, a rather greater distance than is typically the case with intra-oceanic island arcs.

The sequence of continental-margin orogen development begins with the subduction of oceanic lithosphere at, or close to, a continental margin. Frictional heat generated along the upper edge of the underthrusting lithospheric slab at depths exceeding 100 km leads to the upwelling of magma in an expanding dome within the growing orogenic belt which eventually rises above sea level. As the elevation of the orogen increases, the presence of hot, mobile rock within its core promotes gravitational spreading and sliding (see Section 7.3) through upwelling and lateral flowage into the more rigid outer crust. Eventually granite intrusions are emplaced at high levels within the mountain mass and volcanic activity develops. Extensive stretching of the crust above these granite intrusions produces a faulted terrain with active volcanoes. Crustal thickening, due in large measure to the vast volume of granite emplaced above the subduction zone, is considerable in some continental-margin orogens. For instance, crust up to at least 70 km thick occurs below parts of the Andes .

Some continental-margin orogens show evidence of deformation attributable to contraction in the back-arc region. They appear to be jammed against the adjacent cratonic block and are known as **contracted continental-margin orogens**. In other orogens, such as the coastal ranges of the north-west USA, this feature is absent and the term **noncontracted continental-margin orogen** is applied. In contracted continental-margin orogens the back-arc region, known as a **retroarc basin**, is separated from the rear of the volcanic arc by a system of down-thrusted and folded masses of rock (forming a **back-arc fold-thrust belt**) (Fig. 3.8). These features develop if the back-arc lithosphere behind the orogen is moving towards it relative to the underlying asthenosphere, since this convergence will probably lead to partial subduction of the back-arc lithosphere. As the lithosphere carrying the rigid cratonic crust moves towards, and underthrusts, the continental-margin orogen, the stripping of its sedimentary cover gives rise to the back-arc fold-thrust belt. This part of continental-margin orogens can therefore be viewed as a kind of subduction zone, albeit one that involves orders of magnitude less convergence that oceanic trench subduction zones.

The presence of anomalously thin crust, widespread extensional faulting and high rates of heat flow has led some researchers to interpret the Basin and Range Province of the western USA and northern Mexico as an area of back-arc spreading developed on continental crust behind the subduction zone along the Pacific coast. The region, which as a whole has been elevated by 1000–2000 m, has a fault-generated relief of up to 4000 m comprising uplifted mountain blocks and intervening basins (Fig. 3.10). Such back-

Fig. 3.10   *A snow-capped range separating sediment-filled basins in the faulted Basin and Range Province, Nevada, south-west USA.*

arc spreading is analogous to that encountered behind some intra-oceanic island arcs. However, the Pacific–North American plate boundary has undergone significant changes over the past few million years with the subduction of part of the East Pacific Rise spreading ridge and the evolution of a predominantly transform margin. It is possible that these changes have been critical in generating the considerable crustal extension (possibly up to 100 per cent) calculated for the Basin and Range Province.

3.3.1   The Andes

3.3.1.1 General characteristics
The Andes constitute by far the most spectacular present-day example of a continental-margin orogen. Stretching from the Toco Peninsula of Trinidad in the north to Tierra del Fuego in the south, they form a varied and complex mountain system some 9000 km long with elevations in excess of 5000m. Three major divisions can be recognized (Fig. 3.11):

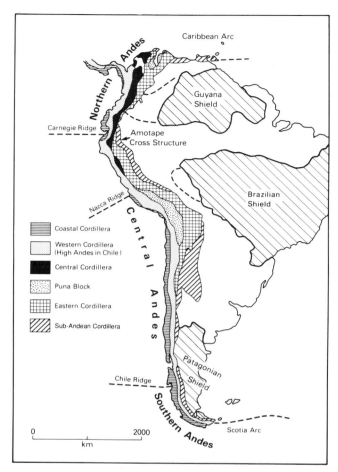

Fig. 3.11 *Major structural and morphological divisions of the Andes. (Modified from A. Gansser (1973)* Journal of the Geological Society London *129, Fig. 2, p. 96. Reproduced by permission of the Geological Society.)*

(1) the Southern (Patagonian) Andes, extending from Tierra del Fuego northwards to the Gulf of Penas at latitude 47 °S where the actively spreading Chile Ridge extends westwards into the Pacific; (2) the Central (Chilean–Peruvian) Andes reaching from the Gulf of Penas northwards to the Amotape cross structure at the Peru–Ecuador border where the north-western trend of the mountain belt changes to a north-easterly orientation; and (3) the northern (Colombian–Venezuelan) Andes, extending from the Amotape cross structure northwards and then eastwards to eventually link with the Caribbean Arc.

Since the development of the plate tectonics model, the Andes have sometimes been regarded as representing a classic example of the orogenic consequences of the sub-duction of oceanic lithosphere beneath an overriding continental plate. This, however, is an over-simplified view as the current phase of uplift and volcanicity represents only the most recent episode in a complex history stretching back to the Mesozoic; moreover, the history of this vast mountain system differs greatly from one part to another.

In the range as a whole block faulting is significant and vertical movements predominate over thrusting and other effects involving laterally induced stress. Any plate tectonics model of the Andes must in fact account for the uplift essentially in terms of vertical tectonics. Another feature which has to be explained is the emplacement of massive granite intrusions of various ages which cover about 464 000 km^2 (about 15 per cent) of the surface of the range; this probably represents a volume in excess of 2×10^6 km^3 – the greatest mass of granite intrusions on Earth. They are concentrated in the Coastal Cordillera of the central Andes but also occur in the High Cordillera (Fig. 3.11). Volcanic activity is localized within the range; in the high plateaus of the central Andes in the border area of Chile, Argentina and Bolivia where the range reaches its greatest breadth, layers of lithified volcanic ash from 500 to 1500 m thick produced by explosive rhyolitic eruptions, cover an area in excess of 200 000 km^2. Quaternary lavas have erupted through these deposits to form over 900 volcanoes ranging in elevation from 5000 to 7000 m (Fig. 3.12).

Fig. 3.12 *The volcanic peak of Cotopaxi reaching an elevation of 5896 m in the Ecuadorian Andes. (Photo courtesy D. Munro.)*

3.3.1.2   Evolution of the central Andes

In the central Andes there has probably been more or less continuous subduction since the Mesozoic. Consumption of the Nazca Plate along the nascent Peru–Chile Trench had begun by at least the Late Triassic – Early Jurassic and marked the conversion of the previously 'passive' continental margin (see Chapter 4) of western South America to an active convergent plate boundary (Fig. 3.13 (A) and (B)). The start of subduction may have been associated with the initiation of spreading on the East Pacific Rise, but in any case it certainly preceded the splitting of South America away from Africa 135–140 Ma ago. This early phase of plate interaction was probably marked by the development of a volcanic arc off the north Chile coast and was apparently contemporaneous with minor emplacement of granite intrusions in the Eastern Cordillera.

A

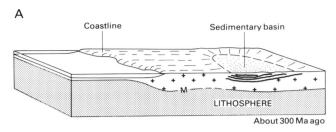

B

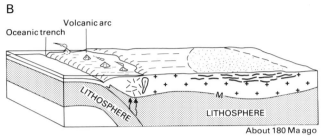

C

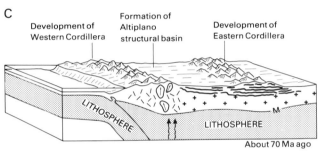

D

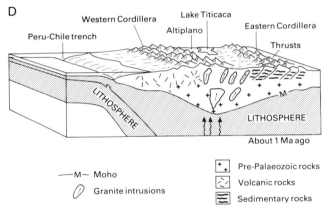

— M — Moho

⊘ Granite intrusions

$\boxed{\begin{smallmatrix}+\\+\end{smallmatrix}}$ Pre-Palaeozoic rocks

$\boxed{\sim}$ Volcanic rocks

$\boxed{\equiv}$ Sedimentary rocks

Fig. 3.13 *Model of the evolution of the central Andes. (Based partly on D. E. James, (1971) Geological Society of America Bulletin* ***82***, *Fig. 10, p. 3341.)*

A major episode of mountain building occurred in the Late Cretaceous–Early Cenozoic to the east of the earlier orogeny (Fig. 3.13(C)). This was the major period of emplacement of granite intrusions which resulted in the growth of the Western Cordillera. A further consequence of this large addition of granite to the core of the mountain mass was the transmission of compressive stresses to the east where the Eastern Cordillera were folded and uplifted. Between these

two ranges a large structural basin was created which now forms the high plateau of the Altiplano. The modern phase of orogeny began in the Miocene with volcanism and the further emplacement of granite intrusions. A more active period of igneous activity in the Pliocene – Quaternary, involving massive emplacements of magma, caused intense folding and the formation of large thrust sheets in the Eastern Cordillera. A series of high, narrow ranges was formed as the Eastern Cordillera were pressed against the resistant block of the Brazilian Shield to the east (Fig. 3.13 (D)). This folded and thrusted belt probably represents the partial subduction of the western fringe of the Brazilian Shield, and the major change of the orientation of the Andes at the Amotape cross structure seems to be the result of the movement of separate basement blocks in the back-arc zone.

There appears to be a clear distinction between the causes of uplift on the western and eastern sides of the central Andes. In the Western Cordillera evidence of tensional deformation indicates that crustal stretching has been associated with the emplacement of magma and volcanic activity related to subduction zone melting. In the Eastern Cordillera, however, compressional stresses apparently generated by the growth of the volcanic arc to the west have caused crustal shortening, and uplift here is attributable to folding, thrusting and crustal thickening. Nevertheless, for the Andes as a whole, compressive stresses arising from plate convergence have not been the major cause of uplift.

3.3.2 The Sunda Arc

The islands which make up the East Indies archipelago form a region of fascinating complexity. For the present, however, we will confine our attention to the Sunda Arc comprising Sumatra, Java and the chain of smaller islands extending eastwards as far as longitude 123 °E (Fig. 3.14). To the west of this point oceanic lithosphere of the Indian Plate is currently being subducted below the Sunda Arc but to the east the Arc is in collision with the continental shelf of northern Australia (this is marked on Figure 3.14). There is thus a transition from a steady-state boundary in the west to a collision boundary in the east. Since the western part of the Sunda Arc is located on the edge of a block of continental crust (at present largely below sea level) it is a continental-margin island arc rather than an intra-oceanic island arc.

Across the Sunda Arc there is a distinct sequence of oceanic trench, outer-arc ridge, forearc basin, volcanic-arc and back-arc zone (Fig. 3.9). The volcanic arc is located about 125 km above the Wadati-Benioff zone representing the seismic activity along the downgoing Indian Plate. This maintains a shallow angle of subduction beneath the outer arc but steepens below the main volcanic arc to about 60°. Beneath Sumatra the subduction zone does not exceed a depth of 200 km, but north of Java it reaches 400 km and north of Flores, at the extreme eastern end of the zone of

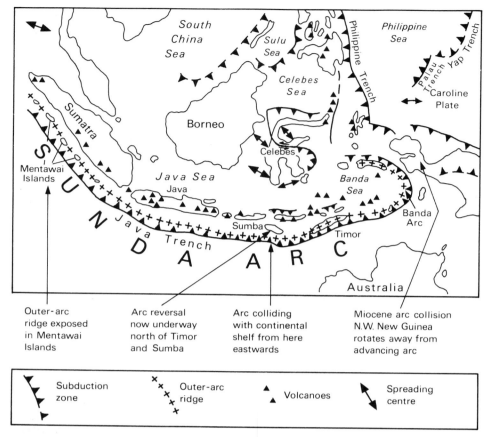

Fig. 3.14 *The tectonic and morphological elements of the Sunda Arc and adjacent region. (Modified from W. Hamilton (1977) in: M. Talwani and W.C. Pitman III (eds)* Island Arcs, Deep Sea Trenches and Back-Arc Basins. *Published by the American Geophysical Union, Washington, DC, Fig. 3, p. 18 and W. B. Hamilton (1988)* Geological Society of American Bulletin, **100**, *Fig. 2, p. 1511.*

steady-state subduction, it attains a depth of 700 km. Sumatra is composed of old, thick continental crust comprising volcanic rocks of Permian, Cretaceous and Cenozoic age. Recent volcanic rocks, including extensive sheets of volcanic ash, are predominantly acidic and intermediate in composition since the magmas have risen through continental crust. By contrast in Java, where the crust is relatively young and thin, the volcanic rocks are mostly intermediate in composition and the large ash sheets indicative of highly explosive eruptions are scarce (see Chapter 5).

The outer-arc ridge of the Sunda Arc is of particular morphological significance. Although it is from 1 to 3 km below sea level off Java, along most of the coast of Sumatra parts of it are exposed as the Mentawai and other islands (Figs 3.9 and 3.14). The height of the ridge is apparently related to the amount of sediment being moved into the adjacent subduction zone where lithosphere is being subducted at a rate of about 65 mm a^{-1}. Large quantities of sediment are present on the ocean floor off Sumatra because of its position with respect to the Bengal Fan which represents the vast submarine extension of the Ganges delta on the Indian subcontinent some 2000 km to the north (Fig. 3.9).

3.4 Intercontinental collision orogens

Although crustal collisions may involve intra-oceanic as well as continental-margin island arcs, it is the convergence and eventual collision of continental crust that gives rise to intercontinental collision orogens (Fig. 3.15). Such mountain belts develop when the oceanic lithosphere originally lying between two continents is eventually consumed. The previously existing subduction zone is converted into a suture zone marking where the two continents are welded together. An embayment of oceanic crust not consumed before the boundary becomes 'locked' is called a **remnant-ocean basin**. **A peripheral foreland basin** is developed on the margin of the underthrusting plate roughly parallel to the strike of the suture belt, either on a remnant-ocean basin covered by a thick layer of sediment, or on true continental crust. A **retroarc foreland basin** occurs on the overriding plate behind the volcanic arc previously developed in association with the earlier subduction of oceanic lithosphere.

3.4.1 General sequence of development

The initial impact occurs between the outer-arc ridge of the

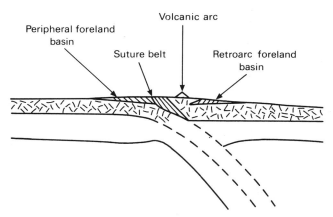

Fig. 3.15 *Major structural and morphological elements of an intercontinental collision orogen.*

overriding continental margin and promontories on the leading edge of the continent entering the subduction zone. The first sections of the converging continental margins to collide suffer the most intense deformation. Wedges of oceanic crust are thrust up on to the overlying sediments of

the subduction zone and uplift ensues (Fig. 3.16). Mountain ranges formed at the points of initial continental collision provide the sediment which accumulates in adjacent remnant-ocean basins (Fig. 3.16(B)). On reaching the subduction zone this sediment is scraped off and stacked into thrust sheets. Final closing of a remnant-ocean basin results in the thrusting of remaining sediments deposited along the continental margin on to the overriding plate and the crustal thickening produced leads to isostatic uplift (Fig. 3.16(C)).

Large-scale crustal shortening is often present in intercontinental collision orogens and this may be accommodated by folding (Figs 3.17, 3.18), or by the limited subduction of continental crust during the initial phase of collision following the previous episode of steady-state subduction. There appear to be two possible mechanisms to account for this. Since the mantle part of the descending lithosphere is cold relative to the asthenosphere it may be sufficiently dense to pull the lower part of the crust down if the upper part is detached. Alternatively, subducted oceanic crust may be able to pull adjacent continental crust down into the asthenosphere. Depending on the assumptions applied this process could lead to the subduction of several hundreds of kilo-

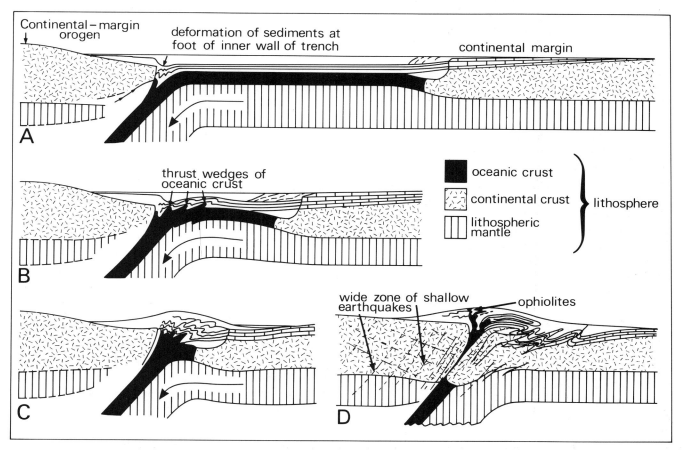

Fig. 3.16 *Schematic representation of the development of an intercontinental collision orogen. (After J. F. Dewey and J. M. Bird, (1970)* Journal of Geophysical Research *75, Fig. 13, p. 2642. Copyright by The American Geophysical Union.)*

metres of continental crust; more likely, however, is rather more limited subduction, but even this could involve the complete consumption of peninsulas and microcontinents.

Partial subduction of the underthrusting plate continues until frictional forces and the buoyancy of the continental crust exceed the driving force of plate motion (Fig. 3.16 (D)). Cessation of subduction is eventually followed by a halt to volcanic activity on the overriding plate. Further thrusting may develop at this stage through sliding of large rock masses under the force of gravity along the flanks of the uplifted orogen (see Section 7.3). Accumulation of sediment through the stacking of thrust sheets in the suture zone is probably the major cause of subsidence of the underlying subducted plate. Plate descent itself does not seem to be a major factor as isostatic rebound does not appear to occur after subduction has stopped. A peripheral foreland basin begins to develop as the continental crust of the underthrusting plate is bent just before entering the subduction zone.

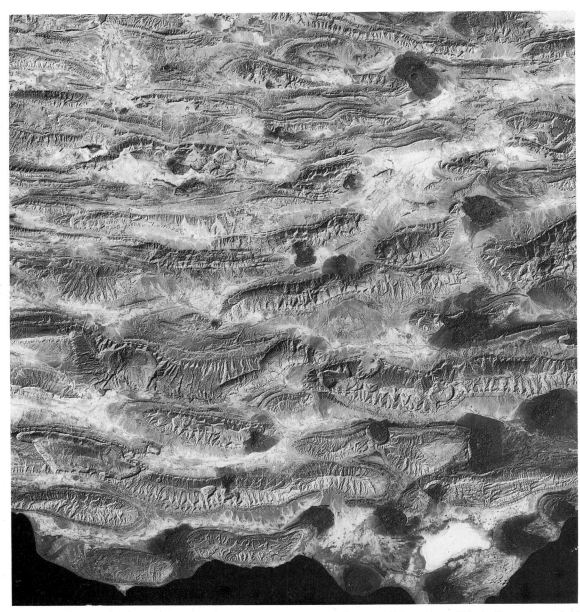

Fig. 3.17 *Landsat scene of part of the southern Zagros Mountains, Iran. Sediments of Mesozoic and Early Cenozoic age laid down on a continental shelf have been folded into a succession of plunging anticlines as a result of compression arising from intercontinental collision. The sediments contain thick salt beds which in places have intruded overlying strata as a result of their lower density and reached the surface to form salt fountains up to 1 km high (small, dark patches on image) The salt spreads under its own weight at rates of a few m a⁻¹ forming salt flows. (Image courtesy N. M. Short.)*

After the phase of partial subduction and the accumulation of a considerable thickness of sediment, possibly up to 5000 m, the peripheral foreland basin may become a site of marked uplift.

Once the two continents have been welded together and active volcanism and tectonic uplift have ceased, a new plate is created in which the orogenic belt formed becomes an intra-plate feature no longer associated with an active convergent plate boundary. A number of ancient suture zones representing previous plate collisions have been recognized but only a few of these, such as the Urals and the Appalachians formed 300 to 250 Ma BP, are major relief features at the present day, most having been extensively eroded and covered by later sediments. Such ancient suture zones may be identified by various lines of evidence such as contrasting pre-collision geological histories either side of the postulated collision zone, the presence of deep ocean floor sediments and associated volcanic rocks, and preserved remnants of andesitic lavas erupted from the continental-margin orogen located on the originally overriding plate prior to collision.

In inter-continental collision orogens major sustained horizontal stresses can be generated which give rise to recumbent folds. Moreover, the back limbs of such folds may become detached along major thrust faults to produce nappes. The classic region for these kinds of structures is the European Alps. Here the convergence of the African and Eurasian Plates has led to the development of successive thrusts and nappes carrying huge masses of rock northwards towards

Fig. 3.18 *Limb of the Saidmarreh anticline, southern Zagros Mountains, Iran. This plunging anticline is experiencing active uplift and although breached by a major stream has retained much of its primary form. (Photo courtesy Aerofilms Ltd.)*

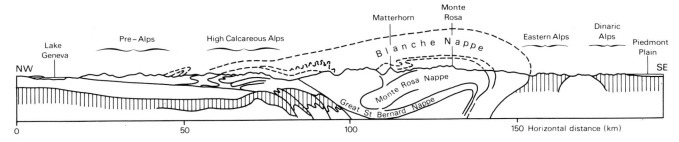

Fig. 3.19 *Highly simplified cross-section from the piedmont plain of northern Italy north-westwards to Lake Geneva illustrating the relationship of major nappes to Alpine topography. (Modified from L. W. Collet (1935)* The Structure of the Alps *(Edward Arnold, London) Plate XI after E. Argand, 1916,* Ecologae Geologicae Helveticae *24, 145–91).*

the European foreland (Fig. 3.19). Horizontal movements along these major thrusts can exceed 100 km and rock thicknesses of several kilometres can be involved. The thrust plane itself represents a zone of detachment (or **décollement**) along a bed of salt or other incompetent lithology. However, detailed investigations of these structures suggest that much of this large-scale lateral movement is accomplished by the spreading or sliding of rock masses under gravity (see Section 7.3).

3.4.2 Pre-collision history and configuration of converging continental margins

Although it is possible to make some generalizations about the likely sequence of development of an intercontinental collision orogen, the pre-collision history of the active margin of the overriding plate and its subduction zone has an important influence on the subsequent evolution of an orogenic belt. The margin of the overriding plate is constantly being modified by the subduction of oceanic lithosphere beneath it; volcanic material is added, sediments accumulate and the margin may be broken by strike-slip faults. The most important of these variables is probably the heating of the plate margin since this will affect its thickness and strength. Thick, strong continental-margin crust will be associated with either slow or recently initiated subduction since in neither case would there be sufficient heating for melting of the upper surface of the downgoing lithosphere. A marked contrast in behaviour would be expected during continental collision between the very thick crust of the Andes, where relatively rapid subduction has occurred throughout at least the Cenozoic, and the Alpine region where the closure of the ocean separating North Africa and southern Europe has been more gradual.

The configuration of continental margins and the angle at which they converge also affects significantly the nature of the orogenic belt that develops (Fig. 3.20). One plate may simply override the other, but another possibility is that a 'flake' of continental crust from the upper surface of the downgoing plate may be thrust over the adjacent plate for a

distance of perhaps 100 km or more. Separation of the upper third of the continental crust on a plate has the effect of reducing the overall buoyancy of the subducted crust by nearly one-half, and this in turn allows more subduction than would otherwise have occurred. The eastern Alps have been suggested as an example of such 'flake tectonics' since this mechanism accounts for the existence of a 'platelet' in the Alps which is much thinner than normal and dips in a southerly direction (Fig. 3.20(B)).

Such a structure is particularly likely to form where a promontory on the advancing edge of the continent on the subducting plate collides with the opposing continent well before the rest of the continental margin (Fig. 3.20(E), (F)). Convergence associated with the continuing subduction of the remaining oceanic lithosphere along the plate margin would be maintained until the plate boundary had been fixed at a sufficient number of points to resist further subduction. Salients of oceanic crust may be left along the suture zone and parts of the Mediterranean Sea represent remnant ocean basins formed in this way.

3.4.3 The Himalayas

3.4.3.1 Morphological and structural elements
The Himalayan intercontinental collision orogen extends in a southward-bending arc, some 200–250 km across, for over 2500 km from the Indus River in the west to the Brahmaputra River in the east (Fig. 3.21). It is separated from the Trans-Himalayan zone and the Tibetan Plateau to the north by the structurally controlled valleys of the Indus and the Brahmaputra (in its upper reaches called the Tsangpo) and to the south it is bounded by the sediment-filled peripheral foreland basin of the Indo-Gangetic plain. Distinct morpho-tectonic zones can be recognized across the orogen along its entire length, although their detailed interpretation is disputed (Fig. 3.22). The two major ranges, the Lesser Himalayas and the Higher Himalayas, both of which exceed 8000 m in height, are bounded by three major linear structures: from south to north these are the Main Boundary

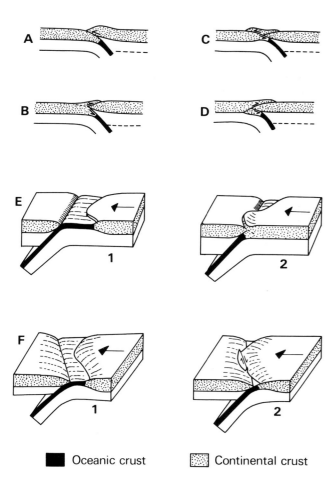

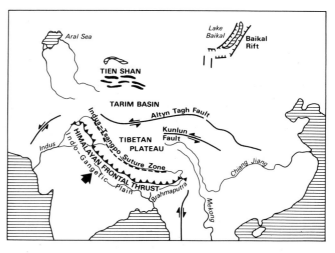

Fig. 3.21 *The Himalayas and Tibetan Plateau in relation to the major structural elements of central Asia (based on various sources).*

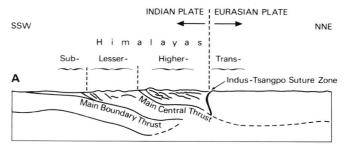

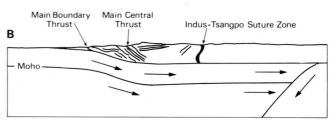

Fig. 3.22 *Two interpretations of the tectonics of the Himalayas according to Le Fort (1975) (A), and Powell and Conaghan (1973) (B). (Modified from P. Le Fort (1975)* American Journal of Science *275A, Fig. 15, p. 38).*

Fig. 3.20 *Various ways in which the collision between the opposing margins of two continents may be accommodated. Shown are: (A) simple overriding; (B) flake formation (e.g. eastern Alps); (C) a detached flake; (D) a deep flake with separating fractures extending below the Moho (e.g. central Alps); (E) a three-dimensional view of flake formation along an irregular continental margin; (F) flake formation associated with oblique subduction and consequential development of strike-slip faulting. (Modified from E. R. Oxburgh (1972)* Nature *239, Figs 2 and 3, p. 204 (E), and E. R. Oxburgh (1974)* Proceedings of the Geologists Association *85, Fig. 26, p. 344 (F) and Fig. 27, p. 345 (A–D).)*

Thrust, the Main Central Thrust and the Indus-Tsangpo suture zone. The belt of ophiolites following the Indus and Tsangpo valleys marks the boundary between the Eurasian Plate and the Indian Plate and to the north the presence of acidic volcanic rocks indicates the subduction of oceanic lithosphere prior to the collision of the Indian and Eurasian land masses.

3.4.3.2 Models of development

From the pattern of ocean floor magnetic anomalies it is possible to trace the movement of India northwards over the past 80 Ma or so (Fig. 3.23). During the period 80 to 50 Ma BP it drifted rapidly at a rate of between 100 and 180 mm a⁻¹. Subsequently, movement has been much slower at only about 50 mm a⁻¹. This reduction in velocity seems to be related to the timing of the contact between India and Eurasia. Collision along what is now the Indus–Tsangpo suture zone seems to have occurred from about the Late Paleocene until the Early Eocene, or possibly a little later. Only limited continental subduction appears to have occurred at this stage. Significant vertical uplift probably began in the Oligocene about 35 Ma ago and has continued to the present time, but at varying rates.

Although clearly an intercontinental collision orogen, the

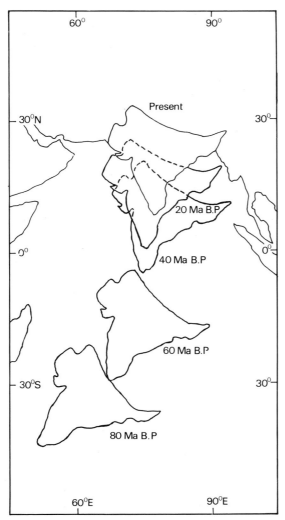

Fig. 3.23   *The northward drift of India over the past 80 Ma reconstructed from palaeomagnetic data. The position of Eurasia is assumed to be fixed in its present location and the configuration of the Indian sub-continent is drawn in its present form (although the shape of the northern boundary prior to collision is unknown). (Based on A. G. Smith, A. M. Hurley and J. C. Briden (1981)* Phanerozoic Paleocontinental World Maps. *Cambridge University Press, Cambridge Map 2, p. 9, Map 10, p. 17, Map 14, p. 21, Map 18, p. 25 and Map 22, p. 29.)*

The most significant development during this period was the formation of the Central Gneiss Zone, some 10–20 km broad and dipping at an angle of 30–40° to the north and bounded to the south by the Main Central Thrust (Fig. 3.22). The development of this zone appears to be related to the major underthrusting and associated metamorphism of local rocks that has apparently occurred along the Main Central Thrust some 100–200 km to the south of the Indus–Tsangpo suture zone. This underthrusting began in the Early Miocene and contributed to the doubling of crustal thickness (up to 80 km in places) and the impressive subsequent uplift of the Higher Himalayas. Detailed seismic evidence indicates that the underthrusting has been at an angle of about 15° over a distance of about 300 km; perhaps significantly this roughly corresponds to the average width of the Himalayas.

The precise relationship between the Main Central Thrust and the intense metamorphism of the Central Gneiss Zone of the Higher Himalayas remains somewhat uncertain. Some researchers have suggested that the metamorphism preceded thrusting and in fact actually promoted it by locally weakening the crust. Another interpretation is that metamorphism occurred at the same time, or post-dated thrusting, and that the frictional heating produced during thrusting contributed to the partial melting of the crust and the intrusion of granitic rocks. A further explanation involves the decoupling of the crust from the underlying lithosphere along the northern margin of the Indian Plate and the exposure of the base of the crust to hot asthenosphere. This could have led to the upwelling of magma, granite emplacement and isostatic uplift through the substitution of hot, low density mantle from the asthenosphere for the cold, dense mantle of the upper lithosphere. A further consequence of this heating could have been a weakening of the upper layers of the crust and the formation of the northward dipping Main Central Thrust along its southern edge.

The extent of subduction of continental crust below the Himalayas is also in dispute. At one extreme underthrusting is seen as proceeding for up to 1500 km at the base of the continental crust through the peeling off of the lower part of the lithosphere of the Eurasian Plate (Fig. 3.22(B)). It is more generally accepted, however, that the buoyancy of the continental crust has limited the extent of continental underthrusting and that this is why the more recent Main Boundary Thrust dipping under the Lesser Himalayas has now taken over the role played earlier by the Main Central Thrust (Fig. 3.22(A)). It seems likely that yet another northward-dipping thrust will develop even further to the south in a few million years.

Although the rate of movement between India and Eurasia fell markedly after their initial collision some 50 Ma ago, convergence of the two land masses since then may have amounted to as much as 2500 km. This represents a major problem because such an amount of continued convergence between two continents after collision is not easily accom-

present elevated form of the Himalayas is not a direct consequence of continental collision; rather it reflects developments that have occurred since the initial impact some 50 Ma BP. Subduction of the oceanic lithosphere of the leading edge of the Indian Plate prior to collision may have been associated with the emplacement of large granite intrusions as well as volcanic activity in southern Tibet. Little relative movement between the Indian and Eurasian Plates seems to have occurred between the Middle Eocene and Early Miocene, but isostatic readjustment of the presumably thickened crust resulted in a withdrawal of the sea on the adjacent Indian continental lowland.

modated within a simple plate tectonics model. The extent of crustal shortening required implies a vast degree of underthrusting by continental lithosphere, extensive crustal compression, crustal deformation over an extensive area behind the suture zone, or a combination of two or more of these effects. Critical to an evaluation of these various hypotheses is an understanding of the development of the high Tibetan Plateau lying to the north of the Himalayas.

3.4.4 The Tibetan Plateau

3.4.4.1 Morphology and structure

The Tibetan Plateau, located between the Himalayan and Karakorum ranges to the south, and the Kunlun and Altyn Tagh ranges to the north, is a roughly triangular area some 1000 km from north to south and 1700 km from east to west (Fig. 3.21). The local relief is relatively subdued but the mean elevation is some 5000 m with isolated masses of probable volcanic origin rising to between 6000 and 7000 m; the Tibetan Plateau is thus by far the highest plateau of significant extent on the Earth's surface. To the east the elevation falls and several major rivers such as the Mekong, Salween, Chiang Jiang (Yangtze) and Huang-He (Yellow) drain in an easterly and southerly direction from the Tibetan Plateau eventually reaching the Yellow and South China Seas.

The Tibetan Plateau has been an unstable area since at least the Late Palaeozoic. It seems to have formed through the successive accretion of continental and island arc fragments (see Section 3.6) from then until the middle Cretaceous and at the time of its collision with the Indian Plate it was probably an area of relatively warm and weak lithosphere. The presence of shallow marine sediments of Late Cretaceous age indicate that the region was below sea level at this time and it apparently did not emerge until the Early Cenozoic. Late Cenozoic volcanism, which is extensive over much of the area, and low seismic wave velocities together suggest that the lithosphere below the Tibetan Plateau is still unusually hot and is therefore probably weaker than in adjacent areas. The crust is now also abnormally thick (about 70 km) but gravity measurements indicate that it is in approximate isostatic equilibrium.

3.4.4.2 Models of development

A number of explanations have been put forward to account for the thick crust and high mean elevation of the Tibetan Plateau which also help to explain the great crustal shortening indicated by the post-collision convergence of the Indian and Eurasian Plates. One model proposes crustal thickening through large-scale underthrusting by the leading edge of the Indian continent (Fig. 3.22(B)), an idea first put forward as long ago as 1924 by the Swiss geologist, Emile Argand. A major problem with the extent of underthrusting required (in excess of 1000 km) is the very low angle of subduction

that it implies (5° or less). This contrasts with the observed northwards dip of the Main Central Thrust at about 15°. Moreover, there is little seismic evidence for the intermediate and deep earthquakes below the Himalayas that would be expected if active subduction is, in fact, occurring.

Of the estimated post-collision convergence of up to 2500 km about 500 km could be accounted for by underthrusting beneath the Himalayas and southern Tibet, another 200–300 km by thrusting and crustal thickening in the various ranges bordering the northern perimeter of the Tibetan Plateau (such as the Pamir, Tien Shan, Altai and Nan Shan ranges) and perhaps a further 300–400 km by the crustal shortening arising from gentle folding within the Tibetan Plateau itself. This leaves around 1400 km of post-collision convergence to be accounted for, much of which might be explained by a doubling of the thickness of the Tibetan crust during the past 50 Ma from a 'normal' value of 35 km to the present 70 km. But not all the remaining convergence can be explained by this process and it has been suggested that the rest has been accommodated by the lateral movement of blocks of continental lithosphere along east–west trending

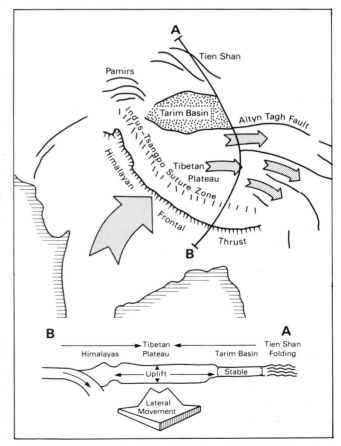

Fig. 3.24 *Schematic representation of the possible effects of post-collision convergence between the Indian and Eurasian land masses based on the model of P. Molnar and P. Tapponnier.*

strike-slip faults to the north of the Tibetan Plateau in Mongolia and western China (Fig. 3.24). Between 500 and 1000 km of north–south crustal shortening could be accounted for by an equivalent amount of lateral crustal movement arising from the 'ploughing' motion of the Indian Plate as it moved northwards and displaced lithospheric blocks in the Eurasian Plate. Supporting evidence for extensive east–west movement comes from the identification of a number of major faults clearly visible on satellite imagery, most notably the Altyn Tagh Fault (Fig. 3.25) which is comparable in length to the San Andreas Fault in California. The amount of displacement along these faults is unknown, but several major earthquakes associated with horizontal strike-slip movements have recently been recorded.

The hypothesized eastward movement of China towards the Pacific would be expected because the continental lithosphere of Eurasia would provide more resistance to lateral movement than the subduction zones of the western Pacific margin. Deformation of the weak Tibetan crust and flow in the upper mantle may have resulted from the north–south squeezing generated by plate convergence. The Tibetan Plateau can perhaps be viewed as the 'pressure gauge' of Asia, with the pressure applied by the Indian continent against Eurasia maintaining its considerable elevation. Moreover, the Tibetan Plateau transmits further pressure to the stable Tarim Basin to the north and this in turn pushes

Fig. 3.25 *Landsat image of the Altyn Tagh Fault north of the Kunlun Shan Mountains. Left-lateral movement along the fault is evident from the offsetting of stream channels which cross it (arrowed). (Image courtesy N. M. Short.)*

northwards to produce thrusting and uplift in the Tien Shan range (Fig. 3.24). The association of thrust faulting on the margins of the Tibetan Plateau with normal faulting in its highest regions has been interpreted as suggesting that it has attained its maximum elevation, and that consequently it is tending to grow outwards rather than increase in altitude. This may be because lower stresses are required for the formation of new thrust faults on its margin than are needed to elevate the entire Tibetan Plateau.

Whatever explanation is applied to the post-collision convergence of the Indian and Eurasian land masses and the formation of the Tibetan Plateau, it is clear that it cannot be accommodated in any simple plate tectonics model. To consider the hypothesized eastward moving segments of lithosphere in Mongolia and China as involving 'micro-plates' is to employ the term 'plate' in quite a different sense from its original definition. The apparent extension of the effects of the collision of the Indian and Eurasian Plates to much of central and eastern Asia, and possibly as far as the Baikal Rift (Fig. 3.21), in fact indicates that the consequences of plate interaction need not necessarily be confined to plate margins.

3.5   Oblique-slip margins

So far we have confined our attention to convergent margins where two plates are in motion towards each other. We now need to consider margins where the movement is predominantly transform and plates meet along transform faults rather than at subduction zones or along collision boundaries. However, it will have become clear from the preceding discussion that purely convergent or transform movement is unlikely to occur except along short sections of a plate boundary. So-called convergent boundaries almost invariably have some element of lateral movement (see, for instance, Figure 3.20(F)), while most transform boundaries possess some convergent or divergent component. It is more realistic, therefore, to think in terms of a continuum of plate boundary types related to various degrees of convergence and divergence. Where a convergent or divergent margin has a significant transform component it is appropriate to refer to it as an **oblique-slip margin**.

Some idea of the variation that may occur along an individual plate boundary can be gained by considering the eastern margin of the Indian Ocean. Here oceanic lithosphere of the Indian Plate is being subducted beneath oceanic and continental lithosphere of the Eurasian Plate (Fig. 3.9). The Indian Plate is moving north-north-eastwards relative to the Eurasian Plate and the margin as a whole is one of low-angle oblique convergence. However, when examined in more detail it is possible to identify significant variations along the margin. For instance, although the forearc zone is predominantly one of convergence and compression, there are also transform elements, such as north of Sumatra. In the

volcanic arc, both in the Sunda Arc and to the north in Burma, movement is predominantly along transform faults. These extend into the back-arc basin of the Andaman Sea where there is evidence of some divergent motion. Where the transform faults occur on land they are associated with significant topographic features such as the valley of the Irrawaddy River in Burma.

In order to emphasize the pattern of stress in oblique-slip margins, the terms **transpression** and **transtension** have been applied to, respectively, the convergent and divergent varieties. Although these two types of margin are associated with distinct types of landscape, they may occur in close proximity owing to the tendency of transform faults to contain offset segments or to have a sinuous form rather than being purely straight (Fig. 3.26). Divergent oblique-slip

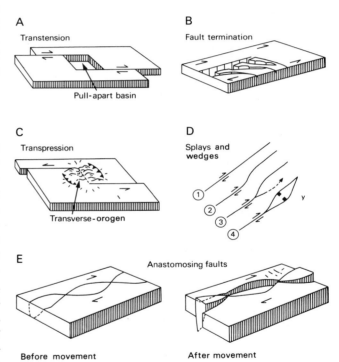

Fig. 3.26   *Fault structures and major landform features associated with oblique-slip margins: (A) pull-apart basin; (B) graben and horst structures caused by fault termination; (C) folding and faulting on a side-stepping fault and creation of transverse orogen; (D) development of fault splays and wedges through the progressive bending (2–4) of an originally straight fault (1); and (E) movement along anastomosing faults leading to upthrusting and the formation of down-sagged pods. (Based largely on J.C. Crowell (1974) in: W. R. Dickinson (ed.) Society of Economic Paleontologists and Mineralogists Special Publication 22 Tectonics and Sedimentation. Figs. 5–7, p. 194; H. G. Reading (1980) in: P. F. Ballance and H. G. Reading (eds) Sedimentation in Oblique-Slip Mobile Zones. Special Publication of the International Association of Sedimentologists 4, Fig. 3, p. 12; and J. T. Kingma, (1958) New Zealand Journal of Geology and Geophysics, 1, Fig. 1, p. 270).*

motion leads to extensional (normal) faulting and subsidence and if this is of sufficient magnitude a **pull-apart basin** may be formed. Oblique convergence, on the other hand, generates compression and results in folding, uplift and the formation of a **transverse orogen** astride the fault.

Complexity is added to this simple relationship because the predominant direction of stress along oblique-slip margins may change through time as a consequence of adjustments to the overall direction of plate movement. The San Andreas Fault System in California, for instance, changed from mainly transtensile in the Miocene to predominantly transpressive in the Pliocene. Very complex patterns of fracture are characteristic of major transform faults and these give rise to local zones of compression and extension over comparatively short distances. Because of the resulting close juxtaposition of areas of uplift and subsidence, considerable local relief can be generated.

3.5.1   The San Andreas Fault System

This situation is illustrated in classic fashion along the southern section of the San Andreas Fault System (Fig. 3.27). Overall the fault system has a length of 1200 km, but the average breadth of the fault zone is some 500 km and it forms a very diffuse boundary between the Pacific and North American Plates. It is estimated that there has been about 1000 km of lateral movement along the fault system over the past 25 Ma. Of the large number of individual faults (Figure 3.27 just shows some of the major ones) only a few are currently regarded as being active, although others have been active in the recent past.

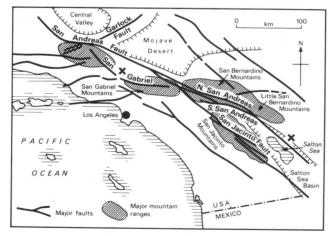

Fig. 3.27   *Simplified map of the main structural and morphological features associated with the San Andreas Fault System in southern California. The points marked by an X indicate areas of similar terrane considered to have been originally adjacent but subsequently displaced laterally along the fault system. (Based on J. C. Crowell (1975)* San Andreas Fault in Southern California. *California Division of Mines and Geology Special Report 118, Fig. 1, p. 11.)*

Fig. 3.28   *Landsat mosaic showing the main structural and morphological features along the San Andreas Fault System in southern California. The mosaic covers approximately the same area as Figure 3.27. (Part of mosaic of ERTS-1 (Landsat) imagery prepared for NASA Goddard Space Flight Center by the USDA Soil Conservation Service.)*

The individual faults which branch, join, bend and side-step each other, give rise to numerous separate areas of uplift and subsidence (Fig. 3.28). A major subsidence feature is the Salton Sea trough, an excellent example of a pull-apart basin. Although it contains a considerable depth of sediment shed from the adjacent mountain ranges its floor still has a maximum elevation of just 74 m below sea level. By contrast an east–west trending zone of compression to the north-east of Los Angeles has led to the formation of a transverse orogen, comprising the San Gabriel and San Bernardino Mountains and collectively known as the Transverse Ranges. These attain elevations in excess of 3000 m and much of this uplift has probably occurred as recently as the Late Quaternary.

3.5.2   The Southern Alps

The Southern Alps of New Zealand provide another example of dramatic uplift and mountain construction along an oblique-slip margin. They run parallel to the Alpine Fault, a remarkably straight transform fault representing the boundary between the Indian and Pacific Plates. It is estimated that the present net movement along the fault involves an average of 40 mm a^{-1} of transform motion and about 22 mm a^{-1} of orthogonal convergence (Fig. 3.29).

Significant compression has only developed comparatively recently (the Alpine Fault was transtensile until the Pliocene) but has generated rapid uplift with the highest peaks in the Southern Alps now exceeding an altitude of 3000 m. The compression has caused thrusting on an eastward-dipping plane along which the continental crust of the Pacific Plate is being heaved up (Fig. 3.30). The western slopes of the Southern Alps are, in effect, an enormous self-perpetuating fault scarp; as rapidly as material is eroded it is replaced by

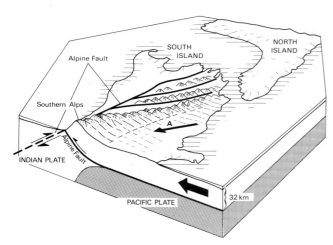

Fig. 3.29 *Schematic representation of the formation of the Southern Alps through the oblique convergence of the Pacific Plate along the Alpine Fault. The approximate direction of movement of the Pacific Plate is indicated by the arrow labelled A. (Based partly on J. Adams (1985) in: M. Morisawa and J. T. Hack (eds)* Tectonic Geomorphology. *Allen and Unwin, Boston, Fig. 5, p. 120.)*

further uplift and crustal shortening against the Alpine Fault (see Section 18.2.4).

3.6 Displaced terranes

As our knowledge of both ancient and modern mountain belts has developed, it has become increasingly evident that not all the complexities observed can simply be explained by the collision of large continental blocks. Many mountain belts situated along present or past plate boundaries contain numerous individual slivers of crust some of which, on the basis of palaeomagnetic and structural evidence, together with their assemblage of rock types, appear to have travelled great distances before arriving in their present positions. Other crustal fragments, while appearing to be of more local origin, have, none the less, experienced significant horizontal displacement and rotation with respect to the plate of which they form a part. These anomalous crustal fragments have been given various names–suspect terrane, allochthonous terrane–but we will use the term **displaced terrane** (Note that in this book the spelling 'terrane' is used to refer to structural units whereas 'terrain' is applied in a topographic sense.)

Several recent studies have indicated the significant role played by displaced terranes in the morphology and structure of both intercontinental collision orogens and continental-margin orogens. For instance, it has been estimated that over 70 per cent of the North American Cordillera is made up of displaced terranes (Fig. 3.31). Most of these seem to have travelled thousands of kilometres and have been accreted to the North American cratonic margin during the Mesozoic and Cenozoic. The terranes are of several types and include slivers of oceanic crust, fragments of intra-oceanic island

Fig. 3.30 *Schist (S) thrust over Quaternary glacial deposits (Q) on the western flank of the Southern Alps, New Zealand. Figure at the bottom of the photograph indicates scale. Fault plane is indicated by arrow.*

arcs and slices detached from unknown continental margins.

Numerous displaced terranes have also been identified in the Alpine–Himalayan chain (Fig. 3.32). Most of these are thought to have rifted away from the margins of Gondwana and moved northward before colliding with the southern margin of the Eurasian land mass and causing extensive deformation and thrusting. Indeed, it has been suggested that much of the structural and morphological complexity of the Alps may be due to the accretion or partial subduction of such crustal fragments rather than the effects of simple collision of two large land masses. Similarly, the morphology of the Himalayan orogen, in particular that of the Tibetan Plateau, may be partly attributable to the accretion of small crustal fragments prior to the main collision between India and Eurasia.

The composition of displaced terranes ranges from that of typical oceanic crust to significantly less dense granitic rock with clear continental affinities. It has been suggested that such terranes are equivalent to oceanic plateaus, which

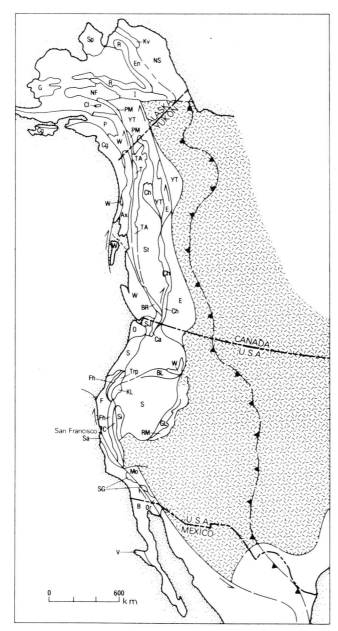

are to be found in the present-day ocean basins and which often rise several kilometres above the adjacent ocean floor (Fig. 3.32). Many of these plateaus have anomalous crustal thicknesses of between 20 and 40 km and an upper layer 10 to 15 km thick with P-wave velocities in the range 6.0–6.3 km s⁻¹. These values are in the range of granitic rocks in the continental crust and so it is inferred that many oceanic plateaus (such as the Seychelles Bank) are drowned continental fragments originating from the edges of ancient land masses and destined to be swept towards a subduction zone in the future. As these plateaus are more buoyant than normal oceanic crust they would tend to resist subduction and therefore be accreted to continents lying along plate margins.

The implications of displaced terranes for the investigation of the relationships between tectonics and landform development have yet to be assessed. Nevertheless, the juxtaposition of terranes with often dramatically different lithologies and structural characteristics is likely to be a significant factor in the long-term development of landscapes.

Fig. 3.31 *Map of displaced (suspect) terranes in western North America. The labelled blocks represent individual terranes which differ significantly in their geological characteristics. A proportion have clear continental affinities and of these some apparently originated in other continents at more southerly latitudes. Other terranes have rock associations indicative of an oceanic origin and represent island arcs and slivers of oceanic crust. The shaded area indicates the extent of North American cratonic basement and the arrowed line marks the eastern limit of Mesozoic–Cenozoic deformation in the North American Cordillera (From P.J. Coney et al. 1980 Nature **288**, Fig. 1, p. 330.)*

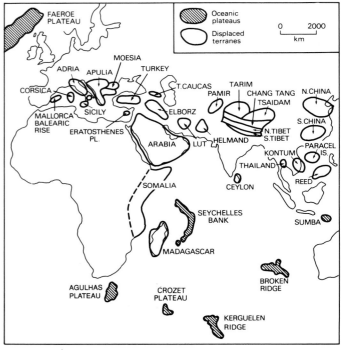

Fig. 3.32 *Map of the probable displaced terranes of the Alpine–Himalayan chain. Several major oceanic plateaus in the Indian Ocean are also shown. (Modified from A. Nur and Z. Ben-Avraham (1982) Journal of Geophysical Research **87**, Fig. 1, p. 3649. Copyright by the American Geophysical Union.)*

3.7 Mesoscale and microscale landforms associated with faulting

Plate margins are invariably regions of active faulting, and microscale and mesoscale landforms directly related to crustal movements along individual faults often provide important evidence of the nature of tectonic activity at a broader scale. Although it must be emphasized that fault-related landforms can be important in areas remote from plate boundaries it will be useful to consider them briefly here. Faulting takes place through sudden, rapid movements over distances ranging from a few centimetres to several metres, each individual movement being associated with an earthquake. Here we will focus on the primary tectonic landforms created by faulting and leave the question of the mode of subsequent landscape development of faulted terrains until Chapter 16.

Both normal and reverse faulting can create a **fault scarp**, the initial angle of which will reflect the dip of the fault plane. But weathering and slumping will very rapidly destroy the original form and reduce the slope to an angle of stability of typically between 20 and 40°. Fault scarps will only be formed where a fault breaks the surface; they can either die out laterally or merge into a monocline (Fig. 3.33(A)). Where faults die out laterally and are replaced by further offset faults an *en echelon* scarp can be formed (Fig. 3.33(B)). Vertical movement along a fault may lead to the impoundment of streams if they are flowing towards the

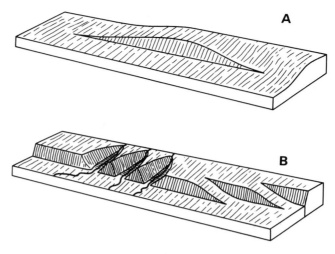

Fig. 3.33 *Block diagram showing various forms of fault scarp and their associated erosional features: (A) a fault scarp dying out laterally but merging into a monocline to the right; (B)* en echelon *scarps to the right and the development of gullies and triangular facets to the left.*

uplifted block, whereas if they are flowing towards the downthrown side they will rapidly cut down through the fault scarp. This leads to the formation of gullies separated along the scarp by characteristic triangular facets (Fig. 3.33 (B)). Eventually the scarp will retreat from the original position of faulting and it may be buried by sediment. A scarp formed

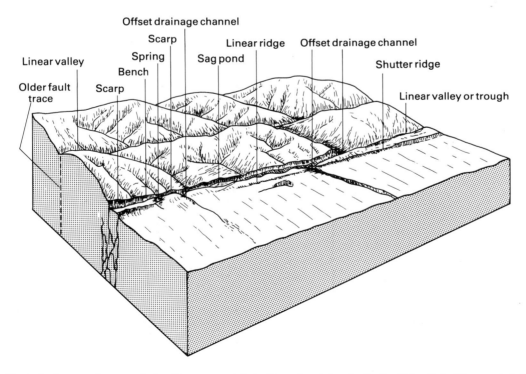

Fig. 3.34 *Typical micro- and mesoscale landforms associated with major strike-slip faults. (After R. L. Wesson* et al. *(1975) United States Geological Survey Professional Paper **914A**, Fig. 11, p. A21.)*

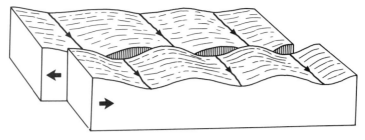

Fig. 3.35   *Block diagram showing a shutter ridge formed along a left-lateral strike-slip fault.*

at the site of a fault through the differential erosion of two rock types juxtaposed along it or through the exposure by erosion of a previously buried fault scarp is termed a **fault-line scarp**. It is important to distinguish such erosional features from true tectonic scarps since only the latter indicate recent or current fault activity.

Strike-slip faulting gives rise to an intriguing range of landforms at the micro- and mesoscale in addition to the major topographic features associated with oblique-slip margins (Fig. 3.34). Where there is no element of convergent or divergent movement along a strike-slip fault it may generate no significant landforms. If, however, faulting occurs across a gullied terrain the resulting offsetting of truncated gullies produces what are known as **shutter ridges** (Fig. 3.35). Localized zones of compressional and tensional stress are common along strike-slip faults and they give rise to a number of distinctive landforms. Divergence across the fault leads to subsidence and the formation of shallow, elongated troughs or sags, typically a few tens of metres across and a few hundred metres long, which, many contain a **sag pond**. In the zones of convergence, on the other hand, compression generates positive topographic features through the formation of ridges and linear and *en echelon* scarplets. Strike-slip faulting also usually leads to the displacement of stream channels (Fig. 3.25) and this feature is often useful in determining the magnitude and frequency of individual movements along the fault (see Section 16.3.2).

Further reading

The literature on the structure and tectonic development of plate margins is vast, but unfortunately very little of it takes a specifically geomorphic perspective. An excellent summary is provided by Miall (1984, pp. 384–440) which, although focusing on sedimentological aspects, also has much relevant material on the morphology of plate margins. Dickinson and Seely (1979) give a more specific treatment of forearc regions with excellent diagrams of their morphology, structure and evolution. A good coverage of the major processes of orogenesis, containing a number of detailed case studies, can be found in Hsü (1983) and Schaer and Rodgers (1987), while Von Huene (1984) and Molnar (1986) provide useful

introductory reviews, and Molnar (1988) considers the problem of reconciling the plate tectonics model with the problem of orogenesis. More details on the complexities of convergent margins are contained in articles by Cross and Pilger (1982), Dewey (1980) and Uyeda (1982) while the possibility of the reversal of subduction polarity is assessed by Johnson and Jaques (1980) and Mattson (1979). Various aspects of island arcs are discussed in the collection of papers edited by Talwani and Pitman (1977), with a more specifically geomorphic perspective being provided by Nunn (1988), Woodroffe (1988) and Yoshikawa *et al.* (1981).

Continental-margin orogens are considered in several of the references mentioned above. Useful introductions to the tectonics and structure of the Andes are provided by Gansser (1973), James (1971) and Dalziel (1986). Garner (1983) discusses the tectonic geomorphology of the Ecuadorian Andes and Jordan *et al.* (1983) consider the effects on Andean topography of variations in the angle of plate subduction. The North American Cordillera, which has a tectonic history of bewildering complexity, has not been examined in detail in this chapter but various aspects of its evolution are considered by Dickinson (1976), Eaton (1987) and Smith and Eaton (1978). The Sunda Arc is discussed by Hamilton (1979) in a comprehensive study of the whole Indonesian region, while Hamilton (1977) provides a more concise analysis.

Articles considering specific aspects of intercontinental collision include those by Burke *et al.* (1977) on suture zones, Molnar and Gray (1979) on the question of limited continental subduction and Şengör (1976) on the effects of the irregular configuration of converging continental margins. There has been a surge of field studies in the Himalayan–Tibetan region since the late 1970s and this has generated a large body of published work. General overviews include those by Allègre *et al.* (1984), Gupta and Delany (1981), Molnar (1984, 1986) and Tapponnier *et al.* (1986), and Selby (1988) provides a specifically morphological study of the Nepal Himalaya. Two contrasting models of the evolution of the Himalayas are presented by Le Fort (1975) and Powell and Conaghan (1973), while the tectonic development of the Tibetan Plateau is discussed by Bingham and Klootwijk (1980) and Ni and York (1978). The model of the Tibetan

Plateau uplift associated with the lateral displacement of lithospheric blocks is introduced by Molnar and Tapponnier (1975), but there is also a more accessible and well-illustrated version (Molnar and Tapponnier, 1977).

Oblique-slip margins are considered in detail in various papers in Ballance and Reading (1980); the introductory article by Reading (1980) is particularly useful as it presents a general review of sedimentation patterns in relation to topography. The evolution of the San Andreas Fault System is concisely covered by Crowell (1974, 1979). The Southern Alps have been the subject of a number of studies examining the relationship between tectonics and landforms; these include those by Adams (1985), Basher *et al.* (1988), Whitehouse (1988) and articles included in Soons and Selby (1982). Berryman (1988) and Kamp (1988) examine landforms associated with oblique convergence in North Island, New Zealand.

Displaced terranes are starting to generate a significant literature and this has been reviewed by Howell (1985) and Schermer *et al.* (1984). The large number of displaced terranes in the North American Cordillera are considered by Coney *et al.* (1980) and the possible role of oceanic plateaus in orogenesis is assessed by Nur and Ben-Avraham (1983). Ollier and Pain (1988) draw attention to the neglected problem of how displaced terranes relate to landscape development in the context of Papua New Guinea. Examples of specific studies of micro-and mesoscale fault-related landforms include those by Gerson *et al.* (1985), Nash (1981) and Wallace (1977), while Armijo *et al.* (1986) and Molnar *et al.* (1987) provide detailed examples of how such landforms can assist in the interpretation of major tectonic features.

References

Adams, J. (1985) Large-scale tectonic geomorphology of the Southern Alps. In: M. Morisawa and J. T. Hack (eds) *Tectonic Geomorphology*. Allen and Unwin, Boston, and London 105– 28.

Allègre, C. J. *et al.* (1984) Structure and evolution of the Himalaya-Tibet orogenic belt. *Nature* **307**, 17–22.

Armijo, R., Tapponnier, P., Mercier, J. L. and Tong-Lin, H. 1986. Quaternary extension in southern Tibet: field observations and tectonic implications. *Journal of Geophysical Research* **91**, 13803–72.

Ballance, P. F. and Reading, H. G. (eds) 1980. *Sedimentation in Oblique-Slip Mobile Zones*. International Association of Sedimentologists Special Publication **4**.

Basher, L. R., Tonkin, P. J. and McSaveney, M. J. (1988) Geomorphic history of a rapidly uplifting area on a compressional plate boundary: Cropp River, New Zealand. *Zeitschrift für Geomorphologie Supplementband* **69**, 117–31.

Berryman, K. (1988) Tectonic geomorphology at a plate boundary: a transect across Hawke Bay, New Zealand. *Zeitschrift für Geomorphologie Supplementband* **69**, 69–86.

Bingham, D. K. and Klootwijk, C. T. (1980) Palaeomagnetic constraints on Greater India's underthrusting of the Tibetan Plateau. *Nature* **284**, 336–8.

Burke, K., Dewey, J. F. and Kidd, W. S. F. (1977) World distribution of sutures; the sites of former oceans. *Tectonophysics* **40**, 69–99.

Coney, P. J., Jones, D. L. and Monger, J. W. H. (1980) Cordilleran suspect terranes. *Nature* **288**, 329–33.

Cross, T. A. and Pilger, R. H. Jr (1982) Controls of subduction geometry, location of magmatic arcs, and tectonics of arc and back-arc regions. *Geological Society of America Bulletin* **93**, 545–62.

Crowell, J. C. (1974) Origin of Late Cenozoic basins in southern California. In: W. R. Dickinson (ed.) *Tectonics and Sedimentation.* Society of Economic Paleontologists and Mineralogists Special Publication **22** (Tulsa) 190–204.

Crowell, J. C. (1979) The San Andreas fault system through time. *Journal of the Geological Society London* 136, 293–302.

Dalziel, I. W. D. (1986) Collision and cordilleran orogenesis: an Andean perspective. In: M. P. Coward and A. C. Ries (eds) *Collision Tectonics*. Geological Society Special Publication **19**, 389–404.

Dewey, J. F. (1980) Episodicity, sequence, and style at convergent plate boundaries. In: D. W. Strangway (ed.) *The Continental Crust and Its Mineral Deposits*. Geologists Association of Canada Special Paper **20**, 553–73.

Dickinson, W. R. (1976) Sedimentary basins developed during evolution of Mesozoic–Cenozoic arc-trench system in western North America. *Canadian Journal of Earth Sciences* **13**, 1268–87.

Dickinson, W. R. and Seely, D. R. (1979) Structure and stratigraphy of forearc regions. *American Association of Petroleum Geologists Bulletin* **63**, 2–31.

Eaton, G. P. (1987) Topography and origin of the southern Rocky Mountains and Alvarado Ridge. In: M. P. Coward, J. F. Dewey and P. L. Hancock (eds) *Continental Extensional Tectonics*. Geological Society Special Publication **28**, 355–69.

Gansser, A. (1973) Facts and theories on the Andes. *Journal of the Geological Society London* **129**, 93–131.

Garner, H. F. (1983) Large-scale tectonic denudation and climatic morphogenesis in the Andes Mountains of Ecuador. In: R. Gardner and H. Scoging (eds) *Mega-Geomorphology*. Clarendon Press, Oxford and Oxford University Press, New York., 1–17.

Gerson, R., Grossman, S. and Bowman, D. (1985) Stages in the creation of a large rift valley–geomorphic evolution along the southern Dead Sea Rift. In: M. Morisawa and J. T. Hack (eds) *Tectonic Geomorphology*. Allen and Unwin, Boston and London, 53–73.

Gupta, H. K. and Delany, F. M. (eds) (1981) *Zagros, Hindu Kush, Himalaya-Geodynamic Evolution* American Geophysical Union, Washington, DC.

Hamilton, W. (1977) Subduction in the Indonesian region. In: M. Talwani and W. C. Pitman III (Eds) *Island Arcs, Deep Sea Trenches and Back-Arc Basins* American Geophysical Union, Washington, DC, 15–32.

Hamilton, W. (1979) Tectonics of the Indonesian region. *United States Geological Survey Professional Paper* **1078.**

Howell, D. G. (1985) Terranes. *Scientific American* **253**(5), 90–103.

Hsü, K. J. (ed.) (1983) *Mountain Building Processes*. Academic Press, London and New York.

James, D. E. (1971) Plate tectonic model for the evolution of the central Andes. *Geological Society of America Bulletin* **82**, 3325–46.

Johnson, R. W. and Jaques, A. L. 1980. Continent–arc collision and reversal of arc polarity: New interpretations from a critical area. *Tectonophysics* **63**, 111–24.

Jordan, T. E., Isacks, B. L., Allmendinger, R. W., Brewer, J. A.,

Ramos, V. A. and Ando, C. J. (1983) Andean tectonics related to geometry of subducted Nazca plate. *Geological Society of America Bulletin* **94**, 341–61.

Kamp, P. J. J. (1988) Tectonic geomorphology of the Hikurangi Margin: surface manifestations of different modes of subduction. *Zeitschrift für Geomorphologie Supplementband* **69**, 55–67.

Le Fort, P. (1975) Himalayas: the collided range; Present knowledge of the continental arc. *American Journal of Science* **275A**, 1–44.

Mattson, P. H. (1979) Subduction, buoyant braking, flipping, and strike-slip faulting in the northern Caribbean. *Journal of Geology* **87**, 293–304.

Miall, A. D. (1984) *Principles of Sedimentary Basin Analysis.* Springer-Verlag, New York.

Molnar, P. (1984) Structure and tectonics of the Himalayas: constraints and implications of geophysical data. *Annual Review of Earth and Planetary Sciences.* **12**, 489–518.

Molnar, P. (1986) The structure of mountain ranges. *Scientific American* **255**(1), 70–9.

Molnar, P. (1988) Continental tectonics in the aftermath of plate tectonics. *Nature* **335**, 131–137

Molnar, P., Burchfiel, B. C., Liang K'Uangyi and Zhao Ziyun (1987) Geomorphic evidence for active faulting in the Altyn Tagh and northern Tibet and qualitative estimates of its contribution to the convergence of India and Eurasia. *Geology* **15**, 249–53.

Molnar, P. and Gray, D. (1979) Subduction of continental lithosphere: Some constraints and uncertainties. *Geology* **7**, 58–62.

Molnar, P. and Tapponnier, P. (1975) Cenozoic tectonics of Asia: effects of a continental collision. *Science* **189**, 419–26.

Molnar, P. and Tapponnier, P. (1977) The collision between India and Eurasia. *Scientific American* **236**(4), 30-41.

Nash, D. B. (1981) Morphologic dating of degraded normal fault scarps. *Journal of Geology* **88**, 353–60.

Ni, J. and York, J. E. (1978) Late Cenozoic tectonics of the Tibetan Plateau. *Journal of Geophysical Research* **83**, 5377–84.

Nunn, P. D. (1988) Plate boundary tectonics and oceanic island geomorphology. *Zeitschrift für Geomorphologie Supplementband* **69**, 39–53.

Nur, A. and Ben-Avraham, Z. (1983) Displaced terranes and mountain building. In: K.J. Hsü (ed.) *Mountain Building Processes.* Academic Press, London and New York, 73–84.

Ollier, C. D. and Pain, C. F. (1988.) Morphotectonics of Papua New Guinea. *Zeitschrift für Geomorphologie Supplementband* **69**, 1–16.

Powell, C. McA. and Conaghan, P. J. (1973) Plate tectonics and the Himalayas. *Earth and Planetary Science Letters* **20**, 1–12.

Reading, H. G. (1980) Characteristics and recognition of strike-slip fault systems. In: P. F. Ballance and H. G. Reading (eds) *Sedimentation in Oblique-Slip Mobile Zones.* International Association of Sedimentologists Special Publication **4**, 7-26.

Schaer, J.-P. and Rodgers, J. (eds) (1987) *The Anatomy of Mountain Ranges.* Princeton University Press, Princeton.

Schermer, E. R., Howell, D. G. and Jones, D. L. (1984) The origin of allochthonous terranes: perspectives on the growth and shaping of continents. *Annual Review of Earth and Planetary Sciences* **12**, 107–31.

Selby, M. J. (1988) Landforms and denudation of the High Himalaya of Nepal: results of continental collision. *Zeitschrift für Geomorphologie Supplementband* **69**, 133–52.

Şengör, A. M. C. (1976) Collision of irregular continental margins: Implications for foreland deformation of Alpine–type orogens. *Geology* **4**, 779–82.

Smith, R. B. and Eaton, G. P. (eds) (1978) *Cenozoic Tectonics and Regional Geophysics of the Western Cordillera.* Geological Society of America Memoir **152**.

Soons, J. M. and Selby, M. J. (eds) (1982) *Landforms of New Zealand.* Longman Paul, Auckland.

Talwani, M. and Pitman, W. C. III (eds) (1977) *Island Arcs, Deep Sea Trenches and Back-Arc Basins.* American Geophysical Union, Washington, DC.

Tapponnier, P., Peltzer, G. and Armijo, R. (1986) On the mechanics of the collision between India and Asia. In: M. P. Coward and A. C. Reis, (eds) *Collision Tectonics.* Geological Society Special Publication **19**, 115–57.

Uyeda, S. (1982) Subduction zones: an introduction to comparative subductology. *Tectonophysics* **81**, 133–59.

Von Huene, R. (1984) Tectonic processes along the front of modern convergent margins - Research of the past decade. *Annual Review of Earth and Planetary Sciences* **12**, 359–81.

Wallace, R. E. (1977) Profiles and ages of young fault scarps, north central Nevada. *Geological Society of America Bulletin* **88**, 1267–81.

Whitehouse, I. E. (1988) Geomorphology of the central Southern Alps, New Zealand: the interaction of plate collision and atmospheric circulation. *Zeitschrift für Geomorphologie Supplementband* **69**, 105–16.

Woodroffe, C. D. (1988) Vertical movement of isolated oceanic islands at plate margins: evidence from emergent reefs in Tonga (Pacific Ocean), Cayman Islands (Caribbean Sea) and Christmas Island (Indian Ocean), *Zeitschrift für Geomorphologie Supplementband* **69**, 17–37.

Yoshikawa, T., Kaizuka, S. and Ota, Y. (1981) *The Landforms of Japan.* University of Tokyo Press, Tokyo.

4

Landforms and tectonics of plate interiors

4.1 Landscapes of plate interiors

We saw in Chapter 3 how a number of major morphological features, notably orogenic mountain belts and island arcs, coincide with convergent plate boundaries; a superficial glance at the Earth's morphology might even suggest that all significant sub-aerial topographic features are confined to such boundaries. But the fact that the effects of plate interactions can extend for thousands of kilometres away from plate boundaries, as we have seen in the case of the Tibetan Plateau and central Asia (see Section 3.4.4.2), shows that this is certainly not the case. Moreover, a closer inspection of global morphology reveals a range of major landform features which are located in the interior of plates and which are unrelated to even distant plate convergence events (Fig. 4.1).

Some continents, most notably Africa, have extensive plateaus which lie at elevations (2000 m or more) which far exceed the average height of the continental platforms. **Basins** represent another major type of landform of plate interiors. These are commonly 1000 km or more across and are either entirely enclosed and therefore drained internally, such as the Lake Eyre Basin of Australia and the Chad and Kalahari Basins of Africa, or they are breached by one or more major river system, such as the region drained by the Zaire (Congo) river system (Fig. 4.1). Although volcanoes and associated volcanic landforms are clearly concentrated along plate margins, significant occurrences are also to be found in plate interiors. On the continents some of this intra-plate volcanic activity is closely related to rift valleys which themselves represent sites of **continental rifting** where the crust has been stretched and faulted.

The break-up of the supercontinent of Pangaea, which began about 180 Ma ago (Fig. 2.16), created many new continental margins. This event was of great significance for the development of landscapes on a continental scale because it led to the establishment of new base levels for continental erosion. These margins exhibit a much lower level of tectonic

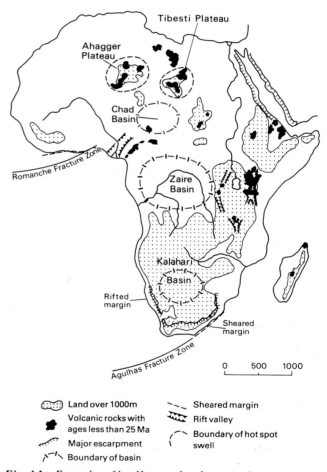

Fig. 4.1 *Examples of landforms related to tectonic processes operating in plate interiors on the African continent. Note that for clarity only some of the basins and hot-spot swells are shown.*

activity than the active margins located along convergent plate boundaries and are consequently termed **passive continental margins** (or simply **passive margins**). Where formed

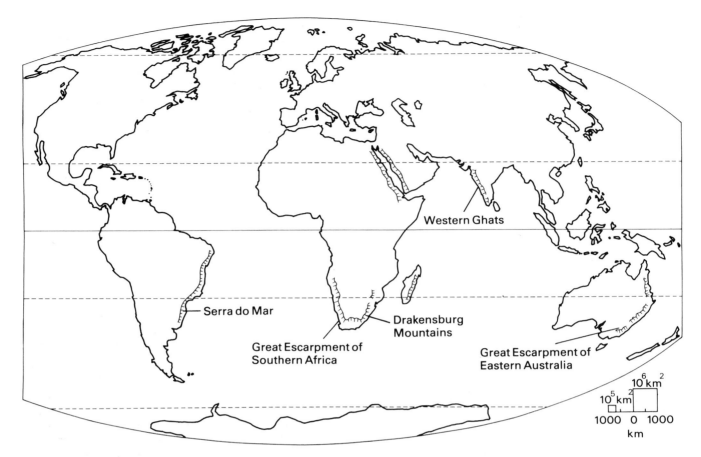

Fig. 4.2 *Great escarpments on passive margins. Although lacking major escarpments, a number of other passive margins possess low amplitude marginal upwarps flanked by a significant break of slope. An example is the Fall Line of the eastern seaboard of North America which marks a significant increase in stream gradients and in places this develops into a distinct escarpment.*

by divergent plate movement they are described as **rifted margins,** but where the motion between two adjacent continental blocks has been transform they are called **sheared margins** (but note that the term rifted margin is often applied rather loosely to passive margins of any type). Many, though certainly not all, passive continental margins are characterized by major escarpments (sometimes called **great escarpments**) (Fig. 4.2). By any standards these are significant morphological features – the Great Escarpment in southern Africa has a relief exceeding 1000 m in places – and they typically separate distinct geomorphic environments. Below great escarpments the topography is usually highly dissected, but inland on the plateau surface the relief is generally subdued.

Along plate margins the horizontal motion of plates is the primary force driving the uplift which occurs in orogens (although as we saw in Section 3.3 vertical movement resulting from thermal effects is of great importance in continental-margin orogens). In plate interiors horizontal plate movements play a less direct role in creating landforms and it is the processes which generate the broad warping and

vertical movements of the crust, that is, the mechanisms of epeirogeny, that are of most concern.

4.2 Mechanisms of epeirogeny

While we now have a reasonably good understanding of how interactions at plate boundaries give rise to major morphological features on the continents our knowledge of the tectonic processes operating in plate interiors is by comparision rather poor. Numerous mechanisms have been suggested to explain the often significant uplift that has occurred in plate interiors but none seems to be applicable to all cases. Although the various models defy any simple classification, for convenience we will consider them under the headings of thermal, phase change and mechanical models.

4.2.1 Thermal models

There are many occurrences of volcanic activity in areas remote from plate margins; the volcanoes of the East African

Fig. 4.3 *Old Faithful geyser, Yellowstone National Park, Wyoming, USA, representing hydrothermal activity associated with a hot spot beneath the North American continent. The surrounding region is characterized by active volcanism and uplift.*

Rift system, such as Kilimanjaro and Mount Kenya, the volcanic and associated hydrothermal activity of the Yellowstone region in Wyoming, U S A (Fig. 4.3), and the chain of volcanic peaks making up the Hawaiian Islands are examples from both continental and oceanic regions. There are also anomalous areas of volcanism on, or near, mid-oceanic ridges; Iceland, for instance, is formed of unusually thick oceanic crust produced by very high localized rates of volcanic activity on the Mid-Atlantic Ridge. Such an area of anomalous volcanism, which is not related to processes of subduction or normal mid-oceanic ridge spreading, is termed a **hot spot**. In addition to volcanic activity hot spots are in many cases associated with significant crustal uplift. The resulting hot-spot swells are typically several hundred kilometres across and may rise 1 km or more above the surrounding terrain.

Hot spots appear to be rather irregularly distributed over the Earth's surface (Fig. 4.4) but the real pattern is difficult

to establish because of the problems of identifying specific hot spots and incomplete data in some regions. Along midoceanic ridges and subduction zones it is often a problem distinguishing between 'normal' and anomalous volcanism. Since it is also difficult in some cases to determine whether neighbouring centres of active volcanism are associated with one or more hot spots, estimates of the total number of hot spots vary significantly from around 40 to well over 100.

A hot spot arises from the presence of unusually hot mantle at the base of the lithosphere (Fig. 4.5(A)). Such a **sublithospheric thermal anomaly** may be generated by a **mantle plume**, an upwelling of hot mantle originating deep in the Earth's interior, possibly at the core-mantle boundary; alternatively, it may be caused by localized heating at shallower depths in the mantle. In either case the presence of hot mantle leads to melting at the base of the lithosphere and the lithosphere as a whole becomes thinner (Fig. 4.5(B)). Since part of the lithosphere is replaced by hot mantle of lower density from the asthenosphere, the lithosphere as a whole experiences isostatic uplift. As the change in density is the result of a change in temperature this kind of isostatic adjustment is termed **thermal isostasy**. Eventually uplift may be sufficient to stretch the brittle overlying crust to the point where it fractures and a rift valley forms (Fig. 4.5(C)).

The time scale over which uplift takes place depends upon the way in which heat is transferred through the lithosphere. Conduction of heat occurs only very slowly through rock, and if this is the only mechanism operating uplift will occur over a very long period – something of the order of 100 Ma. In contrast **penetrative magmatism**, whereby hot and partially molten material from the asthenosphere forces its way up towards the surface, is a much more rapid mechanism of heat transfer and is capable of thinning 100 km thick lithosphere and producing uplift at the surface in only about 20 Ma.

As several million years are required for the transfer of a significant amount of heat to the surface of continental lithosphere, it has been suggested that hot spots are most likely to develop where the lithosphere is more or less stationary with respect to sub-lithospheric thermal anomalies as this would allow time for sustained heating to occur. This idea has been applied to the African Plate which, according to some analyses of plate motions, appears to have been almost stationary with respect to the underlying asthenosphere for the past 25–35 Ma. This period coincides with a well-established increase in volcanic activity in Africa and, less certainly, with a phase of uplift.

Although the idea of linking volcanism and uplift to sublithospheric thermal anomalies is attractive there are problems with this hypothesized relationship. One difficulty is that there is evidence indicating that hot spots can migrate with respect to each other at velocities of up to 20 mm a^{-1}. This means that both hot spots can move under 'stationary' lithosphere and that the lack of motion of the African Plate is

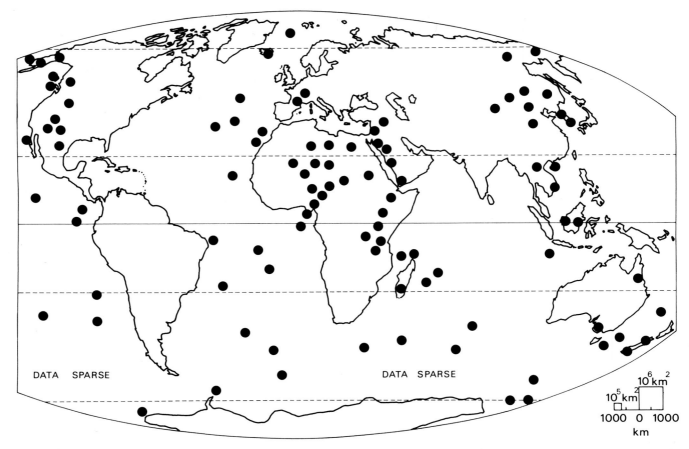

Fig. 4.4 *Global distribution of hot spots. (Based on P. R. Vogt, (1981)* Journal of Geophysical Research **86**, *Fig. 1, p. 951. Published by The American Geophysical Union.)*

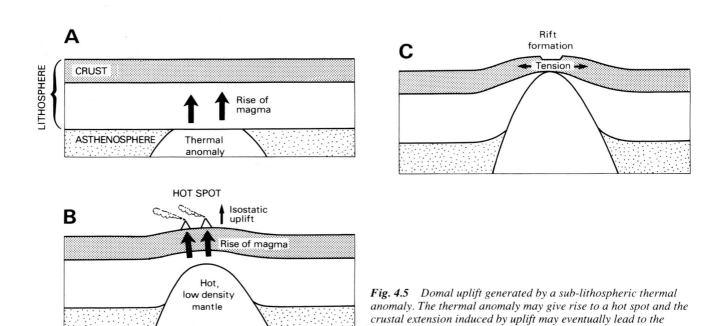

Fig. 4.5 *Domal uplift generated by a sub-lithospheric thermal anomaly. The thermal anomaly may give rise to a hot spot and the crustal extension induced by uplift may eventually lead to the development of a rift (see Section 4.3.2).*

thrown into doubt because its past positions have been in part determined with respect to a reference frame of supposedly fixed hot spots.

4.2.1.1 Lithospheric vulnerability

For a more realistic analysis of the occurrence of hot spots and their associated uplifts we need to consider the thickness of the lithosphere as well as its velocity with respect to sublithospheric thermal anomalies. This is because thick lithosphere will tend to be more resistant to the effects of heat conduction and penetrative magmatism. By relating variations in the thickness of the lithosphere over the Earth's surface to variations in the velocity of plate movement it is possible to determine global variations in **lithospheric vulnerability** (Fig. 4.6; Box 4.1). Lithosphere which is thin and slow moving is predicted to be the most vulnerable to penetrative magmatism and therefore the most likely to experience hot-spot volcanism and associated uplift.

On a global basis there is a fairly strong relationship between lithospheric vulnerability and the distribution of hot spots; few hot spots are to be found in areas of low lithospheric vulnerability whereas they are common in areas of high vulnerability. When specific regions are examined, however, this relationship does not always seem to apply. By carefully examining Figure 4.6 you can see that on the

Box 4.1   Lithospheric vulnerability

The lithospheric vulnerability index (v) is defined in terms of lithospheric thickness (l) and plate velocity (u) as

$$v = \frac{K}{lu^{1/2}}$$

where K is a constant for a property of the lithosphere known as thermal diffusivity which expresses the rate of propagation of a temperature change through time. This expression demonstrates that vulnerability is the same for 200 km thick lithosphere moving very slowly at 10 mm a⁻¹ as it is for lithosphere which is only 63 km thick but which is moving rapidly at 100 mm a⁻¹.

African Plate the relationship is, in fact, the opposite of that predicted with several hot spots located in areas of low lithospheric vulnerability. There are several possible reasons for this including inaccuracies in the estimates of lithospheric thickness and the over-zealous identification of hot spots on the African Plate. Nevertheless it is probable that factors other than lithospheric vulnerability, such as the actual global distribution of sub-lithospheric thermal anomalies, are important in influencing the occurrence of hot-spot volcanism and associated uplift.

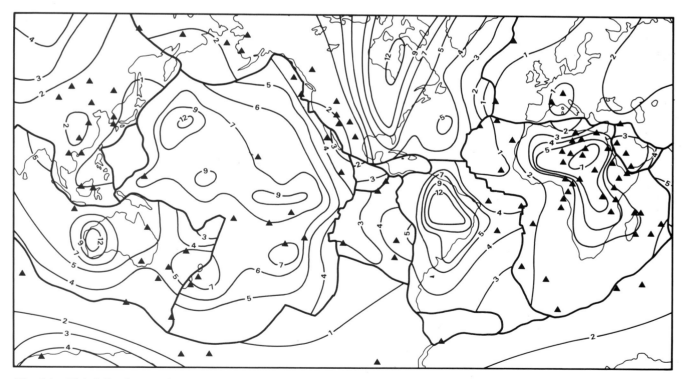

Fig. 4.6 *Global distribution of hot spots (triangles) and contours of the dimensionless lithospheric vulnerability index; lower values of the index indicate areas more vulnerable to penetrative magmatism. Major plate boundaries are also shown. Note that the set of hot-spot locations shown is slightly different from that illustrated in Figure 4.4. (After H. N. Pollack et al., (1981) Journal of Geophysical Research 86, Fig. 3, p. 965. Copyright by the American Geophysical Union.)*

4.2.1.2   *Effects of hot-spot migration*

One intriguing aspect of the effects of sub-lithospheric thermal anomalies on continental topography arises from the consequences of plates moving over such regions of unusually hot asthenosphere. As yet there has been no detailed attempt to assess the possible effects of such hot-spot 'migration' on landscape development, but there is some evidence from the stratigraphic record of the movement of topographic swells across continents which may be related to the passage of hot spots or similar thermal phenomena. For example, from about 160 to 100 Ma BP, North America drifted over a hot spot which is now located near the Great Meteor Seamount in the eastern Atlantic Ocean. Patterns of erosion, sedimentation and igneous activity recorded over the area during this period seem to be consistent with a migrating region of uplift related to the track of this hot spot.

On the basis of an estimated 10 per cent of the Earth's surface currently being part of a hot-spot swell it has been calculated that on average hot-spot epeirogeny should affect a particular area of crust about every 600 Ma. We would expect the slow migration of hot-spot swells across continental interiors to have a significant impact on landscape development through, for instance, the disruption and diversion of drainage systems. Geomorphologists have yet to assess this possibility, although the relationship between migrating hot-spot swells and landscape development provides an intriguing subject for future research.

4.2.2   Phase changes

A major problem with hot spots as a general explanation for uplift in plate interiors is that many elevated continental regions in such areas show no evidence of volcanism. Large parts of southern and eastern Africa stand at over 1500 m yet recent volcanism has been very localized. This might be explained by sub-lithospheric thermal anomalies causing uplift without penetrative magmatism developing to a point where volcanic activity occurs at the surface. Another difficulty with the hot-spot model is that unrealistically large increases in heat flow are apparently required to explain the magnitude of uplift recorded in some regions.

An alternative explanation of such uplifts involves the effects of density changes in minerals in the upper mantle. Such **phase changes** occur when minerals adopt different atomic configurations as temperatures and pressures change. The temperatures experienced by a mineral in the crust or mantle can be affected by heat flow from below while the pressure can be altered by loading and unloading of the crust above. Consequently the depth at which particular phase changes (and therefore density changes) occur can migrate up and down within the lithosphere. The surface above will rise or fall in response to the changes in volume brought about by these alterations in mineral density. A number of possibly significant phase changes have been suggested but

our lack of detailed knowledge of the abundance of specific minerals in the lower crust and upper mantle makes their evaluation rather uncertain.

A phase change, triggered by a small amount of heating at the base of the lithosphere, has been proposed to account for the uplift evident in a number of regions in plate interiors, including southern and eastern Africa, south-east Australia and the Transantarctic Mountains of Antarctica. Each of these areas of continental lithosphere are thought to have over-ridden regions of hot asthenosphere associated with former mid-oceanic spreading ridges. It is thought that this would have increased the flow of heat to the base of the lithosphere sufficiently to cause a phase change in the uppermost continental mantle after a delay of up to 70 Ma (the time taken for heat to be conducted from the base of the lithosphere to just below the Moho). The conversion of eclogite, a rock with a density of 3400 kg m^{-3}, to basalt (density 3000 kg m^{-3}) is the kind of phase change envisaged. If this is so a 11.25 km thick layer of eclogite would need to be converted to a thickness of 12.75 km of basalt to produce an uplift of 1500 m.

The attractiveness of this model is that it requires only a modest increase of temperature of around 100 °C at the base of the lithosphere. This seems much more reasonable than the very large increases of up to 1000 °C which would be required to produce an uplift of around 1500 m by purely thermal effects. A major problem with the model, however, is that it predicts a rather uniform and extensive pattern of uplift which does not accord with the typical morphology of continental plate interiors. Africa, for instance, is characterized by areas of domal uplift, from 500 to 2000 km across and by a prevalence of high elevations along its continental margins.

4.2.3   Mechanical models

The most straightforward mechanical models of uplift are those that involve the isostatic adjustment that takes place when a load is removed from the crust (see Section 2.2.4). The term epeirogeny was in fact first coined by G. K. Gilbert in 1890 to describe the isostatic uplift which occurred as a result of the evaporation of Lake Bonneville in Utah in the western U S A (leaving the much smaller Great Salt Lake of the present day). During a period of high lake levels 15 000–25 000 a BP the load provided by the average water depth of about 145 m depressed the crust. After unloading by evaporation the crust rose and shoreline features are now elevated up to 64 m above the present water level.

Far more spectacular examples of isostatic uplift are to be found in the areas of the northern hemisphere covered by great ice sheets during the Pleistocene. Detailed records of post-glacial uplift have been established for such areas as the Fenno-Scandian Shield and the Canadian Shield from both raised and warped shorelines and gravity data. Uplift of the central region of the Fenno-Scandian Shield began about

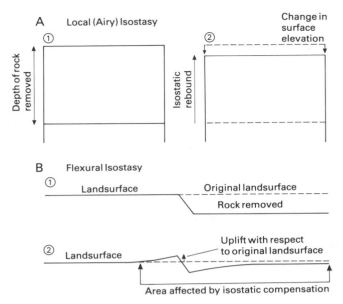

A   Local (Airy) Isostasy

Change in surface elevation

Isostatic rebound

B   Flexural Isostasy

Landsurface   Original landsurface

Rock removed

Landsurface   Uplift with respect to original landsurface

Area affected by isostatic compensation

Fig. 4.7   Models of isostasy showing the contrasting modes of vertical adjustment between local (Airy) isostasy (A), and flexural isostasy (B).

13 000 a BP and the total uplift since then has exceeded 100 m.

Though involving less spectacular rates of uplift, crustal unloading by denudation is a far more pervasive process which extends over long periods of time, in contrast to the rather transient effects of deglaciation and evaporation of lakes. The latter two phenomena essentially restore the crust to its approximate elevation before loading by ice or water, but denudation can lead to continued uplift of the crust for as long as continental crust is sub-aerially exposed. If isostatic compensation is local, that is if vertical adjustments of the lithosphere in response to the removal of a load are confined just to the area covered by the original load, then isostatic uplift simply reduces the rate at which a land-surface is lowered by denudation (Fig. 4.7(A)). Such local compensation, however, cannot occur at the small scale because the lithosphere has a finite strength and so changes in load result in regional isostatic adjustments over a greater area than that actually affected by the change in load (Fig. 4.7(B)).

Such flexural isostasy (see Section 2.2.4) is a familiar phenomenon to geomorphologists who study the uplift and warping of shorelines as a result of crustal loading by nearby ice sheets. In this case flexing of the lithosphere leads to the development of a **forebulge** at some distance from the ice margin which experiences an increase in surface elevation. It is possible that a similar flexural effect is associated with great escarpments along passive continental margins. In this case there is active erosion (and therefore unloading) along the escarpment edge but relatively little erosion on the plateau surface immediately inland of the escarpment. Consequently it is possible that the slowing eroding plateau summit could

actually increase in elevation as a result of the erosional un-loading along the escarpment (Fig. 4.7(B)). A similar effect can lead to the summits of mountains in orogenic belts in-creasing in elevation as the incision of deep valleys between peaks causes unloading of the crust. The extent to which isostatic compensation occurs regionally through flexure, rather than locally, depends on the **flexural rigidity** of the lithosphere, a property which indicates its resistance to bend-ing in response to a change in load.

Another proposed mechanism of epeirogenesis rests on the observation that continental lithosphere is in an unstable mechanical equilibrium because its mantle is denser than the underlying asthenosphere. Consequently if some process, such as cracking or 'erosion' by mantle plumes, punctures the continental lithosphere as far as the crust the relatively dense mantle layer of the lithosphere could become detached and sink (Fig. 4.8). This process has been termed **delamination**

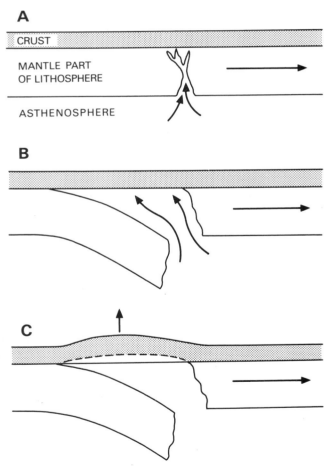

A

CRUST

MANTLE PART OF LITHOSPHERE

ASTHENOSPHERE

B

C

Fig. 4.8   Schematic representation of delamination showing: (A) initial cracking of the mantle part of the lithosphere by penetrative magmatism; (B) the sinking of this portion of the lithosphere into the less dense asthenosphere; (C) crustal uplift associated with the replacement of lithosphere by warmer and less dense sub-lithospheric mantle. Plate motion is to the right in relation to the underlying asthenosphere.

and may be capable of causing epeirogenic uplift through the exposure of the cold base of the crust to hot mantle from the asthenosphere and the replacement of relatively dense lithospheric mantle by less dense mantle from the asthenosphere. Two such delamination 'events' may account for the episodes of uplift of the Colorado Plateau in the south-west U S A. around 30 Ma and 5 Ma BP.

4.3   Continental rifts

4.3.1   Rift structure and location

Rift valleys are important landforms on continental crust in plate interiors where tensional stresses predominate in the lithosphere. They typically have widths of 30–60 km, but some are up to 100 km across. Pioneers of rift valley geology such as J. W. Gregory were strongly influenced in their interpretations of rift structure by the apparently simple rift morphology that they encountered in areas such as East Africa. The conventional image of a rift valley is of a broad, flat-floored and symmetric trough flanked by steep escarpments. The most obvious structural interpretation of this morphology is that of a graben with the rift floor being formed on a downthrown block bounded by normal faults which create the steep escarpments (Fig. 4.9(A)). But a closer examination of both the morphological and structural evidence shows that this model is inadequate. In particular the application of **seismic stratigraphy** to the study of rift valleys during the 1980s has led to a considerable revision of ideas about their structure.

Seismic stratigraphy is a technique whereby seismic waves generated by small explosions set off at the surface are reflected or refracted from discontinuities in the underlying sediments which represent changes in sediment properties. Such changes thus enable discrete sedimentary units to be identified and subsurface faults to be mapped (see Section 17.2.2.2. for a more detailed explanation). Seismic data which show the deep structure of rift systems have revealed a much more complex pattern of faulting than was suspected in the classic rift model, with the presence of listric as well as normal faults. Most significantly, the structure of many rifts examined by seismic methods has been shown to be asymmetric with most of the downthrow occurring along a major boundary listric fault on one side of the rift. Such rifts are said, therefore, to have a **half-graben** structure with the main boundary listric fault forming a **footwall,** and the opposing downwarped crust forming a **hanging wall** leading down from a **roll-over** zone (Fig. 4.9(B)). The presence of an **antithetic fault** on the hanging wall margin can give the impression of a symmetric rift valley if it is exposed and forms an escarpment, even though the overall structure is asymmetric.

Further evidence contradicting the traditional symmetric rift valley model comes from observations of their morphology and surface structure. Careful examination of fault

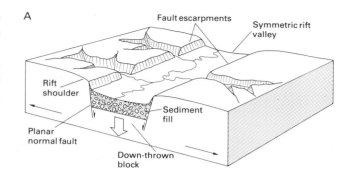

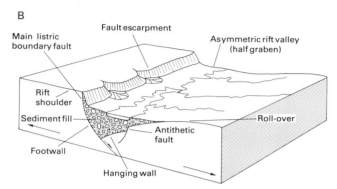

Fig. 4.9   *Schematic representation of contrasting rift structures: (A) classic symmetric graben structure with a downthrown block bounded by normal faults; (B) asymmetric, half-graben structure. In both cases the number of faults in real rifts and the complexity of their structure is much greater than is indicated here.*

patterns, for instance, reveals that major faults tend to occur on only one side of a rift along any one section of the valley. Moreover, these major faults tend to be discontinuous and curved in plan and it seems that main boundary listric faults alternate along rift valleys and are separated by **transfer faults** (Fig. 4.10). Some rift valley lakes, such as Lake Tanganyika, Lake Malawi and Lake Turkana in East Africa

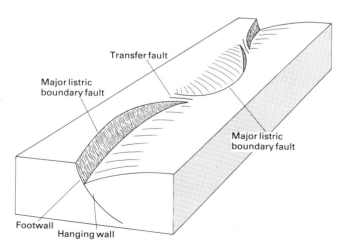

Fig. 4.10   *Highly schematic illustration of alternating half-graben along a rift valley.*

have a slightly sinuous plan form which may reflect the structural control exerted by these alternating curved boundary faults.

Even a cursory examination of global structure and morphology shows the generally close relationship between continental rifts and areas of mid-plate volcanism and crustal doming and uplift. The location of rift systems also seems to be closely related to the distribution of ancient structures representing zones of weakness between stable cratons. The East African Rift System, for instance, parallels the trend of old mobile belts (regions of highly deformed rocks formed during early phases of folding and orogenesis) and tends to follow the margins of Precambrian cratons (Fig. 4.11). Such ancient lines of weakness are apparently more readily activated than adjacent crust and consequently are favoured sites for later faulting and rift development. Some recently active rifts, however, do cut across ancient Precambrian structures, suggesting that the stress pattern initiating rifting, as

well as the position of zones of crustal weakness, is important in determining the location and orientation of continental rifts.

The major differences in rift characteristics relate to their position with respect to plate boundaries and the intensity of volcanic activity associated with them. Some rifts appear to be clearly related to divergent crustal movements with lateral extension being normal to the rift axis. The East African Rift System is of this type and its interpretation as an incipient site of plate rupture is further supported by its continuity with the divergent plate boundaries of the Red Sea and Gulf of Aden. In other cases, such as the Dead Sea Rift, there is a significant transform component in the extensional movement and this feature can in fact be regarded as a large pull-apart basin (see Section 3.5). The other main rift type is to be found orientated approximately at right angles to the strike of intercontinental collision orogens. The development of this type of rift appears to be related to the crustal extension that occurs, as a result of continental collision, beyond the immediate collision zone itself. Examples of this type are the Rhine Rift in the foreland zone of the Alps, and the Baikal Rift System which lies some 3000 km to the north of the Himalayas (see Section 3.4.4.2).

There are significant differences in the morphology and degree of volcanic activity associated with these two types of rift. True intra-plate rifts, such as the East African Rift System, are characterized by prolonged volcanism and typically exhibit 1–2 km of down-faulting along the crest of substantial crustal upwarps. In contrast orogen-related rifts, such as the Rhine and Baikal structures, experience generally less intense volcanism and much more marked down-faulting, amounting to 5–6 km in the case of the Baikal Rift. In this kind of rift, however, the marginal uplift is either narrow or lacking altogether.

4.3.2 Rift formation

The commonly close relationship between rifting and crustal doming in rifts not associated with orogens suggests a causal relationship. There are two possibilities: one involves **active rifting** in which rifting develops as a response to the tensional stresses induced in the crust by uplift resulting from upwelling of the asthenosphere (Fig. 4.5(C)); the other involves **passive rifting** in which rifting is initiated by extensional stresses in the lithosphere, and this permits the subsequent upwelling of hot mantle which in turn induces thermal uplift (Fig. 4.12).

Clearly the timing of uplift, volcanism and rifting is crucial in the assessment of the validity of the active and passive rifting models. The typical sequence for passive rifting is rifting followed by volcanism, which may in turn be succeeded by uplift. Active rifting, on the other hand, would be expected to begin with volcanism which is then followed by uplift before rifting itself is finally initiated; but in some

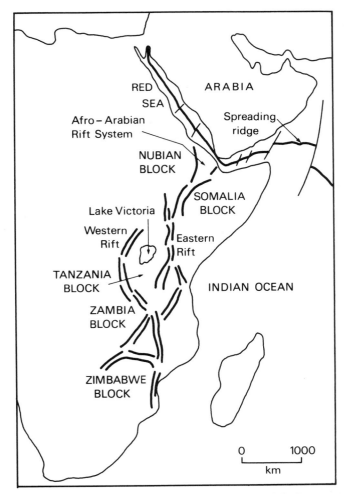

Fig. 4.11 *Simplified map of the main components of the East African Rift System. Note the continuity with the spreading ridge leading from the Red Sea through the Gulf of Aden and into the Indian Ocean. (Based on various sources.)*

A

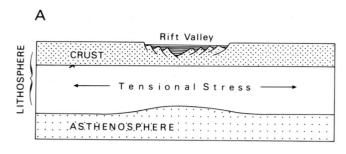

B

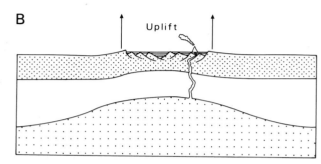

Fig. 4.12 *Formation of a rift through extension of the lithosphere. In contrast to the active rifting model, in this passive rifting mechanism extension and thinning of the lithosphere (A) precedes uplift of the rift flanks (B).*

cases uplift seems to precede volcanism. In practice some rifts seem to be of the active type whereas others seem to have been initiated by passive rifting, although in many cases there is no consensus as to whether specific rifts are active or passive.

4.3.3 Rifting and uplift in Africa

The domal uplifts and rift structures of Africa are at various stages of development and thus provide a useful basis for examining the evolution of rift systems. Two major periods of uplift and rift formation have been recognized in Africa; a Jurassic-Cretaceous rifting phase between 180 and 130 Ma BP associated with the break-up of Gondwana and a more recent period initiated some 35 to 25 Ma BP and continuing up to the present.

There is some evidence in Africa for the idea that uplift may precede volcanism since the Adamawa dome in the Cameroon region, which rises to elevations in excess of 2000 m, lacks evidence of recent volcanic activity. Volcanism and uplift without rifting is exemplified in North Africa where Neogene volcanic rocks are to be found on the Tibesti and Ahaggar Plateaus (Fig. 4.1). Domal uplifts with rifts cutting across them are particularly common in East Africa where the East African Rift System comprises a whole series of such structures (Fig. 4.11). The crests of the domal uplifts typically have three rifts meeting at angles of about 120°; this geometrical arrangement appears to arise from the way

in which the tensional stresses within the lithosphere are most readily accommodated. Eleven such triple junctions, each possibly associated with a separate hot spot, have been identified on the continental part of the African Plate and an intriguing problem is the way in which they may subsequently promote continental rupture and develop into spreading centres.

Spreading can apparently only occur when continental separation can be accommodated into the motions of the global plate system. Where this is not possible rift development is halted, rifts become filled with sediment and any associated uplift subsides. Although the triple junctions identified in Africa, including those along the continental margin as well as those in the interior, have variously evolved by spreading along one, two or all three rift arms, the most common sequence has been for one arm to remain inactive and form an **aulacogen,** with spreading occurring along the other two (Fig. 4.13). Of the 29 rifts identified in Africa, five have led to continental separation and a further three have been associated with the movement of Madagascar away from the African mainland. The two rifts in the Benue Trough in Nigeria went through a phase of opening followed by later closing, and of the remaining rifts ten became inactive before the spreading stage and nine are currently active and may spread in the future.

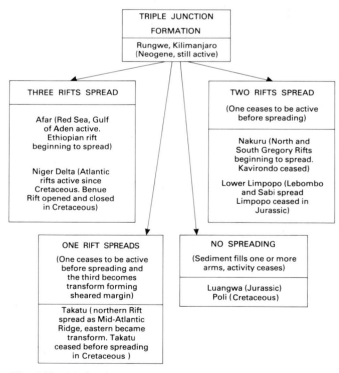

Fig. 4.13 *Mode of evolution of triple junctions illustrated from rift structures in Africa. (Adapted from K. Burke and A. J. Whiteman (1973) in: D. H. Tarling and S. K. Runcorn (eds)* Implications of Continental Drift for the Earth Sciences, *2 vols. Academic Press, London, Table 3, p. 749.)*

4.4 Continental basins

4.4.1 Basin morphology

In addition to upward vertical movements of the crust which give rise to broad upwarps or swells, downward movements also occur creating basins. Basins located in continental interiors are sometimes referred to as **intra-cratonic basins** while those located along continental margins in intra-plate settings are called **passive margin basins.** The latter are primarily submarine features and will not be discussed here although they are certainly an important component of passive margin development (see Section 4.5).

Areas of crustal subsidence usually form topographic features with minimal local relief because they become partly or wholly filled by sediment transported from surrounding highlands; examples include the Lake Eyre Basin in southern Australia, the Kalahari Basin in southern Africa and the Ob Basin in the USSR. The volume of sediment that accumulates in a basin depends on the rates of subsidence, sediment supply and sediment removal. Evacuation of basin sediments is primarily by rivers for basins with an outlet to the ocean (such as the Zaire Basin) but wind action is important in enclosed basins (such as the Kalahari Basin) (Fig. 4.1). Basins in continental interiors are a few hundred to a thousand or more kilometres across and the sedimentary units, with which they are partially or completely filled, generally thicken towards the centre. This sediment thickening indicates the role of sediment loading in sustaining basin subsidence but it does not explain how basins originate. Continental basins generally seem to be long-lived features with a subsidence history often extending over a period of 100 Ma or more.

4.4.2 Mechanisms of subsidence

Basins form as a result of downwarping of the crust, as a consequence of uplift of the surrounding region, or through a combination of both of these effects. Epeirogenic uplift and subsidence are closely related phenomena and a number of the mechanisms of uplift discussed in Section 4.2 may be relevant to crustal subsidence. For example, while heating of the lithosphere may account for uplift, subsidence may result from lithospheric cooling. In this case the connection between uplift and subsidence is especially close as there must first be a heating event for the subsequent cooling to give rise to a basin. On its own heating and cooling of the lithosphere will not create a basin as the crust will simply return to its original elevation. What is required is a mechanism for thinning the lithosphere during the heating phase so that when it cools and subsides a region of negative relief is formed.

One possible way of accomplishing this is through the erosional thinning of the crust when it is elevated during the period of uplift. The amount of thinning and consequently the amount of eventual subsidence will depend on the depth of denudation that has occurred. In most cases, however, this appears to be insufficient to account for the amount of subsidence observed.

If denudation is insufficient as a thinning mechanism then sub-crustal processes must be considered. Phase changes provide a possible mechanism which could account for subsidence as well as uplift, depending on whether the crystalline transformations involved give rise to more, or less, compact mineral structures. A further potential thinning mechanism which has received considerable support since the late 1970s is the stretching of the lithosphere as a result of sustained tensional stress (Fig. 4.12). Lithospheric extension will promote faulting and an initial phase of subsidence associated with upwelling of hot mantle from the asthenosphere into the base of the lithosphere. This is followed by a slower phase of subsidence promoted by cooling of this mantle material as heat is conducted up to the surface.

Irrespective of the mechanism initiating basin formation the amount of subsidence will be amplified by the weight of sediments which accumulate. This is undoubtedly a highly significant process where sediment thicknesses of several kilometres occur. It is also a major reason for the thickening of sedimentary units towards the centre of basins. The initial depression will be the first to accumulate sediment and this will lead to a positive feedback cycle of further subsidence and sediment accumulation extending outwards from the initial focus of subsidence. Because of the rigidity of the lithosphere, down-flexure may extend for 150 km or more beyond the area of sediment loading. While subsidence is maintained by sediment loading, the surrounding upland areas supplying sediment will rise isostatically through unloading. Once initiated such a complementary system of uplift and subsidence is probably self-sustaining for long periods of time in the absence of major tectonic disturbance. In enclosed basins water-loading by lakes may further contribute to subsidence, though their effect will be far more ephemeral than sediment loading as they are likely to undergo repeated phases of growth and desiccation due to changes in climate.

4.5 Passive continental margins

The sequence of uplift, rifting and continental rupture which leads to the formation of passive continental margins is of considerable importance in any attempt to understand the large-scale geomorphology of the present-day continents. Moreover, passive margins have assumed great economic significance since the 1960s as their hydrocarbon potential has begun to be extensively exploited by the oil industry. The break-up of Pangaea (Fig. 2.16) generated a considerable length of passive continental margins and analysis of the palaeomagnetic record of the continents and the subsequently created ocean floor has given us a fairly good idea of the timing of most of these rifting episodes.

A feature common to many, but certainly not all, passive

continental margins is a broad upwarp which separates the coast from interior basins. Such upwarps consist of broad swells running parallel to the coast and are in most cases flanked on their oceanward side by a great escarpment. When rifting leads to the rupture of a continent the original elevation of the landsurface above sea level prior to rifting inevitably creates some relief once a new base level is established. But a number of passive margins appear to have experienced uplift either during or soon after rifting which has led to the development of continental-margin upwarps which stand higher than landsurfaces further inland. Clearly models of passive margin evolution must be able to account for the development of these upwarps.

Our understanding of the vertical (as opposed to the horizontal) movements of the lithosphere during continental rupture is largely derived from the interpretation of the sediments laid down in passive margin basins. Knowledge of these sedimentary sequences, which are now almost entirely located below sea level, has been greatly enhanced by the application of seismic stratigraphy (see Section 4.3.1). Together with evidence from boreholes drilled through continental-margin deposits, investigations using seismic stratigraphy have shown that most passive margins support a wedge of sediment which becomes gradually thinner as it passes seawards from continental to oceanic lithosphere (Fig. 4.14). Deposits laid down during the rifting phase are

termed **synrift sediments** while those laid down on the margin once continental separation has occurred are termed **postrift sediments.** Interpretations of these sedimentary sequences in conjunction with morphological and structural evidence has led to the development of numerous models of the tectonic evolution of passive margins.

4.5.1 Active rifting

In the active rifting model (see Section 4.3.2) rifts propagate between a series of domal uplifts formed above sub-lithospheric thermal anomalies (Fig. 4.15(A)). This gives rise to a sequence of domes and intervening saddles which, as the break-up of the continent proceeds and a new ocean is created, are split to form a sequence of ruptured domes and troughs along the newly formed passive margins. At this early stage of development these new continental margins are termed **nascent passive margins** (Fig. 4.15(B)).

As sea-floor spreading continues the margin ages and subsidence begins to predominate over uplift, at least along its oceanward flank (Fig. 4.15(C)). Such subsidence eventually leads to the submergence of the oceanward part of the margin and the formation of a continental shelf. Subsidence during this **mature passive margin** stage is driven by both **thermal subsidence** arising from the cooling of the margin as it moves away from the region of mantle upwelling

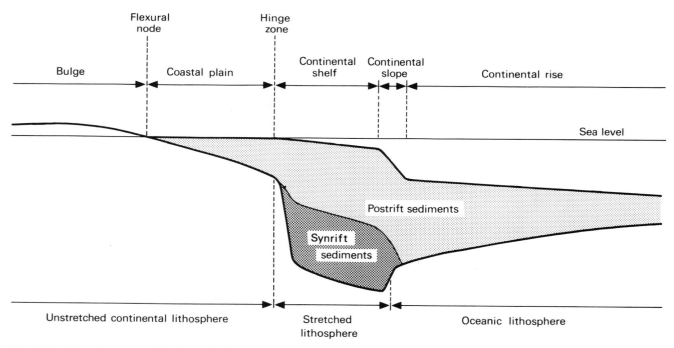

Fig. 4.14 *The major structural and morphological elements of passive (rifted) continental margins. Synrift sediments are those deposited in a rift prior to continental rupture, whereas postrift sediments date from the initiation of sea-floor spreading. The **hinge zone** marks the line along the margin dividing continental lithosphere of normal thickness from lithosphere which has been thinned and stretched during rifting and subsequently subsided under the load of overlying sediments. The **flexural node** separates the (oceanward) portion of the margin subject to subsidence from that experiencing uplift. (Modified from A. B. Watts (1982) Nature **297**, Fig. 2, p. 470.)*

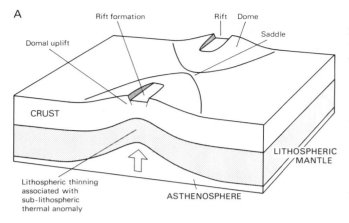

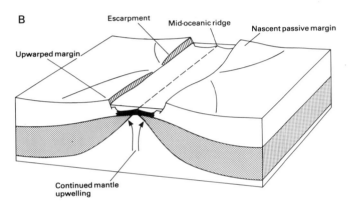

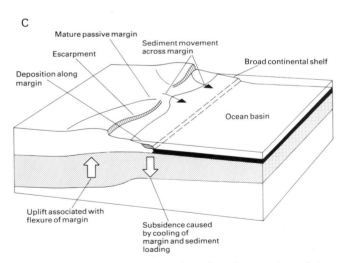

Fig. 4.15 *Schematic representation of passive margin evolution through active rifting showing the initial uplift and rifting (A), nascent passive margin (B) and mature passive margin (C) stages.*

located at the site of rifting, and by the isostatic loading caused by the growing wedge of sediment accumulating offshore. This sediment loading is coupled with isostatic unloading on the landward part of the margin as material is transported offshore.

Some models of passive margin tectonics assume that the isostatic compensation in response to these changes in load is local rather than regional, in which case no net uplift of the landsurface occurs (see Section 4.2.3). But if the lithosphere along the margin has sufficient rigidity the resulting imbalance in load will be accommodated through flexural isostasy. This will lead to a net uplift of landsurfaces along the least eroded part of the margin, that is, inland of major escarpments. The magnitude of this effect is difficult to estimate, but it will depend in part on the flexural rigidity of the lithosphere. Nevertheless, flexural isostasy provides a possible uplift mechanism for passive margins during their mature stage of development when the effects of the initial thermal uplift have diminished.

Unless the zones of sub-lithospheric mantle upwelling which precipitate uplift and rifting are in a line, sheared margins will develop along zones of transform plate motion linking offset domal upwarps (Fig. 4.16(A)). Modelling by geophysicists indicates that a 200 km diameter convective upwelling could produce a domal uplift about 1000 km across with a maximum elevation of around 1000 m. Rift arms at triple junctions which fail to spread will form aulacogens striking at a high angle into the continental margin (Fig. 4.16(B), (C)).

The extensive series of domal uplifts traversed by the East African Rift System has been widely regarded as a clear example of active rifting with uplift preceding the development of rift structures. Some recent work, however, has cast doubt on this interpretation and suggests instead that rifting may have been initiated before uplift. The evolution of the nascent rifted margins of the Red Sea has also been interpreted both in terms of active and passive rifting mechanisms.

4.5.2 Passive rifting

The idea that simple extension of the lithosphere can produce the structural features observed along passive margins was first presented by D. McKenzie in 1978. In this initial presentation of the passive rifting model it was assumed that the amount of extension was uniform throughout the lithosphere (Fig. 4.17(A)). Calculations based on this assumption, however, indicated that any uplift would be confined to the region over which stretching occurs and that no uplift at all would occur if the thickness of the crust were less than about 20 per cent of the thickness of the lithosphere. In fact this **uniform extension** model predicted that in most geologically reasonable situations subsidence caused by thinning of the lithosphere through stretching would outweigh any thermal isostatic uplift arising from the partial replacement of cool lithosphere by hot asthenosphere below the zone of extension.

Since this simple model was unable to explain the kind of uplift observed along rifted margins there were immediate attempts to modify it. One way the model was altered was to assume that the the mantle part of the lithosphere stretches

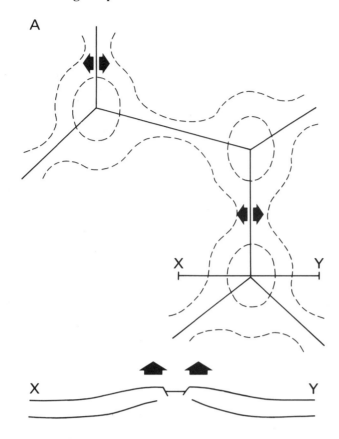

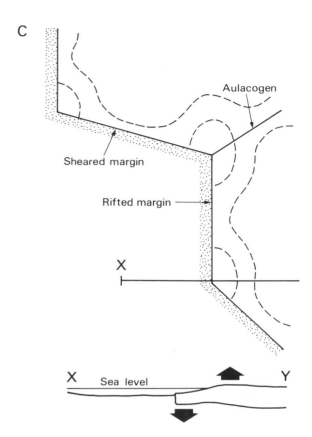

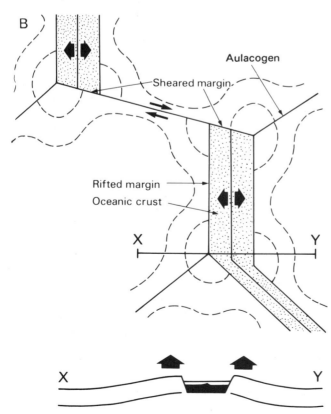

Fig. 4.16 *Plan view of the sequence of continental break-up and passive margin formation under active rifting: (A) rifting stage; (B) nascent passive margin stage with the formation of a rifted margin where the motion along the new plate boundary is largely divergent, and the creation of a sheared margin where the motion is largely transform; (C) mature passive margin stage with failed rift arms (aulacogens) striking into the new continental margin. X–Y cross-sections indicate the general nature of associated vertical movements across the margin and the dashed lines schematically indicate contours.*

by a greater amount than the crust during extension (Fig. 4.17(B)). In this **non-uniform extension** (or **depth-dependent extension**) model uplift will take place because relatively dense lithosphere is replaced by warmer and less dense asthenosphere below relatively less stretched and thinned crust. Uplifts of 1–2 km can apparently be produced if the crust experiences very little or no extension and the lithosphere is stretched by a large amount (Box 4.2).

Another modification of the model assumes that uniform extension occurs but takes account of the fact that stretching of the lithosphere is likely to occur over a long period of time, probably several million years. In his original model McKenzie assumed, for ease of calculation, that the lithosphere is stretched instantaneously. If we make the more realistic assumption that it takes a finite period of time then we have to take into account the fact that replacement of the lower lithosphere by hot asthenosphere during extension will lead to heat being conducted laterally into unextended,

A

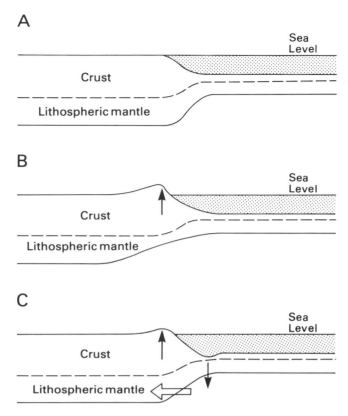

B

C

Fig. 4.17 Response of rifted margins to different modes of passive rifting: (A) (instantaneous) uniform extension; (B) non-uniform (depth-dependent) extension in which the sub-crustal lithosphere stretches by a greater amount than the crust and uplift occurs in the zone of relatively unthinned crust; (C) uniform extension over a finite period of time with uplift promoted by lateral heat flow into adjacent unthinned lithosphere. (Modified from C. Beaumont et al. (1982) Philosophical Transactions of the Royal Society London *A305, Fig. 10, p. 310.)*

Box 4.2   Uplift by non-uniform extension

We can calculate the maximum uplift to be expected during non-uniform, or depth-dependent, extension if we know how much the sub-crustal lithosphere has been stretched in regions where the overlying crust has experienced little or no extension. It is in fact often difficult to determine these stretching factors empirically, but we can make some calculations using geologically reasonable values. The maximum uplift (Δh) is given by

$$\Delta h = \frac{0.5\rho_m\alpha(T_a - T_s)\,l_1}{\rho_c}\left(1 - \frac{1}{\beta_L}\right)$$

where l_1 is the initial thickness of the lithosphere, β_L the proportionate amount of stretching experienced by the sub-crustal lithosphere, ρ_m and ρ_c the densities of the mantle (3300 kg m^{-3}) and crust (2700 kg m^{-3}) respectively, $T_a - T_s$ the difference in temperature between the base and the top of the lithosphere (1300 °C) and α the volumetric coefficient of thermal expansion (that is, the proportionate change in volume accompanying a change in temperature) ($\alpha = 3.4 \times 10^{-5}$°C^{-1}).

If we assume that the initial lithospheric thickness is 150 km and the sub-crustal lithosphere is stretched by a factor (β_L) of 1.1. (that is, its 'length' is increased by 10 per cent) then the maximum uplift in unstretched overlying crust will be about 0.37 km. But if we increase the stretching factor we obtain greater changes in surface elevation; thus for $\beta_L = 1.2$ we get 0.68 km and for $\beta_L = 1.5$ we get 1.35 km. These higher figures are comparable to the amount of uplift observed on a number of passive margins. The initial lithospheric thickness also affects the ultimate amount of uplift generated in unstretched overlying crust. For example, if the initial lithospheric thickness is only 100 km where the stretching factor (β_L) is 1.5, the uplift is reduced to 0.9 km.

and therefore unthinned, lithosphere, as well as vertically into the overlying upper lithosphere (Fig. 4.17 (C)). Depending on the amount of heat involved this mechanism could lead to a significant uplift through thermal isostasy due to the heating of this unthinned lithosphere. Since this heat transfer will occur only slowly the maximum uplift of the passive margin would probably occur as long as 60 Ma after rifting; this contrasts with the non-uniform extension model in which uplift takes place concurrently with stretching and rifting.

A further approach to explaining the development of significant continental-margin upwarps as a consequence of passive rifting has been to invoke a process known as **secondary convection**. In this model it is proposed that thinning of the lithosphere and its partial replacement by hot asthenosphere might allow convective upwelling to become established below the zone of extension. It is envisaged that this active mantle upwelling would lead to a significant heating of the overlying lithosphere and could generate uplifts of up to around 1 km. The difference between this model and

active rifting is that the convective upwelling is a consequence, not a cause, of lithospheric thinning and rifting. The secondary convection model has been applied to the Gulf of Suez at the northern end of the Red Sea. Here significant uplift of rift flanks has occurred although there is convincing evidence (from the deposition of marine sediments at the time of rifting) that rifting preceded uplift.

The sequence of rifted margin evolution to be expected from passive rifting is illustrated schematically in Figure 4.18. This figure is non-committal as to the precise mechanism likely to lead to the generation of passive margin upwarps as it illustrates secondary convection as well as non-uniform (depth-dependent extension) and the effects of lateral heat flow into unthinned lithosphere. As is evident from comparing Figures 4.15 and 4.18, by the mature stage continental margins formed by both active and passive rifting tend to assume a similar mode of evolution, with subsidence, at least in the offshore zone, predominating over uplift as the margin both cools and is loaded by sediment shed from the adjacent land mass.

A further component of passive rifting that needs to be considered is whether a rifting event is accompanied by

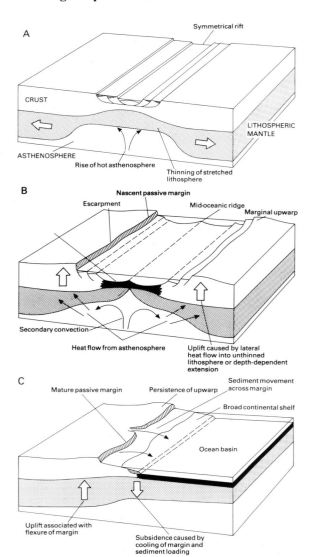

A

Symmetrical rift

CRUST

LITHOSPHERIC MANTLE

ASTHENOSPHERE

Rise of hot asthenosphere

Thinning of stretched lithosphere

B

Nascent passive margin

Escarpment

Mid-oceanic ridge

Marginal upwarp

Secondary convection

Heat flow from asthenosphere

Uplift caused by lateral heat flow into unthinned lithosphere or depth-dependent extension

C

Mature passive margin

Persistence of upwarp

Sediment movement across margin

Broad continental shelf

Ocean basin

Uplift associated with flexure of margin

Subsidence caused by cooling of margin and sediment loading

Fig. 4.18 *Schematic representation of passive margin evolution through symmetric passive rifting showing the initial rifting (A), nascent passive margin (B) and mature passive margin (C) stages.*

abundant volcanism. It has been pointed out that enormous thicknesses of basaltic lavas erupted during the early stages of rifting characterize some passive margins, such as those of eastern Brazil, western India and the south-eastern part of southern Africa. A number of other passive margins, however, show no sign of such major volcanism. This observation has led to the suggestion that whether a passive margin is volcanic or non-volcanic depends on whether the site of rifting happens to coincide with the location of a mantle plume and its associated hot spot.

Where rifting occurs over a mantle plume the sub-lithospheric mantle will be up to 200 °C hotter than normal and large quantities of magma will be generated. This occurs because rifting of the lithosphere allows hot mantle

rock to move towards the surface. The resulting reduction in pressure, or decompression, experienced by this material causes it to melt. Some of the magma generated is erupted on to the surface to form enormous lava flows (see Chapter 5), but by far the largest proportion is accreted to the base of the crust, a process known (somewhat misleadingly) as **underplating**. According to the underplating model the addition of volcanic rock thickens the crust and the resulting isostatic adjustment leads to the formation of a broad hot-spot swell up to 2000 km across and an increase in surface elevation of up to 2000 m. It is indeed interesting to note that the highest passive margins also seem to be those characterized by rift-related volcanism, just as the model predicts. Although mantle plumes do not drive the rifting in this model, the surface uplift caused by underplating would assist the rifting process as diverging plates would tend to slide down either side of a hot-spot swell.

All the models of passive rifting discussed so far assume that rifting is more or less a symmetric process; that is, we would expect the opposing passive margins formed through continental break-up to have a similar structure and morphology because they have experienced a similar tectonic history. But does symmetric rifting accord with the evidence? We saw in our discussion of continental rifts (see Section 4.3) that asymmetric rather than symmetric rifting appears to be the norm, at least at the scale of individual rift structures. Moreover, if we compare opposing passive margins there seem to be a number of cases where their morphology and structure are very different. For example the western margin of southern Africa has a relatively narrow continental shelf and a well-developed marginal upwarp, whereas the opposing margin of the eastern coast of southern South America to which it was joined prior to the break-up of Gondwana has a very broad continental shelf and no significant marginal upwarp. Another instance of apparent passive margin asymmetry is provided by the eastern margin of Australia, which has a marked upwarp which forms the Great Dividing Range, and the Lord Howe Rise which represents a now submerged fragment of continental crust which rifted away from eastern Australia around 95 Ma BP.

Studies of extensional terranes, such as the Basin and Range Province of the south-western USA, using seismic methods have revealed shallow-dipping faults which appear to extend through the entire lithosphere. These have been interpreted as shear zones representing surfaces of lithospheric detachment. If lithospheric extension occurs primarily along such **detachment faults** rather than by thinning throughout the entire zone of lithospheric extension, then we would expect opposing passive margins to display a marked complementary asymmetry (Fig. 4.19).

Such detachment models predict that two types of passive margin will be produced by continental rupture. An **upper-plate margin** is formed by crust lying above the detachment

A

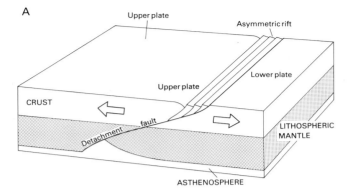

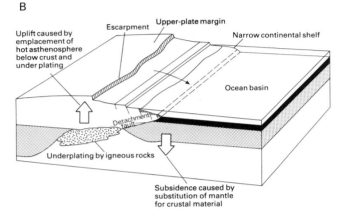

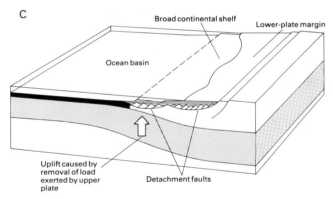

Fig. 4.19 *Schematic representation of the detachment model of asymmetric passive margin evolution. Asymmetric rifting (A) leads to the development of morphologically and structurally distinct upper-plate (B) and lower-plate (C) margins.*

fault while a **lower-plate margin** is developed on deeper crustal rocks lying below the detachment zone and which may be overlain by faulted fragments of the upper plate. Upper- and lower-plate margins will differ significantly in their uplift and subsidence history.

As displacement occurs along the detachment fault the margin of the lower plate will be upwarped as the load previously exerted by the upper plate is progressively removed (Fig. 4.19(C)). The predominant effect, however, will be sub-

sidence due to the considerable thinning of the lower plate lithosphere. The resulting continental margin morphology is likely to consist of a broad continental shelf and a rather modest amount of uplift inland.

The behaviour of the upper plate is quite different. The pulling away of the lower plate along the detachment fault exposes the base of the lithosphere on the upper plate to hot, rising asthenosphere (Fig. 4.19(B)). This substitution of less dense asthenosphere for denser lithosphere will result in uplift of the landsurface on the upper-plate margin. Oceanward of this uplift zone, however, there will be subsidence since here movement along the detachment fault results in the substitution of mantle for less dense lower crust. The coupling of this offshore subsidence and onshore uplift will induce a flexure of the margin which will accentuate uplift inland. The resulting upper-plate margin morphology will comprise a broad marginal upwarp which drops abruptly down to a relatively narrow continental shelf.

Superimposed on all these models of passive margin rifting is the possible contribution to uplift and subsidence made by flexural isostasy as material is eroded from the margin and transported offshore. As with the active rifting model this effect could generate uplift of the margin, even during its mature phase when other mechanisms are largely inoperative.

4.5.3 Passive margins: a key research focus

The modelling of passive margin evolution is still in its infancy. A wide range of mechanisms have been proposed by geophysicists attempting to model the broad patterns of uplift and subsidence observed but modelling has run well ahead of empirical testing. Figure 4.20 summarizes some of the key processes that appear to be involved. To date relatively little attention has been paid to the detailed morphological evidence that might indicate which, if any, of the proposed models is applicable to any particular passive margin. Nevertheless, it is impossible to underestimate the

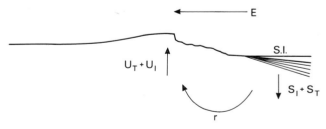

Fig. 4.20 *Summary of some of the major tectonic factors controlling the morphological evolution of rifted passive margins: U_T = thermally-driven uplift; U_I = isostatic uplift associated with denudational unloading; S_T = thermally driven subsidence; S_I = isostatic subsidence associated with sediment loading; r = rotation of the margin resulting from U_I and S_I; E = erosion by escarpment retreat; S.l. = sea level.*

importance of passive margins to an understanding of long-term landform development, a topic we will be returning to in Chapter 18. It seems probable that many of the models outlined above will be found to be applicable to at least some passive margins, but much work remains to be done in this field. This is especially true of sheared margins which have been neglected by geophysicists. Their attention has been focused on rifted margins largely because of their potential oil and gas reserves.

4.6 The break-up of supercontinents

Discussion about the way continental rifts form and how, in some cases, they develop into sites of continental rupture and sea-floor spreading raises the much broader question of why supercontinents break up. At least part of the answer to this question may lie in the contrasting properties of continental and oceanic crust. Continental crust is only about one-half as efficient at conducting heat as oceanic crust, so if a **supercontinent**, such as Pangaea, covers a significant part of the Earth's surface heat will build up in the mantle below it. After perhaps 80 Ma or so the increase in the temperature in the sub-lithospheric mantle below the supercontinent will lead to uplift, rifting and continental fragmentation. New ocean basins are created which become wider through time but eventually subduction zones will become established, either along previously passive margins, or within ocean basins. This establishment of subduction zones results in a change from continental dispersal to continental aggregation which eventually leads to the reassembly of a large land mass.

This idea of a repetitive sequence of continental fragmentation and aggregation is termed the **Wilson cycle** after its originator J. T. Wilson. The average duration of each cycle seems, on the basis of geological evidence, to be very roughly around 500 Ma. This is well beyond the time scale generally relevant to landform development but it is still important to be aware of how present landscapes, especially in plate interiors, relate to the most recent episode of supercontinent fragmentation. For instance, dispersal patterns of the continents following the break-up of Pangaea show that the now separate continents have been moving away from Africa, which itself appears to have remained relatively stationary. In other words Africa may have remained close to the region of major sub-lithospheric heating which preceded the break-up of Pangaea and this may explain why it has a high mean elevation today in spite of lacking significant orogenic belts. Australia, on the other hand, has moved rapidly away from its original location on the edge of Pangaea and it has the lowest mean elevation of the major continents. There is obviously much uncertainty in these broad generalizations, but geomorphic evidence has a key role to play in relating such global tectonic models to the geological history of the continents.

Further reading

The large-scale relationships between landforms and tectonics in plate interiors have received comparatively little attention from geomorphologists so the literature from this perspective is rather limited. King (1967) is still useful for its descriptions of the continental landscapes of plate interiors although his interpretations of the tectonic processes involved have now been superseded by more recent work. Ollier (1981) provides an introduction to several topics discussed in this chapter, and my own brief reviews in *Progress in Physical Geography* (Summerfield, 1986, 1988, 1989) try to relate some of the work by geophysicists to problems of landscape development. The morphology of passive margins is discussed in general terms by Ollier (1985), and with respect to specific regions by Ollier (1982) (eastern Australia), Summerfield (1985a) (Africa), Ollier and Marker (1985) (southern Africa) and Ollier and Powar (1985) (India).

There is an extensive and rapidly growing literature on the tectonics of plate interiors, but unfortunately much of it is written from a geophysical rather than a geomorphic perspective and it is not readily accessible to readers lacking a good grounding in physics and maths. Two general reviews of the field presented at an introductory level are Bally *et al.* (1980) and Burchfiel *et al.* (1980). Both of these edited volumes contain a number of relevant and clearly written papers on a variety of specific topics. A more detailed treatment of epeirogenic uplift is to be found in McGetchin and Merrill (1979); in spite of its general title it is almost solely concerned with the Colorado Plateau, but it does contain a brief but wide-ranging review of uplift mechanisms. An interesting assessment of the differential uplift of the continents through the Late Cenozoic is provided by Bond (1978).

A clear introduction to hot spots is provided by Burke and Wilson (1976), while a more recent assessment of their global distribution is that by Stefanick and Jurdy (1984). Hot spots as a cause of uplift are discussed by Crough (1983) and Gass *et al.* (1978), while Theissen *et al.* (1979), Summerfield (1985b) and Sahagian (1988) examine the relationship between epeirogeny and hot spot activity in Africa. The important, but neglected, problem of the range of scales of uplift associated with thermal mechanisms is addressed by Le Bas (1980), and the idea of lithospheric vulnerability is presented by Pollack *et al.* (1981). Crough (1979) explores the possible morphological consequences of continents migrating over hot spots and also cites a possible example from North America (Crough, 1981).

Turning to other possible uplift mechanisms, the phase change model is considered by Smith (1982) with particular reference to Africa and Australia, and by Smith and Drewry (1984) with respect to the Transantarctic Mountains. The mechanism of delamination is described by Bird (1979). Flexural isostasy is discussed by Karner and Watts (1982) and Watts *et al.* (1982), and at a more introductory level by Watts

(1981), but unfortunately these treatments focus on offshore subsidence rather than uplift inland.

Rift systems have been studied for many decades but have received renewed attention since the early 1970s because of their association with crustal upwarps and their role in continental rupture. An accessible introduction is provided by Pàlmason (1982); this is an edited compilation containing a brief but stimulating review of continental rifts by Mohr (1982). More recent ideas about the half-graben structure of many rifts are introduced by Frostick and Reid (1987) and examined in depth by Rosendahl (1987) and in Frostick *et al.* (1986). The influence of pre-rift structures on eventual rift form is considered by Versfelt and Rosendahl (1989) and the relationship between rifts and domal uplift in Africa is discussed by Burke and Whiteman (1973).

Much of the research into the tectonic evolution of basins overlaps with work on passive margin development, and although written for students of sedimentology the book by Miall (1984) provides a valuable introduction for geomorphologists on both of these themes. Bally and Snelson (1980) provide an extremely comprehensive review of basin types and origins and Turcotte (1980) presents one of the more accessible reviews of the geophysics of basin subsidence.

A good introduction to the tectonics of passive continental margins is provided by Scrutton (1982) while Courtillot and Vink (1983) and Bonatti (1987) look at the processes of continental rifting and break-up. The extent and magnitude of uplift to be expected from active and passive rifting mechanisms are compared by Keen (1985), while Beaumont *et al.* (1982) and Buck *et al.* (1988) assess the patterns of uplift likely to result from different models of passive rifting. The uniform extension model is presented by McKenzie (1978) and the idea of non-uniform stretching is clearly described by Rowley and Sahagian (1986). Steckler (1985) proposes secondary convection as an explanation of the uplift in the Gulf of Suez region, White and McKenzie (1989) outline the underplating model, and the asymmetric detachment model of continental break-up is presented by Lister *et al.* (1986). The eastern Australian passive margin is one region where there has been a fruitful interchange of ideas and data between geomorphologists and geophysicists, although this has not resulted in a consensus as to either the history of uplift or the mechanisms that have caused it. Veevers (1984) provides a comprehensive background to the history of the eastern Australian margin while Bishop (1988) reviews the various models that have been proposed to account for uplift. Lambeck and Stephenson (1986) and Wellman (1987, 1988) provide contrasting interpretations of the uplift mechanisms that have operated in the region.

Finally, the possible mechanisms underlying the cyclic fragmentation and reassembly of continents are discussed at an introductory level by Nance *et al.* (1988) and in detail by Gurnis (1988). Both these discussions derive ideas from earlier work by Anderson (1982).

References

Anderson, D. L. (1982) Hotspots, polar wander, Mesozoic convection and the geoid. *Nature* **297**, 391–3.

Bally, A. W., Bender, P. L., McGetchin, T. R. and Walcott, R. I. (eds) (1980) *Dynamics of Plate Interiors*. American Geophysical Union, Washington D C; Geological Society of America, Boulder.

Bally, A. W. and Snelson, S. (1980) Realms of subsidence. In: A. D. Miall (ed.) *Facts and Principles of World Petroleum Occurrence*. Canadian Society of Petroleum Geologists Memoir **6**, 9–94.

Beaumont, C., Keen, C. E. and Boutilier, R. (1982) A comparison of foreland and rift margin sedimentary basins. *Philosophical Transactions of the Royal Society London* **A305,** 295–317.

Bird, P. (1979) Continental delamination and the Colorado Plateau. *Journal of Geophysical Research* **84**, 7561–71.

Bishop, P. (1988) The eastern highlands of Australia: the evolution of an intraplate highland belt. *Progress in Physical Geography* **12**, 159–82.

Bonatti, E. (1987) The rifting of continents. *Scientific American* **256**(3), 74–81.

Bond, G. (1978) Evidence for Late Tertiary uplift of Africa relative to North America, South America, Australia and Europe. *Journal of Geology* **86**, 47–65.

Buck, W. R., Martinez, F., Steckler, M. S. and Cochran J. R. (1988) Thermal consequences of lithospheric extension: pure and simple. *Tectonics* **7**, 213–34.

Burchfiel, B. C., Oliver, J. E. and Silver, L. T. (eds) (1980) *Continental Tectonics*. National Academy of Sciences, Washington D C.

Burke, K. and Whiteman, A. J. (1973) Uplift, rifting and the break-up of Africa. In: D. H. Tarling and S. K. Runcorn (eds) *Implications of Continental Drift for the Earth Sciences* (2 vols). Academic Press, London 735–55.

Burke, K. C. and Wilson, J. T. (1976) Hot spots on the Earth's surface. *Scientific American* **235**(2), 46-57.

Courtillot, V. and Vink, G. E. (1983) How continents break up. *Scientific American* **248**(7), 43–9.

Crough, S. T. (1979) Hotspot epeirogeny. *Tectonophysics* **61**, 321–33.

Crough, S. T. (1981) Mesozoic hotspot epeirogeny in eastern North America. *Geology* **9**, 2–6.

Crough, S. T. (1983) Hotspot swells. *Annual Review of Earth and Planetary Sciences* **11**, 165–93.

Frostick, L. and Reid, I. (1987) A new look at rifts. *Geology Today* **3**, 122–6.

Frostick, L .E., Renaut, R. W., Reid, I. and Tiercelin, J. J. (eds) (1986) *Sedimentation in the African Rifts* Geological Society Special Publication **25**.

Gass, I. G., Chapman, D. S., Pollack, H. N. and Thorpe, R. S. (1978) Geological and geophysical parameters of mid-plate volcanism. *Philosophical Transactions of the Royal Society London* **A288,** 581–97.

Gurnis, M. (1988) Large-scale mantle convection and the aggregation and dispersal of supercontinents. *Nature* **332**, 695–9.

Karner, G. D. and Watts, A. B. (1982) On isostasy at Atlantic-type continental margins. *Journal of Geophysical Research* **87**, 2923–48.

Keen, C. E. (1985) The dynamics of rifting: deformation of the lithosphere by active and passive driving forces. *Geophysical Journal of the Royal Astronomical Society* **80**, 95–120.

King, L. C. (1967) *The Morphology of the Earth: A Study and Synthesis of World Scenery* (2nd edn). Oliver and Boyd, Edinburgh.

Lambeck, K. and Stephenson, R. (1986) The post-Palaeozoic uplift history of south-eastern Australia. *Australian Journal of Earth Sciences* **33**, 253–70.

Le Bas, M. J. (1980) Alkaline magmatism and uplift of continental crust. *Proceedings of the Geologists Association* **91**, 33–8.

Lister, G. S., Etheridge, M. A. and Symonds, P. A. (1986) Detachment faulting and the evolution of passive margins. *Geology* **14**, 246–50.

McGetchin, T. R. and Merrill, R. B. (eds) (1979) *Plateau Uplift: Mode and Mechanism.* Elsevier, Amsterdam.

McKenzie, D. (1978) Some remarks on the development of sedimentary basins. *Earth and Planetary Science Letters* **40**, 25–32.

Miall, A. D. (1984) *Principles of Sedimentary Basin Analysis.* Springer-Verlag, New York.

Mohr, P. (1982) Musings on continental rifts. In: G. Pàlmason (ed.) *Continental and Oceanic Rifts.* American Geophysical Union, Washington DC; Geological Society of America, Boulder, 293–309.

Nance, R. D., Worsley, T. R. and Moody, J. B. (1988) The supercontinent cycle. *Scientific American* **259**(1), 72–9.

Ollier, C. D. (1981) *Tectonics and Landforms.* Longman, London and New York.

Ollier, C. D. (1982) The Great Escarpment of eastern Australia: tectonic and geomorphic significance. *Journal of the Geological Society of Australia* **29**, 13–23.

Ollier, C. D. (1985) Morphotectonics of continental margins with great escarpments. In: M. Morisawa and J. T. Hack (eds) *Tectonic Geomorphology.* Allen and Unwin, Boston and London, 3-25.

Ollier, C. D. and Marker, M. E. (1985). The Great Escarpment of southern Africa. *Zeitschrift für Geomorphologie Supplementband* **54**, 37–56.

Ollier, C. D. and Powar, K. B. (1985) The Western Ghats and the morphotectonics of Peninsular India. *Zeitschrift für Geomorphologie Supplementband* **54**, 57–69.

Pàlmason, G. (ed.) (1982) *Continental and Oceanic Rifts.* American Geophysical Union, Washington DC; Geological Society of America, Boulder.

Pollack, H. N., Gass, I. G., Thorpe, R. S. and Chapman, D. S. (1981) On the vulnerability of lithospheric plates to mid-plate volcanism: Reply to comments by P. R. Vogt. *Journal of Geophysical Research* **86**, 961–6.

Rosendahl, B. R. (1987) Architecture of continental rifts with special reference to East Africa. *Annual Review of Earth and Planetary Sciences* **15**, 445–503.

Rowley, D. B. and Sahagian, D. (1986) Depth-dependent stretching: a different approach. *Geology* **14**, 32–5.

Sahagian, D. (1988) Epeirogenic motions of Africa as inferred from Cretaceous shoreline deposits. *Tectonics* **7**, 125–38.

Scrutton, R. A. (ed.) (1982) *Dynamics of Passive Margins.* American Geophysical Union, Washington DC; Geological Society of America, Boulder.

Smith, A. G. (1982) Late Cenozoic uplift of stable continents in a reference frame fixed to South America. *Nature* **296**, 400–4.

Smith, A. G. and Drewry, D. J. (1984) Delayed phase change due to hot asthenosphere causes Transantarctic uplift? *Nature* **309**, 536–8.

Steckler, M. S. (1985) Uplift and extension at the Gulf of Suez: indications of induced mantle convection. *Nature* **317**, 135–9.

Stefanick, M. and Jurdy, D. M. (1984) The distribution of hotspots. *Journal of Geophysical Research* **89**, 9919–25.

Summerfield, M. A. (1985a) Plate tectonics and landscape development on the African continent. In: M. Morisawa and J.T. Hack (eds) *Tectonic Geomorphology.* Allen and Unwin, Boston and London, 27–51.

Summerfield, M. A. (1985b) Tectonic background to long-term landform development in tropical Africa. In: I. Douglas and T. Spencer (Eds) *Environmental Change and Tropical Geomorphology.* Allen and Unwin, London and Boston, 281–94.

Summerfield, M. A. (1986) Tectonic geomorphology: macroscale perspectives. *Progress in Physical Geography* **10**, 227–38.

Summerfield, M. A. (1988) Global tectonics and landform development. *Progress in Physical Geography* **12**, 389–404.

Summerfield, M. A. (1989) Tectonic geomorphology: convergent plate boundaries, passive continental margins and supercontinent cycles. *Progress in Physical Geography* **13**, 431–41.

Theissen, R., Burke, K. and Kidd, W. S. F. (1979) African hotspots and their relation to the underlying mantle. *Geology* **7**, 263–6.

Turcotte, D. L. (1980) Models for the evolution of sedimentary basins. In: A. W. Bally, P. L. Bender, T. R. McGetchin and R. I. Walcott (eds) *Dynamics of Plate Interiors.* American Geophysical Union, Washington DC; Geological Society of America, Boulder, 21–6.

Veevers, J. J. (ed.) (1984) *Phanerozoic Earth History of Australia.* Clarendon Press, Oxford and New York.

Versfelt, J. and Rosendahl, B. R. (1989) Relationship between pre-rift structure and rift architecture in Lakes Tanganyika and Malawi, East Africa. *Nature* **337**, 354–7.

Watts, A. B. (1981) The U. S. Atlantic continental margin: Subsidence history, crustal structure and thermal evolution. In: A. W. Bally *et al.* (eds) *Geology of Passive Continental Margins: History, Structure and Sedimentologic Record (With Special Emphasis on the Atlantic Margin).* Education Course Note Series #19. American Association of Petroleum Geologists, Tulsa, Section 2.

Wellman, P. (1987) Eastern Highlands of Australia; their uplift and erosion. *BMR Journal of Australian Geology and Geophysics* **10**, 277–86.

Wellman, P. (1988) Tectonic and denudational uplift of Australian and Antarctic highlands. *Zeitschrift für Geomorphologie* **32**, 17–29.

White, R. S. and McKenzie, D. P. (1989) Volcanism at rifts. *Scientific American* **261** (1), 44–55.

5

Landforms associated with igneous activity

5.1 Extrusive and intrusive igneous activity

Igneous activity involves the movement of **magma** (molten rock), or solid rock undergoing slow deformation, towards, or on to, the Earth's surface. **Extrusive igneous activity** is more commonly referred to as **volcanism** and occurs where magma erupts on to the surface either as flowing **lava**, or as fragmental material thrown into the air by explosive volcanic activity. Volcanism directly gives rise to predominantly constructional landforms, although not all volcanic activity results in the development of volcanoes since some types of eruption create extensive sheets of lava or fragmental material.

There are five major types of volcanic activity. One very obvious form is the volcanism of island arcs and continental-margin orogens associated with plate convergence. More extensive, however, are the vast sheets of basaltic lavas which have formed on the continents at various times in the past and are referred to as **continental flood basalts.** Yet volcanism is predominantly a submarine rather than a continental phenomenon since the outpouring of magma associated with submarine eruptions far exceeds anything witnessed on the continents. This submarine volcanic activity is of three types. In addition to the volcanic activity marking the creation of new lithosphere at mid-oceanic spreading centres, there are also volcanoes in the ocean basins representing hot spots, as well as flood basalts covering large parts of the ocean floor.

Intrusive igneous activity involves the movement of rock bodies towards the Earth's surface. Such rock bodies may be either molten, in the strict sense, or they may be solid and move very slowly at elevated temperatures and pressures through a form of plastic flow. A rock body emplaced in the surrounding country rock in this way is called an **intrusion** or, alternatively, a **pluton**. The formation of intrusions can affect the landsurface in two ways. During emplacement the overlying strata may be significantly uplifted and deformed.

Secondly, the rocks of intrusions are often more resistant to erosion than the surrounding country rock so they frequently form prominent landforms once exposed at the surface.

The various types of igneous rock associated with different forms of extrusive and intrusive activity, and the different types of magma from which they originate, play an important role in influencing the ultimate landforms created. Igneous petrology, the subdiscipline concerned with the classification and origin of igneous rocks, is a complex field. None the less, for the purposes of understanding the development of landforms associated with volcanism and intrusive igneous activity, we need to appreciate a few key properties of igneous rocks. Figure 5.1 presents a simple classification of

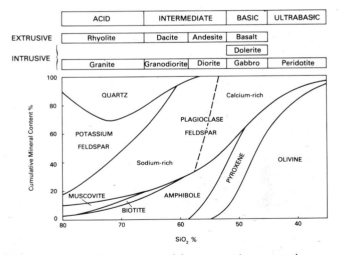

Fig. 5.1 *Mineral composition of the common igneous rocks. Note that dolerite is a medium-grained rock of basic composition whereas gabbro is coarse-grained. The terms 'acid' and 'basic' originate from the idea current in the nineteenth century that silica is derived from silicic acid, although we now know that the silica content of a rock is not a measure of acidity.*

igneous rocks based on their mineral and chemical composition and on differences in grain size, or texture.

Igneous rocks can be classified on the basis of the proportion of silica (SiO_2) that they contain. **Acid** rocks have more than 66 per cent SiO_2, **intermediate** rocks between 52 and 66 per cent, **basic** rocks between 45 and 52 per cent, and **ultrabasic** rocks less than 45 per cent. The proportion of silica is roughly related to the relative abundance of light-coloured **felsic** minerals (quartz and feldspar) and dark-coloured **mafic** minerals (silicates rich in magnesium and iron). The grain size of igneous rocks is strongly in-fluenced by the rate at which they cooled when formed, and not surprisingly we find that intrusive igneous rocks nearly always have a coarser texture than volcanic rocks.

5.2   Volcanism

In addition to its importance in directly creating landforms, volcanism is also of geomorphic significance for other reasons. The material erupted during episodes of volcanic activity can be dated by radiometric techniques and can

Fig. 5.2   *Volcanoes of central Java, Indonesia. Several large strato-volcanoes can be seen in this Landsat image which covers an area 185 km across. The peaks labelled are Muria (Mu) (1602 m), Lawu (L) (3265 m), Merapi (Me) (2911 m) and Sumbing (S) (3371 m). Murjo is now inactive and, as the image shows, is more deeply eroded than the other volcanoes. The volcanoes of this region are among the most active and explosive in the world. Merapi, for instance, erupts frequently producing nueés ardentes and lahars. (Image courtesy N. M. Short.)*

therefore provide a minimum age for the landsurface over which it lies. Since volcanoes often have a rather regular, symmetric form when first created, their original form can often be reconstructed with some confidence even after significant erosion has occurred. Together with radiometric age determinations of the most recent eruptive episode, such reconstructions enable estimates to be made of the amount of material eroded over a known period of time. Volcanoes can thus provide important test cases for estimates of rates of denudation (see Section 15.4.1.2). Since volcanoes composed of similar rock types are also found in a range of climatic environments they also provide the opportunity to compare the effects of climate on the rate and nature of denudation.

5.2.1   Distribution of volcanic activity

There are presently around 600 active volcanoes on the continents or exposed above the sea as islands. A further several thousand extinct volcanoes on land are known from their form, structure or characteristic rock types. But these totals are dwarfed by the immense numbers of submarine volcanoes in the world's ocean basins; at least 50 000 are known to exist on the floor of the Pacific Ocean.

Confining our attention here to sub-aerial volcanic activity, we have already considered some aspects of the distribution of volcanoes at the global scale in the three previous chapters. Most present-day volcanic activity is associated with convergent plate boundaries and there is a considerable concentration of volcanoes in the island arc systems (Fig. 3.6) and continental-margin orogens bordering the Pacific Ocean. Around 60 per cent of active volcanoes are, in fact, to be found around the rim of the Pacific Ocean, with a third of these concentrated in Indonesia (Fig. 5.2). But volcanoes are also found in the interiors of plates where they are associated with hot spots and tend to occur either as clusters or lines.

Volcanic clusters occur both on continental and oceanic lithosphere, and are particularly common on the African Plate which apparently has a high concentration of hot spots (see Section 4.2.1). Clusters range in size from fairly restricted features such as Tristan da Cunha in the South Atlantic, to much more extensive volcanic groups covering a considerable area such as the Azores in the North Atlantic and the Galapagos Islands in the eastern Pacific. Active volcanism at any one time is normally confined to a limited number of centres within a particular cluster. **Volcanic lines** are most abundant in the Pacific Ocean and comprise series of volcanic islands (or seamounts in the case of submarine volcanoes) which typically become progressively younger along a line and often terminate in an active volcano. The Hawaiian Islands provide a classic example of a volcanic line, and the generally accepted explanation of this type of feature involves the passage of part of a plate over a sub-lithospheric thermal anomaly which, through the periodic

generation of magma in the overlying crust, promotes a sequence of volcanic activity at the surface (Fig. 5.3).

Within volcanic arcs individual volcanoes tend to be aligned along essentially straight lines over distances of up to 1000 km, the arcuate form not being evident at this

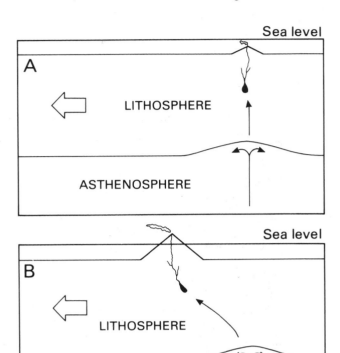

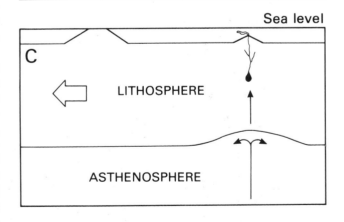

Fig. 5.3 *Formation of a volcanic line through the motion of lithosphere over a sub-lithospheric thermal anomaly. A volcano marking the hot spot grows from the ocean floor as melting of the crust generates magma (A) and eventually emerges above sea level (B). Plate motion finally moves the new volcano away from the underlying sub-lithospheric thermal anomaly (C); the original volcano is eroded and a new volcano develops above the zone of magma generation (see Section 17.6.3).*

smaller scale. In fact, volcanic arcs appear to consist of series of slightly off-set linear elements which have been interpreted as representing the subduction of discrete segments of lithosphere. Volcanic lines and clusters also exhibit patterning at the regional and local scale with both radial and rectilinear arrangements being found. Such patterns appear to reflect the way in which the lithosphere fractures as magma forces its way up to the surface. In any one cluster the spacing of volcanoes is usually consistent and is generally close to the thickness of the lithosphere at that locality (Table 5.1). The inference here is that the volcano spacing is controlled by fracture patterns in the lithosphere, with increasingly thick lithosphere exhibiting progressively more widely spaced fracture patterns through which magma is extruded.

Table 5.1 Relationship between volcano spacing and estimated lithospheric thickness in eastern Africa

LOCATION	AGE	VOLCANO SPACING (km)	ESTIMATED LITHOSPHERIC THICKNESS (km)
Erta-ali, Afar	Quaternary	10 ± 3	16
Dubbi, Afar	?Pliocene–Quaternary	19 ± 6	?25
Ethiopian rift	Pliocene–Quaternary	43 ± 13	35–50
Gregory rift	Pliocene–Quaternary	42 ± 11	35–50
Eastern Uganda	Miocene	72 ± 9	75–80
Addis Ababa	Miocene–Pliocene	70 ± 10	80
Ethiopian plateau	Oligocene–Miocene	109 ± 22	?80–120

Source: From C. M. Clapperton (1977) *Progress in Physical Geography* 1, Table 1b, p.390 based on data in P.A. Mohr and C. A. Wood (1976) *Earth and Planetary Science Letters* 33, 126–144.

5.2.2   Volcanic activity through time

The fact that volcanic activity is highly episodic in nature is evident even from the elementary classification of volcanoes into active, dormant and extinct. Although widely used, these categories are often difficult to apply in practice. **Extinct** volcanoes are those that have not erupted in historic time, whereas **active** volcanoes have been seen to erupt. The term **dormant** volcano is applied during the period between eruptions to those volcanoes thought to be potentially active.

The most obvious manifestation of volcanic activity comes from the infrequent but spectacular eruption of volcanoes on land. Such individual explosive events can indeed involve impressive quantities of material. Around 10 km³ of volcanic debris was hurled into the atmosphere during the famous 1883 eruption of Krakatau (Krakatoa) in Indonesia, but more than ten times this quantity was involved in the 1815 eruption of Tambora, another Indonesian volcano. Going back into prehistoric time even these volumes were far exceeded by the eruption of Toba, yet another volcano in the very active Indonesian region. When it erupted around

75 000 a BP an estimated 2000 km³ of material was dispersed over a wide area of the East Indies.

Interesting as these figures are, they do not give an accurate indication of the average rate of volcanic activity over time. Estimates based on volumes of erupted material in fact show that the discharge from volcanoes forming island arcs and continental-margin orogens averages a very modest 1 km³ a⁻¹. This rate has no doubt varied somewhat through geological time as the rate of lithospheric subduction has changed. It is, in fact, the much less spectacular, but more consistent, activity of volcanic centres related to hot spots that have more impressive rates of production. The island of Hawaii alone seems to have a long-term average rate of construction of over 0.4 km³ a⁻¹, while Iceland has sustained a rate of about 0.13 km³ a⁻¹ during historic time, and a rather lower average of 0.06 km³ a⁻¹ over the past 16 Ma.

Although the volume of continental flood basalts amounts to around 1×10^7 km³, the long-term average rate of accumulation is very modest. There have been periods in the Earth's history, however, when this rate has increased dramatically. It has been estimated, for example, that around 600 000 km³ of basalt was erupted within just 1 Ma to form the Deccan flows in India. On the basis of their vast volume, flood basalts on the ocean floor have erupted at a much faster long-term average rate than their continental equivalents, although the major contribution of volcanic material on the ocean floor comes from mid-oceanic ridge spreading centres. These create between 5 and 6 km³ a⁻¹ of new volcanic material (that is, new oceanic crust). This production is, however, episodic as is illustrated by the volcanic record of Iceland, which is both a hot spot and part of a mid-oceanic ridge. In the major eruption of Laki in Iceland in 1783 over 10 km³ of lava was produced in just 50 days, and it spread to eventually cover an area of 370 km².

Rates of volcanic activity through time are especially critical in determining whether intra-plate volcanoes formed in ocean basins (as opposed to volcanoes associated with subduction zones) will reach sea level to form a volcanic island, or remain submerged to form a seamount. Once it has been formed at a mid-oceanic ridge new oceanic lithosphere subsides as it moves away from a spreading centre and becomes cooler, thicker and more dense (see Section 17.6.3). Consequently, if a volcanic island is to be formed the rate of growth of a volcano must exceed the rate of ocean floor subsidence. Since intra-plate volcanicity is linked to effectively stationary sub-lithospheric thermal anomalies, there is only a limited amount of time available for a volcano to develop before plate motion removes it from its underlying region of crustal melting and source of magma generation (Fig. 5.3).

Dating of the volcanic rocks of the Hawaiian Islands has shown that each of the main islands was formed in about 1 Ma. These volcanoes are large landforms which load the underlying crust and are partly isostatically supported by deep roots (Fig. 5.4). Taking into account the volume of

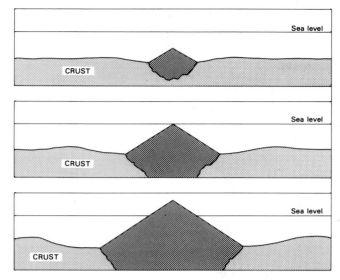

Fig. 5.4 *Growth of a volcanic island associated with a hot spot showing the growth of the compensating root to maintain isostasy. This root is approximately twice as thick as the volcanic pile built upon the ocean floor. Since the lithosphere has a certain degree of rigidity some of the isostatic compensation will be achieved by flexure which creates a 'moat' and peripheral 'bulge' around the volcano (see Sections 4.2.3 and 17.6.3).*

5.2.3 Products of volcanic activity

Volcanic eruptions can involve both lava which flows over the surface, and fragmented particles, or **pyroclasts,** thrown into the air by violent explosions. All volcanic material which falls through the air is described as **tephra,** while **pyroclastic flow deposits** result from the rapid 'flow' of pyroclasts mixed with abundant hot gas.

5.2.3.1 Lava forms

Lava is formed from magma by cooling and the loss of gas. The behaviour of lava once erupted is significantly affected by its viscosity, that is, its resistance to flow (see Section 8.2.1). This in turn is related to the composition and temperature of the lava. Acid lavas, such as those composed of rhyolite, which are derived from magma with a high silica content, are the most viscous and are generally erupted at the lowest temperatures (Fig. 5.5). They are consequently able to flow only very short distances before they solidify. Basic magmas, however, which are predominantly of basaltic composition, have a much lower silica content and are erupted at high temperatures. Their low viscosity consequently allows them to flow over considerable distances. Low-viscosity basaltic lavas produce extensive flows, whereas acidic lavas tend to produce flows which are convex in cross-section, the degree of convexity increasing with lava viscosity. Once erupted the rate at which a lava flow cools and solidifies also depends on its depth. While a 1 m thick flow of basalt will

erupted volcanic material lying both on and below the surface of the ocean floor, it can be estimated that a minimum construction rate of between 0.005 and 0.01 $km^3\,a^{-1}$ is required for a volcanic island to be produced in 1 Ma on lithosphere between 20 and 40 Ma old. The rate is lower on younger lithosphere underlying shallow ocean, but greater on older lithosphere underlying the deeper parts of ocean basins. The Hawaiian Islands have attained elevations well above sea level so their rates of growth have been much higher, with 0.44 $km^3\,a^{-1}$ being estimated for the island of Hawaii itself. Interestingly, this is about ten times the rate of eruption of lava recorded over historic time. The reason for this discrepancy is not known; it might be because there are short-term fluctuations in the rate of lava eruption and that recently the rate has been well below the long-term average. An alternative explanation is that much of the growth of the island has occurred through intrusive igneous activity rather than the extrusion of lava on the surface.

A revealing comparison can be made between the growth of Hawaii and Iceland. Hawaii has exhibited a much faster rate of construction, but its volcanism will be short-lived as it will be carried away from its magma source by plate motion. Iceland, on the other hand, is located on a mid-oceanic ridge and consequently has not moved significantly with respect to its hot spot since the opening of the North Atlantic Ocean over 50 Ma BP. It has accumulated a pile of volcanic material 10 km thick and it has five times the volume of Hawaii in spite of a long-term rate of volcanic activity less than one-sixth as high.

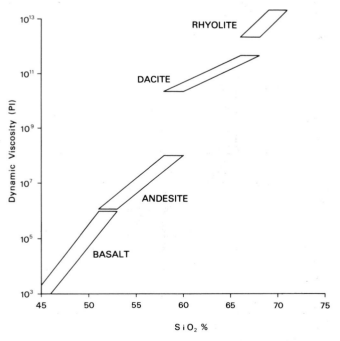

Fig. 5.5 *Relationship between SiO_2 content and viscosity for various magma types. Note that there is a difference of 10 orders of magnitude between the most viscous rhyolitic magma and the least viscous basaltic magma.*

take around 12 days to cool from 1100 to 750 °C, the same amount of cooling will take 3 a for a 10 m flow and 30 a for a 100 m flow.

Eruptions of lava give rise to a diverse range of surface forms depending on composition, viscosity and gas content. **Pahoehoe** has a smooth, ropy surface as a result of the partial solidification of a thin skin which is dragged into folds by continuing lava flow underneath. Pahoehoe is formed when lava has a low viscosity, but when it cools, or loses some of its gas content, it is transformed into a much more viscous form. This is called **aa** and its surface is composed of jagged, angular fragments resembling clinkers (Fig. 5.6).

Where there is abundant gas in erupting magma the resulting lava contains numerous gas-filled vesicles, or cavities, and has a sponge-like form known as **scoria**. If such vesicles are extremely abundant **pumice** is formed. This has such a low density that it can float in water and blocks of pumice may travel thousands of kilometres across the world's oceans. Lava erupted underwater cools very rapidly with a plastic skin forming around lumps of still molten material. These lumps can become somewhat rounded through rolling and they are piled up on top of each other to create **pillow lava.**

The shrinkage of lava as it cools produces joints. These are usually irregular, but in extensive flows of basalt a highly regular hexagonal jointing pattern is sometimes produced giving rise to the kind of vertical columnar basalt outcrops which form the Giant's Causeway in Northern Ireland, UK. Such regular joint patterns appear to develop when the centres of contraction are evenly spaced. The lines joining these centres represent the directions of greatest tensile stress in the lava flow as it cools, and the hexagonal joint patterns are generated by cracks forming at right angles to these maximum stress lines.

A common feature of some flows, particularly of the pahoehoe type, are **lava tunnels** or **lava tubes**. These form as a result of the continuing movement of lava below a surface crust. A main feeder lava tunnel often branches towards the terminus of a flow rather like the distributaries of a river delta. A deep and often sinuous **lava channel** can form if the roof of a lava tunnel subsequently collapses. In low-viscosity lavas features known as **lava levees** may also develop. These consist of solidified fragments of lava which are piled up on either side of a flow by the hotter lava moving more rapidly in the centre of the flow.

5.2.3.2   Tephra

Tephra can be categorized on the basis of the size of the pyroclasts of which it is composed. The finest material (< 2 mm across) is termed **ash** when freshly deposited, and **volcanic tuff** when compacted, while material of intermediate size (2–64 mm across) is called **lapilli.** The largest calibre pyroclasts (> 64 mm across), which usually show some degree of rounding, are known as **volcanic bombs**. Volcanic bombs become partially streamlined and rounded as they are thrown through the air or pass rapidly through the magma in the neck of a volcano when they are violently erupted. Volcanic bombs can be composed of either lava, or of pre-existing rock which has become incorporated into the magma in the neck of a volcano. Most volcanic bombs are relatively small, being about the size of a football, but much larger bombs can be produced in very violent eruptions. For instance, in one eruption of the Andean volcano Cotopaxi a 200 t volcanic bomb was hurled a distance of 14 km.

Volcanic eruptions invariably generate pyroclasts of a range of sizes. The resulting tephra is deposited predominantly downwind and forms well-bedded deposits which become progressively finer upwards. This size-grading occurs because the coarser material generated by an eruption is heavier and is therefore deposited first. For the same reason large-calibre pyroclasts are concentrated close to the site of eruption, whereas ash can be dispersed over distances of hundreds, or even thousands, of kilometres. Such ash forms a progressively thinner deposit as the distance from the site

Fig. 5.6   *Recent flows of pahoehoe (left) and aa (right) on the flanks of Kilauea, Hawaii.*

of eruption increases. In regions of highly active volcanism tephra can periodically blanket the landscape.

Since it can be dated radiometrically tephra provides a valuable stratigraphic marker and a means both of determining the minimum age of the landsurface on which it is deposited, and of recording the activity of particular volcanoes. So valuable has the dating of tephra become that it has been given a specific name – **tephrochronology.** In Iceland each major volcanic eruption has resulted in a layer of tephra being deposited, and in many cases these have covered much of the country. Tephra horizons are also widespread in New Zealand where ashes of Late Quaternary age cover about half of the North Island.

5.2.3.3 *Pyroclastic flow deposits*

In some kinds of violent volcanic eruptions pyroclasts combine with large quantities of hot gases to form a fluid-like flow which is capable of moving very rapidly even over very low gradients. The great mobility of these **pyroclastic flows** is attributable to the volume of hot gas present and the heating of engulfed air which renders the solid material highly buoyant. An alternative term for such flows is **nuées ardentes** (literally burning clouds).

As a result of their great mobility pyroclastic flows can generate pyroclastic flow deposits, or **ignimbrites,** which cover a large area. The ignimbrites of the Lake Toba region in Sumatra extend over 25 000 km², while those of the Taupo-Rotorua area in the North Island of New Zealand cover more than 26 000 km². Although ignimbrites can be unconsolidated deposits, the pyroclasts are frequently so hot when deposited that the contained gas is expelled and the individual particles are welded together to form a compact, impervious rock.

5.2.3.4 *Volcanic products associated with water*

Magma contains up to 4 per cent water by weight, and some volcanic activity takes the form of eruptions of super-heated water and steam from **geysers** (Fig. 4.3). Such **hydrothermal activity** is most common during the later stages of volcanism. High concentrations of dissolved material, especially silica and calcium carbonate, are often contained in hydrothermal solutions, and on cooling this material is precipitated to form **sinter.**

Where volcanic debris is mixed with water a rapidly moving mud flow, or **lahar,** can be produced. Lahars can be a direct result of an eruption, where for instance freshly ejected lava or pyroclasts melt ice or snow, or they can occur as a result of the destabilization of the unconsolidated volcanic debris on the slope of a volcano by an earthquake, or as a result of heavy rain. In some cases lahars travel large distances and form thick deposits up to tens of metres deep. The 1980 eruption of Mount St Helens, Washington, U S A, led to a lahar which flowed some 27 km down the valley of the Toutle River and filled the valley bottom to a depth of 60 m.

Equally catastrophic results can arise when volcanic ac-

tivity occurs beneath a glacier. This has been a relatively frequent event in Iceland and it can lead to the rapid melting of large volumes of ice and the release of enormous quantities of water, creating a cataclysmic flood known as a **jökulhlaup** (although such floods can also be produced by other means (see Section 11.4.1)).

5.3 Volcanoes

5.3.1 Types of eruption

Volcanic eruptions involve the rapid release of enormous quantities of energy. A typical violent eruption from an individual volcano has an energy of between 10^{12} and 10^{15} J. This compares with the 10^{16} J released by a one megaton hydrogen bomb. But major eruptions involve much larger amounts of energy. For instance, the 1980 eruption of Mount St Helens was equivalent to about 30 one megaton hydrogen bombs, while the cataclysmic eruption of Laki in Iceland in 1783 had an estimated energy of 10^{20} J (equivalent to 10 000 one megaton hydrogen bombs).

The initial form of a volcano is very closely related to the nature of the eruptive activity that produces it. Eruptions can be of three primary types, each associated with a particular form of ejecta - **exhalative** (gas), **effusive** (lava) and **explosive** (tephra). All volcanic activity involves the expulsion of some gas but major landforms are only formed through the eruption of lava or tephra. None the less, gas is crucial in the actual mechanisms which control the kind of eruptive activity that occurs (Fig. 5.7). Various attempts have been made to categorize types of volcanic activity on the basis of the nature of the material erupted, and the violence of eruptive events. A simplified version of such a classification is presented in Table 5.2.

As magma rises towards the surface it exerts a considerable pressure on the overlying rock. Of particular importance is the pressure exerted by the gas (especially CO_2, SO_2 and water vapour) which is formed through the release of volatiles and by the conversion of ground water to steam. The first manifestation of volcanic activity at the surface is normally the exhalation of gas from tensional fractures. The character of the ensuing volcanism is largely a function of the viscosity of the magma which is, in turn, dependent on its chemical composition and temperature. As the magma rises towards the surface the confining pressure drops and gas bubbles begin to form and move upwards. In fluid magmas, such as those of basaltic composition, these gas bubbles can expand freely. Consequently, on reaching the surface they have a comparatively low residual pressure and explosive activity is very limited. This contrasts with the situation for a viscous magma such as rhyolite in which gas expansion is constrained in the lower levels of the magma column. On reaching the surface the residual gas pressure is consequent-

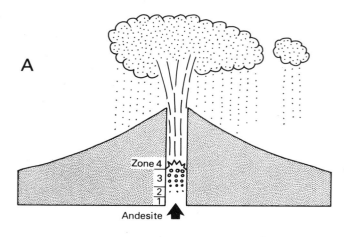

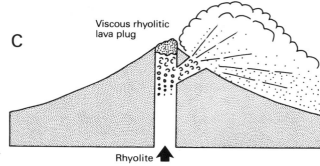

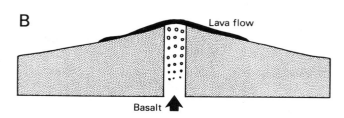

Fig. 5.7 *Schematic representation of three major forms of eruptive activity related to magma type and associated mode of gas expansion. In (A) the andesitic magma in zone 1 may be either saturated or undersaturated with volatiles, but as it rises it enters a zone of lower pressure (2) and bubbles of gas begin to form (3). The viscosity of andesitic magma is sufficient to restrain gas expansion up to the explosion zone (4) where the gas pressure is able to overcome the resistance of the overburden. In (B) magma of basaltic composition is fluid enough to allow the unrestricted expansion of gas bubbles and their relatively quiescent exhalation at the surface. In (C) the rhyolitic magma is so viscous that the bursting of gas bubbles does not occur until very near the surface. Violent lateral explosions producing ignimbritic eruptions consisting of dense mixtures of gases, liquids and pyroclasts are caused by the periodic blocking of the main vent by a viscous lava plug. (After C. M. Clapperton (1977)* Progress in Physical Geography *Fig. 7, p. 394.)*

ly very high and this results in the violent bursting of gas bubbles in the magma and the generation of highly explosive eruptions which generate large volumes of tephra.

The emission of hydrothermal solutions, steam and other gases without significant quantities of fragmental material or lava can give rise to minor landforms. Vents through which steam or gas are emitted are known as **fumaroles** or **solfataras,** while the precipitation of minerals such as silica and calcium carbonate around a vent can create a **sinter mound. Mud volcanoes** are rather more impressive features which may attain elevations of up to 100 m, eruptions of mud arising from the heating of shallow ground water. If the confining

Table 5.2 Classification of volcanic eruptions

TYPE OF ERUPTION	TYPE OF MAGMA	NATURE OF EFFUSIVE ACTIVITY	NATURE OF EXPLOSIVE ACTIVITY	STRUCTURES FORMED AROUND VENT
Icelandic	Basic, low viscosity	Thick, extensive flows from fissures	Very weak	Very broad lava cones; lava plains with construction of cones along fissures in terminal phase
Hawaiian	Basic, low viscosity	Normally thin, extensive flows from central vents	Very weak	Very broad lava domes and shields
Strombolian	Moderate viscosity; partly basic, partly acid	Flows absent, or thick and moderately extensive	Weak to violent	Cinder cones and lava flows
Vulcanian	Acid, viscous	Flows frequently absent; thick if present	Moderate	Ash cones, explosion craters
Vesuvian (strong Vulcanian)	Acid, viscous	Flows frequently absent; thick if present	Moderate to violent	Ash cones, explosion craters
Plinian (extremely strong Vulcanian	Acid, viscous	Flows may be absent; variable in thickness where present	Very violent	Widespread pumice and lapilli; generally no cone construction
Peléan	Acid, viscous	Domes and/or short, very thick flows; may be absent	Like Vulcanian but with nuées ardentes	Domes; cones of ash and pumice
Krakatauan	Acid, viscous	Absent	Cataclysmic	Large explosion caldera

Source: Based largely on G.A. MacDonald (1972) *Volcanoes*. (Prentice-Hall, Englewood Cliffs) Table 10 – 1, p. 211.

pressure at the surface is reduced by some mechanism, such as the draining of a pool, steam may be generated and any overlying fine-grained sediments liquefied.

The complexity of some volcanic eruptions is illustrated by the detailed observations made of Mount St Helens in 1980. After 123 a of quiescence, noticeable seismic activity began on 20 March, 1980. A week later the first steam and ash began to erupt from a newly formed crater on the summit. During April the movement of magma within the volcano was indicated by a bulge which began to develop on its north-eastern flank; by 23 April this had grown to 100 m and was expanding at a daily rate of 1.5 m. On 18 May a magnitude 5 earthquake triggered a failure of the bulge and a huge landslide crashed down the mountain. This caused the instantaneous release of gas and steam which had been confined under high pressure and the whole northern flank of the volcano was removed by an enormous lateral blast. Hot gas, steam and tephra devastated an area of 600 km² and pyroclastic flow deposits covered several square kilometres. A vertical eruption carried volcanic debris to an altitude of 25 km and overall the elevation of Mount St Helens was reduced by 350 m.

5.3.2 Volcano morphology

The form of a volcano is controlled by several variables. The most fundamental is the type of magma present since this influences both the type of eruption and the nature of the erupted material. Eruptions can occur through either fissures or vents, again depending on magma type. Andesitic eruptions typically occur from a single central vent, while both basaltic and, to a lesser extent, rhyolitic eruptions are more often associated with fissures, in part because the major phase of expansion in these forms of volcanism occurs near to the surface. The most violent eruptions in fact leave some of the least impressive landforms; in extreme cases these consist simply of a broad depression surrounded by a plateau of ignimbrite.

The eruption of highly viscous lava, such as that formed by the degassing of rhyolite or andesite, can create large **domes** which may reach a height of several hundred metres and a width of several kilometres. Small domes composed of viscous lava may form within craters. These can be either rounded **tholoids,** cylindrical **plug domes** or irregular **cumulo-domes.** Highly explosive eruptions of gas-charged magma originating in the upper mantle produce small, shallow craters called **maars.**

A change in magma composition may occur during a single eruption as different parts of the magma chamber are tapped, but more significant compositional changes, often leading to the production of more acidic ejecta, may arise during the overall lifespan of a volcano as a result of chemical changes in the magma reservoir. The duration of both an individual eruption and the overall lifespan of a volcano will also affect its form, and its size. The length of single eruptive episode, which can range from a few days to several years, will depend on the eruptive energy available once the ground surface is broken. The total active life of a volcano, possibly involving hundreds of individual eruptions and spanning up to hundreds of thousands of years, will be related to much more deep-seated processes associated with the heat source promoting crustal melting and magma generation.

Volcanoes are highly variable in their morphology and are therefore very difficult to classify satisfactorily. Moreover, size itself is an important factor since greater quantities of lava or tephra do not simply produce landforms of a similar form but larger size. Rather volcanoes can be said to have a **morphological capacity** representing the maximum size attainable by a particular type of eruption and associated ejecta. For instance, **cinder cones,** which are composed entirely of tephra, are structurally too weak to attain a large size (Fig. 5.8). Consequently all large volcanoes are composed of lava, or a mixture of lava and tephra. This notion of morphological capacity is evident in Table 5.3 and Figure 5.9 in which the main types of volcanic landforms are classified and illustrated.

Once created, denudational processes can rapidly modify the original constructional form of a volcano. The rate of this modification will be a function of the resistance to weathering and erosion of the volcanic materials laid down, the initial relief created by the volcano and the prevailing climatic environment. The effect of erosion on volcano morphology is beautifully illustrated in the Hawaiian Islands where the degree of dissection of each island is proportional to the time elapsed since volcanic activity ceased. The island of Hawaii itself is currently active and shows little sign of erosion (Fig. 5.10), whereas the island of Kauai, where most volcanic activity ceased around 4 Ma BP, is deeply eroded and cut by spectacular valleys (Fig. 5.11).

5.3.3 Basalt domes and shield volcanoes

The low viscosity of basaltic lavas means that extensive flows are more common than distinct cones. Where lava erupts from a single vent a low **exogenous dome** may be formed from a succession of flows, but basaltic lava flows so readily that such features will only develop on nearly level surfaces. Exogenous domes grade upwards in size to **shield volcanoes** which, together with plateaus formed by flood basalts or ignimbrite sheets, rank as the largest volcanic landforms. The smaller versions range in height from 100 m to 1000 m, but because their slopes typically have low gradients (usually < 10°) they may have basal diameters of 2–20 km. The largest examples are to be found on oceanic lithosphere marking the site of hot spots. The greatest group of shield volcanoes on Earth forms the island of Hawaii. The two major volcanoes, Mauna Loa and Mauna Kea, have attained elevations of over 4000 m above sea level, but they rise

Fig. 5.8 *Sunset Crater, Arizona, USA., a small cinder cone formed by an eruption in 1066 AD. Note the fresh appearance of the lava flow in the foreground which issued from the base of the cone during the eruption.*

Table 5.3 Classification of volcanoes and related landforms

TYPE OF MAGMA	TYPE OF ACTIVITY	QUANTITY OF MAGMA PRODUCED			
		Small ←————————————————————————→ Large			
Fluid, very hot, basic in composition	Effusive	Lava flows[1]	Exogenous[2] domes	Basalt domes and shield volcanoes	Icelandic[3A]
					Hawaiian[3B]
Increasing viscosity, gas content, and acidity (high proportion of silica)	Mixed	Scoria cones with flows[4]	Composite or strato-volcanoes[6]		Volcanic fields with multiple domes
		Loose tephra cones with thick flows[5]			
		Endogenous domes (plug domes, tholoids)[7]	Ruptured endogenous domes with thick lava flows[8]		
Viscous, relatively cool, acidic	Explosive	Maars of tephra[9]	Maars with ramparts[10]	Collapse and explosion calderas[13]	Ignimbrite sheets
Extremely viscous, abundant crystals	Explosive, mostly gas	Gas maars[11]	Explosion craters[12]		

Note: The numbers refer to the diagrams in Fig. 5.9.
Source: Adapted from A. Rittmann (1962) *Volcanoes and Their Activity*, Wiley, New York, Table 4, p. 115.

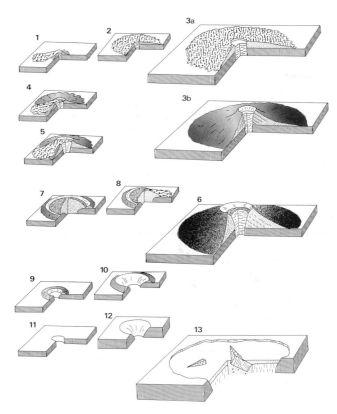

Fig. 5.9 *Schematic block diagrams of simple central volcanoes. The numbers refer to the volcano types classified in Table 5.3. The forms of actual volcanoes are normally far more complex since they usually result from a complicated history of numerous eruptive episodes. (After A. Rittmann (1962) Volcanoes and Their Activity. Wiley, New York, Fig. 57, p.114.)*

more than 9000 m from the floor of the Pacific Ocean from a base well in excess of 200 km across (Fig. 5.12). Nevertheless, these dimensions are dwarfed by the enormous volcanoes now known to exist on Mars, the highest of which, Olympus Mons, rises 26 km from the planet's surface (see Section 19.3.2).

5.3.4 Strato-volcanoes

More common than purely basaltic lava volcanoes are those comprising a mixture of tephra and lava. Such cones are known as **strato-volcanoes** or **composite volcanoes.** They often assume a highly symmetric form with slopes at, or close to, the angle of stability of the ejecta (Fig. 3.12). Large strato-volcanoes invariably contain lava flows which help to support the weight of overlying tephra and in fact most contain a complex network of lava flows together with hori-

Fig. 5.10 *Gently sloping south-eastern flank of Mauna Loa, Hawaii, showing no evidence of significant erosion.*

Fig. 5.11 *The deeply dissected north-west coast of Kauai, the oldest of the five major islands in the Hawaiian group (compare with Figure 5.10).*

zontal and vertical igneous intrusions known respectively as sills and dykes (dikes) (Fig. 5.13) (see Section 5.5).

Rather than having straight slopes, strato-volcanoes typically have concave profiles. Such a slope form has been variously attributed to the effects of lava flows emitted from the summit crater, the changing angle of slope stability in response to a change in the particle-size characteristics of the tephra ejected, and the effects of slumping. But such a form can also be anticipated from the pattern of deposition of tephra as an ash cloud spreads out from the summit crater during an eruption over a wedge-shaped sector of the cone, the rate of ash fall-out being the reciprocal of the distance from the vent. Although this effect may be instrumental in creating the characteristic concave profile, it is the subsequent formation of lava flows and development of intrusions within the cone that support the structure of the volcano and allow its form to persist.

5.3.5 Calderas

Calderas are sometimes simply defined as large craters, the lower size limit normally being a diameter of 2 km. Although arbitrary, this size limit can be related to the idea of morpho-

logical capacity since craters below this size are normally a constructional component of strato-volcanoes, whereas those above are nearly always formed by subsidence and collapse following catastrophic eruptions. Such **explosion calderas** formed by rapid collapse after the evacuation of underlying magma chambers are relatively common on the summits of large strato-volcanoes and should be distinguished from the **subsidence calderas** of large shield volcanoes which are associated with the non-explosive eruption of basaltic lavas.

Calderas created by violent explosions can be of enormous size. One on the Yellowstone Plateau in Wyoming, U S A, covers an area of around 2500 km² and forms a large basin partly filled by massive sheets of ignimbrite. Smaller, but still impressive, calderas are to be found at the summits of strato-volcanoes. An example often cited is Crater Lake in Oregon, U S A. (Fig. 5.14). Here a 600 m deep lake occupies a 9 km diameter crater (Fig. 5.15). The frequent association of massive quantities of ignimbrite with large calderas suggests a genetic link. A favoured hypothesis is that the persistence of a large body of magma allows it to become vertically stratified with the more viscous and gas-rich component accumulating towards the top (Fig. 5.16). An unknown mechanism which causes the magma to subside leads

Fig. 5.12   *The island of Hawaii, the largest shield volcano complex on Earth, seen in a Landsat mosaic. It reaches a height of 4135 m above sea level and descends 4900 m to the ocean floor. The more recent eruptive activity is evident from the dark, fresh lava flows. Most of those visible on the flanks of Mauna Loa, the prominent summit just to the south of the centre of the island, have been formed in the past 150 a. The area covered by the image is 185 km across. (Image courtesy N. M. Short.)*

to a drop in pressure which promotes a rapid expansion of gas and tephra through fissures in the cone above. This initial eruption further lowers the pressure in the magma chamber and more eruptions are generated.

5.4   Other extrusive igneous landforms

We have already noted how certain kinds of highly explosive eruptions can produce massive pyroclastic flows which create sheets of ignimbrite. In some cases these can completely

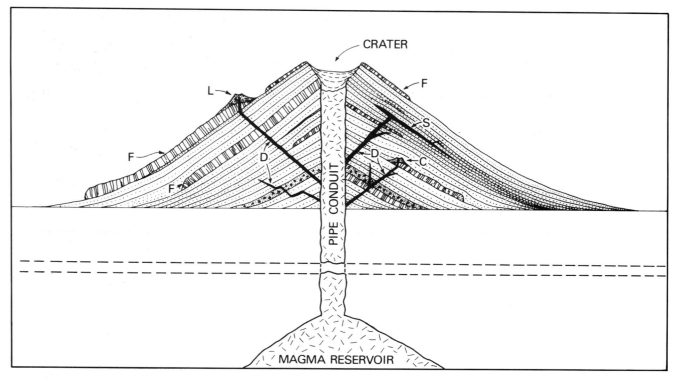

Fig. 5.13 *Schematic cross-section of a typical strato-volcano showing dykes (D), a dyke conduit feeding a lateral cone (L), lava flows (F), a buried cinder cone (C) and a sill (S). Ash deposits are dotted, layers of coarser tephra are marked with small triangles and lava flows are irregularly cross-hatched. (After G. A. MacDonald, (1972) Volcanoes. Prentice-Hall, Englewood Cliffs, Fig. 2.1, p. 23.)*

cover the topography over an extensive area and form an ignimbrite plateau. It is the eruption of low viscosity basaltic lavas, however, that is responsible for the construction of the most impressive and extensive volcanic landforms.

Basaltic magmas are normally expelled at the surface in a very hot and fluid state, and the resulting lavas are often able to flow over considerable distances. Thus, in addition to forming shield volcanoes and domes, basaltic lavas erupting from an extensive regional system of fissures can generate flood basalts and create massive basalt plateaus. Small-scale basalt eruptions can fill valleys and in some cases over-top interfluves: but large-scale flood basalts can completely bury a pre-existing topography as great numbers of superimposed individual flows build up a basalt plateau.

The Columbia Plateau, which covers 130 000 km^2 of the states of Washington and Oregon in the north-west USA, is composed of hundreds of separate basalt flows of Miocene age which in places reach a total thickness of 2000 m and have buried a pre-existing relief of more than 1500 m. Even more extensive examples are the Deccan Plateau of India which covers more than 500 000 km^2 and the vast Parana Plateau which extends over some 750 000 km^2 of Uruguay and southern Brazil. The location and age of these massive outpourings of basalt – Cretaceous to Eocene

in India and Early Jurassic in South America - suggest that they were associated with the break-up of Gondwana (see Section 4.5.2). The Early Jurassic Karoo basalts of southern Africa have been reduced by erosion to an area of 50 000 km^2, but originally they were probably ten times as extensive. This is suggested by the widespread occurrence in southern Africa of thick dolerite intrusions which are contemporaneous with the Karoo flood basalts and which themselves have created significant thickening of the sedimentary sequences into which they are intruded.

Since highly fluid flows have an initial slope of only around 1° they form monotonously flat constructional landscapes. Nevertheless the erosion of such landscapes can expose the successive lava flows which may then form low, but laterally persistent, cliff-like faces in which the hexagonal joint patterns that are sometimes developed in basalt are visible. More generally, the great weight of basalt erupted at the surface may, in some cases, be sufficient to promote significant isostatic subsidence in the underlying bedrock.

5.5   Landforms associated with igneous intrusions

Igneous intrusions form through the penetration of country rock by a mobile body of igneous rock which need not be

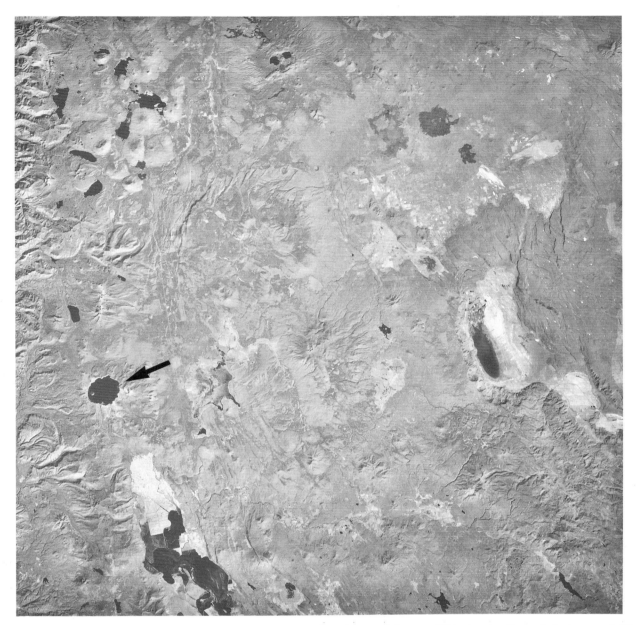

Fig. 5.14 *Landsat image covering an area about 180 km across of south-central Oregon, USA, showing Crater Lake (arrowed). The left third of the image covers part of the Cascade Range which includes several volcanic peaks, while the right half consists of lava plains which fringe the southern edge of the Columbia Plateau. Crater Lake itself was formed by a catastrophic eruption of Mount Mazama, which originally had an elevation of at least 3600 m, around 6500 a BP. Sufficient ejecta were generated to cover a surrounding area of nearly 13 000 km² with tephra at least 0.15 m deep. (Image courtesy N. M. Short.)*

molten. If subject to stress for a sufficiently long period of time all 'solid' materials are capable of flow. The length of time required for a solid to flow rather than fracture is known as its **rheidity** and the type of flow produced is termed **rheid flow.** The rheidity of ice is a few weeks, that of salt about a year, but igneous rocks will only exhibit rheid flow after a stress has been applied for hundreds of years, in the case of deeply buried rocks subject to elevated temperatures, or tens of thousands of years for rocks at the surface. Consequently, intrusions can be formed either by molten rock or by rock experiencing rheid flow.

Intrusions can be composed of a wide range of rock types, although very large intrusions with a surface exposure of over 100 km² known as **batholiths,** often have an acidic granitic composition (Fig. 5.17). Granite masses originate at depth and rise very slowly to the surface as **diapirs** over

Fig. 5.15   *Crater Lake, Oregon, USA. Wizard Island, a small cinder cone, can be seen emerging from the lake (see Figure 5.14).*

millions of years. The process involved is known as **dia-pirism,** and diapirs rise because the hot rock of which they are composed is less dense than the cooler surrounding country rock. Diapirism involves both the moving aside of overlying strata and its partial reincorporation into the rising mass of rock. Pods of country rock surrounded by intrusive material create **roof pendants** when subsequently exposed by erosion (Fig. 5.17). Processes along continental-margin subduction zones often lead to the formation of massive granite batholiths which play an important part in orogenesis. They frequently underlie and support the most elevated sectors of continental-margin orogens, as is the case in the Andes (see Section 3.3.1).

An important factor in the mode of landscape develop-ment on granite batholiths once they are sub-aerially exposed is the joint structure that they form. Initially, three sets of largely orthogonal joints develop, but as the overlying rock is unloaded by erosion the release of pressure in the upper 100 m or so of the batholith generates a secondary set of joints aligned roughly parallel with the surface. These joint systems play an important role in the development of weath-ering forms (see Section 6.3.1) and of drainage patterns (see Section 16.2.2).

In contrast to the irregular form characteristic of acidic granite batholiths, large basic intrusions which typically have a gabbro-type composition are normally layered and have an overall saucer shape (Fig. 5.17). Such intrusions, which rarely attain the dimensions of batholiths, are known as **lopoliths** and their layered structure tends to lead to the formation of series of outward-facing scarps as they are exposed by erosion.

Small intrusions occur both in association with larger forms, and with extrusive igneous phenomena. They can be conveniently categorized as **concordant** or **discordant** de-pending on whether or not they cut across the bedding of the pre-existing strata (Fig. 5.17). The form they take is a func-tion of both the viscosity of the magma and the configur-ation of fractures and other lines of weakness in the country rock. Where exposed by erosion even quite small intrusions can be significant features in the landscape, especially if they are composed of a rock type which is markedly more re-sistant to weathering than the adjacent strata.

Dykes are vertically discordant intrusions often composed of dolerite and they frequently exist in swarms. They are typically 1–20 m across although much greater dimensions occur; the Great Dyke of Zimbabwe averages 6–8 km across

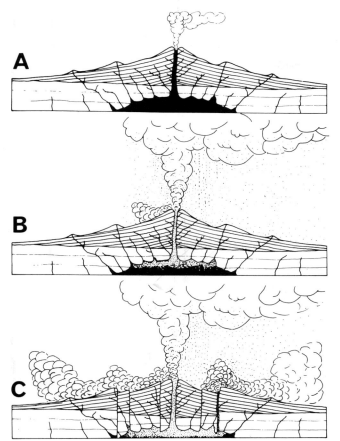

*Fig. 5.16   The formation of Crater Lake caldera, Oregon, USA: (A) before the eruption; (B) during an early stage of the caldera-forming eruption, showing an ash cloud and a small ash flow issuing from the central vent; (C) during the climax of the eruption, with large ash flows erupting from the central vent and from ring fractures on the volcano slopes while the summit starts to sink as a series of blocks; (D) after the eruption; (E) in its present state with new eruptions on the floor and the caldera partly filled with water. The small volcano marked X and clearly visible in Figure 5.15 is Wizard Island, a cinder cone formed since the Pleistocene. (From G. A. MacDonald (1972) Volcanoes. Prentice-Hall, Englewood Cliffs, Fig. 12–5, p.301, modified after H. Williams (1942) Carnegie Institute, Washington, Publication **540**.)*

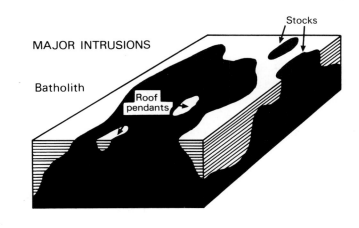

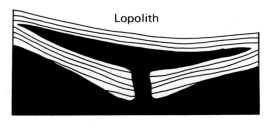

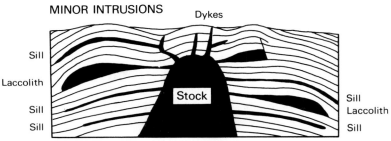

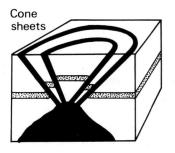

Fig. 5.17   *Various types of major and minor intrusion. Lopoliths, dykes, stocks and cone sheets are discordant structures, whereas batholiths, sills, laccoliths are concordant. Note that the diagrams are not to scale and that many other varieties of intrusion exist. (Modified from B. W. Sparks (1971) Rocks and Relief. Longman, London, Fig. 3.1, p. 68, Fig. 3.2, p. 70, Fig. 3.7, p. 90 and Fig. 3.17, p. 101.)*

Fig. 5.18 *A large contorted dolerite dyke cutting across the unconsolidated sediments of a shallow basin (pan), northern Cape Province, South Africa.*

and is over 500 km in length. In some cases curved concentric series of dykes radiate outward and upward from a common magma chamber at depth; such features are described as **cone sheets.** As with other intrusions, dykes form either positive or negative features depending on whether they are more or less resistant than the surrounding rock (Fig. 5.18).

The most areally extensive small intrusion is a sheet-like form called a **sill.** Sills can be hundreds of metres thick but 10 m–30 m is a more typical range. If formed from viscous magma they are normally limited in lateral extent but where composed of basic rocks they may extend for thousands of square kilometres. The vast series of dolerite sills which have intruded the Karoo sediments of southern Africa underlie an area of over 500 000 km² and probably constitute a total volume of in excess of 200 000 km³. In this region, as in many others, the intruded rock is more resistant than the surrounding country rock and gives rise to prominent topographic features (Fig. 7.26).

A **laccolith** is a type of sill which has thickened to produce a dome which causes the overlying rock to be upwarped. Perhaps the most noted examples of this kind of structure are provided by the Henry Mountains in Utah, U S A (Fig. 5.19). These formed the topic of a classic study by G. K. Gilbert published in 1877. Subsequent investigations have led to some revisions of Gilbert's original findings and it is now thought by some researchers that the laccoliths radiate from central discordant intrusions known as **stocks** (Fig. 5.17). Uplift associated with the intrusion of these stocks has given rise to several peaks rising some 1500 m above the surrounding Colorado Plateau.

Further reading

There are a number of texts on the nature of volcanic activity and the development of volcanic landforms. Excellent general introductions to volcanic processes are provided by Bullard (1976), Decker and Decker (1989), Francis (1976), MacDonald (1972), Rittmann (1962) and Williams and McBirney (1979). Volcanic landforms are considered more specifically in the classic work by Cotton (1944) and in the more recent books by Ollier (1988) and Green and Short (1971). Examples of more detailed analyses of volcanic landforms in particular regions are to be found in Soons and Selby (1982) (New Zealand) and Yoshikawa *et al.* (1981) (Japan).

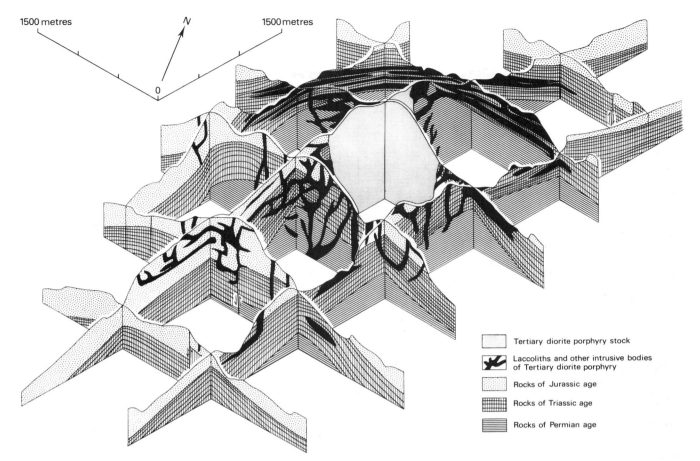

1500 metres   N   1500 metres

0

Tertiary diorite porphyry stock

Laccoliths and other intrusive bodies of Tertiary diorite porphyry

Rocks of Jurassic age

Rocks of Triassic age

Rocks of Permian age

Fig. 5.19 *Fence diagram showing the structural relations of the intrusions of Mount Ellsworth, Henry Mountains, Utah, USA. (from C. B. Hunt (1953)* United States Geological Survey Professional Paper **228***, Plate 16.)*

Ollier (1988) discusses the various types of volcanic products and their associated landforms in some detail. Spry (1962) considers the development of columnar joints in basalts while Walker (1973) examines the factors that determine the length of lava flows. Self and Sparks (1981) contains papers on various aspects of tephra, and Fisher and Schmincke (1984) and Sparks *et al.* (1973) deal with pyroclastic flow deposits.

Factors controlling the global occurrence of volcanism are evaluated by Clapperton (1977) while more detailed analyses of volcano spacing are presented by Mohr and Wood (1976) and Vogt (1974). Menard (1986) discusses volcanism with particular reference to oceanic islands. The characteristics of the various types of volcanic eruption are considered in the general texts mentioned above. For examples of particular eruptions see Self and Rampino (1981) and Simkin and Fiske (1983) on the cataclysmic Krakatau eruption of 1883 and the extremely detailed and fully illustrated collection of papers on the 1980 Mount St Helens eruption edited by Lipman and Mullineaux (1981). Chester *et al.* (1985) provide a detailed analysis of Mount Etna, while Decker *et al.* (1987) present

a comprehensive study of volcanism in Hawaii. The catastrophic collapse of volcanoes is discussed by Francis and Self (1987) and the global occurrence of different volcano types is assessed by Suzuki (1977).

Landforms associated with intrusions are considered by Sparks (1971), while the classic work by Gilbert (1877) on laccolith structures in the Henry Mountains, Utah has been reassessed by Hunt (1953). Ollier and Pain (1981) consider an interesting case of an apparent diapir of gneiss that has risen to create a major domal landform.

References

Bullard, F. M. (1976) *Volcanoes of the Earth*. Texas University Press, Austin.

Chester, D. K., Duncan A. M., Guest, J. E. and Kilburn, C. R. J. (1985) *Mount Etna: The Anatomy of a Volcano*. Stanford University Press, Stanford.

Clapperton, C. M. (1977) Volcanoes in space and time. *Progress in Physical Geography* **1**, 375–411.

Cotton, C. A. (1944) *Volcanoes as Landscape Forms*. Whitcombe and Tombs, Christchurch.

Decker, R. and Decker, B. (1989) *Volcanoes.* (2nd edn) W. H. Freeman, San Francisco.

Decker, R.W., Wright, T.L. and Stauffer, P.H. (eds) (1987) Volcanism in Hawaii. *United States Geological Survey Professional Paper* **1350.**

Fisher, R.V. and Schmincke, H.U. (1984) *Pyroclastic Rocks.* Springer-Verlag, New York.

Francis, P. (1976) *Volcanoes.* Penguin Books, Harmondsworth.

Francis, P. and Self, S. (1987) Collapsing volcanoes. *Scientific American* **256** (6), 73–9.

Gilbert, G.K. (1877) *Report on the Geology of the Henry Mountains.* United States Department of the Interior, Washington, D C.

Green, J. and Short, N.M. (1971) *Volcanic Landforms and Surface Features.* Springer-Verlag, New York.

Hunt, C.H. (1953) Geology and geography of the Henry Mountains region, Utah. *United States Geological Survey Professional Paper* **228.**

Lipman, P.W. and Mullineaux, D.R. (eds) (1981) The 1980 eruptions of Mount St Helens, Washington. *United States Geological Survey Professional Paper* **1250.**

MacDonald, G.A. (1972) *Volcanoes.* Prentice-Hall, Englewood Cliffs and London.

Menard, H.W. (1986) *Islands.* Scientific American Books, New York.

Mohr, P.A. and Wood, C.A. (1976) Volcano spacings and lithospheric attenuation in the eastern rift of Africa. *Earth and Planetary Science Letters* **33,** 126–44.

Ollier, C.D. (1988) *Volcanoes.* Blackwell, Oxford.

Ollier, C.D. and Pain, C.F. (1981) Active gneiss domes in Papua New Guinea: new tectonic landforms. *Zeitschrift für Geomorphologie* **25,** 133–45.

Rittmann, A. (1962) *Volcanoes and Their Activity* (Translated by E.A. Vincent). Wiley, New York and London.

Self, S. and Rampino, M.R. (1981) The 1883 eruption of Krakatau. *Nature* **294,** 699–704.

Self, S. and Sparks, R.S.J. (eds) (1981) *Tephra Studies as a Tool in Quaternary Research.* Reidel, Dordrecht.

Simkin, T. and Fiske, R.S. (1983) *Krakatau 1883 – The Volcanic Eruption and its Effects.* Smithsonian Institute Press, Washington., DC.

Soons, J.M. and Selby, M.J. (eds) (1982) *Landforms of New Zealand.* Longman Paul, Auckland.

Sparks, B.W. (1971) *Rocks and Relief.* Longman, London and New York.

Sparks, R.S.J., Self, S. and Walker, G.P.L. (1973) Products of ignimbrite eruptions. *Geology* **1,** 115–18.

Spry, A. (1962) The origin of columnar jointing, particularly in basalt flows. *Journal of the Geological Society of Australia* **8,** 191–216.

Suzuki, T. (1977) Volcano types and their global population percentages. *Bulletin of the Volcanology Society of Japan* **22,** 27–40.

Vogt, P.R. (1974) Volcano spacing, fractures, and thickness of lithosphere. *Earth and Planetary Science Letters* **21,** 235–52.

Walker, G.P.L. (1973) Lengths of lava flows. *Philosophical Transactions of the Royal Society London* **A274,** 107–18.

Williams, H. and McBirney, A.R. (1979) *Volcanology.* Freeman Cooper, San Francisco.

Yoshikawa, T., Kaizuka, S. and Ota, Y. (1981) *The Landforms of Japan.* University of Tokyo Press, Tokyo.

Part III

Exogenic processes and landforms

6

Weathering and associated landforms

6.1 The weathering system

Material carried to the sea by rivers, or transported by glaciers or the wind, experiences some degree of chemical decomposition or physical breakdown prior to being eroded. Weathering is therefore an appropriate place to begin our look at the operation and effects of exogenic geomorphic processes. Geomorphologists are concerned with the rates at which different weathering processes operate as a function of environmental conditions, and with the nature of the weathered material that is produced. But they are also especially interested in how weathering gives rise to specific landforms. This contrasts with, for instance, the soil scientist's interest in weathering which stems from a concern for the way it contributes to differences in soil characteristics and the release and movement of nutrients.

6.1.1 The nature of weathering

Weathering can be divided into those processes involving chemical reactions and the formation of new minerals (**chemical weathering**) and those that involve only physical changes (**physical weathering**). Although the differences between these two types of weathering are distinct in theory, in practice they rarely operate separately; rather, the effects of one aid the operation of the other. For instance, a rock shattered through physical weathering will be more liable to chemical weathering because of the increased surface area made available for chemical reactions. Conversely, chemical weathering along microfractures in a rock will weaken it and help physical processes break it down more rapidly.

Weathering can be defined as the adjustment of the chemical, mineralogical and physical properties of rocks in response to environmental conditions prevailing at the Earth's surface. For some igneous and metamorphic rocks

formed at great depths and at high temperatures and pressures this adjustment can involve a complete transformation of their constituent minerals. Weathering occurs through complex interactions between the lithosphere, the atmosphere, the hydrosphere and the biosphere, and gives rise to three major types of product. Chemical processes lead to the release of compounds in solution and the creation of new mineral products, while physical processes cause the breakdown of the original rock into smaller particles. Dissolved material may subsequently be reprecipitated or be reincorporated into other minerals, but the great proportion is carried by rivers to the ocean.

6.1.2 Water in rocks and soils

Water plays a vital role in nearly all mechanisms of physical and chemical weathering. This arises in part from the fact that it is a polar solvent; that is, the covalently bonded H_2O molecule has a positive charge at each end balanced by a negative charge in the middle. The positive and negative parts of water molecules become attached, respectively, to the anions and cations of solids and free them by neutralizing their charge; water is thus a highly effective solvent. A second significant property of water is its ionized state. A proportion of H_2O molecules are always decomposed into **hydrogen ions** (H^+) and **hydroxyl ions** (OH^-) (a state known as **dissociation**), the concentration of H^+ ions being expressed as **pH**. This is defined as the negative logarithm to the base 10 of the H^+ concentration in grams per litre. At the standard temperature of 25 $^{\circ}$C there are 10^{-7}g of H^+ ions per litre of pure water, giving a neutral pH of 7. Lower pH values indicate **acidity** (higher concentration of H^+ ions), and higher values represent **alkalinity** (lower concentration of H^+ ions). It is important to bear in mind that a change in pH of one unit represents a tenfold change in hydrogen ion concentration.

Water may enter the soil or bedrock simply through percolation through interconnected voids between particles under the force of gravity. But gravity is not the only factor controlling the distribution and movement of water in rock or soil, as is apparent from the ability of water to move laterally and upward as well as downward. The most important mechanism whereby horizontal and upward movements of moisture can occur is the **capillary suction** which affects the water films attached to soil and rock particles. This suction effect can be readily illustrated by imagining the rise of water in a thin (capillary) tube through surface tension effects. If such a tube is placed vertically into a tank of water the level of the water in the tube will rise above that in the tank. The rise is a result of the capillary suction which acts against the force of gravity. The suction is relatively more effective in comparison with the effects of gravity in a thin tube because the circumference of the tube (along which the suction effect operates) is large relative to its cross-sectional area (to which the effect of gravity is proportional). Interconnected pores in soils and rock possess a suction in a manner analogous to a capillary tube and this can be measured by the upward movement of water against the force of gravity. Water in fine-grained materials with small pores experiences a greater capillary suction in just the same way that the height to which water rises in a capillary tube increases as the diameter of the tube decreases.

Capillary suction, together with the force of gravity, therefore influence both the movement and distribution of water in soil and rock, and the interaction of these two factors gives rise to four zones of moisture storage (Fig. 6.1). Below the **water table** rock and soil pores are saturated; water movement is controlled by gravity and is subject to **hydrostatic pressure**, that is pressure exerted by the weight of overlying water. Above the water table capillary suction provides an additional control. A zone of **capillary saturation** exists immediately above in which the moisture content, which does not vary with depth, is present in the form of continuous films of water around particles with entrapped air between. Above this zone the moisture content decreases upward to a point where the continuous film of moisture eventually separates into discrete droplets. The two zones characterized by continuous moisture films are referred to collectively as the **capillary fringe**, the depth of which depends on pore size. In materials composed of sand-sized particles this depth is just a few millimetres, but in clay-sized materials in which the voids are 1 μm or less across it can be tens of metres. The rate of capillary movement in very fine pores is, however, so slow that the theoretical equilibrium height over which **capillary rise** should occur is rarely attained in such cases.

6.2   Chemical weathering

6.2.1   Chemical characteristics of rock-forming minerals

There are two main types of chemical bond existing between atoms contained in the compounds constituting the Earth's rock-forming minerals – **ionic bonds** and **covalent bonds**. Atoms with eight electrons in their outermost (valence) shell are chemically stable, but all the elements with which we are concerned have either more or less than this stable number. Those with one additional electron readily lose it and become positively charged ions (**cations**) (for instance, K^+, Na^+) and those with one less tend to gain an electron and become negatively charged ions (**anions**) (for example, F^-, Cl^-). For other elements the loss or gain of two electrons similarly creates ions with a double charge (for instance, Ca^{2+}, O^{2-}). Such ions are stable on their own but only form stable compounds when their electrostatic charge is neutralized by ionic bonding. In those elements which gain or lose three or more electrons in becoming ions, adjacent atoms can share electrons through covalent bonding. Some elements, such as oxygen, are able to form both covalent and ionic bonds.

This is a simplified picture because in reality most chemical bonds in the majority of rock-forming minerals are intermediate between ionic and covalent (Fig. 6.2). As the bonds with oxygen are particularly important in rock-forming minerals Table 6.1 gives estimates of the pro-

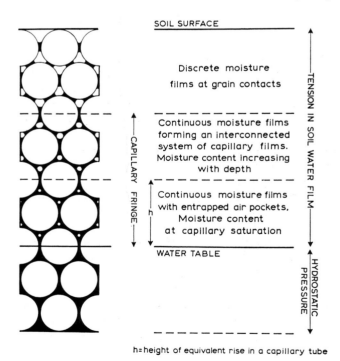

SOIL SURFACE

Discrete moisture films at grain contacts

Continuous moisture films forming an interconnected system of capillary films. Moisture content increasing with depth

Continuous moisture films with entrapped air pockets. Moisture content at capillary saturation

h

WATER TABLE

CAPILLARY FRINGE

TENSION IN SOIL WATER FILM

HYDROSTATIC PRESSURE

h = height of equivalent rise in a capillary tube

Fig. 6.1   *The zones of capillary moisture in regolith and soils. (From M. A. Carson (1969) in R. J. Carson (ed.)* Water, Earth and Man. *Methuen, London, Fig. 4.11.4, p. 100.)*

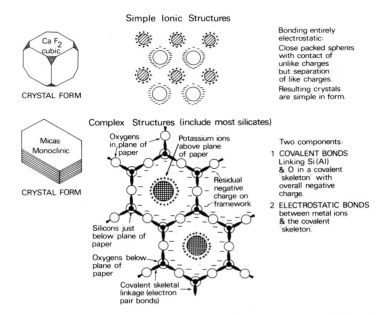

Fig. 6.2 *Ionic and covalent bonding in minerals. Silicate minerals have a basic structure consisting of silicon atoms covalently bonded between oxygen atoms to form tetrahedra with an overall negative charge. This is neutralized through electrostatic bonds with various cations, most commonly potassium, calcium, sodium and magnesium (from C. D. Curtis, (1976) in: E. Derbyshire (ed.)* Geomorphology and Climate. *Wiley, London, Fig. 2.2, p. 34.)*

Table 6.1 Bonding characteristics and ionic properties of some common elements

ELEMENT	ION	APPROXIMATE IONIC CHARACTER OF BOND WITH OXYGEN (%)	IONIC RADIUS (Å) (O^{2-} = 1.40)
Predominantly ionic bonding			
Potassium	K^+	87	1.60
Sodium	Na^+	83	1.16
Calcium	Ca^{2+}	79	1.12
Magnesium	Mg^{2+}	71	0.72
Iron (ferrous)	Fe^{2+}	69	0.77
Intermediate ionic-covalent bonding			
Aluminium	Al^{3+}	60	0.53
Iron (ferric)	Fe^{3+}	54	0.65
Titanium	Ti^{4+}	51	0.61
Predominantly covalent bonding			
Silicon	—	48	—
Phosphorus	—	35	—
Carbon	—	23	—
Sulphur	—	20	—

Source: Based on data in C.D. Curtis (1976) in: E. Derbyshire (ed.) *Geomorphology and Climate*. Wiley, London, Table 2.2, p. 33.

portion of bonding of an ionic type for a number of common elements. Although the relative dimensions of ions (indicated in Table 6.1 by ionic radius) can be used to predict the general form of crystalline compounds where the bonds are largely ionic (such as the cubic structure of fluorite (CaF^2, Fig. 6.2)), most rock-forming minerals are built from complicated silicate structures with complex charge patterns (Fig. 6.3).

6.2.2 Chemical reactions: thermodynamics and kinetics

Although chemical weathering reflects the tendency for new minerals to be formed which are stable under conditions prevailing at the Earth's surface, the rate at which these stable forms are produced is often very slow. This necessitates two complementary approaches to the study of chemical weathering: **thermodynamics** considers the ultimately stable forms by analysing the energy changes involved in chemical reactions, while **kinetics** focuses on rates and mechanisms of change.

The thermodynamic approach can be illustrated using a simple mechanical analogy (Fig. 6.4). All substances contain various amounts of energy within their chemical structures in rather the same way that the ball in Figure 6.4 possesses a potential energy proportional to its height above some datum. When chemical reactions occur energy is usually liberated as heat (although in some kinds of reactions heat is absorbed) just as the ball loses potential energy as it moves downslope. This liberation of heat represents a net release of free energy (ΔR) (more strictly referred to as **Gibbs free energy**). Whether a particular chemical reaction is likely to occur is related to the change in free energy involved. This can be calculated by subtracting the sum of all the free energies of the reacting

ARRANGEMENT OF Si O₄ TETRAHEDRA	NO. OF OXYGENS SHARED WITH ADJACENT TETRAHEDRA	EXAMPLES
A 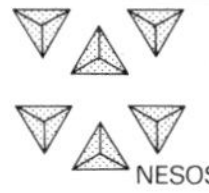 ![NESOSILICATES]	None	Olivine
B  ![SOROSILICATES]	1	Melilite
C 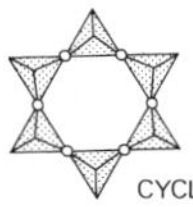 ![CYCLOSILICATES]	2	Beryl
D 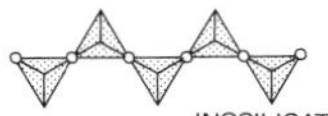 ![INOSILICATES]	2	Pyroxenes e.g. hypersthene diopside augite
E 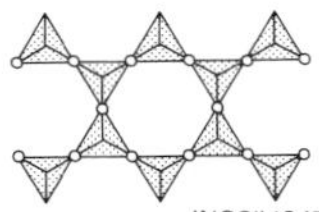 ![INOSILICATES]	2 and 3	Amphibolite e.g. hornblende glaucophane
F 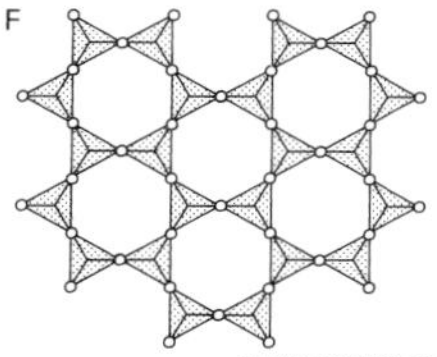 ![PHYLLOSILICATES]	3	Micas e.g. muscovite biotite

Fig. 6.3 *The primary structural forms of silicate minerals (shared oxygen atoms shown by open circles): (A) discrete tetrahedra; (B) a tetrahedra pair sharing one oxygen; (C) a ring of six tetrahedra each sharing two oxygens; (D) a single tetrahedra chain; (E) a double tetrahedra chain with the outward-facing tetrahedra sharing two oxygens, the inward-facing tetrahedra sharing three oxygens and the creation of hexagonal 'holes' between chains sufficiently large to accommodate ions such as OH⁻ (hydroxyl) or F⁻ (fluorine); (F) a tetrahedra sheet with a continuous network of hexagonal holes (this is the structure shown in Figure 6.2). A final structural type of great importance (tectosilicates) consists of a continuous three-dimensional framework of tetrahedra in which all four oxygens are shared; examples are quartz and feldspar). (Based partly on A. Holmes, (1978)* Principles of Physical Geology *(Nelson, Sunbury-on-Thames) Fig. 4.10, p. 52.)*

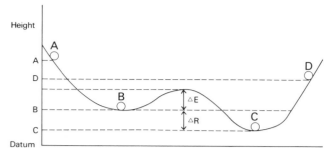

Fig. 6.4 *Types of stability illustrated by a ball on an undulating surface. The potential energy at each position is proportional to its height; the vertical axis is therefore an energy scale. The ball is stable at B and C since it has no spontaneous tendency to move, but A and D are unstable positions since the ball will spontaneously move down the slope towards the nearest trough losing potential energy (height) as it does so. Position B is not as stable as C (it is metastable) since with the addition of a small amount of energy (ΔE) it could roll over the hump to the right to position C where it would have a lower potential energy (height) (ΔR) than at its original position. In this model the most stable state is the one with the least potential energy (lowest height). (based on C. D. Curtis (1976) in: E. Derbyshire (ed.)* Geomorphology and Climate. *Wiley, London, Fig. 2.3, p.38, and Fig. 2.4, p.43.)*

substances from the sum of the free energies of the reaction products. If the calculated free energy change is negative this indicates that the reaction will occur spontaneously; the larger the negative value the more readily the reaction will occur and the more stable the reaction products will be in comparison with the original reactants.

Unfortunately, weathering processes and products cannot be understood simply in terms of the formation of stable mineral forms since many minerals are abundant on the Earth's surface even though thermodynamic considerations tell us that they are unstable. We therefore have to consider the kinetics of weathering by examining those factors which affect the *rate* of chemical reactions. A fundamental control is the degree of instability of the reacting system, or in our mechanical analogy (Fig. 6.4), how far up the slope our ball is located. Another crucial factor is the concentration of reactants; in the case of a solution (such as soil water) reacting with a solid (such as a rock) this will be related to the surface area of contact between the water and the constituent minerals of the rock. A fine-grained, permeable rock, for instance, will provide a much greater surface area than a densely cemented, massive rock. Temperature is another significant factor; this is expressed in the **Arrhenius equation** which indicates that reaction rates approximately double for each 10 °C rise in temperature. **Catalysts** can also greatly speed up chemical reactions, as is illustrated by the familiar oxidation of iron. In completely dry air this reaction proceeds very slowly, but with the addition of water as a catalyst the rate of oxidation increases dramatically. A further important control is the intensity of **leaching**, the downward movement of water through the weathering zone which leads to the removal of soluble products.

6.2.3   Chemical weathering processes

6.2.3.1   Solution

Solution (or **dissolution**) is the simplest process whereby minerals can be decomposed and involves water acting as a solvent. The dissolution of quartz (a crystalline form of silica) provides an example:

$$SiO_2 \quad + \quad 2H_2O \quad \rightarrow \quad Si(OH)_4^0$$
quartz   water   silica in solution
                                  (silicic acid)

The **equilibrium solubility** of a mineral represents the extent to which it will dissolve in water; it is usually expressed in ppm (parts per million by volume) or mg l^{-1}. Some minerals, such as halite (NaCl), are highly soluble in earth surface environments but others, such as quartz, have a very low equilibrium solubility and dissolve in pure water at an exceedingly slow rate. Equilibrium solubility is affected by the temperature and pH of the environment (Fig. 6.5) while the rate of throughput of water is important in controlling the rate of dissolution. This last variable is particularly significant as the film of water in direct contact with the mineral surface eventually becomes saturated with solutes, thereby hindering further dissolution. Consequently the saturated zone must be constantly flushed by under-

saturated water if dissolution is to be an effective mechanism.

Some minerals have a marked ability to absorb water into their crystal structure through a reversible reaction known as **hydration**. This can be illustrated by the hydration of iron oxide:

$$2Fe_2O_3 \quad + \quad 3H_2O \quad \rightleftharpoons \quad 2Fe_2O_3 . 3H_2O$$
iron oxide   water   hydrated iron oxide

This reaction is significant in chemical weathering because it aids other chemical processes by introducing water molecules deep into crystal structures.

6.2.3.2 Hydrolysis

In addition to acting as a solvent, water may react directly with minerals through **hydrolysis**. This involves the replacement of metal cations (most commonly K^+, Na^+, Ca^{2+} and Mg^{2+}) in a mineral lattice by H^+ ions and the combining of these released cations with hydroxyl (OH^-) ions. The effects of this process can be illustrated by crushing different minerals in pure water. The resulting **abrasion pH** indicates the amount of exchange between H^+ ions in the water and cations in the mineral, the increase in pH being due to the abstraction of the H^+ ions from the water (Table 6.2).

The basic hydrolysis reaction can be illustrated by the weathering of the silicate mineral albite (a sodium-rich plagioclase feldspar) to the clay mineral kaolinite:

$$4NaAlSi_3O_8 + 6H_2O \rightarrow Al_4Si_4O_{10}(OH)_8 + 8SiO_2 + 4Na^+ + 4OH^-$$
albite   water   kaolinite   silica   dissolved hydroxyl
                                                              sodium ions

Note that some of the silicon is retained in kaolinite and that sodium is removed in solution. This reaction leads to the production of hydroxyl ions and would make pore and surface waters alkaline, yet in most environments these waters are neutral to slightly acid. Consequently, we can conclude that hydrolysis as such is not a realistic weathering reaction.

6.2.3.3 Carbonation

The **bicarbonate ion** (HCO_3^-) is invariably present in weathering solutions and is easily the most abundant anion in most surface waters. It is formed from the dissolution and dissociation of carbon dioxide in water in a reversible reaction:

$$H_2O \; + \; CO_2 \; \rightleftharpoons \; H_2CO_3 \; \rightleftharpoons \; H^+ \; + \; HCO_3^-$$
water   carbon   carbonic acid   hydrogen   bicarbonate
         dioxide   ion   ion

Carbon dioxide is fixed from the atmosphere by photosynthesis and enters the weathering system through the intermediary of respiration by plant roots and the breakdown of plant debris by bacteria. Carbon dioxide is

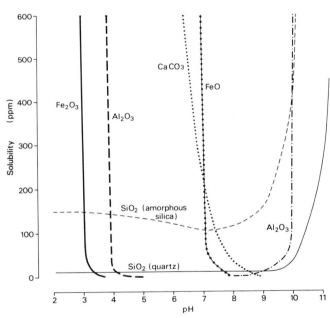

Fig. 6.5 *Equilibrium solubility (expressed in parts per million (ppm)) in water of some components involved in chemical weathering as a function of pH. Note that: (1) the equilibrium solubility of quartz (SiO₂) (crystalline silica) is only about one-tenth of that of amorphous silica; (2) ferrous iron (FeO) is soluble below a pH of about 7, whereas ferric iron (Fe₂O₃) is insoluble except under highly acidic conditions (pH<3.5); (3) alumina (Al₂O₃) is insoluble in the pH range 4–10; and (4) the solubilities of calcium carbonate (CaCO₃) and silica (SiO₂) are inversely related over the pH range 6.5–9. (Based on various sources including H. Blatt et al. (1972) Origin of Sedimentary Rocks. Prentice-Hall, Englewood Cliffs Fig. 10-12, p.364; R. K. Iler (1979) The Chemistry of Silica. Wiley, New York, Fig. 1.6, p.42; and F. C. Loughnan (1969) Chemical Weathering of the Silicate Minerals. Elsevier, New York, Fig. 15, p.32.)*

Table 6.2 Abrasion pH of some common rock-forming minerals

MINERAL	GENERAL FORMULA	ABRASION pH
Olivine	$(Mg, Fe)_2 SiO_4$	10–11
Augite (pyroxene)	$Ca(Mg, Fe, Al) (AlSi)_2O_6$	10
Hornblende (amphibole)	$(Ca, Na)_2 (Mg, Fe, Al)_5 (AlSi)_8O_{22} (OH)_2$	10
Dolomite (carbonate)	$CaMg (CO_3)_2$	9–10
Albite (feldspar)	$NaAlSi_3O_8$	9–10
Biotite (mica)	$K(MgFe)_3 (AlSi_3) O_{10}(OH)_2$	8–9
Anorthite (feldspar)	$CaAl_2Si_2O_8$	8
Hypersthene (pyroxene)	$(Mg, Fe)_2 Si_2 O_6$	8
Orthoclase (feldspar)	$KAlSi_3O_8$	8
Calcite (carbonate)	$CaCO_3$	8
Muscovite (mica)	$K_2Al_4 (Si_6 Al_2 O_{20} (OH)_4$	7–8
Quartz	SiO_2	7

consequently abundant in the atmosphere of soils, especially those characterized by high rates of organic activity; it may reach a concentration of 10 per cent (in comparison with 0.035 per cent in the free atmosphere). The dissolution of carbon dioxide in precipitation provides an additional source of bicarbonate ions.

Surface waters are in fact weak carbonic acid solutions and we can write a more realistic reaction for the weathering of albite in which the release of metal cations (in this case sodium) is matched by the production of bicarbonate ions, a reaction termed **carbonation**:

$$4NaAlSi_3O_8 + 6H_2O + 4CO_2 \rightarrow 8SiO_2 + Al_4Si_4O_{10}(OH)_8$$

albite   water   carbon   silica   kaolinite
                         dioxide

$$+ 4Na^+ + 4HCO_3^-$$

dissolved   bicarbonate
sodium   ions

In humid tropical environments, where organic activity is very high and leaching intense, we would expect silica to be removed in solution as silicic acid:

$$4NaAlSi_3O_8 + 22H_2O + 4CO_2 \rightarrow Al_4Si_4O_{10}(OH)_8$$

albite   water   carbon   kaolinite
                           dioxide

$$+ 4Na^+ + 4HCO_3^- + 8Si(OH)_4^0$$

dissolved   bicarbonate   silicic
sodium   ions   acid

In general it appears that the leaching of metal cations from silicate minerals is controlled by the supply of acids; this not only involves carbonic acid, although this is the most pervasive, but also sulphuric acid and, more significantly, a range of organic acids.

Carbonation plays a particularly important role in the weathering of calcareous rocks. Limestones contain a high proportion of calcium carbonate (nearly always in the form of the mineral calcite) and its weathering involves complex reversible reactions with carbon dioxide in the soil or subterranean atmosphere and carbonic acid in natural waters

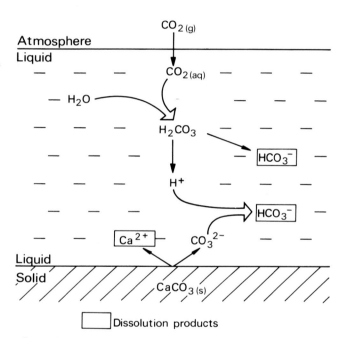

Fig. 6.6 *General scheme of the reactions involved in the chemical weathering of calcium carbonate. (After S. Trudgill, 1985,* Limestone Geomorphology. Longman, London, *Fig. 2.8, p.17.)*

(Fig. 6.6). In the weathering of calcite, half of the bicarbonate is derived from the calcite itself:

$$CaCO_3 + H_2O \rightarrow Ca^{2+} + 2HCO_3^-$$

calcite   water   calcium   bicarbonate
                         ion   ions

The factors controlling the efficacy of limestone solution are complex (Fig. 6.7) but the role of temperature is of particular interest. Whereas the rate of chemical reaction between carbonic acid and calcite increases, as we would expect, with temperature, the equilibrium solubility of carbon dioxide decreases with temperature (at 20 °C it is only half that at 0 °C). This has the effect that high

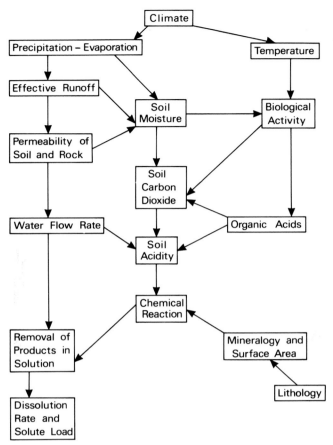

Fig. 6.7 *Key variables in the control of limestone solution. (After S. Trudgill (1985)* Limestone Geomorphology. *Longman, London, Fig. 3.1, p. 27.)*

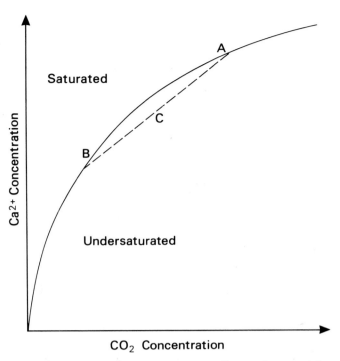

Fig. 6.8 *Schematic representation of the effect on the potential dissolution of calcite of mixing waters with different bicarbonate concentrations arising from the non-linear relationship between CO_2 and Ca^{2+}. The curve represents the equilibrium concentration of Ca^{2+} in solution at given CO_2 levels. Mixing of waters at A and B occurs along the dashed line C in the zone of undersaturation. (After S. Trudgill (1985)* Limestone Geomorphology. *Longman, London, Fig. 2.16, p. 21.)*

concentrations of carbonic acid can be attained in cold regions even though the rate of production of carbon dioxide by organic activity in such environments is relatively low. The mixing of saturated waters with different equilibrium solubility concentrations can also cause further decomposition of calcite (Fig. 6.8).

The concentration of dissolved $CaCO_3$ in surface waters varies considerably. Tropical waters are frequently found to be supersaturated with respect to the prevailing pH levels and the equilibrium concentration of dissolved carbon dioxide. This suggests that the solution of limestone in such areas must be controlled by components other than carbonic acid and it seems likely that organic acids play a major role.

6.2.3.4 *Oxidation and reduction*

Oxidation is the process whereby an atom or ion loses an electron and thus acquires an increase in its positive charge or decrease in its negative charge. Oxygen dissolved in water is by far the most common oxidizing agent. The reaction can be reversed by **reduction** which involves the gaining of an electron.

Oxidation acts as a weathering process in two distinct ways. Various elements, such as iron, titanium, manganese and sulphur can be oxidized to form oxides or hydroxides. For iron this reaction is written:

$$4Fe^{2+} + 3O_2 \rightarrow 2Fe_2O_3$$
$$\text{iron} \quad \text{oxygen} \quad \text{iron oxide}$$

Iron usually exists in the bivalent ferrous state (Fe^{2+}) in most rock-forming minerals, but it can be converted by oxidation to the trivalent ferric form (Fe^{3+}) with the effect that the neutral charge of the crystal structure is upset and can only be regained by the loss of other cations. This mechanism can lead to the collapse of the mineral lattice (as in the formation of the clay mineral vermiculite from biotite) and it also renders a mineral more vulnerable to the operation of other weathering processes.

The tendency for oxidation or reduction to occur is indicated by the **redox potential** (**Eh**) of the environment. This is measured in units of millivolts (mV), with positive values registering an oxidizing potential and negative values a reducing potential. Abundant oxygen is dissolved in most surface waters and the Eh is predominantly positive in weathering environments; that is, oxidation occurs spon-

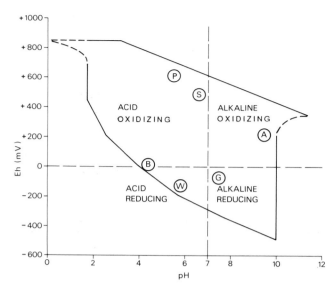

Fig. 6.9   *Range of Eh–pH values encountered in the environment. The enclosed area represents the extreme limits of natural Eh–pH values as indicated by the extensive collation of data by Baas Becking* et al. *(1960). The letters represent typical values for various types of environments: (P) precipitation, (S) stream water, (B) bog water, (W) waterlogged soils, (G) ground water and (A) aerated alkaline environments. (Based on R. M. Garrels and C. L. Christ (1965)* Solutions, Minerals, and Equilibria. *Freeman, Cooper, San Francisco, Fig. 11.1, p. 380 and Fig. 11.2, p. 381 after L. G. M. Baas Becking* et al. *(1960)* Journal of Geology **68** *Fig. 31, p. 276.)*

taneously, though not necessarily rapidly. However, in some localities, such as below the water table and in waterlogged soils, reducing conditions can prevail. Eh also varies with pH, becoming generally lower as alkalinity increases. This relationship can be seen clearly by plotting pH and Eh values measured in various Earth surface environments (Fig. 6.9). Where a particular environment is located on such an Eh–pH diagram provides a useful indication of the nature of chemical reactions that are likely to occur.

6.2.3.5   *Cation exchange*

Cation exchange is the substitution of one cation for another of a different element in a mineral structure. This can occur with any mineral, but it is by far the most common in clay minerals which typically have rather loosely bonded cations on their surfaces which can be readily exchanged for cations in solution. Each type of clay mineral has a different propensity for the exchange of cations which can be measured by its **cation exchange capacity (c.e.c.)** in units of milliequivalents per 100 g (meq $100 \, g^{-1}$) of clay. The mechanism is extremely important in the alteration of one clay mineral to another. The two cations most readily absorbed on to clay minerals are H+ and Ca^{2+}, and the cation most easily released is Na^{2+}.

Cation exchange capacity is affected by the pH of the surrounding solution since under acidic conditions H+ readily replaces metal cations. Where the environment is alkaline, however, metal cations are less easily detached and may even replace H+ adsorbed on to a mineral surface.

6.2.3.6   *Organic processes*

We have already referred to the role of organic acids in chemical weathering, the anions of which may be important in the removal of metal cations from silicate minerals and in the dissolution of carbonates. However, it is the formation of **chelating agents** by organic processes that is of particular significance in weathering. These are able to mobilize metal cations, such as Fe^{3+} and Al^{3+}, which are virtually insoluble under normal Eh–pH conditions, a process known as **chelation**. Chelating agents can come from various sources, but most are organic compounds either secreted directly by organisms such as lichen, or formed through the decomposition of humus in the soil. Although the effects of chelation are well known its detailed mechanisms are not. It appears that a stable ring structure is formed between complexing agents and metal cations. The structure may be subsequently broken down by microbial activity and the complexed cation precipitated.

The dark staining characteristic of rock surfaces in arid regions, which is also observed to a limited extent in more humid environments, appears to be primarily a result of biogenic processes. **Desert varnish** (alternatively **rock varnish**), as it is termed, is composed of clay minerals, oxides and hydroxides of manganese and/or iron, together with detrital particles, which form a layer typically around 10–30 μm thick. It probably takes many thousands of years for a characteristic black to brown desert varnish to develop, apparently through the action of bacteria which preferentially oxidize and concentrate certain elements (especially manganese) supplied from dust and surface water.

6.2.4   Products of chemical weathering

6.2.4.1   *The weathering mantle*

Chemical processes assisted by the physical disintegration of the bedrock combine to produce a **weathering mantle** or **regolith** which, if differentiated into identifiable horizons, constitutes a **weathering profile**. The interface between weathered material and unweathered bedrock is known as the **weathering front**. The thickness of a weathering mantle at a particular locality represents a balance between the rate of bedrock weathering and the rate of removal of weathered material by denudational agents (Fig. 6.10). Depths of weathering may exceed 100 m and exceptionally reach 300 m or more.

Thick weathering mantles will only form in regions of

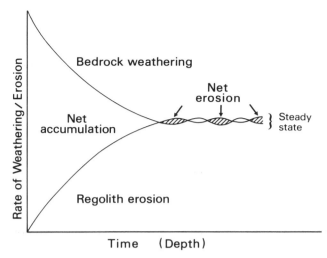

Fig. 6.10 *Schematic representation of the achievement of a steady-state weathering profile thickness dependent upon the rate of weathering and rate of erosion. (After M. J. Crozier (1986)* Landslides: causes consequences and Environment. *Croom Helm, London, Fig. 4.8, p. 138.)*

minimal local relief where rates of erosion have been low for a prolonged period. High rates of weathering are not a prerequisite for the formation of deep weathering profiles if the rate of removal of weathered material is sufficiently slow. This observation is well supported by the presence of very great depths of weathering on the Russian Platform. The general absence of deep weathering mantles in many regions of low relief in the mid-latitudes of the northern hemisphere can probably be attributed to the effects of glacial scouring during the repeated ice-sheet advances of the Pleistocene.

Deep weathering profiles are widespread in the humid tropics, where high temperatures and abundant rainfall create conditions in which the potential for chemical weathering is great. However, when the weathering mantle has developed to a depth of several metres the rate of water movement at the weathering front becomes very slow. Consequently, under these conditions the rate of chemical decomposition at the weathering front is minimal, an effect which is clearly evident in the typically very low concentrations of solutes recorded in streams draining highly leached deep weathering profiles in lowland humid tropical regions (see Section 15.3).

The situation on mountain slopes in the humid tropics is, however, quite different as here high erosion rates lead to the rapid removal of weathered material. Deep weathering mantles cannot develop even though the actual rate of weathering is probably very high as a result of the proximity of the weathering front to the surface and the associated rapid downslope movement of soil water. We need, therefore, to distinguish carefully between *potential* rates of weathering related to climatic parameters, and *actual* rates

of weathering related to conditions at the weathering front.

A further complication in the interpretation of regolith thickness is climatic change. In presently arid central Australia, for instance, deep weathering profiles are to be found which have clearly been inherited from a previously more humid climatic regime. Their partial preservation has been aided by the presence of siliceous crusts which are themselves highly resistant to weathering and erosion (see Section 6.5).

The physical characteristics of weathering profiles depend upon the rock type and its structural properties and mineralogy as well as the intensity and nature of the chemical weathering processes. A number of geomorphic studies have focused on weathering profiles developed on granite under humid tropical and subtropical climates. Although considerable differences exist in profile characteristics on other rock types and under different climatic regimes, deep weathering profiles on granite exhibit features which are common to many (Fig. 6.11). There is typically a gradation over tens of metres from fresh bedrock, through weathering material incorporating **corestones** of relatively unweathered rock, to clay-rich horizons capped by a soil.

The mineralogical composition of the various horizons will be influenced by the mineralogy of the bedrock, the hydrological regime and, related to this, the intensity of leaching. For instance, ferrous iron taken into solution below the water table where reducing conditions prevail may be drawn upwards by capillary suction and precipitated as ferric iron in the upper part of the profile. Under a seasonally dry climate this leads to the formation of a characteristically red-coloured, iron-rich horizon overlying a bleached **pallid zone** which has been effectively leached of ferric iron.

In some weathering mantles original bedrock structures are preserved. The term **saprolite** is given to this type of regolith and it reflects the operation of **isovolumetric weathering**, that is, weathering accomplished without any change in volume. This is probably a very common phenomenon, but in many cases its occurrence cannot be proved either because the original bedrock lacked well-defined structures or because any such structures have been destroyed by organic activity.

6.2.4.2 *Mineral stability and the formation of secondary minerals*

The types and proportions of the various minerals in a weathering profile are usually quite different from the original bedrock. Some minerals seem to survive more or less unaltered even after being subject to prolonged weathering, whereas others decompose very rapidly. A classic study by S. S. Goldich published in 1938 showed a consistent variation in the stability in the weathering environment of common silicate minerals occurring in a range of igneous and metamorphic rocks (Fig. 6.12). Goldich pointed

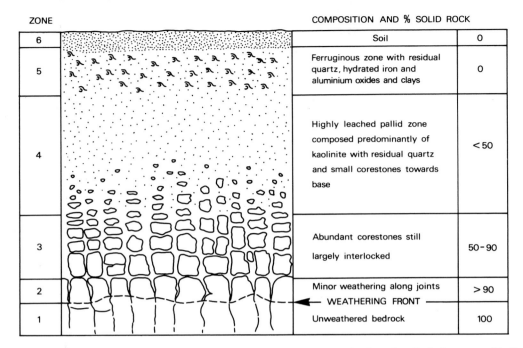

Fig. 6.11 *Features occurring in a fully developed weathering profile on granitic rocks. Based on B. P. Ruxton and L. Berry, 1957, Bulletin of the Geological Society of America **68**, Fig. 2, p. 1266.)*

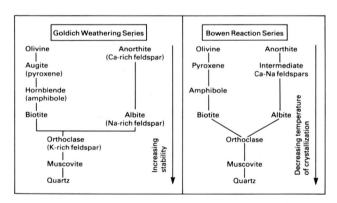

Fig. 6.12 *The weathering series of Goldich and the reaction series of Bowen.*

out that this weathering series closely matched Bowen's reaction series which listed various silicate minerals in terms of their order of crystallization from a silicate melt. Olivine, for instance, is the most unstable mineral in the weathering series and crystallizes at the highest temperature in the reaction series. Quartz, on the other hand, is the most stable and crystallizes at the lowest temperature.

Part of the reason for this relationship between susceptibility to weathering and temperature of crystallization appears to lie in the relative strength of the bonds between oxygen and the cations in each mineral. But the structure of the bonding also seems to be significant, with minerals having few silicon–oxygen bonds (such as the neosilicates and inosilicates) generally being less stable than those with more bonds (such as the phyllosilicates and tectosilicates) (Fig. 6.3). The relationship can also be viewed thermodynamically, there being a larger change in free energy accompanying the decomposition of an unstable mineral like olivine than a comparatively stable mineral like muscovite.

In addition to quartz, which can at least partially survive prolonged weathering in most environments, the mineralogical composition of the weathering mantle includes a range of **secondary minerals** (Table 6.3). These form by the decomposition of primary minerals and the incorporation of dissolved constituents into new mineral forms, by far the most important of which are **clay minerals**. Clay minerals occur in a bewildering variety of forms; only the major groups are listed in Table 6.3. In most cases they are formed through the recombination of silica, alumina and metal cations released during weathering into layered phyllosilicate-type structures (Fig. 6.13). Other secondary minerals formed during weathering include those composed of oxides and hydroxides of iron and aluminium and, less significantly, of titanium oxide, as well as amorphous or finely crystalline silica.

The type of clay mineral produced during weathering depends on the composition of the circulating pore waters (specifically the concentration of dissolved silica and the type and concentration of cations), the mineralogy of the

Table 6.3 Major secondary minerals formed in weathering environments

MINERAL GROUP	COMPOSITION	GENERAL FORMULA	MAJOR VARIETIES	COMMENTS
Oxides and hydroxides				
Oxides of silicon	Silicon dioxide	SiO_2 SiO_2 $SiO_2 \cdot nH_2O$	Quartz Amorphous silica Opaline silica	Water content usually < 10%
Hydroxides of aluminium	Hydrous alumina	$Al_2(OH)_6$ $AlO(OH)$	Gibbsite Boehmite	Has a clay-like layered structure
Oxides and hydroxides of iron	Iron oxide and hydrous iron oxide	Fe_2O_3 $FeO \cdot OH$	Hematite Goethite	
Oxide of titanium	Titanium dioxide	TiO_2	Anatase	
Clay minerals				
Kaolinite group	Hydrous aluminium silicate	$Al_2Si_2O_5(OH)_4$		Kaolinite is the most common variety
Illite group	Hydrous potassium silicate	$KAl_2(AlSi_3)O_{10}(OH)_2$ (K content varies and there is partial substitution of Al by Mg and Fe)		Layered structure similar to mica. Occurs in 1-, 2- and 3-layer forms
Smectite group	Complex hydrous magnesium aluminium silicate	$Al_4Si_8O_{20}(OH)_4 \cdot nH_2O$ (variable replacement of Mg for Al and Al for Si with attachment of Ca and Na)		Readily hydrated with expansion of up to 10 times: montmorillonite is a common variety.
Chlorite group	Hydrous silicate of aluminium iron and magnesium	$(Mg, Fe)_5 Al(AlSi_3)O_{10}(OH)_9$		
Vermiculite group	Hydrous magnesium silicate	$Mg_3[Si_4O_{10}](OH)_2 \cdot nH_2O$ (some Al replacement of Si and introduction of Mg, Fe and Al may occur)		Structure intermediate between smectite and chlorite
Mixed-layered group	Hydrous silicate			Regular or irregular interstratification of two clay minerals. Illite–smectite and chlorite-vermiculite mixed layers are the most common
Palygorskite and sepiolite group	Hydrous magnesium silicate	$Mg_4[Si_6O_{15}](OH)_2 \cdot 6H_2O$		Modified amphibole-type double-chain structure rather than layered phyllosilicate structure of other clay minerals

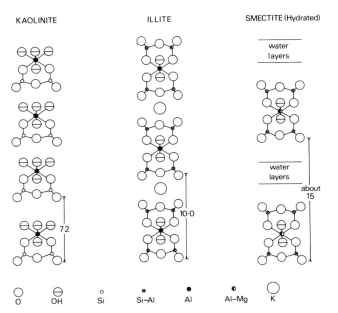

bedrock, the intensity of leaching and the prevailing Eh–pH conditions (Table 6.4). Primary minerals with an existing layered structure, such as biotite, can be fairly readily altered into a clay mineral (often illite which is similarly potassium-rich), whereas most other silicate minerals require a more fundamental reorganization of their crystal structure.

Thermodynamics can indicate which mineral species will be in equilibrium with conditions prevailing in the weathering environment; but the rate at which this equilibrium is achieved is often extremely slow and consequently it is

Fig. 6.13 *Basic structures of some major clay mineral groups. All the clay minerals illustrated have a layered phyllosilicate structure comprising successions of tetrahedral (Si–O) and octahedral (Al–O or Al–OH) layers with, in some cases, intervening layers of absorbed water. The various clay minerals have a characteristic spacing of these layers (shown in the diagram in ångstroms (Å) (10^{-10} m). (Modified from C. Ollier, (1984)* Weathering. *(2nd edn) Longman, London, Fig. 6.5, p. 73.)*

Table 6.4 Mineralogy of weathering products in relation to environmental controls

ENVIRONMENT	pH	Eh	BEHAVIOUR OF MAJOR ELEMENTS	MINERALOGY OF WEATHERING PRODUCTS
Non-leaching. Mean annual precipitation 0–300 mm. Hot	Alkaline	Oxidizing	Some loss of Na^+ and K^+. Iron present in ferric state	Partly decomposed parent minerals. Illite, chlorite, smectite and mixed-layered clay minerals. Hematite, carbonates, secondary silica and salts. Organic matter absent or sparse
Non-leaching below water table	Alkaline to neutral	Reducing	Some loss of Na^+ and K^+. Iron present in ferrous state	Partly decomposed parent materials. Illite, chlorite, smectite and mixed-layered clay minerals. Siderite (iron carbonate) and pyrite (iron sulphide). Organic matter present
Moderate leaching mean annual precipitation 600–1300 mm. Temperate	Acid	Oxidizing to reducing	Loss of Na^+, K^+, Ca^{2+} and Mg^{2+}. Some loss of SiO_2. Concentration of Al_2O_3, Fe_2O_3 and TiO_2	Kaolinite with or without degraded (K-deficient) illite. Some hematite present. Organic matter generally present
Intense leaching. Mean annual precipitation > 1300 mm. Hot	Acid	Oxidizing	Loss of Na^+, K^+, Ca^{2+}, Mg^{2+} and SiO_2. Concentration of Al_2O_3, Fe_2O_3 and TiO_2	Hematite, goethite, gibbsite and boehmite with some kaolinite. Organic matter absent or sparse
Intense leaching. Mean annual precipitation > 1300 mm. Cool	Very acid	Reducing	Loss of Na^+, K^+, Ca^{2+}, Mg^{2+} and some iron and Al_2O_3. SiO_2 and TiO_2 retained	Kaolinite, possibly with some gibbsite or degraded illite. Organic matter abundant

Source: Modified from F. C. Loughnan (1969) *Chemical Weathering of the Silicate Minerals*. Elsevier, Amsterdam, Table 21, p. 73.

important to know what factors control the rate at which more stable weathering products are created. The intensity of leaching is probably the major such kinetic control in

most weathering environments (Fig. 6.14). Clay minerals such as kaolinite, which have been stripped of metal cations together with iron and aluminium oxides and hydroxides, are prevalent in environments characterized by intense leaching since here the large throughput of water removes cations in solution and prevents their concentration in pore waters within the regolith. Conditions of extreme leaching can even lead to a significant removal of iron and silicon to leave an aluminium-rich residue in the form of the mineral gibbsite.

Where leaching is of only moderate intensity, cations released during weathering can build up in the solutions moving through the weathering mantle and the formation of cation-bearing clays such as illite and smectite is favoured.

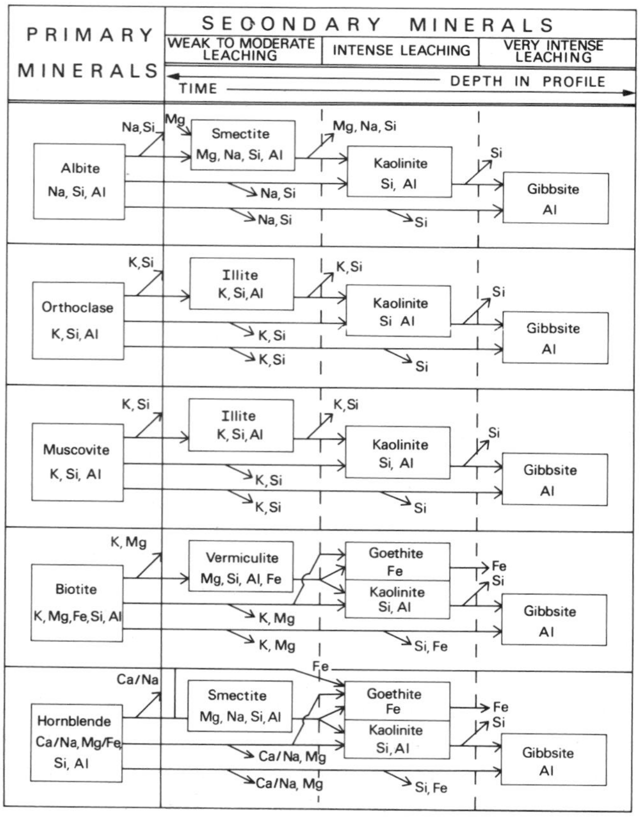

Fig. 6.14 *Schematic representation of the relationship between leaching intensity and types of clay minerals formed. The main constituent elements of each mineral are indicated together with the progressive loss in solution of soluble components at each stage of weathering. A range of primary rock-forming minerals are shown in order to illustrate the role of bedrock mineralogy in controlling the types of clay minerals produced under conditions of weak leaching. Under intense leaching such differences are gradually eliminated. Since the weathering of a unit of bedrock and the leaching of soluble constituents will be progressive the sequences of clay minerals may, in some cases, represent a temporal succession. However, irrespective of the duration of weathering, gibbsite is only likely to form where there is free drainage and intense weathering. Similarly, leaching intensity is likely to vary down profile and this may be reflected in the changing proportions of various secondary minerals*

In arid and semi-arid regions leaching may be minimal and solutions in the weathering mantle can attain high concentrations of dissolved constituents. This condition commonly leads to the accumulation of calcium carbonate at, or near, the surface, while under more extreme conditions of aridity soluble salts such as gypsum ($CaSO_4 \cdot 2H_2O$), halite ($NaCl$) and thenardite (Na_2SO_4) may be precipitated.

The intensity of leaching is never uniform throughout an entire weathering profile since the throughput of water normally decreases with depth due to compaction and the downward translocation of fine particles which reduces the size of voids. The resulting reduction in the intensity of leaching is usually reflected in mineralogical changes with depth (Fig. 6.15). Clay minerals such as illite and smectite, which are readily altered in the zone of active leaching towards the top of a profile, may be relatively stable at lower levels where leaching is less intense. Vertical differentiation in mineralogy in weathering profiles may also reflect the stage-by-stage alteration of primary rock minerals. Smectite and illite may form preferentially near the weathering front only to be eventually altered to kaolinite, and perhaps gibbsite, as gradual lowering of the weathering front and erosion at the top of the profile effectively causes individual clay mineral particles to move up through the profile.

The relative abundance of dissolved constituents released through weathering reactions into surface waters can be compared to their relative abundance in the bedrock in order to provide information on the types of chemical reactions that are occurring. A number of studies have attempted to rank the relative mobilities of the major constituents released during weathering, and although the precise order varies depending on the environmental conditions the generally agreed sequence is

$$Ca^{2+} > Mg^{2+} > Na^+ > K^+ > Fe^{2+} > Si^{4+} > Fe^{3+} > Al^{3+}$$

6.2.5 Factors influencing chemical weathering

It has long been recognized that five factors control the nature of soil development: climate, parent material, topography, organic activity and time. The factors controlling chemical weathering, and indeed weathering as a whole, can be viewed in a similar way. Figure 6.16 illustrates the way in which the four environmental factors influence both the thermodynamics and kinetics of weathering reactions by indicating the conditions required for maximum rates of weathering to be attained.

These factors clearly operate at different scales. Apart from montane environments, where climatic gradients are typically sharp, the effects of climatic differences on weathering tend to be most apparent at the continental and global scale. Indeed attempts have been made to produce world maps indicating the distribution of different weathering products (Fig. 6.17). Although there is a reason-

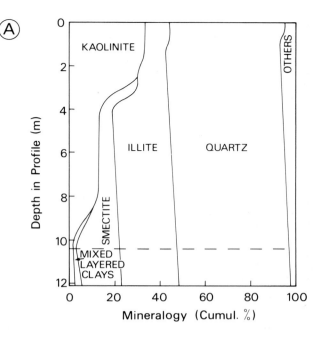

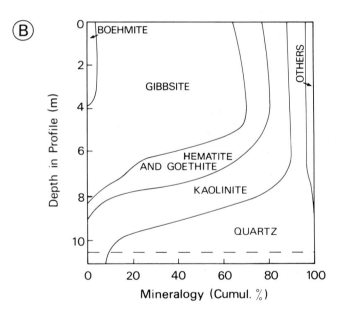

Fig. 6.15 *Variation in mineralogy for weathering profiles developed on clay slates at Goulburn, New South Wales, Australia, under a humid, warm temperate climate (A) and on a sandstone at Weipa, Queensland, Australia, under a hot, monsoonal climate (B). The approximate transition between bedrock and weathered material is indicated by a dashed line. (Adapted from P. S. Bayliss and F. C. Loughnan, (1964), Clay Mineral Bulletin **5**, Fig. 3, p. 358, and F. C. Loughnan and P. Bayliss, 1961, American Mineralogist **46**, Fig. 4, p. 212.)*

ably good correlation between the major climatic regions and particular weathering zones, the correspondence becomes very weak in more mountainous areas where topographic factors can become predominant. This is

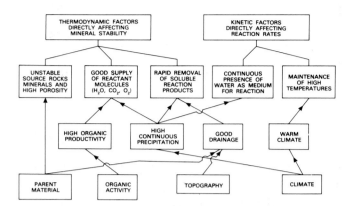

Fig. 6.16 *Schematic representation of the way in which environmental factors can contribute to high rates of chemical weathering. (Modified from C. D. Curtis (1976) in: E. Derbyshire (ed.)* Geomorphology and Climate. *Wiley, London, Fig. 2.5, p. 50.)*

especially the case when the depth of weathering is considered; although a fairly systematic relationship with climate and vegetation is evident as a consequence of variation in precipitation and temperature (Fig. 6.18), this only applies in the absence of marked relief.

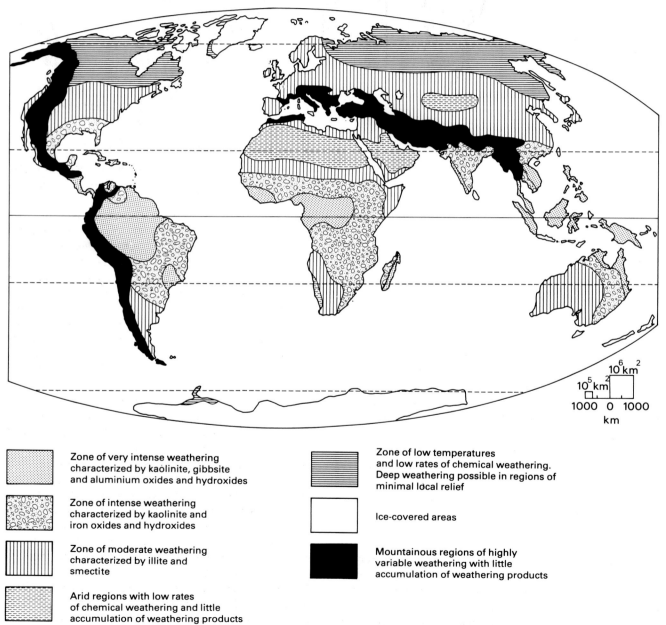

Legend:

- Zone of very intense weathering characterized by kaolinite, gibbsite and aluminium oxides and hydroxides
- Zone of intense weathering characterized by kaolinite and iron oxides and hydroxides
- Zone of moderate weathering characterized by illite and smectite
- Arid regions with low rates of chemical weathering and little accumulation of weathering products
- Zone of low temperatures and low rates of chemical weathering. Deep weathering possible in regions of minimal local relief
- Ice-covered areas
- Mountainous regions of highly variable weathering with little accumulation of weathering products

Fig. 6.17 *Global distribution of major weathering zones. Modified from N. M. Strakhov (1967)* Principles of Lithogenesis *Vol. 1. Oliver and Boyd, Edinburgh, Fig. 1, p. 5.)*

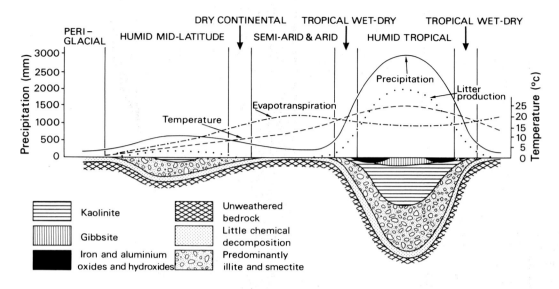

Fig. 6.18   *Variation in weathering mantle depth and composition in relation to climatic and biotic variables. (Modified from N. M. Strakhov, (1967)* Principles of Lithogenesis *Vol. 1. Oliver and Boyd, Edinburgh, Fig. 2, p. 6).*

Temperature is an important factor in rates of chemical weathering, both through the direct effect it has on the rate of chemical reactions (see Section 6.2.2), and indirectly through its influence on rates of organic activity and hence the production of both soil carbon dioxide and organic acids – both critical components in chemical weathering. As long as there is an abundant supply of water high temperatures tend to be associated with high rates of weathering, although this does not necessarily result in deep weathering profiles for, as we have already noted, where slope gradients are steep the products of weathering may be removed almost as soon as they are created.

Rates of weathering are potentially highest where temperatures are high and rainfall is abundant. As indicated in Table 6.4 intensity of leaching, and hence the mineralogy of weathering deposits, is largely a function of temperature, precipitation and drainage. Although the precise mineralogical composition of weathering mantles is influenced by bedrock mineralogy, the effects of variations in precipitation are often evident with the relative abundance of kaolinite, gibbsite and iron and aluminium hydroxides showing a positive correlation with mean annual precipitation (Fig. 6.19).

Vegetation influences weathering, as we have already indicated, through the release of organic acids and in the supply of carbon dioxide to soil waters. Important to both these factors is the production of litter. This varies enormously not only between desert and forest ecosystems but also between temperate forests with a typical range of 0.1–$0.3 \times 10^6 \, kg \, km^{-2} \, a^{-1}$ and tropical rain forests which produce 0.4–$1.3 \times 10^6 \, kg \, km^{-2} \, a^{-1}$. Organic activity is closely

related to climatic controls, but vegetation type also varies at a local scale as a result of topographic factors and soil properites.

Topography affects weathering largely through the way it influences the movement of water through the regolith. Weathering rates and the intensity of leaching depend critically on the total throughput of water and this is likely to be far higher on a steep, well-drained slope than in flat terrain with poor drainage. If drainage is too efficient, however, little water may be present in the regolith for much of the time and rates of weathering will consequently be reduced.

Parent material influences the nature and rate of weathering in two major ways. Its mineralogical composition, particularly in terms of the proportion of unstable mineral phases present, will affect both the rate of chemical decomposition and the kinds of secondary minerals that are most likely to form. Although we might expect bedrock mineralogy to exert a pervasive control over the products of chemical weathering this is not the case, at least at the broad scale, both because the rocks and minerals exposed at the Earth's surface are dominated by just two or three types (Table 6.5), and because prolonged weathering tends to lead to a convergence of secondary mineral types irrespective of parent material mineralogy. The second factor is the physical nature of the parent material, especially its particle size and permeability, which influences the rate of weathering though not the weathering products.

The final factor of time is important because of the slowness of most chemical reactions at the Earth's surface. There is always a significant lag between the establishment of a particular set of conditions in the weathering environ-

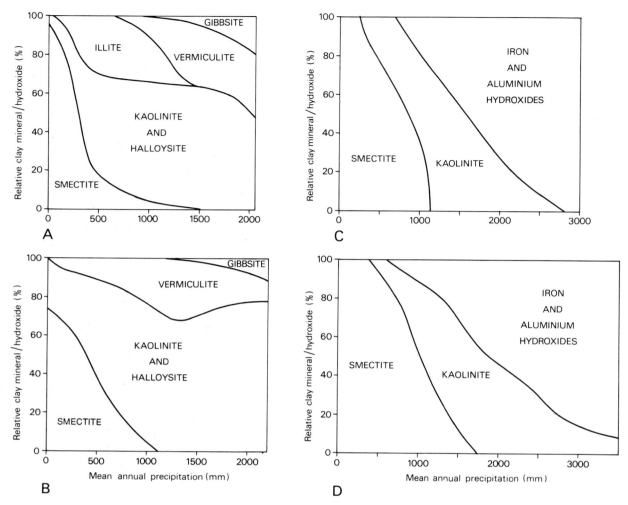

Fig. 6.19 *Variations in the clay mineral and residual hydroxide composition of soils in relation to mean annual precipitation: surface soil samples on feldspar and quartz rich igneous rocks (A) and an olivine, amphibole and pyroxene-rich igneous rocks (B) (both from California); soils developed on basalt under an alternating wet and dry climate (C) and a continuously humid climate (D) (both from Hawaii). (Constructed from data and figures in I. Barshad (1966)* International Clay Conference Proceedings *1, Figs 1 and 2, p. 167 and G. D. Sherman (1962)* Problems of Clay and Laterite Genesis. American Institute of Mining and Metallurgical Engineering, New York, Fig. 2, p. 158 and Fig. 3, p. 159.)*

ment and the adjustment of the mineralogical and physical properties of the regolith to these conditions. Consequently, weathering profiles are rarely in full equilibrium with environmental conditions; in most cases the weathering mantle adjusts to long-term average conditions rather than to conditions at a specific time. We always have to be aware of the kinetics of weathering reactions when looking at the mineralogical composition of weathering profiles. As we shall see in Section 6.5, some weathering products show a remarkable resilience to changing environmental conditions to the extent that in some circumstances they can persist in the landscape essentially unaltered for tens of millions of years.

6.3   Physical weathering

Physical weathering encompasses a range of mechanisms, the relative effectiveness of which are not accurately known but clearly vary significantly as a function of environmental conditions. The physical breakdown of rock is always associated with some kind of volume change and it is useful to categorize the various processes into those involving an overall volumetric change in the rock mass and those related to changes in volume of material introduced into voids or fissures in the rock.

6.3.1   Volumetric changes of the rock mass

As noted in Chapter 5 in the context of granite intrusions (Section 5.5), rocks formed at depth or located beneath a

Table 6.5 Relative abundance of rock types and minerals exposed to weathering at the Earth's surface

Proportion of area covered by major rock types (%)	
Shale	52
Sandstone	15
Granite and granodiorite	15
Limestone and dolomite	7
Basalt	3
Others	8
Total sedimentary	~ 75
Total igneous and metamorphic	~ 25
Proportion of common minerals in bedrock exposed to weathering at the Earth's surface (%)	
Feldspars	30
Quartz	28
Clay minerals and micas	18
Calcite and dolomite	9
Iron oxide minerals	4
Pyroxenes and amphiboles	1
Others	10

Source: Data from L. B. Leopold *et al.* (1964) *Fluvial Processes in Geomorphology*. W. H. Freeman, San Francisco, p. 100.

thick overburden are under considerable internal stress. As the overlying strata are gradually removed by erosion and these rocks reach the surface they undergo expansion or dilation – a process known as **pressure release** – which promotes the development of joints. At the smaller scale micro-fissures and incipient joints related to the original pattern of mineral crystallization in the rock provide lines of weakness along which **exfoliation** (the spalling off of thin sheets of rock) (Fig. 6.20) and **granular disintegration** (the disaggregation of individual crystals or particles) can occur. In some kinds of rock, such as granite, the creation of steep, bare rock faces can lead to significant lateral expansion into the valley side as well as vertical dilation and this may give rise to **exfoliation domes** (Fig. 6.21).

Fig. 6.20 *Exfoliation in granite-gneiss in north-western Cape Province, South Africa. The hat towards the right-hand edge of the photograph indicates the scale.*

Fig. 6.21 *An exfoliation dome formed in granite in the Yosemite Valley, Sierra Nevada, USA.*

Insolation weathering, that is, the breakdown of rock as a result of volume changes arising from thermal expansion and contraction, is another potentially important physical weathering process. Fluctuations in temperature experienced by rocks at the Earth's surface due to day-time heating and night-time cooling certainly cause them to expand and contract. Nevertheless it has proved difficult to establish whether the resulting volume changes are sufficient to shatter rock, even in desert environments where diurnal temperature ranges on rock surfaces can easily exceed 30 °C. Since rock is a poor conductor of heat a thermal gradient is created when the surface is warmed. The exterior of the rock consequently expands more than the interior and stresses are set up. Data on bedrock surfaces in desert regions show that on dark rocks temperatures of 80 °C can be attained by solar radiation and diurnal ranges of 50 °C are not uncommon. Further (intergranular) stress might be expected to arise from the differential thermal expansion of individual minerals in a rock related to differences in colour, specific heat and coefficient of thermal expansion.

In spite of the abundance of shattered rock in hot desert regions and elsewhere, experimental work in the 1930s cast considerable doubt on the efficacy of insolation weathering. Small rock samples subjected to repeated cycles of heating and cooling over temperature ranges well in excess of those produced naturally by solar radiation failed to shatter. The presence of moisture was found to be necessary to promote chemical weathering and weaken the rock sufficiently for disintegration to occur. Subsequently these conclusions have been challenged, it being pointed out that experiments involving the rapid heating and cooling of small individual rocks do not accurately re-create field conditions. In particular, stresses set up in such samples can be relieved in all directions whereas stresses in bedrock are much more concentrated since the rock is confined both laterally and vertically.

The occurrence of scrub and forest fires provides another mechanism whereby rocks can be subjected to significant thermal expansion and contraction. It is certainly important in semi-arid regions characterized by vegetation communities, such as chaparral (North America) and eucalyptus (Australia), which require fires for regeneration. Such areas are frequently seen to be characterized by spalling rock outcrops and shattered boulders.

Chemical weathering frequently gives rise to minerals which are less dense than their precursors. The associated volume increase can exert pressure on the surrounding rock and the weathering rinds produced peel off to expose fresh rock to further weathering. The chemical processes involved are usually associated with hydration (see Section 6.2.3.1) and the addition of water plays a significant role in the physical breakdown of rock. Such **hydration weathering** can occur simply through wetting and drying which causes expansion and contraction. The process is most active in some clay minerals, notably those of the smectite group, which are capable of absorbing large quantities of water into their crystal structures.

A further hydration weathering mechanism, restricted to environments in which freezing occurs, is **hydration shattering**. This can occur in very fine-grained materials such as clays which are capable of retaining significant quantities of unfrozen water at temperatures well below 0 °C. At these low temperatures the dipolar molecules of this supercooled water become ordered in such a way that a repulsive force is set up across voids. It is thought that while this force is insignificant in large voids it can be sufficiently strong in the small pores in very fine-grained material to cause rock fracture. Chemical weathering is also probably associated with the mechanism which so far has only been investigated on clay-rich rocks, and it may be restricted to such lithologies.

6.3.2   Volumetric changes within rock voids and fissures

Under this heading we need to consider two main sets of processes: the effects of crystal growth and expansion, primarily of ice and salt, and the stresses induced by biological activity. The latter are induced by both fauna and flora. Earthworms, for instance, ingest and excrete huge volumes of soil, as so graphically reported in a classic study by Charles Darwin in the last century. The resulting mixing of the soil, or **bioturbation**, probably averages about a 5 mm depth annually over the more humid parts of the continents. Other soil fauna contribute to this effect, with termites being particularly significant in moving large quantities of soil in the tropics and subtropics.

The growth of tree roots and their penetration and enlargement of incipient rock fractures can contribute significantly to the break-up of rock which may have experienced relatively little weakening through chemical alteration. Large tap roots can penetrate rock to a depth of several metres and the process commonly works in association with other mechanisms, particularly ice crystal growth. It is also likely that chemical weathering is enhanced in the vicinity of roots where carbon dioxide concentrations are likely to be higher and where the extraction of metal cations is occurring. Although such organic activity can be significant, especially locally, it is the physical processes of crystal formation and expansion that are of more fundamental importance in rock weathering.

6.3.2.1   Frost weathering

In arctic and alpine environments the surface is often seen to be composed of a layer of angular rock fragments commonly described by the term **felsenmeer** and attributed to the operation of **frost weathering**. This process, which is also referred to as **frost shattering** and **frost wedging**, involves the breakdown of rock or other solid materials as a result of stresses induced by the freezing of water.

Early explanations of frost weathering alighted on the simple and seemingly obvious effect of the 9 per cent volume expansion which accompanies the phase change from water to ice. Under ideal conditions it was estimated that ice formation could exert a maximum pressure of around 200 MPa (at which point the freezing temperature is −22 °C), but it was appreciated that this theoretical maximum pressure could never be attained under natural conditions, not the least because it far exceeded the tensile strength of most rocks (around 25 MPa). Even to build up moderately high pressures sufficient to shatter rock it was realized that a closed system was required in order for the pressure not to be relieved by the expulsion of water into adjacent voids. Such a closed system might be produced if freezing proceeds rapidly from the surface downwards since initial ice formation at the surface can then seal in water contained within pores in the rock.

Subsequent experimental work has failed both to clarify fully the exact mechanisms involved in frost weathering and to define precisely the climatic conditions under which the process is likely be most effective. It has been suggested that a freezing rate of at least 0.1 °C per minute and temperatures of −5 to −10 °C are required for rock shattering to occur. Most experimental studies have emphasized the importance of rapid freezing, a minimum temperature of −5 °C, the frequency of freeze–thaw cycles and the overriding significance of the moisture content of the rock. This last factor is crucial since if a rock is not saturated ice formation will simply cause the expulsion of water into air-filled cavities, thereby inhibiting a build-up of pressure. Some theoretical work, however, has questioned the importance of water freezing in sealed cracks and the frequency of freezing events, suggesting instead that crack growth will be greatest when the temperature ranges from approximately −4 to −15 °C. Other experimental studies have supported the idea that frost weathering is likely to be more effective in alpine and marine arctic environments, where freeze–thaw may occur diurnally, than in polar regions where freezing and thawing are predominantly seasonal phenomena.

Uncertainties about the efficacy of volume expansion on freezing in rock shattering have encouraged the examination of other possibilities. One of these is the pressure exerted by ice-crystal growth, the idea being that this is directional and that if the direction of growth is resisted by the wall of a void or crack then pressure will be exerted which may be sufficient to wedge the crack apart. As applied to the crystallization of ice this mechanism is largely conjectural at present since little experimental work has been undertaken to estimate the pressures likely to be generated. A further possibly important mechanism in rock breakdown associated with freezing, at least in very fine-grained lithologies, is hydration shattering (see Section 6.3.1) and it is likely that some of the effects of physical weathering previously attributed to frost weathering in fact result from this process.

6.3.2.2 Salt weathering

Salt weathering involves three major processes: the precipitation of salt in voids and the expansion of salt crystals through either hydration or heating. It is most active in arid environments where rates of evaporation are high relative to precipitation, and consequently surface and soil waters can become saturated with respect to a variety of salts. Although frequently associated with hot desert regions, the effects of salt weathering have also been observed in high-latitude arid regions such as Antarctica. Moreover, the high salinity of sea water makes it likely that salt is an active weathering agent in coastal environments (see Section 13.3.1.1). In inland areas salt is derived from a number of sources. These include surface and soil waters containing cations released during bedrock weathering which have subsequently been concentrated by evaporation, precipitation with a high salt content (especially in regions near coasts), saline ground waters and saline deposits blown inland from the coast or originating in inland saline basins. A number of salts are found in the soils of arid regions but only a few are generally quantitatively significant – $CaSO_4(.2H_2O)$, $NaCl$, Na_2SO_4, Na_2CO_3, K_2SO_4, $MgSO_4$ and KCl.

When saline solutions in rock pores become saturated as a result of a temperature change or evaporation, salt crystals begin to form and considerable pressures are generated. Whether these lead to rock fracture will depend on both the stresses produced by crystal growth and the tensile strength of the rock. Salt crystal growth in unconsolidated sediments can cause considerable ground heave, and where salt solutions enter desiccation cracks various types of **patterned ground** may be formed (Fig. 6.22). A second cause of stress arises from the capacity of several common salts to absorb significant quantities of water into their crystal structure. Such hydration expansion commonly occurs in response to changes in relative humidity which, since they are closely related to temperature, may be diurnal. The importance of thermal expansion, the third mechanism of salt weathering, has yet to be fully evaluated. Its possible significance derives from the observation that the coefficients of thermal expansion of many salts are greater than that of most rocks. Sodium chloride (common salt), for instance, can expand by up to 1 per cent under the very large diurnal temperature ranges typical of continental deserts.

The rather specific environmental requirements of salt weathering mean that its action is spatially highly variable. It is likely to be intense in west coast deserts, such as the Atacama (Chile/Peru) and Namib (Namibia), where frequent fogs supply both salt and moisture. It is also active in areas adjacent to nearly enclosed seas in hot deserts, for example the Red Sea, where both humidity and salinity are high. Other favourable environments include the margins of salt-covered basins, along river channels and in other areas of low-lying topography where shallow saline ground waters can be drawn to the surface by capillary action.

Different salts also vary significantly in their efficacy as weathering agents, and several laboratory investigations have indicated the particular effectiveness of sodium sulphate. This can be attributed to its various physical properties. First, it experiences a large increase in volume when hydrated (from Na_2SO_4 to $Na_2SO_4. 10H_2O$). Secondly, it has a high equilibrium solubility so that large quantities can be precipitated from a given volume of solution. Thirdly, its equilibrium solubility is particularly sensitive to temperature, reaching a maximum at 32.3 °C. This is within the range of diurnal temperature change in many hot deserts and precipitation will occur as the temperature rises

Fig. 6.22   *Polygonal patterned ground developed in the saline bed of the southern margin of Lake Eyre, South Australia.*

or falls across this value. Finally, crystallization occurs preferentially along a single crystal axis and large pressures can be generated within rock fractures.

6.4   Lithology and weathering forms

Just as variations in the physical and mineralogical properties of bedrock can influence the mineralogical products of weathering so different lithologies give rise to a range of weathering forms. On most lithologies landforms related directly to weathering tend to be minor features, but on rock types such as limestones, in which a large proportion of the products of chemical weathering processes are removed in solution, major landforms can be produced. In this section we will focus on weathering forms in limestone, although it should be pointed out that similar weathering forms can develop on other lithologies.

6.4.1   Karst weathering forms

Karst is the German form of a Slovene word meaning 'bare stony ground' and is used to describe limestone terrain characterized by a lack of surface drainage, a discontinuous or thin soil cover, abundant enclosed depressions and a well-developed system of underground drainage including caves, all features attributable to the solubility of limestone. Limestone is defined as a rock which contains at least 50 per cent $CaCO_3$, which nearly always occurs as the mineral calcite. Other carbonate minerals include aragonite (a rare form of $CaCO_3$) and dolomite ($CaMg(CO_3)_2$). Rocks in which more than 50 per cent of the carbonate content occurs as the mineral dolomite are given the rock name dolomite. The term **karstification** is used to refer to the process of karst landscape development. Karst landforms are classically exemplified by the landscape of northern Yugoslavia and a number of terms used to describe karst landforms originate from this region.

The term **pseudokarst** is applied to landforms in non-carbonate rocks which are morphologically similar to those characteristic of limestone terrains. In most cases pseudokarst develops as a result of processes analogous to those operating on true karst; such is the case on some pure siliceous rocks subject to prolonged weathering where the slow dissolution of quartz creates karst-like forms. In other cases quite different processes can give rise to topography superficially resembling true karst, an example being the creation of 'underground drainage' through the formation of lava tunnels in volcanic terrains.

Karst scenery is not entirely, or in many cases even primarily, a product of weathering since fluvial processes play a crucial role in its development. Nevertheless, it is appropriate to consider here those components of karst topography that owe much of their origin to weathering processes. Karst landforms related more specifically to fluvial activity are discussed in Chapter 9.

The fullest development of karst scenery is attained where the limestone is relatively pure (in excess of 80 per cent $CaCO_3$), very thick, mechanically strong and massively jointed (so that water flow is concentrated along joints rather than passing through the mass of the rock). In addition a humid climate is required together with sufficient relief for extensive vertical water flow down to a regional water table deep below the surface. In the karst region of Yugoslavia there are pure limestones up to 4000 m thick

and part of the area is more than 2000 m above sea level. Although chalk is a very pure limestone it is mechanically too weak to allow the development of large underground cavities. Similarly, thin beds of limestone do not permit the development of underground drainage while in arid regions there is insufficient water for significant limestone weathering to occur. There is an extensive vocabulary of special terms to describe karst landforms which can be confusing to the uninitiated, and even to the expert.

6.4.1.1 Minor forms

The German term **karren** and the French term **lapies** refer to the small-scale solutional forms developed on limestone. (Note that the terms 'solution' and 'dissolution' are used widely in discussions of karst landforms, but this is just a form of shorthand since, as was discussed in Section 6.2.3.3, limestone weathering involves complex chemical reactions.) There is no English equivalent for these terms although some specific forms have acquired local names. A

Table 6.6 Classification of solutional microforms developed on limestone

		FORM	TYPICAL DIMENSIONS	COMMENTS
Forms developed on bare limestone	Developed through areal wetting	**Rainpit**	<30 mm across, <20 mm deep	Produced by rain falling on bare rock. Occurs in fields on gentle rather than steep slopes. Can coalesce to give irregular, carious appearance
		Solution ripples	20–30 mm high; may extend horizontally for >100 mm	Wave-like form transverse to downward water movement under gravity. Rhythmic form implies that periodic flows or chemical reactions are important in their development
		Solution flutes (rillenkarren)	20–40 mm across, 10–20 mm deep	Develop due to channelled flow down steep slopes. Cross-sectional form ranges from semi-circular to V-shaped but is constant along flute
		Solution bevels	0.2–1 m long, 30–50 mm high	Flat, smooth elements usually found below flutes. Flow over them occurs as a thin sheet
		Solution runnels (rinnenkarren)	400–500 mm across, 300–400 mm deep, 10–20 m long	Down runnel increase in water flow leads to increase in cross-sectional area. May have meandering form. Ribs between runnels may be covered with solution flutes
	Developed through concentration of runoff	**Grikes** (kluftkarren)	500 mm across, up to several metres deep	Formed through the solutional widening of joints or, if bedding is nearly vertical, of bedding planes
		Clints (flackkarren)	Up to several metres across	Tabular blocks detached through the concentration of solution along near-surface bedding planes in horizontally bedded limestone
		Solution spikes (spitzkarren)	Up to several metres	Sharply pointed projections between grikes
Forms developed on partly covered limestone		**Solution pans**	10–500 mm deep, 0.03–3 m wide	Dish-shaped depressions usually floored by a thin layer of soil, vegetation or algal remains. CO_2 contributed to water from organic decay enhances dissolution.
		Undercut solution runnels (hohlkarren)	400–500 mm across, 300–400 mm deep 10–20 m long	Like runnels but become larger with depth. Recession at depth probably associated with accumulation of humus or soil which keeps sides at base constantly wet
		Solution notches (korrosionkehlen)	1 m high and wide, 10 m long	Produced by active solution where soil abuts against projecting rock giving rise to curved incuts
Forms developed on covered limestone		**Rounded solution runnels** (rundkarren)	400–500 mm across, 300–400 mm deep 10–20 m long	Runnels developed beneath a soil cover which become smoothed by the more active corrosion associated with acid soil waters.
		Solution pipes	1 m cross, 2–5 m deep	Usually become narrower with depth. Found on soft limestones such as chalk as well as mechanically stronger and less permeable varieties

Note: The commonly encountered German terms are given in parentheses
Source: Based largely on discussion in J. N. Jennings (1985), *Karst Geomorphology*. Blackwell, Oxford, pp. 73–82.

bewildering variety of karren have been described and the most important of these are classified and described in Table 6.6

Several factors influence the form of karren. Probably of most importance is the presence or absence of a cover of soil or vegetation during its formation. This is probably because the solution process operates differently on covered and bare surfaces. On covered surfaces soil and vegetation maintains a continuous presence of moisture on the rock surface while organic decay increases the quantity of organic acids and the abundance of carbon dioxide; both increase the aggressiveness of the water percolating on to the rock surface. (**Aggressivity** refers to the propensity of water to dissolve calcium carbonate.) The resulting solutional forms tend to be rounded and less sharp-edged than those formed on bare rock.

A second factor is lithology. Karren are best developed on relatively massive and uniform, mechanically strong and impermeable limestones in which there is a sharp contact between soil and rock. Chalk, for instance, is too porous and mechanically weak for karren development. Climate is also important since adequate moisture must be present for effective solution. In arid climates there is generally insufficient surface water, whereas in periglacial regimes frost action may be so effective that solutional forms have little time to develop. Snow patches, with their enhanced concentrations of CO_2, which experience seasonal melting may, nevertheless, be favourable environments for karren development. Time is the final significant factor since karren developed under a relatively humid climate may survive for a period after a change to a more arid climatic regime. Moreover, forms developed under a soil cover may subsequently be exposed by erosion, in some cases promoted by human activities such as forest clearance.

The role of organisms in directly creating both erosional and depositional microforms on limestone has probably been underestimated. Micro-organisms may be instrumental in many commonly observed microforms in limestones, but the precise nature of their role is as yet poorly known. The terms **biokarst** and **phytokarst** are used to describe forms attributable in large part to organic processes.

6.4.1.2   *Major forms*

If one landform can be regarded as typifying karst scenery it is the **doline**. A doline is a closed depression which may range in shape from bowl-shaped to cylindrical and in size up to 100 m deep and 1 km across. In some limestone terrains the surface is pitted with large numbers of dolines which each provide separate foci for surface drainage, in contrast to the integrated valley systems of fluvially eroded landscapes. In fact it is likely that doline formation may be a kind of contagious process whereby the establishment of an initial depression will tend, through its effect on lowering the water table in its immediate vicinity, to promote the

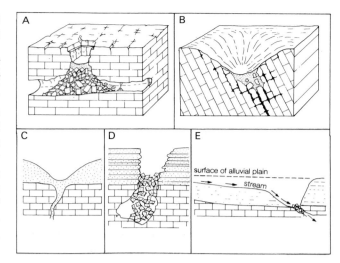

Fig. 6.23   *Major types of doline: (A) collapse doline; (B) solution doline; (C) subsidence doline; (D) subjacent karst collapse doline; (E) alluvial stream sink doline. (From J. N. Jennings (1985)* Karst Geomorphology. *Blackwell, Oxford, Fig. 37, p. 107).*

development of further depressions near by.

Various types of doline are recognized in terms of their mode of formation (Fig. 6.23). A **collapse doline** is created through the fall of the roof of a cave to leave a steep-walled cavity. These steep walls are, however, likely to be subject to rapid degradation through dissolution and physical weathering processes in all but arid climates, and their gradients may be consequently reduced in a short period of time to the extent that a collpase doline may be indistinguishable in form from other types. Collapse into a water-filled cave or a subsequent rise in the local water table can lead to the formation of a lake, such as in the case of the **cenotes** which are common in the Yucatan Peninsula, southern Mexico.

Solution dolines are formed when localized dissolution usually concentrated at the intersection of major joints lowers the bedrock surface. The initial development of the depression leads to the capture of more surface drainage and this in turn promotes further doline enlargement. Solution dolines are one of the few karst landforms which are capable of forming in relatively soft limestones such as chalk. **Subsidence dolines** form in a similar manner, but in this case there is a covering of superficial deposits which collapse either suddenly or progressively into the cavity developing in the underlying limestone. More dramatic collapse characterizes **subjacent karst collapse dolines** in which a covering non-calcareous rock unit collapses into a void in the underlying limestone. The final generally recognized type is the **alluvial stream sink doline** which consists of a stream flowing into a doline before disappearing underground.

Dolines, especially of the solution type, may be so numerous that they merge into each other to form **uvalas**. Where dissolution is significantly controlled by joint systems uvalas may be elongated forms, but in other cases they are irregular depressions. In the humid tropics dolines characteristically have an irregular to star-shaped plan form due to their mutual interference as they have grown and completely consumed the original surface. This produces a polygonal pattern of ridges surrounding individual dolines which are sometimes described as **cockpits** following the local name given to them in Jamaica. Residual hills are often present on the divides separating depressions. They are usually rounded in form and are termed **cones** or **kegel**, and the whole landscape assemblage is called **cockpit karst** (or alternatively **cone karst** or **kegelkarst**). Many cockpits contain stream sinks and the morphology of cockpits is probably attributable to the abundance of surface runoff in humid tropical environments which leads to the creation of minor valley-like solutional forms on their slopes. In some localities in the humid tropics the residual hills on doline divides have a remarkable sharp-edged, pinnacle form and this has given rise to the term **pinnacle karst** (Fig. 6.24).

Large flat-floored depressions ranging in size up to 200 km² or more are found in some karst terrains and are termed **poljes**. Polje floors are commonly covered by alluvium and one or more sides of the depression has a steep angle in excess of 30°. Water flows on to the polje floor via springs along its edge or from impermeable strata which commonly form one side and leaves the polje either through stream sinks called **ponors** or via gorges penetrating one of the polje walls. Early attempts to explain the formation of poljes regarded them as a late stage in a developmental sequence of depression enlargement beginning with dolines, but current ideas now focus on structural controls. Many poljes appear to be aligned with structural trends represented by fold axes, faults or junctions between limestone and non-calcareous rocks; some may indeed be graben but in the majority of cases it appears that structure controls the location of lateral planation necessary for polje development. Most poljes are located at the junction of limestone and impermeable beds, and runoff from the latter will be undersaturated with calcium carbonate and will therefore be aggressive when entering the limestone terrain.

In humid tropical and subtropical environments some limestone landscapes are dominated by spectacular tower-like hills, or **mogotes**, up to 100 m or more high separated by broad, alluvial valley floors (Fig. 6.25). This type of terrain is described as **tower karst** and has been thought to represent the effects of enhanced limestone dissolution in regions of high precipitation. Acid waters in the alluvial valleys cause rapid lateral planation and undercutting of rock faces while high temperatures promote the evaporation of water flowing across the exposed rock leading to the deposition of a protective layer of calcite. This climatic interpretation of tower karst, however, has been revised in the light of the recognition of tower karst in the far from tropical environment of the Mackenzie Mountains in northwest Canada. In this region the limestone is massive and very thick with widely spaced joints and the tower karst appears to represent the last stage of karst landscape development. This apparently began with the formation of deep, joint-controlled dolines and progressed through the creation of long, narrow gorges called **karst streets** and then the development of a rectilinear network of deep gorges forming **labyrinth karst** to a final phase of lateral planation of the gorge rock walls to form towers.

6.4.2 Other weathering forms

Exposures of bare rock are found in many environments and result from differential weathering of bedrock and the removal of the weathered debris by slope processes. Rock outcrops which stand out on all sides from surrounding slopes are known as **tors**. They are particularly common on crystalline rocks, but also occur on other resistant lithologies such as quartzites and some sandstones. Some researchers have regarded deep weathering as a prerequisite for the development of tors, with a phase of intense chemical weathering preferentially focused along joints being followed by stripping of the weathered material in response to a change in conditions more favourable to erosion. Other workers, however, have argued that tors can develop in the absence of deep weathering where weathering and stripping operate simultaneously on rocks of variable resistance.

Fig. 6.24 *Pinnacle karst in the Mulu area of Sarawak. The development of this extraordinary topography is probably attributable to preferential dissolution focused along joints or bedding planes in steeply dipping metamorphosed limestone. The apparently rapid rate of limestone dissolution is related to high rainfall totals (in excess of 5000 mm a⁻¹), frequent mists and high rates of organic activity. (Interpretation by M. J. Day, and photo courtesy of the Royal Geographical Society and M. J. Day.)*

Fig. 6.25   *Tower karst near Guilin, Guangxi Province, China. (Photo courtesy D. Munro.)*

Nearly all exposed rock outcrops have irregular surfaces which appear to be related to the effects of weathering. These include rills, pits and cavernous forms. They are found on essentially all rock types and in all climates, but in view of the extent of exposed bedrock they are most apparent in arid and semi-arid environments. The origin of the wide range of forms that occurs is poorly understood; suggested mechanisms include salt weathering, hydration shattering, insolation weathering and frost action.

Two forms which have attracted particular attention are **honeycomb weathering** and **tafoni**. Honeycomb weathering consists of numerous small pits a few millimetres or centimetres in width and depth and the honeycomb form is produced when they coalesce to create a network structure. Tafoni are larger features ranging up to several metres in size and are cut into steep, bare rock faces (Fig. 6.26). In some cases selective chemical disintegration of the rock is apparent but in others evidence of chemical decomposition seems to be quite absent and mechanical processes such as salt weathering appear to be more probable processes. Organic activity may also be instrumental through the localized enhancement of acidity.

Case-hardening and core softening may also be import-

Fig. 6.26   *Tafoni developed on the side of Ayers Rock, Northern Territory, Australia.*

ant. **Case-hardening** appears to result from the localized mobilization and reprecipitation of minerals on the rock surface, thereby strengthening it and rendering it generally less permeable. Areas of the rock surface less effectively case-hardened would then be susceptible to weathering processes which could then selectively penetrate into the rock interior producing cavities. Similarly, there is evidence that the cores of boulders may be weathered more rapidly than the surface and again this could lead to the selective development of cavities.

Differential heave in unconsolidated deposits arising from changes in volume can create a range of patterned ground phenomena (Fig. 6.22). Such changes in volume are most commonly associated with wetting and drying and freeze–thaw cycles and patterned ground is most evident in arid and periglacial regions (see Chapter 12).

6.5 Duricrusts

Duricrusts are hard layers formed in the weathering zone at, or near, the landsurface as a consequence of the absolute or relative accumulation of particular components through the replacement or cementation of pre-existing rock, soil, weathering materials or other unconsolidated deposits. The most important components in duricrust formation are iron and aluminium oxides and hydroxides, silica, calcium carbonate and gypsum. They are of interest not only in terms of their origin but also because of their role in landscape development and the evidence they provide of past climates. Duricrusts may reach thicknesses in excess of 50 m although 1–10 m is a more usual range. The ferruginous, calcareous and especially siliceous forms can be exceedingly hard, and particularly where they cap less resistant materials they form prominent and highly resistant elements of the landscape. Duricrusts may form several metres below the landsurface but they can be subsequently exposed by erosion of less resistant overlying materials.

Iron- and aluminium-rich duricrusts are known as, respectively, **ferricrete** and **alcrete**. Both occur in association with deep weathering profiles in the humid to subhumid tropics although alcretes are more prevalent where rainfall totals are particularly high. The term **laterite** has long been used to describe iron- and aluminium-rich weathering deposits while the term **bauxite** refers to deposits containing economically extractable concentrations of aluminium. Many laterites and bauxites are, however, relatively weak materials and the terms ferricrete and alcrete are reserved for the indurated forms.

Siliceous duricrusts, or **silcretes**, are commonly composed of more than 95 per cent SiO_2 and are found in both humid and arid tropical environments. They are particularly prominent in central Australia and parts of southern and northern Africa. In some cases they occur in weathering profiles in close association with ferricretes, while in more arid regions they are found in conjunction with calcium carbonate crusts or **calcretes**. Calcretes have an average $CaCO_3$ content of around 80 per cent and their distribution in general broadly coincides with areas with a current mean annual precipitation of between 200 and 600 mm (Fig. 6.27).

Fig. 6.27 *A calcrete profile more than 10 m thick exposed in the Molopo gorge in the southern Kalahari Basin on the border between Botswana and South Africa. The calcrete has formed through the replacement of quartzites and tillites.*

They cover a significant proportion of the world's semi-arid environments and they possibly underlie up to 13 per cent of the global landsurface area. Gypsum crusts, or **gypcretes**, by contrast have a much more limited distribution and appear to be largely confined to very arid regions where the mean annual precipitation is below 250 mm. The gypsum ($CaSO_4 . 2H_2O$) content is highly variable but may exceptionally reach 95 per cent. Gypcrete reaches a maximum thickness of about 5 m, somewhat less than other duricrust types.

6.5.1 Models of duricrust formation

The formation of the different types of duricrust can be largely understood in terms of the relative mobility of different elements in the weathering environment. Ferricretes and alcretes in most cases represent the relative accumulation of iron and aluminium oxides and hydroxides as more mobile components are removed from the weathering mantle (Fig. 6.28). Such relative accumulation occurs under conditions of intense leaching under climates where rainfall is abundant for a significant part of the year. Locally absolute accumulation may occur as iron and aluminium are mobilized to a limited extent within the weathering mantle or are transported either mechanically or in solution from high to low topographic positions.

Silcrete, calcrete and gypcrete form through absolute accumulation of silica, calcium carbonate and gypsum respectively. Their formation requires a source for these constituents, a means of transferring them to the site of formation and a mechanism for precipitation. Possible sources include the weathering of bedrock or sediments, inputs from rainfall or dust, plant residues and solutes in ground water. Translocation of solid or dissolved source materials can be either lateral or vertical, and in the latter

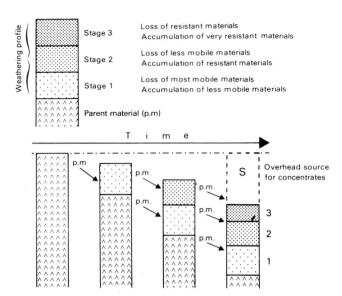

Fig. 6.28 *Stages of weathering leading to the relative accumulation of iron or aluminium to form ferricrete or alcrete. The profile in which the iron and aluminium accumulates experiences a continuous development whereby stage 1 weathering effectively occurs at progressively lower levels as the weathering front moves downward. Similarly, the landsurface experiences a progressive lowering and the iron and aluminium gradually becomes concentrated in the zone of stage 3 weathering. (From M. J. McFarlane (1983) in A. S. Goudie and K. Pye (eds)* Chemical Sediments and Geomorphology: Precipitates and Residua in the Near-Surface Environment. *Academic Press, London, Fig. 2.3, p. 22.)*

case may be either upward as a result of capillary rise (*per ascensum* model), or downward through percolation (*per descensum*) model (Fig. 6.29). Downward movements are probably more important in most cases since capillary rise will lead to a concentration of precipitation close to the surface and this will reduce the efficiency of surface evaporation and thus reduce the rate of subsequent capillary rise.

Precipitation can arise from a variety of factors in addition to the evaporation of weathering solutions: these include pH changes, reactions with other cations in solution and organic activity. Changes in pH appear to be of particular importance in the case of silcrete and may explain why this type of duricrust is found in both predominantly alkaline arid and semi-arid environments, and in acidic humid tropical weathering profiles (Fig. 6.30). The solubility of silica is more or less constant up to a pH of around 9, but above this point its solubility increases dramatically (Fig. 6.5). In alkaline environments, such as those commonly found around ephemeral lake beds in semi-arid and arid regions, pH may vary between 8 and 10 over short distances. Silica in solution moving from a locality where the pH is above 9 to one where it is below 9 will therefore be precipitated. In such circumstances

silcrete can develop within calcrete while under different conditions calcrete can be seen replacing silcrete. A contrasting situation exists in the highly acidic conditions associated with abundant organic activity in humid tropical regions. At a pH of below 4 aluminium becomes more mobile than silica so clay minerals can be silicified as they lose aluminium, even though the concentration of silica in solution is invariably very low.

There are few firm data on how quickly duricrusts can form although it is clear that gypcrete and calcrete generally develop much more quickly than the other types. The rapid formation of some calcretes is demonstrated by the cementation of human artefacts, such as gravestones and even Coca-Cola tins! The slow development of silcretes, ferricretes and alcretes is probably in part explained by the very low concentrations of silica, iron and aluminium in most surface and regolith waters.

6.5.2   Environmental controls

As with other weathering products the factors important in the development of the different types of duricrust are scale dependent. At the broadest scale climate (and vegetation in so far as it affects pH) is the predominant control; but the overlap in the climatic conditions under which different types of duricrust develop (Fig. 6.30) demonstrates that other factors operate at the regional and local scale. Topography is of considerable significance. Low local relief is an essential requirement for virtually all duricrust formation since the rate of development must exceed the rate of denudation for a duricrust to be created. Secondly, topography is the key factor in determining local drainage conditions. Silica will not be retained in weathering profiles to form silcrete unless the drainage is very poor whereas alcrete develops preferentially where efficient drainage encourages intense leaching. Bedrock also plays a role since it influences the availability of source constituents – silica, iron, aluminium, calcium carbonate, gypsum – but, as with the development of other weathering products, lithological variations tend to become progressively masked after prolonged duricrust development.

Although the climatic parameters for calcrete and gypcrete genesis are known fairly well, the same is not true for other types of duricrust. This is because ferricretes, alcretes and silcretes are chemically and mechanically resistant so that once formed under a certain set of climatic conditions they can persist in the landscape for long periods even though the prevailing climate may have changed dramatically. Consequently, these types of duricrust may frequently not be in equilibrium with prevailing environmental conditions; indeed, they are likely to reflect in their chemical, mineralogical and physical characteristics a history of changes in climate and other variables affecting their development rather than a single equilibrium state.

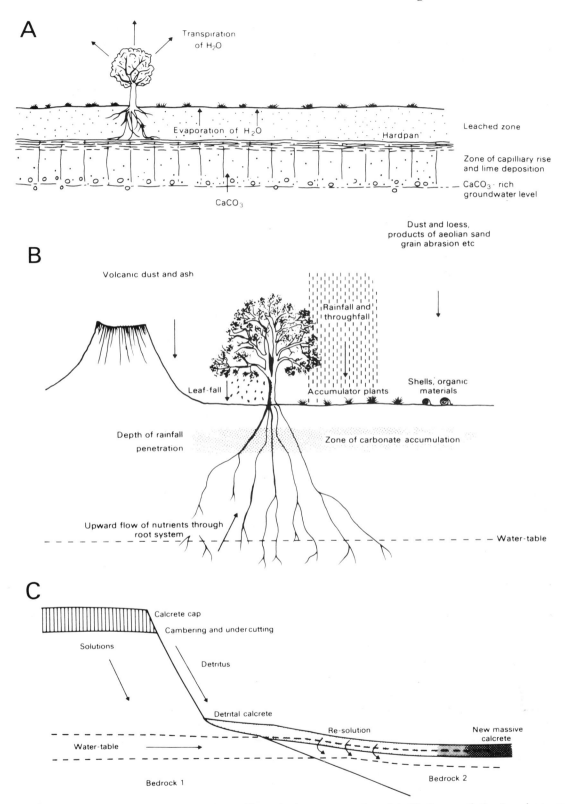

Fig. 6.29 *Models of the formation of calcrete: (A) the capillary rise* (per ascensum) *model; (B) the percolation* (per descensum) *model with inputs of carbonate from above; (C) the detrital model in which a secondary calcrete is formed through the solution and disintegration of an existing calcrete horizon at a higher level in the landscape. Calcium carbonate as well as silica may also be translocated in river waters, eventually being precipitated at some distance from the original source. (From A. S. Goudie (1983) in A. S. Goudie and Pye, K. (eds)* Chemical Sediments and Geomorphology: Precipitates and Residua in the Near-Surface Environment. *Academic Press, London, Fig. 4.4, p. 113 and Figs. 4.5 and 4.6, p. 115.)*

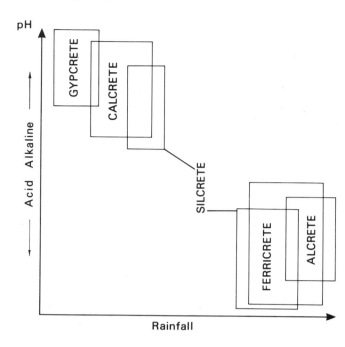

pH

Acid ← Alkaline →

GYPCRETE

CALCRETE

SILCRETE

FERRICRETE

ALCRETE

Rainfall

Fig. 6.30 *A schematic representation of the relationship between the formation of different types of duricrust and prevailing conditions of climate and pH. Note that the chemical and physical resistance of some kinds of duricrust mean that many occurrences may now exist outside the climatic and pH zones in which they were formed. In many instances local factors such as topography and drainage outweigh broad climatic controls.*

Calcretes and especially gypcretes are by contrast much more responsive to changing conditions; the relatively high solubility of gypsum, for instance, would lead to the rapid destruction of a gypcrete should the climate become humid. This contrasting 'survivability' in the face of environmental changes is largely responsible for the fact that few gypcretes and calcretes which have been continuously exposed at the surface are older than Late Cenozoic while some ferricretes and many silcretes date back to the Early Cenozoic, and even beyond.

Duricrusts are normally significantly harder and more resistant to erosion than the materials which they overlie. Consequently they tend to armour, or protect, landsurfaces from denudational processes (Fig. 6.31). Although some

Fig. 6.31 *Flat-topped hills (koppies) capped by silcrete near Riversdale, in southern Cape Province, South Africa. It is likely that the silcrete originally formed at a low elevation in the landscape and that subsequently topographic inversion has occurred (see Figure 6.32).*

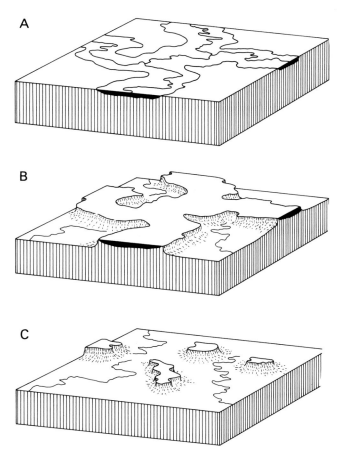

Fig. 6.32 *Schematic representation of stages in the development of inverted topography. A resistant duricrust, such as silcrete, is initially formed preferentially at low points in the landscape (A). Subsequently, differential erosion lowers the surrounding, less resistant, terrain (B) to leave the duricrust capping residual hills (C).*

types of duricrust seem to preferentially develop at low points in the landscape where ground water flows and surface runoff converge, this armouring can subsequently lead to the more rapid lowering of surrounding areas and the creation of **inverted topography** (Fig. 6.32). Even where a duricrust horizon has been broken up by prolonged erosion the duricrust fragments produced may still remain on the surface and perform a protective role for a long period of time. This is the case in the **gibber plains** which cover extensive areas of central Australia and which consist of silcrete boulders scattered on the surface (Fig. 6.33).

Further reading

There is an extensive literature on weathering though not all of it is written from a specifically geomorphic perspective. The books by Birkeland (1984) and Ollier (1984) provide good introductions to the processes and products of weathering in the context of landforms, while the reviews

by Whalley and McGreevy (1985, 1987, 1988) are excellent sources for references, especially on physical weathering processes. The book by Gerrard (1988) is a valuable summary of the role of lithology in weathering and landscape development. The behaviour of water in the soil and regolith which is so important to an understanding of both physical and chemical weathering processes is discussed in detail by Williams (1982).

The best introduction to chemical weathering for the geomorphologist is probably still the short book by Loughnan (1969) although it is confined to silicate minerals. Drever (1985) provides a more detailed treatment of processes while Colman and Dether (1986) consider various aspects of chemical weathering rates. Curtis (1976a) and Ross (1987) provide brief and accessible introductions to energy considerations in chemical weathering while the complex sequence of reactions involved in the weathering of limestone is discussed by Gunn (1986) and Trudgill (1985). The pH and Eh conditions in natural environments are assessed by Baas Becking *et al.* (1960). The role of earthworms in bioturbation is examined in the classic study by Darwin (1881) and Goudie (1988) looks at the important role played by both earthworms and termites in the tropics. The probable role of organic processes in the formation of desert varnish is discussed by Dorn and Oberlander (1982) and Whalley (1983).

Weathering products are covered in the book by Ollier (1984) and in the volumes edited by Goudie and Pye (1983) and Wilson (1983). The deep weathering profiles of the humid tropics are discussed by Thomas (1974) and in the classic paper by Ruxton and Berry (1957). Mineral stability is treated from a thermodynamic perspective by Curtis (1976b), while Williams *et al.* (1986) look in detail at the chemical weathering of granite in terms of mineral stability and element mobility. The classic study of granite weathering by Goldich (1938) is still worth consulting. The formation of secondary minerals is examined in general terms by Loughnan (1969) and with specific reference to clays by Millot (1979). Loughnan (1969) also provides an excellent discussion of the environmental factors controlling chemical weathering, while Pye *et al.* (1986) show how apparently minor differences in bedrock chemistry can influence weathering rates.

Turning to physical processes Rice (1976) reassesses the importance of insolation weathering while McGreevy (1985) looks at the implications of the thermal properties of rock for physical weathering. Valuable field data on temperatures attained by rocks in the field are presented by Kerr *et al.* (1984), Peel (1974) and Smith (1977), while Ollier and Ash (1983) look at the role of fire in rock weathering. The importance of cracks within rocks for the effective operation of frost and salt weathering is emphasized by Whalley *et al.* (1982), while McGreevy and Whalley (1985) assess the importance of rock moisture to frost weathering both in

Fig. 6.33   *A gibber plain east of Lake Eyre, South Australia.*

the laboratory and under field conditions. Fifteen years of experimental work on frost weathering is summarized in Lautridou and Ozouf (1982), and Thorn (1979) reassesses the efficacy of freeze–thaw action in alpine environments. An important recent trend has been the greater application of basic physical principles to studies of frost weathering and this approach is exemplified by Walder and Hallet (1985). Salt weathering has attracted increasing attention from geomorphologists as its importance, especially in arid environments, has become appreciated. Sperling and Cooke (1985) and Smith and McGreevy (1983) report experimental and simulation studies of salt weathering, while Goudie and Day (1980) and Goudie and Watson (1984) provide evidence of its potency in the field.

There is a very large literature on weathering forms developed on bedrock, especially in the case of limestone. Both minor and major karst forms are comprehensively treated in Jennings (1985), Sweeting (1972), Trudgill (1985) and White (1988), while Viles (1988) provides a useful review of biokarst. Kemmerly (1986) presents a general model for the development of closed depressions, while Williams (1972) shows the value of applying morphometric techniques to karst terrains and Brook and Ford (1978) and Williams (1987) consider the long-standing problem of the development of tower karst. A variety of cavernous weathering forms, including tafoni and honeycomb weathering, occur on non-calcareous rocks and the origin of these features is considered by Conca and Rossman (1985) and Mustoe (1982, 1983). Pseudokarst microforms on granite are recorded by Watson and Pye (1985) while Young (1986) reports karst-like tower forms developed in sandstone.

Good summaries of the characteristics and origin of the various forms of duricrust are given in the volume edited by Goudie and Pye (1983). Of particular relevance are the chapters by Goudie (1983), McFarlane (1983), Summerfield (1983a) and Watson (1983). Other useful articles on particular duricrust occurrences are contained in Wilson (1983). Additional information can be found in Watson (1985) on gypsum crusts in Tunisia and Namibia, in Summerfield (1983b) and Young (1985) on contrasting interpretations of the climatic conditions under which different types of silcrete form and in Goudie (1985) on the role of duricrusts in landscape development.

References

Baas Becking, L. G. M., Kaplan, I. R. and Moore, D. (1960) Limits of the natural environment in terms of pH and oxidation–reduction potentials. *Journal of Geology* **68**, 243–84.

Birkeland, P. W. (1984) *Soils and Geomorphology* (2nd edn.). Oxford University Press, New York and London.

Brook, G. A. and Ford, D. C. (1978) The origin of labyrinth and tower karst and the climatic conditions necessary for their development. *Nature* **275**, 493–6.

Colman, S. M. and Dether, D. P. (eds) (1986) *Rates of Chemical Weathering of Rocks and Minerals*. Academic Press, Orlando.

Conca, J. L. and Rossman, G. R. (1985) Core softening in cavernously weathered tonalite. *Journal of Geology* **93**, 59–73.

Curtis, C. D. (1976a) Chemistry of rock weathering: fundamental reactions and controls. In: E. Derbyshire (ed.) *Geomorphology and Climate*. Wiley, London and New York, 25–57.

Curtis, C. D. (1976b) Stability of minerals in surface weathering reactions: a general thermo-chemical approach. *Earth Surface Processes* **1**, 63–70.

Darwin, C. (1881) *The Formation of Vegetable Mould, through the Action of Worms, with Observations on their Habits*. Murray, London.

Dorn, R. I. and Oberlander, T. M. (1982) Rock varnish. *Progress in Physical Geography* **6**, 317–67.

Drever, J. I. (ed.) (1985) *The Chemistry of Weathering*. Reidel, Dordrecht.

Gerrard, A .J. (1988) *Rocks and Landforms*. Unwin Hyman, London and Boston.

Goldich, S. S. (1938) A study of rock weathering. *Journal of Geology* **46**, 17–58.

Goudie, A. S. (1983) Calcrete. In: A. S. Goudie and K. Pye (eds) *Chemical Sediments and Geomorphology: Precipitates and Residua in the Near-Surface Environment*. Academic Press, London and New York, 93–131.

Goudie, A. S. (1985) Duricrusts and landforms. In: K. S. Richards, R. R. Arnett and S. Ellis (eds) *Geomorphology and Soils*. Allen and Unwin, London and Boston, 37–57.

Goudie, A. S. (1988) The geomorphological role of termites and earthworms in the tropics. In: H. A. Viles (ed.) *Biogeomorphology*. Blackwell, Oxford and New York, 166–92.

Goudie, A. S. and Day, M. J. (1980) Disintegration of fan sediments in Death Valley, California, by salt weathering. *Physical Geography* **1**, 126–37.

Goudie, A. S. and Pye, K. (eds) (1983) *Chemical Sediments and Geomorphology: Precipitates and Residua in the Near-Surface Environment*. Academic Press, London and New York.

Goudie, A. S. and Watson, A. (1984) Rock block monitoring of rapid salt weathering in southern Tunisia. *Earth Surface Processes and Landforms* **9**, 95–8.

Gunn, J. (1986) Solute processes and karst landforms. In: S. T. Trudgill (ed.) *Solute Processes*. Wiley, Chichester and New York, 363–437.

Jennings, J. N. (1985) *Karst Geomorphology*. Blackwell, Oxford and New York.

Kemmerly, P. R. (1986) Exploring a contagion model for karst-terrane evolution. *Geological Society of America Bulletin*, **97**, 619–25.

Kerr, A., Smith, B. J., Whalley, W. B. and McGreevy, J. P. (1984) Rock temperatures from southeast Morocco and their significance for experimental rock weathering studies. *Geology* **12**, 306–9.

Lautridou, J. P. and Ozouf, J. C. (1982) Experimental frost shattering: 15 years of research at the Centre de Géomorphologie du CNRS. *Progress in Physical Geography* **6**, 215–32.

Loughnan, F. C. (1969) *Chemical Weathering of the Silicate Minerals*. Elsevier, New York.

McFarlane, M. J. (1983) Laterites. In: A. S. Goudie and K. Pye (eds) *Chemical Sediments and Geomorphology: Precipitates and Residua in the Near-Surface Environment*. Academic Press, London and New York, 7–58.

McGreevy, J. P. (1985) Thermal properties as controls on rock surface temperature maxima and possible implications for rock weathering. *Earth Surface Processes and Landforms* **10**, 125–36.

McGreevy, J. P. and Whalley, W. B. (1985) Rock moisture content and frost weathering under natural and experimental conditions: a comparative discussion. *Arctic and Alpine Research* **17**, 337–46.

Millot, G. (1979) Clay. *Scientific American* **240**(4), 76–84.

Mustoe, G. E. (1982) The origin of honeycomb weathering. *Geological Society of America Bulletin*, **93**, 108–15.

Mustoe, G. E. (1983) Cavernous weathering in the Capitol Reef desert, Utah. *Earth Surface Processes and Landforms* **8**, 517–26.

Ollier, C. (1984) *Weathering* (2nd edn). Longman, London and New York.

Ollier, C. D. and Ash, J. E. (1983) Fire and rock breakdown. *Zeitschrift für Geomorphologie* **27**, 363–74.

Peel, R. F. (1974) Insolation weathering: some measurements of diurnal temperature changes in exposed rocks in the Tibesti region, central Sahara. *Zeitschrift für Geomorphologie Supplementband* **21**, 19–28.

Pye, K., Goudie, A. S. and Watson, A. (1986) Petrological influence on differential weathering and inselberg development in the Kora area of central Kenya. *Earth Surface Processes and Landforms* **11**, 41–52.

Rice, A. (1976) Insolation warmed over. *Geology* **4**, 61–2.

Ross, S. M. (1987) Energetics of soil processes. In: K. J. Gregory (ed.) *Energetics of Physical Environment: Energetic Approaches to Physical Geography*. Wiley, Chichester and New York, 119–43.

Ruxton, B. P. and Berry, L. (1957) Weathering of granite and associated erosional features in Hong Kong. *Bulletin of the Geological Society of America* **68**, 1263–92.

Smith, B. J. (1977) Rock temperature measurements from the northwest Sahara and their implications for rock weathering. *Catena* **4**, 41–64.

Smith, B. J. and McGreevy, J. P. (1983) A simulation study of salt weathering in hot deserts. *Geografiska Annaler* **65A**, 127–33.

Sperling, C. H. B. and Cooke, R. U. (1985) Laboratory simulation of rock weathering by salt crystallization and hydration processes in hot, arid environments. *Earth Surface Processes and Landforms* **10**, 541–55.

Summerfield, M. A. (1983a) Silcrete. In: A. S. Goudie and K. Pye (eds) *Chemical Sediments and Geomorphology: Precipitates and Residua in the Near-Surface Environment*. Academic Press, London and New York, 59–91.

Summerfield, M. A. (1983b) Silcrete as a palaeoclimatic indicator: Evidence from southern Africa. *Palaeogeography, Palaeoclimatology, Palaeoecology* **41**, 65–79.

Sweeting, M. M. (1972) *Karst Landforms*. Macmillan, London.

Thomas, M. F. (1974) *Tropical Geomorphology: A Study of Weathering and Landform Development in Warm Climates*. Macmillan, London.

Thorn, C. E. (1979) Bedrock freeze–thaw weathering regime in an

Alpine environment, Colorado Front Range. *Earth Surface Processes* **4**, 211–28.

Trudgill, S. (1985) *Limestone Geomorphology.* Longman, London and New York.

Viles, H. A. (1988) Organisms and karst geomorphology. In: H. A. Viles (ed.) *Biogeomorphology.* Blackwell, Oxford and New York, 319–50.

Walder, J. and Hallet, B. 1985. A theoretical model of the fracture of rock during freezing. *Geological Society of America Bulletin* **96**, 336–46.

Watson, A. (1983) Gypsum crusts. In: A. S. Goudie and K. Pye (eds) *Chemical Sediments and Geomorphology: Precipitates and Residua in the Near-Surface Environment.* Academic Press, London and New York, 133–61.

Watson, A. (1985) Structure, chemistry and origins of gypsum crusts in southern Tunisia and the central Namib Desert. *Sedimentology* **32**, 855–75.

Watson, A. and Pye, K. (1985) Pseudokarstic micro-relief and other weathering features on the Mowati Granite (Swaziland). *Zeitschrift für Geomorphologie* **29**, 285–300.

Whalley, W. B. (1983) Desert varnish. In: A. S. Goudie and K. Pye, (eds) *Chemical Sediments and Geomorphology: Precipitates and Residua in the Near-Surface Environment.* Academic Press, London and New York, 197–226.

Whalley, W. B., Douglas, G. R. and McGreevy, J. P. (1982) Crack propagation and associated weathering in igneous rocks. *Zeitschrift für Geomorphologie* **26**, 33–54.

Whalley, W. B. and McGreevy, J. P. (1985) Weathering. *Progress in Physical Geography* **9**, 559–81.

Whalley, W. B. and McGreevy, J. P. (1987) Weathering. *Progress in Physical Geography* **11**, 357–69.

Whalley, W. B. and McGreevy, J. P. (1988) Weathering. *Progress in Physical Geography* **12**, 130–43.

White, W. B. (1988) *Geomorphology and Hydrology of Karst Terrains.* Oxford University Press, New York and Oxford.

Williams, A. G., Ternan, L. and Kent, M. (1986) Some observations on the chemical weathering of the Dartmoor granite. *Earth Surface Processes and Landforms* **11**, 557–574.

Williams, P. J. (1982) *The Surface of the Earth: An Introduction to Geotechnical Science.* Longman, London and New York.

Williams, P. W. (1972) Morphometric analysis of polygonal karst in New Guinea. *Geological Society of America Bulletin* **83**, 761–96.

Williams, P. W. (1987) Geomorphic inheritance and the development of tower karst. *Earth Surface Processes and Landforms* **12**, 453–65.

Wilson, R. C. L. (ed.) (1983) *Residual Deposits: Surface Related Weathering Processes and Materials.* Geological Society Special Publication **11**. Blackwell Scientific Publications, Oxford and Boston.

Young, R. W. (1985) Silcrete distribution in eastern Australia. *Zeitschrift für Geomorphologie* **29**, 21–36.

Young, R. W. (1986) Tower karst in sandstone: Bungle Bungle massif, northwestern Australia. *Zeitschrift für Geomorphologie* **30**, 189–202.

7

Slope processes and forms

7.1 Properties and behaviour of slope materials

Slopes constitute the basic element of the landsurface and so it is not surprising that they have long formed a focal point for landform studies. The main emphasis in this chapter is on the characteristics and behaviour of valley-side slopes or hillslopes which extend from drainage divides to stream channels, although much of the discussion is also relevant to other kinds of slopes such as those along coasts. Until the 1950s much of the research carried out on slopes was focused on the question of how the history of a landscape might be reconstructed from the slope forms of which it is composed. Such studies rarely considered in detail the specific processes operating on slopes and it has been only in the past two decades or so that major advances have been made in this regard. These advances have been due in large part to the application of research carried out by engineers into the problems of slope stability to slopes in the natural landscape.

At any point on the landsurface the form of the landscape is dependent upon the nature, frequency and intensity of the geomorphic processes acting upon it and the strength, or resistance to deformation, of the surface materials of which it is composed. The properties of these surface materials are clearly important in understanding the form and mode of development of the slopes which make up the landscape. Two fundamental types of material can be distinguished – rock and soil. Since our concern here is with mechanical properties which are relevant to the behaviour of slope materials, it is necessary to define rock and soil in this specific context.

Rock is a hard, coherent material comprising individual particles or crystals. It is discontinuous in the sense that it is broken to a greater or lesser extent by joints and fractures but it is not significantly weakened when saturated with water. **Soil**, (or regolith), by contrast, is a weak, un-consolidated deposit which forms an essentially continuous mass lacking significant joints or fissures, but which is further weakened when saturated with water. This definition of soil includes unconsolidated weathering and sedimentary deposits as well as materials we would more normally regard as soils. Many slopes are, of course, composed of a mixture of rock and soil, but the distinction between the two is still important when analyzing the behaviour of slope materials.

Unconsolidated material transported across, and deposited on, slopes is termed **talus** when composed of relatively large rock fragments, and **colluvium** when composed predominantly of finer material. The term **talluvium** is sometimes used for material that is a mixture of fine and coarse material. Colluvium may contain fossil soil layers representing periods of relative slope inactivity characterized by low rates of erosion or deposition. Boulders dislodged from cliffs or free faces accumulate at the cliff foot to form a **talus slope**, (or **scree slope**). Where the rock fragments are funnelled down a notch or gully in an exposed rock face the material accumulates as a **debris cone**.

7.1.1 Factors determining the strength of slope materials

The response of slope materials to stress is determined by their strength which we can define as the ability to resist deformation and fracture without significant failure. In the context of slope materials experiencing stresses generated by gravity, it is strength in relation to shear stress that is most important. The **shear strength** of slope materials as they occur in the field is usually rather variable over time and space because rocks and soils are generally complex mixtures of mineral particles, water and air; nevertheless, the factors determining shear strength are well known.

One controlling factor, which is particularly important

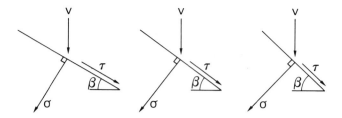

V is the vertical stress
τ is the shear stress
σ is the normal stress
β is the angle of the shear plane

Fig. 7.1 *Change in shear stress and normal stress with change in angle of the shear plane (After M. J. Crozier (1986)* Landslides: Causes, Consequences and Environment. *Croom Helm, London, Fig. 3.2, p. 41.)*

for soils, is the frictional resistance between the constituent particles of a material. This is related to the size of particles, their shape and arrangement, their resistance to crushing and the number of contact points per unit volume. These inherent frictional properties collectively determine the **angle of internal friction** of the material. The greatest angles are achieved by materials composed of angular particles of a range of sizes which are capable of close packing.

The effects of gravity on a slope are expressed by the weight per unit area of the slope materials. This load contributes to both the frictional resistance of the material and to its propensity to move downslope and it can be resolved into two components (Fig. 7.1). One acts parallel to the **shear plane**, or surface of rupture along which movement occurs, and constitutes the shear stress. The other is termed the **normal stress** because it acts at right angles to the shear plane. It contributes to frictional resistance because the load exerted forces the constituent particles of the slope material together. Frictional resistance is therefore a function of both the inherent frictional properties of slope materials and the normal stress acting on them. Clearly, as the shear plane angle becomes steeper the shear stress becomes larger and the normal stress smaller. The maximum angle attained when failure of the slope materials occurs is known as the **threshold angle of stability**.

This analysis is in fact over-simplified because normal stress is exerted only at points of contact between individual particles and not along the entire surface area of the shear plane. Below the water table the voids between particles are filled by water and this gives rise to a positive (greater than atmospheric) **pore-water pressure** which has a buoyancy effect on the overlying material and thus acts in opposition to the normal stress. Consequently, in this case the **effective normal stress** is less than the total normal

stress. This accounts for the common observation that slope failures often occur after heavy rain when pore-water pressures in slope materials are high and effective normal stresses are low. Where slope materials are completely dry and voids are entirely occupied by air the pore-water pressure is zero (atmospheric pressure) and effective normal stress and total normal stress are identical.

The second factor affecting shear strength is **cohesion** which can be defined as the inherent strength of a material which is present irrespective of any load imposed directly on it normal to the surface along which movement tends to occur. Cohesion includes the chemical bonding of rock and soil particles and the adhesion of clay-sized material as a result of electromagnetic and electrostatic forces. Most rocks are highly cohesive but the chemical bonding of soils can also occur through the presence of cements commonly composed of silica, carbonates or iron oxides. In partially moist soil, water is drawn over particle surfaces by capillary forces and pore-water pressure is negative. In these circumstances thin water films on particles contribute to adhesion by creating **capillary cohesion**. This effect is most apparent in soils with a high silt or clay content which are able to retain some moisture permanently in very small pores. It is also illustrated by the ability of sand to stand at a much higher angle when it is damp than when it is completely dry and cohesionless.

We can now take into account the role of pore-water pressure in either moderating normal stress or contributing to cohesion by defining shear strength as effective normal stress plus cohesion, a relation often referred to as the **Coulomb–Terzaghi shear strength equation** (Box 7.1). The relative contribution of normal stress and cohesion to shear strength will vary greatly depending on the state of the material being considered (Fig. 7.2).

The shear strength of soils can be measured by using either a shear box, in which the maximum force that can be transmitted through the sample before failure is established, or by triaxial apparatus which has the advantage of being able to monitor and control the pore-water pressure in the sample. Such tests show that natural soils show rather more complex responses to shear stress than would be anticipated from the factors we have discussed so far.

Box 7.1   The Coulomb–Terzaghi shear strength equation

The total shear strength of a slope material (s) is given as

$$s = c + \sigma'.\tan \phi$$

where c is cohesion, σ' the effective normal stress and $\tan \phi$ the coefficient of plane sliding friction which characterizes the packing, surface roughness and hardness of the constituent particles.

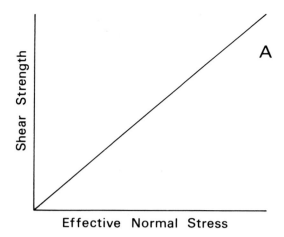

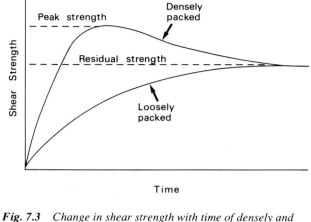

Fig. 7.3 *Change in shear strength with time of densely and loosely packed soils. The behaviour of most soils lies somewhere between the two curves depicted.*

Figure 7.3 illustrates curves showing changes in shear stress over time for both densely packed and loosely packed soils. Although both ultimately attain the same residual shear strength their initial behaviour differs. On the application of stress the densely packed material rapidly attains a peak shear strength which then declines to the residual value. This behaviour can apparently be explained by the rearrangement of soil particles as stress is applied. This rearrangement eventually leads to a decline in shear strength to the residual value. During this rearrangement under stress the material as a whole may expand, an effect known as **dilatancy**, probably due in part to the riding of particles over each other.

7.1.2 Rock properties

Having considered the general factors determining the strength of slope materials we will now look at rock and soil more specifically. Different rock types vary enormously in their **intact strength**, that is, the strength of the rock excluding the effects of fractures and joints (Table

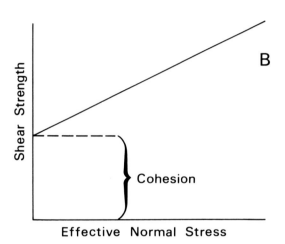

Fig. 7.2 *Relationship between shear stress and normal stress for two types of slope material: (A) a dry sand in which there is only a frictional component directly related to effective normal stress; (B) the usual situation where the slope material has some initial strength from cohesion irrespective of effective normal stress, but its shear strength increases as effective normal stress increases.*

Table 7.1 Intact strength of various types of rock

CHARACTERISTICS	SCHMIDT HAMMER REBOUND VALUE	INTACT STRENGTH CLASSIFICATION	EXAMPLES
Requires severe blows from hammer to break intact sample	100–60	Very strong	Quartzite, dolerite, gabbro, basalt
Hand-held sample can be broken with hammer	60–50	Strong	Marble, granite, gneiss
Shallow indentations can be made by firm hammer blow	50–40	Moderately strong	Sandstone, shale, slate
Deep indentations can be made by firm hammer blow	40–35	Weak	Coal, siltstone, schist
Crumbles under sharp hammer blow — can be cut with knife	35–10	Very weak	Chalk, rock salt

Source: Modified from M. J. Selby (1982) *Hillslope Materials and Processes*. Oxford University Press, Oxford, Table 4.3, p. 65.

7.1). Intact strength can be rapidly assessed using a Schmidt hammer, an instrument which measures the rebound from the rock surface of a known impact. The Coulomb–Terzaghi relation is capable of describing the shear strength of intact rock, but in a fractured and jointed rock mass factors other than the cohesion and frictional properites of the rock are largely responsible for its shear strength. These factors include the degree of weathering, the width, continuity and spacing of joints and other fissures, the orientation of fissures (especially joints and bedding planes) with respect to the ground slope, the presence of infilling material in partings, and the movement of water through the rock mass. Of these factors the distance between individual fissures is in general the most important, there being no cohesive strength along fissures. In some cases, however, the orientation of fissures can be critical, especially where the dip of bedding planes is close to the slope angle.

It is possible to assess the presence of these factors in the field and thereby provide a rating of rock strength in relation to shear stress (Table 7.2). Each variable is given a weighting in proportion to its estimated importance and its value can be assessed for each variable for any particular rock type. This enables an estimate of **rock mass strength** to be made.

7.1.3 Soil properties

Whereas joint spacing is the most significant variable in determining rock strength, it is the clay content that is the most critical factor in influencing the strength of the majority of soils. With an increasing proportion of clay in the soil there is a greater potential for swelling and shrinking through hydration, for lowering of permeability due to increased water retention, for an increase in cohesion and for a lowering of the angle of friction.

Soils are complex materials containing voids filled with air or water, in addition to solid particles of various sizes. Whether a soil behaves as an elastic, plastic or viscous material under stress depends on the proportions of these components and especially the water content. A fine-grained cohesive soil with a low water content will exhibit essentially elastic behaviour, failing by brittle fracture with minimal plastic deformation. As the moisture content increases the behaviour of the soil changes from elastic to plastic, the point of transition being known as the **plastic limit**. If the moisture content is increased further, a stage is eventually reached where the soil consists of a suspension of particles in water and exhibits viscous behaviour. This transition is referred to as the **liquid limit**. It has also been noted that when fine-grained cohesive soils are .dried beyond the plastic limit the soil continues to decrease in volume down to a certain moisture content, known as the **shrinkage limit**. Beyond this limit no further volume change occurs in spite of further moisture losses. These transitions between different forms of soil behaviour are known as **Atterberg limits**, after the Swedish soil scientist who first devised a means of quantifying the effects of

Table 7.2 Classification of rock mass strength

VARIABLE	WEIGHTING %	VERY STRONG	STRONG	MODERATE	WEAK	VERY WEAK
Intact rock strength (Schmidt hammer rebound value)	20	100–60 $r = 20$	60–50 $r = 18$	50–40 $r = 14$	40–35 $r = 10$	35–10 $r = 5$
Weathering	10	Unweathered $r = 10$	Slightly weathered $r = 9$	Moderately weathered $r = 7$	Highly weathered $r = 5$	Completely weathered $r = 3$
Joint spacing	30	>3 m $r = 30$	3–1 m $r = 28$	1–0.3 m $r = 21$	300–50 mm $r = 15$	<50 mm $r = 8$
Joint orientations	20	Very favourable. Steep dips into slope, cross joints interlock $r = 20$	Favourable. Moderate dips into slope $r = 18$	Fair. Horizontal dips or nearly vertical dips (hard rocks only) $r = 14$	Unfavourable. Moderate dips out of slope $r = 9$	Very unfavourable. Steep dips out of slope $r = 5$
Joint width	7	<0.1 mm $r = 7$	0.1–1 mm $r = 6$	1–5 mm $r = 5$	5–20 mm $r = 4$	>20 mm $r = 2$
Joint continuity and infill	7	None, continuous $r = 7$	Few, continuous $r = 6$	Continuous, no infill $r = 5$	Continuous, thin infill $r = 4$	Continuous, thick infill $r = 1$
Ground-water outflow	6	None $r = 6$	Trace $r = 5$	Slight <40 ml s^{-1} m^{-2} $r = 4$	Moderate 40–200 m s^{-1} m^{-2} $r = 3$	Great >200 ml s^{-1} m^{-2} $r = 1$
Total rating		100–91	90–71	70–51	50–26	<26

Source: Modified from M. J. Selby (1980) *Zeitschrift für Geomorphologie* **24**, Table 6, pp. 44 and 45.

Table 7.3 Atterberg limits for various clay minerals

MINERAL	SURFACE AREA ($m^2 kg^{-1}$)	PLASTIC LIMIT (%)	LIQUID LIMIT (%)	SHRINKAGE LIMIT (%)	VOLUME CHANGE
Smectite	800 000	50–100	100–900	8.5–15	High
Illite	80 000	35–60	60–120	15–17	Medium
Kaolinite	15 000	25–40	30–110	3–15	Low

Source: Modified from M. J. Selby (1982), *Hillslope Materials and Processes*. Oxford University Press, Oxford, Table 4.8, p. 79; based on data in R. E. Grim, 1968, *Clay Mineralogy*. McGraw-Hill, New York.

water content on soil behaviour. They are expressed in terms of the weight of contained water as a percentage of the weight of dry soil.

When subject to the same stress different soils with an identical moisture content may fail by brittle fracture, deform plastically or behave as a viscous fluid. This is because of the varying abilities of different clay minerals to absorb water (Table 7.3). Smectite generally has the highest plastic and liquid limits because its crystal structure enables it to provide an enormous surface area for water absorption. Consequently, smectite-rich soils swell and shrink significantly when they are wetted and dried. Kaolinite, by contrast, has a much more limited water absorption capacity.

An important parameter derived from Atterberg limits is the **plasticity index**. This is defined as the liquid limit minus the plastic limit and indicates the range of moisture content over which a soil exhibits plastic behaviour. It is an important indicator of the potential instability of the soil since the higher the plasticity index the less stable slope materials will be.

Two further aspects of soil properties need to be mentioned briefly. Certain soils have an open 'honeycomb' structure which allows them to retain water at proportions in excess of the liquid limit. The structure of these soils, termed **sensitive soils**, is potentially unstable. If they are subject to high shear stresses (such as those induced by an earthquake), or high compressive stresses (such as those arising from loading by burial), they can collapse catastrophically as the water is squeezed out and the soil becomes a fluid. Such soils are sometimes called **quick clays** and are often associated with major and rapid flows of slope materials.

Sands can also act in a fashion similar to sensitive soils under certain conditions. In a saturated mass of sand most of the strength arises from point to point contacts between the solid sand grains. If the sand is shaken violently, by a seismic shock for instance, all the effective stresses can be transferred from grain-to-grain contacts to the pore water, an effect known as **liquefaction**. All the strength from interparticle friction is thereby lost, the sand mass consequently has no resistance to shear stress and liquid deformation occurs.

7.2 Mass movement

Mass movement is the downslope movement of slope material under the influence of the gravitational force of the material itself and without the assistance of moving water, ice or air. The distinction between mass movement and the transport of material by other denudational processes is, however, not always clear-cut in practice since mass movements involving material with a high water content grade into fluvial transport where streams carry very large loads of fine sediment. In a very real sense glacier flow itself is a form of mass movement, involving as it does the downslope movement of coherent masses of ice; nevertheless, glaciers have particular characteristics which merit special consideration and they are examined separately in the context of glacial landscapes (see Chapter 11). Very large-scale movements of rock which are transitional to tectonic processes can also occur under gravity; we discuss these briefly in Section 7.3. The term **mass wasting** is often regarded as synonymous with mass movement, but it is also used in a broader sense to encompass all processes involved in the lowering of the landscape. Before we consider the various mechanisms of mass movement we will look briefly at the conditions which give rise to them.

7.2.1 Slope stability

The stability of a slope can be expressed in terms of the relationship between those stresses tending to disturb the slope material and cause it to move and those forces tending to resist these driving stresses. Clearly, movement will occur where driving forces exceed resisting forces and this relationship is represented as the **safety factor** for a slope. This is expressed as the ratio between shear strength and shear stress (Box 7.2).

Slopes can exist in one of three states. Where shear strength is significantly larger than shear stress the slope is described as **stable** (safety factor > 1.3). Where shear stress exceeds shear strength (safety factor < 1) there will be continuous or intermittent movement and the slope is described as **actively unstable**. Since shear strength can vary over time, especially in response to changes in the water content of slope materials, the third stability category is the

Box 7.2   Safety factor for a slope

The safety factor (F) is defined as

$$F = \frac{s}{\tau}$$

where s is the total shear strength along a specific shear plane, and τ the total amount of shear stress developed along this plane. For shallow, translational slides F is defined as

$$F = c + \frac{(\gamma z \cos^2 \beta - u) \tan \phi}{\gamma z \sin \beta \cos \beta}$$

where c is cohesion, γ the unit weight of regolith, z the vertical depth to the shear plane, β the angle of the shear plane, u the pore-water pressure at the shear plane and ϕ the angle of internal friction.

conditionally stable slope which has a safety factor of 1–1.3 and fails on occasion in response to transient changes in shear strength. Numerous factors contribute to the occurrence of mass movements, and these are listed in Table 7.4. They can be categorized as either preparatory factors or triggering factors. Preparatory factors make the slope susceptible to movement without actually initiating failure by transforming it into a conditionally stable state. Triggering factors transform the slope from a conditionally stable to an actively unstable state.

Although the slope stability approach to analyzing mass movements provides a good theoretical understanding of the factors which promote movement, it has limited applicability to specific situations. This is because both cohesion and pore-water pressure are highly variable on most natural slopes, even over short distances and brief periods of time. For instance, fissures may traverse slope materials leading to drastic variations in pore-water pressure from place to place.

7.2.2   Mass movement processes

There have been numerous attempts to classify the diverse modes of mass movement, none of them universally satisfactory. Here we identify six fundamental types of movement – creep, flow, slide, heave, fall and subsidence. Each of these can be subdivided into more specific forms of mass movement (Table 7.5). Classifications of the various processes of mass movement are valuable in indicating the range of mechanisms and forms of motion, but it must be appreciated that most movements in reality involve a combination of processes. **Debris avalanches**, for example, may begin as slides consisting of large masses of rock but then rapidly break up to form flows as the material is pulverized in transit. The compound nature of many forms of mass movement is illustrated in Figure 7.4, which also indicates how the different types of movement vary in their moisture content and velocity. Flows tend to be wet and slides dry, while heave can occur over a fairly broad range of moisture conditions. Heave processes are invariably slow, whereas both flows and slides tend to be rapid.

7.2.2.1   Creep
Creep is the slow, plastic deformation of rock or soil in

Table 7.4 Factors contributing to the occurrence of mass movement

FACTOR	EXAMPLES
Factors contributing to increased shear stress	
Removal of lateral support through undercutting or slope steepening	Erosion by rivers and glaciers, wave action, faulting, previous rock falls or slides
Removal of underlying support	Undercutting by rivers and waves, subsurface solution, loss of strength by extrusion of underlying sediments
Loading of slope	Weight of water, vegetation, accumulation of debris
Lateral pressure	Water in cracks, freezing in cracks, swelling (especially through hydration of clays). pressure release
Transient stresses	Earthquakes, movement of trees in wind
Factors contributing to reduced shear strength	
Weathering effects	Disintegration of granular rocks, hydration of clay minerals, dissolution of cementing minerals in rock or soil
Changes in pore-water pressure	Saturation, softening of material
Changes of structure	Creation of fissures in shales and clays, remoulding of sand and sensitive clays
Organic effects	Burrowing of animals, decay of tree roots

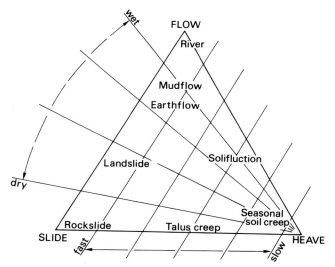

Fig. 7.4 *Classification of mass movements in terms of pure flow, slide and heave. Note the compound nature of some movements. (After M. A. Carson and M. J. Kirkby (1972)* Hillslope Form and Process. *Cambridge University Press, Cambridge, Fig. 5.2, p. 100.)*

Table 7.5 Classification and characteristics of the major types of mass movement

PRIMARY MECHANISM		MASS MOVEMENT TYPE	MATERIALS IN MOTION	MOISTURE CONTENT	TYPE OF STRAIN AND NATURE OF MOVEMENT	RATE OF MOVEMENT
LATERAL COMPONENT PREDOMINANT	Creep	Rock creep	Rock (especially readily deformable types such as shales and clays)	Low	Slow plastic deformation of rock, or soil producing a variety of forms including cambering, valley bulging and out-crop bedding curvature	Very slow to extremely slow
		Continuous creep	Soil	Low		
	Flow	Dry flow	Sand or silt	Very low	Funnelled flow down steep slopes of non-cohesive sediments	Rapid to extremely rapid
		Solifluction	Soil	High	Widespread flow of saturated soil over low to moderate angle slopes	Very slow to extremely slow
		Gelifluction	Soil	High	Widespread flow of seasonally saturated soil over permanently frozen subsoil	Very slow to extremely slow
		Mud flow	>80% clay-sized	Extremely high	Confined elongated flow	Slow
		Slow earthflow	>80% sand-sized	Low	Confined elongated flow	Slow
		Rapid earthflow	Soil containing sensitive clays	Very high	Rapid collapse and lateral spreading of soil following disturbance, often by an initial slide	Very rapid
		Debris flow	Mixture of fine and coarse debris (20–80% of particles coarser than sand-sized)	High	Flow usually focused into pre-existing drainage lines	Very rapid
		Debris (rock) avalanche (sturzstrom)	Rock debris, in some cases with ice and snow	Low	Catastrophic low friction movement of up to several kilometres, usually precipitated by a major rock fall and capable of overriding significant topographic features	Extremely rapid
		Snow avalanche	Snow and ice, in some cases with rock debris	Low	Catastrophic low friction movement precipitated by fall or slide	Extremely rapid
		Slush avalanche	Water-saturated snow	Extremely high	Flow along existing drainage lines	Very rapid
	Slide / Translational	Rock slide	Unfractured rock mass	Low	Shallow slide approximately parallel to ground surface of coherent rock mass along single fracture	Very slow to extremely rapid
		Rock block slide	Fractured rock	Low	Slide approximately parallel to ground surface of fractured rock	Moderate
		Debris/earth slide	Rock debris or soil	Low to moderate	Shallow slide of deformed masses of soil	Very slow to rapid
		Debris/earth block slide	Rock debris or soil	Low to moderate	Shallow slide of largely undeformed masses of soil	Slow
	Slide / Rotational	Rock slump	Rock	Low	Rotational movement along concave failure plane	Extremely slow to moderate
		Debris/earth slump	Rock debris or soil	Moderate	Rotational movement along concave failure plane	Slow
	Heave	Soil creep	Soil	Low	Widespread incremental downslope movement of soil or rock particles	Extremely slow
		Talus creep	Rock debris	Low		
VERTICAL COMPONENT PREDOMINANT	Fall	Rock fall	Detached rock joint blocks	Low	Fall of individual blocks from vertical faces	Extremely rapid
		Debris/earth fall (topple)	Detached cohesive units of soil	Low	Toppling of cohesive units of soil from near-vertical faces such as river banks	Very rapid
	Subsidence	Cavity collapse	Rock or soil	Low	Collapse of rock or soil into underground cavities such as limestone caves or lava tubes	Very rapid
		Settlement	Soil	Low	Lowering of surface due to ground compaction usually resulting from withdrawal of ground water	Slow

Source: Based largely on D. J. Varnes (1978) in: R. L. Schuster and R. J. Krizek (eds) *Landslide Analysis and Control*, Transportation Research Board Special Report 176. National Academy of Sciences, Washington, DC, 11–33.

response to stress generated by the weight of overburden. It begins once the yield stress of the slope material is exceeded. In rock it can extend to hundreds of metres below the surface. It occurs at very slow rates, typically 1 mm to 10 m a^{-1} and is likely to be especially active where weakly competent materials, such as clays, are overlain by more competent beds. Creep is often a precursor of slide-type movements, but it can also cause specific observable effects. The bending of the lower parts of tree trunks is often cited as evidence of creep, but this phenomenon can also be caused by other mechanisms. More substantial evidence of creep is provided by the downslope curvature of strata near the surface. Another consequence of creep is **cambering** which involves the extrusion of weak rocks (usually clays), either lying below, or interbedded with, more rigid strata which causes valley sides to bulge.

It is important to distinguish the type of creep described here from soil creep and talus creep. The former acts solely under gravity, whereas the latter involve heave and are consequently considered in Section 7.2.2.4.

7.2.2.2   Flow

In a pure **flow**, shear occurs throughout the moving mass of material and there is no well-defined shear plane (Fig. 7.5(A)). Flow is distinguished from creep by having discrete boundaries or narrow peripheral zones experiencing shear. Shear is at a maximum at the base of the flow, but here the rate of flow is relatively slow and nearly all the movement occurs as turbulent motion within the body of the flowing mass. **Dry flows** can occur, but abundant water is usually present. They are often initiated by falls or slides, becoming flows when the moving soil or rock mass breaks up. Flows are categorized as **avalanches**, **debris flows**, **earthflows** or **mudflows** depending on whether they consist of predominantly snow and ice, rock fragments, sand-sized material or clay (Fig. 7.6).

Where the flow has a high water content it may extend as a long, narrow tongue well beyond the base of the slope from which it originated. Such flows are usually more or less confined to existing drainage lines (Fig. 7.7) and there is in fact a transition between mudflows and streams laden with abundant fine sediment. Earthflows involve the extrusion of lobes downslope and are usually slow moving. Where the slope material is composed of sensitive soils, however, an initial disturbance can cause an instantaneous loss of shear strength and promote a rapid earthflow. The slowest type of flow is **solifluction** which involves the downslope movement of saturated soil. Solifluction can occur at slope angles as low as 1° and is particularly active in periglacial environments. Here abundant moisture is made available by seasonal thawing of soil above a frozen subsurface, and this form of solifluction is termed **gelifluction** (see Section 12.2.3). Although solifluction can be regarded as a distinct mass movement process it

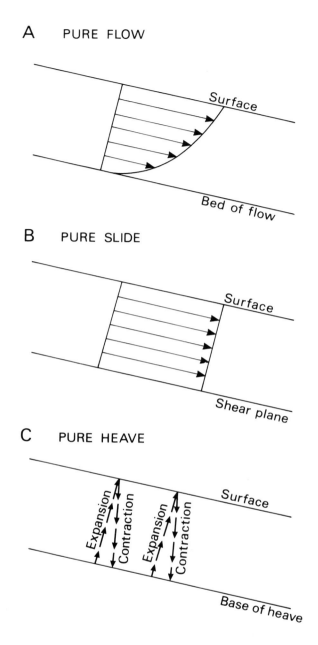

Fig. 7.5   *Velocity profiles for ideal types of mass movement: (A) pure flow, (B) pure slide and (C) pure heave. (Modified from M. A. Carson and M. J. Kirkby (1972) Hillslope Form and Process. Cambridge University Press, Cambridge, Fig. 5.1, p. 100).*

frequently occurs in close association with soil creep. Many hillslopes show the combined effects of both processes (Fig. 7.8).

In addition to the transport of material from higher to lower elevations, flows can be effective erosional agents, especially the more energetic varieties. This is particularly true for debris flows and for debris and **snow avalanches** which travel over the ground, rather than predominantly

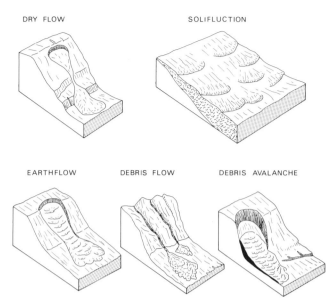

DRY FLOW   SOLIFLUCTION

EARTHFLOW   DEBRIS FLOW   DEBRIS AVALANCHE

Fig. 7.6   *Varieties of flow-type mass movement. (Based on D. J. Varnes (1978) in R. L. Schuster and R. J. Krizek (eds)* Landslides: Analysis and Control. *Transportation Research Board Special Report 176. National Academy of Sciences, Washington, DC, Fig. 2.1.)*

Fig. 7.7   *Debris flow on western flank of the Southern Alps, South Island, New Zealand.*

Fig. 7.8   *Valley cut into a chalk escarpment in Kent, UK. Such valleys on the chalk are known as coombes, and solifluction deposits, known as coombe rock, are often present on valley floors (evident in photograph from the flat bottom of the valley). These solifluction deposits probably accumulated at the end of the last glacial 10 000–15 000 a BP under periglacial conditions (and are therefore more accurately described as gelifluction deposits). Terracettes can be seen on the valley-side slopes and are probably related to soil creep.*

through the air. The major geomorphic effects of avalanches are the removal of debris from gullies and slope faces, the excavation of rock to form **avalanche chutes**, the erosion and redistribution of unconsolidated slope deposits, and the deposition of snow and/or rock debris.

Major avalanches can be one of the most violent and destructive forms of geomorphic activity. Two particularly catastrophic avalanches, both triggered by earthquakes, crashed from the the mountain peak of Huascaran in the Peruvian Andes in 1962 and 1970. In the 1962 event 3 Mt of ice and 9 Mt of rock were transported at speeds well in excess of $100 \, \text{km h}^{-1}$ over a horizontal distance of 20 km. An estimated 3500 people were killed. In 1970 a second avalanche occurred which initially moved along the same route as the 1962 event, but after travelling some 16 km at an average speed of around $300 \, \text{km h}^{-1}$ part of the turbulent flow of rock debris and ice jumped across a 300 m high ridge and buried the town of Yungay beyond. On this occasion the death toll was possibly as high as 40 000.

The extremely high velocities achieved by some avalanches clearly require explanation. Some may ride on a layer of compressed air trapped between the avalanche debris and the ground surface. This has the effect of greatly reducing the frictional drag on the moving mass. Other possible explanations of high rates of movement include **fluidization** where fine particles are kept in suspension by a flow of air, and cohesionless grain flow in which particle motion is sustained by continuous collisions as in a fluid.

Although spectacular, avalanches are not of widespread significance. They are largely confined to terrain characterized by high local relief and steep slopes and are thus most common in active orogenic belts. On a global basis the more ubiquitous but less spectacular slower forms of flow are more significant denudational agents.

7.2.2.3   Slide

Slide is an extremely widespread form of mass movement, and the term landslide is part of our everyday vocabulary. This presents problems when using it in a specific technical sense since landslide in general usage simply means the rapid downslope movement of slope material. Applied in this sense many landslides also involve fall and flow. In a pure slide failure occurs along a well-defined shear plane (Fig. 7.5(B)). Resistance to movement falls sharply immediately the initial failure takes place, and downslope movement continues until there is a sufficient increase in resistance, often related to a decrease in slope angle, to halt it.

Slides are nearly always long in relation to their width and depth, their length–width ratio typically being 10:1. They can be subdivided into **translational slides**, which have predominantly planar shear surfaces, and **rotational slides** in which the shear plane is concave-up (Fig. 7.9). Rotational slides are most common where slopes consist of thick, homogeneous materials, such as clays. The rotational movement can result in the upper part of the slumped mass being back-tilted towards the failure surface (Fig. 7.10). The material can move as a single block, but usually it is broken into several discrete segments separated by transverse fissures. Movement at the base of rotational slides in clay or similar cohesive material is often transformed into that of an earthflow and this gives rise to a chaotic, hummocky

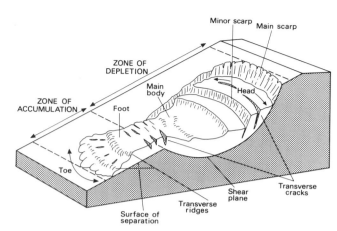

Fig. 7.10   *Major features of a rotational slide (Modified from D. J. Varnes (1978) in: R. L. Schuster and R. J. Krizek (eds)* Landslides: Analysis and Control. *Transportation Research Board Special Report* **176**. *National Academy of Sciences, Washington, DC, Fig. 2.1.)*

surface. Both rotational and translational slides are precipitated by a temporary excess of shear stress over shear strength within the slope (Box 7.3). The difference between the two types of movement is that in thick, relatively homogeneous material the depth at which the ratio between

Box 7.3   Analysis of rotational slides

The shear plane of a rotational slide is curved and therefore the stability analysis outlined in Box 7.2 must be modified. One way the stability of a rotational slide can be evaluated is to divide the slide into a number of 'slices' of length L and aggregate the forces acting of each of these slices (Fig. B7.3). The weight (W) is taken as operating through the centre of each slice. The angle of the shear plane (α) is calculated for each slice from the centre of rotation (O). The effective normal stress (σ') at the base of each slice is $W \cos \alpha$ and the shear strength (s) is $W \sin \alpha$. The safety factor (F) can then be defined as

$$F = \frac{\sum_{B}^{A} [cL + (W \cos \alpha - uL) \tan \phi]}{\sum_{B}^{A} W \sin \alpha}$$

where c is cohesion, u the pore-water pressure at the base of the slice and $\tan \phi$ the angle of internal friction.

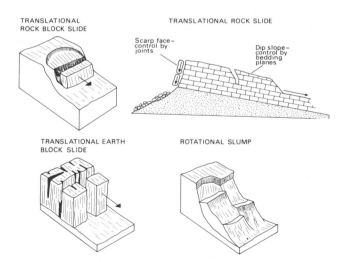

Fig. 7.9   *Varieties of slide-type mass movements. (Based on D. J. Varnes, 1978, in: R. L. Schuster and R. J. Krizek (eds)* Landslides: Analysis and Control. *Transportation Research Board Special Report* **176**, *National Academy of Sciences, Washington, DC, Fig. 2.1.)*

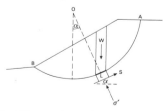

Fig. B7.3   Stability analysis of a deep-seated slide using the method of slices. (After M. A. Carson and M. J. Kirkby (1972) Hillslope Form and Process. *Cambridge University Press, Cambridge, Fig. 7.11, p. 167.)*

shear strength and shear stress is at a minimum (that is, the potential shear plane) forms an arc rather than a straight line.

The great majority of slides are small and shallow with lengths of a few tens of metres and depths of 2–3 m. Slides in bedrock are less common, but may attain enormous dimensions and involve the movement of millions of cubic metres of material. Very large slides usually break up to form debris avalanches but in some cases the rock travels a significant distance as a coherent mass. This is most likely to occur where competent beds slide over incompetent strata (often clay or mudstone) dipping steeply roughly parallel with the ground slope. Such conditions contribute to low shear strength and enhance the probability of slides occurring.

Probably the largest slide on Earth is the Saidmarreh slide located in south-western Iran. Although it occurred more than 10 000 a BP it has suffered only superficial modification by subsequent erosion. The deposits are, crudely stratified, indicating that the movement was not predominantly one of a turbulent debris avalanche. A mass of limestone some 15 km long, 5 km wide and at least 300 m thick slid off the underlying interbedded marl and limestone which dips at an angle of around 20° out of the slope. The initial vertical component of movement was only about 1000 m, but the slide travelled a total distance of 18 km, crossing an 800 m high ridge en route.

7.2.2.4   Heave

In pure **heave** the slope material experiences cycles of expansion and contraction (Fig. 7.5(C)). Downslope movement arises from the fact that while expansion occurs normal to the sloping ground surface contraction under gravity tends to be more nearly vertical. Cohesion between particles usually prevents a purely vertical return movement under gravity. Two types of heave can be distinguished on the basis of the size of the constituent particles – **soil creep** and **talus creep**; the latter involves coarser material than the former. Expansion and contraction can be caused by wetting and drying, freezing and thawing (in which case the process is described as **frost creep** – see Section 12.2.3), temperature changes and the burrowing activity of worms and other organisms.

The rate of soil or talus creep on a slope depends on a number of factors. It will become greater with increasing slope angle since this increases the downslope component of movement. It will also be high in soils containing abundant quantities of clays, such as smectite, which expand significantly on wetting, or in silt-sized material which is capable of substantial ice accumulation. Soil and talus creep will, however, decrease with depth below the slope surface, both because of the moderation of changes associated with wetting and drying and freeze–thaw, and the increase in the weight of the overburden. There is some support from experimental and field observations for these expected relationships, although there is an upper limit to

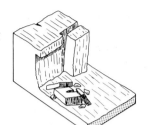

Fig. 7.11 *Varieties of fall-type mass movements. (Based on D. J. Varnes (1978) in R. L. Schuster and R. J. Krizek (eds)* Landslides: Analysis and Control. *Transportation Research Board Special Report* **176**. *National Academy of Sciences, Washington, DC, Fig. 2.1.)*

the operation of soil creep because above an angle of around 25° the soil cover is thin or absent. On steep, grass-covered slopes flights of narrow steps, called **terracettes**, are commonly present and it is likely that these are related to soil creep, although shallow landslides may also be instrumental in their development (Fig. 7.8).

7.2.2.5   Fall

Fall involves the downward motion of rock or, more rarely, soil through the air (Fig. 7.11). Soil is not frequently involved for the simple reason that free fall can only occur from very steep slopes or cliff faces which, of course, have very little soil cover. An exception is the toppling of slabs of earth along river banks, a process often referred to as **bank calving**. This arises from the undercutting of banks by streams and is a very common phenomenon which contributes large quantities of sediment to river channels. **Topples** are distinguished from other types of fall by the rotation of the block of material as it falls away. Topple can occur in rock especially where joints are vertically extensive in relation to their width and where they dip out of a slope.

Rock can become detached as a result of various physical weathering processes, including pressure release and joint widening by frost action, and the fragments produced are rapidly removed under gravity. **Rock falls** are common phenomena in terrain characterized by high, steep rock slopes and cliffs (Fig. 7.12). Where the dislodged fragments are large they accumulate a significant amount of kinetic energy by the time they impact on the slope below and they can therefore be an active erosive agent by detaching other fragments. Large rock falls originating from a considerable height above the ground spread their debris over an extensive area unless the dispersal of material is confined by topography. As already mentioned, large rock falls are often transformed into debris avalanches once they have made their initial impact. In situations where deep valleys are cut into hard rocks, such as granites and some sandstones, by glaciers or rapidly incising river channels, the release of

Fig. 7.12   *Rock fall in a recently glaciated valley in the Southern Alps, South Island, New Zealand. The area of fresh rock face from which the large blocks of rock have fallen is more than 20 m high.*

also occur naturally where the volume of poorly compacted materials is decreased by the addition of water (**hydro-compaction**) or by vibrations such as those generated by earthquakes.

7.3   Gravity tectonics

Gravity tectonics is a useful term which covers a range of processes extending from the very large-scale movements of rock masses involved in the development of thrusts and nappes to smaller scale downdip translocations of material which are transitional to landslides. Many of the massive nappes occurring in intercontinental collision orogens such as the Alps are now considered to be vast **gravity slides** moving away from their high, axial zones (see Section 3.4.1).

Gravity tectonics can involve both spreading and sliding, the two processes being closely associated, but not identical. Rocks located high up in mountain masses and bounded by steep slopes gradually yield and move downslope under gravity. This kind of motion can be accommodated by internal movements within the rock (**gravity spreading**) or it may occur primarily through **gravity sliding** of the rock mass over a few well-defined planes composed of incompetent strata. Although the movement is essentially downslope, internal deformation and the rotation of blocks along small-scale faults can create chaotic patterns at the local scale (Fig. 7.13).

Such structures clearly have a combined endogenic and exogenic origin. The initial energy input is endogenic, involving as it does orogenic uplift. But the subsequent sliding, which may take place over gradients as low as 0.5°,

lateral confining pressure along the valley walls can give rise to pressure release and generate tension joints running roughly parallel with the ground slope. These can promote slab failure when the progressive widening of these joints eventually leads to the detachment of thick slices of rock (Fig. 6.21).

7.2.2.6   Subsidence

Subsidence can occur either as the more or less instantaneous collapse of material into a cave or other cavity (**cavity collapse**) or as a progressive lowering of the ground surface (**settlement**). Cavity collapse is largely confined to limestone terrains where the roofs of underground cavities occasionally collapse. More rarely lava tubes within lava flows may experience a similar fate. Cavity collapse can also occur as a result of human activites such as mining. Settlement usually arises from the lowering of water tables and is most dramatically illustrated in areas where there has been oil drilling or large-scale abstraction of ground water for irrigation. Settlement can

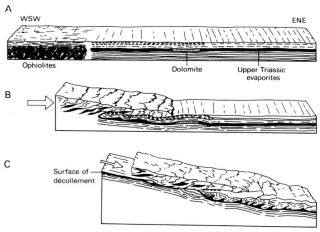

Fig. 7.13   *Schematic representation of gravity tectonics in the northern Apennines, near Florence, Italy: (A) the original depositional basin; (B) thrusting from the west; (C) gravity sliding over Upper Triassic evaporites (solid shading) acting as a décollement surface. (After P. Elter and L. Trevisan (1973) in: K. A. De Jong and R. Scholten (eds) Gravity and Tectonics. Wiley, New York, Fig. 15, p. 187.)*

is dependent on stresses arising from the gravitational instability of the rock mass. Unloading of mountain ranges by gravity spreading and sliding, a process sometimes described as **tectonic denudation**, occurs in continental-margin arc orogens such as the Andes as well as in inter-continental collision belts. Slide rates are slow, reaching a maximum of about 100 m a^{-1} over low-friction décollement layers such as salt, but are orders of magnitude lower in most cases.

7.4   Water erosion and solute transport on slopes

In addition to being affected by mass movement processes, slope material can be transported by water. Three mechanisms are involved: **rainsplash erosion**, in which particles on the soil surface are dislodged by the impact of raindrops; **slope wash**, which is the process whereby sediment is entrained and transported by a thin sheet of water flowing over the slope surface; and solute transport where soil materials taken into solution during weathering reactions are transported downslope as solutes. These mechanisms are sometimes collectively referred to as **wash processes**. Rainsplash and slope wash are effective only on soil-covered slopes, whereas solution can operate on both soil and rock slopes. In order to understand how materials are transported downslope by water we first need to look at the way water moves over, and within, hillslopes.

7.4.1   Hillslope hydrology

Precipitation falling on a slope either runs off the surface, is held in surface depressions (**surface detention**), or infiltrates through the slope surface. Except on impermeable rock slopes some proportion of the precipitation reaching the slope almost invariably infiltrates the surface and either percolates down to the water table to contribute to ground-water storage, or moves laterally down through the slope more or less parallel to the surface (Fig. 7.14). This lateral movement occurs because compaction, the infilling of voids by fine particles washed down from above, and, in some cases, the precipitation of iron oxide, silica, calcium carbonate and other compounds reduces the ability of the soil to transmit water at depth. Water that is diverted late-rally is termed **throughflow**, although the term **interflow** is also sometimes applied to water moving laterally below the soil proper but above the water table. The route taken by water that has infiltrated the slope surface and the rate at which it is transmitted through the slope are particularly important in terms of the transport of solutes.

The proportion of precipitation that flows down a slope surface rather than infiltrating can depend on both the intensity and duration of precipitation as well as the pro-perties of the slope surface. The latter determine the **infiltration capacity** of the surface – that is, the rate at which it can absorb water. Many factors influence the infiltration capacity of natural surfaces. The most important are particle size and the abundance of organic matter and intensity of faunal activity. Each of these factors promotes the development of an open soil structure which is capable of absorbing water efficiently from a slope surface. Slope surfaces with low infiltration capacities are most likely to be encountered in arid and semi-arid environments where soils have a low organic content, but they are also common in all environments where human activities lead to an artificial removal of the vegetation cover or compaction of the surface.

Where the intensity of precipitation exceeds the infil-tration capacity of the surface a proportion of the precipi-tation will flow over the surface as **infiltration-excess overland flow** (or **Hortonian overland flow** as it is sometimes termed after the pioneer in this aspect of hydrological research, R. E. Horton). In other cases the infiltration capacity of the slope surface may not be exceeded by the rate of incoming precipitation but the combined intensity and duration of precipitation may be sufficient locally to saturate the soil and raise the water table to the surface. These conditions will give rise to **saturation overland flow** which is most likely to occur in areas of a drainage basin where the water table is already relatively close to the surface, such as close to stream channels. This is the predominant form of overland flow in humid environments.

7.4.2   Rainsplash erosion

Raindrops possess kinetic energy by virtue of their mass and velocity. Although the impact velocity of raindrops varies depending on droplet size, wind speed and turbu-lence, raindrops of the maximum size under normal conditions of around 6 mm diameter have an impact velocity of about 9 m s^{-1}. At this speed rain drops can directly move particles more than 10 mm across and coarser material can be dislodged by the removal of downslope support provided by finer sediment.

Rainsplash erosion can occur wherever vegetation does not entirely cover the ground, although it is a more potent erosive agent in environments where there is little or no vegetation cover. Both slope gradient and surface charac-teristics influence the effectiveness of rainsplash erosion. Experimental studies have shown that on low angle slopes of 5° only about 60 per cent of the particles dislodged by raindrop impacts move downslope but this percentage increases with gradient reaching 95 per cent on 25° slopes. It also appears that rainsplash erosion is more effective on sandy surfaces than those containing a high proportion of clay and silt-sized material, apparently because the pres-ence of finer particles contributes to cohesion.

Although rainsplash alone can cause significant erosion,

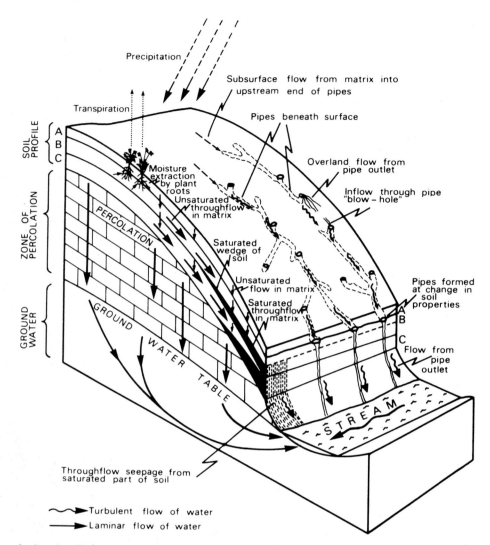

Fig. 7.14 *Routes of subsurface flow on hillslopes. (From T. C. Atkinson (1978) in M. J. Kirkby (ed.)* Hillslope Hydrology, Wiley, *Chichester, Fig. 3.1, p. 74.)*

especially on very steep, highly permeable slopes where it may be as important as slope wash, it is most effective when it is combined with slope wash. If slope wash occurs without rainsplash the loose surface material is rapidly removed, but the rate of erosion quickly declines once the more compacted material below is exposed. But where rainsplash is combined with slope wash a high rate of erosion can be more effectively sustained because the surface is constantly being disrupted by raindrop impacts and particles are continuously being made available for entrainment and transport. Raindrop impacts also increase the turbulence of the water flow and this enhances its ability to entrain and transport sediment. Probably the main role of rainsplash is the dislodgement of particles from small ridges and bumps on hillslopes which are not covered by overland flow. These particles move into adjacent microtroughs where surface-water movement is concen-

trated. Finally, rainsplash can contribute to slope wash by lowering infiltration capacity through the creation of a thin surface crust formed through the infilling of near-surface voids by fine particles displaced by raindrop impacts.

7.4.3   Slope wash

The movement of water across a slope surface, irrespective of how it is generated, is termed **sheet flow**, although this is a rather misleading description since the water flow is never of uniform depth because of the microtopography of hillslope surfaces. Sheet flow can in fact grade into channelled fluvial flow as the water movement becomes progressively more concentrated into particular downslope routes, and the distinction between the two is sometimes difficult to make.

The mechanisms whereby particles are entrained and

transported by flowing water will be considered in Chapter 8. Nevertheless, an important distinction which needs to be made here is that between true sheet flow, which can only move very fine particles, and concentrated flow where the greater depth of flow allows larger material to be transported. The erosional effectiveness of sheet flow is largely controlled by the characteristics of the surface. These include particle size and degree of particle cohesion, the extent and nature of the vegetation cover, and slope gradient. By contrast, the rate of erosion for concentrated flow is determined more by the depth and velocity of flow. In addition to material dislodged by rainsplash, slope wash readily removes particles disturbed by the growth of ice crystals or salts, or by animal activity.

Sustained concentrated flow can eventually produce **rills**, microchannels a few centimetres in depth and width (Fig. 7.15). In humid environments the presence of vegetation means that rills usually develop only on artificially disturbed surfaces, but in arid and semi-arid environments they can occur naturally. Although rills may be destroyed between rainfall events by other slope processes, especially soil creep, those favourably located may eventually be enlarged into **gullies** and form a permanent part of a

channel network (see Section 8.1.3). Contour curvature has a significant effect on slope wash since where contours are convex in plan sheet flow will be dispersed downslope and erosion will be minimized. Conversely, at valley heads and other locations where contours are concave in plan, the flow will be concentrated downslope and rill erosion will consequently be more effective (Fig. 7.16).

7.4.4 Soil erosion

The rate at which material is eroded from a slope is a function of both **erodibility**, or the resistance of slope materials to entrainment and transport, and **erosivity**, the potential of slope processes to cause erosion. A multitude of factors influence erodibility and erosivity, and it is extremely difficult to quantify these in order to predict the rate of erosion on a particular slope under a given set of conditions. None the less, there have been several attempts to express the rate of soil erosion as a function of specific variables and these have been applied to the prediction of soil loss, especially where this is of importance to agriculture.

The most comprehensive index of soil erosion is provided by the **Universal Soil-Loss Equation**. This was

Fig. 7.15 *Rills feeding into gullies on an unvegetated natural surface in the Painted Desert of northern Arizona, USA. The bag in the bottom-left corner of the photograph indicates the small size of the rills.*

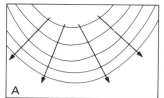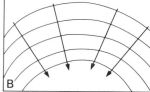

Fig. 7.16 *Effect of contour curvature on sheet flow and rill erosion: (A) contours convex in plan with sheet flow dispersed; (B) contours concave in plan with sheet flow concentrated.*

specifically developed for cropland and it incorporates a range of erodibility factors. These include the resistance of the soil to rainsplash and slope wash, slope length and gradient, the proportion of the ground surface that is covered by crops, and the presence or absence of soil conservation measures such as contour ploughing. In addition it contains a measure of erosivity derived from the kinetic energy of rainfall events. Although extensively used to predict soil erosion on agricultural land, the Universal Soil-Loss Equation is only of limited applicability in studies of erosion on slopes under natural conditions.

7.4.5 Solute transport

The transport of slope materials in solution has been the least studied aspect of slope denudation, and there are comparatively few quantitative data by which direct comparisons can be made with the effectiveness of other slope processes. The loss of weathered material in solution leads to the rearrangement and settling of the remaining particles, although it is difficult to calculate the effect of such loss on the form of the slope itself. The rate of solute transport can certainly be estimated by measuring the discharge of subsurface flow on a slope and relating this to its solute concentration. But this cannot be converted to a volume change unless we know what alterations in bulk density have accompanied the weathering reactions which gave rise to the release of the solutes. Indeed, in the extreme case of isovolumetric weathering (see Section 6.2.4.1), the release of material in solution is exactly compensated by a decrease in the bulk density of the weathered material, so there is no change in volume and, therefore, no effect on the configuration of the slope surface.

The clearest impact of solution is on slopes in limestone terrain. Here the release of solutes occurs without the accumulation of a significant weathering residue so solute transport is converted more directly into a change in surface form, even allowing for a significant proportion of the solution occurring below the surface. Rates of solute release are likely to be most rapid during the initial phase of percolation or throughflow and decrease thereafter as the soil solutions move towards equilibrium with the minerals present within the slope materials. Solute transport will be greatest in those parts of a slope where subsurface flow is at a maximum, such as at the base of slopes and where contour curvature encourages the convergence of throughflow.

7.5 Rates of slope processes

Numerous estimates have been made of the rate of operation of individual slope processes and of the overall rate of slope denudation. A range of techniques have been employed (Table 7.6), but virtually all measurements are limited to a period of a few months or a few years, and it is uncertain whether such short-term estimates are representative of average rates over the long term.

Table 7.6 Techniques for estimating present rates of slope processes

PROCESSES	TECHNIQUES
Soil creep	1. Young pits – rods or plates buried in a pit which is re-excavated after a period of years.
	2. Surface rods and buried and re-excavated cylinders, deformation of plastic tubes
	3. Tilt of cylinders monitored from above by viewing through cross-wires
Solifluction	1. Buried plastic tubes with deformation measured by inclinometers or strain gauges
	2. As for soil creep
Slope wash and rainsplash	1. Recording of ground loss using rods marked to show position of the ground surface (erosion pins)
	2. Collection of sediment moving over slope in a downslope wash trap
Solute transport and loss	1. Monitoring discharge and solute concentration in streams to estimate solute loss from slopes
	2. Monitoring of solute concentration in throughflow on individual slope sections.
Slides	Estimation of volume of debris moved in individual slides related to their estimated frequency of occurrence
Cliff and slope retreat	1. Repeated surveying or comparison of large-scale maps constructed at different times over periods of several decades
	2. Erosion pins and the direct recording of individual rock falls and talus accumulation

Source: Based on I. Saunders and A. Young (1983) *Earth Surface Processes and Landforms* **8**, 475–87.

Few studies have attempted simultaneously to monitor various slope processes in a specific area. One notable example, however, is the classic investigation by A. Rapp in the Kärkevagge area of northern Sweden. Rapp monitored a range of processes over a period of 9 a on slopes in mica-schist and amphibolite with gradients ranging from 15 to 45°. Rock fall was estimated from the accumulation of debris on seasonal snow patches, on areas of matting laid on the ground and in wire netting slung at the base of rock faces. The rates of the various processes measured are given in Table 7.7. Two comments need to be made on these results. First, the considerable proportion of movement accomplished by earth slides was largely achieved in a

Table 7.7 Relative importance of slope processes at Kärkevagge, Sweden

Rapid movements (%)	
Rock falls	7
Debris avalanches	8
Earth slides	34
Total	49
Slow movements (%)	
Solifluction	2
Scree movement	1
Solute loss	48
Total	51

Source: Data from A. Rapp (1960)
Geografiska Annaler **42**, 73–200.

single major storm event; this underscores the problem of assessing the true long-term significance of geomorphic events which have long recurrence intervals. Secondly, the impressive figure for solution initially suggests the importance of chemical weathering in ground loss on slopes in the area. But a more appropriate assessment of its significance would have to make allowance for any decrease in the bulk density of the regolith associated with the weathering reactions leading to the release of solutes.

The rates at which different slope processes operate on an individual slope vary greatly, but there is also a considerable variation in the activity of particular processes under different morphoclimatic regimes (Fig. 7.17). The latter arise from the influence of climate on erosivity and its indirect effects on the erodibility of slope materials, especially in terms of the abundance of vegetation and moisture content. We consider the rates of overall landscape denudation further in Chapter 15.

7.6 The slope system

The ability of processes on hillslopes to fashion slope form is determined by the capacity of these processes to transport the available slope material. In 1877 G. K. Gilbert identified two kinds of situation for slope development. In one, erosion is limited by the rate at which material is made available through weathering. In the other, there is no effective limit to the availability of weathered material and slope erosion is therefore controlled by the capacity of the transport processes. This distinction has subsequently developed into the idea that slopes can be viewed either as **weathering-limited** or as **transport-limited**. But how valid is this widely accepted categorization? It misleadingly implies that on slopes where there is a ready supply of erodible particles transport processes will always operate at capacity. Moreover, it underplays the contribution of mass movement processes in determining slope form and over emphasizes the role of water erosion on slopes.

A more useful basis for understanding the relationship between slope processes and slope form is the idea of **detachment control**. All material forming hillslopes can be regarded as having a range of **detachability** with respect to particular hillslope processes (Fig. 7.18). At one end of this continuum, surface materials may be completely undetachable by a particular process, and erosion is limited by the rate at which weathering creates particles which *are* detachable by the process. At the other end of the continuum surface materials may, in effect, be infinitely detachable by a particular process and the rate of erosion on the hillslope is controlled by the transporting capacity of this process.

The fashioning of slope form by slope processes can be influenced by both external conditions and by the nature of the slope system itself. An example of the role of an external factor is provided by the situation where the moisture content of slope materials is significantly increased by a period of intense rainfall. As the moisture content rises and pore pressures increase, the effective normal stress acting on the slope will fall until a point is reached where shear stress exceeds shear strength and failure occurs. A similar situation occurs when surface runoff is established on a slope since the depth and velocity of flow must attain a critical level before the detachment and transport of slope materials can begin. Both these cases provide examples of slope behaviour being influenced by thresholds. And since in both cases it is external (climatic) factors that are promoting the changes in slope conditions, these are instances of the operation of **extrinsic thresholds**.

A contrasting situation occurs where internal changes within slope materials themselves lead to an adjustment of slope form. Where a slope is being actively undercut and thereby maintained close to its threshold angle of stability, it will experience successive periods of progressive weathering and weakening of slope materials interspersed with episodes of slope failure when the regolith becomes incapable of sustaining the existing slope angle. In this case slope failure occurs as a result of the breaching of an **intrinsic threshold** since there are no changes in external variables. The slope failure itself exposes less-weathered material which is able to maintain a higher slope angle, and failure will not occur again until the shear strength of the slope materials has been sufficiently reduced by further weathering. It is, of course, possible that changes in external variables could also promote a slope failure, but the important point is that this is not necessary for a discontinuous adjustment in slope form to occur.

7.6.1 Slope form

Slope form is most often represented in terms of two-dimensional **slope profiles**. Slope profiles run from drain-

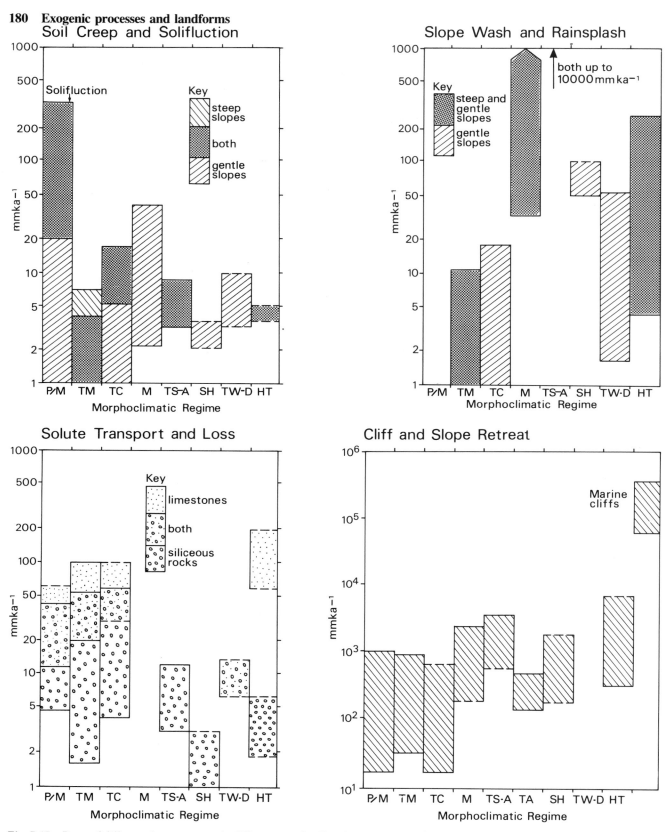

Fig. 7.17   *Rates of different slope processes in different morphoclimatic environments based on short-term field measurements. Key to morphoclimatic regimes: P/M – periglacial and montane environments: TM – temperate maritime including eastern USA and western Europe; TC – temperate continental including the humid interior of the USA and eastern Europe; M – Mediterranean and regions with similar climate elsewhere such as southern California; TS–A – tropical semi-arid; TA – tropical arid; SH – subtropical humid environments such as the south-east USA; TW–D – tropical wet–dry; HT – humid tropical. (Modified from A. Young and I. Saunders (1986) in A. D. Abrahams (ed.)* Hillslope Processes. *Allen and Unwin, Boston, Fig. 1.1, p. 7.)*

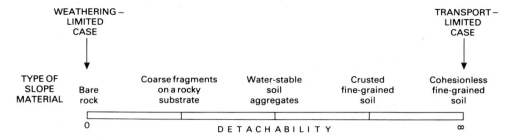

Fig. 7.18   *The detachability continuum for slope processes. (After A. J. Parsons (1988)* Hillslope Form. *Routledge, London, Fig. 7.1, p. 108.)*

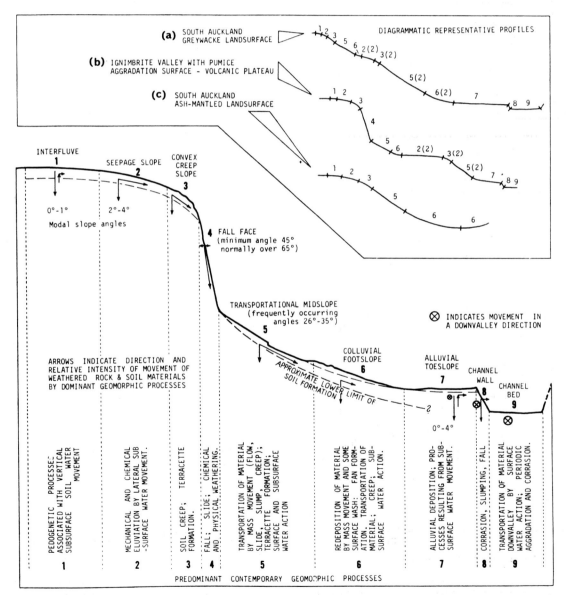

Fig. 7.19   *A theoretical slope profile containing a comprehensive range of slope units related to particular types of slope processes. (From J. B. Dalrymple* et al. *(1968)* Zeitschrift für Geomorphologie *12, Fig. 1, p. 62).*

age lines up the steepest slope to drainage divides. They are usually measured using simple surveying equipment, such as an Abney level or clinometer, which provides slope angle measurements over a known distance on the ground. Slope profiles can be subdivided into individual components or units consisting of convex and concave **elements** and straight, or rectilinear, slope **segments**.

Slope forms vary enormously, but in many cases they comprise an **upslope convexity** leading down to a rectilinear **main slope** which terminates in a **basal concavity**. The main slope can consist of either a single segment or a more complex sequence of segments at different angles. In some cases bare rock is exposed, most often in the upper part of the main slope, and this part of the slope is termed a **free face**. Where active vertical or lateral undercutting is present at the slope foot the basal concavity will be absent. Figure 7.19 illustrates the complete range of slope units that can be encountered, although very few individual slopes contain all these components, and not all the units always occur in the sequence illustrated. This diagram also indicates how different slope processes tend to predominate on different slope units.

Although slope form is conventionally thought of in two-dimensional terms, it must be remembered that slopes are components of the three-dimensional surface that constitutes the landscape. The plan form of slopes is important because contour curvature controls the routes taken by water, sediment and solutes moving downslope (Fig. 7.16). Since slope units can be straight, convex or concave in plan as well as in profile there are nine possible three-dimensional slope forms (Fig. 7.20).

Numerous factors control slope form, but since there are important contrasts between those influencing rock slopes and those affecting soil-mantled slopes we will consider each separately.

7.6.1.1   Rock slopes

Rock slopes generally lie towards the weathering-limited end of the continuum of detachability, their form being controlled by the weathering resistance and shear strength of the rock rather than by the activity of transport processes. The detailed form of rock slopes depends on variations in their rock mass strength as defined in Table 7.2 (see Section 7.1.2). A rock slope may therefore have several units but each unit will tend towards an angle which is in equilibrium with rock strength on that part of the slope (Fig. 7.21). We would expect these equilibrium angles to be sustained through time for as long as the rock strength characteristics remain constant. In rock types with a very high rock mass strength vertical, or even overhanging, free faces can develop (Fig. 7.22).

Rock slopes are not common in humid tropical morphoclimatic environments because rock strength is usually significantly reduced by weathering. In such environments

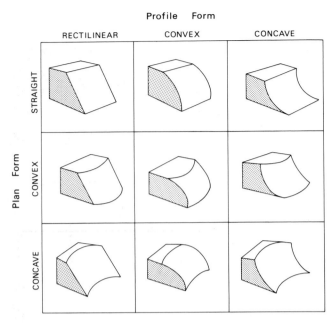

Fig. 7.20   *The nine possible shapes of three-dimensional hillslope forms. (Modified from A. J. Parsons (1988)* Hillslope Form. *Routledge, London, Fig. 2.5, p. 16.)*

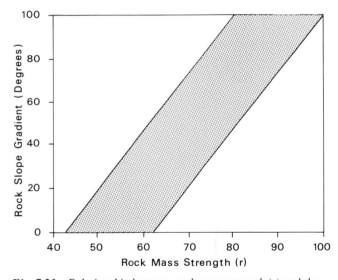

Fig. 7.21   *Relationship between rock mass strength* (r) *and the gradient of rock slopes based on over 250 measurements. (Modified from A. D. Abrahams and A. J. Parsons (1987)* Earth Surface Processes and Landforms *12, Fig. 2, p. 634.)*

rock slopes are generally confined to coasts and regions of high local relief where steep slopes associated with active basal undercutting allow the removal of weathered material as rapidly as it is produced.

Talus will accumulate below rock slopes if basal removal occurs at a slower rate than debris is supplied. Moreover, the thickness of talus that accumulates reflects the balance

Fig. 7.22 *Overhang (slope gradient > 90°) developed in a rock slope in Clarens Sandstone, a thick unit at the top of the Karoo sedimentary sequence in southern Africa.*

between the supply and removal of material through time. In their initial stage of development talus slopes tend to be concave in form. This is due in part to the rolling and bouncing of debris falling from the rock face over the talus surface, although slumping and sliding in the upper part of the talus slope can also contribute to the development of a concave profile. As the talus slope builds up and the rock face becomes progressively buried the distance that debris can fall from the rock face becomes gradually less. Falling rock fragments consequently have, on average, less kinetic energy and are less likely to move beyond the talus toe, so the talus slope becomes gradually straighter. Talus slopes formed of fine material can also be significantly modified by overland flow.

The form of rock slopes is influenced a great deal by the lithological and structural properties of the rock. Tensional joint systems generated in response to pressure release can induce a degree of slope convexity (Fig. 6.21), while alternating beds of varying resistance can give rise to

compound slopes comprising a series of free faces separated by talus slopes.

7.6.1.2 Soil-mantled slopes

Rectilinear units on soil-mantled slopes are developed by those mass movement processes which transport debris downslope to a relatively constant depth. Extensive straight main slopes are common where stream incision is active. The slope is steepened by basal undercutting until a shallow debris or earth slide restores the slope to the angle at which it is stable. As mentioned earlier (see Section 7.1.1) the gradient at which failure occurs is termed the threshold angle of stability (alternatively, **limiting angle of stability**) and is related to the shear strength of the slope material. Numerous field measurements of the angle of straight slope segments appear to show that some angles are rather more frequent than others in the landscape (Table 7.8). These peak angle frequencies are most clearly evident when the data for particular lithologies are examined (Fig. 7.23) and appear to be linked to the threshold angle of stability of different kinds of slope materials.

Convex slope segments most frequently occur on slope crests. The cause of this slope convexity was attributed to soil creep by G. K. Gilbert in a paper published in 1909. He argued that if the interfluves between drainage channels are to be lowered there must be increasing rates of transport of material at increasing distances from the divide. Since rates of soil creep increase with gradient, but not with distance from the divide, the increasing rate of transport can only be accomplished by a downslope increase in slope angle, thereby giving rise to slope convexity. This argument also applies to rainsplash erosion and solifluction. Whereas soil creep is probably the main cause of the convex divides so common in humid temperate environments, rainsplash erosion and solifluction may equally be responsible for convex crests in, respectively, semi-arid and periglacial regions.

Concave slope segments are normally associated with either slope wash or, as already mentioned, the deposition of talus. In the case of slope wash there is an increase in

Table 7.8 Typical threshold slope angles for various types of slope material

SLOPE ANGLE (°)	SLOPE MATERIAL
43–45	Jointed and fractured rock that is virtually cohesionless but with a high packing density
33–38	Same material as above but with looser packing
25–28	Taluvial slopes in which high pore-water pressures can be attained
19–21	Sandy material
8–11	Clays

Source: Based on M. A. Carson and M. J. Kirkby (1972), *Hillslope Form and Process*. Cambridge University Press, Cambridge, pp. 183–184.

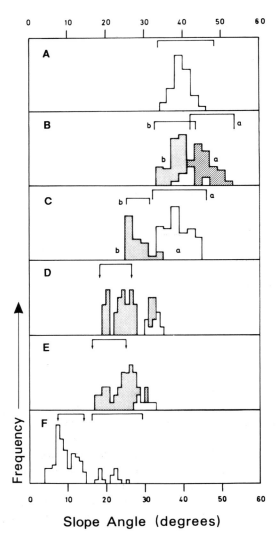

Fig. 7.23 *Frequency distribution of angles of straight slopes in six areas illustrating the presence of threshold slopes: (A) slopes in shale, Mesa Verde, Colorado, USA; (B) undercut (a) and protected (b) slopes, Verdugo Hills, California, USA; (C) slopes in Queenstown Shale, Ontario, Canada, with rubble (a) or clay (b) regolith; (D) slopes in sandstone, Pennines and Exmoor, UK; (E) slopes in Laramie Mountains, Wyoming, USA (unshaded histogram denotes dry talus slopes); (F) slopes in clay–shale, Derbyshire, UK. Threshold angles are divided into semi-frictional (arrowed) and frictional. (After M. A. Carson (1976) in: E. Derbyshire (ed.) Geomorphology and Climate, Wiley, London, Fig. 4.11, p. 123, based on various sources.)*

depth of flow downslope because the contributing area increases away from the drainage divide. Consequently the velocity of flow can be maintained at progressively lower slope angles. In addition the continuous weathering and sorting of particles as they move downslope means that the average grain size generally decreases away from the slope crest. Since a given load of fine material requires less power to transport it than an equivalent load of coarser material the rate of sediment transport can be maintained

over a lower gradient. Slopes dominated by the effects of slope wash are therefore concave since they become progressively less steep downslope.

Field measurements of rates of various slope processes can be incorporated into mathematical models which predict the approximate equilibrium form that should arise given the operation of specific processes. One approach which has been widely applied uses a **continuity equation**. This expresses the transporting capacity of the slope as a function of slope gradient and distance from the drainage divide, and indicates the form which the profile will tend to approach over time (Fig. 7.24). Since this equilibrium profile depends on the processes operating and not the initial slope geometry it is described as a **characteristic form** (see Section 18.2.3).

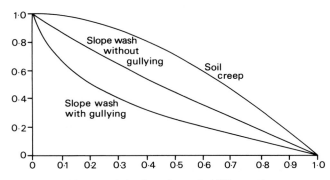

Fig. 7.24 *Characteristic slope forms for different slope processes. (Modified from M. J. Kirkby (1971)* Institute of British Geographers Special Publication **3**, *Fig. 5, p. 26.)*

In the absence of basal undercutting many soil-mantled slopes are convexo-concave in profile as the dominance of rainsplash or soil creep in the upper part of the slope gives way at some point downslope to the prevalence of slope wash. Where active stream incision removes material from the slope base as quickly as it is supplied the basal concavity is replaced by a rectilinear slope standing at, or close to, the threshold angle of stability of the slope material.

7.6.2 Slope evolution
The manner in which slope form changes through time provided a central focus for geomorphic research until the 1950s. Various models of slope evolution were proposed, notably by W. M. Davis, W. Penck and L. C. King, which formed a framework within which the development of the landscape as a whole was considered. Davis presented a model of **slope decline** in which there is a progressive decrease in overall slope angles through time as the rate of basal downcutting by streams decreases and slopes become mantled with weathered material of ever finer calibre which can be transported across ever lower gradient slopes (Fig. 7.25(A)).

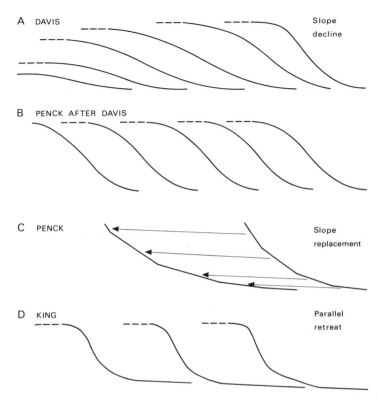

Fig. 7.25 *Classic models of slope development through time: (A) Davis's model of slope decline; (B) Davis's misinterpretation of Penck's model indicating parallel retreat; (C) Penck's model of slope replacement; (D) King's model of parallel retreat.*

Penck's model was originally presented in German, and it was introduced to the English-speaking world by Davis who seriously misrepresented Penck's ideas. Rather than advocating **parallel retreat** of the major part of a slope, as implied by Davis (Fig. 7.25(B)), Penck in fact considered that where rates of denudation are declining hillslopes evolve through a process of flattening from the base upwards (see Section 18.1.2). In effect each part of the slope profile is replaced by a slope of lower gradient as it retreats, and through this process of **slope replacement** a broadly concave profile is produced (Fig. 7.25 (C)).

In rejecting Davis's assertion that slope gradients decline through time, King pointed to the widespread occurrence of escarpments, notably in southern Africa, which have apparently experienced prolonged parallel retreat (Fig. 7.25 (D)). King argued that the free face retreats parallel to itself as material is weathered and then removed by rill erosion, and a low angle slope, or **pediment**, grows at its base (see Section 14.1). Such slope retreat eventually creates isolated, residual hills, known as **inselbergs**.

Although they have been extremely important in influencing previous views about landscape development (see Chapter 18) none of these classic models is based on detailed empirical observations. Moreover, they do not yield specific quantitative predictions as to how slope form may be expected to change over time. Studies of slope processes since the 1960s have clearly demonstrated that the evolution of slopes is, in most cases, likely to be far more complex than implied in these classic models, and that the mode of development will depend on structural and lithogical properties as well as the slope processes operating.

A rock slope will experience parallel retreat if the strength of the rock mass remains constant and basal debris is continuously removed, but lithological variations and climatic changes mean that over an extended period of time there will inevitably be some change in slope profile form. Parallel retreat is also likely to predominate in situations where a flat-lying resistant lithology overlies less resistant strata. Such situations are common where weakly resistant regolith or other types of unconsolidated deposits are overlain by duricrusts (Fig. 6.31), or where igneous intrusions, such as dolerite sills, cap less resistant sedimentary strata (Fig. 7.26). Once the resistant cap rock is finally removed by back-wearing, subsequent slope evolution may occur through slope decline.

Where rates of basal undercutting decline we might expect a reduction in slope gradient as material can be more intensively weathered and the angle of threshold slopes is reduced. This is suggested by the presence of distinct groupings of threshold slopes, the number and angle of which apparently depends on the weathering sequence of

Fig. 7.26 *An example of parallel slope retreat where a resistant dolerite sill caps less resistant Karoo sedimentary rocks, near Steynsburg, South Africa.*

the bedrock (Fig. 7.23). We can envisage slope development in a well-jointed rock in which there is an initial phase of scree formation, a second stage marked by the production of taluvium stable at a lower angle, and finally the development of a cover of colluvium with the lowest angle of threshold stability. Whether the associated changes in slope gradient occur by an overall decline in slope angle or by a replacement of individual slope elements by segments standing at a lower angle is probably dependent on the nature of the predominant slope processes and the way these interact with the slope materials. These factors are summarized in Figures 7.27, 7.28 and 7.29 which illustrate the probable course of profile development on different lithologies and under contrasting climatic conditions. The profiles show the likely change in form for slopes initially subject to active basal downcutting, but which subsequently develop in the absence of active basal erosion.

Further reading

Given the nodal position of slope studies in geomorphology it is not surprising that a large literature has been produced.

Since the 1970s much of the research generated on hillslopes has reflected the growing links being developed between engineers concerned primarily with questions of slope stability, and geomorphologists attempting to apply physical principles to slope behaviour in the natural landscape. These links are evident in the pioneering book by Carson and Kirkby (1972) which remains a valuable guide to the broad field of slope studies. Although it emphasizes physical principles it is firmly focused on the problem of how real slopes develop. The excellent book by Parsons (1988) provides a more up-to-date treatment of the factors controlling hillslope form. Finlayson and Statham (1980) also give a useful introduction to slope processes, while the book by Selby (1982a) is particularly strong on rock slopes. A clear appreciation of some of the major issues currently being addressed can be gained from the volume edited by Abrahams (1986).

The books by Statham (1977) and Williams (1982) provide an excellent background on the properties and behaviour of slope materials, and Selby (1980) describes the classification and application of rock mass strength to slope analysis. The papers included in the two books edited by Brunsden and Prior (1984) and Anderson and Richards

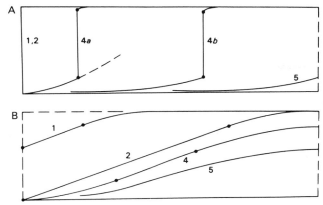

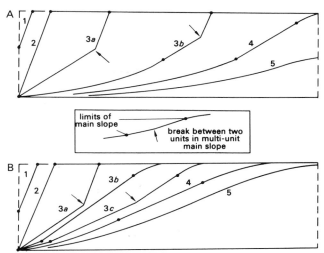

Fig. 7.27 *Slope profile development in massive sandstone. (A) Semi-arid environment: (1) Slope development during initial phase of active stream incision. The initial slope is assumed to be vertical because slope processes act very slowly relative to the rate of stream downcutting. Slope processes predominate thereafter. There is no immediate replacement of the initial slope by another unit because loose sand grains which accumulate at the cliff base are easily removed by slope wash. (4) Modification of the stable main slope predominantly by retreat due to active slope wash of loose material that is only poorly, if at all, bound together by vegetation. Shortening of the main slope is due mostly to rainsplash. (5) Final elimination of rock wall leaves extensive but gently inclined concave slope or pediment. (B) Humid temperate environment: (1) As for (A) but slope gradient is less (20–30°) and the summit convexity is more fully developed. Lower main slope angle assumes rapid weathering of the bedrock under humid conditions and loss of cohesion of the surface soil cover. Development of convexity during stream incision is due to soil creep and rainsplash. (2) There is no immediate replacement of the initial slope by another unit (3) because the sandy soil-mantled slope is assumed to be already at the angle of ultimate stability, since once the rock has weathered to a sandy mantle little further weathering and therefore change in shear strength is likely to occur. (4) Modification of main slope mainly through shortening as a result of both soil creep and rainsplash. Slope wash is much less effective than in semi-arid regions because of the abundance of vegetation. Retreat is consequently much less, although solution may be important in some instances. (5) Further shortening through downslope expansion of summit convexity produces a predominantly convex slope. (From M. A. Carson and M. J. Kirkby, (1972) Hillslope From and Process. Cambridge University Press, Cambridge, Fig. 15.7, p. 380, and caption based on same source, pp. 379–82.)*

Fig. 7.28 *Slope profile development in strong, closely jointed rock. (A) Semi-arid environment: (1) Slope development during the initial phase of active stream incision. Initial slopes are assumed to be steep (45–75°) but not vertical since weakening of the jointed rock mass accompanies stream downcutting and the rock mass effectively acts as a densely packed cohesionless system. (2) Slope at instant when active stream incision ceases. Slope processes dominate thereafter. (3) The rock wall is transformed into a talus-mantled slope. It is assumed that the removal of fine debris from the talus slope by surface water prevents replacement by a lower angle main slope. Consequently, the rock wall–talus slope retreats as one slope leaving behind a concave basal slope or pediment. (4) Main slope progressively consumed by upslope extension of basal concavity predominantly through slope wash. (5) Further extension of basal concavity obliterates main slope. (B) Humid temperate environment: (1) As for (A). (2) As for (A). (3) Three phases of instability related to the weathering sequence (bedrock–talus–taluvium) are indicated, although there may be more. The rock wall (3a) is replaced by a talus-mantled slope (3b) and this, in turn, is replaced by the taluvial slope (3c). (4) Conversion of the taluvium to finer colluvium could result in further replacement or decline (3c–4) before the ultimate angle of stability of the regolith is reached. (5) Upslope extension of basal concavity and downslope extension of summit convexity obliterates the main slope and produces a convexo-concave profile. (From M. A. Carson and M. J. Kirkby (1972) Hillslope form and Process. Cambridge University Press, Cambridge, Fig. 15.8, p. 380, and caption based on same source, pp. 379–82.)*

(1987) provide a very broad and detailed coverage of slope stability analysis, while Selby (1982a) and Crozier (1986) present useful introductions to this topic.

The terminology of mass movement is a quagmire of confusion with different researchers using different terms for the same phenomenon. Various classifications have been proposed including those by Sharpe (1938) and Skempton and Hutchinson (1969), but the categorization used in this chapter is based on that of Varnes (1978). Innes (1983) provides a detailed review of debris flows, while Hsü (1975) looks at catastrophic debris avalanches and

Browning (1973) assesses the 1970 Huascaran avalanche. Slides are discussed by Crozier (1986), and Watson and Wright (1969) describe the Saidmarreh landslide. Soil creep is considered by Kirkby (1967) and in the classic paper by Gilbert (1909).

Hillslope hydrology is comprehensively covered in Kirkby (1978) and reviewed concisely by Kirkby (1985a, 1988) while the movement of solutes on slopes is discussed by Burt (1986). Morgan (1986) presents an accessible treatment of soil erosion on slopes and Dunne and Aubry (1986) provide a detailed analysis of slope wash and rain-splash erosion. Data on slope process rates are presented by

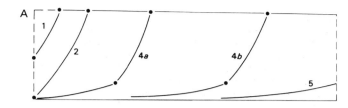

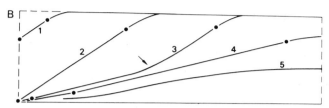

Fig. 7.29 *Slope profile development in a clay mass. (A) Semi-arid environment: (1) Slope development during initial phase of active stream incision. The slope is steep as a result of true cohesion together with capillary cohesion, and slightly concave due to rill and gully erosion which accompanies sliding during stream incision. (2) Slope at instant when active stream incision ceases. Slope processes dominate thereafter. There is no immediate replacement of the initial slope by another unit as there is little change in shear strength or cohesion at this stage. (4) Continuing retreat of main slope with little change in shear strength or cohesion. (5) Main slope is eventually eliminated to leave a concave profile. (B) Humid temperate environment: (1) As for (A) except that slope gradient is less (20–30°) and summit convexity is more fully developed. The lower-angle main slope reflects the more rapid weathering of the clay under humid conditions and loss of cohesion. Slope convexity develops through the action of soil creep and rainsplash. (2) As for (A). (3) One phase of slope replacement or decline may occur associated with a change from a temporary to an ultimate angle of stability. (4) Continued reduction in angle of the main slope. (5) Downslope extension of summit convexity produces a predominantly convex profile. (From M. A. Carson and M. J. Kirkby (1972)* Hillslope Form and Process. *Cambridge University Press, Cambridge, Fig. 15.9, p. 381, and caption based on same source, pp. 379–82.)*

Saunders and Young (1983) and Young and Saunders (1986) while Leopold *et al.* (1966) and Rapp (1960) provide examples of the detailed field monitoring of slope processes, and Reneau *et al.* (1989) employ dated colluvial deposits.

Turning to the slope system as a whole, the distinction between weathering-limited and transport-limited movement is clearly drawn by Carson and Kirkby (1972), and reassessed by Parsons (1988). Young (1972) provides a useful survey of slope profile forms, while Dalrymple *et al.* (1968) present a nine-unit landsurface model, and Parsons (1979) demonstrates the importance of considering slopes in three-dimensions as well as in profile. The development of rock slopes is considered by Moon (1986) and Selby (1980, 1982b, 1987), while Carson and Petley (1970) apply the concept of threshold stability angles to slope development, an approach examined further by Anderson *et al.* (1980) and Francis (1987). Examples of the mathematical

modelling of slope form are contained in Ahnert (1976, 1987), Armstrong (1987) and Kirkby (1985b), while Douglas (1988) discusses the problems of applying such models to slope development in real landscapes.

References

Abrahams, A. D. (ed.) (1986) *Hillslope Processes.* Allen and Unwin, Boston, and London.

Ahnert, F. (1976) A brief description of a comprehensive three-dimensional process–response model of landform development. *Zeitschrift für Geomorphologie Supplementband* **25**, 29–49.

Ahnert, F. (1987) Process–response models of denudation at different spatial scales. *Catena Supplement* **10**, 31–50.

Anderson, M. G. and Richards, K. S. (eds) (1987) *Slope Stability: Geotechnical Engineering and Geomorphology.* Wiley, Chichester and New York.

Anderson, M. G., Richards, K. S. and Kneale, P. E. 1980. The role of stability analysis in the interpretation of the evolution of threshold slopes. *Transactions of the Institute of British Geographers* NS **5**, 100–12.

Armstrong, A. C. (1987) Slopes, boundary conditions, and the development of convexo-concave forms – some numerical experiments. *Earth Surface Processes and Landforms* **12**, 17–30.

Browning, J. M. (1973) Catastrophic rock slide, Mount Huascaran, north-central Peru, May 31, 1970. *American Association of Petroleum Geologists Bulletin* **57**, 1335–41.

Brunsden, D. and Prior, D. B. (eds) (1984) *Slope Instability.* Wiley, Chichester and New York.

Burt, T. P. (1986) Runoff processes and solutional denudation rates on humid temperate hillslopes. In: S. T. Trudgill (ed.) *Solute Processes.* Wiley, Chichester and New York, 193–249.

Carson, M. A. and Kirkby, M. J. (1972) *Hillslope Form and Process.* Cambridge University Press, Cambridge.

Carson, M. A. and Petley, D. J. (1970) The existence of threshold hillslopes in the denudation of the landscape. *Transactions of the Institute of British Geographers* **49**, 71–95.

Crozier, M. J. (1986) *Landslides: Causes, Consequences and Environment.* Croom Helm, London and Dover.

Dalrymple, J. B., Blong, R. J. and Conacher, A. J. (1968) A hypothetical nine-unit landsurface model. *Zeitschrift für Geomorphologie* **12**, 60–76.

Douglas, I. (1988) Restrictions on hillslope modelling. In: M. G. Anderson (ed.) *Modelling Geomorphological Systems.* Wiley, Chichester and New York, 401–20.

Dunne, T. and Aubry, B. F. (1986) Evaluation of Horton's theory of sheetwash and rill erosion on the basis of field experiments. In: A. D. Abrahams (ed.) *Hillslope Processes.* Allen and Unwin, Boston, and London, 31–53.

Finlayson, B. and Statham, I. (1980) *Hillslope Analysis.* Butterworths, London.

Francis, S. C. (1987) Slope development through the threshold concept. In: M. G. Anderson and K. S. Richards (eds) *Slope Stability: Geotechnical Engineering and Geomorphology.* Wiley, Chichester and New York, 601–24.

Gilbert, G. K. (1909) The convexity of hilltops. *Journal of Geology* **17**, 344–50.

Hsü, K. J. (1975) Catastrophic debris streams (sturzstroms) generated by rock falls. *Geological Society of America Bulletin,* **86**, 129–40.

Innes, J. L. (1983) Debris flows. *Progress in Physical Geography* **7**, 469–501.

Kirkby, M. J. (1967) Measurement and theory of soil creep. *Journal of Geology* **75**, 359–78.

Kirkby, M. J. (ed.) (1978) *Hillslope Hydrology*. Wiley, Chichester and New York.

Kirkby, M. J. (1985a) Hillslope hydrology. In: M. G. Anderson and T. P. Burt (eds) *Hydrological Forecasting*. Wiley, Chichester and New York, 37–75.

Kirkby, M. J. (1985b) A model for the evolution of regolith mantled slopes. In: M. J. Woldenberg (ed.) *Models in Geomorphology*. Allen and Unwin, Boston, and Chichester, 213–37.

Kirkby, M. (1988) Hillslope runoff processes and models. *Journal of Hydrology* **100**, 315–39.

Leopold, L. B., Emmett, W. W. and Myrick, R. M. (1966) Channel and hillslope processes in a semi-arid area, New Mexico. *United States Geological Survey Professional Paper* **352G**, 153–253.

Moon, B. P. (1986) Controls on the form and development of rock slopes in fold terrane. In: A. D. Abrahams (ed.) *Hillslope Processes*. Allen and Unwin, Boston and New York, 225–43.

Morgan, R. P. C. (1986) *Soil Erosion and Conservation*. Longman, London and Wiley, New York.

Parsons, A. J. (1979) Plan form and slope profile form of hillslopes. *Earth Surface Processes and Landforms* **4**, 395–402.

Parsons, A. J. (1988) *Hillslope Form*. Routledge, London and New York.

Rapp, A. (1960) Recent development of mountain slopes in Kärkevagge and surroundings, northern Scandinavia, *Geografiska Annaler* **42**, 71–200.

Reneau, S. L., Dietrich, W. E., Rubin, M., Donahue, D. J. and Jull, A. J. T. (1989) Analysis of hillslope erosion rates using dated colluvial deposits. *Journal of Geology* **97**, 45–63.

Saunders, I. and Young, A. (1983) Rate of surface processes on slopes, slope retreat and denudation. *Earth Surface Processes and Landforms* **8**, 473–501.

Selby, M. J. (1980) A rock mass strength classification for geomorphic purposes: with tests from Antarctica and New Zealand. *Zeitschrift für Geomorphologie* **24**, 31–51.

Selby, M. J. (1982a) *Hillslope Materials and Processes*. Oxford University Press, Oxford and New York.

Selby, M. J. (1982b) Rock mass strength and the form of some inselbergs in the central Namib Desert. *Earth Surface Processes and Landforms* **7**, 489–97.

Selby, M. J. (1987) Rock slopes. In: M. G. Anderson and K. S. Richards (eds) *Slope Stability: Geotechnical Engineering and Geomorphology*. Wiley, Chichester and New York, 475–504.

Sharpe, C. F. S. (1938) *Landslides and Related Phenomena*. Columbia University Press, New York.

Skempton, A. W. and Hutchinson, J. N. (1969) Stability of natural slopes and embankment foundations. *State-of-the-Art Report: Proceedings of the 7th International Conference on Soil Mechanics and Foundation Engineering, Mexico*, 291–355.

Statham, I. (1977) *Earth Surface Sediment Transport*. Clarendon Press, Oxford and New York.

Varnes, D. J. (1978) Slope movement and types and processes. In: R. L. Schuster and R. J. Krizek (eds) *Landslides: Analysis and Control*. Transportation Research Board Special Report **176**. National Academy of Sciences, Washington DC, 11–33.

Watson, R. A. and Wright, H. E. Jr (1969) The Saidmarreh landslide, Iran. *Geological Society of America Special Paper* **123**, 115–39.

Williams, P. J. (1982) *The Surface of the Earth: An Introduction to Geotechnical Science*. Longman, London and New York.

Young, A. (1972) *Slopes*. Longman, London and New York.

Young, A. and Saunders, I. (1986) Rates of surface processes and denudation. In: A. D. Abrahams (ed.) *Hillslope Processes*. Allen and Unwin, Boston and New York, 3–27.

8

Fluvial processes

8.1 Drainage basin hydrology

In Chapter 7 we looked briefly at the movement of water on and within slopes, but it is now necessary to extend our discussion of water movements to the scale of entire drainage basins. All water in rivers ultimately originates as precipitation, although there can be a considerable lag before this water enters the fluvial system. The main components of drainage basin hydrology are illustrated in Figure 8.1, and are represented schematically in terms of a system of inputs, outputs, storages and transfers in Figure 8.2.

Runoff, or more strictly **basin channel runoff**, is the quantity of water which enters stream channels in a drainage basin over a specified period of time, and can be determined by a **water-balance equation**. This expresses runoff in terms of precipitation, losses through evapotranspiration and changes in the amount of soil moisture and ground water storage. In environments where much of the precipitation falls as snow the water-balance equation is complicated by having to take into account water released during melting. A further output which has to be considered, in addition to runoff and evapotranspiration, is **deep out-**

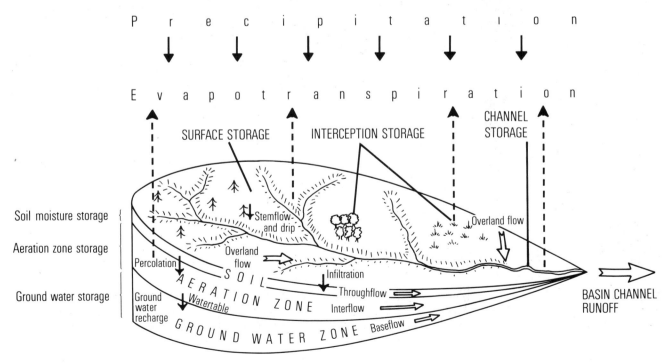

Fig. 8.1 *Primary storages and transfers of water within drainage basins.*

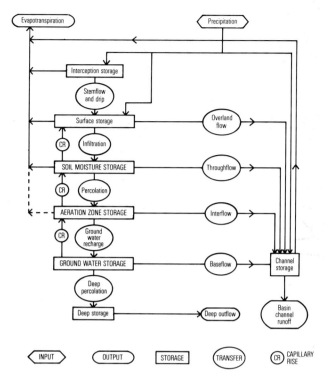

Fig. 8.2 *Major elements of the drainage basin hydrological cycle.*

flow from ground water. In most cases this happens so slowly that it can be ignored in calculating the water balance for a basin. In limestone terrains, however, movements of water at depth may have little relation to surface patterns of water flow and a complex hydrological system can result.

8.1.1   Channel discharge

Water discharge in river channels varies over both time and space at a variety of scales. Changes in discharge through time are represented by a **hydrograph**. In a **discharge hydrograph** changes in discharge are plotted against time, but the direct measurement of stream discharge is time-consuming and more often discharge is estimated from a **stage hydrograph**. This shows variation in **stage**, or elevation of the water surface, through time. These changes in stage can be converted to a record of discharge variations using a **stage-discharge rating curve** which plots changes in discharge against changes in water depth. The curve is initially calibrated by taking discharge measurements in a well-defined channel cross-section, often a specially constructed weir. Thereafter a **stage recorder** can be used to monitor subsequent variations in discharge.

Although discharge is usually expressed in units of $m^3 s^{-1}$, represented in this way it is a function of drainage basin area. If we wish to relate discharge to the rate of operation of geomorphic processes, such as the transport of sediments or solutes, over a basin as a whole, it may be appropriate to

represent discharge as an equivalent mean depth of water over the basin. A common index of this kind is obtained by dividing the mean annual discharge by the drainage basin area. This mean annual depth-equivalent measure of discharge ranges from around 1000 mm for the Amazon River to 177 mm for the Mississippi and 31 mm for the Colorado. In terms of absolute discharge the Amazon has by far the highest mean runoff in the world estimated at 230 000 $m^3 s^{-1}$,, not a surprising figure when we consider that its channel at Obidos, 700 km upstream from its mouth, is nearly 2.5 km across and attains a depth of 60 m. The Amazon is followed by the Zaire River with a mean discharge of around 40 000 $m^3 s^{-1}$. These totals can be compared with the mean discharge from the Mississippi Basin which is a mere 18 000 $m^3 s^{-1}$; even the greatest ever recorded flood of the Mississippi achieved a discharge of only 57 000 $m^3 s^{-1}$.

8.1.1.1   Spatial variations

In the great majority of river systems discharge increases downstream as tributaries progressively add more runoff to the trunk channel. In arid regions, however, losses through evaporation and seepage can lead to a downstream decrease in discharge. At the global scale river discharge is closely related to the balance between precipitation and evapotranspiration, and the highest discharges are achieved under humid climatic regimes (Fig. 8.3). Since precipitation also tends to increase with elevation, the greatest depth-equivalent discharges are found in the mountainous regions of south-east Asia and along the mid-latitude mountain ranges, such as those of Chile and Scandinavia, which intercept rain-bearing weather systems.

8.1.1.2   Temporal variations

In environments with a clear seasonal variation in precipitation or temperature, stream flow will tend to vary fairly systematically throughout the year, being highest during the wet season or during the period when lower temperatures reduce water losses through evapotranspiration. The term **river regime** is applied to these average annual variations in discharge. River systems in humid wet-dry and monsoonal environments experience dramatic annual fluctuations in discharge as do basins experiencing heavy winter snowfalls. But not all channel systems maintain a constant flow of water. In **intermittent stream channels** flow occurs for at least one month per year in response to seasonally generated runoff. In arid environments many channels carry only occasional flow after storms and are consequently described as **ephemeral stream channels**.

Much attention in fluvial geomorphology has been focused on the factors that determine variations in the discharge of streams draining comparatively small drainage basins. It is only over relatively small areas that it is realistic to monitor changes in soil moisture and groundwater storage, and to track the movement of water from

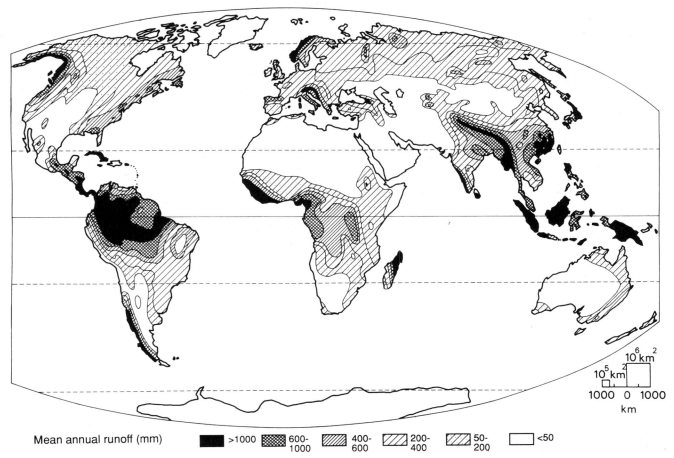

Fig. 8.3 *Global variation in mean annual runoff expressed as depth-equivalent discharge. (Based on Unesco (1978)* Atlas of the World Water Balance, *Unesco, Paris.)*

where precipitation arrives at the surface to the point at which it leaves the drainage basin. Since it is the peak flows resulting from storms which are often of most significance in terms of the geomorphic work that they accomplish, particular attention has been paid to how run-off is generated by periods of heavy rainfall.

Hydrographs for small basins in humid environments typically show a relatively constant discharge, termed **base flow**, punctuated by flood events which are represented by a rapidly rising limb, a short-lived peak and a slowly declining falling limb (Fig. 8.4). The precise form of such a **storm hydrograph** depends on the intensity, duration and areal extent of precipitation, and on the hydrological properties of the drainage basin.

The mean frequency, or recurrence interval, of floods of a specified magnitude can be estimated if there is a record of stream discharge extending over many years, preferably several decades. Recurrence intervals of floods can be expressed in two ways. The usual procedure is to rank the highest discharge attained in each year to produce an **annual series**. The recurrence interval of a flood which is

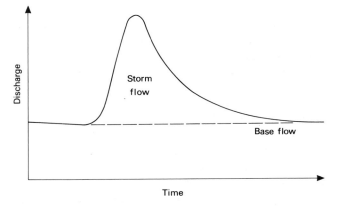

Fig. 8.4 *Typical form of a storm hydrograph.*

equal or greater in magnitude than one of a particular size in the annual series is then given as $(n + 1)/r$, where n is the number of years of record, and r is the rank order of the flood of the specified magnitude. For instance, if we have a 50 a annual series the recurrence interval of a flood of the same or greater magnitude than the tenth largest annual flood in the series is $(50 + 1)/10 = 5.1$ a.

The alternative means of expressing the recurrence interval of a flood is through a **partial duration series**. This is constructed by listing and ranking all discharges which exceed a given magnitude over a specified period of time. The recurrence interval for a partial duration series is calculated using the same formula as for the annual series except that n represents the total number of flood events exceeding the specified magnitude. In a partial duration series there may be more than one flood event recorded in some years, and none in others. A recurrence interval calculated using this approach gives the mean interval of time between discharges of a size which equal, or exceed, the specified flood magnitude.

A flood which is at least equal in magnitude to the mean flood in an annual series is termed the **mean annual flood**. This has a recurrence interval of about 2.33 a; that is, the maximum discharge reached each year will equal, or exceed, the mean annual flood on average once every 2.33 a. The **most probable annual flood** in an annual series has a recurrence interval of about 1.58 a, this being equivalent in magnitude to the flood in the partial duration series which has a recurrence interval of 1 a. In general, in alluvial channels the most probable annual flood seems to be roughly equivalent in magnitude to **bankfull discharge**, that is, the discharge at which the channel is filled to capacity. Floods of this magnitude appear to be of special significance because, in many cases, they show a close relationship with certain channel form variables and are therefore presumably important in shaping channel morphology.

Rare, 'instantaneous' increases in channel discharge can occur as a result of the catastrophic failure of natural dams. Such dams can be formed by glacial ice or moraines (see Chapter 11), but may also be created by landslides. Many landslide dams fail soon after they are formed, usually by overtopping. On occasion, however, massive landslides can dam major rivers and lead to the formation of large lakes. In 1911 an earthquake in the USSR triggered a $2.0–2.5 \times 10^9$ m^3 rock avalanche which dammed the Murgab River. This natural dam is 550 m high, nearly twice the height of the world's largest artificial dam. The failure of such dams can, of course, create enormous, short-term discharges.

8.1.2 Runoff generation

Water contributing to stream flow can be divided into two types in terms of the rapidity with which it enters the stream channel after a rainfall event. In most streams there is a relatively low level of base flow which is maintained between major rainfall events and this is supplied by **delayed flow** from the ground water or from slow per-

colation of water through the soil. By contrast, water that enters stream channels soon after a storm is termed **quick flow** and is generated in different ways depending on the nature of the topography, vegetation and soils in a basin.

In drainage basins lacking a significant vegetation cover, such as in semi-arid and arid regions, or where vegetation has been cleared and the surface compacted, soil infiltration capacities may be sufficiently low to be exceeded by rainfall intensities during storms. In such circumstances infiltration-excess overland flow will be generated once the storage capacity of the surface has been exceeded; this is known as the **partial area model** of stream flow generation. Large volumes of water can be transmitted to river channels by this means because of the high velocities achieved by overland flow – up to 500 m h^{-1} – and its simultaneous occurrence over large parts of an individual basin.

In well-vegetated basins soils characteristically have infiltration capacities far in excess of likely rainfall intensities, yet high peak discharges are still observed during storms. In such basins peak flows are generated from rain falling directly on stream channels and from saturation overland flow contributed from zones adjacent to channels where the soil has become completely saturated. Areas contributing saturation overland flow change during a storm as the saturated zones initially expand with the onset of rain, and then contract once the rain stops. This is known as the **variable source area model** of stream-flow generation. An important distinction between infiltration-excess and saturation overland flow is that whereas the former is influenced by soil infiltration capacity the latter is related to antecedent soil-moisture conditions and location within a basin.

A second important route for quick flow in humid environments where thick soils are present is the lateral movement of water through the soil itself. This lateral movement is most pronounced immediately above soil horizons with a lower **hydraulic conductivity**, that is, a lower permeability or capacity to transmit water. Under such conditions a higher proportion of the water percolating through the soil will be deflected laterally downslope. Such subsurface flow, or throughflow, is generally much slower than overland flow, and usually only reaches a maximum velocity of 0.4 m h^{-1}. In some soils, however, much more rapid subsurface flow, at velocities of up to 200 m h^{-1}, can occur through natural **pipes** formed by the rotting of roots or the burrowing of animals. Such pipes can develop to form well-integrated conduits enabling the rapid downslope transmission of water. Where there is a downslope thinning of permeable soil horizons subsurface flow may emerge at the surface as **return flow** towards the slope base. Although most subsurface flow is too slow to contribute to the main peak flows accompanying storm events, it is important in priming areas adjacent to the stream channel and thereby enhancing the area of saturation overland flow in subsequent storms.

Various properties of the drainage system and morphological characteristics of the basin also influence the runoff response to storms. These include the density of channels, and the relief and shape of the basin (see Section 9.2). Such factors affect the arrival time of runoff from tributaries at the main channel.

8.1.3   Channel initiation

An obvious question to ask about stream channels is how do they originate? Channels may be created on a newly exposed surface or develop through the expansion of an existing channel network, but in order to understand how they are initiated we must look at the conditions under which water flowing on a slope becomes sufficiently concentrated for channel incision to occur. It is also necessary to establish how, once they are established, channels are maintained and enlarged to form 'permanent' features in the landscape.

In Chapter 7 we saw how both surface and subsurface flows converge in areas of contour concavity, and such convergence is an important factor in channel development. Moreover, it was also pointed out how infiltration-excess overland flow can lead to the development of rills, although the precise mechanism that brings this about is far from clear. Microtopography on slopes tends to disrupt sheet flow and promote the concentration of water movement and rill formation. But such rill development can be counteracted by the lateral shifting of flow lines, or by rainsplash erosion which tends to even out the surface.

In the model put forward by R. E. Horton, before erosion by overland flow can occur on a hillslope it must reach a critical depth at which the eroding stress of the flow exceeds the shear resistance of the soil surface (Fig. 8.5). Horton therefore thought that a 'belt of no erosion' is present on the upper part of slopes because here the flow depth is insufficient to cause erosion. Subsequent work has shown that some surface wash is possible even on slope crests, although here it does not lead to rill development because the rate of incision is slow and incipient rills are infilled as a result of rainsplash.

Although the infiltration-excess overland flow model provides a reasonable framework for understanding channel initiation in semi-arid and arid environments, in humid regions we have to look for an alternative mechanism. In such areas channel initiation is less related to a critical distance of overland flow than to the location of surface and subsurface flow convergence, usually in slope concavities and adjacent to existing drainage lines. In one case rills have been seen to develop as a result of the sudden outburst of subsurface flow at the surface close to the base of a slope. In humid environments channel development is

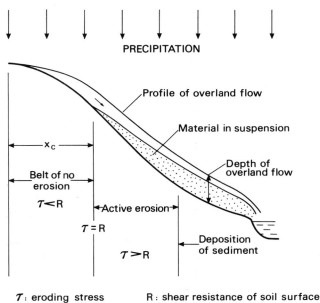

PRECIPITATION

T: eroding stress   R: shear resistance of soil surface

x_c: critical distance of overland flow

Fig. 8.5   *Horton's model of surface erosion by infiltration-excess overland flow. (After R. E. Horton (1945) Bulletin of the Geological Society of America **56** Fig. 14, p. 316.)*

likely to be especially favoured where subsurface pipes are present. Pipe networks appear capable of promoting channel development either through roof collapse or by the concentration of runoff and erosion downslope of pipe outlets. Although infiltration-excess overland flow is usually regarded as dominant in semi-arid regions, piping is also frequently important in such environments.

Channels may also be initiated through **spring sapping** where ground-water flow is concentrated in more permeable zones within the bedrock. This encourages chemical weathering or dissolution of the bedrock which, in turn, leads to an increase in hydraulic conductivity and a further increase in the rate of water movement. This positive feedback mechanism can lead to channel development by headward erosion as the water flow emerges at the surface along a zone of enhanced permeability. Spring sapping is likely to be especially active where a permeable lithology, such as sandstone or limestone, overlies an impermeable lithology, such as clay.

8.2   Open channel flow

Two opposing forces act on water flowing in an open channel. The driving force is gravity which acts in a downslope direction and is determined by gravitational acceleration and channel gradient. The resisting force arises from friction both within the water body and between the flowing water and the channel surface. The ability of flowing water to entrain and transport material, and hence its capacity to do geomorphic work, is essentially determined by the relationship between these two forces. However, before looking at the processes operating in natural river channels we need to consider the fundamental characteristics of the behaviour of flowing water.

8.2.1   Resistance to flow

Water is a fluid – that is, its shape is changed continuously by the smallest applied external stress. This change in shape is sustained for as long as the force is applied. Resistance of a fluid to a change in shape is represented by its **viscosity**. One type of resistance to deformation provided by viscosity arises from internal friction caused by cohesion and collisions between molecules as they move past each other, and consequently this type of viscosity is termed **molecular viscosity** or **dynamic viscosity**. Around 97 per cent of the energy of rivers is expended as frictional heat generated by molecular impacts, leaving only about 3 per cent for the transport of sediment. As with all liquids the molecular viscosity of water increases with a decrease in temperature. Moreover, in the natural environment we are concerned not only with pure water but also with water containing fine sediment or dissolved constituents which has a greater dynamic viscosity.

8.2.2   Laminar and turbulent flow

A fluid moving over a flat solid surface can act as a series of thin 'layers' sliding over one another, the resistance to movement resulting from molecular viscosity. This form of motion is described as **laminar flow** and while it is relatively common in highly viscous fluids, such as lava flows, it is extremely rare for water moving in natural channels (Fig. 8.6). In stream channels water movement nearly always occurs as **turbulent flow**, that is, the velocity of flow fluctuates in all directions within the fluid. Water is constantly interchanged in eddies between adjacent zones of flow, and local changes in velocity occur which work against the mean velocity gradient and lead to a loss of energy. The resulting additional resistance to shear is termed **eddy viscosity**. In most channels there is a thin **laminar sublayer** within which the velocity of flow initially increases in a roughly linear fashion. Above this zone the rate of increase in flow velocity is approximately logarithmic (Fig. 8.6).

Whether flow is laminar or turbulent is determined by the mean flow velocity, the molecular viscosity and density of the fluid and the dimensions of the flow section. For stream channels the dimensions of the flow section used are either the depth of flow or the **hydraulic radius**. The hydraulic radius of a stream channel is defined as the cross-sectional area of the flow in a channel divided by the

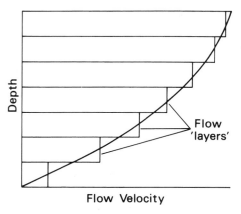

LAMINAR FLOW

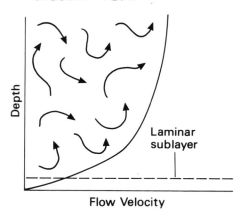

TURBULENT FLOW

Fig. 8.6   *Schematic representation of velocity profiles for laminar and turbulent flow in a river channel.*

wetted perimeter, the latter being the length of the boundary along which water is in contact with the channel (Fig. 8.7). In broad, shallow channels the hydraulic radius is closely approximated by the flow depth.

Conditions under which laminar or turbulent flow occur are defined by a **Reynolds number** (*Re*) which is essentially a dimensionless measure of flow rate. A Reynolds number is calculated by multiplying the mean flow velocity and hydraulic radius, and dividing by the **kinematic viscosity** which represents the ratio between molecular viscosity and fluid density (Box 8.1). In stream channels the maximum Reynolds number at which laminar flow is sustained is around 500. Above a value of around 2000 flow is turbulent, and at transitional values of between 500 and 2000 elements of both laminar and turbulent flow are present.

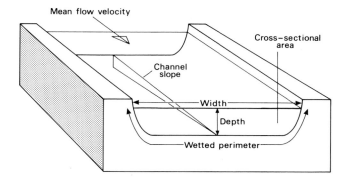

Discharge = Cross–sectional area x Mean flow velocity
Hydraulic radius = Cross-sectional area/wetted perimeter

Fig. 8.7 *Properties of a stream channel relevant to flow characteristics.*

Box 8.1 Reynolds number

The Reynolds number (*Re*) is defined as

$$Re = \frac{vR}{v}$$

where *v* is the velocity of flow, *R* the hydraulic radius and *v* the kinematic viscosity.

8.2.3 Flow regimes

In natural channels local variations in the depth of flow caused by irregularities on the channel bed create waves which exert a weight or gravity force. The ratio of the mean flow velocity to the velocity of these gravity waves, or ripples, defines the **Froude number** (*F*) of the flow and this index can be used to distinguish different flow states (Box 8.2). When the Froude number is less than 1 the wave velocity is greater than the mean flow velocity and the flow is described as **subcritical** or tranquil. Under such flow conditions ripples propagated by a pebble dropped into a stream can travel upstream. When the Froude number is 1 the flow is termed **critical** and when it is greater than 1 it is described as **supercritical** or rapid.

The reason for the existence of these different types of flow is that changes in discharge can be accomplished by

Box 8.2 Froude number

The Froude number (*F*) is defined as

$$F = \frac{v}{\sqrt{gd}}$$

where *v* is the flow velocity, *g* the acceleration of gravity and *d* the depth of flow.

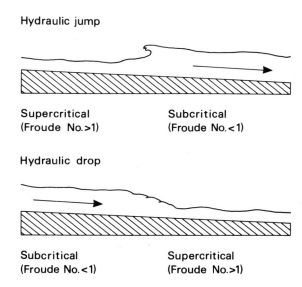

Fig. 8.8 *Transitions between supercritical and subcritical flow resulting in a hydraulic jump or hydraulic drop.*

changes in both the depth and velocity of flow; in other words a given discharge can be transmitted along a stream channel either as a deep, slow-moving, subcritical flow or as a shallow, rapid, supercritical flow. Transitions between subcritical and supercritical flow are determined by the velocity of flow. A sudden transition from supercritical to subcritical flow is called a **hydraulic jump** and it gives rise to a stationary wave and an increase in water depth (Fig. 8.8). When a flow changes from subcritical to supercritical there is a decrease in the water depth, a situation described as **hydraulic drop**. Sudden transitions in flow regime can occur where there is an abrupt change in channel bed form. This is a common feature in mountain streams where there are large boulders or bedrock obstructions in the channel.

In natural channels mean Froude numbers rarely exceed 0.5 and supercritical flows are only temporary since the large energy losses associated with this kind of flow promote bank erosion and channel enlargement. This erosion in turn leads to a lowering of flow velocity and a consequential reduction in the Froude number of the flow (an excellent example of negative feedback). Froude and Reynolds numbers can be combined to specify four distinct flow regimes (Fig. 8.9). As we will see in Section 8.2.4 the Froude number associated with a particular flow regime has important consequences for bedforms in sandy, alluvial channels.

8.2.4 Velocity of flow

The velocity of stream flow is influenced by the gradient, roughness and cross-sectional form of the channel. Natural

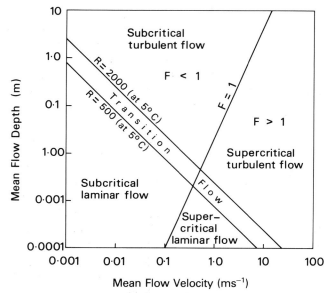

Fig. 8.9 *Boundary values for Reynolds (Re) and Froude (F) numbers specifying distinct flow regimes related to mean flow velocity and depth of flow. (Modified from J. R. L. Allen (1970) Physical Processes of Sedimentation. Allen and Unwin, London, Fig. 4.3, p. 126.)*

channels almost invariably have rough surfaces which induce significant frictional energy losses and cause a reduction in flow velocity especially close to the channel boundary. The direct measurement of stream flow velocity is time-consuming and consequently empirical equations have been developed in order to estimate mean flow velocities (Box 8.3). In the **Chezy equation** velocity is estimated in terms of the hydraulic radius and gradient of the stream channel, and a coefficient expressing the gravitational and frictional forces acting on the water. The **Manning equation** is a more widely applied estimator

Box 8.3   Estimates of flow velocity

The Chezy equation defines the mean flow velocity ($\bar{v}$) as

$$\bar{v} = C \sqrt{Rs}$$

where R is the hydraulic radius, s the channel gradient and C the Chezy coefficient which represents gravitational and frictional forces.

The Manning equation defines the mean flow velocity ($\bar{v}$) as

$$\bar{v} = \frac{R^{2/3} s^{1/2}}{n}$$

where R is the hydraulic radius, s the channel gradient and n the Manning roughness coefficient.

which incorporates an index of channel bed roughness. This **Manning roughness coefficient** (n) is usually estimated from tables (see Table 8.1), or by comparison with photographs illustrating channels of known roughness.

The use of the Manning equation is complicated by the fact that resistance to flow varies with discharge and flow depth. The Manning roughness coefficient decreases as flow depth increases up to bankfull discharge. Once channel capacity is exceeded the flow spreads over a much larger area and the Manning roughness coefficient increases. A further complication arises from the adjustments in bed form, and hence channel roughness, that accompany changes in flow regime in alluvial channels (Fig. 8.10).

8.3   Fluvial erosion and sediment entrainment

8.3.1   Erosion of bedrock channels

The reduction of the landscape through the action of fluvial processes can involve the incision of stream channels into

Table 8.1 Values of Manning's roughness coefficient for various types of natural channel

CHANNEL TYPE	NORMAL VALUE	RANGE
Small channels (width <30 m)		
Low-gradient streams		
Unvegetated straight channels at bankfull stage	0.030	0.025–0.033
Unvegetated winding channels with some pools and shallows	0.040	0.033–0.045
Winding vegetated channels with stones on bed	0.050	0.045–0.060
Sluggish vegetated channels with deep pools	0.070	0.050–0.080
Heavily vegetated channels with deep pools	0.100	0.075–0.150
Mountain streams (with steep unvegetated banks)		
Few boulders on channel bed	0.040	0.030–0.050
Abundant cobbles and large boulders on channel bed	0.050	0.040–0.070
Large channels (width >30 m)		
Regular channel lacking boulders or vegetation	—	0.025–0.060
Irregular channel	—	0.035–0.100

Source: Based on data in V. T. Chow (ed.) (1964) *Handbook of Applied Hydrology*. McGraw-Hill, New York.

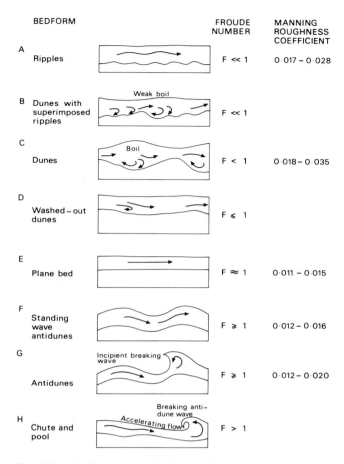

BEDFORM		FROUDE NUMBER	MANNING ROUGHNESS COEFFICIENT
A	Ripples	$F \ll 1$	0·017 – 0·028
B	Dunes with superimposed ripples	$F \ll 1$	
C	Dunes	$F < 1$	0·018 – 0·035
D	Washed – out dunes	$F \leqslant 1$	
E	Plane bed	$F \approx 1$	0·011 – 0·015
F	Standing wave antidunes	$F \geqslant 1$	0·012 – 0·016
G	Antidunes	$F \geqslant 1$	0·012 – 0·020
H	Chute and pool	$F > 1$	

*Fig. 8.10 Bedforms in sandy alluvial channels in relation to flow regimes expressed by Froude numbers. At low flow velocities ripples are formed (A), but as the flow velocity increases ripples are transformed into larger forms called **dunes** (B and C), both being out of phase with waves on the water surface. With a further increase in velocity bed undulations are planed off, resistance to flow is lowered and sediment transport rates increase (D and E). This is a transitional state between subcritical and supercritical flow. With a further increase in velocity, supercritical flow gives rise to **antidunes** which because they are in phase with standing waves at the water surface present a low resistance to flow (F and G). Antidunes move upstream since sediment is lost from their downstream side more rapidly than it is deposited. At the highest flow velocities fast – flowing shallow chutes alternate with deeper pools (H). (Based on D. B. Simons and E. V. Richardson (1963) Transactions of the American Society of Civil Engineers, **128**, Fig. 2, p. 289.)*

bedrock as well as the entrainment and downstream transportation of sediment. The erosion of bedrock channels must be of considerable significance in mountainous regions, and although the mechanisms involved are poorly understood three major processes appear to operate.

Corrosion is the chemical weathering of minerals in contact with stream water and the removal of soluble products downstream, and the weathering processes involved are discussed in Chapter 6. The key factors controlling rates

of corrosion are bedrock mineralogy, the solute concentration of the stream water, the stream discharge and velocity of flow. Maximum rates of corrosion are achieved where fast-flowing, undersaturated stream waters pass over lithologies with a high proportion of reactive minerals; for instance, corrosion is an important process in bedrock channels in mountainous limestone terrains in humid environments.

A second mechanism is **abrasion**, or **corrasion**, and consists of the wearing away or detachment of bedrock by particles moved by the water flow. The particles involved can be of any size that can be transported at prevailing flow velocities, and large boulders several metres across may be in motion in fast-flowing, deep river channels. The effectiveness of abrasion depends on the concentration, hardness and kinetic energy of the impacting particles and the resistance of the bedrock surface. Since kinetic energy is proportional to the square of velocity, rates of abrasion increase rapidly as flow velocities increase.

A third mechanism involves hydraulic action, that is the movement of water alone. One way this can occur is through the detachment of loose rock fragments by the force of moving water. Another process which is almost certainly more important, but which is poorly documented in natural channels, is **cavitation**. This is a well-known effect on ship propellers, dam spillways and other artificial structures which are subject to rapid flow. An acceleration of flow in a fluid causes a drop in pressure which, if of sufficient magnitude, leads to the formation of air bubbles. Cavitation occurs when these bubbles implode and emit tiny jets of water at velocities as high as 130 m s^{-1}. Such velocities can generate stresses sufficient to fracture solid rock. Although the high initial mean flow velocities of 10 m s^{-1} or so required for cavitation to operate are certainly attained in some stream channels, especially in rapids and at the base of waterfalls, its effects usually appear to be concealed by features produced by abrasion.

8.3.2 Sediment entrainment

The majority of rivers do not cut directly into bedrock but flow in alluvial channels formed in unconsolidated sediments. These sediments may range in calibre from boulders to clay-sized material. Alluvial channels are 'self-formed' equilibrium or quasi-equilibrium landforms in that their morphology arises from the mobilization, transportation and deposition of sediment and represents an adjustment to prevailing hydrological and sedimentological conditions. Alluvial channels can generally adjust rapidly to changes in the balance between the stresses generated by the flowing water and the resistance of the channel bed sediments to movement. In this respect they differ significantly from bedrock channels which can usually change only slowly and whose morphology is dominated by structural and

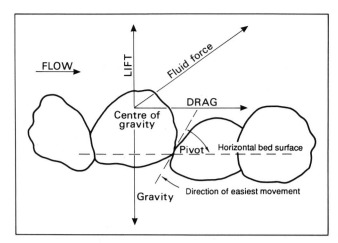

Fig. 8.11 *Schematic representation of the forces acting on a grain resting with others of similar size and shape on a channel bed and subject to fluid flow. (After G. V. Middleton and J. B. Southard (1984)* Mechanics of Sediment Movement. *Society of Economic Paleontologists and Mineralogists Short Course No. 3 Fig. 6.1, p. 6–3.)*

lithological controls. Sediment in alluvial channels includes particles previously carried downstream as well as material contributed directly from valley-side slopes and from bank erosion.

The initial setting into motion of a solid particle in a fluid is called **entrainment** and this occurs when the stresses acting on a particle exceed the resisting forces. The resisting forces are the immersed weight of the particle in water acting normal to the channel bed and the constraining effects of neighbouring grains. The driving forces are the downslope component of the immersed weight of the particle and the fluid forces, the latter being of two types (Fig. 8.11). A particle on a channel bed is subject to a **drag** or **tractive force** largely as a result of the difference in the fluid pressure on its upstream and downstream sides. The stress becomes larger as the velocity of flow increases, but in turbulent flow temporary increases in velocity in eddies can generate local stresses three or four times the mean value. Small particles experience relatively low drag forces since they are located within the laminar sublayer. As the flow velocity increases this zone becomes thinner, and particles are subject to markedly increased stresses as they begin to protrude through the laminar sublayer and are exposed to higher velocity turbulent flows and eddies. Particles are also subject to a **lift force** arising from the acceleration in the flow of water over projecting grains. This increase in velocity leads to a drop in pressure above the grain and is known as the **Bernoulli effect**. If this lift force is sufficient to overcome the resistance to movement the grain will rise from the channel bed. As it does so the lift force rapidly declines, the drag force quickly increases and the particle is carried downstream in the water flow.

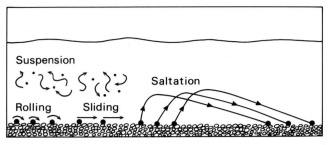

Fig. 8.12 *Schematic representation of the modes of sediment transport in flowing water.*

Unless the flow is highly turbulent and energetic larger particles move on a trajectory converging with the channel bed at a low angle (Fig. 8.12). This kind of particle motion is termed **saltation**. When large numbers of grains are in motion under rapid flow conditions the ideal saltation trajectory is not attained because there are frequent collisions between particles. In this situation there is a concentrated dispersion of particles near the channel bed dominated by interparticle collisions and deflections. Larger particles which cannot be lifted from the channel bed may simply move across it by either rolling or sliding. Impacting saltating grains returning to the channel bed may help to precipitate this movement.

In addition to rolling, sliding and saltating, solid particles may experience suspended motion in which their trajectories are more irregular and more prolonged than for saltation (Fig. 8.12). The weight of fine particles in true **suspension** is entirely supported by the upward pulses of flow generated by eddies. Grains descending during saltation may be temporarily buoyed up by upward movements in turbulent flows and this condition is more appropriately described as **incipient suspension**.

8.4   Fluvial transport and deposition

8.4.1   Modes of fluvial transport

Material can be transported by rivers either as solid particles or in solution. **Solute load**, or **dissolved load**, which is derived largely from bedrock weathering (see Chapter 6), is dispersed throughout the flow. Solid load is of two main types. **Bed load**, or **traction load**, encompasses all material rolling, sliding or saltating along the channel bed. **Suspended load** is invariably of fine calibre and includes all particles prevented from falling to the channel bed by the upward momentum imparted by eddies within turbulent flows. The finest fraction of suspended load, consisting of very small clay-sized particles, is termed **wash load** and is able to stay essentially in permanent suspension as long as some flow is maintained.

Solute and solid loads can be estimated by relating solute and sediment concentrations to stream discharge (see

Section 15.3.1). The concentration of solutes and suspended load can be measured from water samples collected by inserting sampling devices into the stream at various depths and at different stages of flow, although such devices also collect some saltating grains which are a component of bed load. These measurements can then be multiplied by discharge to provide an estimate of suspended sediment and solute transport rates. Bed load is much more difficult to estimate because most of the devices used interfere to some extent with the channel bed. Techniques employed have ranged from a variety of traps to acoustic and pressure-difference recording apparatus. The movement of very coarse bed material can be monitored by painting individual boulders and tracking their movement along the channel bed over a period of time, while fluorescent dyes and radioactive tracers can be used to monitor the movement of finer material of a specific size or mineralogy.

Sediment in transit in river channels is subject to abrasion and weathering. Although there is evidence for significant rounding of pebbles as they are transported downstream, it is less certain to what extent sand and finer material experiences attrition during fluvial transport. A downstream reduction in mean particle size is a widespread feature of river channels, but the great majority of this appears to be due to sorting (see Section 8.4.3).

8.4.2 Sediment transport

Although the general principles whereby grains are set in motion in a fluid are fairly well understood, the problem of predicting the rate of particle movement in natural river channels under different flow conditions and with varying sediment characteristics has proved extremely difficult. Advances in this field have been achieved largely through research employing laboratory flumes to measure rates of sediment movement under a range of flow conditions. In such experimental work the variables involved can be fairly closely controlled, but the complexity present in natural channels means that it is not possible to extrapolate these laboratory findings directly to field situations.

It is also possible to approach the problem of sediment transport from fundamental physical principles. A particle lying on a channel bed possesses potential energy by virtue of its elevation above base level. Once set in motion a particle moves by expending kinetic energy and by overcoming frictional resistance. An important concept in relating fluvial energy to sediment transport is that of **stream power** (Box 8.4). In order to transport sediment, work must be performed. Work is defined as the product of force and distance and power is the rate of doing work. Stream power is therefore the power per unit length of stream. In other words, it is the rate of energy supply at the channel bed which is available for overcoming friction and

---

Box 8.4   Stream power

Stream power (Ω), or the power per unit length of stream, is measured in W (watts) or J s^{-1} (joules per second) and is defined as

$$\Omega = \rho_W g Q s$$

where ρ_W is the density of water, g the acceleration of gravity, Q discharge and s channel gradient.

The energy available per unit area of the channel bed (w) can be related to stream power (Ω) to give the specific stream power (ω):

$$\omega = \frac{\Omega}{w} = \tau_0 \bar{v}$$

where w is channel width, τ_0 channel bed shear stress and $\bar{v}$ mean flow velocity. The specific power (ω) of rivers in Britain varies over three orders of magnitude from 100 to 1000 W m^{-2} in steep gradient, high runoff streams, to less than 100 W m^{-2} in low gradient, low runoff streams.

---

for transporting sediment. Stream power is at a maximum in streams with high discharge, high gradient and large hydraulic radius. **Specific stream power** is defined as stream power per unit area of channel bed and equals bed shear stress times mean flow velocity.

The concept of stream power can be applied to conditions under which erosion or deposition will predominate through the **threshold of critical power**. This is defined as the state in a specific channel reach when the power available is exactly sufficient to transport the mean available sediment load. If the available power is greater than that required to transport this mean sediment load then erosion and downcutting will occur; if it is less than that required, deposition will predominate.

In order to predict the overall rate of transport of sediment in a channel rather than the conditions which lead to the entrainment of individual particles, it is convenient to use a readily measured average property of the water flow. One approach, pioneered by F. Hjulström, uses mean flow velocity to define conditions of sediment entrainment and transport. Hjulström collated data from some 30 experimental studies relating flow velocity to sediment transport to construct the curves illustrated in Figure 8.13. These curves, which have been broadly confirmed by later research, show the **competence** of different flow velocities – the competence of a flow being the maximum particle size that can be entrained at a specific flow velocity. It is important to note that flow velocity in this case refers to the mean velocity in the channel and not the much lower velocity on the channel bed where the sediment is entrained. A major limitation of this approach is that there is no unique mean flow velocity at which particles of a particular size are set in motion. The velocity on the channel bed is the more important controlling variable, and for a given

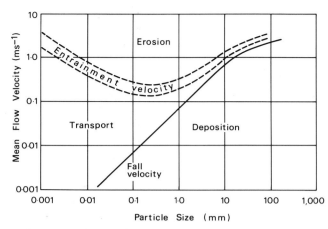

Fig. 8.13 *Velocity thresholds for entrainment and deposition as a function of particle size. Once in motion a particle is not deposited until the flow velocity falls below the fall threshold velocity; this is less than the velocity required to set it in motion because more force is required to entrain a particle than to keep it in motion. Note that these curves relate to data for well-sorted material and do not indicate threshold velocities for poorly sorted sediments. (After F. Hjulström (1935) Bulletin of the Geological Institute, University of Uppsala 25.)*

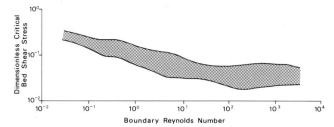

Fig. 8.14 *A Shields diagram showing the relationship between the dimensionless critical bed shear stress and the boundary Reynolds number. The scatter evident is a result of a wide range of experimental conditions and the problem of deciding precisely when a threshold is reached. Note that the dimensionless critical bed shear stress is almost constant, at a mean value of around 0.06, for a broad range of particle dimensions above a boundary Reynolds number of around 1.0. At lower boundary Reynolds numbers the dimensionless critical bed shear stress increases progressively down to a value of about 0.3. This is thought to be due to the presence of a smooth boundary to flow with particles lying entirely within the laminar sublayer. (Based on M. C. Miller et al. (1977) Sedimentology 24, Fig. 2, p. 511.)*

threshold bed velocity the mean flow velocity increases with water depth.

An alternative approach uses the mean bed shear stress necessary to initiate particle movement. This depends not only on particle size but also on the roughness of the channel bed. The critical shear stress on the channel bed can be represented in a dimensionless form and related to a boundary Reynolds number which defines the degree of channel bed roughness and is proportional to the ratio between grain size and the thickness of the laminar sublayer. This relationship can be represented in a **Shields diagram** (Fig. 8.14) and shows a broadly similar form to the Hjulström curve relating mean flow velocity to sediment entrainment.

The curves indicate that the particle size which can be moved at the lowest velocity (about 0.2 m s⁻¹) is roughly 0.25–0.5 mm (medium sand). Larger, heavier grains require higher threshold velocities to initiate movement, while silt and clay-sized material also requires higher velocities because such small particles are at least partially protected by lying within the laminar sublayer. For clay-sized material cohesion between individual particles further enhances their resistance to entrainment. Once set in motion fine particles can continue to be transported in suspension even if the flow velocities fall significantly, but coarser grains are more rapidly deposited when flow velocities are reduced.

Although useful in indicating some of the fundamental controls on sediment transport rates, the relationships evident in the Hjulström curves and the Shields diagram are

difficult to apply to natural channels for a number of reasons. First, they relate to well-sorted sediments, that is, sediments of a relatively uniform grain size. In poorly sorted, or heterogeneous sediments, small particles may be sheltered by larger grains from the drag stress exerted by the water flow. The extreme example of this is the production of a channel armour where coarse grains cover the channel bed and conceal the finer sediment below. Secondly, the curves refer to smooth channels whereas most natural channels are irregular. Thirdly, the curves assume flow to be steady but in natural channels the velocity of flow is often highly variable. Fourthly, the way particles are packed together can significantly affect the ease with which they are entrained. Finally, experimentally or theoretically derived sediment transport equations are especially difficult to apply to gravel-bed channels. Sediment transport occurs at almost all flows in sandy, alluvial channels, but significant transport in gravel-bed rivers only occurs at relatively high flows, normally where bankfull discharge is approached. Consequently, in channel beds composed of gravel or boulders. there are relatively long periods of base flow with little or no transport during which sediment can settle and interlock thereby rendering it less susceptible to entrainment in subsequent high flows than would be expected from standard transport equations.

8.4.3 Fluvial deposition

Just as there are threshold flow velocities for the entrainment of particles of different sizes so there are thresholds for sediment deposition. The velocity at which a particle settles to a channel bed, known as its **fall velocity**, is a function of both its density, size and shape and of the viscosity and density of the transporting fluid. Since viscosity and density change with the concentration of sediment in a stream flow, deposition is not related simply to flow velocity. As flow velocity decreases the coarser sediment begins to be deposited while the finer particles remain in motion, and this differential settling of the material in transit gives rise to sediment sorting.

Conditions on a channel bed can change rapidly over time and space, so whether sediment is entrained or deposited depends on local rather than average conditions. Indeed it is possible for deposition and entrainment to be occurring simultaneously in the same channel reach and coarse sediment can be deposited at the same time that finer particles are being entrained. Nevertheless, as discharge fluctuates between high and low stages, episodes of degradation during which erosion predominates will alternate with periods of aggradation during which deposition prevails.

Further reading

Fluvial processes are discussed in a number of texts on fluvial geomorphology including those by Knighton (1984), Morisawa (1985) and Petts and Foster (1985), but the pioneering work by Leopold et al. (1964) still remains valuable as a clear introduction to the topic. A somewhat more advanced treatment of processes in alluvial channels is provided by Richards (1982). Other useful material is to be found in Richards (1987) and Schumm et al. (1987).

The hydrology of drainage basins has generated an enormous literature, but the important concepts are summarized in several general texts including those by Dunne and Leopold (1978) and Ward (1975) and in the book edited by Kirkby (1978). Numerous papers on specific aspects of drainage basin hydrology are to be found in the periodicals *Journal of Hydrology, Hydrological Processes* and *Water Resources Research*. The initiation of erosion on slopes subject to infiltration-excess overland flow is considered in the classic paper by Horton (1945), while more recent ideas on the role of both surface and subsurface flow in establishing networks of open channels are reviewed by Jones (1987). Catastrophic discharges resulting from the failure of natural dams are considered by Costa and Schuster (1988).

The book by Leeder (1982), although directed towards sedimentologists, provides an excellent treatment of open channel flow, a topic which is also well covered by Knighton (1984) and Richards (1982). A detailed discussion of bedform – flow interactions is to be found in Simons and Richardson (1966). The erosion of bedrock channels is considered by Foley (1980). Leeder (1982) is also a good starting-point for acquiring a more detailed understanding of the physical principles controlling sediment entrainment and transport. Advanced treatments of this topic include those by Graf (1971) and Yalin (1977), while Bagnold (1966, 1977) demonstrates how general physical principles can be applied to the problem of sediment transport. Bull (1979) introduces the idea of a threshold of critical power for sediment transport and deposition. Sundborg (1956) develops the approach pioneered by Hjulström of relating sediment entrainment to a threshold flow velocity, while the problems involved in defining a critical threshold parameter for particle entrainment are reviewed by Miller *et al.* (1977).

Turning to more specific topics, sediment transport in river bends, as opposed to straight channels, is examined by Dietrich (1987), and the difficulties of predicting sediment transport rates in coarse-bed channels are discussed by Baker and Ritter (1975), Carling (1983), Bathhurst (1987) and in numerous contributions in the book edited by Thorne *et al.* (1987). Channel bed armouring is of particular significance in modifying sediment transport rates and this phenomenon is discussed by Gomez (1983). Techniques for monitoring sediment movement in natural channels are described by Crickmore (1967) and Kennedy and Kouba (1970) for sand, and by Sayre (1965) and Reid *et al.* (1985) for bed load, while the paper by Ergenzinger and Conrady (1982) illustrates an approach to the measurement of the overall rate of bed-load transport. Sediment deposition and the complex relationships between erosion and deposition are examined by Colby (1964).

References

Bagnold, R. A. (1966) An approach to the sediment transport problem from general physics. *United States Geological Survey Professional Paper* **422–I**.

Bagnold, R. A. (1977) Bed load transport by natural rivers. *Water Resources Research* **13**, 303–12.

Baker, V. R. and Ritter, D. F. (1975) Competence of rivers to transport coarse bedload material. *Geological Society of America Bulletin* **86**, 975–8.

Bathhurst, J. C. (1987) Measuring and modelling bedload transport in channels with coarse bed materials. In: K. Richards (ed.) *River Channels: Environment and Process*. Institute of British Geographers Special Publication **18**. Blackwell, Oxford, pp. 272–94.

Bull, W. B. (1979) Threshold of critical power in streams. *Geological Society of America Bulletin* **90**, 453–64.

Carling, P. A. (1983) Threshold of coarse sediment transport in broad and narrow natural streams. *Earth Surface Processes and Landforms* **8**, 1–18.

Colby, B. R. (1964) Scour and fill in sand-bed streams. *United States Geological Survey Professional Paper* **462D**.

Costa, J. E. and Schuster, R. L. (1988) The formation and failure of natural dams. *Geological Society of America Bulletin* **100**, 1054–68.

Crickmore, M. J. (1967) Measurement of sand transport in rivers with special reference to tracer methods. *Sedimentology* **8**, 175–228.

Dietrich, W. E. (1987) Mechanics of flow and sediment transport in river bends. In: K. Richards (ed.) *River Channels: Environment and Process*. Institute of British Geographers Special

Publication **18**. Blackwell, Oxford, pp. 179–227.

Dunne, T. and Leopold, L. B. (1978) *Water in Environmental Planning*. W. H. Freeman, San Francisco.

Ergenzinger, P. and Conrady, J. (1982) A new technique for measuring bedload in natural channels. *Catena* **9**, 77–80.

Foley, M. G. (1980) Bed-rock incision by streams. *Geological Society of America Bulletin* **91** Part II, 2189–213.

Gomez, B. (1983) Temporal variations in bedload transport rates: the effect of progressive bed armouring. *Earth Surface Processes and Landforms* **6**, 235–50.

Graf, W. H. (1971) *Hydraulics of Sediment Transport*. McGraw-Hill, New York.

Horton, R. E. (1945) Erosional development of streams and their drainage basins: hydrophysical approach to quantitative morphology. *Bulletin of the Geological Society of America* **56**, 275–370.

Jones, J. A. A. (1987) The initiation of natural drainage networks. *Progress in Physical Geography* **11**, 207–45.

Kennedy, V. C. and Kouba, D. L. (1970) Fluorescent sand as a tracer of fluvial sediments. *United States Geological Survey Professional Paper* **562–E**.

Kirkby, M. J. (ed.) (1978) *Hillslope Hydrology*. Wiley, Chichester and New York.

Knighton, D. (1984) *Fluvial Forms and Processes*. Edward Arnold, London and Baltimore.

Leeder, M. R. (1982) *Sedimentology: Process and Product*. Allen and Unwin, London and Boston.

Leopold, L. B., Wolman, M. G. and Miller, J. P. (1964) *Fluvial Processes in Geomorphology*. W. H. Freeman, San Francisco.

Miller, M. C., McCave, I. N. and Komar, P. D. (1977) Threshold of sediment motion under unidirectional currents. *Sedimentology* **24**, 507–27.

Morisawa, M. (1985) *Rivers: Form and Process*. Longman, London and New York.

Petts, G. and Foster, I. (1985) *Rivers and Landscape*. Edward Arnold, London and Baltimore.

Reid, I., Frostick, L. E. and Layman, J. T. (1985) The incidence and nature of bedload transport during flood flows in coarse-grained alluvial channels. *Earth Surface Processes and Landforms* **10**, 33–44.

Richards, K. (1982) *Rivers: Form and Process in Alluvial Channels*. Methuen, London and New York.

Richards, K. (ed.) (1987) *River Channels: Environment and Process*. Institute of British Geographers Special Publication **18**. Blackwell, Oxford.

Sayre, W. W. (1965) Transport and dispersion of labeled bed material, North Loup River, Nebraska. *United States Geological Survey Professional Paper* **433–C**.

Schumm, S. A., Moseley, M. P. and Weaver, W. E. (1987) *Experimental Fluvial Geomorphology*. Wiley, New York and Chichester.

Simons, D. B. and Richardson, E. V. (1966) Resistance to flow in alluvial channels. *United States Geological Survey Professional Paper* **422J**.

Sundborg, A. (1956) The river Klaralven. A study in fluvial processes. *Geografiska Annaler* **38**, 125–316.

Thorne, C. R., Bathurst, J. C. and Hey, R. D. (eds) (1987) *Sediment Transport in Gravel-Bed Rivers*. Wiley, Chichester and New York.

Ward, R. C. (1975) *Principles of Hydrology* (2nd edn). McGraw-Hill, London and New York.

Yalin, M. S. (1977) *Mechanics of Sediment Transport* (2nd edn). Pergamon, Oxford and New York.

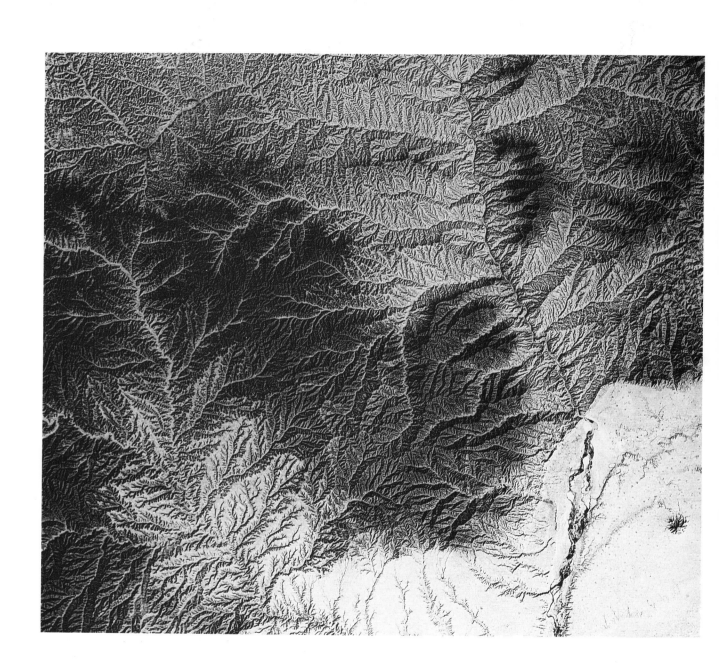

9

Fluvial landforms

9.1 The fluvial system

Rivers and the landforms they create can be considered at an enormous range of scales. In Chapter 8 we focused on the microscale aspects of how sediment is entrained, transported and deposited by rivers. At this scale we can examine the movement of individual particles on a river bed and the relationships between such factors as the roughness of the channel floor, the velocity of water flow and the rate of sediment transport. In this chapter we are primarily concerned with the way fluvial processes create fluvial landforms. These range from the processes operating in a single bend in a river, to the different channel patterns arising from contrasting conditions of water flow, sediment transport and channel gradient, and ultimately to the morphology of entire drainage basins .

Rivers and streams can be simply defined as bodies of water flowing in an open channel. They have three important roles in landscape creation: they erode the channels in which they flow, they transport sediments and solutes provided by weathering and slope processes as well as by the other denudational agents of ice and wind, and they produce a wide range of erosional and depositional landforms. River systems are the primary agents of erosion, transportation and deposition in most landscapes, including many where surface water is not present for most of the time. Over much of the Earth fluvial processes determine the overall form of the landsurface.

Fluvial systems can for convenience be regarded as consisting of three main elements: a zone of sediment production, a zone of sediment transfer and a zone of sediment deposition. This categorization is, of course, over-simplified because some erosion, transport and deposition occurs in all three zones; nevertheless, within each zone one of these three processes is usually dominant. In large basins the upstream zone in which sediment production predominates is usually a mountainous or upland region. The zone dominated by deposition is generally located along a coast and takes the form of a delta or lowland coastal plain. The depositional zone may, however, be located at the centre of an interior basin, and local sites of significant deposition may occur in the piedmont environment where rivers emerge at a mountain front. In the intervening zone in which transport predominates, inputs and outputs of sediment may be roughly in balance, at least over the medium term. Sediment may, however, be stored in this zone for long periods of time before being removed downstream into the zone of final deposition.

9.2 The drainage basin

A **drainage basin** is an area within which water supplied by precipitation is transferred to the ocean, a focus of internal drainage, such as a lake, or to a larger stream. (Note that the term drainage basin is synonymous with **catchment** and, in the USA, with **watershed**. In the UK the term watershed refers to the drainage divide between basins.)

For a number of reasons the drainage basin is the fundamental unit of fluvial geomorphology. Drainage basins are usually well-defined areas, clearly separated from each other by **drainage divides**, within which surface or near-surface flows of water and associated movements of sediment and solutes are contained. Since it is the transfer of material that causes changes in the elevation and form of the landsurface over time, drainage basins constitute the natural unit for the analysis of fluvially-eroded landscapes. The outlet of a drainage basin provides a very convenient point at which to monitor these movements of water, sediment and solutes. There are important exceptions to these generalizations – such as the partially subsurface flow of water in limestone terrains which may be unrelated to surface topography (see Section 9.6) – but they are valid for the great majority of landscapes. Another important property of drainage basins is their hierarchical nature; each tributary in a drainage system has its own basin area contributing runoff, and so larger basins consist of a hierarchy of smaller ones.

A final feature of drainage basins which makes them such important units of analysis in geomorphology is that, following the pioneer work of R. E. Horton and A. N. Strahler, many of their important properties can be expressed quantitatively in a way which allows one basin to be compared with another. Such quantitative description is termed **drainage basin morphometry** and can be applied to the areal and relief properties of basins as well as the characteristics of their river channel systems.

9.2.1   Channel network characteristics

River systems are a type of **network**; that is, they consist of a series of links which connect nodes. Networks can be analyzed with respect to two main sets of properties: the **topological** aspects of stream networks concern the interconnections of the system, whereas the geometrical aspects involve length, area, shape, relief and orientation properties.

The basic element of stream networks is the **stream segment**, or link. This is a section of stream channel between two channel junctions or, for 'fingertip' tributaries, between a junction and the upstream termination of a channel. **Stream order** expresses the hierarchical relationship between stream segments. It is a fundamental property of stream networks since it is related to the relative discharge of a channel segment.

Various systems of stream ordering have been proposed, but the two most frequently used are those of A. N. Strahler and R. L. Shreve (Fig. 9.1). In the Strahler system a stream segment with no tributaries is designated a first order segment. A second order segment is formed by the joining of two first order segments, a third order segment by the joining of two second order segments and so on. It is important to note that with the Strahler ordering method there is no increase in order when a segment of one order is joined by another of a lower order. In contrast, the stream ordering system proposed by Shreve defines the magnitude of a channel segment as the number of fingertip tributaries that feed it. As **stream magnitude** is closely related to the proportion of the total basin area contributing runoff, it provides a good estimate of relative stream discharge for small river systems.

Stream order as defined by Strahler has been applied to numerous river systems and has been shown to be statistically related to various elements of drainage basin morphometry (Fig. 9.2). A widely used topological property of stream networks is the **bifurcation ratio**. This is the ratio between the number of stream segments of one order and

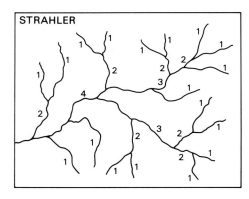

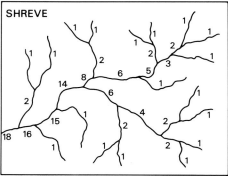

Fig. 9.1   *Schemes of stream ordering proposed by A. N. Strahler (stream order) and R. L. Shreve (stream magnitude) (see text for explanation).*

the number of the next highest order. This number varies slightly between different successive orders in a basin (that is, between first and second order, second and third order and so on), so a mean bifurcation ratio for a whole basin is normally used. Where lithology is relatively homogeneous the bifurcation ratio is rarely more than 5 or less than 3. Nevertheless, a value of 10 or more can be attained in highly elongated basins which can develop where there are narrow, alternating outcrops of soft and resistant lithologies.

Various geometrical properties of stream networks are defined in Table 9.1. Probably the most important of these is **drainage density**. This reflects a balance between erosive forces and the resistance of the ground surface, and, as a consequence, is closely related to climate and lithology. Drainage densities range from less than 5 km km^{-2} on permeable sandstones, to extreme values of more than 500 km km^{-2} on unvegetated clay 'badlands'. The role of climate is indicated by the very high drainage densities in some semi-arid environments which appear to result from the prevalence of surface runoff and the relative ease with which new channels are initiated. Drainage density is related to the **length of overland flow** (Table 9.1), the latter being approximately equal to the reciprocal of twice the drainage density.

Further aspects of channel network geometry include

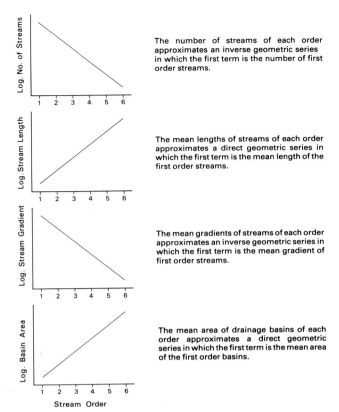

The number of streams of each order approximates an inverse geometric series in which the first term is the number of first order streams.

The mean lengths of streams of each order approximates a direct geometric series in which the first term is the mean length of the first order streams.

The mean gradients of streams of each order approximates an inverse geometric series in which the first term is the mean gradient of first order streams.

The mean area of drainage basins of each order approximates a direct geometric series in which the first term is the mean area of the first order basins.

Fig. 9.2 Statistical relationships between stream order and various drainage basin properties. These relationships are often referred to as 'laws' of drainage basin morphometry, although it must be emphasized that they are only general trends around which there may be significant variability.

Table 9.1 Some morphometric properties of drainage basins

PROPERTY	DEFINITION
Network properties	
Drainage density	Mean length of stream channels per unit area
Stream frequency	Number of stream segments of all orders per unit area
Length of overland flow	Mean distance from channels up maximum valley-side slope to drainage divide
Areal properties	
Circularity ratio	Total drainage basin area divided by the area of a circle having the same perimeter as the basin
Elongation ratio	The diameter of a circle of the same area as the drainage basin divided by the maximum length of the basin measured from its mouth
Relief properties	
Basin relief	Difference in elevation between the highest and lowest point in a drainage basin
Relief ratio	Basin relief divided by the maximum length of the basin
Ruggedness number	Basin relief multiplied by drainaged density

drainage patterns and orientation. In the absence of structural effects tree-like patterns develop, but where structural influences are strong a variety of drainage patterns occur. Drainage patterns are discussed further in Chapter 16.

Initial studies of stream networks indicated that fluvial systems with topological properties similar to natural systems could be generated by purely random processes. Although such random-model thinking has been extremely influential in channel network studies, later research has identified numerous regularities in stream network topology. These systematic, as opposed to random, variations appear to be attributable to various factors including the need for lower order basins to fit together, the sinuosity of valleys and the migration of valley bends downstream, and the length and steepness of valley sides. Although they are most pronounced in large basins, such non-random elements of stream network topology are also present in small catchments.

9.2.2   Areal and relief characteristics

Areal properties express the overall plan form and dimensions of drainage basins, while relief properties express elevation differences (Table 9.1). The **elongation ratio** has important hydrological consequences because, in contrast to more circular catchments, precipitation delivered during a storm in highly elongated basins has to travel a wide range of distances to reach the basin outlet. The resulting delay in the arrival of a proportion of the storm flow consequently leads to a flattening of the storm hydrograph (Fig. 9.3).

The relief, or height differences, in a basin can be expressed quite simply using maximum and minimum elevation values. **Local relief** is the difference between maximum and minimum elevations within a given area (usually of limited extent). Relief is related to the slope and stream gradients in a basin, and so indirectly has an influence on the rates of slope processes and sediment transport by rivers. Applied to large basins, however, an index such as **relief ratio** (Table 9.1) will invariably conceal significant variations in relief within the basin. A much more useful measure of relief can be produced by averaging the local relief over cells of a given size across an entire basin, and such average relief estimates have been found to be closely correlated with fluvial denudation rates (see Section 15.5).

There is a close relationship between drainage density, mean slope angle and relief (Fig. 9.4). If drainage density is constant and stream channels maintain a constant spacing through time, an increase in local relief due to stream incision must, of necessity, cause an increase in mean slope angles in the basin. There is a limit to this effect, however, as a progressive increase in slope angles cannot continue indefinitely. At a certain point rates of erosion will be so

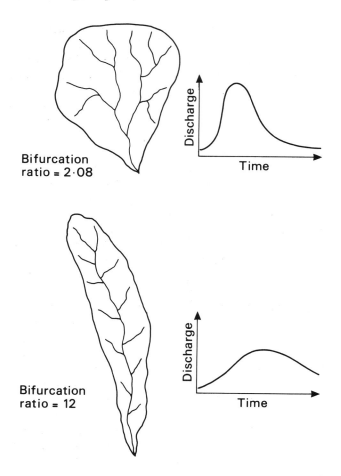

Fig. 9.3 *Schematic representation of the effect of basin shape on the form of the storm hydrograph.*

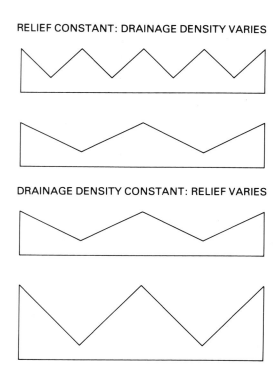

RELIEF CONSTANT: DRAINAGE DENSITY VARIES

DRAINAGE DENSITY CONSTANT: RELIEF VARIES

Fig. 9.4 *Relationships between relief, drainage density and mean slope.*

high that interfluves will be lowered as rapidly as stream channels and local relief will attain a constant value.

Another approach to the description of drainage basin relief involves hypsometric analysis in which elevation is related to basin area (Fig. 9.5). Such analysis allows the calculation of a **hypsometric integral** which summarizes the form of a drainage basin in a single value. Hypsometric integrals lie between 25 and 75 per cent for most drainage basins, high values indicating a relatively large proportion of land at high elevation within the basin and low values indicating a small proportion. Thus basins in which stream channels have incised deep valleys, leaving extensive areas of relatively high elevation, have high hypsometric integrals. The idea that such basins might be modified by denudation through time to a state where little uneroded high land remains has led to the use of the hypsometric integral as an index of the stage of development of a landscape. Although this may be true in a limited sense for relatively small catchments, in large drainage basins tectonic uplift and subsidence as well as sea-level changes can operate in conjunction with denudation to change the areal distribution of elevation over time.

9.3   River channels

Three major types of river channel can be identified. **Bedrock channels** are cut into rock. In general they experience gradual modification but retain their overall form for long periods of time. Significant lateral shifting of channels may, however, occur where the bedrock is only weakly resistant. In **alluvial channels** the bed and banks are composed of sediment being transported by the river. They can undergo dramatic changes in form as weakly resistant alluvium is eroded, transported and redeposited in response to changes in water discharge and sediment load, among other factors. **Semi-controlled channels** are of intermediate type, being only locally controlled by bedrock or resistant alluvium. A semi-controlled channel will be stable where it is cut into bedrock or resistant alluvium, but over time it may migrate laterally into alluvium and be much more responsive to changes in hydrological and sedimentological variables.

9.3.1   Alluvial channels: plan form

Alluvial channels exhibit a great variety of plan form. Numerous channel patterns can be recognized but they all represent variations of just a few basic types. One key

A

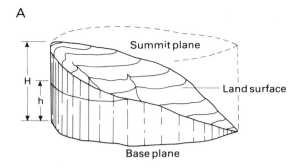

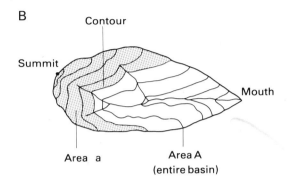

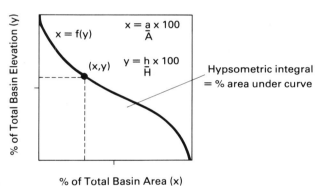

Fig. 9.5 *The components of hypsometric analysis showing how the area–elevation data are calculated (A) and plotted to produce a hypsometric curve (B). The proportion of total basin elevation is calculated for contours at appropriate intervals by dividing the contour elevation by the maximum basin elevation (calculated with respect to the basin outlet). The proportion of total basin area is calculated by dividing the area of the basin above each contour interval by the total basin area. The plot of these two ratios (normally represented as percentages) defines the hypsometric curve for the basin. The hypsometric integral is the area under the hypsometric curve as a percentage of the total area defined by the two graph axes and can be simply estimated by superimposing it on graph paper and counting squares. (Modified from A. N. Strahler (1952) Bulletin of the Geological Society of America 63, Figs 1 and 2, pp. 1119 and 1120.)*

property is **sinuosity** which represents the irregularity of the channel course and is expressed as the ratio of channel length (measured along the centre of the channel) to valley length (measured along the valley axis). The ratio of valley gradient to channel gradient provides an alternative definition. Sinuosity ranges from 1.0 for perfectly straight channels to around 3 for highly tortuous river courses. Channels with a sinuosity greater than 1.5 are usually described as **meandering** (although this is a purely arbitrary break-point). Even straight channels generally have a sinuous **thalweg**, that is, a line of maximum depth along the channel floor.

A second fundamental form that channels may assume involves **braiding**. This represents the extent to which flow in the channel is divided by islands or **bars**, that is exposed accumulations of sediment (Fig. 9.6). Islands are vegetated and are relatively long-lived features, whereas bars are less stable, being composed of unvegetated sands or gravels. The degree of braiding is expressed quantitatively as the percentage of channel length that contains islands or bars. If a channel contains islands whose width is more than three times the width of the water at mean discharge it is described as **anabranching**. The degree of anabranching is expressed as the percentage of channel length occupied by such islands.

A third type of channel pattern is termed **anastomosing**. Anastomosing channels consist of distributaries which branch and rejoin and have a superficial resemblance to braided patterns. Braided channels, however, are single-channel forms in which the flow is diverted around obstructions in the channel itself, whereas anastomosing patterns consist of discrete, interconnected channels which are separated by bedrock or by stable alluvium. While braided channels are primarily depositional forms, anastomosing channels are essentially erosional in nature since material between channels is too resistant to be transported, except by exceptional flows.

Although many alluvial channels can be described as stable in that they are not experiencing a dramatic change in form, *some* change is an inevitable element of the behaviour of all alluvial channels since they are, at least partly, composed of material which is eroded or deposited as the stress exerted on the channel bed and banks by the flowing water changes over time. Such changes can occur in various ways (Fig. 9.8), including the downstream migration of bars, the gradual shifting of meanders and the rapid alteration of course through **cut-offs** or channel diversion. A most dramatic example of the latter process, known as **avulsion**, occurred in China in 1851 when the mouth of the Huang He shifted over 300 km to the north.

Studies of alluvial channels from a wide range of environments have demonstrated that although the size of alluvial channels is controlled largely by the water discharge flowing through them, the channel pattern and shape are related primarily to the quantity and size of sediment being transported and the valley floor gradient. A genetic classification proposed by S. A. Schumm identifies the three fundamental channel patterns and relates these to the nature

Fig. 9.6 *Braided channel of the Waimakariri, just north of Christchurch, South Island, New Zealand. (Photo courtesy S. J. Smith.)*

of the transported sediment among other factors (Fig. 9.9).

Although the proportions of bed load and suspended load invariably change over time it is useful to distinguish between **suspended-load channels** transporting less than 3 per cent of total sediment as bed load, **bed-load channels** transporting more than 11 per cent as bed load, and **mixed-load channels** with 3–11 per cent of sediment being transported as bed load (Table 9.2). Suspended load channels are narrow and deep with a width–depth ratio of less than 10. If the valley gradient is low the channel will be straight (Fig. 9.9, pattern 1), but sinuosity becomes greater with an increase in gradient (Fig 9.9, pattern 3a). Mixed-load channels have a lower width–depth ratio of 10–40 and sinuosity ranges from 1.3 to 2.0 (Fig. 9.9, pattern 3b). Even if the channel is relatively straight the thalweg is nearly always sinuous (Fig. 9.9, pattern 2). Bed-load channels are straight and have a high width–depth ratio of over 40. More than one thalweg tends to develop (Fig. 9.9, pattern 4) and where bed-load transport is very high distinct bars form and a braided channel is created (Fig. 9.9, pattern 5). Overall the mean flow velocity, the drag force on the channel bed and stream power increase, while channel stability decreases from patterns 1 to 5 (in Figure 9.9).

Table 9.2 Classification of stable alluvial channels

TYPE OF CHANNEL	BED LOAD AS % OF TOTAL LOAD	CHANNEL CHARACTERISTICS
Suspended load	<3	Width–depth ratio <10; sinuosity usually >2.0; relatively gentle gradient
Mixed load	3–11	Width–depth ratio 10–40; sinuosity usually 1.3–2.0; moderate gradient
Bed load	>11	Width–depth ratio >40; sinuosity usually <1.3; relatively steep gradient

Source: Modified from S. A. Schumm (1977), *The Fluvial System*. Wiley, New York, Table 5–4, p. 156.

9.3.1.1   Meandering channels

Meandering channels vary in form, but a number of morphological characteristics can be defined which are relatively consistent for a large proportion of rivers (Fig. 9.10). These include the observation that meander wavelength is commonly about ten times channel width and about five times the mean radius of curvature. Natural meanders rarely have a perfectly symmetric and regular form apparently because of variations in channel bed material. In

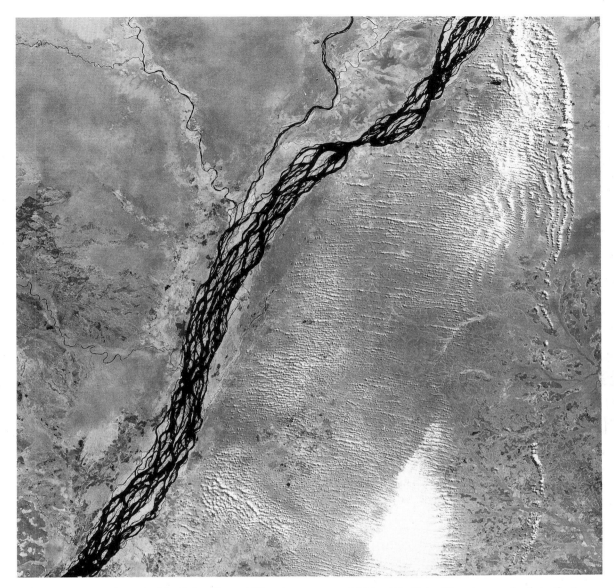

Fig. 9.7 *Anastomosing channel of the Zaire (Congo) River in west-central Africa. The Zaire River is second only to the Amazon in annual discharge, and the anastomosing channel system in its lower reaches shown here is up to 15 km across. The area covered is about 170 km across. (Landsat image courtesy N. M. Short.)*

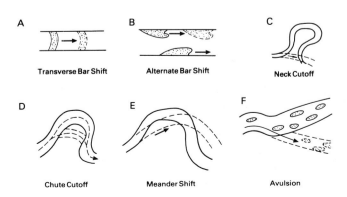

Fig. 9.8 *Types of channel changes. (A) and (B) show the downstream migration of bars or islands within channels; in (B) alternate bars are moving in association with a shifting in position of the deepest part of the channel, or thalweg, with the channel banks alternatively being protected from, and exposed to, erosion as the bars migrate; (C), (D) and (E) illustrate rapid changes in the course of meandering channels; (F) shows the establishment of a new course through avulsion. (After S. A. Schumm, (1985)* Annual Review of Earth and Planetary Sciences *13, Fig. 4, p. 11.)*

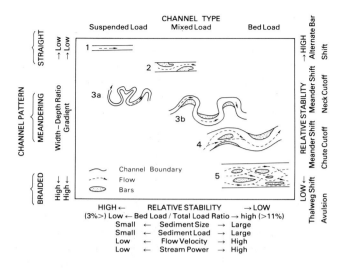

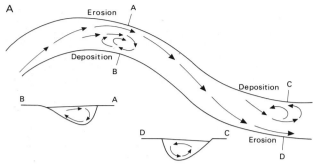

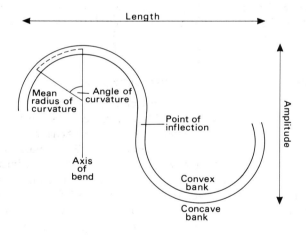

Fig. 9.9 *Classification of channel form based on pattern and sediment load. (After S. A. Schumm (1981) Society of Economic Paleontologists and Mineralogists Special Publication 31, Fig. 4, p. 24.)*

Fig. 9.10 *Principal components of meander geometry.*

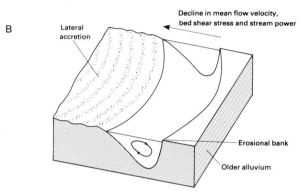

Fig. 9.11 *Nature of water flow in a meandering channel: (A) downstream changes in cross-sectional channel form and sites of erosion and deposition; (B) water movement in a channel bend and associated stresses exerted on channel bed material. ((B) based on J. R. L. Allen (1970) Physical Processes of Sedimentation. Allen and Unwin, London, Fig. 4.5, p. 133(B).)*

rivers with coarse bed material meander forms are often highly distorted.

The fact that straight channels are rare and meandering channels are common raises the question of what causes meandering. The meandering behaviour of rivers may initially seem anomalous since sinuous channels take a longer, lower gradient course when a steeper, shorter course is available. But we simply have to note that air-flows such as jet streams and ocean currents meander to recognize that meandering may be the normal behaviour of fluids in motion.

There is an extensive literature on the origin of meanders suggesting a range of possible mechanisms for initiating and sustaining meandering behaviour. Research by hydraulic engineers has now shown that friction with the channel bed and banks causes shear and turbulence in the water flow and the development of instabilities which promote the formation of alternating bars along the channel. A helical flow is established, with the water surface being elevated on the outer (concave) bank of each curve and return currents at depth directing the flow towards the opposite bank downstream (Fig. 9.11(A)). The outer bank is eroded as a result of the higher flow velocity, whereas deposition takes place along the inner (convex) bank forming a **point bar** (Figs 9.11(B), 9.12).

If meandering is an inherent feature of fluid flow we might ask why all channels do not meander. Experimental studies have shown that at a given discharge water introduced at an angle into a flume can, as expected, generate excellent meanders. If the gradient is reduced, however, there is a critical minimum gradient below which a straight channel will form and be stable even though lateral flow is still introduced into the flume (Fig. 9.13). The explanation for this behaviour seems to be that, although secondary helical flows are established at very low gradients under the prevailing low flow energy conditions they are not suf-

Fig. 9.12   *Point bars clearly evident along a meandering channel, Louisiana, USA.*

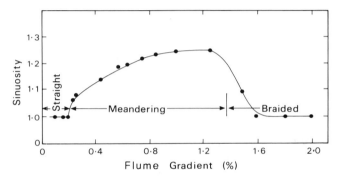

Fig. 9.13   *Relationship between valley gradient and thalweg sinuosity determined from experimental studies employing a flume. (After S. A. Schumm and H. R. Khan (1972)* Geological Society of America Bulletin *83, Fig. 6, p. 1761.)*

ficiently powerful to form alternating bars and cause bank erosion. By contrast, in high-velocity flows in steep gradient channels the downstream momentum of flow prevents significant cross-channel flows and the development of alternating bars, and consequently straight, braided channels are formed.

9.3.1.2   Braided channels
The development of braided channels is favoured by several factors (Fig. 9.9). In addition to a steep channel gradient the most important appear to be a large proportion of coarse material being transported as bed load, and readily erodible bank material which enables channel shifts to occur with relative ease. Once formed, bars in braided channels can become rapidly vegetated and thereby stabilized as islands. This points to the role of the highly variable discharge which is typical of many braided rivers. By promoting alternating channel degradation and aggradation, large fluctuations in discharge help to suppress the establishment of vegetation on braided-channel bars.

As discharge declines after a flood peak the coarse bed load is the first to be deposited in the channel. This material forms the nucleus of bars which grow downstream as the flow velocity is reduced and finer sediment accumulates. With further decreases in discharge the water level progressively falls and the bars are gradually exposed. During subsequent floods some, or all, of the bars in a braided channel may be submerged depending on the discharge attained. During large floods braided channels can experience major diversions of flow.

9.3.1.3   Channel pattern change
Experimental studies, together with numerous field studies, have shown how channel patterns change in response to changes in controlling variables. An important result from these flume studies is the crucial control that appears to be exerted on channel sinuosity by stream power. Since channel gradient influences stream power it is not surprising that gradient is also related to channel sinuosity. There are thus a number of interrelated variables including discharge, flow velocity, sediment concentration and size, and channel gradient that can change in response to tectonic, climatic and anthropogenic factors and promote a change in channel pattern. The effects of climatic change are examined further in Chapter 14, while tectonic influences are discussed in Chapter 16.

9.3.2   Alluvial channels: hydraulic geometry

In addition to changes in plan form, alluvial channels may also adjust their cross-sectional form in response to changes in discharge. These cross-sectional components of channel form were termed **hydraulic geometry** by L. B. Leopold and T. Maddock in their innovative analysis of channel adjustments published in 1953. There are two components of hydraulic geometry that need to be considered. One involves the relationships representing adjustments over time **at-a-station**, that is, at a particular point along a channel. The other concerns downstream changes evident at a specific time.

9.3.2.1   At-a-station adjustments
The analysis of a large number of rivers has demonstrated rather consistent relationships between changes in discharge at-a-station and resulting adjustments in channel width, channel depth and mean flow velocity (Fig. 9.14). (Note that these relationships are plotted on logarithmic scales since velocity, width and depth are power functions of discharge) (Box 9.1). As discharge increases these dependent variables increase at different rates for different river channels depending on a number of controlling factors. Where bank materials are fine-grained and cohesive, channel sides are steep, and depth increases proportionately more rapidly than width as discharge becomes greater. Where channel materials are coarse and non-cohesive the banks slope

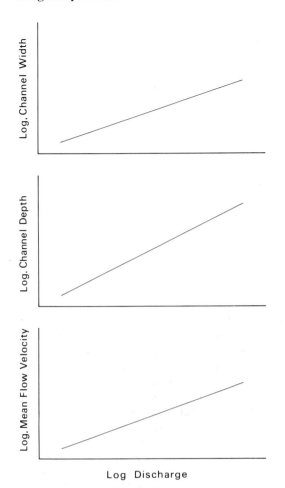

Fig. 9.14 *Schematic representation of at-a-station relationships between stream discharge and channel width, channel depth and mean flow velocity.*

gently towards the channel bed, and width increases rapidly in response to an increase in discharge. The influence of channel bank material helps to explain the observation that channels in semi-arid regions tend to adjust to an increase in discharge primarily by an increase in width rather than depth, whereas the converse is usually the case in humid environments where channels generally have more cohesive banks.

9.3.2.2   Downstream changes

For the majority of rivers, where seepage losses and evaporation are not significant, discharge increases downstream. As with at-a-station hydraulic geometry it is possible to plot corresponding changes in channel width, depth and flow velocity (Fig. 9.15). Because of the difficulty of acquiring data on bankfull discharge these relationships are normally expressed in terms of mean annual discharge. As with at-a-station hydraulic geometry, downstream adjustments are also represented as power relationships (Box 9.1). In general, channel width increases

Box 9.1   Hydraulic geometry

At-a-station adjustments of mean flow velocity, channel width and channel depth to changes in water discharge can be described by power function relations of the form:

$$\bar{v} = kQ^m \qquad w = aQ^b \qquad \bar{d} = cQ^f$$

where Q is the discharge, $\bar{v}$ the mean flow velocity, w the channel width, $\bar{d}$ the mean channel depth, k, a and c are constants, and m, b and f are exponents.

Since $\bar{v}\bar{d}w = Q$ it follows that

$$k + a + c = 1 \quad \text{and} \quad m + b + f = 1$$

Although the values of the exponents m, b and f vary, they seem to approximate to 0.33 for channels in cohesionless sediment. Channels in cohesive sediments have steep banks, so channel width increases only slowly with increasing discharge and the exponent b is low (~0.05). The velocity exponent is high where channel bed roughness decreases rapidly as discharge increases, such as in straight cobble-bed reaches.

The same power functions used to describe at-a-station adjustments in hydraulic geometry can be used to express downstream changes in flow velocity and channel cross-section. In a study of rivers in the mid-west of the USA the exponents m, b and f were found to be 0.16, 0.46 and 0.38 respectively. These values show that, in general, mean flow velocity increases downstream and that width increases more rapidly than depth. This indicates that large rivers are usually slightly wider relative to depth than small rivers.

proportionately more rapidly than channel depth as discharge increases downstream, while in meandering channels an increase in discharge is also accompanied by an increase in meander wavelength.

Until the statistical analysis of downstream hydraulic geometry in the 1950s it was widely believed that mean flow velocity decreases downstream in response to the general decrease in channel gradient. The analysis by Leopold and Maddock, however, suggested that this was not the case and that velocity typically increases slightly downstream. This might be explained by a decrease in frictional effects due to an increase in flow depth and a decrease in channel roughness (associated with a decrease in sediment size) which more than counteracts the effects of a reduction in channel gradient. Nevertheless, such a decrease is not common to all rivers; in some large rivers, for instance, velocity may be more or less constant downstream, while in small streams velocity may either increase or decrease downstream depending on specific channel conditions.

9.3.3   Alluvial channels: longitudinal form

A further component of downstream adjustments in alluvial channels is that of changes in channel gradient, and these can be considered at a range of scales. Looking at long

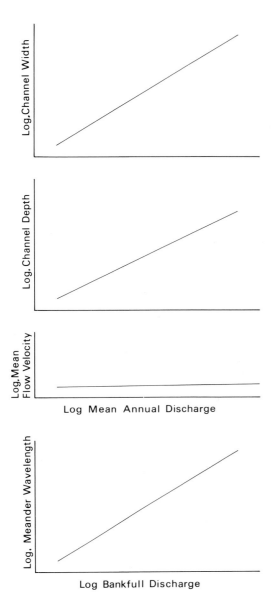

Fig. 9.15 Schematic representation of typical downstream relationships between mean annual discharge and channel width, channel depth and mean flow velocity, and between bankfull discharge and meander wavelength.

sections of a river course, channel gradient is *generally* seen to decrease downstream giving a concave-up **longitudinal profile (long profile)**. This downstream decrease in gradient is accompanied by an increase in discharge, and it can be explained by the ability of a river to transport the same quantity and calibre of sediment load in a lower gradient channel as discharge increases. Average sediment size does, of course, tend to decrease downstream as a result of sorting, attrition and weathering, and at a given flow velocity finer sediment can be transported in lower-gradient channels. None the less, the relationship between channel gradient and sediment size is not a simple causal

one since a progressive decrease in sediment calibre downstream could arise because coarser sediment, once deposited, cannot be transported over lower gradients. Neither channel gradient nor sediment size are independent variables and a feedback relationship operates between them.

The importance of increasing downstream discharge in producing concave profiles in alluvial channels is underlined by observations from rivers in semi-arid environments which experience a net loss of discharge downstream. Such rivers typically have a convex profile as a result of downstream aggradation during rare flood events.

A progressive decline in channel gradient in alluvial channels can be disturbed by either changes in base level downstream or by tectonic activity causing vertical movements of the channel bed itself. A fall in base level can give rise to a **knickpoint**, a discontinuity in the longitudinal profile of a river, and the downstream channel gradient is thereby increased. Alluvial channels will tend to adjust rapidly to such a change through an increase in flow velocity which leads to increased erosion in the steepened reach. The role of active tectonics is discussed in Section 16.4.

In addition to these broad-scale changes in gradient, other much smaller variations in channel slope can be observed in many rivers. Particularly in channel beds composed of material of a wide range of sizes, shallow and deep reaches are seen to alternate downstream. The shallows are formed by high points in the channel bed known as **riffles**. They are composed of coarser material and are characteristically spaced at about five to seven times the channel width. The deep reaches are called **pools** and have a bed of finer calibre than the riffles on either side. At low stage the water surface is steeper over riffles and the flow is shallower and more rapid, whereas over pools the water surface has a lower slope and the flow is deeper and slower.

Theoretical work has shown that even in straight, uniform channels, turbulence along the channel boundary will generate large-scale roller eddies associated with alternating acceleration and deceleration of flow (Fig. 9.16(A)). The dimensions of these eddies are a function of channel size, and their spacing averages about six channel widths. It seems likely that these downstream alternations of fast and slow flow lead to the development of zones of erosion (pools) and zones of sediment accretion (riffles) which eventually assume an equilibrium form. It is the riffles that are the sites of accretion and this may seem to contradict the faster flow observed over riffles. This faster flow, however, occurs during low stage when little sediment is in transit. At high stage when the flow velocity is sufficient for significant erosion and deposition to occur, it is the pools that experience higher bed velocities and greater shear stresses as a result of their lower roughness and the convergence of flow from the riffles upstream.

In Section 9.3.1.1 it was mentioned that lateral com-

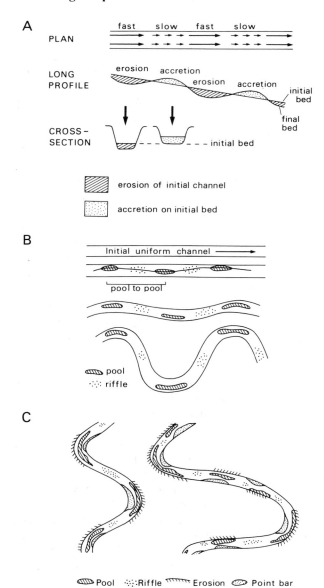

Fig. 9.16 *Development of pool and riffle sequences in channels and their relationship to meander morphology: (A) erosion and accretion of a channel bed corresponding to alternate zones of fast and slow flow; (B) transformation of a straight to a meandering channel in relation to pool spacing; (C) the development of additional pools and riffles with the lengthening of a meandering channel. (After K. Richards (1982) Rivers: Form and Process in Alluvial Channels. (Methuen, London, Fig. 7.2C, p. 184; and G. H. Dury, (1969) in: R. J. Chorley (ed.) Water, Earth and Man (Methuen, London) Fig. 9.II.4, p. 180.)*

ponents of flow are present even in straight channels which are capable of initiating channel sinuosity by selective bank erosion under certain hydraulic and sedimentological conditions. As a result of the greater shear stress exerted on the channel boundary in pools during channel-forming high discharges, banks are eroded preferentially in pools. There is consequently a correspondence between the spacing of

pools and riffles and the wavelength of the meanders that eventually develop, meander wavelength being approximately twice the pool spacing (Fig. 9.16(B)). This situation may change, however, if the channel between each meander loop is lengthened in which case new pools and riffles may develop (Fig. 9.16(C)).

9.3.4 The alluvial channel system

Figure 9.17 summarizes the relationships between the numerous components of the alluvial channel system discussed in the preceding sections. It is a speculative attempt to illustrate the way in which alluvial channels can adjust to changes in the independent variables of valley slope, discharge and bed material size and sorting.

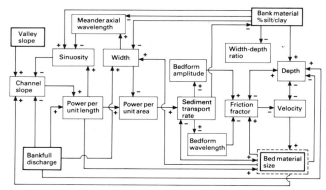

Fig. 9.17 *The alluvial channel system. Independent variables are indicated by a bold outline. Direct relationships are shown by + and inverse relationships by −. Indeterminate relationships are indicated by double-headed arrows. (After K. Richards (1982) Rivers: Form and Process in Alluvial Channels. Methuen, London, Fig. 1.8, p. 26.)*

Under a relatively consistent hydrological regime alluvial channels experience constant adjustments in response to changes in hydraulic and sedimentological variables which maintain a more or less stable channel form. Such a state of equilibrium is usually described as **grade**. The concept of a **graded river** was elegantly expressed by J. H. Mackin in 1948.

A graded river is one in which, over a period of years, slope and channel characteristics are delicately adjusted to provide, with available discharge, just the velocity required for the transportation of the load supplied from the drainage basin. The graded stream is a system in equilibrium; its diagnostic characteristic is that any change in any of the controlling factors will cause a displacement of the equilibrium in a direction that will tend to absorb the effect of the change.

The last sentence provides an excellent description of the process of negative feedback and, indeed, most alluvial channels for most of the time exhibit this form of system behaviour. But changes in controlling variables in most rivers occur very frequently in comparison with the time taken for adjustments in channel properties to occur so a graded channel is most realistically regarded as one in a state of quasi-equilibrium rather than true steady-state equilibrium.

9.3.5 Bedrock channels

Bedrock channels represent a dramatic contrast with alluvial channels in that they are capable of only very slow adjustment of form in response to changes in discharge, sediment load, gradient and other factors. Bedrock channels cannot, therefore, be realistically analyzed in terms of the concept of grade because their response time is so long.

It is generally held that rivers are most commonly cut into bedrock in the upper part of their courses where channel gradients are usually steeper and the coarseness of the load they carry is greater and thus more effective as an agent of abrasion. Although this is undoubtedly true in the majority of cases, there are a number of instances where major rivers flow in alluvial channels for much of their upper course before plunging into bedrock channels as the channel gradient steepens markedly downstream. A number of large rivers in Africa exhibit this characteristic, for example the Orange River in southern Africa (see Fig. 16.24).

Little research has been carried out into bedrock channels in comparison with the abundance of studies of alluvial channels. From the research that has been undertaken it appears that the longitudinal profile is normally much more irregular than in alluvial channels, while the cross-sectional form of bedrock channels is greatly influenced by structural controls with fluvial erosion tending to widen and deepen joints or other zones of weakness.

More attention has been paid to the smaller-scale erosional features evident in bedrock channels. Plucking and cavitation gives rise to irregular channel-bed surfaces, while abrasion and corrosion tend to smooth scour marks. The most characteristic erosional form is the **pothole**. Three types can be recognized: a narrow cleft in a rock face caused by the cutting back of a waterfall; shallow depressions resulting from separated flows which promote a secondary circulation in the water flow on the channel bed; and deep and often almost perfectly circular eddy holes where a vortex-like circulation has become established.

9.3.5.1 *Longitudinal profile*
Irregularities in the longitudinal profile of bedrock channels can arise from a downstream steepening of gradient as a consequence of a fall in base level, but the knickpoint produced may often be difficult to distinguish from a channel bed discontinuity related to the differential erosion of rocks of contrasting resistance. Discontinuities may also be produced by vertical movements along a fault lying across a

channel or by material deposited in the channel by, for instance, a landslide. The most obvious discontinuities in the longitudinal profile of bedrock channels are represented by waterfalls and rapids. The term **waterfall** is applied where there is a vertical fall of water, a succession of waterfalls giving rise to a **cataract**. **Rapids** are less marked irregularities in the channel bed, consisting of short, steep reaches which may be entirely submerged at high stage.

Given their importance in deciphering the history of river systems, waterfalls have been remarkably little studied. It is widely considered that the great majority of waterfalls develop as a result of the erosion of weak rock from beneath a resistant caprock. The classic example are held to be the Niagara Falls not only because of their popularity as a tourist venue but also because G. K. Gilbert carefully considered their development. He emphasized the importance of the undermining of the shale at the base of the falls by the rolling of large limestone boulders detached from the caprock in the cascading water flow. The Niagara model is, however, far from universal since many waterfalls possess no overhanging caprock but instead are buttressed at the base (Fig. 16.24).

Contrary to Gilbert's widely accepted view that falling water has little erosive power unless it carries a coarse sediment load, calculations of both the force exerted by a large mass of falling water and the evidence of rapid scour at the base of dam overflows indicates that the impact of water is probably the predominant mechanism of fluvial erosion operating on waterfalls. Indeed **plunge pools** are commonly found at the base of waterfalls and the overall process of waterfall retreat is also aided by cavitation and the weakening through weathering of rock surfaces exposed to frequent wetting. The rate at which waterfalls retreat and whether the discontinuity in the longitudinal profile that they represent is smoothed over time are important questions which still remain to be satisfactorily answered.

9.3.5.2 *Incised meanders*
Although we noted above that meandering patterns are common in alluvial channels, meandering channels are also found incised into bedrock. **Incised meanders** are particularly well developed in horizontally bedded strata and form when a river cuts down through its alluvium and into the underlying bedrock. Two types of incised meander can be identified. **Intrenched meanders** are symmetric forms and develop where downcutting is sufficiently rapid to minimize the lateral migration of the meanders. This can occur where a significant fall in base level leads to the upstream migration of a knickpoint. **Ingrown meanders**, by contrast, are asymmetric as a result of lateral meander migration occurring concurrently with relatively slow incision. This slow downcutting is initiated by regional warping rather than rapid base level changes. Lateral migration of meanders can proceed to the point where two meander loops cut

Fig. 9.18   *Deformed, intrenched meander, part of the Goosenecks of the San Juan River, Utah, USA.*

through a bedrock spur and form a **natural arch**.

The classic intrenched meanders of the San Juan River, a tributary of the Colorado River in the south-west USA, are highly deformed and compressed and appear not to have been inherited directly from a meandering pattern developed in overlying alluvium (Fig. 9.18). Formation of the meanders seems to have been initiated through rapid downcutting of the Colorado River which lowered the base level of the tributary San Juan. As headward erosion progressed up the initially alluvial channel of the San Juan the underlying bedrock would have been encountered first on the lower downstream limb of each meander. Further downstream, migration of this part of the meander would then be severely retarded, but the upstream limb could continue to shift downstream until it also encountered bedrock thereby compressing the meander loop (Fig. 9.19). Interestingly, incised meanders on the Colorado Plateau exist on the upstream but not the downstream flanks of folds. This seems to be because on the upstream flank intrenching meanders would encounter bedrock, but on the downstream flank the river flows down dip and the increase in channel gradient destroys the meandering channel pattern before it can become incised.

9.4   Fluvial depositional landforms

Deposition by rivers occurs predominantly in the bottom of valleys where gradients are low, at locations where there is a significant change in gradient, or where channelled flow diverges, and is consequently reduced in depth and velocity. It must be emphasized, however, that deposition is not exclusive to any part of the fluvial system. Four major types of fluvial deposition can be distinguished: **channel deposits** and **channel margin deposits** which accumulate within and along river channels; **overbank deposits** formed when bankfull discharge leads to the deposition of fine sediments beyond the confines of the channel itself; and **valley margin deposits** which accumulate at the base of valley slopes (Table 9.3).

9.4.1   Floodplains

Except in mountainous terrain most river channels are flanked by an area of subdued relief termed a **floodplain** formed by deposits laid down when the river floods. Low magnitude – high frequency floods cover only a part of the floodplain, and it is only during rare major floods that the entire floodplain is inundated. Depths of flood water range

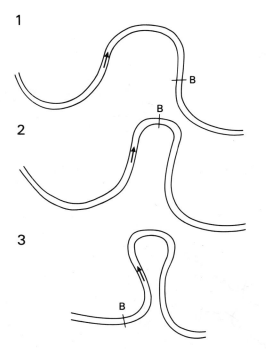

Fig. 9.19 Development of a deformed, intrenched meander (1, 2, 3) when the downstream limb is fixed by bedrock. The letter B indicates the location at which the channel enters bedrock. Upstream of B the channel is in alluvium. (After S. A. Schumm (1977) The Fluvial System. Wiley, New York, Fig. 6–15, p. 200.)

plains, this lies within a much more extensive area of older floodplain deposits. Where closely spaced river channels flow across a region of subdued topography adjacent flood-plains may merge to form extensive **alluvial plains.**

Two modes of sediment deposition can be identified; **lateral accretion deposits** are laid down as the river channel migrates across its floodplain, while **vertical accretion deposits** accumulate beyond the confines of the channel during floods. Lateral accretion deposits are formed by both meandering and braided channels, but in the latter case a large part of the floodplain effectively operates as part of the active river channel at any one time.

Rapid changes in channel courses in braided rivers leave abandoned channels floored by coarse bed-load material which, if they survive reworking during subsequent channel shifts, are eventually filled by organic material and fine sediments laid down during floods. Lateral accretion by meandering channels occurs through the more gradual growth of **point-bar deposits** on the convex side of meander bends (Fig 9.12, 9.20). The overrunning of one meander loop by another produces a **meander cut-off**, or chute, which shortens the channel course (Fig. 9.8). The abandoned meander loop is gradually isolated by the deposition of bed material at each end by the main channel and becomes an **oxbow lake**. Eventually the oxbow lake accumulates channel-fill deposits laid down during floods.

Whereas lateral accretion deposits are constantly being formed and reworked, vertical accretion occurs only during floods when bankfull discharge is exceeded. Morphologically the most prominent depositional landforms in most floodplains are **levees** (Fig. 9.20). These are ridges which lie parallel to the river channel and may rise to a height

from a few centimetres up to several metres for large floods in major rivers. As with the dimensions of river channels, the width of floodplains in most rivers is roughly proportional to discharge. The active floodplain of the lower Mississippi is about 15 km across, but, as with most flood-

Table 9.3 Classification of valley sediments

PLACE OF DEPOSITION	NAME	CHARACTERISTICS
Channel	Transitory channel deposits	Primarily bed load temporarily at rest; part may be preserved in more durable channel fills or lateral accretions
	Lag deposits	Segregations of larger or heavier particles, more persistent than transitory channel deposits
	Channel fills	Accumulations in abandoned or aggrading channel segments; ranging from relatively coarse bed load to fine-grained oxbow lake deposits
Channel margin	Lateral accretion deposits	Point and marginal bars that may be preserved by channel shifting and added to overbank floodplain
Overbank floodplain	Vertical accretion deposits	Fine-grained sediment deposited from suspended load of overbank flood water; including natural levee and backswamp deposits
	Splays	Local accumulations of bed-load materials spread from channels on to adjacent floodplains
Valley margin	Colluvium	Deposits derived chiefly from unconcentrated slope wash and soil creep on adjacent valley sides
	Mass movement deposits	Earthflow, debris avalanche, and landslide deposits commonly intermixed with marginal colluvium; mudflows usually follow channels but also spill overbank

Source: After P. C. Benedict *et al.* 1971, *Journal of the Hydraulics Division, Proceedings of the American Society of Civil Engineers* 97, Table 2 – Q.1, p. 44.

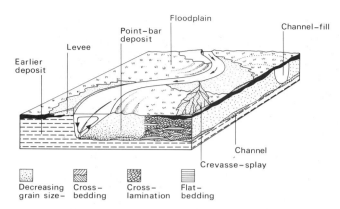

Fig. 9.20 *Depositional forms developed along a meandering channel. Note how levees give rise to steep high banks where they are undercut on the concave side of meander bends. After J. R. L. Allen (1964) Sedimentology 3, Fig. 4, p. 168 and 1970, Geological Journal 7, Fig. 1, p. 131.)*

of several metres above the general level of the floodplain. When bankfull discharge is exceeded water starts to spread rapidly across the floodplain. The decrease in depth in comparison to the river channel rapidly reduces the competence of the flow and the coarser fraction of the suspended load is promptly deposited to form levees. The presence of levees affects the movement of flood waters once bankfull discharge is exceeded. Initially the water is ponded by levees, but as the level rises it breaks through forming **crevasses**. The accelerated flow through crevasses is able to transport a relatively coarse suspended load which is quickly deposited as the flow disperses to form fan-shaped **crevasse-splay deposits** in the **backswamp zone** beyond the levee barrier (Fig 9.20). Flow diverted through a crevasse may travel many kilometres downstream before it is able to regain access to the main channel. This may not occur until a tributary is reached, but tributaries themselves may be diverted downstream by levee development along the main channel. Occasionally the main channel itself may shift its course into the backswamp zone.

Many of the world's floodplains have thick accumulations of vertical accretion deposits, but this is probably a reflection of the particular conditions of the past 10 000 a or so when most river systems have been responding to a rise in base level in response to the post-glacial rise in sea level (see Section 17.6.1). Under conditions of stable base levels we would expect lateral accretion deposits to be quantitatively more significant.

9.4.2 Alluvial fans

An **alluvial fan** is a body of sediment whose surface form approximates to the segment of a cone which radiates downslope from a point on a mountain front, usually where a stream emerges (Fig. 1.10(C)). Most alluvial fans have a radius of less than 8 km, but under certain conditions fan radii may exceed 100 km. In form they have concave-up long profiles, but convex-up cross-profiles. The mean surface slope of fans generally ranges from 1° to 5°, but at the fan apex gradients can exceed 10°. Since the slope of streams emerging at the mountain front is usually similar to the gradient of the upper fan surface, it seems that the deposition that causes fan building occurs primarily as a result of the sudden change from a confined to an unconfined condition as the stream leaves the mountain gorge rather than as a result of reduced channel gradient. The calibre of sediment deposited generally decreases down-fan, but fan deposits are often poorly sorted as they are frequently laid down by torrential floods. In catchments containing abundant, easily erodible debris, sediment concentrations may reach the point where high-viscosity debris flows are generated.

Alluvial fans develop where there is a relatively abundant sediment supply, adequate relief for vertical fan growth and a suitable location for sediment accumulation. Favourable environments include active faulted mountain fronts where the rate of mountain uplift, or adjacent basin subsidence, exceeds the rate of downcutting of the trunk stream feeding the fan. Individual fans along a mountain front may grow laterally to the extent that they coalesce to form a continuous piedmont sedimentary apron. In arid environments such landforms are known as **bajadas** (Figs 1.10(B), 9.21). In humid environments, where alluvial fans tend to be larger, the coalescence of fans can produce immense, gently inclined alluvial slopes, such as those that flank the southern margin of the Himalayas.

Although some alluvial fans in arid environments are simple undissected forms, most are dissected – that is, the trunk channel is entrenched below the upper part of the fan surface and deposition is focused towards the fan toe (Fig. 9.22). **Fan-head trenching** will occur if sediment yield decreases, or channel gradient and flow velocities increase. Such changes can be brought about by climatic changes, increased tectonic activity or even human-induced land use changes in the upstream basin. But we might also anticipate that fan incision will occur in the long term without any change in external factors simply as a result of the reduction in sediment yield as relief, and therefore rates of erosion in the basin, progressively declines through time as the moutain mass is gradually lowered.

Not surprisingly, fan area has been found to be positively correlated with source basin area, although there may be an order of magnitude difference in fan area for a given contributing area. Some of this variation may be explained by lithological factors. Basins in sandstone, for instance, have been found to have smaller associated alluvial fans than those underlain by shales or mudstone. This is presumably

Fig. 9.21 *Bajada formed along the faulted, northern margin of the Turfan Depression, Xinjiang Autonomous Region, China. The area covered by the image is about 180 km across. (Landsat image courtesy A. S. Walker.)*

because sandstone is less easily eroded and gives rise to lower sediment yields. Tectonic factors can also influence fan area. In Death Valley progressive eastward tilting has confined the east-side fans but enabled the west-side fans to extend several kilometres beyond the mountain front (Fig. 1.10(B)).

Alluvial fans in humid environments are distinct in a number of respects from those in arid environments. Apart from being generally smaller and steeper, fans in dry regions are fed by ephemeral stream channels and active deposition tends to move unpredictably from one part of the fan to another. In large arid fans only a very small proportion (usually < 5 per cent) is active during any one flood event. In contrast humid fans are fed by perennial, and often braided, channels which tend to migrate progressively across the fan surface. Such channel migration is well illustrated by the giant Kosi Fan on the southern flanks of the Himalayas (Fig. 9.23). The course of the Kosi River across the fan has moved more than 100 km to the west over the past 250 a. Coarse debris is transported only a few kilometres beyond the fan head and the greater part of the fan is composed of fine- to medium-grained deposits.

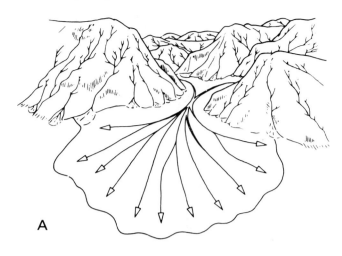

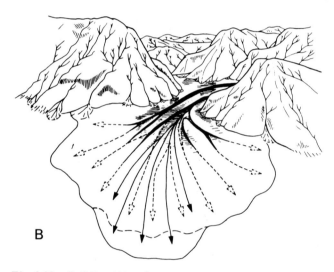

Fig. 9.22 *Building (A) and entrenchment (B) phases of alluvial fan development. (From R. U. Cooke and A. Warren (1973)* Geomorphology in Deserts. *Batsford, London, Fig. 3.8, pp. 186 and 187, after L. K. Lustig (1965) United States Geological Survey Professional Paper **352-F**, Fig. 137, p. 185.)*

9.4.3 River terraces

Changes in channel gradient, discharge or sediment load can lead to a river channel incising into its floodplain. The original floodplain is thereby abandoned and is left as a relatively flat bench, known as a **river terrace**, which is separated from the new floodplain below by a steep slope. As well as forming in the alluvial fill of river valleys, river terraces can also be cut into bedrock and in this case they are covered by only a thin sediment veneer. The term **strath terrace** is sometimes applied to these forms.

River terraces are inclined downstream but not always at the same inclination as the active floodplain. A valley side may contain a vertical sequence of terraces (Fig. 9.24). The

lowest will be the youngest and may retain traces of floodplain morphology, while the highest will be the oldest and will usually be partly degraded. **Paired terraces** form when vertical incision is rapid in comparison with the lateral migration of the river channel (Fig. 9.25). Morphologically similar features can, however, be produced by resistant beds in flat-lying strata. These give rise to **structural benches** and structural controls must be eliminated before a river terrace interpretation is accepted. **Unpaired terraces** form where lateral shifting of the channel is relatively rapid; this results in the river cutting terraces alternately on each side of the valley floor. Since valleys experience phases of aggradation and degradation, once formed terraces can subsequently be covered by sediment. Thus, we would expect complex sequences of buried and partially eroded terraces to lie beneath modern floodplains (Fig. 9.26). Terraces can be formed under a wide range of circumstances. Unpaired terraces do not, in fact, require any change in external conditions since they can form simply through the progressive incision of a river channel migrating from one side of its floodplain to the other.

Many episodes of channel entrenchment and terrace formation probably arise from changes in base level or fluctuations in climate, although tectonic deformation along river courses can also lead to local terrace development (see Section 16.4). A fall in base level can lead to terrace development if it generates a downstream increase in channel gradient. This is then propagated upstream through knickpoint retreat. The past 3 Ma have seen frequent changes in sea level of up to 100 m or so, and this has created conditions highly favourable to terrace formation. Many other river terraces currently present in the landscape are attributable to climatic fluctuations which lead to changes in channel discharge and sediment load. As with base level changes, climatic fluctuations have also been frequent and rapid over the past 3 Ma or so. The response of river systems to such environmental changes are complex and can vary dramatically depending on local channel conditions. For instance, during glacials the Mississippi incised into its floodplain in its lower valley producing a series of terraces (see Section 17.6.1), but in the upper part of its basin the great increase in sediment supplied from the Laurentide ice sheet led to channel aggradation.

The tendency of fluvial systems to aggrade or incise is particularly sensitive to climatic fluctuations in semi-arid regions. This is because in such environments modest changes in annual precipitation can produce significant changes in vegetation cover which are reflected in large changes in the rate of sediment supply to stream channels. In the south-west USA many ephemeral stream channels, known locally as **arroyos**, show evidence of phases of aggradation and entrenchment. Dramatic changes have occurred in a period of a few hundred years with the most

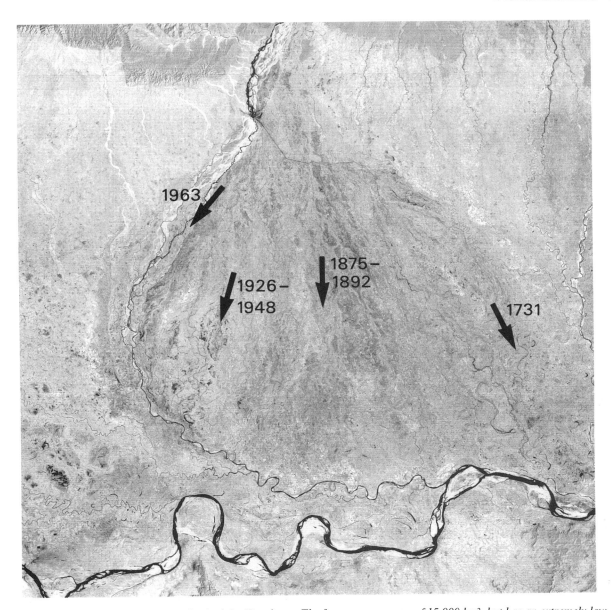

Fig. 9.23 *The Kosi Fan on the southern flank of the Himalayas. The fan covers an area of 15 000 km², but has an extremely low gradient averaging 1 m km⁻¹ at its apex, and only 0.2 m km⁻¹ at its toe near the Ganges River (which flows west to east across the bottom of the image). The Kosi River drains an area of very high relief and monsoonal climate, and these factors combine to generate a high sediment yield. Dates are given for the progressive westward migration of the course of the Kosi River since 1731. The image covers an area about 180 km across. (Landsat image courtesy N. M. Short.)*

recent period of entrenchment and terrace formation occurring between the 1860s and around 1915. The cause of this episode of channel entrenchment is uncertain. In some areas it is apparently due to subtle climatic fluctuations, but in other localities significant changes in land use may also have played a role by influencing the supply of sediment to channel systems. The possibility remains, however, that alternating phases of aggradation and degradation may represent the normal behaviour of fluvial systems even in the absence of changes in external controls.

9.5 Fluvial activity through time

9.5.1 Geomorphic thresholds, complex response and episodic erosion

In Section 7.6 we noted how slope behaviour can be analyzed in terms of the exceeding of extrinsic or intrinsic thresholds. Similar reasoning can be applied to fluvial systems, but in this case changes through time in the morphology of the landform itself can lead to the crossing

Fig. 9.24 *Terrace sequence formed in glacial outwash debris, Cave Stream, Canterbury High Country, South Island, New Zealand.*

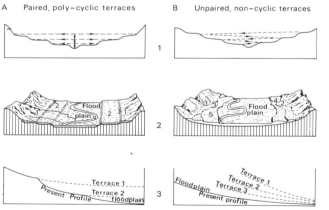

Fig. 9.25 *Paired and unpaired terraces. Note how the vertical spacing of terraces is more or less retained in paired terraces but converges downstream in unpaired terraces. (After B. W. Sparks (1972) Geomorphology (2nd edn). Longman, London, Figs 9.5, 9.7 and 9.8, pp. 296 and 297; and W. D. Thornbury, (1969) Principles of Geomorphology (2nd edn). Wiley, New York and London, Fig. 6.9, p. 157 after Longwell et al.)*

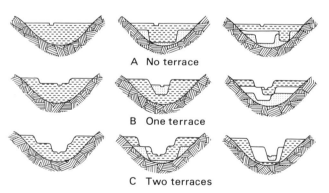

Fig. 9.26 *Valley cross-sections showing some possible terrace and alluvial-fill combinations. (After L. B. Leopold and J. P. Miller (1954) United States Geological Survey Water-Supply Paper 1261, Fig. 2, p. 5).*

of a threshold. An example is provided by ephemeral stream channels. These may experience periods of sediment accumulation on the valley floor until a point is reached where the channel gradient has been so increased that rapid erosion ensues. This is an example of a **geomorphic threshold** since the breaching of the threshold is inherent in the way the morphology of the landform changes through time. (Note, however, that the term geomorphic threshold has been applied more broadly by some writers to encompass both intrinsic changes and external factors – I retain the more specific usage here.) The existence of geomorphic thresholds means that many dramatic adjustments in fluvial systems which have hitherto been ascribed to the effects of external factors may simply be the result of the inherent behaviour of the landform itself. We must clearly be very careful in our interpretations of river channel adjustments, especially in climatically variable and tectonically active environments where we might tend to assume that all dramatic landform changes are the result of external factors.

The appreciation of the importance of thresholds in fluvial systems has important implications for the way in which we conceive how such systems change through time. The existence of thresholds indicates the inability of river channels necessarily to adjust rapidly to new equilibrium conditions, and we would expect different components of the fluvial system to respond at different rates to external changes. Because of the intricate relationships between these different components – alluvial channels, tributary streams, valley-side slopes and divides – we would expect fluvially-eroded landscapes to exhibit a **complex response** as the effects of any change in external factors, such as those arising from tectonic activity or climatic change, diffuse through drainage networks.

This behaviour has been demonstrated experimentally in laboratory drainage systems where a base-level fall at the basin outlet leads to successive phases of incision and aggradation as the system hunts for a new equilibrium (Fig. 9.27). The fall in base level initially promotes incision, but as the increase in channel gradient diffuses throughout the drainage system the sediment supply from the upper tributaries increases to the point where the downstream channel is incapable of transporting all the sediment being delivered and aggradation begins. This aggradation then extends upstream, reduces channel gradients and leads to a decrease in upstream erosion and sediment supply. This in turn allows channel incision to begin again downstream. The response of the drainage basin is characterized by **episodic erosion** punctuating periods of deposition. In this model of drainage basin development, features such as flights of river terraces and alluvial-fill sequences may simply reflect the attempt of the system to re-establish equilibrium after an initial perturbation and need not be indicative of successive changes in external variables. The concepts of geomorphic threshold, complex response and episodic erosion clearly have implications for the way we view the long-term evolution of fluvially-eroded landscapes, and we examine this topic in Chapter 18.

9.5.2 Magnitude and frequency of fluvial activity

A key question addressed by fluvial geomorphologists

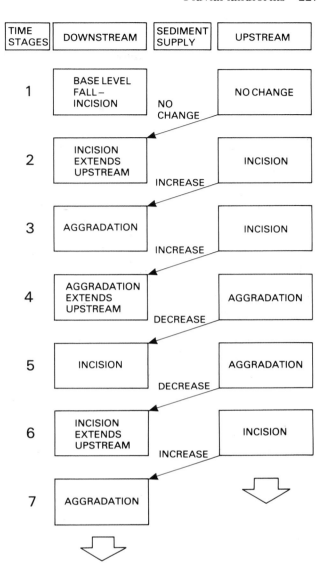

Fig. 9.27 *Schematic representation of complex response in a drainage system as a result of a fall in base level.*

since the early 1960s has been at what relative magnitude of discharge is most sediment transported in rivers and most geomorphic work accomplished. Although sediment transport increases with increasing flow velocity and discharge, high flows occur much less frequently than low discharges. In a highly influential paper published in 1960, M. G. Wolman and J. P. Miller argued that in most river systems about 90 per cent of the total sediment load is transported by relatively modest flows, that is, those that have a recurrence interval of up to 5–10 a. In most cases large floods individually transport large sediment loads, but they do not occur with sufficient frequency to have a major impact on overall sediment transport rates. Although there are some discrepancies between studies of different rivers with contrasting proportions of high discharges, most researchers support the view that bankfull discharge corresponds closely with the most effective sediment-transporting flows. This observation is of particular interest because there is considerable evidence supporting the view that in many rivers bankfull discharge is generally the most significant in determining channel morphology.

Although the prevailing philosophy of fluvial geomorphologists is that most rivers for most of the time are in a state of quasi-equilibrium in which channel morphology is broadly adjusted to prevailing discharge and sediment load, some researchers have highlighted the critical role of major floods in certain kinds of fluvial systems. There are numerous examples of high magnitude floods which have been seen to produce either little, or very short-term, change in fluvial channel form, but there are also instances where spectacular changes have occurred. One such case was associated with a storm over a part of central Texas, USA, on 2 August 1978. At least 790 mm of rain fell in just 24 h in some localities, the most intense rainfall ever recorded in the USA up to that time. At a gauging station on the Medina River the peak flow was 15 m deep and the discharge reached 7920 m^3 s^{-1} from a catchment area of a little over 1200 km^2. The recurrence interval of such a flood cannot be accurately estimated from the brief, 38 a discharge record available, but probably lies in the range 100–500 a. The geomorphic effects of this flood were considerable. Trees lining the low-flow banks of the Medina River were ripped up and the underlying alluvial sediments which had accumulated over centuries suffered extensive erosion. Boulders over 2 m in diameter were detached from jointed bedrock ledges and carried tens of metres downstream. In other channel sections giant ripples were formed; the largest of these were composed of gravel and boulders and their enormous wavelength of up to 80 m indicates a flow velocity of 3–4 m s^{-1} and a flow depth of 10 m.

The role of flood events in shaping fluvial landforms depends in part on the frequency distribution of high discharges. In semi-arid environments, for instance, rare flood events may be relatively important in determining channel characteristics because the much smaller discharges occurring for the great majority of time will be ineffective in reshaping the channel between these major flood events. Another very important factor, however, is the resistance of the materials experiencing the effects of flood discharges. To play a significant geomorphic role the erosional force exerted by floods (expressed as specific stream power) must exceed the resistance of materials, including vegetation, forming the channel bed and banks. The important point here is that there is an enormous range in the resistance of different kinds of channel materials, and consequently a wide range in the critical discharge at which significant channel modification can occur.

Most alluvial channels have a low resistance threshold and are highly sensitive to floods of relatively small magnitude and high frequency. When a major flood occurs there is a rapid response in terms of channel morphology and bed roughness, but there is equally a relatively prompt readjustment to lower flow conditions after a flood event. The situation in bedrock channels, or channels composed of very coarse alluvium, is very different. In highly resistant bedrock, for instance, erosion and channel modification will only occur during extremely high magnitude and very rare floods. Since such floods are usually short-lived, there is insufficient time for bed morphology to achieve 'equilibrium' with flow conditions, and consequently the resulting 'disequilibrium' forms may persist for a long period of time in certain environments (see Section 11.4.1).

9.6   Fluvial systems in limestone terrains

The susceptibility of rocks composed predominantly of calcium carbonate to the weathering process of carbonation gives rise to a distinctive form of fluvial activity on limestone terrains. We have already discussed the development of solutional forms on limestone associated with weathering and the development of karst scenery in general terms (see Section 6.4.1), but we have yet to consider specifically the genesis of fluvial landforms in such environments.

The distinctive nature of fluvial activity on limestone arises from the high permeability of the bedrock and the diversion underground of a variable, but in many cases significant, proportion of runoff. The permeability of rock is in part related to its **primary porosity**, that is, the proportion of the rock occupied by voids. (Strictly it is the **effective porosity** that is important – that is, the proportion of voids that are connected and thus able to transmit water.) In many types of limestones, however, it is their **secondary porosity**, arising from the solutional enlargement of joints, fissures and bedding planes, that makes by far the greatest contribution to their permeability. Ultimately such enlargement can give rise to caves. Since such secondary porosity increases over time as calcium carbonate is removed in solution, the proportion of subsurface flow depends both on the lithological properties of the limestone and on the length of time that it has been subject to active dissolution.

9.6.1   Surface drainage

In addition to a topography characterized by numerous closed depressions, karst landscapes may also contain fluvially eroded valleys. In limestone regions, such as those underlain by chalk, where karst has not developed the landscape may be predominantly one of fluvially eroded valleys. If recently exposed through the removal of overlying impermeable strata these valleys may contain streams, but **dry valleys** with no active stream, or valleys which occasionally support short-term flow following a storm or longer period of heavy rainfall, are much more typical of limestone terrains.

In some cases only the lower section of a valley is dry as a result of the surface flow in its upper part disappearing underground via a **sinkhole**. If this situation persists the floor of the upper part of the valley may be lowered significantly below the level in its lower part due to the drastically reduced potential for erosion down-valley of the sinkhole. The two sections of the valley may eventually be separated by a limestone 'cliff', in which case the upper section is referred to as a **blind valley** (Fig. 9.28(A)). In other cases water may emerge at the surface on a valley floor as a spring or **resurgence**. This frequently occurs at the junction of an impermeable lithology and a thick over-lying limestone, and the marked difference in surface flow up-valley and down-valley of the spring causes a much more rapid incision of the down-valley section forming a **pocket valley** (Fig. 9.28(B)). A steep headwall formed in the limestone may be produced which gradually retreats through spring sapping at its base. An excellent example is provided by Malham Cove in Yorkshire, UK.

Allogenic valleys are formed where river courses cross from an area of strata capable of supporting permanent surface flow on to limestone (Fig. 9.28(C)). The length of such valleys on the limestone outcrop depends both on the discharge of the river and the permeability of the valley floor. Small streams may disappear underground within a few hundred metres, whereas large rivers may be able to sustain flow for several tens of kilometres or more. Allogenic valleys cut through massive limestone can form spectacular gorges, such as the 300 m deep Tarn Gorge in southern France.

Although individual dry valleys can generally be explained by the progressive loss of subsurface flow over time as the permeability of the underlying limestone increases, extensive integrated networks of dry valleys, such as those of the chalklands of southern England and northern France are somewhat more problematic. Numerous hypotheses have been proposed to explain their development. One idea is that a drainage system was initially established on overlying impermeable strata which provided sufficient surface runoff to cut a valley network in the underlying chalk as the latter was gradually exposed. Another possible explanation is that during the glacial episodes of the past few million years the ground became perennially frozen a short distance below the surface and this rendered the chalk effectively impermeable and thus allowed surface flow and valley formation to occur. Yet another possibility is that the water table has been progressively lowered through time at a faster rate than the upper tributary streams could lower their valley floors. This is particularly likely to happen where the strata are gently dipping (Fig. 9.28(D)), but it might also be a normal product of the dissection of highly permeable lithologies where lower lying trunk streams can maintain flow even during periods of low water table.

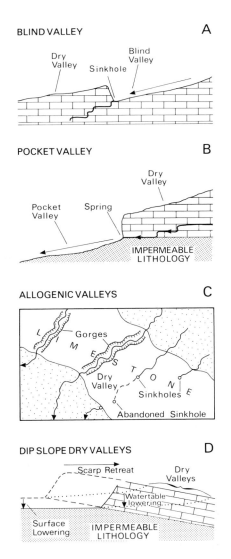

Fig. 9.28 *Formation of different types of valleys in limestone terrain.*

9.6.2 Subsurface drainage

The size and frequency of joints and fissures in limestone have an important influence on the type of subsurface water flow that occurs. At one extreme limestones, such as chalk, possess a high density of small fractures and consequently surface infiltration rates are uniformly high. Chalk is mechanically weak so fissures can only be enlarged to a certain point before they collapse. The rock is essentially permeable throughout (although water movement is still concentrated in the larger fissures) and there is a distinct and continuous water table. At the other extreme are the majority of limestones which are densely cemented and through which the greater amount of water flow occurs along major joints and fissures. This creates discrete systems of subsurface flow with no single continuous water table, although in time a somewhat more integrated system of subsurface drainage may develop as conduits are enlarged to the point where the rock is honeycombed with linked cave systems and there is one general water table level.

In limestone hydrology the term **vadose zone** is usually applied to the region lying above the water table in which voids may contain either air or water. This is equivalent to the soil moisture and aeration zones defined in Figure 8.1. Below the water table is the **phreatic zone** (equivalent to the ground water zone (Fig. 8.1)) in which all voids are completely filled with water. At any point below the water table there is a pressure equal to atmospheric pressure plus the **pressure head** (the product of the depth of water and its unit weight). The **hydraulic head** at this point is the sum of the pressure head and the elevation above some datum. Points of equal hydraulic head define **equipotential surfaces** within the phreatic zone. Water moves from high to low potential along flow paths which are orthogonal to these equipotential surfaces (Fig. 9.29). In lithologies with relatively uniform permeabilities, including limestones such as chalk, water flow in the phreatic zone approximates to **Darcy's law** (Box 9.2). This states that flow is proportional to the **hydraulic gradient**, that is, the difference in elevation between the inflow and outflow points divided by the distance between them. Darcy's law

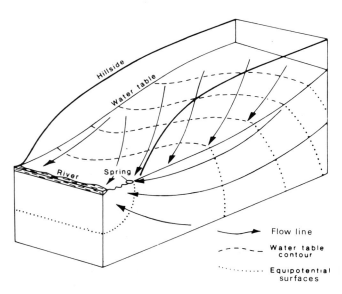

Fig. 9.29 *Water table contours, equipotential surfaces and flow lines. (After D. Ford and P. Williams (1989)* Karst Geomorphology and Hydrology *Unwin Hyman, London. Fig. 5.7, p. 138.)*

Box 9.2   Darcy's law

The flow of water flowing through a porous medium is given by

$$v = -K \frac{dh}{dl}$$

where v is the specific discharge (discharge per unit cross-sectional area), K the hydraulic conductivity and dh/dl the hydraulic gradient. The hydraulic gradient is determined by the length of flow (dl) and the difference in the elevation of the inflow and outflow points (dh). The negative sign on the right-hand side of the equation indicates a loss in hydraulic head in the direction of flow. Note that, strictly, hydraulic conductivity is not synonymous with permeability; the former refers to the ability of a medium to transmit fluids in terms of the properties of both the medium *and* the fluid (specifically its density and dynamic viscosity), whereas the latter refers to material properties only (such as void size and connectivity).

assumes that flow is laminar, and while this assumption is valid for slow rates of flow it is not for high flow rates or in larger conduits (above about 10 mm diameter) where flow is turbulent and frictional effects become important.

A major point of dispute in limestone hydrology is whether subsurface drainage and the forms it produces are best interpreted in terms of water moving through the mass of the rock as **diffuse flow** and establishing an extensive, continuous water table, or whether the concept of water movement largely confined to discrete conduits (**conduit flow**) with no single water table is more realistic. On the basis of the large quantity of field data that has now been collected it is apparent that both diffuse flow and conduit flow occur and that their relative importance varies depending on topography and lithology. Diffuse flow is more likely in low relief situations and where the limestone is horizontally bedded and relatively porous throughout. The conduit flow model is more valid, however, in regions of high relief with densely cemented limestones.

Concentrated subsurface flow is a prerequisite for the development of **caves**, which can be arbitrarily defined as cavities sufficiently large for a person to enter. Rates of solution increase rapidly once conduits have been enlarged to the size where turbulent flow begins to predominate. Although the initial opening of fissures to form larger conduits must occur primarily through solution, abrasion by sediment carried in subsurface flows can be important in the enlargement of caves especially where they are fed by sediment-charged rivers. Caves vary from those that permanently contain water, to those that are inactive or experience only occasional flow. They are preferentially developed in massive well-jointed limestone which has sufficient mechanical strength to resist crushing and roof collapse.

The most active debate about the formation of caves revolves around the level at which they develop relative to the water table. Most caves are presently located in the vadose zone and we would expect the most aggressive waters to be located in this zone. Nevertheless, such caves may initially have formed in the phreatic zone before being drained following a fall in the water table. Many other cave systems lie just below the present water table, and development here would be favoured by the high rates of ground water flow at this level and the presence of relatively undersaturated waters. Formation deep within the phreatic zone is also possible since here large volumes of water can be transmitted through the rock at high rates of flow as a result of high pressures. Although we might expect waters at depth within limestone to be saturated with respect to calcium carbonate and thus incapable of accomplishing further solution, the mixing of different water bodies can produce aggressive, undersaturated solutions (see Section 6.2.3.3). A fall in the water table would elevate the passages produced into the vadose zone. **Phreatic caves** typically have rather regular circular or elliptical cross-sections indicative of uniform rates of solution by flows completely filling the passage. In **vadose caves** vertical incision of passages is important as the lower part of the passage is subject to higher rates of solution than the partly air-filled upper part. If the water table is lowered and a phreatic passage is moved into the vadose zone a channel will begin to cut down into the base of the phreatic passage. Most large caves are probably of multiple origin, having experienced phases of both phreatic and vadose zone development. Structural factors, such as the orientation of bedding planes and fracture systems are certainly crucial in determining the overall architecture of cave systems.

Although the morphological evidence which would help document stages of cave development is usually destroyed during their formation, cave deposits can not only provide information on the most recent phases of cave history but also material which can yield palaeoclimatic data. Calcite deposits precipitated in the vadose zone are known as **speleothems** and include the familiar vertically hanging **stalactites** and upward-growing **stalagmites** as well as other more exotic forms. The dating of such speleothems can provide detailed information on the more recent phases of cave history by indicating periods of limestone solution and precipitation. Such data can be supplemented by sediments which may contain mineralogical and fossil evidence of surface environments as well as hydrological conditions within the cave itself.

Further reading

Several textbooks, each with a different emphasis, provide an excellent coverage of fluvial landforms and develop the topics introduced here. To be recommended are Knighton (1984), Morisawa (1985), Petts and Foster (1985) and Richards (1982), although in spite of being published two decades ago Leopold et al. (1964) still provide a stimulating survey of some of the core concepts of fluvial geomorphology. More recent treatments focusing on concepts and experimental work are Schumm (1977) and Schumm et al. (1987).

The drainage basin as a basic geomorphic unit is considered briefly by Chorley (1969) and in detail by Gregory and Walling (1973). An overview of the quantitative description of drainage basins is provided by Strahler (1964), while Abrahams (1984a) presents a comprehensive review of channel network analysis. The assumption that stream network topology is random is challenged by Abrahams (1987). Other aspects of channel networks are discussed by Abrahams (1984b) and Shreve (1967), and Howard (1967) considers the classification of drainage patterns. The hypsometric analysis of drainage basin form is outlined by Strahler (1952), while Melton (1958) provides a pioneering attempt to relate basin morphometry to geomorphic processes.

The best introduction to alluvial channels is the book by Richards (1982) and this is updated by papers in his later edited volume (Richards, 1987). Schumm (1985) provides an excellent introduction to channel patterns, while Hooke (1984) reviews ideas on the way meanders change through time. The classic paper by Langbein and Leopold (1966) is worth reading as a statement of the equilibrium symmetrical meander model, although this has been challenged by Carson and Lapointe (1983), Howard and Knutson (1984) and Lapointe and Carson (1986) who argue that asymmetry is a natural element of meander development. Carson (1986) looks at gravel-bed meandering rivers, while Callander (1978) represents a hydraulic engineer's approach to understanding meander development. Thompson (1986) points to the role of secondary flows in initiating meander development.

Braided channels are considered by Miall (1977), Parker (1976) and Smith (1974), while Leopold and Wolman (1957) attempt to define the conditions under which meandering, braided and straight channels develop. Schumm and Khan (1972) report experimental work on this problem, while Carson (1984a, 1984b) and Smith and Smith (1984) discuss field examples. The complexity of controls on channel form are emphasized by Ferguson (1987). Hydraulic geometry is defined and introduced in the classic paper by Leopold and Maddock (1953), while Ferguson (1986) provides an excellent review of this topic. Other aspects of hydraulic geometry are considered by Knighton (1975, 1987) and Park (1977). The longitudinal form of alluvial channels is discussed in detail by Richards (1982) and riffle–pool sequences are considered by Lisle (1979) and Richards (1976). Mackin (1948), with his detailed discussion of the concept of grade, and Richards (1982) provide overviews of the alluvial channel system as a whole.

Bedrock channels are a neglected landform and there is little further reading that can be recommended. Of particular value are the re-evaluation of waterfalls by Young (1985), the discussion of incised meanders contained in the book by Schumm (1977) and the discussion of the role of large floods in eroding bedrock channels by Baker (1988).

Turning finally to depositional landforms Lewin (1978) provides a useful review of floodplain forms and in a later paper he discusses the response of floodplains to changing channel patterns (Lewin, 1983). The factors influencing the development of alluvial fans is discussed by Bull (1977), Harvey (1989) and Hooke and Rohrer (1979), and in terms of specific examples in Rackocki and Church (1989). Green and McGregor (1987) and Schumm (1977) assess the range of factors that affect terrace formation, and Ritter (1982) illustrates these with a case study. The entrenchment of ephemeral stream channels is considered in detail by Cooke and Reeves (1976).

The importance of geomorphic thresholds, episodic erosion and complex response in the functioning of fluvial systems, as well as other types of geomorphic systems, is highlighted by Schumm (1979) and illustrated in the collection of papers edited by Coates and Vitek (1980). Some of the original experimental work from which these ideas were developed is presented by Schumm (1977) and Schumm and Parker (1973). The magnitude and frequency of fluvial activity is discussed in the classic paper by Wolman and Miller (1960) and reassessed by Wolman and Gerson (1978). The role of floods as channel-shaping events is discussed from various angles in the volumes edited by Mayer and Nash (1987) and Baker *et al.* (1988).

Both surface and subsurface drainage in limestone terrains is covered in several excellent texts and innumerable journal articles. Jennings (1985) provides a more than adequate introduction, but for a comprehensive coverage of the topic see Ford and Williams (1989) or White (1988).

References

Abrahams, A. D. (1984a) Channel networks: a geomorphological perspective. *Water Resources Research* **20**, 161–88.

Abrahams, A. D. (1984b) Tributary development along winding streams and valleys. *American Journal of Science* **284**, 863–92.

Abrahams, A. D. (1987) Channel network topology: regular or random? In: V. Gardiner *et al.* (eds) *International Geomorphology 1986* Part II. Wiley, Chichester and New York, 145–58.

Baker, V. R. (1988) Flood erosion. In: V. R. Baker, R. C. Kochel, and P. C. Patton, (eds) *Flood Geomorphology*. Wiley, New York and Chichester, 81–95.

Baker, V. R., Kochel, R. C. and Patton, P. C. (eds) (1988) *Flood Geomorphology*. Wiley, New York and Chichester.

Bull, W. B. (1977) The alluvial fan environment. *Progress in Physical Geography* **1**, 222–70.

Callander, R. A. (1978) River meandering. *Annual Review of Fluid Mechanics* **10**, 129–58.

Carson, M. A. (1984a) Observations on the meandering-braided river transition, the Canterbury Plains, New Zealand: Part One. *New Zealand Geographer* **40**, 12–17.

Carson, M. A. (1984b) Observations on the meandering-braided river transition, the Canterbury Plains, New Zealand: Part Two. *New Zealand Geographer* **40**, 89–99.

Carson, M. A. (1986) Characteristics of high-energy 'meandering' rivers: the Canterbury Plains, New Zealand. *Geological Society of America Bulletin* **97**, 886–95.

Carson, M. A. and Lapointe, M. F. (1983) The inherent asymmetry of river meander planform. *Journal of Geology* **91**, 41–55.

Chorley, R. J. (1969) The drainage basin as the fundamental geomorphic unit. In: R. J. Chorley (ed.) *Water, Earth and Man*. Methuen, London, 30–52.

Coates, D. R. and Vitek, J. D. (eds) (1980) *Thresholds in Geomorphology*. Allen and Unwin, Boston and London.

Cooke, R. U. and Reeves, R. W. (1976) *Arroyos and Environmental Change in the American South-West*. Clarendon Press, Oxford.

Ferguson, R. I. (1986) Hydraulics and hydraulic geometry. *Progress in Physical Geography* **10**, 1–31.

Ferguson, R. (1987) Hydraulic and sedimentary controls of channel pattern. In: K. Richards (ed.) *River Channels: Environment and Process*. Institute of British Geographers Special Publication **18**. Blackwell, Oxford, 129–58.

Ford, D. C. and Williams, P. W. (1989) *Karst Geomorphology and Hydrology*. Unwin Hyman, London and Boston,

Green, C. P. and McGregor, D. F. M. (1987) River terraces: a stratigraphic record of environmental change. In: V. Gardiner *et al.* (eds) *International Geomorphology 1986* Part I. Wiley, Chichester and New York, 977–87.

Gregory, K. J. and Walling, D. E. (1973) *Drainage Basin Form and Process: A Geomorphological Approach*. Edward Arnold, London.

Harvey, A. M. (1989) The occurrence and role of arid zone alluvial fans. In: D. S. G. Thomas (ed.) *Arid Zone Geomorphology*. Belhaven Press, London; Halsted Press, New York, 136–58.

Hooke, J. M. (1984) Changes in river meanders: a review of techniques and results of analyses. *Progress in Physical Geography* **8**, 473–508.

Hooke, R. Le B. and Rohrer, W. L. (1979) Geometry of alluvial fans: effect of discharge and sediment size. *Earth Surface Processes* **4**, 147–66.

Howard, A. D. (1967) Drainage analysis in geologic interpretation: a summation. *American Association of Petroleum Geologists Bulletin* **51**, 2246–59.

Howard, A. D. and Knutson, T. R. (1984) Sufficient conditions for river meandering: a simulation approach. *Water Resources Research* **20**, 1659–67.

Jennings, J. N. (1985) *Karst Geomorphology*. Blackwell, Oxford and New York.

Knighton, A. D. (1975) Variations in at-a-station hydraulic geometry. *American Journal of Science* **275**, 186–218.

Knighton, A. D. (1987) River channel adjustment – the downstream dimension. In: K. Richards (ed.) *River Channels: Environment and Process*. Institute of British Geographers Special Publication **18**. Blackwell, Oxford, 95–128.

Knighton, D. (1984) *Fluvial Forms and Processes*. Edward Arnold, London and Baltimore.

Langbein, W. B. and Leopold, L. B. (1966) River meanders – theory of minimum variance. *United States Geological Survey Professional Paper* **422–H**.

Lapointe, M. F. and Carson, M. A. (1986) Migration patterns of an asymmetric meandering river: the Rouge River, Quebec. *Water Resources Research* **22**, 731–43.

Leopold, L. B. and Maddock, T. (1953) The hydraulic geometry of stream channels and some physiographic implications. *United States Geological Survey Professional Paper* **252**.

Leopold, L. B. and Wolman, M. G. (1957) River channel patterns – braided, meandering and straight. *United States Geological Survey Professional Paper* **282B**.

Leopold, L. B., Wolman, M. G. and Miller, J. P. (1964) *Fluvial Processes in Geomorphology.* W. H. Freeman, San Francisco.

Lewin, J. (1978) Floodplain geomorphology. *Progress in Physical Geography* **2**, 408–37.

Lewin, J. (1983) Changes of channel patterns and flood-plains. In: K. J. Gregory (ed.) *Background to Palaeohydrology.* Wiley, Chichester and New York, 303–19.

Lisle, T. (1979) A sorting mechanism for a riffle pool sequence. *Geological Society of America Bulletin* **90**, 1142–57.

Mackin, J. H. (1948) Concept of the graded river. *Bulletin of the Geological Society of America* **59**, 463–512.

Mayer, L. and Nash, D. (eds) 1987. *Catastrophic Flooding.* Allen and Unwin, Boston and London.

Melton, M. A. (1958) Correlation structure of morphometric properties of drainage systems and their controlling agents. *Journal of Geology* **66**, 442–60.

Miall, A. D. (1977) A review of the braided-river depositional environment. *Earth Science Reviews* **13**, 1–62.

Morisawa, M. (1985) *Rivers: Form and Process.* Longman, London and New York.

Park, C. C. (1977) World-wide variations in hydraulic geometry exponents of stream channels: an analysis and some observations. *Journal of Hydrology* **33**, 133–46.

Parker, G. (1976) On the cause and characteristic scale of meandering and braiding in rivers. *Journal of Fluid Mechanics* **76**, 459–80.

Petts, G. and Foster, I. (1985) *Rivers and Landscape.* Edward Arnold, London and Baltimore.

Rackocki, A. H. and Church, M. J. (eds) (1989) *Alluvial Fans: A Field Approach.* Wiley, New York and Chichester.

Richards, K. (1976) The morphology of riffle-pool sequences. *Earth Surface Processes* **1**, 71–88.

Richards, K. (1982) *Rivers: Form and Process in Alluvial Channels.* Methuen, London and New York.

Richards, K. (ed.) 1987. *River Channels: Environment and Process.* Institute of British Geographers Special Publication **18**. Blackwell, Oxford.

Ritter, D. F. 1982. Complex river terrace development in the Nenana valley near Healy. Alaska. *Geological Society of America Bulletin* **93**, 346–56.

Schumm, S. A. (1977) *The Fluvial System.* Wiley, New York and London.

Schumm, S. A. (1979) Geomorphic thresholds: the concept and its applications. *Transactions Institute of British Geographers* NS **4**, 485–515.

Schumm, S. A. (1985) Patterns of alluvial rivers. *Annual Review of Earth and Planetary Sciences* **13**, 5–27.

Schumm, S. A. and Khan, H. R. (1972) Experimental study of channel patterns. *Geological Society of America Bulletin* **83**, 1755–70.

Schumm, S. A., Mosley, M. P. and Weaver, W. E. (1987) *Experimental Fluvial Geomorphology.* Wiley, New York and Chichester.

Schumm, S. A. and Parker, R. S. (1973) Implications of complex response of drainage systems for Quaternary alluvial stratigraphy. *Nature Physical Science* **243**, 99–100.

Shreve, R. L. (1967) Infinite topologically random networks. *Journal of Geology* **75**, 178–86.

Smith, N. D. (1974) Sedimentology and bar formation in the upper Kicking Horse River, a braided outwash stream. *Journal of Geology* **82**, 205–23.

Smith, N. D. and Smith, D. G. (1984) William River: an outstanding example of channel widening and braiding caused by bed-load addition. *Geology* **12**, 78–82.

Strahler, A. N. (1952) Hypsometric (area-altitude) analysis of erosional topography. *Bulletin of the Geological Society of America* **63**, 1117–42.

Strahler, A. N. (1964) Quantitative geomorphology of drainage basins and channel networks. In: V. T. Chow (ed.) *Handbook of Applied Hydrology.* McGraw-Hill, New York, Section 4–II.

Thompson, A. (1986) Secondary flows and the pool-riffle unit: a case study of the processes of meander development. *Earth Surface Processes and Landforms* **11**, 631–41.

White, W. B. (1988) *Geomorphology and Hydrology of Karst Terrains.* Oxford University Press, New York and Oxford.

Wolman, M. G. and Gerson, R. (1978) Relative scales of time and effectiveness of climate in watershed geomorphology. *Earth Surface Processes* **3**, 189–208.

Wolman, M. G. and Miller, J. P. (1960) Magnitude and frequency of forces in geomorphic processes. *Journal of Geology* **68**, 54–74.

Young, R. W. (1985) Waterfalls: form and process. *Zeitschrift für Geomorphologie Supplementband* **55**, 81–95.

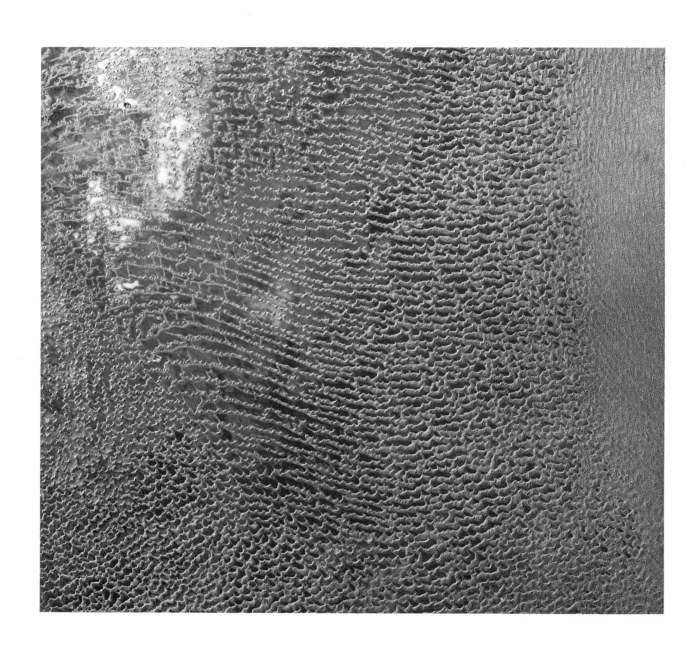

10

Aeolian processes and landforms

10.1 Aeolian activity

Wind is a comparatively feeble geomorphic agent over much of the Earth's surface, but in areas free of vegetation it can have significant effects. In the arid regions of the subtropics vast sand seas and extensive grooved bedrock surfaces are a testament to the power of wind action. Deserts are still some of the least known areas of the globe and until comparatively recently our knowledge of desert landforms rested heavily on observations of explorers involved in pioneering expeditions to these inhospitable environments during the late nineteenth and early twentieth centuries. Increased accessibility associated with economic development, together with the global reconnaissance of deserts made possible by satellite imagery, is rapidly increasing our knowledge of aeolian landforms, while experimental studies and field instrumentation are beginning to improve significantly our understanding of aeolian processes.

The effectiveness of wind action is limited by a number of factors, and on a global basis is a far less potent erosional agent than fluvial activity. Compared with water, air has a low density and viscosity so only very fine particles can be carried in suspension, except at very high wind speeds. Moreover, vegetation greatly reduces wind speeds near the ground, and together with moisture it tends to bind surface particles together and prevent them from being entrained by the wind. Consequently aeolian activity is only effective in areas which lack a relatively complete vegetation cover and where the surface material dries out at least occasionally.

Most important among such areas are the world's arid regions, but more localized wind action also occurs along sandy coasts, over bare fields (especially where soil management techniques are poor) and in river plains containing migrating channels, particularly around the margins of glaciers and ice sheets. In such environments aeolian processes may play an important role in landform development and in areas of massive sand accumulations within the world's major deserts wind is by far the predominant geomorphic agent.

10.1.1 Global distribution of aeolian landforms

The great proportion of wind-formed landscapes are contained within the world's hot deserts and so we will largely confine our attention to these areas in this chapter. Aeolian features formed in periglacial and coastal environments are discussed in Chapters 12 and 13 respectively.

The most striking aeolian landforms are **dunes**. These are accumulations of wind deposited particles and they can assume a bewildering variety of forms. The great majority of dunes are composed of sand, although silt and clay-sized material may accumulate into dune-like features under certain conditions. Sand is not evenly spread throughout desert regions but rather is concentrated into sand seas or **ergs**. It has been estimated that 85 per cent of the sand in active sand bodies is contained in ergs greater than 32 000 km² in area. Large active ergs are more or less confined within the 150 mm mean annual isohyet, but **fixed** or **relict ergs** (which have become inactive due to the stabilizing effects of vegetation) are to be found in the subhumid fringes of the world's arid regions (Fig. 10.1). Some active ergs are of vast extent; the largest is Rub'al Khali which covers 560 000 km² of Saudi Arabia. Very extensive blankets of silt-sized material, known as **loess**, are found in those mid-latitude areas that were marginal to the Pleistocene ice sheets, and, to a more limited extent, at lower latitudes.

Quartz is by far the predominant component of desert sands, both because of its abundance as a rock-forming mineral and because of its resistance to chemical decomposition and abrasion. Ultimately aeolian quartz grains must

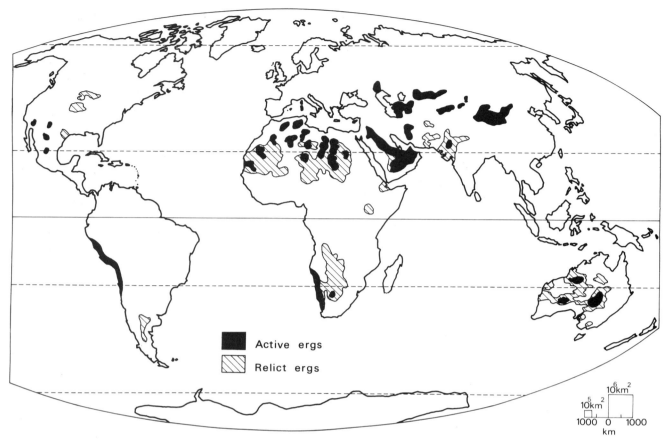

Fig. 10.1 *Global distribution of active and relict (fixed) ergs. The extent of relict ergs, thought to have been active during arid phases during the Pleistocene, is approximate only. (Based partly on M. Sarnthein (1978)* Nature *272, Fig. 1, p. 44 and G. Wells (1989) in: L. Friday and R. Laskey (eds)* The Fragile Environment. *Cambridge University Press, Cambridge, Fig. 8.18, p. 171.)*

have been derived from weathered or abraded rock, either as first-cycle quartz from igneous or metamorphic rocks or, more frequently, as previously cycled grains in sedimentary rocks. Most deserts contain extensive outcrops of quartzose sandstones, some of which themselves originated as aeolian deposits. Present-day desert aeolian sands are mostly derived from alluvium rather than directly from bedrock weathering. This is suggested by the tendency for North African ergs to be located near the centre of basins towards which drainage systems (now largely inactive) are focused.

The distribution of erosional aeolian landforms is more difficult to specify in view of the uncertain origin of many landforms ascribed to aeolian action, such as the depressions of various sizes common in some desert areas. Satellite imagery is now making possible a new evaluation of the origin of some of the larger scale varieties of these possible aeolian features and many do indeed appear explicable in terms of present or past phases of wind action.

10.1.2 Wind characteristics

A knowledge of wind characteristics is vital to an under-

standing of the development of aeolian landforms, yet there is a frustrating paucity of data on the temporal and spatial variations in wind speed and direction in desert areas. Nevertheless, some generalizations are possible. The large-scale properties of desert winds are determined by the general circulation which, at the latitude of the world's hot deserts, is dominated by sub-tropical high-pressure systems. Large areas are subject to fairly constant trade winds, but regional and local factors, such as those arising from contrasts in the thermal properties of desert surfaces, can add to global effects. Differential heating of the surface associated with variations in albedo lead to pressure gradients which in turn generate local winds.

Of particular significance to the movement of particles by the wind is the vertical variation of wind speed above the ground. Close to the surface wind speeds are reduced by friction, and the magnitude of this effect is largely a function of surface roughness. As wind speed and surface roughness increase, the airflow becomes more turbulent and the potential for erosion is consequently increased. Surface friction consumes energy, and as the frictional effect declines away from the surface wind speed increases with height. The rate of increase is rapid close to the surface but increases progressively more slowly at greater heights. The actual rate of increase in any particular instance depends largely on surface roughness and wind speed.

10.1.3 Effect of surface characteristics

The character of desert surfaces merits consideration because surface roughness and the size and cohesion of surface particles significantly affect the ability of the wind to erode. Although sand dunes are perhaps thought of as the land-forms most typical of arid regions, only 25–30 per cent of the total area of hot deserts is covered by sand. The re-mainder is composed of bedrock, coarse weathered debris, fluvial sediments, weathering crusts, poorly developed soils, and deposits of clay, silt or salt. The proportion of each of these types of surface in particular deserts reflects the com-plex interaction of a variety of climatic and geological factors.

Particle size and cohesion, together with vegetation cha-racteristics, are the main properties of desert surfaces which influence their susceptibility to wind erosion. In most cli-matic regions the surface is covered with soils rendered cohesive by the binding effects of moisture, humus and clay-sized particles, but in hot deserts all of these elements are, to a greater or lesser extent, lacking. The susceptibility of soils to erosion by the wind is known to be related to their water content, desiccated soils being much more prone to erosion. Moisture retention and interparticle bonding are more marked in clay- and silt-rich materials and these are consequently more resistant to wind erosion.

10.1.4 Sediment entrainment and transport

Both water and air are fluids and there are close similarities between the mechanisms of sediment entrainment in air-flows and water flows (see Section 8.3.2). Movement is resisted by the weight of particles, and by friction and inter-particle cohesion. Movement is induced by drag and lift forces, and the impact of grains already in motion.

Lift forces can cause particles to jump up into the airflow in much the same way that the wings of aeroplanes generate lift. In the case of a wing, lift occurs because its cross-sectional form is such that air flowing over its upper surface is forced to take a longer path, and therefore moves faster, than air flowing over its lower surface. This causes a pressure difference between the upper and lower wing surfaces which induces lift. Imagine a sand grain pro-truding slightly above the average level of a sand-covered surface. Depending on the exact form that the grain presents to the wind it might also produce an acceleration of air over it. In addition a pressure difference might result from the contrast in wind velocity between the almost stationary air

in the voids around the lower part of the grain and the faster-moving air above.

A much more important mechanism is the drag force resulting from the difference in fluid pressure on the wind-ward and leeward sides of grains in an airflow. This can cause particles to move downwind by rolling or sliding, a process known as **surface creep**. Drag forces can probably initiate movement at lower wind velocities than those necessary for lift and they can also move particles too large to be lifted by the wind. The wind speed needed to promote movement through lift or drag is known as the **fluid threshold velocity**.

This simple picture is greatly complicated by turbulence. Airflows are rarely uniform but rather are characterized by bursts of higher velocity flow which can momentarily pro-duce a pocket of low pressure above the surface and create a high potential for lift and drag. Such variations in wind velocity are particularly important because the drag force varies with the second power of velocity; in other words a doubling of wind velocity will produce a fourfold increase in drag.

Whether a particle will move at a given wind speed depends on many factors in addition to its size; these in-clude its density and shape, the packing of grains around it and the amount of cohesion between particles. Neverthe-less, for dry particles of similar shape and density there is a fairly direct relationship between size and fluid threshold velocity since there is a strong correlation between drag and average wind speed (Fig. 10.2). Grains larger than about 1 mm across will rarely be moved by drag alone since the required wind speeds are hardly ever attained under natural conditions. For grains smaller than about 0.6 mm across, the threshold velocity necessary for movement actually increases as particle size decreases. One reason for this is the reduced surface roughness, and therefore lower level of turbulence, generated by surfaces composed of fine material. Another reason is the greater interparticle cohesion charac-teristic of fine sediments which provides a greater resist-ance to entrainment. The sorting and packing of grains also affects their susceptibility to entrainment as small particles can be sheltered from the airflow by larger grains.

Very small particles less than about 0.1 mm across are normally capable of being carried away in suspension once set in motion by the wind. Somewhat larger grains up to around 0.4 mm in diameter can be temporarily lifted into the airflow before returning to the surface. As a result of this process of saltation, once lift and drag have initiated particle movement grains downwind start to be bombarded by those particles already in motion. This mechanism of **ballistic impact** applies an additional forward momentum to surface particles and means that grains of a particular size can be set in motion at a wind velocity lower than that required to initiate movement. The difference between this **impact threshold velocity** and the fluid threshold velocity

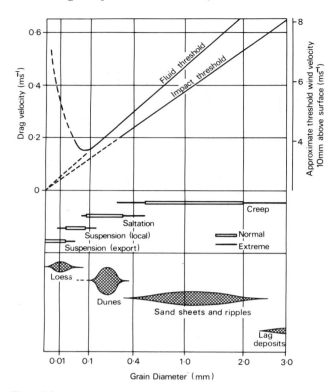

Fig. 10.2 *Relationships between grain size and fluid and impact threshold wind velocities. Also shown are the relationships between grain size and characteristic modes of wind transport and associated depositional forms. (Modified from J. A. Mabbutt (1977) Desert Landforms. Australian National University Press, Canberra, Fig. 65, p. 219, based on R. A. Bagnold (1954), The Physics of Blown Sand and Desert Dunes. Methuen, London, Fig. 28, p. 88 and R. L. Folk (1971) Sedimentology 16, Fig. 16, p. 43.)*

is small for fine particles but is quite significant for larger grains (Fig. 10.2). Once a gust of wind has caused an initial movement in some particles further motion can be temporarily sustained by ballistic impact even after the wind speed has dropped below the fluid threshold velocity.

The proportion of total movement occurring by surface creep becomes less with decreasing grain size, and even for relatively coarse sands it rarely exceeds 25 per cent. Saltation is the major mechanism of sand movement, and in view of its importance we need to examine the process more closely. On rising into the airflow, saltating grains encounter increasingly higher wind speeds and they are carried downwind (Fig. 10.3). They return to the surface along a trajectory determined by a balance between gravity and drag and usually strike the surface at an angle of between 6 and 12°. After impact saltating grains may bounce and

return to the air. They may also cause other grains disturbed by their impact to saltate. This process gives rise to a cloud of saltating grains extending a few centimetres above the surface. Most saltating grains travel within 10 mm of the surface with individual horizontal leaps of around 0.5–1.5 m. The height which saltating grains reach depends on particle size, wind speed and surface characteristics. In general height increases with greater wind speeds and smaller particles, but as larger particles may bounce better they may occasionally rise out of the saltating cloud. Saltating grains also reach greater heights over hard, non-sandy surfaces apparently because they bounce more effectively on such surfaces.

The major factors influencing the rate of aeolian transport are wind speed, air density (which varies slightly with altitude and temperature), particle size and surface characteristics. Experimental investigations have shown the rate of sand transport to be proportional to the third power of wind speed. This relationship suggests that most sand is moved during periods of very high wind velocity. Very strong winds, however, are much less frequent than those of moderate speeds so it seems likely that most sand transport in fact occurs during relatively frequent periods of moderately strong winds of around 12–18 m s[-1] (Fig. 10.4).

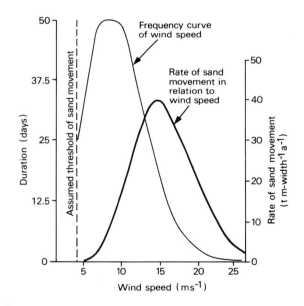

Fig. 10.4 *Relationship between wind speed and sand movement. The frequency of winds of different speeds is based on a hypothetical, but realistic, distribution through an annual cycle. The estimated rate of sand movement (t m-width^{-1} a^{-1}) is based on one of a number of formulae derived from experimental data. (After A. Warren (1979) in: C. Embleton and J. Thornes (eds) Process in Geomorphology. Edward Arnold, London, Fig. 10.8, p. 335.)*

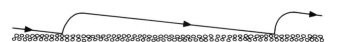

Fig. 10.3 *Trajectory of a single saltating grain.*

10.2   Aeolian erosion

10.2.1   Deflation and abrasion

Aeolian erosion involves two processes; **deflation**, which is the removal of loose particles by the wind, and aeolian abrasion, which is caused by the bombardment of rock and other surfaces by particles carried in the airflow. Deflation involving sand-sized particles is rather localized since sand grains cannot be moved great distances except over long periods of time. Silt, and especially clay-sized material, on the other hand, can be lifted by turbulence and carried in suspension in the atmosphere while the very finest material can be carried great distances in dust storms.

The world's hot deserts are a major source of atmospheric dust (see Section 15.6). Accurate estimates are difficult to obtain but it has been estimated that between 130 and 800 Mt of material is deflated from the continents annually, with the Sahara alone contributing between 60 and 200 Mt. Deflation is a temporally and spatially concentrated process. Even in arid regions it is largely restricted to areas where surface conditions and wind speeds are particularly favourable. Most of the material deflated each year is removed in major dust storms, each usually lasting only a few days, and satellite reconnaissance has provided a potent technique for tracking these events. Some of this material is redeposited elsewhere in desert areas forming loess but much of the finer material is carried out over the ocean and makes a significant contribution to sedimentation in the world's ocean basins.

Rock- and boulder-covered surfaces can be abraded by particles of a range of sizes. Although it has been widely thought that sand-sized material is the most effective, some research has pointed to the role of fine silt-sized particles and dust, especially in producing fluted surfaces. The rate of abrasion is maximized where high velocity winds cross an abundant supply of hard sand grains upwind from soft, friable lithologies. As sand particles are only carried close to the ground, even at high wind speeds, abrasion caused by this calibre of material is confined to within 1 or 2 m of the surface.

10.2.2   Erosional landforms

Although aeolian erosion may be active over alluvial plains and on beaches, wind-eroded landforms are rarely preserved in such environments because of destruction by fluvial processes or wave action. Only in arid areas, where other denudational agents are lacking or weakly active, are aeolian erosional landforms abundant. A significant proportion of many deserts is covered not by sand, but by **lag deposits** comprising particles of gravel size or coarser (Fig. 10.5). Lag deposits usually form a thin layer lying over predominantly finer material (Fig. 10.6). Where the stone cover is continuous such surfaces are known as

Fig. 10.5   *Gravel lag deposits, Skeleton Coast, northern Namibia.*

Fig. 10.6   *Experimental plot in Arizona, USA, where the surficial gravel lag deposit has been removed in order to determine the rate at which it is re-established.*

desert pavements, or **stone pavements**. It was long thought that they form by deflation, with the wind removing finer particles from poorly sorted deposits such as alluvium.

Recently, however, it has been argued that the observed distribution of fine and coarse particles is better explained by the accretion over large areas of aeolian sand and silt which progressively infiltrates into alluvial gravels.

10.2.2.1   Small-scale forms

A characteristic feature of stony desert surfaces is the presence of faceted cobbles and pebbles called **ventifacts**. Their precise mode of development is disputed but abrasion by dust and silt (rather than sand-sized particles) is probably the main mechanism. Facets are produced at angles of between 30 and 60° facing the prevailing wind and are separated from the protected lee side by a sharp edge (Fig. 10.7). More than one side may be wind abraded and three-facet pyramid-shaped ventifacts (called **dreikanter**) are particularly common. Abrasion of more than one side may indicate winds prevailing from more than one direction, but experimental studies demonstrate that ventifacts can be formed by essentially undirectional winds. In this case erosion occurs simultaneously on all sides as dust and silt

particles are carried in vortices in flow lines. Ventifacts may also be disturbed or rolled into a new position during formation, thereby realigning them with respect to a single prevailing wind.

Bedrock surfaces may also exhibit evidence of wind abrasion in the form of small-scale pits, flutes, grooves and polished surfaces, but such forms are often difficult to ascribe with certainty to aeolian action as similar features can be produced by running water and chemical weathering.

10.2.2.2   Intermediate-scale forms

Much larger than these small-scale bedrock features are a variety of grooved forms and shallow depressions with dimensions of tens to hundreds of metres. The most characteristic grooved form is the **yardang**, a streamlined parallel ridge usually less than 10 m high and 100 m or more in length aligned with, and typically tapering away from, the direction of the prevailing wind (Fig. 10.8). Yardangs are most commonly developed in soft lithologies such as lacustrine sediments and are numerous in some desert lake beds (Fig. 10.9), but in areas such as western Egypt they occur in extremely resistant granites and quartzites.

The extent to which yardangs are wind-formed features is not universally agreed. It has been argued that some merely represent a pre-existing bedrock relief only slightly modified by wind action. Nevertheless their consistent orientation, which can frequently be correlated with the dominant wind direction, is strong evidence for a primary aeolian origin. Their formation appears to depend essentially on the excavation of material from the troughs lying between the yardangs. They are frequently undercut at the base and this would accord with a higher rate of wind

Fig. 10.7   *Ventifact on a gravel lag surface in the Namib Desert in a region of strong coastal winds.*

Fig. 10.8   *A mud yardang in the Kharga Depression, Egypt. The unidirectional prevailing wind is from the right. Sand-blasting is confined to the blunt, windward face, and the leeward tapering of the yardang indicates the importance of erosion by fine particles carried in suspension in secondary flows across the long leeside tail. (Photo and interpretation courtesy M.I. Whitney, from M.I. Whitney, (1985)* Journal of Geological Education *33, Fig. 1, p. 94.)*

Fig. 10.9 *Extensive yardang field in the Lut Desert, Iran. The grooving evident on the image, which covers an area about 140 km across, is a result of lines of very large yardangs up to 80 m high and spaced at least 100 m apart. The yardangs are formed in horizontally bedded lacustrine deposits and trend NNW parallel to the prevailing wind. (Landsat image courtesy A. S. Walker.)*

abrasion near the ground. Weathering and surface runoff may be significant in modifying yardangs once they begin to form and recently one study has suggested that deflation of unconsolidated trough material may be of more importance than abrasion in their development. Whatever their exact mode of formation, yardangs seem to be confined to the very arid core regions of deserts where there is comparatively little sand and where erosion by dust and silt particles occurs in a unidirectional prevailing wind.

Similar in scale to yardangs are the shallow depressions found across many desert regions of low relief. These have been given a variety of local names but are generally referred to as **deflation hollows**. They exist in a continuum of sizes from small forms less than 1 m deep and only a few metres across, to large features which grade into the macro-scale basins discussed below. The development of deflation hollows is clearly influenced by those factors which control the process of deflation and they tend to form where high wind velocities are associated with bare surfaces covered by relatively fine-grained, desiccated sediments. The problem is to explain the differential erosion involved and this could be due to a variety of local conditions, such as a lower sediment moisture content or disturbance of the vegetation cover. In some areas, however, deflation hollows are regularly spaced and this suggests some control by waves in the airstream. Much research remains to be done before these features are fully understood.

10.2.2.3   Large-scale forms

The occurrence of large enclosed basins in deserts has been known since the first explorations of these regions, but the extent and regularity of large-scale grooving in bedrock has only been fully appreciated since the advent of satellite imagery. The basins range from landforms a few metres deep and upwards of 100 m across, such as the **pans** which are so abundant in parts of southern Africa, to very large features over 100 m deep and more than 100 km across. The smaller, shallow basins probably represent localized deflation; some are orientated along drainage lines whereas others are located in troughs between dunes. But in both situations their long axes are aligned with the prevailing wind.

Very large basins are much more complex landforms and almost certainly owe their origin to the interplay of several processes. In many cases their gross form appears to be related to tectonic activity, and this can be established by the presence of bounding faults. In such tectonic basins aeolian activity may only slightly modify the existing form. But if surface deposits are susceptible to removal by wind, deflation may be a key process in maintaining their relief, or even increasing it, since this provides the only mechanism capable of completely removing material from an enclosed basin. Where faulting is absent there is frequently much uncertainty about the origin of these features, some of which represent the removal of enormous amounts of material. Perhaps the most remarkable concentration of large basins is to be found in Egypt where they cover more than 70 000 km² and have an average maximum depth of 250 m (Fig. 10.10). The Qattara Depression, the deepest of the basins, reaches a depth of 134 m below sea level, and has a volume of 3200 km³.

Large-scale bedrock grooving clearly visible on satellite imagery has been widely attributed to aeolian action. Such grooves are the major landform over a 90 000 km² area on the south-eastern margin of the Tibesti Plateau in North Africa, and an aeolian origin is strongly suggested by the close match between the regional atmospheric circulation and their gently curving NE–SW trend. The grooves are between 0.5 and 1 km wide, several tens of kilometres long and are spaced 0.5–2 km apart. They are eroded into sandstone, and are generally assumed to have formed in a similar way to yardangs, that is, largely through a combination of

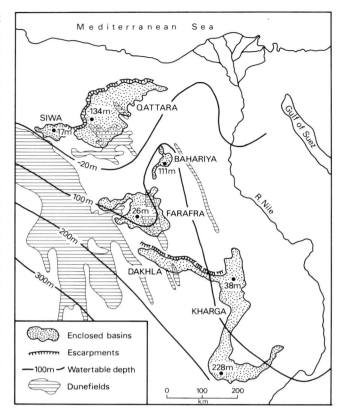

Fig. 10.10 *Large enclosed basins in the western Egyptian Desert. Escarpments rim the northern margins of the basins but there is no evidence of significant faulting in the area, only gentle warping. The southern basins are the oldest and were probably initiated by fluvial erosion as far back as the Eocene, but the deepening of these basins once enclosed, and the development of the northern basins, is attributed largely to prolonged deflation. Judging by the relationship between basin depth (shown by spot heights for low points) and depth of the water table (indicated by contours) deflation has been limited by the presence of ground water. Retreat of the south-facing escarpments has occurred through spring sapping and weathering, the resulting sediment being deflated. This combination of denudational processes is clearly most effective since the Qattara Depression is cut into Pliocene rocks and is therefore primarily of Quaternary age. This implies a mean rate of lowering of around 90 mm ka⁻¹, a rate comparable to fluvial denudation. (Modified from J. A. Mabbutt (1977) Desert Landforms, Australian National University Press, Canberra, Fig. 62, p. 212.)*

wind abrasion and deflation. However, since they also follow the trend of bedrock lineations thought to have a tectonic origin, it is possible that winds may have been in part channelled along pre-existing lines of structural weakness once these had been initially excavated.

10.3   Depositional landforms

10.3.1   Basic depositional forms

Although capable of movement by surface creep and salta-

Fig. 10.11 *Ripples on the surface of a dune, north of Walvis Bay, Namibia. The scale is indicated by the lens cap in the bottom right corner. Regular ripples such as these form in well-sorted sand, whereas irregular ripples are produced in poorly sorted sediments.*

tion, sand grains spend the vast majority of time in storage in sand accumulations which vary enormously in size and form. The smallest depositional features are called **ripples** and consist of regular, wave-like undulations orientated at right angles to the direction of the prevailing wind (Fig. 10.11). The dimensions of ripples increase with particle size, their heights ranging from 1 to 500 mm and wavelengths from 0.01 to 5 m. Dunes are much larger depositional forms having typical heights of 5 to 30 m and wavelengths of 50 to 300 m. Some dunes, however, attain even greater dimensions with heights of up to 400 m and wavelengths up to 4 km (Fig. 10.12). The terms **draa**, or **megadune**, are sometimes applied to these very large depositional forms. Measurements from ergs around the world reveal a close relationship between the height, width and spacing of sand dunes. This suggests that dune forms are closely adjusted to controlling variables.

In the early 1970s I. G. Wilson argued, on the basis of research focused on Saharan dunes, that for a given grain size the spacing of aeolian depositional forms falls into a hierarchy of three discrete size classes consisting of ripples, dunes and megadunes with no transitional forms (Fig. 10.13). He suggested that the contrasting spacing of dunes and draas was related to different scales of secondary airflows. The dimensions of each of these depositional forms were held to be related to grain size, with coarse sands giving rise to the largest forms in each category (Fig. 10.13). This relationship was interpreted as being a result of the minimum drag velocity required to move sand of a given size, which, in turn, constrains the minimum spacing and dimensions of depositional forms of each type.

For a time this idea was widely accepted and seemed to provide a useful basis for classifying aeolian depositional forms. More recent research, however, has led many geomorphologists to question whether this size hierarchy is universally valid. Data from the Namib, Kalahari and Australian Deserts, for instance, have failed to show a clear discontinuity in the gradation from dunes to megadunes (Fig. 10.13). Nevertheless, it is acknowledged that such a size discontinuity may occur in dune types which develop normal to a single prevailing wind direction, and in dune fields where there is a large range in grain size.

10.3.1.1 Ripples

Ripples are asymmetric in cross-section with windward slopes around 10° and lee-side gradients near the angle of repose of dry sand of 30–35°. According to the model presented by R. A. Bagnold they develop from very slight irregularities in the sand surface through a combination of surface creep and saltation. As saltating grains strike the surface at a shallow angle they dislodge more grains on the windward side of an irregularity than on the leeward side which is protected from the flight path of the saltating grains (Fig. 10.14). Ballistic impacts release grains from the windward side which either creep up the windward slope to the crest, or saltate a characteristic distance to add to the sand accumulation forming the next ripple downwind. The most common saltation distance depends on particle size and wind speed and will influence ripple spacing. As sand is constantly being eroded from the windward side of ripples and accreted to the leeward side a series of ripples will migrate downwind while maintaining their characteristic spacing. This process does not lead to an indefinite increase in ripple height because the ripple crest gradually builds into a zone of higher wind speeds until eventually deposition is balanced by erosion.

This simple model works well for sand with a more or less uniform grain size, but most natural sands are only moderately well sorted. In such cases the large grains tend to move slowly by creep and they accumulate just on the leeward side of the ripple crest where they are protected from the impacts of smaller saltating grains. In this way

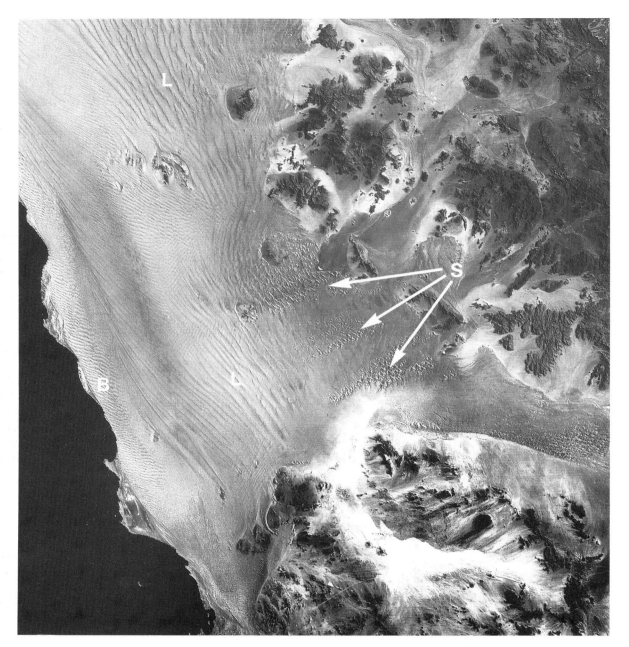

Fig. 10.12   *Megadune scale forms of the southern Namib Sand Sea, Namibia. Three major dune types are represented. Compound barchanoid ridges (B) predominate along the Atlantic coast, while compound and complex linear forms (L) predominate inland. Lines of large star dunes (S) are visible in the centre of the image. The area covered is about 170 km across. (Landsat image courtesy A. S. Walker.)*

ripple height can increase substantially as much higher wind speeds are required to move large grains. The presence of coarse particles on ripple crests provides a better bouncing surface for saltating grains which consequently have longer trajectories, and consequently ripple spacing increases. Where very coarse sand is subject to very strong winds megaripples may form with wavelengths of 5 m and heights of 0.5 m (Fig. 10.15). Low relief ripples may also originate

from the aerodynamic effects of the wind on sand, but such forms are almost invariably obscured by ripples generated by ballistic impact.

10.3.1.2   Dunes

For a dune to form a patch of sand must first begin to accumulate. This occurs where the wind speed is reduced by an increase in surface roughness, or by primary instabilities in

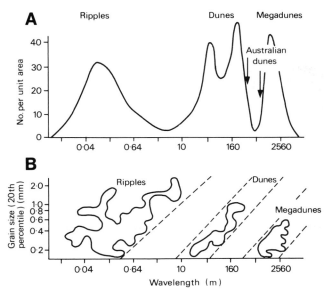

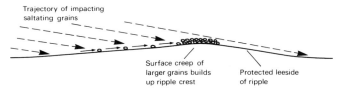

Fig. 10.13   *Relationships proposed by I. G. Wilson between areal frequency and wavelength (spacing) (A), and between grain size and wavelength for aeolian depositional forms (B). Note that ripples, dunes and megadunes form three discrete clusters, but that Australian dunes plot between the dune and megadune groups identified by Wilson. (Based on I. G. Wilson (1972) Sedimentology 19, Figs. 2 and 3, p. 13 and R. J. Wasson and R. Hyde (1983), Earth Surface Processes and Landforms 8, Fig. 1, p. 302).)*

Fig. 10.14   *Mechanism of ripple formation according to the model of R. A. Bagnold.*

Fig. 10.15   *Large ripples with a wavelength of more than 1 m developed in very coarse sediments on the flank of a linear megadune, Skeleton Coast, northern Namibia (lens cap indicates scale).*

the airflow. Once formed the patch of sand grows by trapping saltating grains which are unable to rebound on impact as readily as they are on the surrounding stony surface. For this effect to work the sand body must be broader than the flight lengths of saltating grains. A width of around 1–5 m seems to be a critical lower limit for the growth of sand bodies and thus represents the limiting size for dunes. As a result of the separation and deceleration of the airflow on the lee side of the growing sand body, sand accumulates more rapidly than it is removed and consequently the dune increases in size.

The basic shape of a dune formed by winds from a single prevailing direction is asymmetric in cross-section. Such dunes have a gently inclined windward slope at a typical angle of 10–15° separated by a sharp crest from a much steeper leeward **slip-face** sustained close to the threshold angle of stability of dry sand of between 30 and 35° (Fig.

10.16). Where the airflow descends to the ground there is more sand movement. Deposition occurs in the zones of converging, upward moving air forming dunes which grow vertically until an equilibrium height is attained where the wind velocity is sufficient to remove sand from the dune crest as rapidly as it is supplied from upwind. In addition the wave-like form of the airflow itself may exert a control

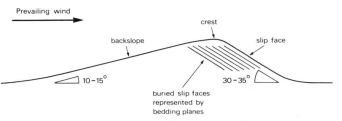

Fig. 10.16   *Basic components of the cross-profile of a transverse type dune with a single slip-face orientation. Linear dunes contain two opposing slip-face orientations and star dunes multiple slip-face orientations.*

on dune height. Sand eroded from the windward side is deposited on the leeward slip-face so dunes move in the direction of the prevailing wind. Rates of dune movement depend both on dune type and size and the frequency and strength of winds necessary to cause sand movement, but 10–20 m a^{-1} is typical.

The overall form of megadunes is similar in cross-section to that of dunes, but in detail is often complicated by the presence of superimposed dunes. The formation of megadunes appears to be more clearly related to primary instabilities in the airflow than is the case for dune-sized features. Since sand movement is confined to the surface of a dune, the area of which decreases as a proportion of volume as dune size increases, large dunes move much more slowly than small dunes. Measurements on a medagune in Peru have indicated a rate of movement as low as 0.5 m a^{-1}, whereas smaller superimposed dunes were found to be moving at 9 m a^{-1}. Indeed these slow rates of movement mean that very large megadunes must take hundreds of years to attain an equilibrium form.

10.3.2   Classification of dune morphology

Our discussion so far has been confined to the general characteristics of ripple and dune forms in cross-section. Adding the third dimension of spatial organization we see that, whereas ripples are comparatively simple features forming a succession of more or less parallel ridges at right angles to the prevailing wind, dunes occur in a great variety of much more complex patterns. The extent and regularity of these patterns at the megadune scale have only been fully appreciated with the advent of satellite imagery and the global overview that it makes possible (Fig. 10.12).

Any categorization of features as variable as dunes is to a certain extent arbitrary, but a reasonably consistent classification can be made on the basis of two key characteristics – the overall shape of the dune and the position and number of slip-faces. Plan shape is an obvious factor to consider, but slip-face characteristics are also important since they provide information on the nature of the formative winds. Wind direction and velocity, sand supply and the presence of vegetation or topographic obstacles are the most important factors influencing dune morphology. None the less, it is convenient to distinguish between **free dunes**, whose form is primarily a function of wind characteristics, and **impeded dunes** whose morphology is influenced significantly by the effects of vegetation, topographic barriers or highly localized sediment sources.

10.3.3   Classification of free dunes

The major types of free dune are classified in Table 10.1. The primary wind regime thought to be associated with each type is given but the nature of the formative airflows

is considered in more detail below. Dunes with a single slip-face orientation are associated with unidirectional winds. Their axes are orientated normal to the prevailing wind, and for this reason they are often referred to collectively as **transverse dunes**. Simple, straight, parallel dunes called **transverse ridges** (Fig. 10.17(A)) frequently grade into more common **barchanoid ridges** (Fig. 10.17(B)), composed of coalesced crescentic forms, and into individual crescent-shaped **barchans** (Figs 10.17(C), 10.18). Highly elongated forms with two, more or less opposing, slip-faces, are called **linear dunes** although the term **longitudinal dune** is used by some researchers for certain types (Figs. 10.17(D), 10.19). In contrast to transverse dunes net sand transport is parallel to the crest line. There is disagreement about the origin of linear dunes, but most researchers now consider that they develop where there are two obliquely converging prevailing winds. Linear dunes can attain lengths of tens of kilometres and they may

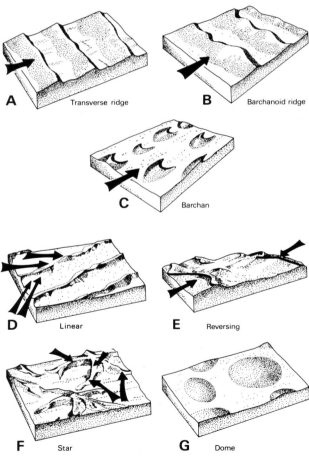

Fig. 10.17 *Morphology of the major types of free dune. Arrows indicate the probable formative prevailing wind direction(s), although note that the origin of linear dunes is controversial (see Section 10.3.5.2). (After E. D. McKee (1979)* United States Geological Survey Professional Paper *1052, Figs. 3–5, 7, 10–12, pp. 11–13.)*

Table 10.1 Classification of basic types of free dune

NUMBER AND GEOMETRY OF SLIP-FACE ORIENTATIONS	INFERRED PRIMARY WIND REGIME		DUNE TYPE	MORPHOLOGY
One; unidirectional	Unidirectional	Transverse forms	**Transverse ridge**	Asymmetric ridge
			Barchanoid ridge	Row of contiguous crescentic forms
			Barchan	Crescentic form
Two; opposing	Bidirectional; opposing at ~180°		**Reversing**	Asymmetric ridge
Two; opposing	Bidirectional; obliquely converging?		**Linear**	Symmetric ridge; straight to sinuous in plan
Three or more; multidirectional	Multidirectional		**Star**	Central peak with three or more arms
None	—		**Dome**	Circular or elliptical mound

coalesce downwind forming Y-shaped junctions (Fig. 10.20).

Reversing dunes (Fig. 10.17(E)) are similar to linear dunes in that they have two slip-face orientations, but they are more appropriately regarded as a type of transverse dune. They are formed where two prevailing winds of similar strength and duration blow normal to the dune axis from opposite directions. The opposing slip-faces develop alter-nately as wind direction changes periodically.

Dunes with several slip-faces orientated in different directions have a roughly pyramidal shape but with elongated and often irregular arms (Figs 10.17(F), 10.21). These forms, which are attributed to strong winds blowing from several different directions during an annual cycle, have a variety of names, but will be referred to here as **star dunes**. Other sand accumulations including **sheets, streaks** and **dome dunes**

Fig. 10.18 *Barchan west of the Salton Sea, California, USA (Photo courtesy K. Mulligan.)*

Fig. 10.19 *Complex linear megadune at the northern extremity of the Namib Sand Sea. The vegetated area in the middle foreground marks the site of the Kuiseb River which flows occasionally and limits northward dune migration. (Photo courtesy J. T. Teller.)*

Fig. 10.20 *Y-junction between linear dunes, South Australia.*

exhibit no external slip-face. Sheets and streaks rarely give rise to significant topographic features but domes (Fig. 10.17(G)) are true dune forms, in some cases attaining significant dimensions.

In addition to these basic types are **compound dunes** comprising two or more of the same basic dune type which have either coalesced or are superimposed. The latter can occur when, for instance, a dune-sized barchan rides up the back of a much larger megabarchan. **Complex dunes** can be formed when different basic dune types are associated. For example, star dunes are commonly seen surmounting linear dunes while barchans may occupy the troughs between them.

10.3.4   Global occurrence of free dune types

Only since the advent of the satellite reconnaissance of dune fields has it been possible to gain any idea of the relative abundance of the different types of dune on a global basis. Previously our knowledge of the distribution of dune types was provided only by rather patchy air photograph coverage and by even more limited ground observations. Although satellite imagery is extremely valuable it has limitations since some dune types cannot be readily differentiated at the resolutions available.

Some idea of the relative abundance of the principal dune types in most of the world's major deserts, including some fixed ergs, can be gained from Table 10.2, which is based on an extensive survey conducted by the United States Geological Survey. It should be noted that the dune types listed are confined to those that can be distinguished on Landsat imagery and that major dune fields in Australia and central Asia are omitted. Nevertheless the data indicate that sand sheets and streaks, transverse dunes and linear dunes are all quantitatively significant, with star and dome

dunes being much less common. Regional contrasts between deserts are interesting; for instance linear dunes are particularly predominant in the Kalahari, whereas star dunes are numerous in the north-east Sahara.

10.3.5   Development of free dunes

The two most significant factors determining free dune morphology are wind regime and sand supply (Fig. 10.22). The importance of sand supply is illustrated by the occurrence of barchans which appear only to develop on hard, rocky surfaces where sand is scarce. Under a similar undirectional wind regime abundant sand typically gives rise to series of barchanoid ridges. Sand availability is also crucial in determining whether megadunes develop. Australian deserts are remarkable for their lack of megadunes, the Simpson Desert, for instance, being characterized by low linear dunes separated by relatively sand free inter-dune troughs. It has been estimated that if all the sand of the Simpson Desert were spread out evenly it would be only 1 m thick. Similar calculations for the Great Erg Oriental of Algeria, where

Fig. 10.21 *Star dune, United Arab Emirates. Note the orientation of the ripples and slip-face indicating a left-to-right prevailing wind shaping the arm of the star dune in the foreground.*

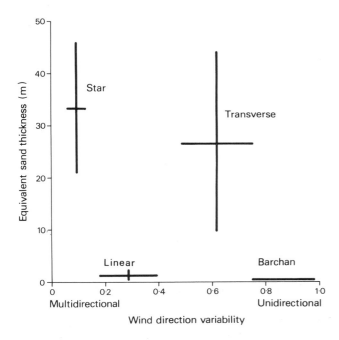

Fig. 10.22 *Relationship of major dune types to measures of sand availability (equivalent sand thickness) and variability in wind direction. Equivalent sand thickness is the thickness that would be attained by dune sand in an area if it were spread evenly. Wind direction variability is measured by calculating the frequency and strength of winds blowing from different directions. High values indicate a single predominant wind which accounts for a large proportion of sand movement, whereas low values indicate a highly variable wind regime. The majority of examples of each dune type plot within the indicated ranges. (After R. J. Wasson and R. Hyde (1983)* Nature, **304**, *Fig. 4, p. 339.)*

The morphology of dunes represents a response to the complex interaction between secondary flows in the wind and the form presented by the sand surface. Cause and effect are so intimately related that it is frequently difficult to establish whether particular types of airflow around dunes are instrumental in their formation or merely a result of their existence. Empirical data on the configuration of airflows associated with dune forms are sparse, especially for larger-scale features, and there is still much to be learnt about the detailed mechanisms of dune formation.

Examination of the relationships between present-day wind regimes and dune patterns can provide valuable insights

megadunes are abundant, give a much greater average depth of 26 m.

Table 10.2 Relative abundance of free dune types in some major world deserts

	THAR	TAKLA MAKAN	NAMIB	KALAHARI	SAUDI ARABIA	ALA SHAN	SOUTH SAHARA	NORTH SAHARA	NORTH-EAST SAHARA	WEST SAHARA	AVERAGE
Linear type dunes (total)	13.96	22.12	32.55	85.55	49.81	1.44	24.08	22.84	17.01	35.49	30.54
Simple and compound	13.96	18.91	18.50	85.85	26.24	1.44	24.08	5.74	2.41	35.49	23.26
Feathered	—	—	—	—	4.36	—	—	3.56	1.13	—	0.91
With crescentic superimposed	—	3.21	—	—	—	—	—	4.02	7.32	—	1.46
With stars superimposed	—	—	14.34	—	19.21	—	—	9.52	6.15	—	4.92
Transverse type dunes (total)	25.61	36.91	11.80	—	14.91	27.01	28.37	33.34	14.53	19.17	24.09
Single barchanoid ridges	8.96	3.21	11.80	—	0.59	8.62	4.08	0.06	—	0.65	3.80
Megabarchans	—	—	—	—	—	—	—	7.18	1.98	—	0.92
Complex barchanoid ridges	16.65	33.70	—	—	14.32	18.39	24.29	26.10	12.55	18.52	16.45
Star dunes	—	—	9.92	—	5.34	2.87	—	7.92	23.92	—	5.00
Dome dunes	—	7.40	—	—	—	0.86	—	—	0.80	—	0.90
Sheets and streaks	31.75	33.56	45.44	13.56	23.24	67.82	47.54	35.92	39.25	45.34	38.34
Undifferentiated	—	—	—	—	6.71	—	—	—	4.50	—	1.12

Figures are for percentage area covered of sample Landsat images by the particular dune type.
Source: Modified from S. Fryberger and A. S. Goudie (1981) *Progress in Physical Geography* **5**, Table 1, p. 423, based on analysis of maps in C. S. Breed *et al.* (1979) *United States Geological Survey Professional Paper* **1052**, 305–97 by A.S. Goudie.

into the genesis of particular dune forms. Nevertheless, this approach is not without difficulties. Some dune systems are so large that they have been developing over periods of thousands of years. It is unlikely that wind regimes have remained constant over such long periods and today we may just be witnessing the slow modification of patterns formed by earlier wind regimes. This problem is most relevant to megadune scale features since larger forms will take longer to adjust to changing wind patterns that smaller dune-sized forms. In some areas, such as the western Sahara, dunes are orientated quite differently from the megadunes on which they lie and it is difficult to explain this without invoking a change in wind direction.

A second problem is the existence of **relict dunes**. Large areas marginal to presently arid regions are covered by inactive dunes stabilized by vegetation. The patterns exhibited by these relict dune systems may be quite unrelated to present-day wind regimes and their interpretation may have to rest on uncertain analogies with morphologically similar, but active, forms elsewhere. Relict ergs, such as that covering much of the Kalahari Desert of southern Africa, develop in response to major climatic changes and thus provide important evidence of phases of greater aridity and higher wind speeds in the past (see Section 14.5.2).

10.3.5.1 Transverse-type dunes

One possible explanation for the development of transverse ridges involves wave-like motions in the wind. These could be initiated by some pre-existing minor surface irregularity or by temperature gradients in the atmosphere. Sand moves most rapidly where such waves descend to the ground, and consequently it will tend to accumulate in zones of slower movement where airflows are ascending. An undulating surface transverse to the wind direction will begin to form which, once initiated, will enhance the wave-like airflow. Movement of sand up the windward slope of the developing dune will eventually produce a slip-face and an equilibrium transverse ridge morphology.

Simple transverse ridges are comparatively rare apparently because 'corkscrew'-like motions, or vortices, usually develop in the airflow probably as a result of surface irregularities. These vortices distort the transverse ridge into the sinuous form called a barchanoid ridge (Fig. 10.23). Once formed a barchanoid ridge will itself influence the airflow so that further vortices are propagated which shape the next ridge downwind (Fig. 10.24). Linguoid elements of one ridge usually coincide with barchanoid elements of the next ridge downwind due to a complex interaction of transverse and longitudinal components in the airstream. This gives rise to a series of nearly linked ridges trending obliquely at 10–20° to the prevailing wind direction which, with the transverse elements, form what is termed a **dune network** or **aklé pattern**. The distance over which airflows are disturbed in the lee of dunes may provide another possible mechanism

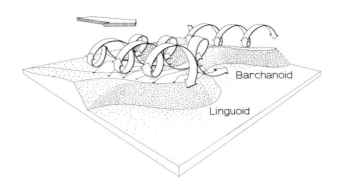

Fig. 10.23 *A transverse ridge distorted by longitudinal vortices. These cause variations in wind speed along the ridge crest which in turn lead to variations in the height of the ridge. Since lower dunes move more rapidly than higher dunes, the lower sections of the undulating ridge crest move more quickly than the higher sections. This effect eventually produces a sinuous barchanoid ridge composed of protruding linguoid and recessed barchanoid elements. (After A. Warren (1979) in: C. Embleton and J. Thornes (eds)* Process in Geomorphology. *Edward Arnold, London, Fig. 10.11, p. 339.)*

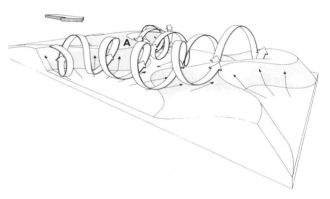

Fig. 10.24 *A barchanoid ridge (A) propagating vortices which shape a second ridge downwind. (From A. Warren, 1979,* Process in Geomorphology. *Edward Arnold, London, Fig. 10.12, p. 340.)*

for producing the relationship observed between dune size and spacing. Field data indicate that wind velocities return to their upwind value within a horizontal distance equivalent to 12 to 15 times dune height.

Where the supply is sparse, sand is entirely removed from the zone represented by the linguoid elements of barchanoid ridges and is swept downwind to form elongated arms thereby creating barchans. Indeed, transitions from transverse ridges to barchans may be observed where there is a marked diminution in sand supply downwind. Once formed a barchan generates its own secondary airflows which maintain its equilibrium form while it migrates downwind (Fig. 10.25). The elongation of barchan arms is due to the higher rate of sand movement on the lower, outer parts of the dune. Eventually the arms extend sufficiently downwind

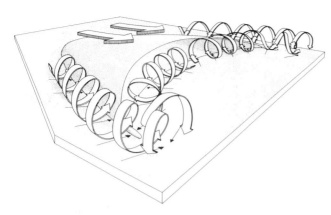

Fig. 10.25   *The pattern of airflow around an isolated barchan reconstructed from field data collected near In Salah in Algeria by P. Knott. (After A. Warren (1979)* Process in Geomorphology. *Edward Arnold, London, Fig. 10.13, p.341.)*

to receive some protection in the lee of the main body of the barchan. Like transverse and barchanoid ridges, barchans often show a regularity in width and spacing and this may

similarly reflect the effects of airflow patterns set up by barchans upwind.

10.3.5.2   Linear dunes

Although all linear dunes consist of elongated ridges, two subtypes can sometimes be distinguished. One type, sometimes referred to as a **seif dune**, is sinuous, relatively short and has a sharp crest and a pointed downwind terminus. The other subtype is long and straight and consists of a narrow crest surmounting a broad plinth (Fig. 10.26). There are widely divergent interpretations of the formation of linear dunes. A minority view is that they are created by the removal of sand from interdune troughs and that they are therefore erosional features. The great majority of researchers, however, believe them to be, like other free dune types, predominantly depositional forms. The main point of dispute is whether they develop in response to unidirectional winds or whether they are formed under bidirectional wind regimes.

Vortices in a unidirectional airflow might originate from convection cells in interdune troughs as a result of localized surface heating (Fig. 10.27). According to this model Y-

Fig. 10.26   *High altitude oblique aerial view of linear dunes in the Simpson Desert, central Australia. Y-junctions can be seen between individual dunes, many of which extend for tens of kilometres. Sand volumes here are low, as is evident from the broad, sandless interdune troughs.*

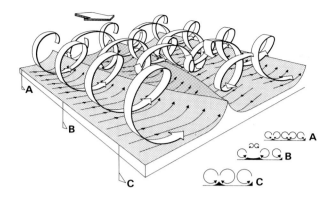

Fig. 10.27 *Hypothetical reconstruction of roll-vortices that might shape linear dunes. Note the change in vortex diameter downwind (cross-sections A, B and C); this provides a possible mechanism of Y-junction formation. As discussed in the text, this model of linear dune development is not now widely favoured. (From A. Warren (1979)* Process in Geomorphology. *Edward Arnold, London, Fig. 10.14, p.342.)*

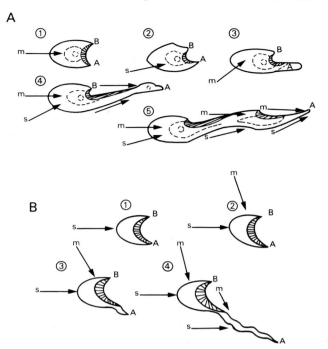

Fig. 10.29 *Two models for the development of seif-type linear dunes from barchans. In the model proposed by Bagnold (A) a symmetric barchan is initially formed by unidirectional, moderately strong prevailing winds (m) (1). High velocity storm winds (s) blowing obliquely across the barchan lead to a build-up of sand on one horn (2). This horn is elongated by storm winds until it enters the flow of sand moving from the other horn (3–4). Subsequently the horn is elongated by winds from both directions to form a sinuous linear, or seif, dune (5). By contrast, in the model proposed by Tsoar (B) the orientation of a barchan is considered to be related to the prevailing direction of storm winds (s) (1). The elongation of one of the barchan horns into a seif dune is thought to result from sand transport by more common gentle winds blowing from a different direction (2–4). (After H. Tsoar, (1984)* Zeitschrift für Geomorphologie *28, Figs. 1 and 2, pp. 100 and 101.)*

junctions would form where the axes of vortices rise from the surface, possibly as a result of heating. The major shortcoming of this explanation of linear dune formation is that the dimensions of vortices created by atmospheric instability are much larger than average dune spacings.

It has been noted that in many regions linear dunes are aligned obliquely to winds blowing from two prevailing directions. The two principal winds may be either seasonal or diurnal, but in either case slip-faces develop in two directions (Fig. 10.28). Under such conditions we would expect a dune to extend in the resultant direction of sand movement. Seif dunes, for instance, can apparently develop from barchans through the extension of one of the two arms by a second principal wind direction (Fig. 10.29).

Detailed field monitoring of a linear dune in the Namib Desert has demonstrated that its form changed in response to a seasonal wind regime. In summer westerly winds were

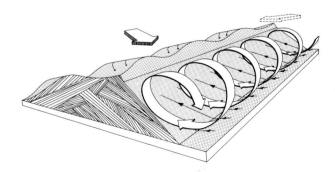

Fig. 10.28 *Internal structure of a linear dune formed by winds blowing from two principal directions. Note that the slip faces dip in opposing directions. (From A. Warren, (1979)* Process in Geomorphology. *Edward Arnold, London, Fig. 10.15, p.342.)*

found to transport sand from the western flank of the dune and to deposit it on its easterly lee side. In winter the situation was reversed. The crest of the dune was seen to move 15 m back and forth over each annual cycle but the base of the dune appeared to be fixed. Nevertheless, sand was being transported across the dune which was extending along a resultant of the easterly and westerly winds.

Although many researchers now believe that linear dunes are a product of bidirectional wind regimes, problems remain. For instance, it is not clear how compound linear dunes can be accounted for by this model. Moreover, it does not explain the regular spacing of linear dunes.

The limiting case for dune extension is when the two prevailing winds are each orientated normal to the dune crest. This type of wind regime produces a reversing dune which grows by vertical accumulation rather than horizontal extension. Reversing dunes are therefore strictly speaking a type of transverse dune. They have periodically opposing slip-faces and typically develop under a strong seasonal wind regime. Reversing dunes in Death Valley, California, advance towards the north-east in spring and summer and retreat to the south-east in the autumn.

10.3.5.3 Other types of free dune

Star dunes are most common in areas with complex wind regimes and appear to form in response to strong winds from several different directions. Some megadune-scale examples, such as those in the Grand Erg Oriental of the Sahara, reach an immense size and appear to be preferentially located at the nodes of crossing trends formed by linear dunes. It has been suggested that some isolated star dunes form at the nodes of thermal convection cells, but it seems unlikely that such airflows could be strong enough to pile up large quantities of sand. Star dunes along the eastern and southern margin of the Rub'al Khali erg in Saudi Arabia appear to fan out from the mouths of intermittent streams and this suggests an association with localized sand supplies.

Dome dunes probably form in a variety of ways. Some reveal internal slip-faces dipping in one direction and this suggests that they may have evolved from barchans through truncation of the crest and deflation of the arms by strong winds. Other types of dome dune appear to have evolved from star dunes or to be controlled by vegetation and moisture distribution.

10.3.6 Development of impeded dunes

A variety of dune forms are related to vegetation, topographic barriers or localized sources of sediment. The major varieties of these impeded dunes are listed and described in

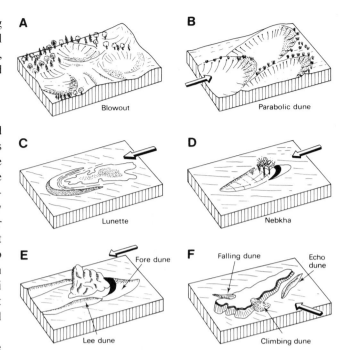

Fig. 10.30 *Types of impeded dune. Prevailing wind direction indicated by arrows.*

Table 10.3 and illustrated in Figure 10.30.

In sand accumulations which have been stabilized by vegetation, localized disturbance of the vegetation cover, by for instance overgrazing, can give rise to **blowouts**. These are circular or elliptical depressions, usually only a few metres across, formed by the deflation of sand (Fig. 10.30(A)). Higher wind speeds are attained over the unvegetated surface and this dries the sand and makes it more susceptible to entrainment. The deflated sand is

Table 10.3 Classification of dune types controlled largely by vegetation, topographic features or localized sediment sources

TYPE	FORM AND POSITION	MODE OF DEVELOPMENT
Blowout (A)	Circular rim around depression	Localized deflation
Parabolic dune (B)	'U' or 'V' shape in plan view with arms opening upwind to enclose a blowout	Deposition of sand locally deflated upwind; arms are usually fixed by vegetation
Lunette (C)	Crescent-shaped opening upwind	Accumulation downwind of localized sediment source such as desiccated lake basin or pan
Shrub-coppice dune (nebkha) (D)	Roughly elliptical to irregular in plan, streamlined downwind	Accumulation around and downwind of vegetation clump
Lee dune (E)	Elongated downwind from topographic obstruction	Accumulation on protected lee side of obstacle
Fore dune (E)	Roughly arcuate with arms extending downwind either side of obstruction	Accumulation in zone of disrupted airflow immediately windward of obstacle
Climbing dune (F)	Irregular accumulation rising up windward side of large topographic obstruction	Accumulation in zone of disrupted airflow on windward side of obstacle
Falling dune (F)	Irregular accumulation descending leeward side of large topographic obstruction	Accumulation in zone of disrupted airflow on upwind side of obstacle
Echo dune (F)	Elongated ridge roughly parallel to, and separated from, windward side of topographic obstruction	Accumulation in zone of rotating airflow upwind from large obstacle

Note: Letters A–F refer to illustrations in Figure 10.30

carried downwind generally as far as the edge of the originally disturbed area.

If deflation is intense the blowout becomes larger and its leeward rim migrates downwind leaving trailing arms on either side which are usually stabilized by vegetation. This forms a U or V-shaped **parabolic dune** (Fig. 10.30(B)). High wind speeds may destroy the transverse component of the parabolic form and leave only two elongated arms orientated downwind. Superficially, parabolic dunes may resemble barchans, but they can be readily distinguished since their arms extend upwind whereas in barchans they point downwind. Parabolic dunes are common where sand has been stabilized by vegetation, such as along coasts and in semi-arid areas. Patchy reduction in vegetation cover can eventually give rise to extensive fields of parabolic dunes. They cover nearly 30 per cent of the Thar Desert in north-west India, a region which has experienced considerable climatic fluctuations and consequently changes in vegetation density.

Crescent-shaped **lunettes** are found on the downwind side of some saline ephemeral lakes and pans and along tidal lagoons in coastal areas (Fig. 10.30(C)). In general, clay-sized particles constitute a much more important constituent than sand. Periodically exposed salts and clays are dried and carried as sand-sized aggregates downwind until vegetation is encountered. The crescent shape may be enhanced by wave action at times of lake or tidal flooding. Lunettes are common in the semi-arid region of south-east Australia and around pans in the north–central Kalahari Desert in Botswana. They appear to be confined to areas with precipitation sufficiently high to introduce clays and salts in runoff, but sufficiently low to permit desiccation and deflation. An optimum mean annual precipitation of about 380 mm has been suggested for Australian lunettes, but it is likely that many are associated with fluctuating semi-arid climates.

Since vegetation reduces wind speeds near the ground we would expect its presence to promote sand accretion. This is particularly likely to occur where vegetation lies downwind of a large sand supply. This situation is found not only in inland deserts, where localized vegetation sustained by near-surface ground water may be present, but also along coasts in most climatic zones where sand moving inland encounters vegetation. Isolated shrubs or vegetation clumps in deserts generate localized sand accumulations which taper downwind and are known as **shrub-coppice dunes** or **nebkhas** (Figs 10.30(D), 10.31). Especially along temperate coasts, grasses may colonize and stabilize dunes.

Fig. 10.31 *Nebkha, or shrub-coppice, dunes near Swakopmund, Namibia (hammer indicates scale).*

Such dunes are frequently sensitive to vegetation disturbance or short periods of high intensity winds associated with storms, both of which can generate numerous blow-outs and initiate dune deflation.

Topographic features can significantly affect near-surface airflows through the reduction of wind speeds and the creation of eddies. If they lie astride sand-carrying winds, dunes will form where secondary airflows converge. This usually occurs close to the upwind side of the obstruction, where a **fore dune** develops, and on the downwind side, where a streamlined **lee dune** is formed (Fig. 10.30(E)). A single lee dune can develop behind small topographic features, but a large obstruction characteristically generates two parallel lee dunes tapering downwind from its edges.

Where the wind encounters a high, steep slope, such as along an escarpment, a large roller vortex can develop which separates the fore dune from the topographic barrier. These may extend for several kilometres and because they reflect the form of the adjacent escarpment are often called **echo dunes** (Fig. 10.30(F)). They may develop into huge forms and the highest recorded dune in the world at over 400 m is an echo dune in Algeria. Where gradients along a topographic feature are gently inclined, such as along gullies, sand may be funnelled up the slope to form a **climbing dune** and if sand is carried over the crest of the barrier a **falling dune** may develop on the lee side (Fig. 10.30(F)).

10.3.7 Fine-grained deposits

Silts and especially clays can be transported considerable distances by the wind. Loess, consisting of well-sorted, very fine-grained deposits, cover large areas particularly in the mid-latitudes of the northern hemisphere, but they are also found on the fringes of hot deserts. Loess blankets substantial areas of central Europe, the southern USSR and China, and the USA (Fig. 10.32). Typically 80–90 per cent of the particles are between 0.005 and 0.5 mm across. On a global basis it has been estimated that some 10 per cent of the Earth's total land area is covered by loess from 1 to 100 m thick. These deposits appear to have originated largely in the extensive depositional plains formed on the margins of the Pleistocene ice sheets. Sediments were deflated, transported and redeposited during glacial episodes which were characterized by high wind speeds. This accounts for the concentration of loess in areas close to the equatorward limits of the Pleistocene ice sheets, since it is too coarse to

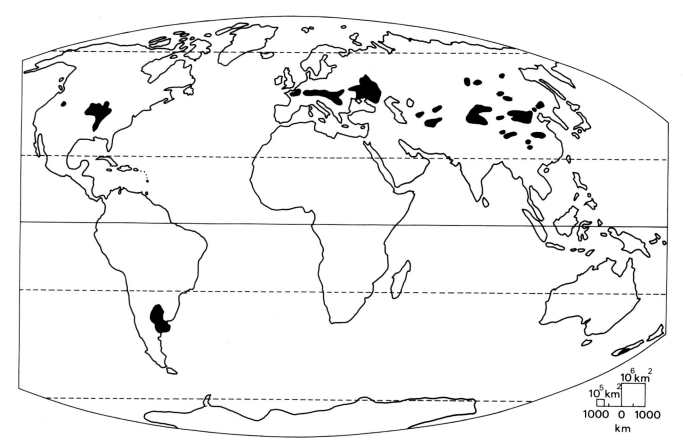

Fig. 10.32 *Global distribution of major loess occurrences. (After K. Pye, (1984) Progress in Physical Geography 8, Fig. 1, p. 178).*

travel very large distances and in fact can be seen to thin rapidly away from assumed source regions. These loess deposits are considered further in Section 12.2.6.

In most areas loess simply blankets the existing landscape and its main topographic effect is to smooth out minor surface irregularities since it accumulates more deeply in depressions. Rare occurrences of loess ridges, similar in form to sand dunes, are known, but it is not clear whether these are primarily of depositional or erosional origin. Unusually thick deposits known as **loess-lips**, are found downwind of major river valleys, such as the Mississippi and Missouri, which provided major sources of silt during glacial phases. A high degree of sorting, angularity of constituent particles, and small particle size associated with relatively strong interparticle bonds and moisture-retention capability makes loess a relatively cohesive deposit which in some cases forms steep cliffs. A high clay content or the precipitation of calcium carbonate in voids in the more calcareous forms of loess enhance this property.

Surprisingly, dune forms in clay are more common that those developed in silt. **Clay dunes** are mostly found near coastal lagoons and downwind of exposed clay-rich saline lake deposits and often take the form of lunettes. Clay dune slopes are rarely above 15° and the steeper gradient occurs on the windward side. The origin of these features composed of such fine material is more explicable when it is realized that the original constituents were sand-sized aggregates of clay particles which were entrained by the wind in the same way as ordinary sand grains. Such aggregates can form through the desiccation and cracking of clay-rich deposits, perhaps aided by the binding effect of crystallized salts. Once deposited, rain leaches out the salt and the clay aggregates decompose into fine particles to form a solid clay mass.

Further reading

Aeolian processes and landforms are discussed in various books on desert geomorphology, while additional information on littoral aeolian forms is to be found in texts on coastal geomorphology (see Chapter 13). Cooke and Warren (1973) and Mabbutt (1977) provide useful introductions to most of the topics covered in this chapter, but there have been major advances in the understanding of aeolian landforms since these books were written. This more recent research is discussed in the volumes edited by Nickling (1986) and Thomas (1989). McKee (1979) is an important global survey of depositional forms drawing heavily on remote sensing data and El-Baz (1984) similarly emphasizes the value of satellite imagery in the interpretation of aeolian features in desert landscapes. Other general sources include the book on aeolian processes by Greeley and Iverson (1985), the collection of papers edited by Brookfield and Ahlbrandt (1983) on aeolian sediments and the pion-

eering treatment of aeolian processes by Bagnold (1941) which is still worth consulting. Annual reviews of arid geomorphology in *Progress in Physical Geography* frequently contain valuable updates of research on aeolian processes and landforms.

The literature on the mechanisms of particle movement and deposition has grown rapidly over the past decade with the increasing use of wind-tunnel experiments. An overview of the mechanisms of sand entrainment and transport by the wind is provided by Sarre (1987), while Pye (1987) considers the mobilization of finer particles. Specific factors controlling sediment entrainment by the wind are examined by Logie (1982) and Willetts (1983). For more detail on the effect of turbulence on grain entrainment see Lyles and Krauss (1971), and for quantitative estimates of critical threshold velocities for particle movement over a wide variety of surface types see Gillette *et al.* (1980). A general model of aeolian sediment transport by both saltation and suspension is presented by Anderson and Hallet (1986), while saltation is considered specifically by Gerety and Slingerland (1983), Greeley *et al.* (1983) and Iversen (1983).

Deflation is discussed in the context of the distribution and frequency of dust storms by Goudie (1978, 1983a) and Middleton *et al.* (1986). Anderson (1986) considers the variation in rate of grain abrasion with height above the ground, and the formation of ventifacts is discussed by Whitney and Dietrich (1973). The development of yardangs is discussed by McCauley *et al.* (1981) and Whitney (1985), while Ward and Greeley (1984) provide a detailed field study from Rogers Lake, California. Whitney (1978) examines the role of vortices in generating wind-eroded lineations. The origin of pans is discussed by Lancaster (1978) with respect to the Kalahari and by Goudie and Thomas (1985) in the context of southern Africa as a whole.

Turning to depositional forms, the development of ergs is discussed by Fryberger and Ahlbrandt (1979) and Wilson (1971, 1973), while Wasson and Hyde (1983a) emphasize the importance of wind regime and sand supply in controlling dune morphology. The existence of a well-defined dimensional hierarchy of aeolian depositional forms is proposed by Wilson (1972, 1973) and challenged with respect to particular localities and dune types by Lancaster (1989a), Thomas (1988) and Wasson and Hyde (1983b). The formation of ripples is considered by Brugmans (1983), Ellwood *et al.* (1975) and Sharp (1963). Useful discussions on the development of free dunes include those by Cooke and Warren (1973), Fryberger (1979), Warren (1979) and Wilson (1973). Breed *et al.* (1979) provide a global survey of major sand seas based largely on the interpretation of satellite imagery, while studies of specific ergs include those by Lancaster (1989a) and McKee (1982) on the Namib, Anton (1983) on the eastern Arabian Desert, Wasson *et al.* (1983) on the Thar Desert, Wasson (1983) on the Strzelecki and Simpson Deserts in Australia and Mainguet and Chemin (1983) on the Sahara.

Important analyses of the origin of particular types of dune include those by Hastenrath (1967), Howard *et al.* (1978) and Warren and Knott (1983) on barchans; Folk (1971), Lancaster (1982) and Tsoar (1983) on linear dunes; Lancaster (1980), Tsoar (1984) and Warren (1976) on the relationship between barchans and seif dunes; and finally Lancaster (1989b) on star dunes. In addition, the report of the detailed monitoring of a linear dune in the Namib by Livingstone (1989) illustrates the value of field measurements in complementing laboratory and theoretical investigations of aeolian processes. Ideas on the formation of obstruction and vegetation-related dunes are reviewed by Cooke and Warren (1973), while Breed and Grow (1979) and Hack (1941) provide a more detailed examination of parabolic dunes, and Campbell (1968) discusses Australian lunettes. Relict dunes are briefly discussed by Goudie (1983b), and Bowler (1976), Grove and Warren (1968) and Talbot and Williams (1978) provide interesting regional studies. At the global scale Sarnthein (1978) presents evidence of the considerable extension of ergs during the Late Pleistocene.

A brief review of the voluminous literature on loess is provided by Pye (1984), but a much more detailed treatment is available in Pye (1987). The collection of papers edited by Péwé (1981) contains much useful material on the erosion, transport and deposition of desert dust. Smalley and Krinsley (1978) discuss the origin of desert loess and Yaalon and Dan (1974) consider its distribution and accumulation on the desert fringe in Israel. Clay dunes remain a little studied phenomenon, but Bowler (1973) provides a useful review of their occurrence and development.

References

Anderson, R. S. (1986) Erosion profiles due to particles entrained by wind: application of an eolian sediment-transport model. *Geological Society of America Bulletin* **97**, 1270–8.

Anderson, R. S. and Hallet, B. (1986) Sediment transport by wind: Toward a general model. *Geological Society of America Bulletin* **97**, 523–35.

Anton, D. (1983) Modern eolian deposits of the Eastern Province of Saudi Arabia. In: M.E. Brookfield and T.S. Ahlbrandt (eds) *Eolian Sediments and Processes.* Elsevier, Amsterdam, 365–78.

Bagnold, R. A. (1941) *The Physics of Blown Sand and Desert Dunes.* Chapman and Hall, London.

Bowler, J. M. (1973) Clay dunes: their occurrence, formation and environmental significance. *Earth Science Reviews* **9**, 315–38.

Bowler, J. M. (1976) Aridity in Australia: age, origins and expression in aeolian land forms and sediments. *Earth Science Reviews* **12**, 279–310.

Breed, C. S., Fryberger, S. G., Andrews, S., McCauley, C., Lennartz, F., Gebel, D. and Horstman, K. (1979) Regional studies of sand seas using Landsat (ERTS) imagery. *United States Geological Survey Professional Paper* **1052**, 305–97.

Breed, C. S. and Grow, T. (1979) Morphology and distribution of dunes in sand seas observed by remote sensing. *United States Geological Survey Professional Paper* **1052**, 253–302.

Brookfield, M. E. and Ahlbrandt, T. S. (eds) (1983) *Eolian Sediments and Processes.* Elsevier, Amsterdam.

Brugmans, F. (1983) Wind ripples in an active drift sand area in the Netherlands: a preliminary report. *Earth Surface Processes and Landforms* **8**, 527–34.

Campbell, E. M. (1968) Lunettes in southern South Australia. *Transactions of the Royal Society of South Australia* **92**, 85–109.

Cooke, R. U. and Warren, A. (1973) *Geomorphology of Deserts.* Batsford, London.

El-Baz, F. (ed.) (1984) *Deserts and Arid Lands.* Martinus Nijhoff, The Hague.

Ellwood, J., Evans, P. and Wilson, I. G. (1975) Small-scale aeolian bedforms. *Journal of Sedimentary Petrology* **45**, 554–61.

Folk, R. L. (1971) Longitudinal dunes of the northwestern edge of the Simpson Desert, Northern Territory, Australia. Part 1. Geomorphology and grain size relationships. *Sedimentology* **16**, 5–54.

Fryberger, S. G. (1979) Dune forms and wind regime. *United States Geological Survey Professional Paper* **1052**, 137–69.

Fryberger, S. G. and Ahlbrandt, T. S. (1979) Mechanisms for the formation of eolian sand seas. *Zeitschrift für Geomorphologie* **23**, 440–60.

Gerety, K. M. and Slingerland, R. (1983) Nature of the saltating population in wind tunnel experiments with heterogeneous size-density sands. In: M. E. Brookfield and T. S. Ahlbrandt (eds) *Eolian Sediments and Processes.* Elsevier, Amsterdam, 115–32.

Gillette, D. A., Adams, J., Edno, A. and Smith, D. (1980) Threshold velocities for input of soil particles into the air by desert winds. *Journal of Geophysical Research* **85**, 5621–30.

Goudie, A. S. (1978) Dust storms and their geomorphological implications. *Journal of Arid Environments* **1**, 291–310.

Goudie, A. S. (1983a) Dust storms in space and time. *Progress in Physical Geography* **7**, 502–30.

Goudie, A. (1983b) The arid Earth. In: R. Gardner and H. Scoging (eds) *Mega-Geomorphology.* Clarendon Press, Oxford and New York, 152–71.

Goudie, A. S. and Thomas, D. S. G. (1985) Pans in southern Africa with particular reference to South Africa and Zimbabwe. *Zeitschrift für Geomorphologie* **29**, 1–19.

Greeley, R. and Iversen, J. D. (1985) *Wind as a Geological Process.* Cambridge University Press, Cambridge and New York.

Greeley, R., Williams, S. H. and Marshall, J. R. (1983) Velocities of windblown particles in saltation: preliminary laboratory and field measurements. In: M. E. Brookfield and T. S. Ahlbrandt (eds) *Eolian Sediments and Processes.* Elsevier, Amsterdam, 133–48.

Grove, A. T. and Warren, A. (1968) Quaternary landforms and climate on the south side of the Sahara. *Geographical Journal* **134**, 194–208.

Hack, J. T. (1941) Dunes of the western Navajo Country. *Geographical Review* **31**, 240–63.

Hastenrath, S. L. (1967) The barchans of the Arequipa region, southern Peru. *Zeitschrift für Geomorphologie* **11**, 300–31.

Howard, A. D., Morton, J. B., Gad-El-Hak, M. and Pierce, D. B. (1978) Sand transport model of barchan dune equilibrium. *Sedimentology* **25**, 307–38.

Iversen, J. D. (1983) Saltation threshold and deposition rate modelling. In: M. E. Brookfield and T. S. Ahlbrandt (eds) *Eolian Sediments and Processes.* Elsevier, Amsterdam, 103–14.

Lancaster, I. N. (1978) The pans of the southern Kalahari, Botswana. *Geographical Journal* **144**, 81–98.

Lancaster, N. (1980) The formation of seif dunes from barchans –

supporting evidence for Bagnold's model from the Namib desert. *Zeitschrift für Geomorphologie* **24**, 160–7.

Lancaster, N. (1982) Linear dunes. *Progress in Physical Geography* **6**, 475–504.

Lancaster, N. (1989a) *The Namib Sand Sea: Dune Forms, Processes and Sediments*. Balkema, Rotterdam.

Lancaster, N. (1989b) Star dunes. *Progress in Physical Geography* **13**, 67–91.

Livingstone, I. (1989) Monitoring surface change on a Namib linear dune. *Earth Surface Processes and Landforms* **14**, 317–32.

Logie, M. (1982) Influence of roughness elements and soil moisture on the resistance of sand to wind erosion. *Catena Supplement* **1**, 161–73.

Lyles, L. and Krauss, R. K. (1971) Threshold velocities and initial particle motion as influenced by air turbulence. *Transactions of the American Society of Agricultural Engineers* **14**, 563–6.

Mabbutt, J. A. (1977) *Desert Landforms* Australian National University Press, Canberra.

Mainguet, M. and Chemin, M. C. (1983) Sand seas of the Sahara and Sahel: an explanation of their thickness and sand dune type by the sand budget principle. In: M.E. Brookfield and T. S. Ahlbrandt (eds) *Eolian Sediments and Processes*. Elsevier, Amsterdam, 353–63.

McCauley, J. F., Grolier, M. J. and Breed, C. S. (1981) Yardangs. In: D. O. Doerhing (ed.) *Geomorphology of Arid Environments*. Allen and Unwin, London, 233–69.

McKee, E. D. (ed.) (1979) A study of global sand seas. *United States Geological Survey Professional Paper* **1052**.

McKee, E. D. (1982) Sedimentary structures in dunes of the Namib Desert, South West Africa. *Geological Society of America Special Paper* **188**.

Middleton, N. J., Goudie, A. S. and Wells, G. L. (1986) The frequency and source areas of dust storms. In: W. G. Nickling (ed.) *Aeolian Geomorphology*. Allen and Unwin, Boston, and London, 237–59.

Nickling, W. G. (ed.) (1986) *Aeolian Geomorphology*. Allen and Unwin, Boston, and London.

Péwé, T. (ed.) (1981) Desert dust: origin, characteristics, and effect on man. *Geological Society of America Special Paper* **186**.

Pye, K. (1984) Loess. *Progress in Physical Geography* **8**, 176–217.

Pye, K. (1987) *Aeolian Dust and Dust Deposits*. Academic Press, London and Orlando.

Sarnthein, M. (1978) Sand deserts during glacial maximum and climatic optimum. *Nature* **272**, 43–6.

Sarre, R. D. (1987) Aeolian sand transport. *Progress in Physical Geography* **11**, 157–82.

Sharp, R. P. (1963) Wind ripples. *Journal of Geology* **71**, 617–36.

Smalley, I. J. and Krinsley, D. H. (1978) Loess deposits associated with deserts. *Catena* **5**, 53–66.

Talbot, M. R. and Williams, M. A. J. (1978) Erosion of fixed dunes in the Sahel, Central Niger. *Earth Surface Processes* **3**, 107–13.

Thomas, D. S. G. (1988) Analysis of linear dune sediment-form relationships in the Kalahari dune desert. *Earth Surface Processes and Landforms* **13**, 545–53.

Thomas, D. S. G. (ed.) (1989) *Arid Zone Geomorphology*. Belhaven, London.

Tsoar, H. (1983) Dynamic processes acting on a longitudinal (seif) dune. *Sedimentology* **30**, 567–78.

Tsoar, H. (1984) The formation of seif dunes from barchans – a discussion. *Zeitschrift für Geomorphologie* **28**, 99–103.

Ward, A. W. and Greeley, R. (1984) Evolution of the yardangs at Rogers Lake, California. *Geological Society of America Bulletin* **95**, 829–37.

Warren, A. (1976) Morphology and sediments of the Nebraska Sand Hills in relation to Pleistocene winds and the development of aeolian bedforms. *Journal of Geology* **84**, 685–700.

Warren, A. (1979) Aeolian processes. In: C. Embleton and J. Thornes (eds) *Process in Geomorphology*. Edward Arnold, London, 325–51.

Warren, A. and Knott, P. (1983) Desert dunes: a short review of needs in desert dune research and a recent study of micrometeorological dune-initiation mechanisms. In: M. E. Brookfield and T. S. Ahlbrandt (eds) *Eolian Sediments and Processes*. Elsevier, Amsterdam, 343–52.

Wasson, R. J. (1983) The Cainozoic history of the Strzelecki and Simpson dune fields (Australia), and the origin of the desert dunes. *Zeitschrift für Geomorphologie Supplementband* **45**, 85–115.

Wasson, R. J. and Hyde, R. (1983a) Factors determining desert dune type. *Nature* **304**, 337–9.

Wasson, R. J. and Hyde, R. (1983b) A test of granulometric control of desert dune geometry. *Earth Surface Processes and Landforms* **8**, 301–12.

Wasson, R. J., Rajaguru, S. N., Misra, V. N., Agarwal, D. P., Dhir, R. P., Singhoi, A. K. and Kameswara Rao, K. (1983) Geomorphology, late Quarternary stratigraphy and palaeoclimatology of the Thar dune field. *Zeitschrift für Geomorphologie Supplementband* **45**, 117–51.

Whitney, M. (1978) The role of vorticity in developing lineation by wind erosion. *Geological Society of American Bulletin* **89**, 1–18.

Whitney, M. I. (1985) Yardangs. *Journal of Geological Education* **33**, 93–6.

Whitney, M. I. and Dietrich, R. V. (1973) Ventifact sculpture by wind-blown dust. *Geological Society of American Bulletin* **84**, 2561–82.

Willetts, B. (1983) Transportation by wind of granular materials of different grain shapes and densities. *Sedimentology* **30**, 669–80.

Wilson, I. G. (1971) Desert sandflow basins and a model for the development of ergs. *Geographical Journal* **137**, 180–97.

Wilson, I. G. (1972) Aeolian bedforms – their development and origins. *Sedimentology* **19**, 173–210.

Wilson, I. G. (1973) Ergs. *Sedimentary Geology* **10**, 77–106.

Yaalon, D. H. and Dan, J. (1974) Accumulation and distribution of loess-derived deposits in the semi-desert fringe areas of Israel. *Zeitschrift für Geomorphologie Supplementband* **20**, 91–105.

11

Glacial processes and landforms

11.1 Glacier characteristics and dynamics

In some ways it is appropriate to describe the Earth as an ice planet. At present some 10 per cent of the surface of the continents is covered by ice, but as little as 18 000 a BP the figure was nearer 30 per cent. This greater extent of ice sheets has characterized much of the past 2–3 Ma and consequently landforms created by glacial action cover large areas that are now experiencing non-glacial climates. Since the majority of glacial landforms are formed beneath ice sheets and glaciers it is only when these ice bodies retreat that many types of glacial landform are revealed. This situation poses particular problems in understanding the processes whereby glaciers produce landforms since the mechanisms at work can only rarely be observed directly. In this chapter we examine specific landforms created by glaciers, but as many glaciated landscapes are located in regions which are now experiencing quite different temperate conditions we consider the features of glacial landscapes as a whole in Chapter 14.

11.1.1 Glacier distribution and classification

Ice covers some 14.9 million km^2 at the present day. Most of this is accounted for by the two vast ice sheets of Antarctica and Greenland, the remaining 4 per cent comprising ice caps and glaciers located mainly in high latitudes, and innumerable glaciers found at high elevations in all latitudes (Table 11.1). Glaciers can be classified morphologically on the basis of their relationship to the underlying bedrock topography (Table 11.2; Figs 11.1 and 11.2). It is difficult to generalize about the distribution of the different glacier types since, apart from ice sheets, most occur over a broad range of latitudes.

The occurrence of glaciers is determined not only by climate but also by topography since there must be a suitable surface on which ice can accumulate. Glaciers can only

Table 11.1 Present global distribution of glaciers

REGION		APPROXIMATE AREA (km²)
North polar area		
Greenland ice sheet		1 726 400
Queen Elizabeth Islands		106 988
Other Greenland glaciers		76 200
Spitsbergen and Nordaustlandet		58 016
Other		114 012
	Subtotal	2 081 616
Continental North America		
Alaska		51 476
Other		25 404
	Subtotal	76 880
South American Cordillera		26 500
Europe		9 276
Asian continent		115 021
African continent		12
Australasia		1 015
South polar area		
Antarctic ice sheet (excluding ice shelves)		12 535 000
Other glaciers on Antarctic continent		50 000
Sub-Antarctic islands		3 000
	Subtotal	12 588 000
	Total	14 898 320

Source: Data from R. F. Flint, (1971) *Glacial and Quaternary Geology*, Wiley, New York, Table 4–B, pp. 76–7, based on various sources.

form where snow persists from year to year. Whether this occurs is determined by the rate of snow accumulation, which is largely a function of the amount of precipitation falling as snow, and the rate of melting, which is largely a function of temperature (Fig. 11.3). In the mid-latitudes the seasonal distribution of precipitation is more crucial than the total amount. Extensive glaciers are not found, for instance, in the high mountains along mid-latitude west

Table 11.2 Morphological classification of glaciers

BASIC TYPE	COMPONENT OR SUBTYPES	CHARACTERISTICS
Ice sheet and **ice cap** (unconstrained by topography)*	**Ice dome**	A dome-like ice mass with a convex cross-profile formed in response to the basic flow characteristics of ice
	Outlet glacier	Glaciers which radiate out from an ice dome often occupying significant depressions (Fig. 11.1). Within the ice dome they can be distinguished by a zohe of rapidly moving ice termed an ice stream
Ice shelf		A floating ice cap or part of an ice sheet only partially constrained by the coastal configuration and which deforms under its own weight
Glaciers constrained by topography	**Ice field**	A roughly level area of ice distinguished from an ice cap because of the absence of a dome-like form and the control on ice flow exerted by the underlying topography
	Cirque glacier	A small ice mass usually occupying an armchair-shaped bedrock hollow and characteristically wide in relation to its length
	Valley glacier	A glacier which occupies a rock valley and is overlooked by rock cliffs (Fig. 11.2). They may originate in an ice field or a cirque glacier into which they may imperceptibly merge at their upper end. Large valley glaciers may be joined by tributary glaciers, forming a dendritic pattern of ice flow
	Other small glaciers	Glaciers which occur in a wide variety of topographic positions, but all of which are closely controlled by the underlying topography

*Ice sheets and ice caps are essentially distinguished on the basis of size, the former exceeding around 50 000 km^2 in area.
Source: Based on classification and discussion in D. E. Sugden and B. S. John (1976) *Glaciers and Landscape*. Edward Arnold, London.

Fig. 11.1 *Outlet glaciers near Bartholins Brae, Blosseville Kyst, Greenland. Note the medial and lateral moraine. (Photo courtesy of the Geodetic Institute, Copenhagen, Denmark.)*

Fig. 11.2 *Harker Glacier, a valley glacier discharging into a fjord, South Georgia. (Photo courtesy D. E. Sugden).*

coasts in spite of heavy precipitation because most falls as rain in summer and the relatively high summer temperatures lead to substantial melting of winter snows. By contrast the largest ice sheets receive very small amounts of precipitation although this nearly all falls as snow. The mean annual precipitation over the northern part of the Greenland ice sheet is only 150 mm, but rates of summer melting are so low that ice accumulation takes place.

Although there are other sources of heat supplied to glaciers, solar radiation is by far the most significant in

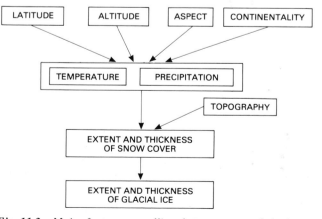

Fig. 11.3 *Major factors controlling the occurrence of glaciers.*

accounting for their global distribution. Because melting can occur only at temperatures above 0 °C it is the mean summer temperature rather than the mean annual value that influences the amount of snow-melt that will occur. Latitude, elevation, aspect and continentality all exert an indirect control over the distribution of glaciers through their influence on precipitation and temperature (Fig. 11.3). The effect of continentality on precipitation is a particularly significant factor in the development of ice sheets. This is illustrated by the pattern of snow accumulation over the Antarctic ice sheet which is closely related to the distance from the nearest area of ice free ocean in summer.

11.1.2 Characteristics of glacier ice

Glaciers are composed not only of ice but also of smaller amounts of air, water and rock debris. A property of ice of considerable geomorphic importance is its ability to deform and flow under its own weight. Although glacier ice can form directly from the freezing of liquid water or water vapour on the glacier surface, snowfall is the most important source. Snow with a density of only 50–70 kg m^{-3} is gradually converted to ice by a complex process of compaction and recrystallization through an intermediate stage known as **firn**. This is composed of a loosely consolidated mass of ice crystals with a bulk density of around 400 kg m^{-3}. The transformation into ice with a density of about 800 kg m^{-3} involves an increase in crystal size and the closing of voids, but the final conversion to true glacier ice (density of about 900 kg m^{-3}) only occurs when the pressure of overlying ice leads to the elimination of most of the remaining air bubbles.

Temperature is a key characteristic of glacier ice as it exerts a profound influence on glacier behaviour. In addition to solar radiation heat can be supplied to a glacier from the surface by the incorporation of firn which is warmer than the existing glacier ice and by the release of latent heat as water is refrozen. A glacier is also warmed from below by geothermal heat and by frictional heat generated by sliding and the deformation of ice at its base. Variations in the supply of heat from these sources give rise to two types of glacier ice which have fundamentally different geomorphic properties.

Cold ice is at a temperature below melting point, but **warm ice** is so close to melting point that it contains liquid water. Strictly we should use the term **pressure melting point** here as the freezing point of water decreases as pressure increases. For example, the melting point at a depth of 2164 m at the base of the Antarctic ice sheet at Byrd Station is −1.6 °C. Cold ice occurs where the glacier surface experiences very low winter temperatures or where low summer temperatures lead to negligible surface melting.

Warm ice occurs where geothermal or frictional heat are sufficient to raise the temperature of at least the basal ice to the pressure melting point. In small glaciers the percolation of surface meltwater may be the most important heat source since the refreezing of 1 g of water releases sufficient heat to raise the temperature of 160 g of ice by 1 °C. The rate of firn accumulation can also be important since a high rate can introduce a significant quantity of cold material into the glacier. This helps to counteract the geothermal heating at the glacier base and so reduces the rate of increase in temperature with depth. Once the base of a glacier reaches pressure melting point all the heat from geothermal and frictional sources can be used in melting rather than simply raising the temperature of the ice.

It has been accepted practice to classify glaciers as temperate or polar depending on whether they are supposed to be composed of warm or cold ice. However, it is now recognized that not only do both types of ice occur in glaciers in both temperate and polar environments but that both types may be present in an individual glacier or ice sheet. The Antarctic ice sheet, for example, is composed predominantly of cold ice but is now known to contain basal layers of warm ice in places. In the following discussion we shall refer to those glaciers or parts of glaciers with a basal layer at pressure melting point as **warm-based** and those with a basal layer below pressure melting point as **cold-based**. A further important distinction is that between **active ice** which is moving downslope and being replenished by fresh accumulations of snow in its source region and **stagnant ice** which is decaying *in situ* and has ceased to experience any significant lateral movement.

11.1.3 The glacier mass balance

The gains and losses of ice experienced by a glacier constitute its **mass balance** or **glacial budget**. The ice, firn and snow added to the glacier constitute its **accumulation**. The losses, or **ablation**, arise largely from melting in most glaciers, but evaporation, **sublimation** (the direct conversion of ice to water vapour), wind erosion and **calving** (the breaking away of blocks of ice into standing water) may also be important in certain situations.

A mass balance is normally calculated over the **balance year**. This is the interval between two successive times at which ablation has reached a maximum value. Although this normally occurs at the end of the summer the balance year may not be exactly 365 days. The total masses added to and lost from a glacier during the balance year are called, respectively, the **gross annual accumulation** and **gross annual ablation**. The difference between these two amounts is the **net annual accumulation** or **net annual ablation**, depending on whether there has been a net gain or loss in mass. The **net specific balance** is the net annual accumulation or ablation at a particular point on a glacier, and the integration of a large number of such point measurements

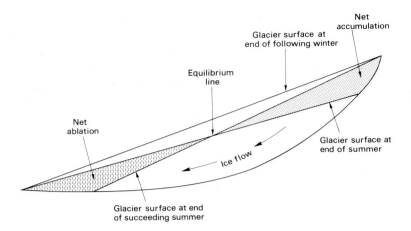

Fig. 11.4 *Variations in accumulation and ablation down an idealized glacier. The distribution of net annual accumulation and ablation over a glacier is of great significance to glacier movement. Net ablation generally increases down-glacier below the equilibrium line while net accumulation tends to increase up-glacier above this point. The rate of increase in the net balance with height up a glacier is important for the rate of ice movement because the higher the rate of increase the faster is the rate of ice movement needed to maintain the same glacier profile.*

can provide an estimate of the total mass balance.

The taking of such measurements is a difficult and time-consuming task and consequently there are few accurate mass balances for real glaciers. The data that do exist clearly show that the annual net balance generally varies over a glacier in a systematic manner with a positive balance (net accumulation) in the upper part of the glacier and a negative balance (net ablation) in the lower part. These two zones meet at the **equilibrium line** where accumulation is exactly compensated by ablation and where the mass balance is therefore zero (Fig. 11.4). It is important to draw a distinction between the situation over an entire balance year and seasonal variations since, depending on the season, much or all of a glacier may be experiencing accumulation or ablation.

11.1.4 Glacier motion

11.1.4.1 Mechanisms of ice movement

The movement of glaciers can in many respects be treated in the same way as other forms of mass movement, although ice masses have special properties which must be taken into consideration. Two, more or less distinct, types of ice movement can be distinguished – **internal deformation** and **basal sliding**. The relative importance of these two mechanisms varies significantly with basal sliding accounting for up to 90 per cent of the movement of warm-based glaciers, but being largely inoperative in cold-based glaciers. Glacier ice deforms internally because it is subject to stress. At any one point in a glacier this stress has two components: hydrostatic pressure, which is exerted in all

directions and is related to the mass of overlying ice, and shear stress, which is related not only to the thickness of overlying ice but also to the surface slope of the glacier. High shear stresses are generated towards the base of thick glaciers with steep slopes whereas lower basal shear stresses are produced by thin, gently sloping glaciers. The variations in shear stress at the base of most glaciers are, however, not large, usually lying between 0.05 and 0.15 MPa with a mean of about 0.1 MPa.

The main mode of internal deformation involves slippage within and between ice crystals and is called **creep**. Although the exact mechanism is not fully understood, various attempts have been made to model ice deformation by this process. In one such formulation, known as **Glen's power flow law**, the predicted rate of deformation depends not only on the shear stress but also on the temperature of the ice (Box 11.1). This model of creep appears to accord with several observations about the behaviour of glaciers. First, deformation rates are at a maximum at the base of the glacier, where both stresses and, in the case of cold ice, temperatures are highest. Secondly, it explains the faster rates of movement of warm ice. Thirdly, it accounts for the way glaciers can regulate their discharge through negative feedback since an increase in glacier thickness will increase the basal shear stress which will in turn accelerate the rate of ice flow and thereby reduce ice thickness. Where stresses within the glacier cannot be accommodated sufficiently quickly the ice may move by fracturing. This is most likely to occur on the margins of glaciers where thrusting can give rise to shear fractures and tensional stresses can produce crevasses (see Section 11.1.5.2).

Basal sliding involves four major processes. One mechanism is the slippage of the glacier bed over a thin

Box 11.1   Glen's power flow law

On the basis of laboratory experiments J. W. Glen established that the strain rate in a block of ice soon attains a
steady value on being subject to a constant stress. The relationship determined by Glen, known as the power flow
law, has been adapted by J. F. Nye to apply to glaciers. The
relationship is

$$e = A\tau^n$$

where e is the strain rate, τ the effective shear stress, A a
constant related to temperature and n an exponent.

The values of n have been determined by several inves-
tigators and vary from 1.7 to 4.5, the mean value being
around 3. This exponent makes the strain rate highly dependent on shear stress since with a value of n of 3 a doubling of the shear stress produces an eightfold increase in
strain rate. In Glen's experiments the value of A was found
to be significantly affected by differences in temperature,
changing from 0.17 at 0 °C to only 0.0017 at –13 °C, a
variation of two orders of magnitude.

layer of water only a few millimetres thick. This can significantly reduce the friction between the glacier and its bed
and considerably increase rates of glacier movement. A
second mechanism, known as **regelation creep**, involves
the movement of a warm-based glacier over minor irregularities in its bed. The higher pressures on the up-glacier side
of an obstacle lead to melting and the water produced
migrates to the zone of lower pressure on the lee side of the
obstruction where it refreezes. The process is most effective where the obstacle is small enough (less than about 1 m
across) to enable the latent heat released by the refreezing
to be effectively conducted to the up-glacier side where it
promotes further melting. Where larger obstacles are present
on the glacier bed a third process known as **enhanced
basal creep** can become important. The increased stress on
the up-glacier side of such an obstruction allows the ice to
flow round it. This mechanism promotes the movement of
basal ice even where it is below pressure melting-point.

Traditionally, the analysis of basal sliding has been
founded on the assumption that glaciers move over a rigid
rock bed. That this is frequently not the case is clear from
the sediment-covered surfaces found immediately in front
of retreating glaciers. Furthermore, some 80 per cent of the
area of Europe and North America covered by ice sheets
during the glacial advances of the Pleistocene is mantled by
unconsolidated sediments rather than bedrock. Many instances of deformation structures attributable to ice sheet
movement are known from these sediments, and recent
experimental work has demonstrated that even quite coarse-
grained sediments can be easily deformed beneath a glacier.
All this suggests that a fourth, and possibly widespread,
mechanism of glacier motion involves a significant part of
ice movement being accomplished through the deformation

of subglacial sediments. By facilitating movement such **bed
deformation** would increase the rate of ice flow or allow
flow to occur at a similar rate over a lower bed gradient or
lead to a combination of these effects. In order for bed deformation to occur the subglacial material must be saturated
with water and maintain a high pore-water pressure. Such
conditions will only be met if the glacier bed is at the
pressure melting-point and if basal water accumulates in
the sediment rather than being drained away.

11.1.4.2   Glacier flow

Average rates of movement vary enormously from one
glacier to another but typically lie in the range 3–300 m a⁻¹.
The velocity of ice flow also varies spatially and temporally within an individual glacier. If we look at a vertical
profile we find that the flow velocity is at a maximum at the
glacier surface and decreases downwards (Fig. 11.5(A),
(B)). This at first sight appears paradoxical since, as we
have seen, most movement occurs in the basal layers of a

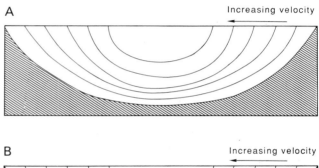

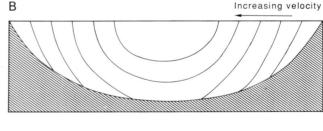

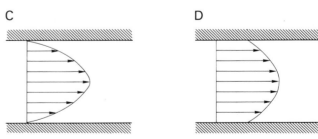

Fig. 11.5 *Schematic representation of vertical and horizontal
variations in the velocity of valley glacier movement: (A) cross-
profile vertical variation in velocity in a glacier with no basal
sliding; (B) vertical cross-profile variations in velocity where
basal sliding is significant (isolines intersect valley floor); (C)
plan view of valley glacier illustrating variations in velocity at
surface where there is no sliding along valley walls. (D) plan
view of velocity variation at surface where there is significant
sliding along valley walls.*

glacier where the shear stress is at a maximum. We have to remember, however, that each imaginary layer of ice is not only moving as a result of the shear stress at that level but is also being carried along by the moving layer below it. The velocities at each level in the glacier are thus cumulative from the base to the surface. There are also significant variations in the velocity across confined glaciers since the increased friction between the ice and rock wall retards ice movement there (Fig. 11.5(C), (D)).

Changes in discharge longitudinally down a glacier are strongly influenced by variations in total accumulation and ablation. Discharge is generally highest around the equilibrium line because the cumulative volume of ice increases from the head of the glacier to reach a maximum at this point. Below the equilibrium line discharge decreases progressively down-glacier with the increasing net loss of ice through ablation (Fig. 11.4).

Because snow is being added from above in the accumulation zone there must be some downward movement of ice from the surface. Similarly, in the ablation zone there must be some upward movement to maintain the surface form by replacing ice lost through ablation. Evidence of these vertical components of movement is provided by the burial of surface debris in the accumulation zone and the emergence at the surface of previously buried debris in the ablation zone. Such longitudinal variations in ice movement are associated with two different types of flow regime within a glacier (Fig. 11.6(A)). Above the equilibrium line **extending flow** predominates since the ice becomes more 'stretched out' down-glacier as the velocity of flow increases with the steeper angle of movement. Conversely, there is a reduction in velocity in the zone of **compressive flow** below the equilibrium line as the upward component of ice movement becomes more marked. In reality the longitudinal zonation of extending and compressive flow is more complex than this since the flow regime is also affected by bedrock topography; extending flow is promoted over convex bedrock surfaces, especially in **ice falls** which occur where a glacier flows down a steep slope, whereas compressive flow prevails over concave surfaces (Fig. 11.6(B)). The upward and downward component of ice movement associated with zones of compressive and extending flow is accommodated along **slip lines** which meet the glacier surface at around 45° and represent trajectories of maximum shear stress. Movement along these slip lines can occur through creep as well as by thrusting along 'faults' within the ice.

In ice sheets and ice caps relatively narrow zones of ice occur in which movement is much more rapid than the adjacent parts of the ice mass. These **ice streams** are responsible for a significant proportion of ice movement in large ice masses and often feed outlet glaciers (Fig. 11.1). Several major ice streams have been identified in the Antarctic ice sheet and seismic measurements indicate that at least one of these in the west Antarctic, known as Ice

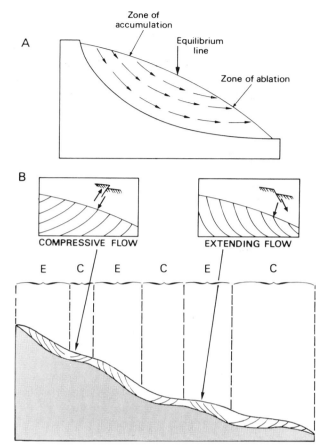

Fig. 11.6 *Flow patterns within a glacier: (A) basic pattern of downward movement of ice in the accumulation zone and upward movement in the ablation zone; (B) zones of extending and compressive flow related to irregularities in valley floor morphology (E = extending, C = compressive). The lines indicating the direction of ice movement with respect to the glacier surface are slip lines which indicate the direction in which the ice has the maximum tendency to shear. (Part (B) modified from J. F. Nye (1952) Journal of Glaciology* **2**, *Fig. 7, p.88 and Fig. 8, p. 90.)*

Stream B, is moving over a bed composed of saturated, deformable sediment.

Changes in ice movement over time occur in response to alterations in the mass balance arising from particular meteorological conditions. Cooler weather promotes accumulation and the glacier thickens and increases its velocity, whereas in warmer conditions increasing ablation induces thinning and a reduction in velocity, both adjustments progressing until a new equilibrium profile is attained. Changes in mass balance can be transmitted down-glacier by **kinematic waves**. These are complex phenomena, but we can grasp a basic understanding of how they function by imagining an increase in accumulation which causes a local raising of the glacier surface. The thicker ice under the bulge will begin to move more rapidly than the surrounding ice because the basal shear stress will be higher. The zone of thicker ice then moves down-glacier as a wave at between two to five times the velocity of the ice itself. Adjustments to the profile of a glacier arising from changes in mass balance can consequently be transmitted to the glacier snout much more rapidly by kinematic waves than by physical movement of the ice itself. Although kinematic waves have been observed moving down glaciers, in exceptional cases involving a surface rise of some 100 m, they need not necessarily have any surface expression since the increase in discharge represented by the wave can occur through adjustments in the ice flow other than those involving alterations in glacier thickness.

The rapidity with which glaciers respond to changes in mass balance and attain a new equilibrium form varies enormously but it is generally related to the size of the ice mass. Relaxation times vary from 3 to 30 a for valley glaciers to thousands of years for ice sheets. Such long and variable relaxation times make it difficult to relate changes in glacier movement to particular climatic or meteorological conditions.

The most spectacular temporal variations in glacier movement are **glacier surges**. These events, during which ice-flow velocities temporarily reach between 10 and 100 times their normal value, are experienced by some, but not all, glaciers. Surging glaciers are fairly common, some 204 having been identified in North America alone, but they seem to have few characteristics in common. In some, surges occur on a more or less regular cycle with a periodicity ranging from 15 to 100 a or more, but in others surging is unpredictable. They also occur in a wide range of glacier types, both warm and cold-based. Surging is initiated when a threshold of instability is reached and ice in the upper ablation zone begins to move rapidly down-glacier. In some cases this may precipitate a rapid movement of the glacier snout, such as the 45 km advance at a rate of up to 5 m h^{-1} recorded during a surge by Brúarjökull, Iceland. In many North American glaciers, however, the effect of a surge is mainly confined to a thickening of ice towards the snout.

Surging is still one of the least understood aspects of glacier movement; it is necessary to explain both the mechanism that triggers a surge and the high flow velocities subsequently attained. The greater abundance of meltwater commonly associated with surges suggests that the rapid movement may be related to greatly enhanced rates of basal sliding through the effect of basal water reducing friction along the glacier bed. The trigger mechanism is much more difficult to explain. It may be linked to external factors such as earthquakes or increased precipitation, but the regularity of surging in many glaciers and the rather similar quantities of ice moved in each surge suggests that it is more likely an intrinsic element of the behaviour of some glaciers. Such glaciers may experience a gradual build-up of basal water as a result of changes in the sub-glacial hydrological system which eventually precipitates

an increase in basal sliding. This could then promote a cycle of positive feedback in which the initial increase in basal sliding generates more meltwater because of the greater frictional heating associated with the higher rate of basal movement.

The concept of glacier movement assisted by the deformation of subglacial sediments has also led to the idea that periodic changes in pore-water pressure could promote surging behaviour. Data collected from boreholes drilled into Trapridge Glacier, Yukon Territory, Canada, indicate that it rests, at least in part, on a bed of potentially deformable sediment. It has been suggested that during non-surging interludes this substrate is efficiently drained through a system of channels and pipes; consequently, pore-water pressures in the sediment are low and significant deformation does not occur. An increase in basal shear stress, perhaps related to an increase in glacier thickness, could promote an initial increase in subglacial sediment deformation leading to the disruption of the subglacial drainage system. Such a reduction in permeability could then lead to a build-up of water within the sediment voids, and the resulting increase in pore-water pressure could induce a considerable increase in bed deformation and thereby glacier movement.

11.1.4.3 Short-term glacier fluctuations

Glaciers advance and retreat in response to changing meteorological and climatic conditions over a range of time scales. Here were are concerned only with the shorter-term fluctuations of up to a few hundred years, since longer-term adjustments are caused by major climatic changes and are therefore more appropriately considered in Chapter 14. The relationships between changing meteorological conditions and short-term climatic fluctuations and glacier behaviour are far more complex than was once assumed. Indeed, changes in weather and climate are so rapid in comparison with typical relaxation times of glacial systems that glaciers in true equilibrium with prevailing climatic conditions are rare. Only over the past 40 a or so have detailed records of mass balance been kept and these for only a few glaciers. Over a longer period less detailed information on glacier fluctuations can be gained from maps, photographs, drawings, historical reports and, over longer periods, botanical evidence.

Variations in rates of glacier movement over weeks and months appear to be related to the abundance of meltwater associated with seasonal changes in ablation rates. Year-to-year fluctuations follow, with a lag, variations in winter snowfall and mean annual temperature, while the widespread and significant phases of glacier advance and subsequent retreat documented over the past 300 a or so appear to be related to the cool climatic interval known as the 'Little Ice Age'. Not all glaciers, however, respond in the same way to a particular climatic fluctuation. Glaciers with a low

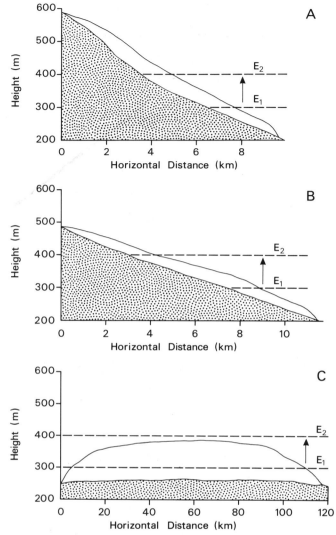

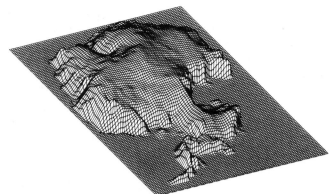

Fig. 11.8 *Morphology of the Antarctic ice sheet illustrated by a computer-generated three-dimensional reconstruction. (From D. J. Drewry (ed.) (1983)* Antarctica: Glaciological and Geophysical Folio. *(Scott Polar Research Institute, Cambridge Sheet 2, Fig. 2b.)*

Fig. 11.7 *Effects of a rise of 100 m in the altitude of the equilibrium line (E₁ to E₂) on a steeply inclined valley glacier (A), a more gently inclined valley glacier (B) and on an ice cap (C). Note that as the slope of the ice surface decreases from A to C the proportion of the glacier affected by the change in the altitude of the equilibrium line increases.*

is largely controlled by the dynamics of ice flow. This can be seen most clearly in the convex form of ice domes (Fig. 11.8). These have a maximum thickness which may exceed 4000 m in Antarctica. Where there is sufficient accumulation the ice builds up until the level of basal shear stress promotes significant ice deformation. As we have already noted the amount of shear stress is related to glacier thickness and surface slope so where the ice is thinner around the margins of the dome a steeper slope is required to maintain the ice flow, and conversely where the ice is thicker below the crest of the dome a less steep gradient is needed to sustain movement. Theoretical models of ice deformation generally accord quite well with the form of most ice domes (Box 11.2), but in reality a smoothly curved convex shape is prevented by factors such as bedrock irregularities underlying the glacier and spatial variations in ice tempera-

surface gradient are more susceptible to such changes than those with a steep slope since a vertical change in the equilibrium line due to, for instance, a change in mean annual temperature, will affect a much larger proportion of the latter and thus produce a much more substantial change in the areas of ablation and accumulation (Fig. 11.7).

11.1.5 Glacier morphology

11.1.5.1 Large-scale forms
While the overall shape of confined glaciers is of course closely related to topography, that of unconfined ice masses

Box 11.2 Estimating the profile of an ice dome

On the basis of a number of assumptions J. F. Nye proposed a simple formula which enables an estimate of the cross-profile of an ice dome to be made. The most important assumptions are that the ice is actively flowing and the ice dome is in equilibrium, but it is also assumed that the ice flow is not affected by variations in temperature, by localized accumulation or by bed irregularities. The elevation (m) of the ice surface at any point (h) is given by

$$h = \sqrt{(2h_0 s)}$$

where h_0 is 11 m, and s the horizontal distance from the margin of the ice dome (m).

This formula provides a fairly good approximation of the form of most ice domes, although it tends to overestimate the maximum elevation. Not only are such calculations useful for examining existing ice domes, but they are also a useful way of estimating the form of the large ice sheets of the Pleistocene.

ture and accumulation. Movement of ice in an ice dome occurs normal to its contours and it can operate through basal sliding as well as internal deformation.

11.1.5.2 Small-scale surface features

Two sets of processes create forms on glacier surfaces – those related to ice movement and those associated with accumulation and ablation. Perhaps the most evident features are **crevasses** formed through ice fracture. They are vertical to subvertical cracks up to several metres across, but do not generally extend to a depth of more than 30 m since below this depth the ice is less rigid and is likely to accommodate to tensional stress by creep rather than fracture. Three types of crevasse can be identified and related to contrasting stress patterns (Fig. 11.9).

A feature common in the ablation zone of many glaciers and attributed to the metamorphism of ice during flow is a banding of white ice containing abundant air bubbles between denser bluish ice. Another form of patterning is that produced by **ogives**, or **Forbes bands**. These are alternating bands of light and dark ice lying across the glacier surface and are developed in some glaciers below ice falls (Fig. 11.10). Where movement is more rapid in the middle of the ice flow the bands are deflected down-glacier. The total width of each pair of light and dark bands corresponds to the distance moved by the glacier in a year. The dark bands are generated during the passage of ice across the ice fall in the summer when melting occurs. Conversely, light bands are produced in winter when fresh snow is incorporated into the ice as it traverses the ice fall.

On the surface of the large ice sheets of Greenland and Antarctica a variety of aeolian features have been recorded. The best documented are dunes called **sastrugi** constructed of hard-packed snow and aligned with the prevailing wind. They most commonly form in the lee of obstacles but are also known in open situations. Wind is also funnelled around rock outcrops protruding above an ice sheet, and snow is excavated from their lee side. On most glaciers the more significant small-scale surface features associated with ablation are produced by meltwater (see Section 11.4.1).

11.2 Glacial erosion

11.2.1 Mechanisms of erosion

Glacial erosion is accomplished by three major processes – abrasion, crushing and fracturing, and **joint-block removal**. Abrasion involves the scratching, grooving and polishing of bedrock by debris carried in the base of a glacier (Box 11.3). Evidence of its efficacy is provided by smoothed bedrock surfaces, often exhibiting parallel sets of fine grooves known as **striations**, and the production of fine particles (usually < 0.1 mm in diameter) known as **rock flour** and occurring in high concentrations in most glacier meltwater

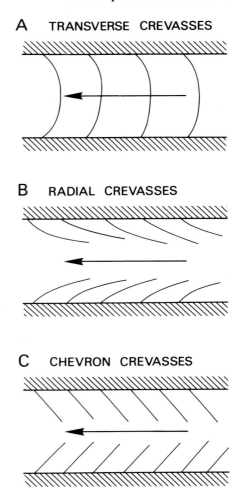

Fig. 11.9 Types of crevasse occurring in valley glaciers: (A) **transverse crevasses** are associated with extending flow and result from expansion normal to the direction of ice flow; (B) **radial** (or **splay**) **crevasses** occur in zones of compressive flow, and although they also arise from expansion normal to the direction of ice flow they differ from transverse crevasses in being concave up-valley; (C) **chevron** (or **en echelon**) **crevasses** associated with the tensional stresses generated by the drag of the valley walls on the margins of a glacier.

streams. The operation of abrasion has been further confirmed by limited direct observations of glacier beds and synthesized in laboratory studies. Its effectiveness is determined by glacier, bedrock and basal debris characteristics and the role of these factors is summarized in Table 11.3.

Evidence for crushing and fracturing comes from series of crescentic cracks and grooves called **chattermarks** found on some bedrock surfaces and thought to represent the effects of pressure exerted by basal glacier debris, and from the presence in glacial deposits of blocks which have clearly been sheared off from fresh bedrock. The typical basal shear stress of around 0.1 MPa is quite inadequate on its own to shear off bedrock protrusions, but where large boulders are being carried along the glacier bed a much more

Fig. 11.10 *Ogives visible on the surface of Austerdalsbreen, a valley glacier in southern Norway. Note the precipitous valley walls.*

Box 11.3 Frictional shear at the bed of a glacier

Ice thickness determines the normal stress at the interface between debris and substrate at the base of a glacier, but for warm-based glaciers this stress is counteracted to some extent by basal water pressure. The basal frictional shear (F) is given by

$$F = A_c \mu (\nu h - p)$$

where A_c is the area of contact, μ the coefficient of friction, ν the unit weight of ice, h the ice thickness, and p the basal water pressure.

As the effective basal pressure increases, the rate of abrasion, for any given sliding velocity and combination of basal debris/glacier bed hardness, will rise (Fig. B11.3). Eventually, however, as the basal normal pressure increases, the friction between a particle and the glacier bed will reach a point where the ice begins to flow over the particle and the rate of abrasion begins to fall. Ultimately, with a further increase in effective basal pressure the particle will stop moving and abrasion will cease.

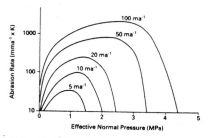

Fig. B11.3 *Relationship between theoretical abrasion rate and effective basal pressure for different rates of glacier movement. The actual abrasion rate will depend on the relative hardness of the glacier bed and the entrained subglacial debris. (Modified from G. S. Boulton (1974), in D. R. Coates (ed.) Glacial Geomorphology. State University of New York, Binghamton, Fig. 7, p. 52.)*

substantial force can be exerted. The process is probably most effective where the bedrock protrusion is much smaller than the transported boulder, and where the shear strength of the protrusion is low.

Joint-block removal involves the 'plucking' or 'quarrying' of large joint-separated blocks by an overriding glacier. Such bedrock joints may originate prior to glaciation through tec-

tonic processes, pressure release or frost wedging. Pressure release, however, can develop as a consequence of rapid glacial erosion since bedrock is replaced by much lower density ice. Moreover, joint widening by frost wedging may possibly occur under warm-based glaciers with abundant meltwater. Subglacial meltwater is, of course, another potent erosional agent and is considered in Section 11.4.2.

Table 11.3 Factors affecting the mechanisms of glacial abrasion

FUNDAMENTAL FACTORS	COMMENTS
Presence of debris in basal ice	Clean ice is unable to abrade solid rock. The rate of abrasion will increase with debris concentration up to the point where effective basal sliding is retarded.
Sliding of basal ice	Ice frozen to bedrock cannot erode unless it already contains rock debris. The faster the rate of basal sliding the more debris passes a given point per unit time and the faster the rate of abrasion.
Movement of debris towards glacier base	Unless particles at the base of a glacier are constantly renewed they become polished and less effective abrasive agents. Thinning of the basal ice by melting or divergent flow around obstacles brings fresh particles down to the rock–ice interface and increases abrasion.

Other factors affecting nature and rate of abrasion

Ice thickness	The greater the thickness of overlying ice the greater the vertical pressure exerted on particles on the glacier bed and the more effective is abrasion. This is the case up to a depth where friction between particles and the bed becomes so high that movement is significantly retarded and abrasion decreases.
Basal water pressure	The presence of water at the glacier base, especially when at high pressure, can reduce the effective pressure on particles on the bed and thus abrasion rates by buoying up the glacier. However, sliding velocities may tend to increase because of the reduced friction.
Relative hardness of debris particles and bedrock	The most effective abrasion occurs when hard rock particles in the glacier base pass over a soft bedrock. If the debris particles are soft in comparison with the bedrock the former are abraded and little bedrock erosion is accomplished.
Debris particle size and shape	Since particles embedded in ice exert a downward pressure proportional to their weight, large blocks should abrade more effectively than small particles. Moreover, angular debris will be a more efficient agent of abrasion than rounded particles.
Efficient removal of fine debris	To sustain high rates of abrasion fine particles need to be removed from the ice–rock interface since they abrade less effectively than larger particles (assuming the latter are continually being supplied from above). Meltwater appears to be the main mechanism for the removal of fine (<0.2 mm) debris.

Source: Based on discussion in D. E. Sugden and B. S. John (1976) *Glaciers and Landscape*. Edward Arnold, London, pp. 153–5.

The thermal regime of a glacier exerts a pervasive influence over processes of glacial erosion. For instance, the potential for abrasion in cold ice is negligible because of the lack of basal slip and the absence of a significant load of basal debris. However, the greater adhesion of cold ice to bedrock would be expected to increase the effectiveness of joint-block removal. Abrasion is probably confined largely to warm-based glaciers since here the presence of basal water both promotes sliding and provides a means for removing fine debris. The fact that a single glacier may have both warm- and cold-based sections and that glaciers may alter from one type to the other in response to climatic change, makes patterns of glacial erosion potentially very complex.

11.2.2 Debris entrainment and transport

Effective glacial erosion can only continue if the eroded material is entrained and transported away by the ice flow (Fig. 11.11). The size of glacially transported debris ranges from very small rock fragments up to huge boulders. **Subglacial debris** is transported along the base of a glacier (Fig. 11.12), but glaciers also acquire **supraglacial debris** through material falling on to the ice surface from rock

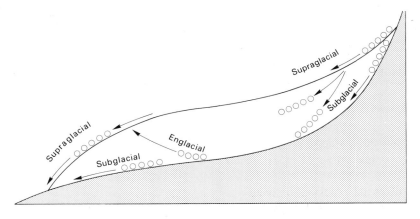

Fig. 11.11 *Schematic illustration of travel paths of debris transported by valley glaciers.*

Fig. 11.12   *Debris exposed at the base of Russell Glacier, Søndre Strømfjord, west Greenland. (Photo courtesy D. E. Sugden.)*

subglacial into englacial debris, and especially in the lower section of the ablation zone, into supraglacial debris.

In some glaciers englacial debris is distributed fairly evenly with depth, but in others it is concentrated in bands separated by relatively clean ice. Such bands run parallel with flow lines and most are probably formed by the freezing of subglacial debris prior to its upward movement through the glacier. Under warm ice, basal melting produces a high concentration of subglacial debris which can be attached on to the base of the glacier by refreezing or **regelation**; but as the regelation layer is only a few centimetres thick this process can only account for thin bands. Creep provides another mechanism for the incorporation of subglacial debris into the base of a glacier, and in warm-based glaciers small amounts of material can probably be squeezed up into sub-glacial cavities. Finally, where a glacier is overriding a large obstruction on its bed, thrusts can develop in the ice which provide a further means of emplacing englacial debris.

walls or other ice-free areas (Fig. 11.1). Naturally, supra-glacial debris is likely to be most abundant in valley and cirque glaciers and absent over large areas of ice sheets. It can continue to be carried on the glacier surface within the ablation zone, but above the equilibrium line it will become progressively buried because of the accumulation of ice from above. Once buried it becomes **englacial debris** and it can travel as such to the glacier snout. Alternatively, it may emerge at the surface in the ablation zone through melting of overlying ice or it can move down to the glacier bed and be trapped by existing subglacial debris. Movement along slip lines in zones of compressive flow tends to transform

11.2.3   Erosional landforms

Glacial erosion produces landforms at a wide range of scales, but irrespective of scale the overriding control is the attempt by glaciers to modify their beds in such a way as to improve the efficiency of ice flow. The extent of the ice cover, the thermal regime of the glacier, the amount of basal debris, the characteristics of the underlying bedrock, the period of time during which glacial erosion has been active and the form of the pre-existing topography all affect

Table 11.4   Landforms and landscapes of glacial erosion

PREDOMINANT PROCESS	ASSOCIATED MORPHOLOGY	LINEAR DIMENSIONS									
		0.01 m	0.1 m	1 m	10 m	100 m	1 km	10 km	100 km	1000 km	10 000 km
Unconfined ice flow	Positive, streamlined			◄——— Whalebacks ———►			Streamlined ► spurs			Landscape of areal scouring	
			◄——— Rock drumlins ———►								
	Positive, partially streamlined	◄——— Roches moutonnées ———►				Flyggbergs ►					
	Negative, streamlined	◄— Striations —►	◄——— Grooves ———►								
	Negative, partially streamlined			◄——— Rock basins ———►							
Channelled ice flow	Negative, streamlined						◄——— Troughs ———		Landscape of linear ice-sheet erosion		
Interaction of glacial and periglacial processes	Negative					◄——— Cirques ———►		Valley glacier landscape			
	Positive					◄— Arêtes —►					
						◄— Horns —►					
					◄— Nunataks —►		Nunatak landscape				

Source: Modified from D. E. Sugden and B. S. John (1976) *Glaciers and Landscape*. Edward Arnold, London, Table 9.2, p. 169.

the nature of glacially eroded landscapes. A useful distinction can be made between those forms associated with largely unconfined ice movement, those produced by channelled glacier flow and those created through the interaction of glacial and periglacial processes. This classification forms the basis for our discussion of glacially eroded landforms (Table 11.4).

11.2.3.1   *Forms associated with unconfined ice flow*
Unconfined ice flow is restricted to ice sheets and ice caps. Although the erosional activity below such ice masses is probably for the most part limited to regions of warm basal ice a range of positive landform features have been attributed to ice-sheet erosion. They include small, smoothly eroded forms a few hundred metres in length known as **whalebacks**, larger streamlined hills sometimes called **rock drumlins** and tapered interfluves and spurs up to several kilometres long. Although there are doubts about the interpretation of some of the larger landforms their alignment with known directions of ice movement and their smoothed surfaces support a glacial origin. There is certainly no doubt about the glacial nature of the asymmetric streamlined form known as a **roche moutonnée**, and its larger variant known as a **flyggberg**. Roches moutonnées range in size up to the dimensions of whalebacks. They have a smoothed end facing the direction of ice flow but a craggy, steeper lee side. Although it is apparent that they are formed by abrasion on their up-glacier side and joint-block removal on their lee side, it is not clear why their down-glacier side is not also smoothed by abrasion as is the case in fully streamlined forms. This may be related to the lower ice pressure on the lee side of a bedrock obstruction in an ice flow. The lower pressure would not only reduce abrasion but would also enable meltwater to migrate to the lee side and possibly aid joint enlargement by frost wedging. Low lee-side ice pressures would be favoured by thin or fast moving ice flowing around a prominent obstruction, since in these circumstances the ice would be less able to mould itself closely around the obstacle.

Negative relief forms attributed at least in part to the effects of unconfined ice flow include grooves aligned with a known direction of ice movement, and shallow **rock basins** often filled by a lake. Grooves up to 30 m deep, 100 m across and 12 km long occur in the Mackenzie Valley in northern Canada while rock basins, from several metres to a few hundred kilometres across, are more ubiquitous and are to be found over much of northern Canada. A glacial origin for rock basins is suggested not only by evidence of abrasion but also by their overdeepening, that is, their erosion to levels below regional base levels related to fluvial systems. Some basins are aligned along faults and major joint systems and one explanation of their development involves preferential joint-block removal where bedrock joints are more closely spaced. Abrasion also probably

plays a role since it might be more effective where basal shear stresses are highest under thicker ice occurring over topographic depressions; in such a case an initially shallow depression might then be prone to overdeepening.

11.2.3.2   *Forms associated with channelled ice flow*
Ice flow concentrated in channels gives rise to steep-sided, **glacial troughs** (Figs 11.1, 11.2, 11.10). These may be formed by valley or outlet glaciers, or by ice streams occurring within ice sheets and ice caps. The walls of glacial troughs, which may exceed 1000 m in height, frequently truncate the tributary valleys and spurs of the pre-existing fluvially-eroded landscape thereby giving rise to **hanging valleys** (Fig. 11.13). The cross-profile of many glacial troughs approximates to a parabola and this form probably results from a combination of higher basal flow rates, and therefore more effective erosion, beneath the middle of the glacier and fluctuations in the height of the surface over time leading to less direct glacial erosion of the upper parts of the trough.

Three varieties of glacial trough can be distinguished. **Alpine troughs** are eroded by valley glaciers whose accumulation zones lie below a mountain mass. **Icelandic troughs**, by contrast, are formed by glaciers flowing from ice sheets or ice caps over the trough head and some enormous examples have been formed by outlet glaciers flowing from the Antarctic ice sheet. Where coastal Icelandic troughs are partially drowned by rising sea level they form **fjords**. A third type, **open troughs**, are so called because they are open at both ends. They frequently breach watersheds and most seem to have been eroded by ice streams within ice sheets or ice caps.

Alpine troughs characteristically have an overdeepened long profile; near the trough head the floor is steeply inclined, but down-valley there is a much lower or even slightly reversed gradient. Icelandic troughs typically have an excessive steepening towards the trough head and flatter trough floors, whereas open troughs normally have a high point roughly in the middle of their long profile. The location of overdeepening in alpine troughs is probably related to the position of the equilibrium line since here the ice is thickest and we would expect this to coincide with the zone of maximum erosion. The overdeepening of Icelandic troughs so near the trough head is more difficult to explain, but may be associated with a localized zone of warm-based ice which promotes more rapid erosion than the surrounding cold-based parts of the glacier.

Superimposed on these characteristic long profile forms are often smaller-scale irregularities consisting of alternating **rock bars** and rock basins which form a series of steps in the trough floor (Fig. 11.13). These features have generated much discussion and explanations have included more effective erosion where ice flow is constricted through a local narrowing of the trough or where tributary glaciers

Fig. 11.13   *Valley below the present location of the snout of Austerdalsbreen, southern Norway. Note the hanging valleys and truncated spurs, and the rock bar in the centre of the photograph.*

enter the main trough, and differential erosion controlled by variations in the spacing of joints in the underlying bedrock or the degree of pre-glacial weathering. If a trough floor contained small initial irregularities then any depressions would tend to be preferentially eroded, especially under warm ice, because of the higher basal shear stress associated with the greater thickness of overlying ice. Deepening of depressions in this way would cause further ice thickening and so irregularities would tend to be accentuated by this positive feedback mechanism. This provides an interesting contrast with fluvial systems in which adjustments between channel bedform and changes in flow depth and velocity are characterized by negative feedback and therefore promote the reduction of irregularities. In glacier flow the development of a stepped trough-floor profile is further aided by the existence of zones of compressive and extending flow since concavities in the glacier bed tend to be eroded to conform to slip lines.

11.2.3.3   *Forms associated with periglacial action*
Cirques (also called **corries** and **cwms**) are the major landforms developed by a combination of periglacial action (see Chapter 12) and glacial erosion and are usually located at the heads of deep valleys. A cirque consists of a bowl-shaped rock basin which extends from a steep **headwall** of shattered rock to a low rim. Fully developed cirques have a fairly consistent ratio of height to length suggesting that they are equilibrium forms which maintain their relative dimensions as they grow in size. They range from modest depressions a few hundred metres across, to massive amphitheatre-like forms several kilometres wide and with headwalls several hundreds of metres in height. Wind-blown snow is important in the mass balance of cirque glaciers, so it is not surprising that they are best developed where prevailing winds can bring in large quantities of snow, or at least where accumulated snow is not readily removed by the wind.

Cirque development starts with the formation of a firn bank in a suitable depression. This is gradually enlarged by a combination of processes known as nivation (see Section 12.2.4) involving active frost weathering and mass movement processes promoted by the presence of meltwater around the firn bank. Deepening proceeds until the firn turns to ice which then begins to flow. Because cirque glaciers have a rather unusual bed configuration their flow characteristics are rather different from other glacier types.

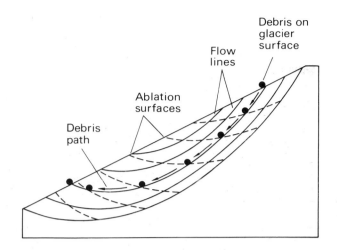

Fig. 11.14 *Schematic longitudinal section through a cirque glacier. Note the form of the flow lines and the path taken by debris which falls on to the glacier surface.*

The regular concave long profile form induces rotational sliding, and the flow lines are inclined away from the glacier surface near the headwall and towards the surface near the terminus. This is illustrated by the movement within the glacier of rock debris which falls on to the surface of the ice at the base of the headwall (Fig. 11.14). As cirque glaciers become larger it is likely that rotational slip becomes less important and internal deformation more so.

In order to maintain the rather constant ratio between height and length, deepening of the cirque floor, largely by glacial abrasion, must be accompanied by retreat of the headwall. This occurs mainly through the removal of large blocks of bedrock, but uncertainty surrounds the mechanism whereby these blocks are produced. It has long been observed that the top layer of actively moving ice on a cirque glacier is usually separated from the headwall by a deep chasm known as a **bergschrund**. The bottom of a bergschrund would be a likely collecting point for meltwater and, being exposed to the atmosphere, it might also experience frequent temperature fluctuations across freezing point and therefore be a site of active enlargement of pressure-release joints by frost weathering. This superficially plausible hypothesis unfortunately lacks supporting evidence since the little data collected do not indicate temperature changes which are either sufficiently frequent or rapid to promote intense frost weathering. A possible alternative mechanism for rock break-up is hydration shattering (see Section 6.3.1) although the effectiveness of this process in a wide range of rock types has yet to be firmly established.

Whatever the process of headwall erosion, cirque growth can consume large areas of upland. The early stages of cirque development focused along plateau margins create deeply notched plateau edges giving rise to so-called **biscuit board topography**. Cirque growth may eventually progress to the point where two encroaching cirque headwalls are only separated by a narrow ridge, known as an **arête**, or where several headwalls converge, by a sharp mountain peak or **horn** (Fig. 11.15). A summary of the sequence of features typically associated with the glaciation of mountain regions and created by the combined action of channelled ice flow and the interaction of glacial and periglacial processes is provided by Figure 11.16.

Periglacial processes also play an important role in the shaping of **nunataks**. These are mountain peaks of limited extent and entirely surrounded by glacial ice. They are most common where an ice sheet or ice cap almost entirely buries a mountain range with only the highest peaks remaining above the ice surface. After wasting of the ice there may be a significant contrast between the irregular, sharp-edged topography characteristic of nunataks and the somewhat smoother and more subdued forms of the surrounding glacially modified terrain.

11.3 Glacial deposition

Before we look at the processes of glacial deposition it is useful to draw a distinction between the deposits laid down by ice, known as **till** or **boulder clay**, and the landforms produced by such deposits, which are termed **moraine**. Although tills are highly variable deposits they usually have certain characteristics in common. These are listed in Table 11.5, but it must be emphasized that any one till deposit is unlikely to exhibit all these features; moreover, deposits originally deposited by ice may subsequently be reworked by glacial meltwater (see Section 11.4.1).

Table 11.5 Characteristics of till

1. Poor sorting – there is a large range in grain size, with large clasts up to boulder size frequently contained in a finer, sometimes clayey, matrix

2. Lack of stratification – laminations and graded bedding (progressive change in grain size with depth) are generally absent except in deposits modified by meltwater which may exhibit stratification

3. Mixture of lithologies – particles may have been derived from widely separated sources, especially in the case of deposits laid down by large ice sheets.

4. Frequent presence of particles with abraded facets and striations.

5. Preferred orientation of particles

6. Compaction associated with pressures developed during deposition

7. Overlies striated rock or sediment basement

8. Predominantly subangular particles due to a combination of fracturing and rounding by abrasion

Fig. 11.15   *The horn of Mount Cook, the highest mountain in the Southern Alps, South Island, New Zealand. (Photo courtesy G. M. Robinson).*

11.3.1   Mechanisms of deposition

Glacial deposition involves a range of processes and any classification is somewhat arbitrary (Table 11.6). An added complication is that researchers have used different terms to describe the same process. Here we make a broad grouping of processes on the basis of whether they operate at the base, on the surface, or around the margins of a glacier. Nevertheless, material deposited in one situation may be subsequently re-entrained and re-deposited elsewhere (Fig. 11.17).

At least three distinct mechanisms of subglacial deposition can be identified. **Undermelt** involves the deposition of material through melting of the underlying ice. Melting may arise from geothermal heat, through frictional heating or as a result of increased pressures around the up-glacier side of obstructions. A second process, called **basal lodgement**, entails the smearing of predominantly fine material

Table 11.6   Classification of mechanisms of glacial deposition

LOCATION RELATIVE TO GLACIER	PROCESS	THERMAL AND DYNAMIC CONDITIONS
Subglacial	Undermelt	Warm-based only, active or stagnant
	Basal lodgement	Predominantly warm-based, active only
	Basal flowage	Warm-based only, active or stagnant
Supraglacial	Meltout	Active or stagnant
	Flowage	Active or stagnant
Marginal	Dumping	Active only
	Pushing	Warm- or cold-based, active only

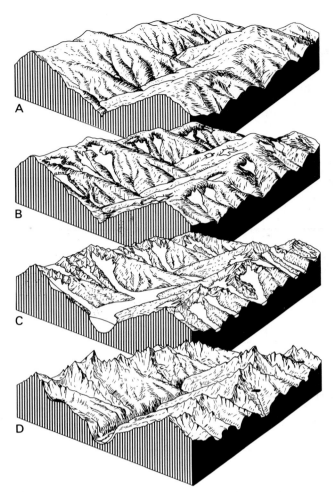

Fig. 11.16 *Development of landforms associated with alpine glaciation: (A) a fluvially eroded mountain landscape prior to glaciation; (B) the initial accumulation of snow and ice; (C) full development of a network of valley glaciers; (D) after deglaciation showing landforms of mountain glaciation including cirques, arêtes, horns, truncated spurs, hanging valleys and a glacial trough. (After R. F. Flint (1971)* Glacial and Quaternary Geology, *Wiley, New York, Fig. 6.3, p. 140.)*

of warm glaciers where up to 20 m of ablation may occur in a single summer. **Supraglacial flowage**, involving the movement down the ice surface of the debris-rich upper layers of a glacier, has also been observed and is particularly common near the snout where intense supraglacial meltout occurs. The movement ranges from slow creep to a rapid liquid flow.

Deposition around a glacier margin can arise from a variety of mechanisms. Water-soaked till can be squeezed from under the ice, and if the glacier margin remains stable for some time significant accumulations of supraglacial and englacial debris can form through dumping of material by meltout. Existing glacial deposits can also be pushed by an actively advancing glacier and deposited when the advance stops, or be overridden and recycled towards the snout as englacial and supraglacial debris. Material deposited at a glacier margin commonly suffers extreme disruption prior to deposition so that any original orientation of particles in the ice is usually lost, except possibly where a frozen mass of till is bulldozed *en masse*.

11.3.2 Depositional forms

Most ice-laid deposits have a definite three-dimensional shape, that is, they form moraines. Few moraines are composed entirely of till – they also contain stratified sediments deposited by meltwater and, in some cases, even a bedrock core. Another characteristic of many moraines is their transience. Those occurring around the margins of a glacier often have an ice core and they collapse into insignificant forms when this melts. Moraines can also be destroyed or drastically modified by meltwater and, in the case of those formed subglacially, can be modified by glacial activity at the ice margin when the glacier retreats.

Perhaps the most helpful way to classify such complex landforms as moraines is to categorize them in terms of their relationship to the direction of ice flow. On this basis there are three major types: those aligned largely parallel to the direction of ice flow, those orientated roughly transverse to the direction of ice flow and those lacking any consistent orientation (Table 11.7).

11.3.2.1 Forms parallel to the direction of ice flow

Moraines orientated roughly parallel with the direction of ice flow can be formed subglacially, supraglacially, or at the ice margin. Most of the subglacially produced forms seem to result from streamlining related to variations in stress across the glacier bed. Often formless sheets of till, known as **ground moraine**, can be modified subglacially by the ice flow into fluted forms up to 10 m high and 1 km long, or exceptionally into megaflutes up to 25 m high and 20 km long. That they can be either predominantly depositional or erosional features is highlighted by the distinction made between **fluted ground moraine**, in which the fluting is

on the glacier bed. Lodgement of subglacial debris occurs where the friction between a particle being transported by the ice and the glacier bed becomes so great that its further movement is retarded. This situation is favoured by thick ice promoting high basal pressures which increases friction on the glacier bed and by low flow velocities which are associated with weak basal shear stresses. **Basal flowage** is a third mechanism of subglacial deposition but it is also partly erosional. It involves both the squeezing of unconsolidated water soaked debris into basal ice concavities and the streamlining of till by overrriding ice (see Section 11.3.2.1).

Supraglacial debris deposition can occur by **meltout** or by **flowage**. Meltout involves deposition of sediment through melting of the glacier surface. It is most active in the snout

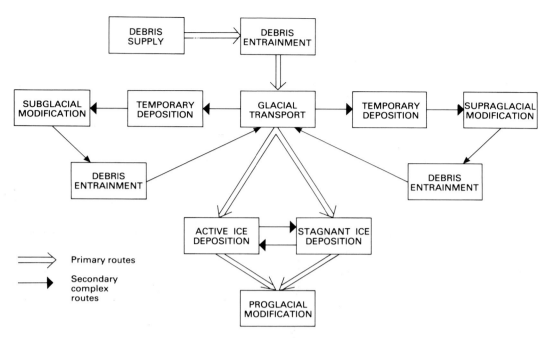

Fig. 11.17 *Schematic representation of the possible routes that debris may follow in the glacial and fluvioglacial system prior to final deposition.*

Table 11.7 Classification of the major types of moraine

PARALLEL TO ICE FLOW	TRANSVERSE TO ICE FLOW	LACKING CONSISTENT ORIENTATION
Subglacial forms with streamlining Fluted and drumlinized ground moraine	*Subglacial forms* Rogen or ribbed moraine	*Subglacial forms* Low-relief ground moraine
Drumlins and drumlinoid ridges	De Geer or washboard moraine	Hummocky ground moraine
Crag-and-tail ridges	Subglacial thrust moraine	
	Sublacustrine moraine	
Ice-pressed forms Longitudinal squeezed ridges	*Ice-pressed forms* Minor transverse squeezed ridges	*Ice-pressed forms* Random or rectilinear squeezed ridges
Ice marginal forms Lateral and medial moraines	*Ice front forms* End moraines	*Ice surface forms* Disintegration moraines
Some interlobate and kame moraines	Push moraines	
	Ice thrust/shear moraines	
	Some kame and delta moraines	

Source: Modified from D. E. Sugden and B. S. John (1976) *Glaciers and Landscape*. Edward Arnold, London, Table 12.1, p. 236, after V. K. Prest (1968) *Geological Survey Papers Canada* 67–57.

due to troughs within the till, and **drumlinized ground moraine** where ridges rise above the general level of the till surface and more clearly represent constructional relief forms. There are numerous theories for the formation of fluted moraine, but several point to the role of large boulders on the glacier bed (Fig. 11.18). Constructional flutings could be formed by unfrozen till being forced into cavities on the up- and down-glacier sides of an obstacle.

The most intensively studied subglacial moraine form is the **drumlin**. Although morphologically somewhat variable,

in plan drumlins tend to be bluntly rounded on their up-glacier margin and sharply pointed on their lee side, while in profile they are usually highest towards their up-glacier end. Typical dimensions are 1–2 km in length, 5–50 m in height and around 500 m in width. While some drumlins appear to be composed entirely of clay-rich till, others have a bedrock core. They rarely occur singly, much more commonly forming **drumlin fields** in which individual drumlins are randomly spaced.

Drumlins remain difficult landforms to explain. The

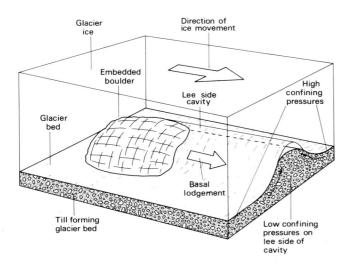

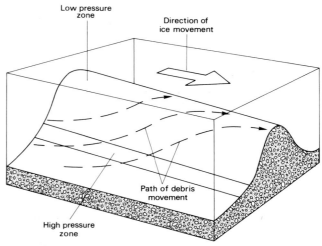

Fig. 11.18 A probable model of fluted moraine formation which may be applicable to most types of fluted moraine ridges. (Based on D. E. Sugden and B. S. John, (1976) Glaciers and Landscape. Edward Arnold, London, Fig. 12.2, p. 239.)

Fig. 11.19 Schematic representation of migration of debris on the flanks of a growing drumlin. (Based on D. E. Sugden and B. S. John (1976) Glaciers and Landscape. (Edward Arnold, London) Fig. 12.4, p.241.)

existence of a bedrock core in some examples suggests that they may form around obstructions. But this is at best an incomplete explanation since it does not account for those forms lacking a bedrock core. The concentration of drumlins in specific areas suggests that the initial accumulation of material is influenced by local conditions. Probably the most significant are variations in bed roughness with subglacial debris being preferentially deposited where the bed is irregular. Bedrock protrusions or large boulders will encourage debris to collect on their up-glacier side while deposition will also occur on the lee side where shear stresses are less. An initial stage of this kind of deposition is represented by **crag-and-tail** forms which consist of a tail of glacial deposits in the lee of a rock obstruction. The drumlin form appears to develop through variations in basal pressure associated with ice flow over an initial irregularity (Fig. 11.19). Drumlin crests would be zones of low pressure while the troughs between would experience high pressures. Particles being transported over the drumlin by the ice flow would tend to move across the gradient of decreasing basal ice pressure and be deposited on the crest. This kind of mechanism accords with the observation that in many drumlin fields there is only a very sparse till cover between drumlins.

Variations in the mechanical properties of till may also be important in drumlin formation, with subglacial deposition occurring where till is more resistant to deformation. This might take place where subglacial meltwater is efficiently drained from till which overlies a permeable bedrock. Another explanation focuses on the role of dilatancy (see Section 7.1.1). Most tills have the property of dilatancy; that is, in very simplified terms, they expand when subject to stress within certain limits. It is envisaged that debris on

a glacier bed will resist movement until the stress builds up to a sufficiently high level to induce the dilatant characteristic. Once this level is reached the till will begin to deform readily and will continue to deform, even when the stress is decreased, until a lower critical threshold is reached and the dilatant property is lost. Imagine debris below thick ice. Here subglacial debris is moving freely in its dilatant state since the basal stress is high. As the ice thins towards the glacier margin the basal stress will gradually decrease until the critical lower limit for dilatancy is reached. The basal debris will now suddenly revert to a compact form and resist further movement by the ice. As it is likely that there will be variations in stress across the glacier bed, zones of compact, stationary till will be surrounded by dilatant material which continues to be transported. The zones of compact till will then start to be moulded by the still-moving, debris-rich basal ice into a streamlined form offering least resistance to ice flow.

Although widely regarded as ice-moulded forms, it has also been suggested that some drumlins, at least, might be formed by subglacial meltwater. One idea is that drumlins can be formed when sedimentation occurs in cavities eroded by meltwater flow directed upwards towards the base of the ice. Another possibility is that drumlins can be remnant erosional ridges created by meltwater flow. These kinds of drumlins are composed of pre-existing bedrock or sediment left protruding as the surrounding material is removed by meltwater erosion.

Moraines deposited on an ice surface or along a glacier margin have a far lower chance of survival than forms of subglacial origin. **Medial moraines** are formed from supraglacial debris concentrated in a thin ribbon in mid-glacier below the confluence of two tributary glaciers (Fig. 11.1).

They can be significant features in the upper part of a glacier but they usually only remain as a thin cover of till near the snout. **Lateral moraines** are formed primarily from frost-shattered debris which has fallen on to the edge of a glacier from the adjacent rock walls and are more likely to survive since they may lie partly on a bedrock substrate. Vertical fluctuations in the level of the glacier can lead to lateral moraines being perched high above the ice surface.

11.3.2.2   Forms transverse to the direction of ice flow
Perhaps the most obvious place to expect moraines aligned perpendicular to the direction of ice movement is along the front of a glacier. **End moraines** form in this position and they mark the maximum extent of a glacier margin. Those associated with continental ice sheets may be up to 100 m high and extend for tens of kilometres. A more or less parallel series of end moraines can often be found representing stages of ice retreat. End moraines are usually composed of both supraglacially and englacially transported debris together with fluvioglacial and lacustrine sediments, and are formed by a combination of dumping, meltout and flowage. Debris brought to the ice surface along flow lines and shear planes developed near the glacier snout can be dumped by sliding off ice-cored ridges on the glacier surface or by collapse due to melting of underlying ice. The rate of melting can significantly affect the detailed form of the moraine. Where melting is rapid and meltwater abundant supraglacial till can flow down even shallow slopes and rather featureless accumulations of till are produced. Where melting occurs slowly debris can collect in troughs lying between ice-cored ridges. As melting progresses these ridges collapse and the originally debris-filled troughs become upstanding ridges as the topography is inverted. Small end moraines, less than 10 m high, can also be formed by the squeezing of water-soaked debris from beneath the ice front or into subglacial crevasses. Moraines developed by this processes of basal flowage may be produced seasonally when summer meltwater percolates to the bed of a warm-based glacier and saturates the underlying till.

Push moraines are another form constructed along an ice front. They are produced by the bulldozing of glacial, and even non-glacial, deposits by an advancing ice margin and can attain considerable dimensions – up to 100 m high and 30 km long. The pushing action of the ice can create complex thrust and fold structures in the till and typically results in a distal slope which is much steeper than the slope facing the glacier front.

Ice-margin processes are not the only mechanisms capable of generating transverse moraines. Transverse ridges up to a few metres in height, in many cases associated with lakes or former lakes, are known as **washboard moraines**, or **De Geer moraines**. Some examples may be subaqueous push moraines, but others may owe their origin to subglacial

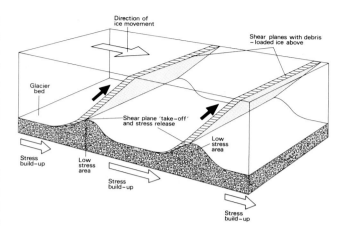

Fig. 11.20   *A model of the formation of ribbed (Rogen) moraine. Debris accumulates at the base of shear planes in zones of compressive ice flow. In addition to particle movement along pressure gradients, the process of formation may also involve pressure melting and lodgement where the shear planes 'take off' from the glacier bed. (Based on D. E. Sugden and B. S. John (1976)* Glaciers and Landscape. *Edward Arnold, London, Fig. 12.8, p.245.)*

thrusting and therefore be allied to **ribbed moraines** or **Rogen moraines**. These consist of fairly regularly spaced ridges, usually arcuate in plan and concave up-glacier, typically 10 m or more in height and up to 1 km long. Many, though not all, have streamlined 'drumlinized' crests. One interpretation is that they form through preferential deposition where shear stresses transverse to the direction of flow are relieved, and experience subsequent streamlining by the ice (Fig. 11.20). Another explanation involves a genetic link between Rogen moraine and fluting, the development of the former being seen as a response to variations in basal stress, or the presence of topographic irregularities, across the glacier bed.

11.3.2.3   Forms lacking consistent alignment
Some moraines have no obvious orientation or linear development and therefore fail to fit into either of the morphological categories discussed above. Ground moraine may occur as virtually featureless sheets of till, but it can also form a gently undulating topography or even a chaotic hummocky terrain with a relief in excess of 100 m. This may develop from debris which has accumulated in basal ice concavities or from the meltout of supraglacial debris spread extensively over an area of wasting ice. On the margins of a retreating ice sheet large masses of ice can become stagnant through being isolated from a source of active ice flow. Such stagnant ice will gradually be reduced by ablation to discrete masses of decaying debris-rich ice and eventually a chaotic assemblage of disintegrating moraines will be created forming **dead-ice topography**.

11.4 Fluvioglacial erosion and deposition

11.4.1 Glacial meltwater

Meltwater is an integral part of the glacial system (Fig. 11.21). It both significantly influences glacier behaviour through its effect on rates of basal sliding and has an erosional and depositional role associated with, but distinct from, that of glacier ice. On most glaciers surface melting is by far the most important source of meltwater, the amount released increasing down through the ablation zone. In temperate regions this surface source may be supplemented seasonally by rainfall and, in valley glaciers, by runoff from valley-side slopes. The chief basal and internal sources of meltwater are basal melting through geothermal and frictional heating, although ground water can be a significant additional source in temperate valley glaciers. Geothermal heat is capable of melting about a 6 mm layer, and heat generated by basal sliding and internal deformation a 10–15 mm layer, of warm ice annually. Further basal melting can arise from the heat brought down by surface

Fig. 11.21 *Meltwater emerging from the snout of Nigardsbreen, a valley glacier in southern Norway.*

water and by the frictional heat generated by the flow of the meltwater itself.

Depending on the thermal regime of the glacier, meltwater may travel over the glacier surface or move englacially and subglacially (Fig. 11.22). Surface runoff may be concentrated into stream channels up to several metres deep, in some cases sunk into 'valleys' on the ice surface. Extensive drainage networks covering hundreds of square kilometres can develop on ice sheets. Stream courses may be influenced by ice surface irregularities, but very regular meandering patterns have also been reported. Internal drainage within a glacier evolves in a manner somewhat analogous to the formation of underground drainage in highly soluble rocks such as limestones. Initially, water permeates warm ice along the boundaries separating individual ice crystals, but eventually a secondary permeability develops with the creation of surface sinkholes, known as **moulins**, and tunnels which tend to form preferentially along cracks and cavities in the ice. Glacier tunnels form dendritic networks both within, and at the base of, the ice, their direction and gradient being controlled both by the general slope of the glacier surface and the subglacial topography. In addition to basal channels, water can collect to form large subglacial lakes. It is suspected that these may be quite extensive below large ice sheets.

Rates of meltwater flow from glaciers vary over a range of time scales. Short-term fluctuations occur in response to diurnal variations in ablation and changing meteorological conditions. Seasonally the lowest flows usually occur in late winter and the highest in early summer – how early depending on the rapidity with which the internal drainage channels are re-established after winter freezing and closure by ice flow. Significant longer-term maxima in meltwater discharge are associated with glacier surges, but the most spectacular peak flows are caused by the catastrophic draining of ice-dammed lakes. Discharges several orders of magnitude greater than normal are attained in a few hours and diminish again even more quickly. Such jökulhlaups are particularly common in Iceland where some are associated with volcanism (see Section 5.2.3.4). One originating from the western part of the Vatnajökull ice cap with a periodicity of about 10 a has reached an estimated peak discharge of 40 000–50 000 m^3s^{-1}, while another in Baffin Island, Canada, involved the draining of a $5 \times 10^6 m^3$ subglacial lake in only 30 h. Such massive discharges could arise in various ways, but one possibility is that a gradually extending internal drainage system eventually taps an ice-dammed, subglacial lake.

Even more cataclysmic jökulhlaups were apparently associated with the draining of Late Pleistocene Lake Missoula in the north-west USA (Fig. 11.23). This **proglacial lake**, which extended across a large area of western Montana, was the largest of a number in the region impounded by ice advancing from the north. It is estimated that at its maxi-

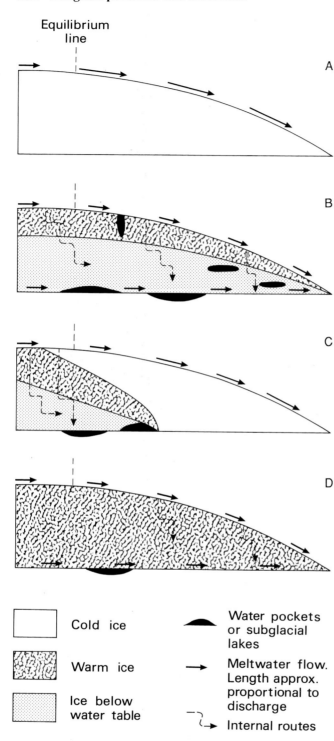

Equilibrium
line

A

B

C

D

Cold ice

Warm ice

Ice below
water table

Water pockets
or subglacial
lakes

Meltwater flow.
Length approx.
proportional to
discharge

Internal routes

Fig. 11.22 Possible meltwater routes in different types of glaciers and glacial ice: (A) cold ice; (B) warm ice; (C) cold ice in the ablation zone and warm ice in the accumulation zone; (D) a high altitude glacier in equatorial latitudes. (After D. E. Sugden and B. S. John (1976) Glaciers and Landscape. *Edward Arnold, London, Fig. 14.18, p. 298.)*

mum Lake Missoula reached a volume of 2000 km³, and much of this water was released during repeated failures of its ice dam. The peak discharge attained during break-out events is thought to have been a staggering 21.3×10^6 m³ s⁻¹, or approximately 20 times the mean discharge of *all* the world's rivers! Several separate flood events occurred between 16 and 12 ka BP which have been reconstructed from the extraordinary landforms they produced. These include a series of enormous streamless canyons, or **coulees**, which form an anastomosing network and which contain giant 'ripples', plunge basins, deep rock basins and vast potholes (Fig. 11.24).

This whole region of anomalous topography is called the Channeled Scabland and its origin was the subject of intense debate after a catastrophic flooding hypothesis was first put forward by J. Harlen Bretz in a paper published in 1923. This explanation was initially rejected by his contemporaries, both because Bretz had no convincing source for such enormous volumes of water (he was unaware of literature documenting the existence of Lake Missoula published in the 1880s), and because they were unwilling, on *a priori* grounds, to accept the notion of catastrophic rather than gradual landscape modification. Bretz's hypothesis was eventually substantiated by his own persistent field work and by other workers who made the link between catastrophic flooding and Lake Missoula. Subsequent work has begun to relate the specific dimensions of the landforms present to flow conditions. Of particular interest are the giant current ripples found in the channels. These are composed of gravel and are typically around 5 m high and have a wavelength of 100 m or so. From these dimensions it is possible to infer mean flow velocities, bed shear stresses and stream power values. It seems that the floods that formed these ripples could transport boulders as large as 10 m across.

11.4.2   Fluvioglacial denudation

The sediment-charged nature of glacial meltwater streams is testament to their transportational role and suggestive of their erosional capabilities. Although data on sediment yields are scarce it is known that the suspended loads of meltwater channels may attain concentrations in excess of 3000 mg l⁻¹. Their characteristic greyish-white colour is due to an abundance of the fine sediment generated by glacial abrasion, but up to 25 per cent or more of the total load may be carried as bed load. Peak yields are usually reached in the early part of the summer presumably due to the flushing out of sediment generated during the preceding winter. The limited information available indicates that meltwater streams also probably carry a significant solute load. The high surface area of unweathered rock provided by fine suspended sediment and the highly turbulent flows in which it is transported might be expected to promote active chemical

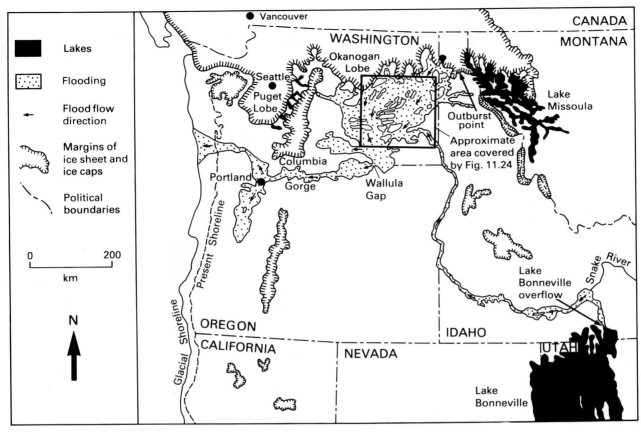

Fig. 11.23 *Area of the north-west USA affected by catastrophic flooding at the end of the last glacial. In addition to catastrophic outbursts from Lake Missoula, flooding also occurred as a result of overspills from Lake Bonneville in Utah and from other ice-dammed lakes in the Columbia Basin. (Modified from V. R. Baker and R. C. Bunker (1985)* Quaternary Science Reviews *4, Fig. 1, p. 2.)*

weathering. Much chemical weathering undoubtedly occurs subglacially and this may indirectly increase erosion rates by weakening the bedrock. During winter low flows the solute load may even exceed the solid load.

High flow velocities and especially high peak discharges suggest that meltwater streams should be potent erosional agents. Although fine rock flour usually represents the major component of the suspended load it probably has a negligible erosional effect. The coarser suspended and saltating load, however, is apparently a highly effective abrasional agent; indeed there is observational evidence of the cutting of meltwater streams into resistant bedrock in just a few years. Since high flow velocities in the range 8–15 m s^{-1} are probably quite common in meltwater streams, and because rough channel beds will be the norm, cavitation may also be an important erosional process especially where stream channels narrow.

Various minor erosional forms consisting of a range of smooth rock depressions have been attributed to the action of meltwater, although for some of these alternative explanations involving glacial erosion have been proposed. Collectively they are known as plastically sculptured forms,

or simply as **p-forms**. Elongated varieties exhibiting striations are certainly likely to be due to glacial abrasion but the origin of another type, known as **sichelwannen**, which occur in resistant crystalline rocks, is less certain (Fig. 11.25). They consist of crescentic depressions up to 5 m or more across and a fluvioglacial origin is suggested by the formation by differential fluvial erosion of similar features in less resistant rock.

The major large-scale erosional fluvioglacial landform is the **meltwater channel**. These can be up to 100 m or more in depth and extend for tens of kilometres. They take various forms but can be broadly grouped into subglacial ice-directed forms, where active ice movement has exerted a major control, and marginal and submarginal forms which tend to run parallel to the glacier margin. Ice-directed channels are generally aligned parallel to the direction of ice movement. Some breach pre-glacial drainage divides while others run downslope parallel with divide crests. They occur singly or as bifurcating or anastomosing networks. A subglacial origin for such channels is supported by their up-and-down long profiles, which are difficult to explain by anything other than subglacial water flow under hydrostatic pressure, and

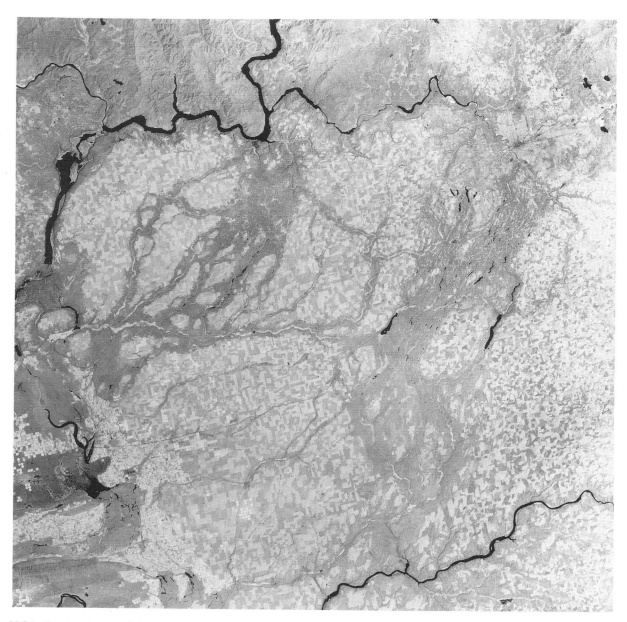

Fig. 11.24 *Landsat image of the Channeled Scabland, Washington, USA. The vast anastomosing channel patterns are clearly visible because they were cut through a cover of loess, thereby exposing the dark-coloured underlying basalt of the Columbia Plateau. The area covered is about 180 km across. (Image courtesy N. M. Short.)*

by the presence of ice moulding on the upper slopes of some examples.

The distinction between subglacial and marginal channels is not always clear since the latter can range from channels cut wholly in lateral moraine or bedrock to submarginal channels located subglacially at the extreme edge of the glacier. The marginal and submarginal environment is certainly a likely focus for meltwater generated in the zone of very active ablation of a valley glacier close to the trough wall, and for non-glacial runoff from adjacent valley slopes.

While submarginal channels are more likely to form in association with warm ice, true marginal channels will develop along the margins of cold glaciers since here there is little opportunity for meltwater to percolate along the frozen ice–rock interface.

11.4.3 Fluvioglacial deposition

Sediment transported by meltwater can be deposited either in contact with the glacier or beyond the ice front in the

Fig. 11.25 *Crescentic scour marks, or sichelwannen, Søndre Stromfjord, west Greenland. These features are probably formed as a result of erosion by high-velocity, sediment-charged subglacial meltwater flows. (Photo courtesy D. E. Sugden.)*

proglacial environment. This distinction provides a convenient basis for classifying fluvioglacial landforms (Table 11.8). Wide temporal variations in meltwater discharge and the potential for reincorporation within a glacier of fluvioglacially deposited debris by refreezing means that many ice-contact forms are transient features, at least in active ice.

Proglacial features are much more likely to survive especially if the ice front is receding.

The basic mechanisms of deposition from meltwater are the same as those for ordinary fluvial systems, except that deposition within a glacier may occur under high hydrostatic pressures if meltwater completely fills an englacial or subglacial tunnel. There are some important differences, however, in the environment of deposition; for instance, tunnels within ice are not usually free to migrate laterally as fluvial channels may do and the rapid temporal fluctuations in meltwater discharge can produce equally rapid vertical changes in the calibre of sediment deposited which may range from large boulders to sand. The key characteristics which distinguish water-laid fluvioglacial sediments from glacial deposits is the absence of fine particles due to sorting and the presence of stratification. Meltwater deposits may be less easy to differentiate from ordinary fluvial sediments, although the glacial sediment source and the typically short distance of transport before deposition means that the former are usually less rounded.

Meltwater can deposit sediment on, in, under and along the margins of glaciers (Fig. 11.26). Although most of the ice-contact landforms to which this deposition gives rise are formed in channels, some may also be constructed either subglacially or marginally in lakes. The terminology used to describe these features is extremely confused. Here we

Table 11.8 A classification of fluvioglacial deposits

DOMINANT SEDIMENT	ENVIRONMENT	GENERAL FORM	RELATIONSHIP TO ICE	GENETIC TERM
Ice-contact deposits Sand and gravel	Fluvial	Ridge	Marginal, subglacial, englacial, supraglacial	**Esker**
		Mound		**Kame** **Kame complex**
		Spread with depressions	Marginal	**Kettled sandur**
Proglacial deposits Sand and gravel	Fluvial	Spread	Proglacial	**Sandur**
Silt and clay	Lacustrine		Proglacial/ marginal	**Lake plain**
Sand and gravel		Terraces, ridges		**Beach**
Clay sand and gravel		Terrace		**Kame delta**
Silt and clay	Marine	Spread		**Raised mud flat**
Sand and gravel		Terraces, ridges		**Raised beach**
Clay, sand and gravel		Terrace		**Raised delta**

Source: After R. J. Price (1973) *Glacial and Fluvioglacial Landforms*. Oliver and Boyd, Edinburgh, Table 3, p. 138.

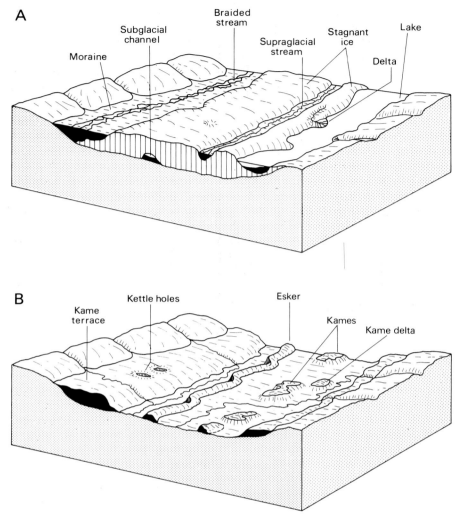

Fig. 11.26   *Development of ice-contact fluvioglacial landforms and deposits: (A) late stage of deglaciation with extensive areas of stagnant ice and abundant meltwater; (B) after deglaciation. (Based on R. F. Flint, (1971)* Glacial and Quaternary Geology. *Wiley, New York, Fig. 8–4, p. 209.)*

will simply distinguish between two main types of feature – **eskers** and **kames**.

Eskers are sinuous, sometimes discontinuous, ridges formed of sand, gravel or boulders and range up to 200 m in height, 3 km in width and 100 km or more in length. Some consist of mounds, lined by narrow ridges and they may form single ridges or interconnected networks. Eskers may initially be deposited in a variety of positions; they can be formed in ice tunnels either subglacially or, much more rarely, englacially, or they may be deposited in channels on the glacier surface. Where initially deposited supraglacially or englacially they can eventually be superimposed on to the subglacial topography as the underlying ice melts. Eskers may also be deposited along the ice margin, either in association with marginal and submarginal channels, or in ice-marginal lakes.

Kame is a broad term describing a mound of sediment formed by the initial deposition of material within a cavity in the ice followed by slumping of this material as the supporting ice walls melt away. Often the term is used as an adjective to describe landforms constructed in particular position within, or along, a glacier. **Kame terraces** for instance, are formed by the lateral or frontal accumulation of fluvioglacial deposits along the ice margin, while a **kame complex** results from the letting down of numerous sediment-filled supraglacial depressions and cavities in stagnant ice on to the subglacial surface. All varieties of kame may be further modified by slumping resulting from the melting of ice cores within the sediments thereby giving rise to **kettle-holes**. Some landscapes described as **kettle-and-kame topography** are dominated by mounds originally representing sites of deposition in depressions and the secondary effects of sediment collapse over melted ice masses.

Fluvioglacial deposition is dominant in the proglacial zone both because of the decreased capacity for transport of meltwater streams once they emerge from the ice front and the common presence of ice-margin lakes in this environment. The rapid dumping of large quantities of coarse debris immediately beyond the glacier margin and the associated shifting of stream channels within this zone produces an extensive depositional plain known as a **sandur** (plural **sandar**). Sandar are broadly analogous to alluvial fans with characteristic entrenchment by the main feeder channels, but they experience much more drastic seasonal fluctuations in discharge. **Valley sandar** form in the laterally confined environment of glacial troughs, whereas **plain sandar** or **outwash plains** can develop along the margins of ice sheets where braided rivers from numerous outlets along the ice front coalesce. Typically the proximal zone of a sandur close to the glacier margin has only a few main meltwater streams discharging from the ice front and a characteristically pitted surface, known as a **kettled-sandur**, produced by the melting of ice masses buried by the fluvio-glacially deposited sediments. Channel braiding becomes accentuated further away from the ice front in the intermediate zone and in the distal zone the channels are so shallow that overflow commonly occurs especially during peak discharges. The distal zone commonly merges into an extensive delta system formed along a proglacial lake. The sandur may extend away from the glacier front, gradually burying the delta as it advances.

Further reading

There are several comprehensive surveys of glacial geomorphology available including Embleton and King (1975), Flint (1971) and Price (1973), while Andrews (1975) provides a brief but stimulating introduction to the subject. Drewry (1986) provides a detailed treatment of glacial processes but these are rather poorly related to landforms. My own favoured treatment is that by Sugden and John (1976). This excellent text combines breadth with depth and has provided much of the framework of discussion on which this chapter is based. Useful supplementary material on a variety of topics is to be found in Coates (1974), while the volume edited by Gurnell and Clark (1987) contains various papers on glacial and fluvioglacial sediment transport with particular reference to Alpine terrains.

Articles on glacial geomorphology are published fairly frequently in the general geomorphology journals, but the other important sources are the *Journal of Glaciology* and *Annals of Glaciology* which emphasize the characteristics of ice and mechanics of glacier flow. Some papers of geomorphic interest are also published in *Arctic and Alpine Research, Quaternary Research, Quaternary Science Reviews, Journal of Quaternary Science* and *Boreas*.

Within the general field of glacier characteristics and dynamics Paterson (1981) is the basic reference, while Armstrong *et al.* (1973) provide illustrated definitions of ice and snow features. Chapter 5 of Sugden and John (1976) contains a discussion of the factors controlling the global distribution of glaciers, and the temperature characteristics of glacier ice are considered in Harrison (1975) and Paterson (1981). The complexity of the mass balances of real glaciers is examined in Mayo *et al.* (1972) and Østrem (1975) looks at the possibilities of the satellite monitoring of glacial budgets.

A general review of glacier movement is provided by Kamb (1964), while a detailed treatment of internal deformation of glacier ice is contained in the classic papers by Glen (1952, 1955) and Nye (1957, 1965a). For a similarly in-depth analysis of the mechanisms of basal sliding see Kamb and La Chapelle (1964), Lliboutry (1968), Morris (1976), Nye (1970) and Weertman (1964). More general reviews can be found in Kamb (1970) and Weertman (1979). The role of bed deformation in glacier movement is assessed by Boulton and Jones (1979) and Boulton and Hindmarsh (1987). Shreve (1984) challenges the conventional distinction between cold and warm-based ice and considers the mechanism of glacial sliding at subfreezing temperatures. Ice flow in the Antarctic ice sheet is considered by Drewry (1983) and Bentley (1987), while Alley *et al.* (1986) and Blankenship *et al.* (1986) discuss the role of substrate deformation in the movement of Antarctic Ice Stream B. It should be pointed out that a full understanding of most of these articles on glacier dynamics requires a fairly sophisticated level of mathematical expertise. The brief but very useful review by Hutter (1982) is, however, readily accessible.

Turning to the movement of glaciers as a whole, extending and compressive flow is treated by Nye (1952a) and applied to a specific glacier by Meier and Tangborn (1965), while the operation of kinematic waves is analysed by Nye (1960, 1965b). Possible mechanisms of surging are considered by Budd (1975), Jarvis and Clarke (1975), Kamb *et al.* (1985) and Robin and Weertman (1973), and Clarke *et al.* (1984) examine the role of bed deformation in the surging behaviour of Trapridge Glacier, Canada. Raymond (1987) and Sharp (1988a) review ideas on the mechanisms of surging and Clarke *et al.* (1986) examine the glacier characteristics common to surging behaviour and Sharp (1988b) looks at its geomorphic consequences. Short-term glacier fluctuations are covered in Chapter 6 of Sugden and John (1976).

The large-scale form of ice masses is discussed by Nye (1952b) and Weertman (1961), and Reeh (1982) presents a model which takes into account the effects of subglacial topography. The properties of the Antarctic ice sheet are documented by Drewry *et al.* (1982) and Drewry (1983). Numerous ice-surface features are illustrated in Post and La Chapelle (1971) while Nye (1952a) considers the generation of crevasses and Hambrey (1976), Kamb (1964) and Lliboutry and Reynaud (1981) discuss the formation of various kinds of ice banding.

The processes of glacial erosion are reviewed by Boulton (1974, 1979) and detailed discussions of particular erosional mechanisms are provided by Kamb and La Chapelle (1964) and McCall (1960) (abrasion), Glen and Lewis (1961) and McCall (1960) (crushing and fracturing) and Lewis (1954), Linton (1963) and Trainer (1973) (joint-block removal). The formation of fluted surfaces at a variety of scales is discussed by Boulton (1974, 1976), Flint (1971), Goldthwait (1979) and Linton (1963). The form of glacial troughs is considered by Graf (1970), Harbor et al. (1988), Hirano and Aniya (1988) and Linton (1963), and their stepped long profiles are examined by Bakker (1965) and, in the broader context of contrasts between glacial and fluvial systems, by King (1970). The morphology of cirques and factors controlling their development are treated by Derbyshire and Evans (1976), Haynes (1968) and Olyphant (1981). Overviews of the creation of glacially eroded landscapes are provided by Linton (1963) and Sugden (1974) while Sugden (1978) provides a more detailed treatment of the morphology generated by the Pleistocene Laurentide ice sheet of North America.

There are several good reviews available on glacial deposition and associated landforms, including Goldthwait (1971), Price (1973) and Schluchter (1979). Boulton (1974) and Kamb and La Chapelle (1964) look at the processes of debris entrainment and transport and Boulton (1972, 1975) discusses mechanisms of deposition. Prest (1968) provides a comprehensive classification of depositional landforms. Fluted moraines are discussed by Baranowski (1970), while drumlins are reviewed by Menzies (1979) and considered in detail in the symposium volume edited by Menzies and Rose (1987). Specific processes of formation are proposed by Evenson (1971), Shaw and Sharpe (1987) and Smalley and Unwin (1968). Embleton and King (1975) and Price (1973) contain extended treatments of end moraines, and Cowan (1968) discusses the formation of ribbed (Rogen) moraine.

For excellent discussions of meltwater in glaciers see Shreve (1972), Stenborg (1969) and Weertman (1972). Nye (1976) provides an additional recent analysis with particular reference to the generation of jökulhlaups, and Thorarinsson (1953) contains a vivid description of the 1934 Grímsvötn jökulhlaup. The Late Pleistocene Lake Missoula cataclysmic flooding is discussed by Baker and Bunker (1985) and Waitt (1985), but the classic papers by Bretz (1923, 1969) are well worth consulting. The relationships between sediment yield and discharge in a meltwater stream are analysed in Østrem et al. (1967). Plastically sculptured forms (p-forms) are discussed by Dahl (1965), but Boulton (1974) and Gjessing (1965) provide alternative explanations for their development. The origin of meltwater channels is dis-

cussed in the classic paper by Mannerfelt (1949) and by Clapperton (1968) and Price (1973). Flint (1971), Parizek (1969) and Price (1973) consider the complex range of depositional fluvioglacial forms and Church (1972) and Krigstrom (1962) provide detailed treatments of the proglacial environment.

References

Alley, R. B., Blankenship, D. D., Bentley, C. R. and Rooney, S. T. (1986) Deformation of till beneath ice stream B, West Antarctica. *Nature* **322**, 57–9.

Andrews, J. T. (1975) *Glacial Systems*. Duxbury Press, North Scituate.

Armstrong, T. E., Roberts, B. and Swithinbank, C. (1973) *Illustrated Glossary of Snow and Ice* (2nd edn) Scott Polar Research Institute, Cambridge.

Baker, V. R. and Bunker, R. C. (1985) Cataclysmic Late Pleistocene flooding from glacial Lake Missoula: a review. *Quaternary Science Reviews* **4**, 1–41.

Bakker, J. P. (1965) A forgotten factor in the interpretation of glacial stairways. *Zeitschrift für Geomorphologie* **9**, 18–34.

Baranowski, S. (1970) The origin of fluted moraine at the fronts of contemporary glaciers. *Geografiska Annaler* **52A**, 68–75.

Bentley, C. R. (1987) Antarctic ice streams: a review. *Journal of Geophysical Research* **92**, 8843–58.

Blankenship, D. D., Bentley, C. R., Rooney, S. T. and Alley, R. B. (1986) Seismic measurements reveal a saturated porous layer beneath an active Antarctic ice stream. *Nature* **322**, 54–7.

Boulton, G. S. (1972) The role of thermal regime in glacial sedimentation. *Institute of British Geographers Special Publication* **4**, 1–19.

Boulton, G. S. (1974) Processes and patterns of glacial erosion. In: D. R. Coates (ed.) *Glacial Geomorphology*. State University of New York, Binghamton, 41–87.

Boulton, G. S. 1975. Processes and patterns of sub-glacial sedimentation: a theoretical approach. In: A. E. Wright and F. Moseley (eds) *Ice Ages: Ancient and Modern*. Seel House Press, Liverpool, 7–42.

Boulton, G. S. (1976) The origin of glacially fluted surfaces - observation and theory. *Journal of Glaciology* **17**, 287–309.

Boulton, G. S. (1979) Processes of glacier erosion on different substrata. *Journal of Glaciology* **23**, 15–38.

Boulton, G. S. and Hindmarsh, R. C. A. (1987) Sediment deformation beneath glaciers: rheology and geological consequences. *Journal of Geophysical Research* **92**, 9059–82.

Boulton, G. S. and Jones, A. S. (1979) Stability of temperate ice caps and ice sheets resting on beds of deformable sediment. *Journal of Glaciology* **24**, 29–43.

Bretz, J. H. (1923) The channeled scabland of the Columbia Plateau. *Journal of Geology* **31**, 617–49.

Bretz, J. H. (1969) The Lake Missoula floods and the channeled scablands. *Journal of Geology* **77**, 505–43.

Budd, W. F. (1975) A first simple model for periodically self-surging glaciers. *Journal of Glaciology* **14**, 3–21.

Church, M. (1972) Baffin Island sandurs. A study of arctic fluvial processes. *Canada Geological Survey Bulletin* **216**.

Clapperton, C. M. (1968) Channels formed by the superimposition of glacial meltwater streams, with special reference to the east Cheviot hills, northeast England. *Geografiska Annaler* **50**, 207–20.

Clarke, G. K. C., Collins, S. G. and Thompson, D. E. (1984) Flow, thermal structure and subglacial conditions of a surge-type glacier. *Canadian Journal of Earth Sciences* **21**, 232–40.

Clarke, G. K. C., Schmok, J. P., Simon, C., Ommanney, L. and Collins, S. G. (1986) Characteristics of surge-type glaciers. *Journal of Geophysical Research* **91**, 7165–80.

Coates, D. R. (ed.) (1974) *Glacial Geomorphology*. State University of New York, Binghamton.

Cowan, W. R. (1968) Ribbed moraine: till fabric analysis and origin. *Canadian Journal of Earth Sciences* **5**, 1145–59.

Dahl, R. (1965) Plastically sculptured detail forms on rock surfaces in northern Nordland, Norway. *Geografiska Annaler* **47**, 83–140.

Derbyshire, E. and Evans, I. S. (1976) The climatic factor in cirque variation. In: E. Derbyshire (ed.) *Geomorphology and Climate*. Wiley, London and New York, 447–94.

Drewry, D. J. (1983) Antarctic ice sheet: aspects of current configuration and flow. In: R. Gardner and H. Scoging (eds) *Megageomorphology*. Clarendon Press, Oxford, 18–38.

Drewry, D. (1986) *Glacial Geologic Processes*. Edward Arnold, London and Baltimore.

Drewry, D. J., Jordan, S. R. and Jankowski, E. (1982) Measured properties of the Antarctic ice sheet: surface configuration, ice thickness, volume and bedrock characteristics. *Annals of Glaciology* **3**, 83–91.

Embleton, C. and King, C. A. M. (1975) *Glacial Geomorphology*. Edward Arnold, London; Halstead, New York.

Evenson, E. B. (1971) The relationship of macro and micro-fabric of till and the genesis of glacial landforms in Jefferson County, Wisconsin. In: R. P. Goldthwait (ed.) *Till: A Symposium*. Ohio State University Press, Columbus, 345–64.

Flint, R. F. (1971) *Glacial and Quaternary Geology*. Wiley, New York.

Gjessing, J. (1965) On 'plastic scouring' and 'subglacial erosion'. *Norsk Geografisk Tidsskrift* **20**, 1–37.

Glen, J. W. (1952) Experiments on the deformation of ice. *Journal of Glaciology* **2**, 111–14.

Glen, J. W. (1955) The creep of polycrystalline ice. *Proceedings of the Royal Society London* **A228**, 519–38.

Glen, J. W. and Lewis, W. V. (1961) Measurements of side-slip at Austerdalsbreen, 1959. *Journal of Glaciology* **3**, 1109–22.

Goldthwait, R. P. (ed.) (1971) *Till: A Symposium*. Ohio State University Press, Columbus.

Goldthwait, R. P. (1979) Giant grooves made by concentrated basal ice streams. *Journal of Glaciology* **23**, 297–307.

Graf, W. L. (1970) The geomorphology of the glacial valley cross-section. *Arctic and Alpine Research* **2**, 303–12.

Gurnell, A. M. and Clark, M. J. (eds) (1987) *Glacio-Fluvial Sediment Transfer: An Alpine Perspective*. Wiley, Chichester and New York.

Hambrey, M. J. (1976) Debris, bubble and crystal fabric characteristics of foliated ice, Charles Rabots Bre, Okstindan, Norway. *Arctic and Alpine Research* **8**, 49–60.

Harbor, J. M., Hallet, B. and Raymond, C. F. (1988) A numerical model of landform development by glacial erosion. *Nature* **333**, 347–9.

Harrison, W. D. (1975) Temperature measurements in a temperate glacier. *Journal of Glaciology* **14**, 23–30.

Haynes, V. M. (1968) The influence of glacial erosion and rock structure on corries in Scotland. *Geografiska Annaler* **50A**, 221–34.

Hirano, M. and Aniya, M. (1988) A rational explanation of cross-profile morphology for glacial valleys and of glacial valley development. *Earth Surface Processes and Landforms* **13**, 707–16.

Hutter, K. (1982) Glacier flow. *American Scientist* **70**, 26–34.

Jarvis, G. T. and Clarke, G. K. C. (1975) The thermal regime of Trapridge Glacier and its relevance to glacier surging. *Journal of Glaciology* **14**, 235–49.

Kamb, B. (1964) Glacier geophysics. *Science* **146**, 353–65.

Kamb, B. (1970) Sliding motion of glaciers: theory and observation. *Reviews of Geophysics and Space Physics* **8**, 673–728.

Kamb, B. and La Chapelle, E. (1964) Direct observation of the mechanism of glacier sliding over bedrock. *Journal of Glaciology* **5**, 159–72.

Kamb, B., Raymond, C. F., Harrison, W. D., Engelhardt, H., Echelmeyer, K. A., Humphrey, N., Brugman, M. M. and Pfeffer, T. (1985) Glacier surge mechanism: 1982–1983 surge of Variegated Glacier, Alaska. *Science* **227**, 469–79.

King, C. A. M. (1970) Feedback relationships in geomorphology. *Geografiska Annaler* **52A**, 147–59.

Krigstrom, A. (1962) Geomorphological studies of sandar plains and their braided rivers in Iceland. *Geografiska Annaler* **44**, 328–46.

Lewis, W. V. (1954) Pressure release and glacial erosion. *Journal of Glaciology* **2**, 417–22.

Linton, D. L. (1963) The forms of glacial erosion. *Transactions of the Institute of British Geographers* **33**, 1–28.

Lliboutry, L. (1968) General theory of subglacial cavitation and sliding of temperate glaciers. *Journal of Glaciology* **7**, 21–58.

Lliboutry, L. and Reynaud, L. (1981) 'Global dynamics' of a temperate valley glacier, Mer de Glace, and past velocities deduced from Forbes' bands. *Journal of Glaciology* **27**, 207–26.

Mannerfelt, C. M. (1949) Marginal drainage channels as indicators of the gradients of Quaternary ice caps. *Geografiska Annaler* **31**, 194–9.

Mayo, L. R., Meier, M. F. and Tangborn, W. V. (1972) A system to combine stratigraphic and annual mass-balance systems: a contribution to the International Hydrological Decade. *Journal of Glaciology* **11**, 3–14.

McCall, J. G. (1960) The flow characteristics of a cirque glacier and their effect on glacial structure and cirque formation. In: W. V. Lewis (ed.) *Investigations of Norwegian Cirque Glaciers*. Royal Geographical Society Research Series **4**, 39–62.

Meier, M. F. and Tangborn, W. V. (1965) Net budget and flow of South Cascade Glacier, Washington. *Journal of Glaciology*. **5**, 547–66.

Menzies, J. (1979) A review of the literature on the formation and location of drumlins. *Earth Science Reviews* **14**, 315–59.

Menzies, J. and Rose, J. (eds) (1987) *Drumlin Symposium*: Proceedings of the Drumlin Symposium: First International Conference on Geomorphology, Manchester 16–18 Sept. 1985. Balkema, Rotterdam.

Morris, E. M. (1976) An experimental study of the motion of ice past obstacles by the process of regelation. *Journal of Glaciology* **17**, 79–98.

Nye, J. F. (1952a) The mechanics of glacier flow. *Journal of Glaciology* **2**, 82–93.

Nye, J. F. (1952b) A method of calculating the thickness of ice sheets. *Nature* **169**, 529–30.

Nye, J. F. (1957) The distribution of stress and velocity in glaciers and ice sheets. *Proceedings of the Royal Society London* **A239**, 113–33.

Nye, J. F. (1960) The response of glaciers and ice sheets to seasonal and climatic changes. *Proceedings of the Royal Society London* **A256**, 559–84.

Nye, J. F. (1965a) The flow of a glacier in a channel of rectangular, elliptic or parabolic cross-section. *Journal of Glaciology* **5**, 661–90.

Nye, J. F. (1965b) The frequency response of glaciers. *Journal of Glaciology* **5**, 567–87.

Nye, J. F. (1970) Glacier sliding without cavitation in a linear viscous approximation. *Proceedings of the Royal Society London.* **A315**, 381–403.

Nye, J. F. (1976) Water flow in glaciers: jökulhlaups, tunnels, and veins. *Journal of Glaciology* **17**, 181–207.

Olyphant, G. A. (1981) Interaction among controls of cirque development: Sangre de Cristo Mountains. Colorado, U.S.A. *Journal of Glaciology* **27**, 449–58.

Østrem, G. (1975) ERTS data in glaciology – an effort to monitor glacier mass balance from satellite imagery. *Journal of Glaciology* **15**, 403–14.

Østrem, G., Bridge, C. W. and Rannie, W. F. 1967. Glaciohydrology, discharge and sediment transport in the Decade glacier area, Baffin Island, N.W.T. *Geografiska Annaler* **49A**, 268–82.

Parizek, R. R. (1969) Glacial ice-contact rings and ridges. *Geological Society of America Special Paper* **123**, 49–102.

Paterson, W. S. B. (1981) *The Physics of Glaciers.* (2nd edn) Pergamon, Oxford.

Post, A. S. and La Chapelle, E. R. (1971) *Glacier Ice.* University of Washington Press, Seattle.

Prest, V. K. (1968) Nomenclature of moraines and ice-flow features as applied to the glacial map of Canada. *Geological Survey Papers Canada* 67–57.

Price, R. J. (1973) *Glacial and Fluvioglacial Landforms.* Oliver and Boyd, Edinburgh.

Raymond, C. F. (1987) How do glaciers surge? A review. *Journal of Geophysical Research* **92**, 9121–34.

Reeh, N. (1982) A plasticity theory approach to the steady-state shape of a three-dimensional ice sheet. *Journal of Glaciology* **28**, 431–55.

Robin, G. de Q. and Weertman, J. (1973) Cyclic surging of glaciers. *Journal of Glaciology* **12**, 3–18.

Schluchter, Ch. (ed.) (1979) *Moraines and Varves: Origin, Genesis, Classification.* Balkema, Rotterdam.

Sharp, M. (1988a) Surging glaciers: behaviour and mechanisms. *Progress in Physical Geography* **12**, 349–70.

Sharp, M. (1988b) Surging glaciers: geomorphic effects. *Progress in Physical Geography* **12**, 533–59.

Shaw, J. and Sharpe, D. R. (1987) Drumlin formation by subglacial meltwater erosion. *Canadian Journal of Earth Sciences* **24**, 2316–22.

Shreve, R. L. (1972) Movement of water in glaciers. *Journal of Glaciology,* **11**, 205–14.

Shreve, R. L. (1984) Glacier sliding at subfreezing temperatures. *Journal of Glaciology* **30**, 341–47.

Smalley, I. J. and Unwin, D. J. (1968) The formation and shape of drumlins and their distribution and orientation in drumlim fields. *Journal of Glaciology* **7**, 377–90.

Stenborg, T. 1969. Studies of the internal drainage of glaciers. *Geografiska Annaler* **51A**, 13–41.

Sugden, D. E. (1974) Landscapes of glacial erosion in Greenland and their relationship to ice, topographic and bedrock conditions. *Institute of British Geographers Special Publication* **7**, 177–95.

Sugden, D. E. (1978) Glacial erosion by the Laurentide ice sheet. *Journal of Glaciology* **20**, 367–91.

Sugden, D. E. and John, B. S. (1976) *Glaciers and Landscape.* Edward Arnold, London.

Thorarinsson, S. (1953) Some new aspects of the Grímsvötn problem. *Journal of Glaciology* **2**, 267–75.

Trainer, F. W. (1973) The formation of joints in bedrock by moving glacial ice. *United States Geological Survey Journal of Research* **1**, 229–36.

Waitt, R. B. (1985) Case for periodic, colossal jökulhlaups from Pleistocene glacial Lake Missoula. *Geological Society of American Bulletin* **96**, 1271–86.

Weertman, J. (1961) Equilibrium profile of ice caps. *Journal of Glaciology* **3**, 953–64.

Weertman, J. (1964) The theory of glacier sliding. *Journal of Glaciology* **5**, 287–303.

Weertman, J. (1972) General theory of water flow at the base of a glacier or ice sheet. *Reviews of Geophysics and Space Physics* **10**, 287–333.

Weertman, J. (1979) The unsolved general glacier sliding problem. *Journal of Glaciology* **23**, 97–115.

12

Periglacial processes and landforms

12.1 The periglacial environment

In the high latitude regions of the northern and southern hemispheres, and in some areas at high elevation elsewhere, prevailing temperatures are so low that the ground remains frozen for much, or all, of the year. In such environments the effects of repeated freezing and thawing and the growth of ice masses in the ground are so pervasive that they give rise to a characteristic range of landforms which merit special consideration. This is the realm of periglacial processes and landforms.

Although their present extent is impressive, the Pleistocene saw the extension of periglacial conditions well into mid-latitudes as the ice sheets of the northern hemisphere advanced southwards. Thus large areas now experiencing comparatively mild climates retain evidence in the form of relict periglacial landforms of much colder conditions as recently as 15 ka BP. Many high latitude regions today have acquired considerable economic and strategic significance and this has served to focus attention on the special characteristics of their landforms.

The term 'periglacial' was introduced in 1909 by the Polish scientist Walery von Lozinski to describe the landforms and processes occurring around the margins of the great Pleistocene ice sheets. Subsequently it was applied more broadly to encompass those processes and landforms (regardless of age) associated with very cold climates in areas not permanently covered with snow or ice (and in many cases located far from glaciers or ice sheets). Such areas of extreme cold are commonly underlain by permanently frozen ground, or **permafrost**, and because many landforms characteristic of these regions owe their existence to permafrost its presence is regarded by some as a prerequisite for the action of periglacial processes. Nevertheless, since many other landforms characteristic of cold regions occur independently of permafrost, most geomorphologists would not use the term in this restricted sense.

Here we will apply the term periglacial to those environments characterized by intense frost action and at least seasonally snow-free ground. Such a qualitative definition makes it difficult to define precisely appropriate climatic parameters in a way which satisfies all researchers. As a general guide the upper limit for mean annual temperature is generally considered to lie between −1 and −3 °C and the mean annual precipitation to be less than 1000 mm. Within this definition there are a great variety of climatic types, ranging from the extremely cold and arid conditions experienced in such areas as northern Greenland and parts of Antarctica, to the comparatively much warmer and humid maritime-influenced climates typified by Spitsbergen and Iceland, and the alpine environments of high mountain ranges in all latitudes. Since an appreciation of this broad climatic spectrum is important for an understanding of the global occurrence of the various periglacial processes and landforms, it is helpful to subdivide the periglacial environment into a number of distinct categories (Table 12.1).

Table 12.1 Classification of periglacial climates

Polar lowlands	Mean temperature of coldest month <−3 °C. Zone is characterized by ice caps, bare rock surfaces and tundra vegetation
Subpolar lowlands	Mean temperature of coldest month <−3 °C and of warmest month >10 °C. Taiga type of vegetation. The 10 °C isotherm for warmest month roughly coincides with tree-line in northern hemisphere.
Mid-latitude lowlands	Mean temperature of coldest month is <−3 °C but mean temperature >10 °C for at least four months per year.
Highlands	Climate influenced by altitude as well as latitude. Considerable variability over short distance depending on aspect. Diurnal temperature ranges tend to be large

Source: Based on the classification presented by A. L. Washburn (1979) *Geocryology*. Edward Arnold, London, pp. 7–8.

12.1.1   Characteristics of permafrost

Although we have indicated that permanently (or more strictly perennially) frozen ground is not an element of all periglacial environments, it plays a key role in many important periglacial processes. Permafrost was first defined by S. W. Muller in 1947 as a thickness of soil, sediment or bedrock at variable depth beneath the surface experiencing a temperature continuouly below freezing for a prolonged period (Muller considered the minumum period to be 2 a). According to this definition cold glaciers are a form of permafrost, but because of their special characteristic of movement they are usually considered separately. Various other definitions have subsequently been offered, a number of these emphasizing the importance of the presence of ice. Although 'dry' permafrost can exist where there is insufficient moisture to permit ice formation, the role of ice is so crucial in the generation of landforms in periglacial environments that we will be almost exclusively concerned here with permafrost containing **ground ice**. The mean surface temperature required to maintain, or develop, permafrost may be less than 0 °C as pressure, salts in solution and the presence of clays can all affect the freezing point of subsurface water. Consequently, most workers now cite a temperature a little below 0 °C in defining permafrost.

Ground ice may be present in a variety of forms ranging from a cement in soil pores to masses of almost pure ice in veins, lenses and wedges (Table 12.2). It is generally confined to the upper part of a permafrost layer, the maximum depth recorded being only about 45 m. In most areas of permafrost, surface temperatures exceed 0 °C for at least a brief period each year. Consequently there is usually a thin **active layer**, up to 3 m in depth, overlying the permafrost which freezes and thaws seasonally (Fig. 12.1). The boundary between the active layer and the permafrost is known as the **permafrost table**. As a result of freezing penetrating downwards at unequal rates in the winter, unfrozen water

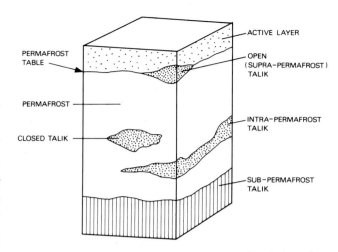

Fig. 12.1 *Nomenclature for features associated with permafrost.*

may be trapped between the permafrost table and the frozen active layer. These water pockets, called **taliks**, develop in part because of the release of latent heat of fusion as water undergoes a change of state from a liquid to a solid. An additional factor is the build-up of **cryostatic pressure**; this is pressure exerted as a result of the volume expansion associated with the freezing of water. Open, or supra-permafrost, taliks may persist for months before finally freezing, while intra-permafrost taliks, occurring within the permafrost layer itself, may survive for much longer periods.

12.1.2   Distribution of permafrost

The very considerable area covered by permafrost, even at the present day, is evident from the data provided in Table 12.3. If ground under ice sheets is included the total exceeds one-quarter of the surface area of the continents. By far the largest proportion occurs in the northern hemisphere, mostly in the USSR, Canada and Alaska (Fig. 12.2). In the very coldest regions permafrost is continuous, but as mean annual temperatures approach 0 °C its occurrence becomes discontinuous and eventually sporadic. Its equatorward limit correlates broadly with a mean annual air temperature of −1 °C. Climates associated with permafrost generally have

Table 12.2 Classification of types of ground ice

MAJOR TYPES	SUBTYPES
Soil ice	Needle ice Segregated ice Ice-filling pore spaces
Vein ice	Single veins Ice wedges
Intrusive ice	Pingo ice Sheet ice
Extrusive ice (formed subaerially e.g. floodplains)	
Sublimation, i.e. formed in cavities by crystallization from water vapour	
Buried ice	Buried icebergs Buried glacial ice

Source: After C. Embleton and C. A. M. King (1975) *Periglacial Geomorphology*, Edward Arnold, London, Table 2.3, p. 34.

Table 12.3 Global occurrence of permafrost

	AREA COVERED (km² × 10⁶)		
	Continuous	Discontinuous	Total
Northern hemisphere	7.64	14.71	22.35
Antarctica	13.21	—	13.21
Mountains	—	2.59	2.59
Totals	20.85	17.30	38.15

Source: Modified from A. L. Washburn (1979) *Geocryology*. Edward Arnold, London, Table 3.1, p. 22, after S. R. Stearns (1966) *US Army Cold Regions Research and Engineering Laboratory, Cold Regions Science and Engineering* **1–A2**.

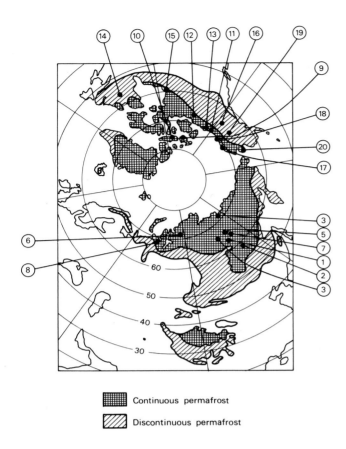

Fig. 12.2 *Distribution of continuous and discontinuous permafrost in the northern hemisphere. Numbers indicate localities in Table 12.4. (Map compiled from various sources.)*

▦ Continuous permafrost

▨ Discontinuous permafrost

daily temperatures below 0 °C for at least nine months of the year, and below −10 °C for at least six months of the year. Temperatures rarely exceed 20 °C at any time and precipitation is typically low – less than 100 mm in winter and less than 300 mm in summer. The large anticyclonic continental polar air masses are largely responsible for producing the required conditions of intense cold and aridity. The distribution of permafrost in Eurasia clearly illustrates the prominent southward depression of its equatorward limit in the continental interior of east-central Siberia and its poleward deflection to very high latitudes in maritime-influenced north-west Europe (Fig. 12.2). The extent of permafrost under the Antarctic and Greenland ice sheets is uncertain but there are strong theoretical grounds for believing it to be present. The minimum elevation at which permafrost is found tends to increase equatorwards but small areas occur on the highest peaks even at very low latitudes; limited occurrences, for instance, exist above an elevation of 4170 m on Mauna Kea, Hawaii, at 19 °N. The most extensive highland occurrences are in western China on the Tibetan Plateau.

At the present day, depths of permafrost range up to at least 1500 m (Table 12.4). Rates of permafrost aggradation

Table 12.4 Approximate depth to base of permafrost and prevailing temperature conditions at various localities in the northern hemisphere.

LOCALITY	DEPTH (m)	MEAN ANNUAL AIR TEMPERATURE (°C)	GROUND TEMPERATURE (°C) AT DEPTH INDICATED (m)	
USSR				
1 Markha River, upper reaches	1450–1500	—	—	—
2 Udokan	900	−12.0	−3.3	500
3 Tiksi	630	−14.0	−3.3	500
4 Mirnyy	550	− 9.0	−1.8	300
5 Vilyuy River, mouth	420	−10.0	−1.0	240
6 Noril'sk	325	− 8.0	−7.5	50
7 Yakutsk	198–250	−10.13	—	—
8 Vorkuta	131	—	—	—
Canada				
9 Winter Harbour	557	−16.0	—	—
10 Resolute	396	−16.2	−5.6	135
11 Mackenzie Delta	18–366	−9.1 to −11.3	−0.4 to −6.0	14
12 Port Radium	107	− 7.1	—	—
13 Inuvik	>91	− 9.6	−3.3	8 to 30
14 Schefferville	>76	− 4.5	−1.1 to −0.3	8 to 58
15 Churchill	30–61	− 7.2	−2.5 to −1.7	8 to 16
16 Dawson	61	− 4.7	—	—
Alaska				
17 Prudhoe Bay	610	—	—	—
18 Barrow	204–405	−12.0	−6.3	179
19 Fairbanks	30–122	− 3.5	−7.0	25
20 Nome	37	− 3.5	—	—

Note: Locations are indicated by numbers on Figure 12.2.
Source: Data from A. L. Washburn, 1979, *Geocryology*. Edward Arnold, London, Table 3.5, pp. 37–40.

can only be of the order of a few centimetres a year so thick permafrost layers must have taken a minimum of several thousands of years to develop. In some areas present-day climates are sufficiently cold for the development as well as the maintenance of permafrost, but elsewhere permafrost depths seem to exceed those expected on the basis of current temperatures. This suggests that permafrost depths may be influenced by past as well as present climatic conditions. Other factors, including the presence and type of soil and vegetation, the existence of water bodies, snow or ice on the surface and exposure to the Sun and wind can influence the depth of permafrost in a particular locality. These, and other factors, can change over time to affect the intimate relationship between air and ground temperatures and thereby lead to the degradation or aggradation of a permafrost layer (Fig. 12.3).

12.2   Periglacial processes

Comparatively few geomorphic processes are solely confined to periglacial environments; rather, a number of processes occurring in other morphoclimatic zones are especially active in periglacial regimes and give rise to some unique landforms. Most of the significant processes arise, directly or indirectly, from the freezing and thawing of water. It is important, however, to emphasize that even though we will be focusing on these 'periglacial processes', the landscapes of periglacial regions as a whole are still dominated by the effects of 'normal' fluvial activity.

12.2.1   Frost action

Frost action encompasses those processes associated with the freezing of water in rock, soil or other material, includ-

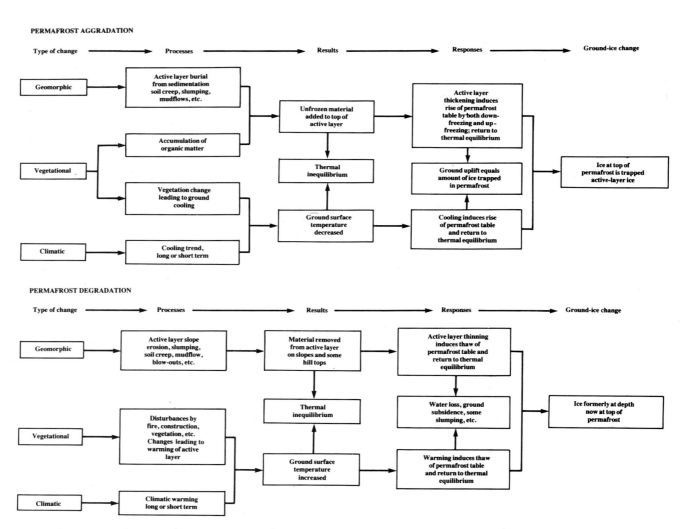

Fig. 12.3   *Factors influencing the growth (aggradation) and decay (degradation) of permafrost. (After A. L. Washburn (1979)* Geocryology. *Edward Arnold, London, Fig. 5.3.31 and 3.32, p. 56, after J. R. Mackay (1971) in: R. J. E. Brown (ed.)* Proceedings of a Seminar on the Permafrost Active Layer, 4 and 5 May 1971. *Canada National Research Council Technical Memo* **103** *Fig. 1, p. 29 and Fig. 2, p. 30.)*

ing frost shattering, heaving and thrusting. Frost cracking (see Section 12.2.1.4) is also usually included although it is caused by the thermal contraction of materials at low temperatures rather than the formation of ice. Although frost action occurs outside the periglacial zone, its significance within it is the crucial element that differentiates periglacial landscapes from those developed under other morphoclimatic regimes.

Many periglacial landforms are associated with the freezing of water in soils and other unconsolidated deposits. At first sight this appears a simple physical mechanism, but in reality the large number of variables at play under natural conditions make it a complex process. Of particular importance in terms of its geomorphic effects is whether water freezes *in situ* in voids, or migrates to form discrete masses of **segregated ice**. The key factors are the rate of freezing, the grain size (and hence pore size) of the material and the potential for capillary suction (see Section 6.1.2) which expresses the ability of water to move towards the point in the soil where it is freezing (often referred to as the **freezing front**).

Coarse deposits, such as gravel, are highly permeable but have a low suction potential, whereas very fine-grained materials with extremely small pores have a low permeability but high suction potential. Consequently it is sediments of intermediate grain size, such as silt, that are most susceptible to the formation of segregated ice since both suction potential and permeability are moderately high in such materials. Grain (pore) size also influences the temperature at which ice will form. As a result of the complex interaction between water molecules and the surfaces of particles the proportion of water remaining unfrozen in sediment at a given temperature varies inversely with grain size since the surface area on to which absorbed water is bound by a variety of physico-chemical effects is greater for small particles. In clays only half of the total water content may be frozen at a temperature of $-2\,°C$, whereas in sands virtually all the water freezes at $0\,°C$. In any material ice forms initially in the larger pores and only as temperature falls is it able to crystallize in smaller voids.

Crucial to an understanding of the role of frost action processes are measurements of the frequency of freeze–thaw cycles experienced by surface materials or bedrock. The frequency of air temperature fluctuations across the freezing point can be readily estimated from conventional meterorological data. But unfortunately such data give little indication of the effectiveness of frost action since many other factors, such as the thermal properties of bedrock and soil and the insulating effects of vegetation and snow, must also be considered. Discrepancies between the number of freeze–thaw cycles in the air and those occurring in the ground are especially marked in climates characterized by many air cycles of small amplitude. In a study of Kerguelen Island in the sub-Antarctic zone no frost was recorded over

a 2 a period at a depth of 50 mm in spite of 441 freeze–thaw cycles at the surface; in many areas there may be only one annual cycle below a depth of a few centimetres. Such findings have important implications for the efficacy of frost action in rock breakdown.

12.2.1.1 Frost weathering

We have already outlined the mechanisms thought to be involved in frost weathering in Section 6.3.2.1, but it is necessary here to reassess the role of this process under present-day periglacial regimes. The important role traditionally attached to frost shattering in producing the angular rock debris so characteristic of periglacial environments rests heavily on the assumed presence of suitable conditions of freeze–thaw and moisture availability, yet field measurements of these crucial parameters are few in number. Moreover, some recent studies suggest that suitable conditions may be less common than previously thought.

A detailed field investigation completed in the Colorado Rockies in the western USA, for instance, involved monitoring temperatures beneath and adjacent to seasonal snow patches over 2 a. This is just the kind of alpine environment (abundant moisture from snow-melt coupled with frequent air temperature fluctuations across the freezing point) in which we would expect frost shattering to be intense. None the less, this study demonstrates that even here the required combination of freezing intensity and moisture availability is absent. Because of the insulating effect of snow, bedrock temperatures at a depth of 10 mm rarely fell below $-5\,°C$, the apparently critical temperature for effective stresses to be generated by the expansion arising from ice formation. On snow-free cliff faces temperatures dropped well below $-5\,°C$ but there was a lack of available moisture.

Such studies question our ideas about conditions assumed to occur in nature and in this case focus attention on other mechanisms of rock breakdown, such as hydration shattering and, at least in arid and coastal locations, salt weathering. In the McMurdo Sound area of Antarctica, for instance, it has been suggested that salts (mainly sodium chloride transported from the sea or from surface accumulations) are incorporated in snow. On thawing, salt solutions enter rock pores and the salt crystallizes out. A variety of surface weathering features occurring in periglacial environments, including tafoni and honeycomb forms (see Section 6.4.2), have been attributed to this process. It may also play an important role in rock breakdown. Alternatively, much of the angular rock debris found in some lower latitude periglacial environments today may be a relic of Pleistocene conditions more favourable to frost shattering.

12.2.1.2 Frost heaving and thrusting

Frost heaving and **frost thrusting** refer respectively to the vertical and horizontal movement of material due to the formation of ice. In nature a combination of vertical and

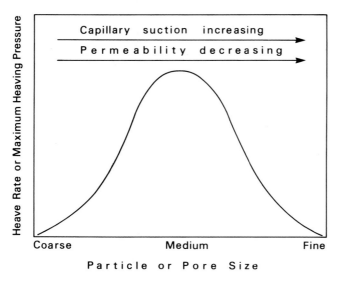

Fig. 12.4 *Relationship between heaving rate and pore size.*

horizontal components is more common, although heaving probably predominates. This is because the pressure generated by ice crystallization, which is at least as crucial as the volume expansion due to freezing, acts predominantly parallel to the direction of the maximum temperature gradient, and this is approximately normal to the ground surface. For the same reason **ice lenses**, which lie parallel to the surface, are a common form of segregated ice, although their precise form is also influenced by the structure and stratification of soils and other deposits.

Since heaving and thrusting are caused by the formation of segregated ice, the factors which control the migration of water to the freezing front also determine the susceptibility of surface materials to such processes (Fig. 12.4). As we have already noted, silts are most prone to the growth of large ice masses. Variations in the particle size of surface deposits can lead to localized ice formation and the development of differential heaving as well as lateral pressures. The nature of the surface cover can also influence the process as heaving has been observed to be most effective where moisture is abundant and where there is little vegetation to insulate the ground. Localized upward movement of surface stones can occur through the formation of **needle ice** which is composed of vertically elongated crystals up to 30 mm in length.

Frost heaving in the active layer appears to be attributable to three main effects: the growth of ice lenses as downward freezing progresses; the development of ice lenses near the base of the active layer by upward freezing from the permafrost layer, and the progressive freezing of unfrozen pore water as the active layer cools below 0 °C. In addition to displacing masses of soil and other materials,

frost heaving is an important mechanism in the differential vertical movement of soil constituents of different sizes.

Numerous field observations confirm the upward movement of stones in soils in periglacial areas, and various mechanisms have been suggested to explain this phenomenon. They may be categorized for convenience into frost-pull and frost-push hypotheses. The frost-pull hypothesis proposes that stones together with the enclosing finer-grained matrix are raised as the ground expands on freezing. On thawing the fine material collapses while the bases of the stones are still supported by ice. Subsequently, as this ice melts, fine particles partly fill the space and support the overlying stone. The frost-push hypothesis is most applicable to stones close to the surface. Soil water flowing around the stone will tend to collect beneath ,it and on freezing the stone is pushed upwards. On melting finer particles collapse into the void and support the stone. There is experimental support for this process which is probably most effective when freezing is rapid; nevertheless the frost-pull mechanism is probably the more important under natural conditions.

12.2.1.3 *Mass displacement*

Mass displacement is the local transfer of material within the soil as a result of frost action. Vertical movements are most marked but horizontal movements also occur. A major cause of mass displacement is probably the cryostatic pressures developed in pockets of unfrozen soil or other sediments trapped between the permafrost table and the freezing front migrating downward from the surface. Localized pockets of unfrozen soil will be produced because the freezing front will advance downward at different rates depending on the grain size and moisture content of the soil. Some recent research, however, has questioned the role of a cryostatic pressure in producing mass displacement since differential heave resulting from the annual freezing and thawing of the ground may produce similar effects. Positive pore-water pressures towards the base of slopes might also generate mass displacement through the injection of water-saturated soil upward through a thin frozen surface layer. This mechanism has been widely cited as a cause of significant displacements within sediments in the active layer giving rise to structures consisting of interpenetrating layers of material which were originally flat-lying. Such forms are called **periglacial involutions**, the qualifying adjective being necessary since non-periglacial processes, such as slumping and other types of mass movement, can produce similar features.

12.2.1.4 *Frost cracking*

Frost cracking is the fracturing of the ground by thermal contraction at sub-freezing temperatures. It is probably the major cause of the polygonal fracture patterns so common in periglacial environments, although similar crack polygons can be formed by **desiccation cracking** and by **dilation cracking** resulting from differential heaving. **Frost crack polygons** are commonly 5–30 m across, crack spacing typically being two to three times crack depth. Although they are found outside the permafrost zone, they are best developed in areas of frozen ground rich in ice. The absence of a thick insulating snow cover appears to be a prerequisite for their development as crack frequency has been observed to diminish markedly with an increase in snow depth. Rate of temperature decrease appears to be more crucial in promoting cracking than the actual minimum temperature attained. Once cracking has been initiated ice films begin to fill the fracture and eventually ice wedges may form (see Section 12.3.2.1).

12.2.2   Chemical weathering

The absence of significant accumulations of chemically weathered rock in most periglacial environments, together with the abundance of largely unweathered angular rock fragments and bare rock surfaces, even in areas of low relief, supports the long-held view that mechanical weathering predominates in such regions. We would expect rates of chemical weathering to be low because of low temperatures, the retention of water as ice for a large proportion of the year and low levels of biotic activity. Nevertheless, there is a lack of detailed investigations directly comparing rates of chemical and mechanical weathering in the periglacial zone and further research is clearly required. One pioneer study carried out by A. Rapp over a 9 a period in northern Sweden in a periglacial environment (but not within the permafrost zone) suggested that material released by chemical weathering and removed in solution by streams in fact accounted for about a half of the total movement of material by all denudational processes (see Section 7.5). Moreover, the absence of thick weathering mantles in most periglacial environments may be partly accounted for by their removal from some areas by glacial erosion during the Pleistocene.

12.2.3   Mass movement

Most of the different types of mass movement are active in periglacial environments to some extent, but two mechanisms – frost creep and solifluction – are of particular significance (see Section 7.2.2). In periglacial environments solifluction frequently occurs in association with permafrost or seasonally frozen ground, and under these circumstances it is more specifically described as gelifluction. In reality it often operates in conjunction with frost creep to the extent that it is frequently difficult to distinguish between the action of the two mechanisms. Downslope movements must realistically be assessed in terms of their combined effect and they are often considered together in assessing rates of movement (Fig. 12.5).

Rates of movement for gelifluction alone are strongly dependent on moisture, and its importance in periglacial environments can be ascribed to two primary factors: saturation of the soil due to restricted drainage promoted by a permafrost layer or seasonally frozen water table, and moisture provided by thawing snow and ice. The overriding importance of soil moisture compared with the influence of vegetation and slope angle in controlling rates of gelifluction has been demonstrated by the observations of A. L. Washburn at an experimental site at Mesters Vig in northeast Greenland. Here the less steep but better vegetated slopes which tended to retain moisture experienced faster rates of movement than steeper but less well vegetated and drier slopes.

Free drainage in gravels and sands on the one hand, and the cohesive properties of clays on the other, means that soils dominated by silt-sized particles are most susceptible to gelifluction. Under ideal conditions of soil type and moisture content gelifluction can occur over gradients as low as 1°. Maximum rates apparently occur at slope angles of around 10–20° since efficient drainage on steeper slopes limits the moisture content of soils. When estimated separately, rates of movement associated with gelifluction and frost creep appear to be of the same order of magnitude. Maximum velocities are generally below 50–100 mm a^{-1}, but may exceptionally exceed 200 mm a^{-1}. Rates decline rapidly with depth and movement seems to be confined to within 1 m of the surface.

12.2.4   Nivation

Nivation is localized denudation by a combination of frost action, gelifluction, frost creep and meltwater flow in association with snow patches. It is most active in subarctic and alpine environments and its major geomorphic effect is the development of **nivation hollows** by the incision of snow patches into hillsides. Once initiated, normally in some existing slight depression, a nivation hollow increases in size as it becomes a collecting site for snow in subsequent years.

Snow provides both meltwater and acts as a ground insulation. The erosional potential of nivation is controlled primarily by snow thickness and the presence or absence of underlying permafrost. Where permafrost is absent an extensive area of freeze–thaw can occur both beneath the snow and around its margins. Where permafrost is present, thawing, and hence frost action, gelifluction and frost creep, are confined to the periphery of the snow patch. As the snow patch gradually thaws the zone of maximum nivation will follow its contracting margin. A thick snow cover insulates the surface and reduces nivation by preventing frequent freeze–thaw cycles. Data on rates of denudation by nivation are lacking, but it appears that features extending 500 m along slopes with a width of 1 m

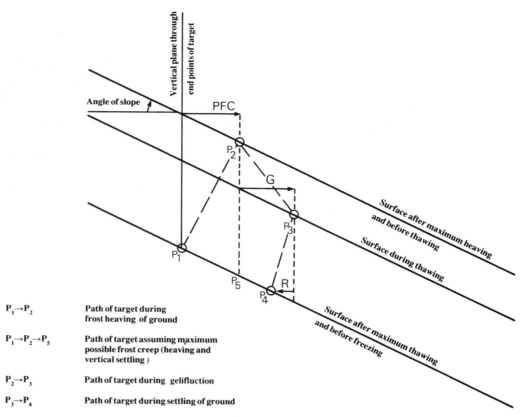

Fig. 12.5   *Operation of frost creep and gelifluction illustrated by the movements of targets inserted in regolith on a slope. PFC indicates the horizontal component of maximum possible frost creep, G is the horizontal component of downslope movement due to gelifluction and R is the apparent retrograde movement due to the tendency for the thawed layer to settle back against the slope rather than purely vertically. (After A. L. Washburn (1979)* Geocryology. *Edward Arnold, London, Fig. 6.8, p. 201.)*

may exceptionally be produced in one winter season in areas of unconsolidated rock. Exceptionally large nivation hollows recorded in Quebec, Canada, may have been excavated at a rate in excess of 1500 m³ a⁻¹.

12.2.5   Fluvial activity

Fluvial action has frequently been regarded as comparatively unimportant in periglacial environments, it being argued that its effectiveness is limited by the freezing of surface water for a large proportion of the year and by the low annual precipitation. Although the hydrological characteristics of stream flow are certainly modified by the long period of winter freezing, the dominance of fluvial erosion is reflected in the largely 'normal' morphometry of periglacial environments.

River regimes are highly seasonal with very large discharges being sustained for short periods during the spring melt (Fig. 12.6). Major lowland rivers may continue to flow throughout the winter under an ice cover, but at greatly reduced rates of flow. In the northward flowing rivers of arctic Canada and Siberia the thaw occurs first in the head-

water region and meltwater can be temporarily dammed behind vast ice accumulations further north before being released and causing widespread flooding.

Since a highly peaked river regime provides far more erosional and transporting potential than a more evenly distributed regime of equivalent total discharge, fluvial action in periglacial environments is far greater than the low annual precipitation and modest annual discharges would suggest. The limited field data available indicate that, during the brief period of high snow-melt discharge, quite small rivers can transport coarse debris and a high total sediment load. The River Mechan in arctic Canada, which lies well within the periglacial zone, provides a clear illustration. Mean annual precipitation over the basin amounts to only around 135 mm, half of which falls as snow. Some 80–90 per cent of the total annual flow is concentrated into a 10-day period during which peak flow velocities reach up to 4 m s⁻¹ and the whole river bed may be in motion.

Although rates of downcutting may be rapid (a rate of 1–3 m ka⁻¹ has been estimated for a river in Spitsbergen) most rivers in the periglacial zone, other than mountain streams, flow on aggraded beds. This may be a result of

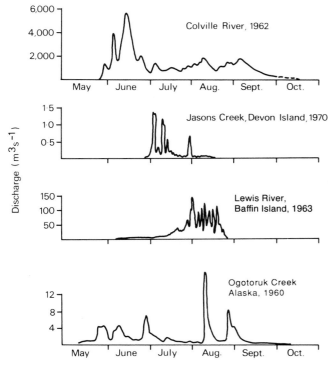

Fig. 12.6 *Runoff regimes of selected rivers in the periglacial zone of North America. (Modified from H. M. French (1976)* The Periglacial Environment. *Longman, London, Fig. 8.1, p. 169, based on M. Church (1974)* Permafrost Hydrology; Proceedings of Workshop Seminar. *Environment Canada, Ottawa 7–20.)*

current rates of weathering and mass movement supplying more debris than can be transported by existing rivers. Some of this material may, however, be a relic of higher rates of rock breakdown and sediment supply during the Pleistocene.

12.2.6 Aeolian activity

Present-day periglacial areas characterized by extreme aridity are favourable environments for aeolian activity, and there is abundant evidence for much more widespread and intense wind action in areas marginal to the great northern hemisphere ice sheets during the Pleistocene. In addition to strong winds and the freeze-drying of sediments, low precipitation and low temperatures are associated with minimal vegetation cover. Aeolian erosion is evident in faceted and grooved bedrock surfaces, deflation forms in unconsolidated fluvial sediments, and in the formation of ventifacts (see Section 10.2.2.1). Wind is also important in the movement and deposition of snow which, together with topographic features, determines areas of snow accumulation and removal and thereby the distribution of an insulating snow cover.

Of great geomorphic significance are various aeolian deposits, especially the loess which blankets much of those areas in North America and Eurasia which were located on the southern margin of the Pleistocene ice sheets. Loess is composed of angular to sub-angular particles and cohesion between grains allows the development of steep, high cliffs where thick blankets have been dissected by fluvial action. Although loess can form in warm deserts, most is believed to have originated in past or present periglacial environments. Glacial grinding, frost cracking and salt weathering appear capable of producing sufficiently fine particles which can be picked up from sites such as glacial outwash plains and carried considerable distances by the wind before being deposited as extensive loess blankets, exceptionally up to 100 m thick and covering tens of thousands of square kilometres (see Section 10.3.7).

During the glacial phases of the Pleistocene, the steep pressure gradient between the high pressure cells over the Laurentide and Fenno-Scandian ice sheets and low pressure to the south would have produced high mean wind velocities and conditions conducive to significant aeolian transport and deposition. Dating of discrete vertically stacked layers of loess interbedded with soils indicative of more temperate and humid conditions in China and Europe has indeed indicated a close correlation between loess deposition and phases of ice sheet extension (see Section 14.3.2).

12.3. Periglacial landforms

12.3.1 Patterned ground

First described more than a century ago, patterned ground is one of the most conspicuous features of periglacial environments. It can vary widely in scale from patterns composed of elements only a few centimetres across to those with dimensions of 100 m or more. Although some forms of patterned ground are found in other environments, especially deserts, the extent of its development is much greater in the periglacial zone than elsewhere since it is most characteristic of surfaces subject to intense frost action.

The most widely used descriptive classification of patterned ground has been developed by A. L. Washburn. This emphasizes both shape and the degree of sorting of the constituent materials. Five basic patterns are recognized – circles, nets, polygons, steps and stripes – although one form may grade into another. Each of these types is subdivided on the basis of whether the pattern results from the sorting of surface materials into fractions of large and small size or whether it has developed without sorting. Slope angle is an additional factor. Circles, nets and polygons are most common on flat surfaces, whereas steps and stripes tend to form on gradients of between 5 and 30°, with transitional forms developing on gentler slopes of between 2 and 5°. Mass movement is generally too active on slopes steeper than 30° to allow patterns to form. A summary of the characteristics and environmental associations of the various types of patterned ground is presented in Table 12.5.

Table 12.5   Characteristics of different forms of patterned ground

Circles	Occur singly or in groups. Typical dimensions 0.5–3 m. Nonsorted type characteristically rimmed by vegetation, sorted type bordered by stones which tend to increase in size with size of circle. Found in both polar and alpine environments but are not restricted to areas of permafrost. Unsorted circles also recorded from non-periglacial environments
Polygons	Occur in groups. Nonsorted polygons range from small features (<1 m across) to much larger forms up to 100 m or more in diameter. Sorted polygons attain maximum dimensions of only 10 m. Stones delimit polygon border and surround finer material. Nonsorted forms are delineated by furrows or cracks. Some types of polygon occur in hot desert environments but most are best developed in areas subject to frost. Ice-wedge polygons only form in the presence of permafrost
Nets	A transitional form between circles and polygons. Usually fairly small (<2 m across). Earth hummocks comprising a core of mineral soil surmounted by vegetation are a common form of unsorted net.
Steps	Found on relatively steep slopes. They develop either parallel to slope contours or become elongated downslope into lobate forms. In unsorted forms the rise of the step is well vegetated and the tread is bare. In the unsorted type the step is bordered by larger stones. Lobate forms are known as stone garlands. Neither type confined to permafrost environments
Stripes	Tend to form on steeper slopes than steps. Sorted stripes composed of alternating stripes of coarse and fine material elongated downslope. Nonsorted variety delineated by vegetation in slight troughs. Not confined to periglacial environments

Source: Based on discussion in A. L. Washburn (1979) *Geocryology*. Edward Arnold, London, pp. 122–56.

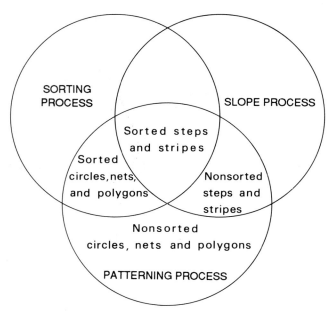

Fig. 12.7   *Relationship between patterning, sorting and slope processes in the formation of patterned ground. (After A. L. Washburn (1979)* Geocryology. *Edward Arnold, London, Fig. 5.41, p. 160.)*

Numerous factors contribute to the form of the great variety of patterned ground including the interactions between sorting, patterning and slope processes (Fig. 12.7). An important distinction is that between those forms of patterned ground in which initial cracking of the surface is essential and those in which it is not (Table 12.6). Causes of cracking include contraction on drying (desiccation cracking), stretching of the surface as a result of heaving (dilation cracking), and thermal contraction of seasonally or perennially frozen ground (frost cracking). Only the latter is restricted solely to periglacial environments. Desiccation and dilation cracking are important in the formation of small non-sorted polygons, whereas frost cracking is instrumental in the development of larger forms. The most important patterning processes which do not depend on initial cracking include differential frost heaving and mass displacement.

Frost heaving is apparently a primary cause of a number of sorted forms of patterned ground; it helps to segregate larger stones, which tend to move upward and outward, from the finer-grained matrix which forms a central core. Continued development at a number of discrete sites eventually leads to coalescence and the formation of polygonal rims formed primarily of stones. Sedimentation and mass

displacement contribute to the development of non-sorted circles while surface wash is probably a factor in the formation of sorted stripes.

The regularity of many forms of patterned ground has suggested to some researchers that convection occurring when frozen ground thaws may be an important mechanism in its development. The idea is that water-saturated soil close to the surface warms up while the temperature of that adjacent to the still frozen subsoil remains close to freezing point. Convection in this situation is possible because water is densest at 4 °C, so water at the thawing front is less dense than the slightly warmer water close to the surface. When the active layer thaws the cooler water at the thawing front rises while descending plumes of warmer water lead to localized melting of the underlying ice. This produces a regular undulating interface between frozen and unfrozen soil which, it is thought, can be reflected in the surface topography. The exact mechanism by which this occurs is not clear, but it probably involves a range of processes including frost heave. Striped forms of patterned ground according to this model are explained by the downslope elongation of convection cells within the soil.

12.3.2   Ground-ice phenomena

12.3.2.1   *Ice wedges*
Ice wedges are downward-tapering bodies of ice (Fig. 12.8). They can be over 1 m across at the top and reach a depth of 10 m or more. They have a polygonal form in plan,

Table 12.6 Processes contributing to various types of patterned ground

| | PATTERNED GROUND MORPHOLOGY | | | | | | | | | |
| | Circles | | Polygons | | Nets | | Steps | | Stripes | |
FORMATIVE PROCESSES	N	S	N	S	N	S	N	S	N	S
Cracking essential										
Desiccation cracking			*	*	*	*			*	*
Dilation cracking			*	*	*	*			*?	*
Salt cracking			*	*						
Seasonal frost cracking			*	*	*	*			*?	*?
Permafrost cracking			*	*	*	*?			*	*
Frost action along joints		*	*	*					*?	*
Cracking not essential										
Primary frost sorting		*		*?		*		*?		*?
Mass displacement	*	*	*?	*?	*	*	*	*	*	*
Differential frost heaving	*	*	*?	*?	*	*	*?	*	*	*
Salt heaving	*	*	*?	*?	*	*	*?	*	*	*
Differential thawing and eluviation			*?			*		*?		*?
Differential mass movement							*	*	*?	*
Rillwork									*?	*

N and S indicate, respectively, nonsorted and sorted types of patterned ground.
Source: Modified from A. L. Washburn (1979) *Geocryology*. Edward Arnold, London, Table 5.1, p. 158.

Fig. 12.8 Ice wedge in the right bank of the Aldan River, 170 km upstream of its junction with the Lena River, Yakutia, USSR. (Photo courtesy A. L. Washburn.)

and **ice-wedge polygons** are associated with the most impressive form of patterned ground in periglacial environments. Frost cracking appears to be the key mechanism necessary for the development of ice wedges. Extreme cooling of ice-rich permafrost in winter leads to thermal contraction and cracking of the surface (Fig. 12.9). As the wedge grows it compresses the surrounding sediments which tend to bulge up around the perimeter of each polygon forming a low ridge. In other cases thawing and erosion of the ice wedge produces a perimeter trough. In dry permafrost, where meltwater is scarce, the cracks may be filled with loess or other coarser sediments to form **sand wedges**.

Recent field studies have revealed that some ice-wedge cracks originate near the top of the permafrost and then propagate both upwards to the surface and downwards into the ice wedge. Only a proportion of wedges tend to crack in any one year, and the frequency of cracking seems to vary inversely with the depth of snow cover. The average amount of ice accreted each year varies enormously so the total volume of ice present is not necessarily a good indicator of how long an ice wedge has been growing. Observations made of the development of ice wedges after the artificial draining of a lake in northern Canada have demonstrated that under ideal conditions they can grow rapidly and reach a significant size after just a few years.

Ice-wedge casts form when an ice wedge is replaced by sediments as a result of the long-term thawing of the permafrost. The form of the ice wedge is preserved by sediment and may survive long after the permafrost in which the original ice wedge developed has completely thawed. Ice-wedge casts are consequently an important indicator of the previous existence of permafrost. Their palaeoenvironmental significance is also enhanced by the fact that ice wedges only seem to form in present-day permafrost environments where the mean annual temperature is below –6 to –8 °C.

12.3.2.2 Pingos

Pingos, named after the Eskimo word for a hill, are large perennial ice-cored mounds. They are roughly circular to elliptical in plan and range in size from 3 to 70 m high and from 30 to 600 m in diameter (Fig. 12.10). Large dilation cracks generated by the progressive growth of the ice core

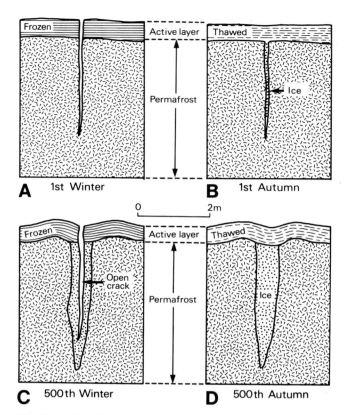

Fig. 12.9 *Hypothesized development of an ice wedge initiated by frost cracking due to thermal contraction. The initial crack, which may be only a few millimetres wide, collects snow and meltwater from thawing of the active layer (A). This water freezes below the permafrost table to form a thin ice vein (B). The initial fracture represents a zone of weakness so will be a favoured site for frost cracking in subsequent years. However, cracking does not necessarily occur each year and expansion during summer heating will tend to close cracks developed during the preceding winter. Further ingress of snow and meltwater adds new ice until an ice wedge is eventually formed (C and D). (After A. H. Lachenbruch (1962) Geological Society of America Special Paper 70, Fig. 1, p. 5.)*

commonly mark the summit. These may widen sufficiently to expose the ice core and initiate thawing, thereby giving rise to a **collapsed pingo**. Various materials can envelop the ice core, including clays, silts, gravels and even bedrock. Drilling has shown that in some cases the core of clear ice extends to depths greater that the pingo height. Vertical growth rates of pingos may reach $0.2 \, \mathrm{m \, a^{-1}}$ but they tend to decline rapidly as the pingo grows. Large pingos certainly take thousands of years to develop, as indicated by the radiocarbon dating of two pingos in the Canadian Arctic which gave approximate ages of 4500 and 7000 a.

The ice core forming a pingo is thought to grow through the freezing of water forced upwards at pressure. Two mechanisms have been proposed to explain the generation of this pressure. In **closed system pingos** cryostatic pressure is involved. This may be brought about in two related ways. If

Fig. 12.10 *Pingo, Wollaston Peninsula, Northwest Territories, Canada. (Photo courtesy A. L. Washburn.)*

a lake in a permafrost zone is infilled by sediment and vegetation, insulation of the underlying ground will be reduced (Fig. 12.11) and freezing will encroach from the base, sides and top. This will trap a body of water which on freezing expands and domes the overlying sediments. Alternatively, diversion of a river or draining of a lake may have a similar effect by reducing ground insulation. Support for this model of cryostatic pressure comes from the distribution of pingos in the Mackenzie Delta area of the Canadian Arctic where 98 per cent of the 1380 pingos recorded are located in, or closely adjacent to, lake basins.

The **open system pingo** model involves the freezing of ground water flowing under hydrostatic pressure as it forces its way towards the surface from below a thin permafrost layer (Fig. 12.12). It has been calculated that very high pressures are required to dome pingos and that these exceed those commonly attained within unconfined ground water. Consequently the open system type may in fact form under temporary closed-system conditions as open taliks are frozen in winter. Open system pingos extend into the zone of discontinuous permafrost where mean annual temperature may be only a little below freezing since here water circulation is freer. They tend to occur in clusters in favourable areas of drainage where the hydrostatic pressure is more constant. In contrast, the closed system type usually occur as isolated features on flat surfaces related to lakes and are confined to the zone of continuous permafrost where minimum mean annual ground temperatures are around $-5 \, °\mathrm{C}$.

12.3.2.3 Palsas

Palsas are mounds or more elongated forms occurring in bogs and containing perennial ice lenses. They differ from pingos in containing peat as a major constituent and a core composed of discrete superimposed ice lenses rather than a

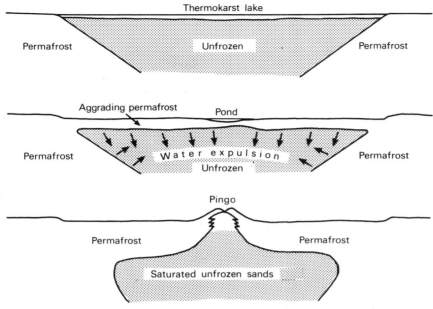

Fig. 12.11 *Schematic representation of the development of a closed system pingo following the infilling of a lake. (Modified from A. L. Washburn, (1979)* Geocryology. *Edward Arnold, London, Fig. 5.49, p. 183.)*

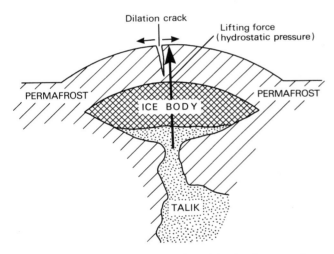

Fig. 12.12 *Schematic representation of the development of an open system pingo. (Modified from F. Muller (1968) in: R. W. Fairbridge (ed.)* Encyclopedia of Geomorphology. *Reinhold, New York, Fig. 2, p. 846.)*

single ice mass. The mode of ice formation is uncertain but probably originates from differential frost heaving with peat playing a crucial insulating role. Palsas are commonly found in areas of discontinuous permafrost.

12.3.2.4 Thermokarst

Thermokarst is a term used to encompass a variety of topographic depressions formed by the thawing of ground ice. It is so called because of the superficial resemblance of the landforms produced to those characteristic of true karst terrain. In addition to the collapse of material into the space previously occupied by ice, thermokarst features may be significantly affected by flowing water released by thawing. High, short-term rates of erosion can occur, especially where ice masses are exposed along cliffs or stream banks. Although some thermokarst features can be attributed to climatic change, many others form as a result of the 'normal' variability of the periglacial environment. Thermokarst most commonly originates from a modification of surface conditions and this can be brought about in a number of possible ways including disturbance of vegetation, cliff retreat and changes in stream courses.

Thaw lakes are perhaps the most ubiquitous thermokarst

Fig. 12.13 *Small thaw lakes, Bathurst Island, Northwest Territories, Canada. (Photo courtesy Geological Survey of Canada.)*

form, and arguably one of the most characteristic landforms of periglacial environments (Fig. 12.13). They are bodies of water which fill small, shallow depressions. The majority are less than 5 m deep, and they are rarely more than 2 km across. The origin of thaw lakes seems to be related to the melting of permafrost which contains a volume of ice that exceeds the normal pore space of the sediment. Subsidence occurs and shallow depressions are formed which fill with water. They are particularly common in poorly drained lowland regions where ground ice is abundant (Fig. 12.14). They are usually filled quite rapidly by sediments and peat, and have a limited life span of only a few thousand years. In some localities, such as in the northern coastal plain of Alaska, thaw lakes are somewhat elongated with length–width ratios of 2 : 1 or 3 : 1 and tend to be orientated in a

specific direction. Their origin is uncertain, but they are probably related to winds prevailing from a particular direction which preferentially deposit sediments along lake banks normal to the wind direction. These deposits insulate the banks from further thawing and protect them from wave erosion.

Thaw slumps are arcuate embayments facing downslope and formed by the exposure and thawing of ground ice. Basal undercutting of river banks or mass movement are sufficient to expose ground ice and initiate the process. Debris saturated by the meltwater produced moves downslope as a mud flow or by gelifluction. The headwalls of thaw slumps may reach 8 m in height and retreat at rates of over 7 m a^{-1}, making this a significant denudational process in some periglacial environments. **Thermocirques** are

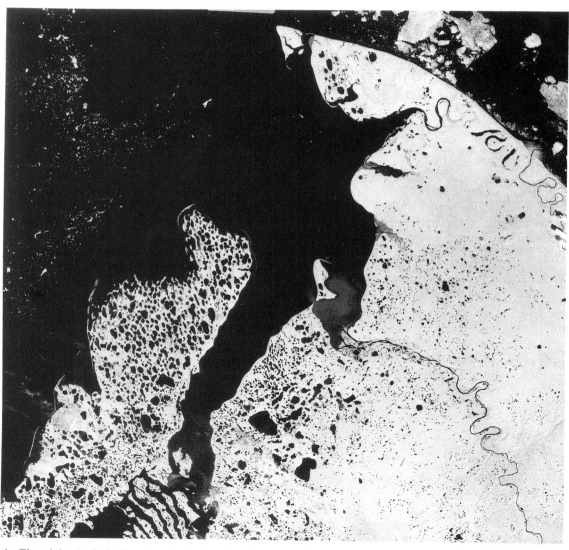

Fig. 12.14 *Thaw lakes in the Tuktoyaktuk Peninsula which lies within the continuous permafrost zone of the Northwest Territories, Canada. The area shown is about 180 km across. (Landsat image courtesy R. S. Williams Jr)*

large-scale variants of thaw slumps formed when retreating slopes intersect ice wedges. The surface thawing of ice-wedge polygons can produce linear and polygonal troughs surrounding a central mound.

Alases are major thermokarst depressions from 3 to 40 m deep and 100 m to 15 km across. They are compound features resulting from climatic change or a widespread disturbance of the surface such as that arising from a forest fire. These environmental perturbations can lead to permafrost degradation which initiates a sequence of surface collapse, lake formation and pingo development (Fig. 12.15). Eventual coalescence of individual alases gives rise to **alas valleys** which may be tens of kilometres long. Alas formation has affected large areas in central Yakutia in eastern Siberia where conditions are particularly favourable. It is estimated that in this area 40–50 per cent of the Pleistocene surface has been subjected to alas development, which apparently

occurred predominantly during a warm phase of the Holocene between 9000 and 2500 a BP.

12.3.3 Depositional forms related to mass movement

The downslope movement of debris in association with periglacial mass movement processes gives rise to a range of locally significant landforms. Deposits associated with gelifluction assume a variety of forms, including sheets, benches and lobes. **Gelifluction sheets** have a smooth, gently sloping surfaces and a bench-like lower margin, whereas **gelifluction benches** are terrace forms. Particularly characteristic are **gelifluction lobes** which consist of tongue-shaped forms, 30–50 m across, which are elongated downslope (Fig. 12.16). They tend to occur on steeper slopes (10–20°) than benches. Where the downslope elongation is very marked the term **gelifluction stream** is used. Both benches and lobes have steep fronts, 1–6 m high, which may be subject to erosion. They appear to form as stones, which have been pushed downslope, accumulate and dam the movement of material above.

Some gelifluction deposits are crudely stratified and most exhibit a preferred downslope orientation of the long axis of their angular constituent particles, which may range in size from boulders down to silt-sized material. Relict gelifluction deposits are difficult to distinguish from other solifluction debris, although they should be less weathered than similar deposits from non-periglacial environments.

Block slopes are slopes mantled with angular boulders, mostly between 1 and 3 m across, covering 50 per cent or more of the ground surface. In the case of **block streams** the boulders are concentrated in valley bottoms or occur as

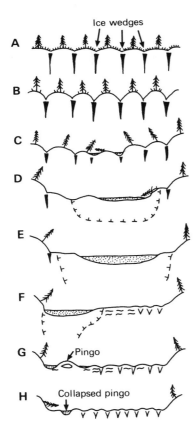

Fig. 12.15 *Schematic representation of the sequence of development and decay of alases: (A) original lowland surface with ice wedges; (B) effect of increase in depth of active layer promoted by climatic warming or ground disturbance; (C) initial thermokarst stage; (D) young alas stage; (E) mature alas stage; (F) old alas stage; (G) phase of possible pingo formation; (H) development of collapsed pingos. (After P. J. Williams (1979)* Pipelines and Permafrost. *Longman, London, Fig. 2.2, p. 21, modified from T. Czudek and J. Demek (1970)* Quaternary Research *1, Fig. 9, p.111.)*

Fig. 12.16 *Gelifluction lobe, Hesteskoen, Mesters Vig, north-east Greenland. (Photo courtesy A. L. Washburn.)*

narrow downslope linear deposits on hillsides. Downslope movement is indicated for both forms by the orientation of the long axes of the boulders which tends to be roughly normal to the slope contours. Gelifluction and frost creep are the favoured transporting mechanisms for most occurrences. **Block fields**, consisting of relatively flat boulder-covered surfaces, are also common. They probably form largely by *in situ* frost shattering and the heaving of jointed bedrock and involve no significant downslope movement.

Rock glaciers are tongue-shaped masses of angular boulders resembling in form a small glacier. They have a steep front, exceptionally exceeding 100 m in height, which stands at the angle of repose of the constituent material. Rock glaciers usually descend from cirques or cliff-faces and may reach a length of 1 km or more. Large boulders and smaller stones typically form the surface which commonly exhibits arcuate ridges and lobes. Active rock glaciers contain ice at depth, either filling voids (ice-cemented type) or forming a core (ice-cored type). Some of the ice-cored variety are probably derived from debris-covered glaciers, whereas others may be transitional to ice-cored moraines. In localities where rock glaciers and ice glaciers are both currently active the distinction between them is essentially one of the quantity of debris carried. Where debris-rich glacier ice becomes stagnant sufficient ice may remain mixed with fine sediment at depth to sustain slow movement, largely by internal deformation and basal shear. Frost creep and gelifluction may also contribute to downslope movement. Irrespective of the actual mechanism of movement, rock glaciers are important transportational agents in highland periglacial environments.

12.3.4 Asymmetric valleys

Asymmetric valleys are valleys with one slope significantly steeper than the other. Virtually all valleys are asymmetric to a certain extent, but some display a degree of asymmetry sufficiently marked to warrant some special explanation. Asymmetric valleys are known from a wide range of environments and can clearly be caused by a number of factors, but geomorphologists working in the periglacial zone have long recognized the prevalence of asymmetric valleys there. Collection of field data has indicated no uniformity in the preferred orientation of the steeper slope in periglacial asymmetric valleys (Table 12.7), although it has been suggested that steeper north-facing slopes predominate in the northern hemisphere below a latitude of about 70 °N.

Various conditions existing in periglacial environments may contribute to valley asymmetry. In the northern hemisphere longer exposure to the Sun on south-facing slopes will result in a more rapid and prolonged thawing and a greater abundance of meltwater, promoting gelifluction and other mass movement processes. Consequently, south-facing slopes will experience a more active reduction of

Table 12.7 Nature of slope asymmetry in the periglacial zone of the northern hemisphere

LOCALITY	ORIENTATION (ASPECT) OF STEEPER SLOPE	VALLEY ALIGNMENT
Central Alaska	N	E–W
North-west Alaska	N	E–W
Southampton Island, Canada	N	E–W
Banks Island, Canada	SW	NW–SE
Caribou Hills, Canada	N, S	E–W
Yakutia, Siberia (3 studies)	N	E–W
Andreeland, west Spitsbergen	S	E–W
Conwayland, west Spitsbergen	W	N–S
Kaffioya-Ebene, west Spitsbergen	S	E–W
Wollaston-Vorland, east Greenland	N	E–W

Source: Modified from H. M. French (1976), *The Periglacial Environment*, Longman, London, Table 8.3 p. 179.

slope angle, while their colluvium will tend to divert streams towards opposing north-facing slopes which will be subject to undercutting and slope steepening. Moreover, as north-facing slopes will be in shade longer during the summer thaw season, permafrost, where present, will be preferentially developed and will help to stabilize unconsolidated sediments at a steeper angle. Prevailing winds may play a role since snow will tend to accumulate on lee slopes and the greater quantity of meltwater produced on thawing will augment mass movement on such slopes. Differences in vegetation cover may also be important. In a study in Alaska, for instance, vegetation was found to be least developed on the colder north-facing slopes which consequently experienced less active mass movement.

Valley asymmetry has also been identified in areas subject to periglacial conditions during the Pleistocene. Although these have commonly been attributed to some of the mechanisms just discussed relevant to periglacial environments, it is difficult to separate any inherited periglacial effects from other processes capable of producing asymmetry. One important factor is the higher angle of inclination of the Sun in mid-latitude Pleistocene periglacial environments in contrast with the present-day high latitude periglacial zone.

12.3.5 Cryoplanation terraces and cryopediments

Cryoplanation terraces (also known as **altiplanation terraces**) are level or gently sloping surfaces found in the periglacial zone which are cut into bedrock on hill summits or upper hillslopes. **Cryopediments** are a similar form developed at the foot of valley sides or on marginal slopes. Unless they transect structure it may be difficult to distinguish them from structurally controlled benches, and in any case lithological and structural factors are important in their formation. Cryoplanation terraces range from 10 m to 2 km across and up to 10 km in length. The risers between terraces may reach a height in excess of 70 m and stand at an angle of 30° or more where debris-covered, and up to 90° where bedrock is exposed. The terraces are mantled by gelifluction debris and may contain patterned ground.

Cryoplanation terraces are thought to be formed by several processes working in conjunction. In essence they appear to form by the combined effects of the break-up of bedrock by frost action and scarp recession. Various stages are involved (Fig. 12.17). According to one model, development begins with the formation of a nivation hollow or bench associated with snow patches. Nivation then proceeds to erode a cliff which recedes as debris is transported away from the cliff base, largely by gelifluction and slope wash. Continued cliff recession on either side of an interfluve finally forms a summit terrace with residual rock masses. Some researchers, however, believe that most descriptions of cryoplanation terraces are of relict forms, and that in many cases the features described may be no more than benches related to the differential erosion of contrasting lithologies.

Cryopediments probably develop in a similar way except that slope wash may be more active than gelifluction in carrying debris away. Although many cryoplanation terraces and cryopediments are associated with present-day permafrost, frozen ground does not seem to be a prerequisite for their formation.

Further reading

Students of periglacial geomorphology have no shortage of excellent texts which they can consult. Washburn (1979) provides a fully illustrated and referenced survey of processes and landforms; a considerable strength of this text is the extensive reference made to the work of Russian and East European scientists who have made very important contributions to periglacial studies. Other general sources include Embleton and King (1975), French (1976) and Williams and Smith (1989). The volumes edited by Church and Slaymaker (1985) and Clark (1988) provide detailed up-to-date treatments of many aspects of periglacial geomorphology, while a number of useful papers are contained in *Zeitschrift für Geomorphologie Supplementband*, 71. Various regional studies are included in Boardman (1987), although here the emphasis is on relict periglacial forms in the British Isles.

The problems involved in the human utilization of high latitude environments have received much attention. Although these are not specifically considered here, discussions of various aspects of applied periglacial geomorphology are to be found in Brown (1970), Harris (1986), Sugden (1982) and Williams (1979). Journals containing articles on periglacial geomorphology include *Biuletyn Peryglacjalny* (a Polish publication but with many articles in English), *Canadian Journal of Earth Sciences, Quaternary Research* and *Arctic and Alpine Research*, in addition to other journals which publish papers on geomorphology generally. Useful annual reviews of periglacial geomorphology are to be found in *Progress in Physical Geography*.

Harris (1986) provides a detailed discussion of the nature and formation of permafrost, while the pioneer work by Taber (1929, 1930, 1943) on frost action is still worth consulting. Thorn (1979) presents some intriguing data on the occurrence of freeze–thaw cycles and the efficacy of frost shattering in the alpine environment of the Colorado Rockies, and White (1976) compares the role of frost shattering and hydration shattering. Both field and laboratory investigations of frost shattering are reviewed by McGreevy (1981), and Pavlik (1980) provides a brief discussion of the factors controlling ground ice formation. Salt weathering in relation to tor formation is considered by Selby (1972).

Mass movement processes are covered by Harris (1987) and Rapp (1986), and also in the more general texts, especially Washburn (1979). Benedict (1976) gives a useful review of frost creep and gelifluction and also provides a detailed discussion on mass movement mechanisms in alpine environments (Benedict, 1970). Fluvial processes in specific periglacial environments are discussed in Church (1972)

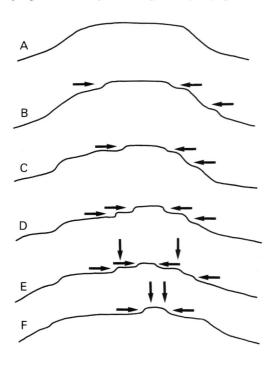

Fig. 12.17 *Stages in the development of cryoplanation terraces in resistant rock: (A) original surface; (B) formation of nivation hollows; (C) initial development of cryoplanation terrace; (D) mature stage of cryoplanation terrace development; (E) initial formation of cryoplanation surface; (F) development of cryoplanation summit surfaces. Arrows indicate the direction of surface modification. (Based on model of J. Demek (1969)* Biuletyn Peryglacjalny *18, 115–25, after H. M. French (1976)* The Periglacial Environment. Longman, London, Fig. 7.11, p. 160.)*

and McCann *et al.* (1972), while Scott (1979) provides a comprehensive annotated bibliography on this topic. Embleton and King (1975) and Thorn and Hall (1980) discuss nivation, while Embleton (1979) and Thorn (1978) consider the action of snow in the landscape.

The classification and origin of patterned ground is covered in the seminal papers by Washburn (1956, 1970), but an extended discussion is also to be found in Washburn's general text. Nicholson (1976) provides some additional material on the characteristics and formation of patterned ground, while Goldthwait (1976) concentrates on the sorting processes. The possible role of convection in the formation of certain types of patterned ground is discussed by Gleason *et al.* (1986) and Krantz *et al.* (1988). The mechanism of frost cracking and the development of ice wedge polygons is considered in the important paper by Lachenbruch (1962), and more recent ideas are discussed by Hamilton *et al.* (1983), Mackay (1974, 1984, 1986) and Mackay and Matthews (1983). The formation of pingos is reviewed in Mackay (1978, 1987) and a detailed field investigation is reported by Mackay (1979).

Thermokarst is assessed in Czudek and Demek (1970) with an emphasis on Siberia, and French (1974) considers contemporary thermokarst development in Arctic Canada. Martin and Whalley (1987) provide a useful review of rock glaciers, while a detailed treatment is to be found in the volume edited by Giardino *et al.* (1987). French (1976) provides a useful discussion of valley asymmetry in both present and past periglacial environments, while Kennedy (1976) takes a broader look at the general factors controlling valley asymmetry. A model of cryoplanation terrace development is presented in Demek (1969).

References

Benedict, J. B. (1970) Downslope soil movement in a Colorado alpine region: rates, processes, and climatic significance. *Arctic and Alpine Research* **2**, 165–226.

Benedict, J. B. (1976) Frost creep and gelifluction: a review. *Quaternary Research* **6**, 55–76.

Boardman, J. (ed.) (1987) *Periglacial Processes and Landforms in Britain and Ireland*. Cambridge University Press, Cambridge.

Brown, R. J. E. (1970) *Permafrost in Canada*. University of Toronto Press, Toronto.

Church, M. (1972) Baffin Island sandurs; a study of Arctic fluvial processes. *Geological Survey of Canada Bulletin* **216**.

Church, M. and Slaymaker, O. (eds) (1985) *Field and Theory: Lectures in Geocryology*. University of British Columbia Press, Vancouver.

Clark, M. J. (ed.) (1988) *Advances in Periglacial Geomorphology*. Wiley, Chichester and New York.

Czudek, T. and Demek, J. (1970) Thermokarst in Siberia and its influence on the development of lowland relief. *Quaternary Research* **1**, 103–20.

Demek, J. (1969) Cryogene processes and the development of cryoplanation terraces. *Biuletyn Peryglacjalny* **18**, 115–25.

Embleton, C. (1979) Nival processes. In: C. Embleton and J. Thornes (eds) *Process in Geomorphology*. Edward Arnold, London, 307–24.

Embleton, C. and King, C. A. M. (1975) *Periglacial Geomorphology* Edward Arnold, London.

French, H. M. (1974) Active thermokarst processes, eastern Banks Island, Western Canadian Arctic. *Canadian Journal of Earth Sciences* **11**, 785–94.

French, H. M. (1976) *The Periglacial Environment*. Longman, London.

Giardino, J. R., Shroder, J. F. Jr and Vitek, J. D. (eds) (1987) *Rock Glaciers* Unwin Hyman, London and Boston.

Gleason, K. J., Krantz, W. B., Caine, N., George, J. H. and Gunn, R. D. (1986) Geometrical aspects of sorted patterned ground in recurrently frozen soil. *Science* **232**, 216–20.

Goldthwait, R. P. (1976) Frost sorted patterned ground: a review. *Quaterny Research* **6**, 27–35.

Hamilton, T. D., Ager, T. A. and Robinson, S. W. (1983) Late Holocene ice wedges near Fairbanks, Alaska, U.S.A.: environmental setting and history of growth. *Arctic and Alpine Research* **15**, 157–68.

Harris, C. (1987) Mechanisms of mass movement in periglacial environments. In: M. G. Anderson and K. S. Richards (eds) *Slope Stability: Geotechnical Engineering and Geomorphology*. Wiley, Chichester and New York, 531–59.

Harris, S. A. (1986) *The Permafrost Environment*. Croom Helm, London.

Kennedy, B. A. (1976) Valley-side slopes and climate. In: E. Derbyshire (ed.) *Geomorphology and Climate*. Wiley, London and New York, 171–201.

Krantz, W. B., Gleason, K. J. and Caine, N. (1988) Patterned ground. *Scientific American* **259**(6), 44–50.

Lachenbruch, A. H. (1962) Mechanics of thermal contraction cracks and ice-wedge polygons in permafrost. *Geological Society of America Special Paper* **70**.

Mackay, J. R. (1974) Ice-wedge cracks, Garry Island, Northwest Territories. *Canadian Journal of Earth Sciences* **11**, 1336–83.

Mackay, J. R. (1978) Contemporary pingos: a discussion. *Biuletyn Peryglacjalny* **27**, 133–54.

Mackay, J. R. (1979) Pingos of the Tuktoyaktuk Peninsula area, Northwest Territories. *Géographie Physique et Quaternaire* **33**, 3–61.

Mackay, J. R. (1984) The direction of ice-wedge cracking in permafrost: downward or upward? *Canadian Journal of Earth Sciences* **21**, 516–24.

Mackay, J. R. (1986) The first 7 years (1978–1985) of ice wedge growth, Illisarvik experimental drained lake site, western Arctic coast. *Canadian Journal of Earth Sciences* **23**, 1782–95.

Mackay, J. R. (1987) Some mechanical aspects of pingo growth and failure, western Arctic coast, Canada. *Canadian Journal of Earth Sciences* **24**, 1108–19.

Mackay, J. R. and Matthews, J. V. Jr (1983) Pleistocene ice and sand wedges, Hooper Island, Northwest Terrritories. *Canadian Journal of Earth Sciences* **20**, 1087–97.

Martin, H. E. and Whalley, W. B. (1987) Rock glaciers part 1: rock glacier morphology: classification and distribution. *Progress in Physical Geography* **11**, 260–82.

McCann, S. B., Howarth, P. J. and Cogley, J. G. (1972) Fluvial processes in a periglacial environment: Queen Elizabeth Islands, N.W.T., Canada. *Transactions of the Institute of British Geographers* **55**, 69–82.

McGreevy, J. P. (1981) Some perspectives on frost shattering. *Progress in Physical Geography* **5**, 56–75.

Nicholson, F. H. (1976) Patterned ground formation and description as suggested by Low Arctic and Subarctic examples. *Arctic and Alpine Research* **8**, 329–42.

Pavlik, H. F. (1980) A physical framework for describing the genesis of ground ice. *Progress in Physical Geography* **4**, 531–48.

Rapp, A. (1986) Slope processes in high latitude mountains. *Progress in Physical Geography* **10**, 53–68.

Scott, K. M. (1979) Arctic stream processes – an annotated bibliography. *United States Geological Survey Water Supply Paper* **2065**.

Selby, M. J. (1972) Antarctic tors. *Zeitschrift für Geomorphologie Supplementband* **13**, 73–86.

Sugden, D. (1982) *Arctic and Antarctic: A Modern Geographical Synthesis*. Blackwell, Oxford.

Taber, S. (1929) Frost heaving. *Journal of Geology* **37**, 428–61.

Taber, S. (1930) The mechanics of frost heaving. *Journal of Geology* **38**, 303–17.

Taber, S. (1943) Perennially frozen ground in Alaska: its origin and history. *Bulletin of the Geological Society of America* **54**, 1433–548.

Thorn, C. E. (1978) The geomorphic role of snow. *Annals of the Association of American Geographers* **68**, 414–25.

Thorn, C. E. (1979) Bedrock freeze–thaw weathering regime in an alpine environment, Colorado Front Range. *Earth Surface Processes* **4**, 211–28.

Thorn, C. E. and Hall, K. (1980) Nivation: an arctic–alpine comparison and reappraisal. *Journal of Glaciology* **25**, 109–24.

Washburn, A. L. (1956) Classification of patterned ground and review of suggested origins. *Bulletin of the Geological Society of America* **67**, 823–65.

Washburn, A. L. (1970) An approach to a genetic classification of patterned ground. *Acta Geographica Lodziensia* **24**, 437–46.

Washburn, A. L. (1979) *Geocryology: A Survey of Periglacial Processes and Environments*. Edward Arnold, London.

White, S. E. (1976) Is frost action really only hydration shattering? *Arctic and Alpine Research* **8**, 1–6.

Williams, P. J. (1979) *Pipelines and Permafrost: Physical Geography and Development in the Circumpolar North*. Longman, New York.

Williams, P. J. and Smith, M. W. (1989) *The Frozen Earth: Fundamentals of Geocryology*. Cambridge University Press, Cambridge and New York.

13

Coastal processes and landforms

13.1 The coastal environment

So far in the preceding chapters we have discussed those processes active in the sub-aerial environment and the landforms to which they give rise. None the less, around 70 per cent of the Earth's surface is covered by water, and the coast, which forms the interface between land and water, experiences a particular array of geomorphic processes and a range of characteristic landforms. Waves and tides involve the movement and dissipation of large amounts of energy and this is capable of causing rapid and spectacular changes in landforms along coastlines.

The coastal environment, or **littoral zone**, can be defined as that area lying at the interface between the land and the sea (or other large water body). It includes both the zone of shallow water within which waves are able to move sediment, and the area landward of this zone, including beaches, cliffs and coastal dunes, which is affected to some degree by the direct or indirect effects of waves, tides and currents. While the terms shoreline or littoral zone refer to the area of frequent, or at least occasional, wave action along the edge of the sea or a lake, the coastal environment itself may extend inland for many kilometres. Figure 13.1 illustrates some of the major components of the coastal environments for both sandy and rocky coasts.

As in other geomorphic environments, coasts are composed of a wide range of landforms, only some of which are in equilibrium, or quasi-equilibrium, with prevailing processes. The Holocene has seen great changes in the position of shorelines and in coastal morphology as a result of predominantly rising sea levels, and there has been a continuous succession of adjustments to these rapidly changing conditions (see Chapter 17 for a detailed discussion of sea-level change). While beaches composed of unconsolidated and readily transportable sand can adjust quickly to changes in the nature and energy of coastal processes, other landforms, either because of their resistance to erosion, or just

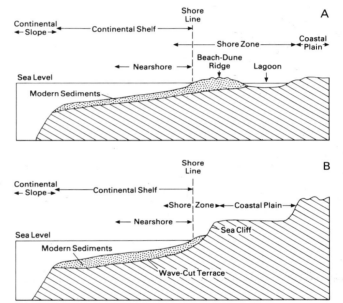

Fig. 13.1 *Schematic representation of some of the important elements of coastal morphology: (A) low elevation coast, and (B) coast backed by cliffs. (After D. L. Inman and C. E. Nordstrom, (1971)* Journal of Geology *79, Fig. 3, p. 6.)*

sheer size, are modified much more slowly. Great care must be taken in relating contemporary forms to current processes especially since important agents of coastal modification, such as storm waves, may change in their frequency of attack over short spans of time.

Many attempts have been made to categorize coastal landforms, either in order to provide a descriptive framework for subsequent investigations or to create a genetic classification which can be applied to the interpretation of coastal morphology. Most such classifications have focused on the consequences of relative changes in sea level, categorizing coasts as either emergent or submergent. Although such classificatory schemes have their uses, the different rates of adjustment of different coastal landforms to changes in sea level, as well as other factors affecting the development of coastal features, make them difficult to apply in many cases. Nevertheless, a growing appreciation of the nature of the global variation of a number of variables affecting coastal development does allow some valuable generalizations to be made about the varying effectiveness of different processes around the world's coasts. Some of this variation is evident in the world maps included in this chapter. Moreover, in keeping with the links between exogenic and endogenic processes which form an important element of this book, it is appropriate here to anticipate the theme of Part IV and illustrate the overall control that tectonics exerts on the broad nature of coastal topography.

It is possible to classify coasts on the basis of their tectonic setting (Fig. 13.2). **Convergent margin coasts**, such as the west coast of South America, lie along convergent plate boundaries, whereas **passive margin coasts**, as their name implies, are located along passive continental margins. Several varieties of passive margin coast can be identified. Young passive margins, such as those flanking the Red Sea, give rise to **nascent passive margin coasts**, and these evolve into **mature passive margin coasts** as the continental margin moves away from the spreading ridge, cools, and subsides. Passive margin coasts lying on the opposite side of a continent from a convergent margin can be described as **passive margin–distal orogen coasts**. Finally, passive continental margins protected by island arc systems offshore are termed **marginal sea coasts**.

There is a fairly clear relationship between the tectonic nature of coastal margins and the width of the continental shelf offshore with, for example, a strong predominance of narrow shelves along convergent margin coasts (Table 13.1). These

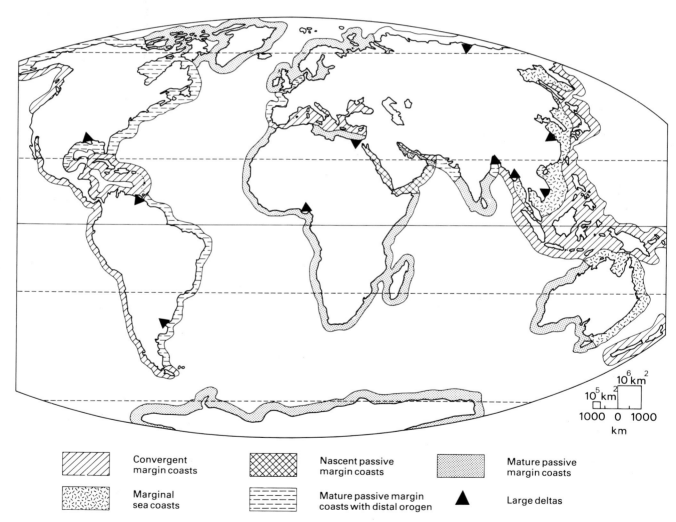

Fig. 13.2 *Tectonic setting of the world's coastline. (Based partly on J. L. Davies (1980)* Geographical Variation in Coastal Development *(2nd edn). Longman, London, Fig. 2, p. 9 and D. L. Inman and C. E. Nordstrom (1971)* Journal of Geology *79, Fig. 4, p. 10.)*

Table 13.1 Relationship of continental shelf width to tectonic setting

	TECTONIC SETTING					% OF WORLD COASTLINE
	CONVERGENT MARGIN COASTS	PASSIVE MARGIN COASTS				
SHELF WIDTH		NASCENT	MATURE	MATURE DISTAL OROGEN	MARGINAL SEA	
Narrow (<50 km)	23.9	6.5	11.2	12.6	3.3	57.5
Wide (>50 km)	1.4	0.1	3.6	29.1	8.3	42.5
Total	25.3	6.6	14.8	41.7	11.6	100.0

Note: All figures are percentages.
Source: Modified from D. L. Inman and C. E. Nordstrom (1971) *Journal of Geology* **79**, Table 6, p. 14.

first order relationships are reflected in second order effects, as is evident in the association between tectonic setting and a range of specific types of coastal features (Table 13.2). Wave erosion, for instance, is important on convergent margin coasts which receive comparatively meagre sediment supplies from continental drainage systems. Deposition by marine action, however, is predominant on passive margin distal orogen coasts, such as those of eastern South America, which are supplied with abundant sediment from an orogenic belt on the other side of the continent. Large deltas in fact appear to be more or less confined to mature passive margin and marginal sea coasts since it is only these that provide the outlets for the world's major drainage systems (Fig. 13.2). Further implications of the relationship between tectonic setting and continental drainage systems are explored in Chapter 16.

Structural and lithological controls are also evident at this second-order scale. Active convergent continental margins in many cases have fold structures running roughly parallel to the coastline, and differential erosion of adjacent litholo-gical units creates a characteristic large-scale coastal morphology (Fig. 13.3). In other cases, especially along passive margin coasts, the structural grain may be oblique or perpendicular to the coastline, and in this case a highly indented coast can be produced, if fluvial valleys or glacial troughs are drowned by a rise in sea level. The resulting submerged valleys are known, respectively, as **rias** and fjords (Fig. 13.4). Finally, we can see in Table 13.3 to what extent it is possible to relate the overall morphology of coasts to their tectonic setting.

13.2 Waves, tides and currents

13.2.1 Waves

Waves are characterized by their length, height (amplitude), velocity (rate of forward motion of the wave peak) and **period** (the interval of time between successive wave peaks passing the same point) (Fig. 13.5). These properties, and the relationships between them, vary greatly depending on

Table 13.2 Relationship of coastal features to tectonic setting

	TECTONIC SETTING					TOTAL	% OF WORLD COASTLINE
	CONVERGENT MARGIN COASTS	PASSIVE MARGIN COASTS					
COASTAL FEATURES		NASCENT	MATURE	MATURE DISTAL OROGEN	MARGINAL SEA		
Wave erosion	47.9	5.6	3.8	4.9	37.8	100	44.7
Wave deposition	15.5	12.4	21.4	33.3	17.4	100	11.6
Fluvial deposition	5.7	6.4	9.9	62.4	15.6	100	3.2
Aeolian deposition	1.9	18.8	79.3	—	—	100	1.2
Biogenic activity	36.1	21.0	11.3	15.8	15.8	100	3.0
Glaciated coasts	6.4	—	7.2	86.4	—	100	30.7
% of world coastline	39.1	4.3	6.8	35.4	8.8		94.4*

*Excluding the Antarctic coastline.
Note: All figures are percentages.
Source: Modified from D. L. Inman and C. E. Nordstrom (1971) *Journal of Geology* **79**, Table 7, p. 14.

Fig. 13.3 *The concordant convergent margin coast of western Pakistan which runs roughly parallel to the tectonically active Makran Coast Range, a region of rapid uplift associated with the convergence of the Arabian and Eurasian Plates. The prominent headlands visible in the image, which covers an area about 180 km across, are recently upthrusted fault blocks. Note the absence of major indentations along the coast and the well-developed zetaform beach to the east of the prominent headland in the bottom left of the image. (Landsat image courtesy N. M. Short.)*

the nature of the mechanism generating the wave, the intensity of this generating mechanism and the environment in which the wave exists.

13.2.1.1 Wind-generated waves

Generation of waves by the wind involves a transfer of energy from moving air to a water surface. Although a very familiar process, the way in which this occurs is still not fully understood. The amount of energy exchanged depends mainly on the velocity, duration and fetch of the wind. The

fetch is the distance over which the wind blows and it has an important influence on wave height and period. The highest waves are produced by strong winds blowing in the same direction over a long distance (Box 13.1). Exceptionally, wave heights of more than 15 m can be generated by the wind. It is possible broadly to identify shoreline environments subject to different levels of wave energy on the basis of prevailing wind speeds, fetch and coastal configuration (Fig. 13.6).

Waves actively being generated by the wind are known

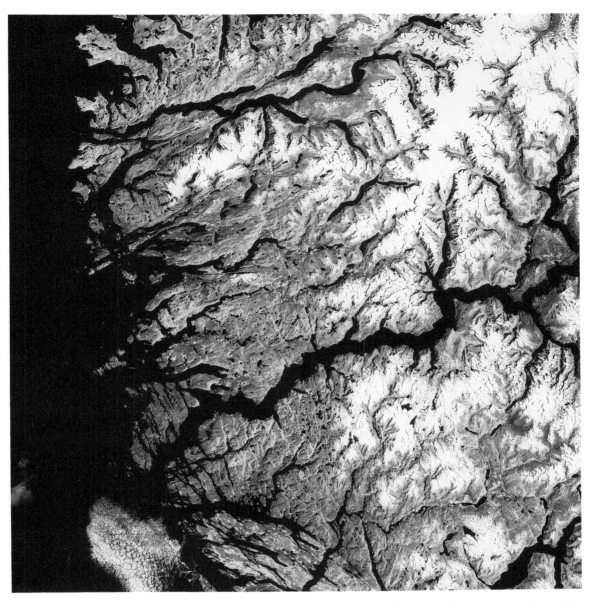

Fig. 13.4 *The mature passive margin coast of south-west Norway where the structural grain is oblique or perpendicular. The discordance between structure and coastal alignment has been emphasized by glacial erosion which has led to the development of fjords extending up to 150 km inland. Sogne fjord, one of the largest, runs roughly east–west across the lower centre of the image. The area covered is about 180 km across. (Landsat image courtesy N. M. Short.)*

Box 13.1   Wave height

Relationships between wind velocity and fetch and wave height have been determined empirically from observational evidence. Wave height is proportional to the square root of fetch and the square of wind velocity:

$$H = 0.36 \sqrt{F} \quad \text{and} \quad H = 0.031 \, U^2$$

where H is the wave height (m), F the fetch (km) and U the wind velocity (m s⁻¹).

as **sea waves**. Their periods vary, so as they disperse from the zone of generation they become separated since longer period waves with a greater wave length travel more rapidly (Box 13.2). This separation of waves of different periods produces **swell waves**. These may travel for thousands of kilometres across whole oceans, gradually losing height and energy. Over the first 200 km of travel a 10 m high wave would be reduced in height to around 2 m with an 80–90 per cent loss in energy, but thereafter the rate of energy loss becomes much less.

Table 13.3 Relationship of coastal morphology to tectonic setting

MORPHOLOGICAL CATEGORY	TECTONIC SETTING				
	CONVERGENT MARGIN COASTS	PASSIVE MARGIN COASTS			
		NASCENT	MATURE	MATURE DISTAL OROGEN	MARGINAL SEA
Mountainous coast	97.2	8.0			2.5
Narrow shelf, hilly coast		75.1	14.1		5.6
Narrow shelf, plains coast		15.9	46.2	1.5	
Wide shelf, plains coast			4.0	89.3	3.1
Wide shelf, hilly coast				2.2	77.4
Deltaic coast		1.0	3.4	1.3	5.8
Reef coast			3.0	1.9	5.6
Glaciated coast	2.8		29.3	3.8	
Total	100.0	100.0	100.0	100.0	100.0
% of world coastline*	39.0	4.6	7.5	35.2	8.1

*Excluding Antarctic coastline.
Note: All figures are percentages.
Source: Modified from D. L. Inman and C. E. Nordstrom (1971) *Journal of Geology* **79**, Table 8, p. 19.

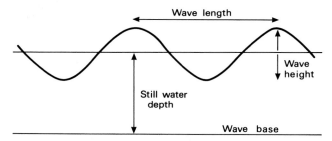

Fig. 13.5 *Major characteristics of waves.*

In the deep water of the oceans there is little actual forward motion of water in waves because it is the waves that move rather than the water. Such deep-water waves are called **oscillatory waves** since the individual water particles move with a virtually circular motion, oscillating around a more or less fixed point. Motion declines rapidly downwards, and at a depth approximately one half that of wave length, known as **wave base**, there is negligible movement (Fig. 13.5). As waves move towards shallower water their mode of movement changes dramatically. Where the water depth decreases to less than that of wave base the sea floor starts to interfere with the oscillatory motion and the orbit of individual water particles become more elliptical. Forward movement of water now becomes important as the oscillatory waves are transformed into **translatory waves**. As the water depth becomes progressively more shallow, wave length and velocity decrease, wave height increases and consequently the wave steepens (Box 13.3). Eventually the wave is over-steepened to the stage where it breaks as its crest crashes forward creating **surf**. This occurs when wave height has built up to the point where it is roughly

Box 13.2 Period and wave length for oscillatory waves

Separation of oscillatory waves as they disperse from the point of generation occurs because of the relationship between wave length, wave velocity and wave period:

$$L = \frac{gT^2}{2\pi} \quad \text{and} \quad U = \frac{gT}{2\pi}$$

where L is the wave length (m), T the period (s), U the wave velocity (m s⁻¹) and g the acceleration of gravity (m s⁻²). Thus waves generated by a storm with a period of 5 s will have a wave length of 39 m and a velocity of 7.8 m s⁻¹, while those with a period of 10 s will have a wave length of 156 m and a velocity of 15.6 m s⁻¹.

equal to water depth, but this relationship is also influenced by wave length and shoreline gradient.

The type of breaking wave formed is related to wave steepness and the gradient of the shore (Fig. 13.7). Once the wave form has been destroyed the remaining water

Box 13.3 Velocity and wave length of translatory waves

As translatory waves enter shallower water their period remains constant but their wave length and velocity changes:

$$U = \sqrt{gd} \quad \text{and} \quad L = T\sqrt{gd}$$

where U is the wave velocity (m s⁻¹), L the wave length (m), d the water depth (m), T the period (s) and g the acceleration of gravity (m s⁻²).

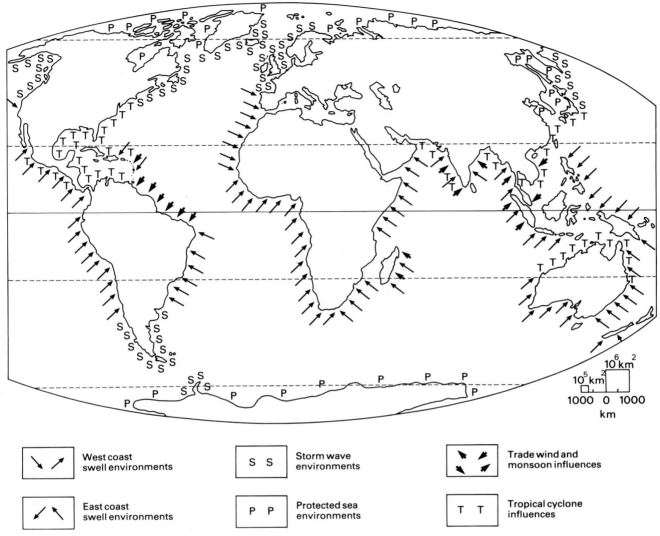

Fig. 13.6 *Global distribution of major wave environments. (Modified from J. L. Davies, (1980)* Geographical Variation in Coastal Development. *(2nd edn) Longman, London, Fig. 27, p. 43.)*

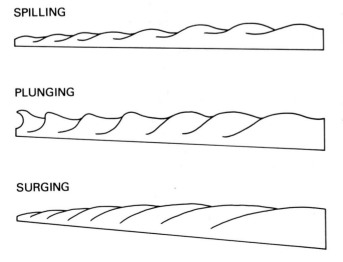

SPILLING

PLUNGING

SURGING

Fig. 13.7 *Major types of breaking wave; in reality breaking waves occur in a continuous spectrum of forms related to the gradient of the shore and wave period. Gentle gradients tend to be associated with* **spilling breakers** *where water from the crest cascades down the wave front.* **Plunging breakers**, *where the crest curls over and crashes down on the advancing wave base, are more characteristic of steeply sloping shores. Where waves of low steepness encounter high gradient shores they may fail to break fully but form instead* **surging breakers** *in which the crest collapses and the base advances up the shore.*

moves up the shore as **swash** and returns under the force of gravity as **backwash**. Breakers, especially those associated with large storm waves, expend large quantities of kinetic energy and are capable of accomplishing considerable geomorphic work (Fig. 13.8). Because swell waves can travel such large distances the energy released by breakers along

Fig. 13.8 *Plunging breakers on a steeply sloping beach composed of fine basalt gravel on the south-east coast of Hawaii. At the time the photograph was taken large waves were being generated by a tropical storm less than 200 km away.*

a coastline may have originated in a storm thousands of kilometres away.

The angle at which waves strike a shoreline is significantly affected by **wave refraction**. This occurs when the water depth as the shoreline is approached falls to less than half the depth of wave base. Beyond this point further decreases in water depth are accompanied by a reduction in wave velocity. Waves are refracted in this situation because they are bent towards the part of the shoreline where they are moving most slowly. If we imagine a wave obliquely approaching a straight coast, those parts of the wave which reach shallow water first will slow down and will tend to be caught by the other parts of the wave still in deep water. This refractive effect means that incoming waves tend to be roughly parallel with the offshore submarine contours as they reach the shoreline, although they are rarely exactly parallel. Wave refraction is important because it causes wave energy to be concentrated at headlands where waves converge and to be dissipated in bays where waves diverge.

13.2.1.2   Storm surges
Storm surges are generated by the combined effects of low atmospheric pressure and very high wind speeds. In the centre of tropical storm systems atmospheric pressure may drop as far as 100 mb below normal, and this can 'suck up' sea level below the centre of the cyclone by up to 1 m. The largest storm surges are produced when winds are blowing towards coastal embayments and are accentuated if they coincide with high tides. By the time the waves generated by the storm surge reach the coast they may have built up to a height of several metres above normal high tides.

Although comparatively uncommon on a global basis, storm surges occur repeatedly along coasts experiencing tropical cyclones, such as the Bay of Bengal, and in mid-latitude areas subject to intense storms where the coastal

configuration is particularly favourable (as in the North Sea) (Fig. 13.6). Storm surges are most destructive along very low-lying coasts where their effects can extend many kilometres inland, but their geomorphic significance arises in large part from the way in which they lead to wave attack at much higher levels along a shoreline than is reached by normal waves.

13.2.1.3   Seismic sea waves
Seismic sea waves (or **tsunami**) are generated by displacement of the ocean floor. They can be produced by volcanic eruptions and catastrophic submarine mass movements, but by far the most significant cause is earthquakes. The displacement of a large mass of ocean water, often at a great depth, generates waves of small amplitude (usually less than 1 m), considerable length (up to 200 km or more) and high velocity. As with translatory waves the wave velocity is related to ocean depth and waves can travel at more than 600 km hr^{-1} over water 3000 m deep (Box 13.3). Seismic sea waves can therefore cross the entire Pacific Ocean in a matter of a few hours. As they near land the decrease in water depth leads to a marked reduction in velocity and an increase in wave amplitude, so by the time the coast is reached wave heights may have reached 15 m. They seem to be most effectively generated by shallow earthquakes associated with oceanic trenches and this probably accounts for their predominance in the Pacific Ocean. They are comparatively rare phenomena, major events being experienced on average only every 25–50 a in Hawaii, for example. Although capable of effecting catastrophic changes in coastal landforms and causing severe flooding and loss of life, seismic sea waves are sufficiently infrequent to be outweighed in their geomorphic significance by less spectacular but more consistent wind-generated wave action.

13.2.2   Tides
Tides result from the gravitational attraction exerted on ocean water by the Moon and the Sun, the former having around twice the effect of the latter. The motions of the Earth, Moon and Sun with respect to each other produce **semi-diurnal tides** along most coasts in which there are two lows and two highs approximately every 24 h. Tides higher than normal, known as **spring tides**, occur every 14–75 days when the Sun and Moon are aligned. In between these periods lower than normal, or **neap tides**, occur when the Sun and Moon are positioned at an angle of 90° with respect to the Earth. Spring and neap tides involve deviations of about 20 per cent above and below normal tidal ranges.

Several factors complicate this simple picture. These include the size, depth and topography of ocean basins, shoreline configuration and meteorological conditions. Much of the Pacific coast, for instance, experiences a regime of

mixed tides in which the highs and lows of each 24 h period are of different magnitudes. Other coastlines, such as those of much of Antarctica, have **diurnal tides** with only one high and one low per 24 h.

Although contrasts between tidal types are important in some coastal processes, of much greater overall geomorphic significance is tidal range. This can vary from a minimal range in virtually enclosed water bodies such as the Mediterranean Sea and Black Sea, to less than 2 m along some open coasts and up to 6 m or more where there is a major semi-diurnal component to the tide. In fact it is useful to consider three categories of tidal range; **microtidal** (less than 2 m), **mesotidal** (between 2 and 4 m) and **macrotidal** (above 4 m) (Fig. 13.9). The most extreme ranges occur where the coastal configuration and submarine topography induce an oscillation of water in phase with the tidal period. This effect is particularly pronounced in the Bay of Fundy in north-eastern Canada where the tidal range is nearly 16 m. In some estuaries with a large tidal range a single wave several metres high can be formed by the incoming tidal flow as it experiences drag in entering shallower water. Such waves, known as **tidal bores**, may reach velocities of 30 km h^{-1} and can be potent erosive agents.

Tidal range and type are of geomorphic importance for several reasons (Fig. 13.10). Tidal type determines the interval between tides and therefore the time available for the shore to dry after high tide. This is important for shoreline weathering processes and biological activity. More significantly, it affects the intensity of tidal currents since, for a given tidal range, the velocity of water movement will be greater in semi-diurnal regimes than for mixed or diurnal types because there is a shorter interval between high and low tide. This effect is particularly important in narrow coastal embayments where tidal flows are concentrated. Tidal range is important because it controls the vertical distance over which waves and currents are effective along the shoreline, and, in conjunction with the shoreline gradient, it determines the extent of the **intertidal zone**, that is, the area between high and low tide.

13.2.3 Currents

The geomorphic importance of currents lies in their role in transporting sediment and, less significantly, in erosion. Currents associated with tides can transport and erode sediment where flow velocities are high. This is usually confined to estuaries or other enclosed sections of coast which

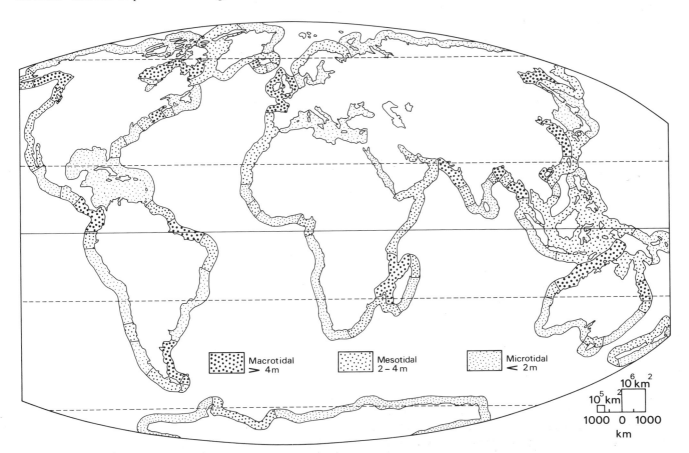

Fig. 13.9 *Global variation in tidal range; the ranges indicated are for spring tides. (Modified from J. L. Davies (1980)* Geographical Variation in Coastal Development. *(2nd end) Longman, London, Fig. 33, p. 51.)*

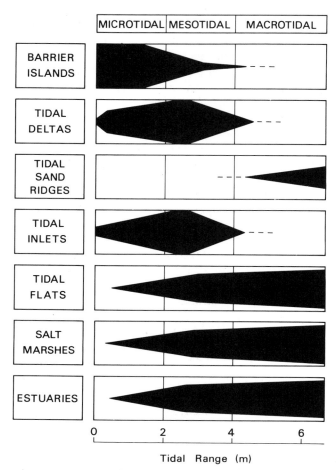

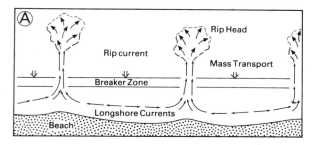

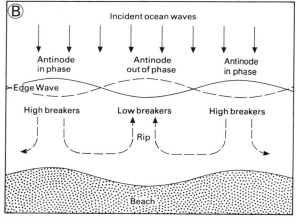

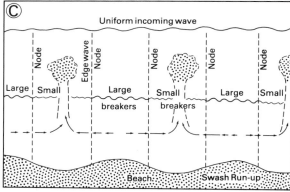

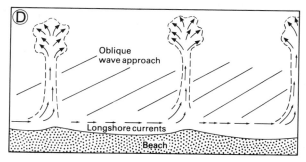

Fig. 13.10 *Relative abundance of different types of coastal landforms in relation to tidal range. (Modified from T. Elliott, 1986, in H. G. Reading (ed.)* Sedimentary Environments and Facies. *(2nd edn) Blackwell Scientific Publications, Oxford, Fig. 7.1, p. 156; and M. O. Hayes (1976) in: L. E. Gronin (ed.)* Estuarine Research: *Vol. 2, Geology and Engineering, Academic Press, London, Fig. 1, p. 5.)*

experience semi-diurnal tides with a high range. In mesotidal and microtidal regimes fast tidal currents are confined to zones of constriction such as around coastal promontories and within narrow channels between islands. The highest velocities usually occur midway between the high tide point and the low tide point, but in long estuaries the timing of maximum tidal current velocities in the tidal cycle may vary with location. This is important since it determines the level in the intertidal zone where tidal currents will be most active.

Currents are also associated with wave action along shorelines (Fig. 13.11). Where waves move in a direction perpendicular to a beach the backwash is normally confined to fairly evenly spaced zones forming **rip currents** (Fig. 13.12) which are fed by water moving parallel to the beach in **longshore currents** (Fig. 13.11 (A)). This cellular circulation pattern is caused by variations in wave height parallel

Fig. 13.11 *Currents in the nearshore zone: (A) feeder longshore currents, rip currents and mass transport returning water to the surf zone; (B) relationship between edge waves and rip currents; (C) position of rip currents with edge waves and incoming waves which are 180° out of phase; (D) longshore currents associated with oblique wave approach. Based on J. L. Davies (1980)* Geographical Variation in Coastal Development *(2nd edn) Longman, London Fig. 76, p. 103 and P. D. Komar and D. L. Inman (1970)* Journal of Geophysical Research **75**, *Fig. 9, p. 5926 Copyright by the American Geophysical Union.*

Fig. 13.12   Rip currents (arrowed) on a sand beach north of Cairns, Queensland, Australia.

with the shore. Water from the higher sections of breakers extends further up the beach and returns at the points where the lower wave sections have broken.

Variations in the height of breaking waves may arise from the configuration of the shoreline, but the presence of cellular circulation along straight beaches suggests another mechanism. Exactly what this mechanism is has been much debated, but it may involve **edge waves** (Fig. 13.11(B)). These are formed by the resonance between waves arriving at the shore (**incident waves**) and those reflected from it. The resonance raises the height of breakers at regular intervals parallel to the shore. Irrespective of the precise mechanism whereby they are initiated, once established rip currents and longshore currents are partially self-sustaining. This is because offshore sediment transport in rip current zones, and onshore transport between them, modifies the form of the beach so as to enhance the cellular circulation pattern (Fig. 13.11(C)). Where waves strike the shore obliquely the longshore currents will be in predominantly one direction (Fig. 13.11(D)). As we will see this has important implications for sediment movement along the shore.

13.3   Coastal processes

13.3.1   Destructional processes

13.3.1.1   Shoreline weathering
A shore zone may be subject to the same range of physical and chemical weathering processes that occur on land, but the presence of sea water and the cycle of wetting and drying produced by tides introduces significant additional factors. The tidal cycle of wetting and drying is instrumental in a variety of weathering processes which are appropriately considered collectively as **water-layer weathering**. The zone affected extends from low water mark to the furthest limit reached by waves and spray at high tide. Its areal extent is therefore controlled largely by tidal range,

but tidal type and meteorological factors are also important since these affect, respectively, the time available for drying between tides and the rate of evaporation. The most aggressive regime for water-layer weathering probably occurs along coasts characterized by high evaporation rates and mixed or diurnal tides.

The most important process in shoreline weathering is probably salt weathering, although its effectiveness depends on the ability of shoreline rocks to absorb sea water and spray. Chemical weathering also plays a role, the nature of the chemical reactions being controlled by rock mineralogy, and their rate being influenced by temperature and variations in micro-environmental factors such as organic activity and pH along the shore. Solution is apparently a significant process on limestones and calcium carbonate-cemented rocks, although there are difficulties in explaining how it occurs since sea water is normally saturated or supersaturated with respect to $CaCO_3$. The lower solubility of CO_2 in warm water would suggest that solution should be a more effective process along cold high latitude coasts, but apparently solutional forms are also found on low latitude coasts. Rock pools on tropical shores may isolate pockets of water which are capable of solution through pH changes associated with photosynthesis. There is still much uncertainty about the effectiveness of sea water in the dissolution of carbonate rocks and recent attention has been focused on the role of shoreline organisms, especially blue-green algae, which may be capable of leaching carbonate through complex biochemical reactions.

In high latitudes frost weathering is a potentially significant shoreline weathering process because of the frequency of wetting of bare rock surfaces in the intertidal zone. Sea water itself is not very effective because, on freezing, the salts it contains become segregated and produce a rather soft ice incapable of transmitting high stresses to the rock. Frost weathering is apparently more effective where fresh water is available from melting snow banks or permafrost.

13.3.1.2   Coastal erosion
Waves are the most important erosive agent along most coasts but their effect varies with wave energy and characteristics, and with the nature of the material exposed to wave attack. Where a coast is formed by steep cliffs which plunge straight into deep water swell waves are not forced to break before they impact. As there is virtually no forward mass displacement of water in such waves they are reflected with little loss of energy and accomplish negligible geomorphic work. Much more commonly coasts are subject to breaking waves. These involve significant mass displacement and a considerable loss of kinetic energy as they break on a shoreline. This energy is dissipated over a short distance where the shore gradient is steep, but over a greater horizontal distance where it is shallow. Of the main types of breaking wave, plunging breakers produce the highest

instantaneous pressures (up to 600 kPa or more) since air can be trapped and compressed between the leading wave front and the shore.

The combined effect of air compression and the impact of a considerable mass of water is capable of dislodging fractured rock and other loose particles, a process known as **quarrying**. Well-jointed rocks and unconsolidated or weakly consolidated sediments are particularly susceptible. Breaking waves may also throw particles against the shore and this leads to the abrasion of shoreline materials. The effectiveness of abrasion is highly dependent on wave energy and on the availability of suitable material, such as pebbles, along the shore. Large boulders can only be moved in the most intense storms, whereas small boulders and pebbles can be moved much more frequently by waves of only moderate energy.

A variety of sea shore organisms, including some molluscs, boring sponges and sea urchins, can destroy rocks by physically boring into them. Their effectiveness is influenced by rock type, most sedimentary rocks being much more susceptible than igneous rocks. The relative importance of biological erosion is much greater along coasts characterized by low wave energies since here abrasion and quarrying operate at only moderate or low intensities.

13.3.2   Constructional processes

13.3.2.1   Sediment movement and deposition

In considering the movement and deposition of sediment along coasts it is necessary to identify the main sources of sediment, its mode of transportation and the zones in which it accumulates or is removed from the littoral zone. If we ignore, for the moment, the movement of sediment parallel to the shore, there are three main sources of sediment: (1) the coastal landforms themselves, including cliffs and beaches; (2) the land area inland from the littoral zone; and (3) the offshore zone and beyond.

The erosion of coastal landforms, especially cliffs, can locally provide abundant sediment in environments with high wave energies (especially where unconsolidated deposits are being eroded), but this is not generally a significant source in the tropics where environments with low wave

Fig. 13.13   *Oblique aerial view showing the effects of longshore drift along the coastline of Tasman Bay, North Island, New Zealand. Abundant sediment is carried down from the Richmond Range by the Motueka River and some of this is transported southward to form small barrier islands. Note the recurved spit towards the bottom centre of the view which clearly indicates the prevailing direction of sediment transport.*

energies are common. This is supported by the relative lack of coasts formed of bedrock in the tropics. Even where present, cliffs formed of well-consolidated strata recede slowly and supply little sediment. In contrast, pre-existing constructional features such as fossil beaches and dunes are much more susceptible to wave attack and sediment mobilization.

Land-derived sediment can be provided by mass movement, especially where cliffs composed of material susceptible to such processes are being actively undercut. In periglacial environments, such as the Arctic coast of the USSR, gelifluction is a particularly important means of transporting sediment into the littoral zone, while in other high latitude environments material can be supplied by glaciers. Nevertheless, by far the most significant source of supply overall is from rivers (Fig. 13.13). It has been estimated that on a global basis rivers contribute about one hundred times as much sediment as marine erosion. This ratio varies latitudinally, with fluvial sources being of even greater relative importance throughout most of the tropics where low wave energies make marine erosion less effective, but even most beaches in the mid-latitudes, where wave energies are on average significantly higher, probably contain less than 5 per cent of material derived from cliff erosion.

Onshore transport may return to the littoral zone previously eroded beach material or fluvial sediments initially deposited well offshore. Particularly severe storm waves, storm surges and seismic sea waves may occasionally bring in sediment from beyond the offshore zone but much of the present-day onshore movement of sediment arises from the post-glacial rise in sea level (see Section 17.6.2). Sediments deposited on previously exposed continental shelves are still being moved inland in response to the Holocene rise in sea level and the adjustment of the littoral zone to present-day tidal and wave regimes. In some localities this sediment supply seems to have been exhausted and some depositional landforms constructed during this period of sea-level rise are now being eroded.

While sediment is constantly being moved more or less perpendicularly towards and away from the shoreline by tidal and wave action, the predominant net movement of sediment along most coasts is parallel with the shore through the effects of longshore currents (Fig. 13.13). This movement is termed **longshore drift** and its rate is dependent on wave energy and the angle at which waves strike the coast (an angle around 30° being the most effective). The effects of longshore drift are clearly illustrated by the accumulation of sand against groynes along a beach and this in fact provides a means of measuring the rate of longshore sediment movement.

Longshore sediment transport occurs below the breaker zone where wave steepness is high, or by **beach drift** where wave steepness is low. Beach drift involves the movement of sediment obliquely up the beach as a result of the angle

of wave attack (Fig. 13.11(D)). As a result of gravity it returns down the beach in the backwash perpendicularly to the shoreline, and so moves a short distance along the beach in each cycle of swash and backwash. The overall significance of longshore drift depends on whether it occurs along coasts of free or impeded transport. Impeded transport is characteristic of coasts with an irregular configuration, the amount of longshore drift being limited by the trapping of sediment against headlands. This contrasts with unimpeded longshore sediment transport along straight coasts. Given a particular coastal configuration the rate of longshore drift will be related to the predominant angle of wave approach and the constancy and velocity of longshore currents.

13.3.2.2 Organic activity

A variety of organisms are either directly or indirectly responsible for the construction of some coastal landforms. The most spectacular example of direct construction is by the corals and other carbonate-secreting organisms which form **coral reefs**. These structures can attain immense sizes, as in the case of the Great Barrier Reef which extends along much of the north-east coast of Australia. Because they require a minimum sea temperature for growth, corals are concentrated in the tropics, as are various calcareous algae which form carbonate encrustations along many tropical shores.

A range of plants are adapted to salt water and these form salt-marsh communities in the intertidal zone along sheltered, muddy coasts. In the tropics mangroves are an important element in coastal vegetation and, together with other halophytic (salt-tolerant) plants, they may play a geomorphic role by trapping sediment within their root systems and thereby aiding the process of deposition. Their precise role, however, is uncertain and we shall return to this in the discussion of tidal flats. More certain is the role of plants in stabilizing coastal dune systems and contributing to sand deposition. Again we will return to this topic in the section on coastal dunes.

13.4 Coastal landforms

13.4.1 Destructional forms

13.4.1.1 Cliffs

Cliffs are steep, often vertical, slopes which rise abruptly from the sea or basal platform. Where they rise sheer from deep water they are termed **plunging cliffs**. Such cliffs are formed by fault scarps, volcanic masses and drowned glaciated valleys and because they are affected largely by swell waves, which are efficiently reflected, they suffer minimal erosion by wave action. Much more commonly the water depth is shallow enough for waves to break and therefore most cliffs are subject to basal erosion by wave action in addition to sub-aerial weathering and mass movement.

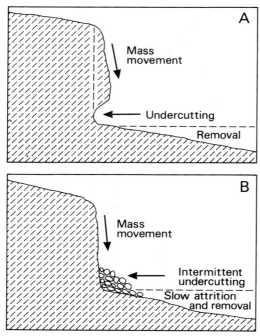

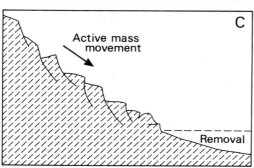

Fig. 13.14 Major processes of cliff retreat: (A) undercutting with rapid removal of collapsed material; (B) undercutting with slow removal of collapsed material; (C) mass movement and removal of material at various rates. (After J. L. Davies (1980) Geographical Variation in Coastal Development. *(2nd edn) Longman, London Fig. 49, p. 76.)*

Fig. 13.15 Shore platform cut in horizontally bedded sandstone, near Sydney, Australia.

Cliff morphology is partly determined by the balance between the efficacy of sub-aerial processes and the rate of basal erosion (Fig. 13.14). Where wave action is capable of removing all the debris accumulating at the base of the cliff, lithological and structural controls become significant in influencing cliff form. The development of long free faces, up to 500 m or more in height, is favoured in mechanically strong, horizontally bedded sedimentary rocks and coarsely jointed igneous rocks. On poorly consolidated materials with a low threshold angle of stability true cliffs may fail to develop even if wave action is aggressive.

A platform at the level of wave action, known as a **shore platform**, is developed as the cliff retreats under the combined effects of quarrying and abrasion (Fig. 13.15). As this platform is extended the base of the cliff starts to be pro-

tected since wave energy is expended in crossing the platform. Wave action gradually becomes less significant in undermining the cliff, and other processes such as salt weathering, solution, frost weathering and the erosive effects of shoreline organisms become more significant. Especially in environments in the tropics where wave energy is low, the combination of organic and inorganic weathering processes is relatively more significant than quarrying and abrasion in cliff development. Well-developed basal notches are characteristic of cliffs along tropical coasts and these seem to develop through water-layer weathering operating within a narrow tidal range. Outside the tropics cliff notches are less well developed, partly because the typically greater tidal range means that wave attack is not concentrated in a narrow vertical zone.

Estimates of the rates of cliff retreat show that most shore platforms cannot have formed in their entirety at present sea levels which have persisted for only about the last 4000 a. Since many shore platforms exist near to present sea levels it can be deduced that they must have developed when cliff retreat was occurring during earlier phases of sea level in the Pleistocene close to that of today. Since a constant sea level would eventually produce a shore platform sufficiently extensive to absorb virtually all the incoming wave energy, periodic rises in sea level are probably necessary to maintain wave attack on the cliff base and thereby sustain cliff retreat and shore platform extension.

13.4.1.2 Shore platforms

Three categories of shore platform can be identified related to contrasting tidal stages and subject to rather different combinations of processes. **Intertidal platforms**, extending between high and low water mark, are formed primarily by quarrying and abrasion and are often referred to as **wave-cut platforms**. They are most common where wave energy is high and are best developed where easily eroded strata are exposed. Once initially formed by cliff retreat, intertidal platforms are subject to modification by weathering and abrasion.

As already mentioned, present-day erosional activity is probably only extending features developed during periods in the recent past when sea level was near to that of the present. Each new episode of wave attack probably has to begin with the removal of superficial material deposited on the platform during the intervening periods of low sea level. Unless there is a gradual rise in sea level, shore platforms can only extend themselves through cliff retreat by being progressively lowered seaward, otherwise sufficient wave action cannot be sustained at the cliff base. Indeed most intertidal shore platforms do exhibit a slight seaward slope termed an **abrasion ramp**. As gradients are typically only around 1°, and as abrasion by waves is under most conditions limited to a water depth of about 10 m, shore platforms can only extend themselves in this way to a width of about 500 m in microtidal environments and perhaps up to 1000 m in macrotidal environments. Any more extensive development than this would require a continuously rising sea level. Platform gradients tend to be less in microtidal than macrotidal environments, since in the former the vertical range of wave action is much smaller.

Shore platforms more extensive than those just discussed must have formed primarily through mechanisms other than quarrying and abrasion. Those developed at approximately high-tide level appear to be mainly attributable to water-layer weathering. They seem to be best developed in permeable rocks along coasts experiencing low to moderate wave energies, microtidal regimes and high evaporation rates. A low tidal range produces a well-defined upper zone experiencing regular saturation. Such **high-tide platforms** may develop through cliff retreat resulting from basal wea-

thering or by the modification of pre-existing platforms.

Low-tide platforms are especially associated with limestones and other calcareous strata. Solution and biological erosion appear to be the key formative processes and they are typical of microtidal, low wave energy environments. They may be partly constructional features where, for instance, carbonate-secreting calcareous algae are active. The action of algae seems to be important in developing the basal cliff notches typically found on their landward margins, but calcareous algae and other organisms contribute to their development through a combination of erosion and deposition. Outside the tropics biological activity is less significant in their development and high-tide platforms formed on hard limestones along mid-latitude and high latitude coasts appear more directly attributable to solution.

13.4.2 Constructional forms

13.4.2.1 Beaches

Beaches constitute quantitatively the most significant accumulations of sub-aerially exposed sediment along coasts. We have already considered some aspects of the supply and movement of sediment to beaches and will now look at beach form and process in more detail. Figure 13.16 provides a useful guide to the main features and processes to be discussed.

Although beaches can be composed of material ranging from fine sand to boulders, most consist of sand or pebbles, intermediate particle sizes being rare. The characteristics of local lithology can account for some of this variation in sediment calibre, but it does not explain the prevalence of pebble beaches in the mid-latitudes and high latitudes, and of sand beaches along tropical coasts. Likely reasons for

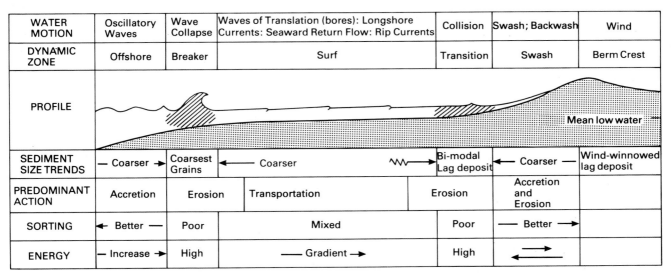

WATER MOTION	Oscillatory Waves	Wave Collapse	Waves of Translation (bores): Longshore Currents: Seaward Return Flow: Rip Currents	Collision	Swash; Backwash	Wind
DYNAMIC ZONE	Offshore	Breaker	Surf	Transition	Swash	Berm Crest
PROFILE						Mean low water
SEDIMENT SIZE TRENDS	← Coarser →	Coarsest Grains	← Coarser	⌇⌇	Bi-modal Lag deposit ← Coarser —	Wind-winnowed lag deposit
PREDOMINANT ACTION	Accretion	Erosion	Transportation	Erosion	Accretion and Erosion	
SORTING	← Better —	Poor	Mixed	Poor	— Better →	
ENERGY	— Increase →	High	—— Gradient →	High	← →	

Fig. 13.16 *Major processes and characteristics of the nearshore zone along a sandy coast. Zones of maximum sand transport are indicated by shading. (After J. C. Ingle, 1966,* The Movement of Beach Sand, *Elsevier, Amsterdam, Fig. 116, p. 181).*

Fig. 13.17   *Beachrock on the island of Aitutaki, Cook Islands, south-west Pacific.*

this contrast include the greater predominance of fine material in fluvial sediments and the smaller relative contribution of material eroded from cliffs to littoral deposits in the tropics. In contrast there is an abundance of coarse material in glacial and periglacial debris in coastal environments at higher latitudes, especially in the northern hemisphere.

Beaches are composed of a variety of organic and inorganic particles, but a major distinction can be made between carbonate and non-carbonate beaches. Carbonate beaches were long thought to be preferentially developed in the tropics where there is a greater abundance of organic carbonate remains. However, it is now realized that many mid-latitude beaches have a comparatively high content of biogenic carbonate. The mineralogical composition of a particular beach will represent a complex balance between local sediment sources, material imported from inland by rivers and material supplied by longshore drift or from offshore.

Under certain conditions beach sediments can be lithified through precipitation of calcium carbonate to form **beachrock** (Fig. 13.17). This is most commonly formed on tropical beaches where calcium carbonate apparently supplied mainly from ground water is precipitated as a result of high rates of evaporation. Micro-organisms may also be involved in the precipitation of calcium carbonate.

The profile form of beaches is determined by the size, shape and composition of beach material, the tidal range and the type and characteristics of incoming waves. The upper section of many beaches consists of a horizontal to slightly landward sloping surface called a **berm** (Fig. 13.16). This feature is not ubiquitous and is usually absent on gravel and pebble beaches. A berm is a zone of vertical accretion formed by backwash deposition and its elevation is therefore limited by the upper limit of swash.

On the seaward side of the berm, and separated from it by the **berm crest**, is the **beach face**. Its gradient is largely determined by the calibre of the beach material, and ranges from around 2° in fine sand to as high as 20° in coarse pebbles. This can be accounted for by the greater permeability of coarser sediments. Since swash percolates more rapidly as it crosses a pebble beach than it does across a sand beach, less water is available in the backwash to transport coarser material back down the beach face. A steeper slope therefore develops to compensate for this. The importance of percolation rates in influencing beach gradients is underlined by the observation that lower gradients are associated with poorly sorted beach sediments (which are more closely packed and therefore less permeable), whereas high gradients characterize beaches with sediments of the same mean size but better sorting. Wave height and steepness are also correlated with beach face slope angle, with steeper and higher waves being associated with gentler gradients as they generate more backwash.

Where the beach gradient is shallow there is typically a submerged **longshore bar** running parallel with the beach and separated from it by a trough. Several bars may be present offshore from beaches with very shallow gradients. They seem to develop in response to the action of breaking waves and apparently migrate to and fro normal to the shoreline as a result of changes in wave characteristics.

The beach profile is not static but rather changes its form over a range of time scales, the most significant of which, at least in the high energy storm wave environments of the mid-latitudes, is a yearly cycle of erosion and deposition. We know a lot about such annual changes because they have been extensively monitored by repeated beach profile surveys. The predominance of swell waves in the summer is associated with deposition from swash and a phase of beach construction, often with the development of a berm just above high-water mark. During the winter, however, storm waves cut back or completely destroy the berm. The sediment eroded is transported just offshore and deposited in longshore bars, before being gradually returned to the beach as the berm is constructed again the following summer. Wave steepness is apparently the critical factor in determining whether a swell or a storm wave beach profile develops. Since the berm is a major dissipator of wave energy, its destruction by storm waves has a major impact on rates of coastal erosion as it can lead to wave attack on backing cliffs.

As well as characteristic profile forms, beaches also exhibit regularities in plan ranging in scale from 1 m or so to features over 100 km in length. These rhythmic features involve a consistent repetition in form along the beach. The smallest significant regularities are **beach cusps**. These are crescent-shaped indentations lying parallel with the shore on the upper beach face and along the seaward margin of the berm. Although not confined to beaches with a particular calibre of sediment they are best developed in a mixture of sand and gravel.

In spite of their simple form there is little agreement as to the origin of beach cusps, several factors seemingly contributing to their development. They form most readily where waves approach the coast perpendicularly and consequently where longshore drift is minimal. Tides have also been suggested as an instrumental factor; under high tidal ranges they frequently form a series of ridges down the beach but they have also been seen to develop along tideless lake shores. It has been observed that cusp spacing is positively correlated with wave height, but a stronger relationship exists with swash distance – the distance between the point at which waves break and the highest point on the beach reached by the swash. Cusps may be initiated by

slight irregularities in form along the beach. These are subsequently enlarged by swash with the eroded material being deposited on either side by the backwash until an equilibrium form is attained and the sediment moves in a closed cycle. Such movement has been documented with the aid of dyed pebbles.

Various other mechanisms contributing to cusp formation have been suggested including seasonal changes in wave energy and the breaching of dune ridges or berms, but more recently the role of standing edge waves with periods equal to, or twice that, of the incident waves has received attention. The wide range of factors and mechanisms briefly mentioned here indicate that we are still some way from a comprehensive and detailed explanation of these intriguing features.

Significantly larger than cusps are **sand waves** and **crescentic bars**. They are submerged features with a longshore wave length usually around 200–300 m. They are found off both straight beaches and along embayed coasts, but are confined to microtidal environments. Standing edge waves with a wavelength twice that of the crescentic bar have been seen as being instrumental in their formation, with cell circulation and rip currents contributing to sediment transport. Longshore drift also seems to play a role in their development, and in this they differ from beach cusps.

At a large scale many coasts display a more or less regular succession of capes and bays. At some localities, along part of the east coast of Australia for instance, there exists an almost continuous series of asymmetrically curved bays linking each headland, with each beach section recessed behind its neighbour. They have been called **headland bay beaches**, but the term **zetaform beaches** is also used because of their similarity in shape to the Greek letter (Fig. 13.3). The form of the Australian examples seems to be related to the incident angle of wave crests refracted around the headlands from the prevailing south-east swell. Beach steepness and sediment size increase northward in response to increasing wave energy and the impeding of longshore transport of sediment by the headlands. Not all zetaform beaches are, however, separated by rocky headlands. Along coasts composed solely of unconsolidated sediments, such as the west coast of Sri Lanka and the eastern coast of the Malay peninsula, river mouths may take the place of headlands in separating individual beaches.

Along some coasts low, linear beach ridges composed of sand or shell debris surrounded by low-lying marshes are found. These features are known as **cheniers** and were first described from the Louisiana coast of the USA. They can be up to 1 km across, 100 km long and 4 m high. Coastal plains formed by alternating ridges and marshes are termed **chenier plains** and have been recorded from a number of coasts including those of eastern China, Australia and New Zealand. Their precise origin is uncertain although it is generally agreed that variations in sediment supply are important in their development. Some workers consider that the most significant factor controlling their formation is rising sea level, but research in China has demonstrated that the cheniers found there have developed since present sea levels were more or less stabilized. Apart from a variable supply of predominantly fine-grained sediment, cheniers appear to develop preferentially where wave energies and tidal ranges are low and longshore drift is active.

13.4.2.2   Barrier island

A **barrier island** (also called a **barrier beach, barrier bar** or **offshore bar**) is an elongated offshore ridge of sand running parallel with a mainland coast and separated from it for almost its entire length by a **lagoon**. Barrier islands range in size from comparatively small features a few metres in width and hundreds of metres long to massive constructional forms up to 1 km or so wide, hundreds of kilometres long and, where large dune systems are developed, up to 100 m high. It has been estimated that barrier islands occur along 13 per cent of the world's coastline so they are truly landforms of global significance. They are most characteristic of very gentle sloping coasts with a low tidal range and are developed particularly extensively along the Atlantic and Gulf of Mexico coasts of the USA (Fig. 13.18).

Tidal inlets are found at intervals along the barrier. They are initially formed by storm breaching and subsequently maintained by the ebb and flow of tidal currents and the escape of river discharge entering the lagoon from its mainland side. They migrate along the barrier as longshore drift leads to removal of sediment from the updrift side and its deposition on the downdrift side. After a certain amount of longshore movement a new breach may form in the wake of the migrating inlet and it too begins its longshore migration. Very large quantities of sediment are moved daily by tidal currents and there are continuous fluctuations in sediment deposition and erosion on different sections of the barrier.

The formation of barrier islands continues to be a controversial topic, although data from boreholes have recently provided vital information on the history of their development and led to a revision of earlier ideas. Many barrier islands were being formed prior to the Holocene rise in sea level, so clearly sea-level changes have played a role in their construction. Some researchers envisage barrier islands beginning as submerged offshore bars, or forming on earlier foundations during the low sea levels of the Late Pleistocene. These were then built up by vertical accretion as sea level rose throughout the early part of the Holocene and large quantities of newly submerged sediment were transported landwards. The present-day progressive erosion of some barrier islands is seen as a consequence of the reduction in this sediment supply with the cessation of the major phase of sea level rise around 4000 a BP.

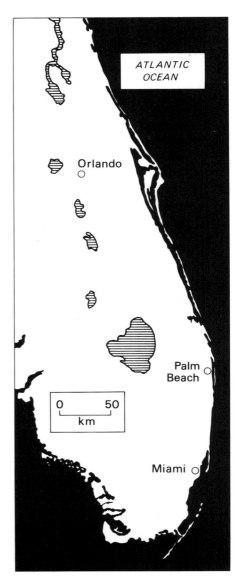

Fig. 13.18   *Barrier islands along the Atlantic coast of Florida, USA.*

An alternative view is that barrier islands initially developed above sea level on beach ridges and coastal dunes, and were subsequently separated from the mainland by flooding and lagoon formation as sea level rose early in the Holocene. There is certainly a lack of evidence for beach processes in lagoonal sediments which would be expected if barrier development had been initiated from submerged features. Yet another explanation sees barrier growth development occurring during the past 4000 a of more or less stable sea level. Each of the models of barrier island formation may be relevant in particular cases, but the extent to which barrier islands may be relict features out of equilibrium with prevailing environmental conditions adds greatly to the difficulties of relating their development to specific processes.

13.4.2.3 Spits, baymouth bars and cuspate forelands

Spits are elongated depositional forms attached at one end to the mainland and usually developed where the coast changes direction. **Baymouth bars** are similar features except that they extend into a bay rather than the open ocean. Longshore drift is the main process involved in their formation, the transported sediment being deposited where it enters a zone of slack water. In addition to elongation they may be widened by deposition on their offshore margin. If baymouth bars continue to extend they may eventually form a lagoon by totally enclosing a bay.

A feature of many spits as well as some barrier islands is their landward curvature at their accreting terminus (Fig. 13.13). Two factors may be important in generating these recurved forms. First, it has been seen simply as a normal consequence of spit development, resulting from the refraction of constructional waves around the accreting terminus of the spit and the consequent landward movement of sediment supplied by longshore drift at this point. This mechanism accords with the lack of recurvature in spits developed in shallow, sheltered localities and its preferential development in macrotidal environments in which there is deep water in the high tide constructional phase. Indeed some spits have a concave plan form with respect to the mainland at low tide and only assume their convex recurved form at high tide. A second possible cause of spit recurvature may arise from the effects of occasional periods of incident waves from a different direction than normal which modify the form of the actively accreting terminus of the spit. This is particularly likely to occur in exposed situations. Where there is a very constant undirectional swell wave regime, concave rather than convex spits can develop, and this situation is exemplified by the southern coast of Australia.

Cuspate forelands are sediment accumulations which are roughly triangular in plan form and have their apex extending offshore. Some seem to be related to special conditions of funnelling of tidal currents, while others, such as Dungeness in southern England, appear to have begun as a spit. In the case of Dungeness the spit extended through an eastwards longshore drift into the path of storm waves from the north. Its location in the narrow English Channel appears to be crucial as constructional waves come from opposite ends of the Channel while cross-Channel waves have a very small fetch and little power to blunt the apex.

13.4.2.4 Tidal flats

Tidal flats are largely depositional forms composed of muddy sediments and characteristically formed in lagoons and tidal estuaries. There is some confusion over the terminology used to describe their various features but three basic units can be recognized (Fig. 13.19): (1) the **high-tide flat** which has a very gently sloping surface, some of which is partly submerged at high tide; (2) the **intertidal slope**, a more steeply inclined though still gently sloping zone

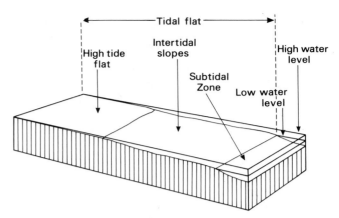

Fig. 13.19 *Relationship of low- and high-tide positions to the morphological elements of tidal flats. (After J. L. Davies (1980)* Geographical Variation in Coastal Development. *Longman, London (2nd edn) Fig. 123, p. 170.)*

between the high tide flat and the lower tidal limit; and (3) a **subtidal zone** which is submerged even at low tide.

Clay and fine silt-sized sediments carried to the coast by rivers tend to flocculate into larger aggregates when they encounter salt water. These particles will tend to settle in quiet coastal waters such as lagoons and sheltered estuaries. This mud is brought in by the incoming tide and is deposited before the tide reverses. If vertical accretion continues, part of the tidal flat will be exposed just above normal high tide level. Once this occurs the mud is colonized by halophytic plants and a salt marsh develops or, in the case of the tropics, commonly a mangrove swamp. Where both occur mangroves are characteristically concentrated on the offshore edge of the high tide flat where a rim with a slightly higher elevation is sometimes developed.

The role of root systems in aiding sediment deposition is uncertain since colonization by vegetation may respond to spatial and temporal variations in sedimentation brought about by channel shifting and other mechanisms. Continued vertical accretion is limited by the reach of the highest tides, a nearly level equilibrium surface developing where the addition of organic debris is more or less balanced by compaction and the decay of organic material. The high tide flat is comparatively stable in comparison with the intertidal flat which is usually subject to successive episodes of erosion and deposition and provides an insufficiently stable stratum for significant vegetation growth to occur.

13.4.2.5 Deltas

Deltas are protuberances extending out from shorelines formed where rivers enter the ocean, partially enclosed seas, barrier-sheltered lagoons or lakes, and supply sediment more rapidly than it can be redistributed by coastal processes (Fig. 13.20). They are normally confined to regions with mature drainage systems where major river channels reach the coast and supply sediment to a limited area; in

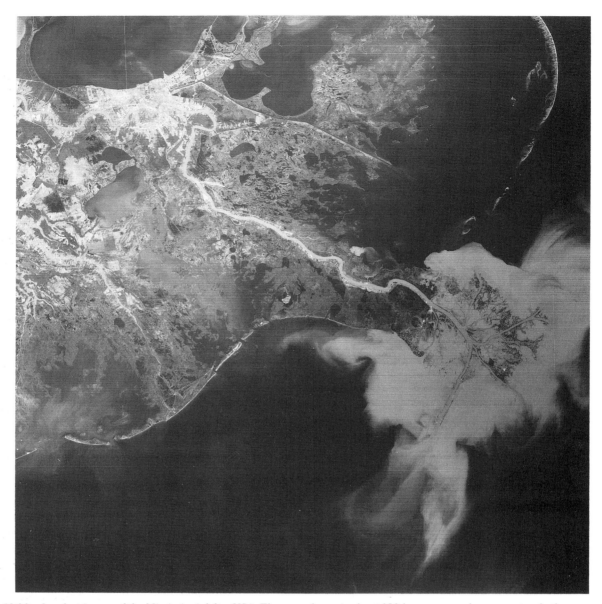

Fig. 13.20 *Landsat image of the Mississippi delta, USA. The area shown is about 180 km across and covers not only the present delta, which can be seen surrounded by a light-coloured halo of suspended sediment, but also some deltas now abandoned (see Figure 13.24). The delicate tracery of distributary channels in the presently active delta shows the overwhelming dominance of fluvial over tidal and wave processes. (Landsat image courtesy of N. M. Short.)*

areas of less organized drainage many small river systems supplying sediment to the littoral zone cause a more or less uniform progradation of the entire coastal plain. Deltas form because when a river enters a lake or the sea the outflow of river water expands and decelerates and its load is deposited. The deposition of bed load occurs rapidly but the behaviour of the suspended load depends significantly on the relative densities of the river, and sea or lake water. Three situations exist; **homopycnal flow** occurs where both waters are of equal density; **hyperpycnal** flow where the

river water is more dense, and **hypopycnal** where it is less dense (Fig. 13.21).

A number of attempts have been made to categorize the broad range of delta types (Table 13.4) in terms of the major processes controlling their development. A continuous spectrum of delta forms is encountered in nature, and one way of taking this variation into account is to categorize deltas using a ternary diagram in terms of fluvial, wave and tidal controls (Fig. 13.22).

The location and configuration of the receiving water

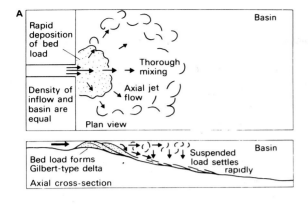

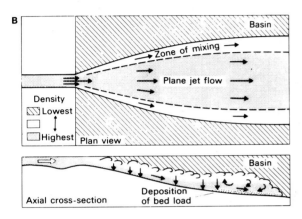

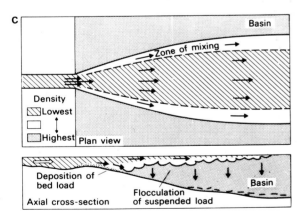

Fig. 13.21 *Various modes of mixing between sediment-laden river waters and receiving water bodies determined by relative water density. Homopycnal flow (A) occurs where both waters are of equal density. There is an immediate vertical and horizontal mixing which promotes rapid sediment deposition. This type of flow is very common in lakes and results in a simple delta structure consisting of **bottomset beds** formed from bed load, **foreset beds** comprising both bed load and suspended load and topset beds formed from overbank deposits and levees. Hyperpycnal flow (B) occurs where the river waters are more dense than the receiving water body and results in the creation of high density basal currents which carry sediment beyond the shore zone and thereby restrict delta development. This can take place when very cold river water enters a warmer lake and may* lead to the erosion of submarine channels on the delta front. Finally, hypopycnal flow (C) occurs where the river waters are less dense than those of the receiving basin. This is generally the case where river waters mix with sea water which is denser by virtue of its salt content. The buoyant river water spreads out near the surface as a jet or plume and may travel a considerable distance out to sea before losing its identity. (After W. L. Fisher et al. (1969) Delta Systems in the Exploration for Oil and Gas. *Bureau of Economic Geology, University of Texas, Austin, Figs. 2–4, based partly on C. C. Bates (1953) American Association of Petroleum Geologists Bulletin 37, Fig. 3, p. 2124.)*

body influences the wave, tidal and associated current processes. The morphology of deltas located on lakes or in lagoons sheltered by barrier islands is dominated by fluvial effects. In partially enclosed seas, such as the Gulf of Mexico and the Mediterranean, wave energy is reduced because of the limited fetch, and tidal effects are minimal. Consequently, deltas located on the shores of such seas experience only limited modification by coastal processes. By contrast, a delta at the head of a narrowing gulf, such as the Ganges–Brahmaputra delta at the head of the Bay of Bengal, may experience a large tidal range which leads to the redistribution of great quantities of sediment. The intensity of fluvial and coastal processes can vary through time, and if a seasonal fluctuation of sediment supply is out of phase with a seasonal variation in wave and tidal energy then periods of delta growth will alternate with periods of erosion.

In essence, delta morphology and sedimentary structure reflect the net effect of fluvial sediment supply and shoreline sediment dispersal. The variables controlling delta structure and morphology relate to the nature of both the receiving water body and the sediment supplied by the drainage system (Table 13.5). Drainage basin characteristics influence the fluvial regime and the nature of the sediment it transports to the delta, while the tidal and wave energy along the shoreline affects the efficacy of sediment removal from the delta area. Climate contributes to both the characteristics of the drainage basin supplying sediment as well as the delta environment itself. Tectonic controls are of particular significance, especially for marine deltas, because deltaic deposition occurs at a land–water boundary which is very sensitive to relative changes in sea level. A delta developing on a stable platform, a common situation for small lake deltas, builds both upward and outward. But where the site of delta formation is subsiding relative to sea level as a result of active tectonic subsidence, the rapid compaction of delta sediments or a sea level rise, a delta progrades by a process of lateral displacement of sediment lobes. Because most coastlines have experienced a relative rise in sea level during the Holocene, delta development in situations where relative sea level is falling is comparatively rare.

Deltas consist of two fundamental components: a **delta front** comprising the shoreline and the gently sloping sub-

Table 13.4 Characteristic delta morphologies and properties

DELTA	COASTLINE AND RIVER MOUTH CONFIGURATION	DELTA FRONT FEATURES	DELTA PLAIN FEATURES
Mississippi (1)	Highly indented coastline, multiple extended digitate distributaries – 'bird foot'	Indented marsh coastline, sand beaches scarce and poorly developed	Marsh, open and closed bays
Danube (2)	Slightly indented with protruding river mouths	Marsh coastline with sand beaches adjacent to river mouths	Marsh, lakes and abandoned beach ridges
Ebro (3)	Smooth shoreline with single protruding river channel	Low sand beaches and extensive spits with some aeolian dunes	Salt marsh with a few beach ridges
Niger (4)	Smooth, arcuate shoreline, multiple river mouths slightly protruding	Sand beaches nearly continuous along shoreline	Marsh, mangrove swamp and beach ridges
Nile (5)	Gently arcuate, smooth shoreline with two slightly protruding distributary mouths	Broad, high sand beaches and barrier formation with aeolian dunes; beach ridges at distributary mouths	Floodplain with abandoned channels and a few beach ridges; hypersaline flats and barrier lagoons near present shoreline
São Francisco (6)	Straight, sandy shoreline with single slightly constricted river mouth	High, broad sand beaches with large aeolian dunes	Large linear beach ridges and dunes
Senegal (7)	Straight coastline with extensive barrier deflecting river mouth but no protrusion	High, broad sand beaches with large aeolian dunes	Large linear beach ridges and swales, aeolian dunes

Note: The numbers refer to delta types shown in Figure 13.22.
Source: After L. D. Wright and J. M. Coleman (1973) *American Association of Petroleum Geologists Bulletin* **57**, Table 1, p. 376.

merged offshore zone, and a **delta plain** which forms an extensive lowland area lying landward of the delta front and is made up of active and abandoned distributary channels. In some deltas, such as the São Francisco Brazil, there is only a single channel, but it is much more common to find a diverging set of channels fanning out across the delta plain. The area between these channels is occupied by floodplains, tidal flats, marshes and lakes. Delta plains are not affected to any significant extent by wave action since they are protected by barrier island systems along wave-dominated deltas. The detailed morphology of delta plains is significantly influenced by climate; mangrove swamps are common in humid tropical and sub-tropical deltas, dunes and shallow salt basins, or **salinas**, are frequently present in arid regions, while pingos, thaw lakes and other permafrost forms are abundant in periglacial environments (Fig. 13.23).

Delta front morphology is highly responsive to the relative effectiveness of fluvial, tidal and wave processes. Delta front form is strongly influenced by fluvial action in lakes because wave and tidal action is usually minimal. The Mississippi is the only significant marine delta in which the delta front morphology is almost solely determined by the mode of sediment deposition by the distributary channels with almost no subsequent modification by wave or tidal processes (Fig. 13.20). Where wave action is relatively more effective a relatively smooth arcuate or cuspate beach shoreline is developed as there is some redistribution of river-borne sediment. Dominant wave action is reflected in a very regular beach shoreline with only slight protuber-ances marking the location of channel mouths. In regions characterized by mesotidal regimes tidal currents are active within the distributary mouths while wave processes operate along the remainder of the delta front. The delta front consequently comprises a series of cheniers separated by tidal channels. In macrotidal settings tidal action is so effective that the delta front is dominated by tidal current ridges, channels and islands.

In deltas which form major protuberances, levees developed along distributary channels may extend out into lakes or seas. These levees are generally lower than the water level attained in the channels during major floods and are consequently fairly frequently breached. Once a breach has been made this may then provide a new route for the main channel, especially if it is shorter and therefore has a steeper gradient than the existing main channel course. In the present Mississippi delta the Atchafalaya River is beginning to divert a significant proportion of the flow from the Mississippi River, since from their point of bifurcation the course of the Mississippi to the Gulf of Mexico is well over twice as long (although this flow diversion is presently being limited by artificial channel controls). The steeper gradient of the Atchafalaya is promoting a process common in deltas, that of channel switching, or avulsion, whereby the existing delta or delta lobe is abandoned and a new delta or lobe is initiated elsewhere. The sediment supply to the previous delta gradually diminishes and eventually it ceases to prograde while the diversion of sediment along the new main distributary channel leads to the construction

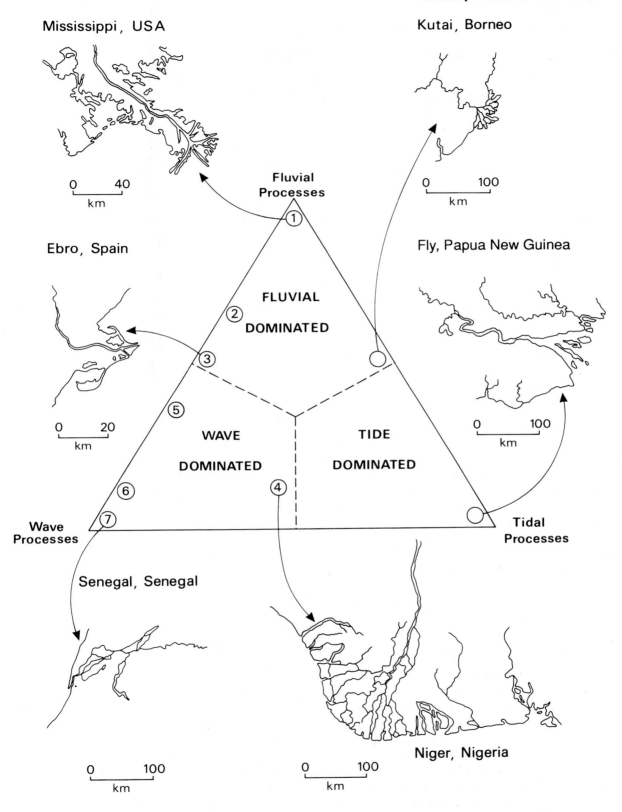

Fig. 13.22 *Ternary diagram of delta types related to fluvial, wave and tidal controls. The numbers refer to the descriptions of delta characteristics in Table 13.4. (Partly based on W. E. Galloway (1975) in M. L. Broussard (ed.)* Deltas, Models for Exploration. *Houston Geological Society, Houston, Fig. 3, p. 92.)*

Table 13.5   Factors influencing the morphology and sedimentary characteristics of deltas

River regime	Flood stage	Sediment load	Quantity of suspended load and bed load (stream capacity) increases during flood
		Particle size	Particle size of suspended load and bed load (stream competence) increases during flood
	Low river stage	Sediment load	Stream capacity diminishes during low river stage
		Particle size	Stream competence diminishes during low river stage
Coastal processes	Wave energy		High wave energy with resulting turbulence and currents erode, rework and winnow deltaic sediments
	Tidal range		High tidal range distributes wave energy across an extended littoral zone and creates tidal currents
	Current strength		Strong littoral currents, generated by waves and tides, transport sediment alongshore, offshore, and inshore
Tectonic factors	Stable area		Rigid basement precludes delta subsidence and forces deltaic plain to build upward as it progrades
	Subsiding area		Subsidence through downwarping coupled with sediment compaction allows delta to construct overlapping sedimentary lobes as it progrades
	Rising area		Uplift of land (or lowering of sea level) causes river distributaries to cut downward and rework their sedimentary deposits
Climatic factors	Wet area	Hot or warm	High temperature and humidity yield dense vegetative cover, which aids in trapping sediment transported by fluvial or tidal currents
		Cool or cold	Seasonal character of vegetative growth is less effective in sediment trapping; cool winter temperature allows seasonal accumulation of plant debris to form delta plain peats
	Dry area	Hot or warm	Sparse vegetative cover plays minor role in sediment trapping and allows significant aeolian processes in deltaic plain
		Cool or cold	Sparse vegetative cover plays minor role in sediment trapping; winter ice interrupts fluvial processes: seasonal thaws and aeolian processes influence sediment transportation and deposition

Source: After J. P. Morgan (1970) in: J. P. Morgan (ed.) *Deltaic Sedimentation: Modern and Ancient*, Society of Economic Paleontologists and Mineralogists Special Publication **15**, Table 1, p. 33.

of a new delta. This process is beautifully illustrated by the sequence of abandoned deltas which lie to the east and west of the modern Mississippi delta and which have formed over the past 7000 a (Fig. 13.24).

Breaches of delta distributary levees need not result in the switching of main channels; more often they feed lakes or bays on the delta plain known as crevasses and the deposits they form which infill the area between extending levees

are known as crevasse-splay deposits. In well-vegetated deltas there is also a significant accumulation of organic debris in these inter-distributary areas which may form thick sequences of peat.

13.4.2.6   Coastal dunes
Coastal dunes are affected by the same basic aeolian processes responsible for the formation of desert dunes (see

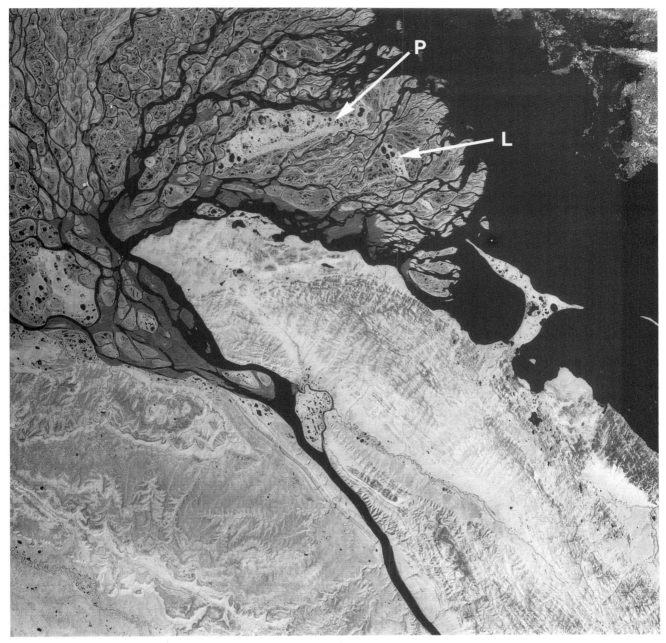

Fig. 13.23 *Part of the delta of the Lena River on the arctic coast of the USSR. River flow is restricted to a short summer season and the delta plain is covered by landforms related to the presence of permafrost such as pingos (P) and thaw lakes (L). The image covers an area about 180 km across. (Landsat image courtesy N. M. Short.)*

Chapter 10) but additional factors must also be considered. The key variable determining whether dunes will develop is the rate of abstraction of sand from the beach. Other important variables are the effects of vegetation and the nature of the backshore zone.

Whether sufficient quantities of sand will be blown landward from a beach to make dune formation possible will depend on several factors. Inland aeolian transportation of beach sand will be favoured by high onshore wind speeds and low humidity and precipitation and, of course, the availability of sediment capable of being moved by the prevailing winds. Tidal range will be crucial since a larger area of beach sand will be exposed to aeolian action where there is a high tidal range, although this will also depend on the beach gradient. Tidal type is also important since semi-diurnal and mixed tides allow less time for beach sediments to dry out and this renders them less susceptible to wind action (although tidal range outweighs this factor). The

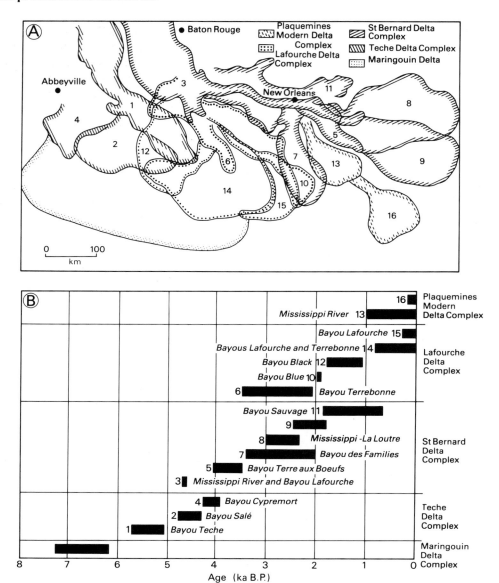

Fig. 13.24 *Distribution of deltas and delta lobes at the outlet of the Mississippi River formed over the past 7000 a (A), and chronology of individual phases of delta construction over this period (B). (After T. Elliott (1978) in H. G. Reading (ed.)* Sedimentary Environments and Facies. *Blackwell, Oxford. Figs. 6.34A and 6.34B, pp. 122 and 123, based on D. E. Frazier (1967)* Transactions of the Gulf-Coast Association of Geological Societies *17, Figs. 11 and 12, pp. 307 and 308.)*

final significant control is sand supply. There is a clear association between those coasts receiving an abundant supply of sand from rivers and those with large coastal dunes. Spectacular examples are the coasts of Oregon and Washington in the USA which have massive dune systems fed ultimately by fluvial sediments originating from high gradient, aggressively-eroding catchments.

A global perspective shows that dunes are preferentially developed in the storm wave coastal environments of the mid-latitudes, notably in north-west Europe and the north-west USA which are characterized by frequent high velocity onshore winds. By comparison coastal dunes are generally poorly developed in the humid tropics. This is not surprising since here fluvial sediments are typically poor in sand, both humidity and precipitation are high and mean wind velocities are low. Such a situation may not apply locally, as along the coast of Queensland, Australia, where there is an unusually abundant sand supply.

Primary dunes fed directly by beach sand can be categorized into free dunes and impeded dunes depending on whether vegetation has been important in initiating their formation (Table 13.6). **Secondary dunes**, which develop through the erosion of impeded dunes, constitute a further major category. Free dunes are most abundant along desert

Table 13.6　Classification of coastal dunes

TYPE	FORMS	ORIENTATION
Primary dunes (sand derived from beach)		
Free dunes (vegetation unimportant)	Transverse ridges barchans, oblique ridges	Wind orientated and generally perpendicular to direction of constructional winds
Impeded dunes (vegetation important)	Frontal dunes, sand beach ridges, dune platforms	Nucleus orientated and generally parallel to rear of source beach
Secondary dunes (sand derived from erosion of impeded dunes)		
Transgressive dunes	Blow-out dunes, parabolic dunes, longitudinal dunes, transgressive sheets	Wind orientated and generally parallel to direction of constructional winds
Remnant dunes (eroded remnants of vegetated primary dunes)		

Source: Adapted from J. L. Davies (1980) *Geographical Variation in Coastal Development (2nd edn)*. Longman, London, p. 157.

coasts, such as those of the Namib and Atacama Deserts and sections of Baja California, where vegetation is extremely sparse due to the hyper-aridity of these regions. They also occur locally in more humid environments where species of sand-binding plants sufficiently vigorous to vegetate and stabilize the dunes are absent. This is especially noticeable along the Washington and Oregon coast where native dune plants are in most localities incapable of establishing themselves in the midst of the very high rates of sand movement. Free dunes are most commonly of the transverse type, but they may evolve into barchans as they migrate inland as a result of the reduced sediment supply

away from the beach (see Section 10.3.5.1).

More abundant globally are the impeded dunes characteristic of humid and subhumid coasts. They take a variety of forms but most have a frontal dune running along the back of the beach. Their detailed form is significantly affected by the type of colonizing plants instrumental in trapping the sand. The eventual height reached by impeded dunes is essentially governed by wind velocity so that it is no surprise to find the highest dunes along the stormy west coasts of the mid-latitudes. Secondary dunes develop as a consequence of dune erosion which can arise from a reduction in sand supply or vegetation cover, or from an increase in wind velocity. Some dunes are particularly sensitive to vegetation removal and blowouts and parabolic dunes can develop rapidly, initiating a new cycle of sand movement inland.

Dunes lithified by calcium carbonate are found on some tropical coasts. They occur in rather similar environments to beachrock, but their distribution is more extensive and reaches into sub-tropical latitudes. They are characteristic of coasts experiencing a warm wet–dry climate in which there is sufficient moisture for $CaCO_3$ to be dissolved from the carbonate fraction of the sand, but sufficiently high evaporation during the dry season to promote its subsequent reprecipitation within the dune. Dunes subject to lithification do not have to be particularly carbonate-rich, lithification having been observed in dunes with as little as 8 per cent detrital carbonate. Lithified dunes are extensively developed along the coasts of Australia, southern Africa and Brazil, among other areas, but direct climatic relationships are difficult to establish since many appear to be relict features.

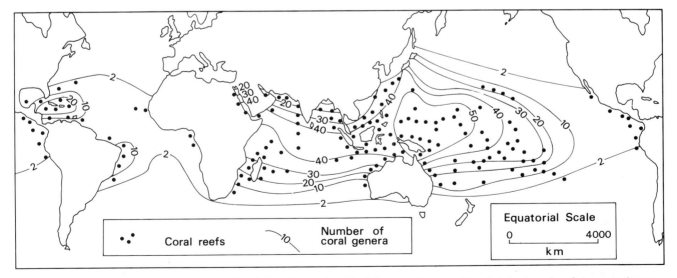

Fig. 13.25　*Global distribution of coral reefs and abundance of reef-building coral genera. The distribution of reefs is approximate only as there are over 400 individual coral reefs. (Based largely on J. L. Davies (1980)* Geographical Variation in Coastal Development, *(2nd edn) Longman, London Fig. 45, pp. 66; and D. R. Stoddart (1969)* Biological Reviews of the Cambridge Philosophical Society *44, Fig. 16, p. 43.)*

13.4.2.7 *Reefs*

Reefs are rocky, shallow-water, submarine ridges. Although they can be composed of inorganic rocks, biogenic forms composed mainly of coral along with the remains of a wide variety of other carbonate-secreting organisms are by far the most common. The environmental requirements of coral growth limit the global distribution of coral reefs broadly to the tropics (Fig. 13.25). A minimum sea temperature of about 18 °C is required with optimum development occurring between 25 and 29 °C. Light requirements limit the depth at which corals grow to around 90 m, but the most vigorous development is confined to a depth of 20 m. A normal ocean water salinity is required along with a moderate amount of water movement and a firm substrate on which the coral can establish itself. Coral growth is adversely affected by sediment deposition on the reef and this contributes to the comparative paucity of reef development in the tropical Atlantic Ocean which receives abundant sediment from several major river systems.

Coral reefs occur in a variety of forms and positions with respect to mainland coasts. Three major types can be identified: **fringing reefs** which are separated from the mainland or island coast by a shallow channel; **barrier reefs** which are separated by a deep channel; and reefs forming islands. Fringing reefs are the most widespread type, but the other varieties are perhaps of more interest in view of their complex histories which reflect the interplay of sea-level changes, tectonic subsidence and coral growth (see Section 17.6.3). Indeed the present-day morphology of most coral reefs is a palimpsest of phases of construction and destruction extending well back into the Pleistocene.

Most coral reefs now exposed above sea level have been subject to a range of geomorphic processes, with wave action and solution and other forms of weathering contributing to the formation of cliffs and shore platforms. Erosion provides sediments which can be built into depositonal forms. Coral reefs can also significantly influence coastal processes along the adjacent mainland, especially through the large amount of wave energy that they are capable of dissipating. Wave-resistant species of coral are concentrated on the oceanward edge of reefs, while organic sediment eroded from this exposed margin is deposited in the calmer zone behind the reef. An intertidal reef flat or sandy lagoon may lie between the reef and island or mainland. Algal limestones are important in coating the very outer rim and contributing to reef stability.

Further reading

There are a number of texts which cover the field of coastal geomorphology or large parts of it. Davies (1980) provides an excellent introduction with a global perspective, while Bird (1984), King (1972) and Pethick (1984) are also good starting points. More detailed treatments of specific topics are to be found in Bird (1985), Clark (1984), Davis (1985) and Komar (1983a) while Bird and Schwartz (1985) provides a global survey of coastal features and Fitzgerald and Rosen (1987) contains papers on glaciated coasts focusing on those regions of North America affected by Quaternary glaciation. Some recent ideas on aspects of coastal erosion are considered by Komar and Holman (1986), while Inman and Nordstrom (1971) provide a stimulating tectonic framework within which to understand coastal morphology and Audley-Charles *et al.* (1977) consider the location of deltas in terms of tectonic setting.

The generation of waves is discussed in Komar (1976), while Suhayda and Pettigrew (1977) look at the relationship between wave height and velocity and Galvin (1968) examines the way in which beach gradient and wave height and wave length together influence the form of breaking waves. Davies (1980) considers the environmental controls of tidal types and Komar (1976) provides a mathematical treatment of the generation of tides and a detailed discussion of currents.

Turning to the processes operating in the coastal environment Robinson and Jerwood (1987) consider the role of frost and salt weathering while the idea of water-layer weathering is introduced by Wentworth (1938). Sediment mobilization and movement in the swash and surf zone is examined by Brenninkmeyer (1976) and Komar (1977), while longshore drift is discussed by Komar (1976). Various aspects of the important role of organisms in coastal processes are assessed by Spencer (1988), Thom (1967) and Stoddart (1969).

The development of cliffs and shore platforms is considered in a detailed treatment of rocky coasts by Trenhaille (1987), while Emery and Kuhn (1982) look specifically at the nature and origin of cliffs and Bradley and Griggs (1976) and Trenhaille (1972, 1978) examine various features of shore platform development. Komar (1976) examines the relative importance of different sediment sources in the formation of beaches, and Thom (1968) points to the importance of the Holocene sea-level rise in the translocation of sediment and the development of contemporary constructional landforms. The formation of beachrock is discussed by Scoffin and Stoddart (1983). DuBois (1972) and McLean and Kirk (1969) consider the relationship between sediment size and sorting and beach gradient and Komar (1976) discusses the role of wave height. General problems in the development of rhythmic topography are discussed by Komar (1976, 1983b) while cusp formation is considered by Guza and Inman (1975) and Dalrymple and Lanan (1976). The characteristics and development of cheniers are examined in general terms by Otvos and Price (1979) and with specific reference to China by Cangzi and Min (1987).

There is a large literature on barrier islands; various views concerning their development are given in Carter and Orford (1984), Cooke (1968), Eddison (1983), Fisher (1968) and

Hoyt (1967, 1968), while the papers in Leatherman (1979) and those in a special issue of the journal *Marine Geology* (1985, vol. 63) give a good idea of the continuing points of debate. The depositional aspects of tidal flats are comprehensively discussed in Ginsburg (1975), while Thom (1967) and Thom *et al.* (1975) consider the relationships between mangrove development and estuarine sedimentation. Elliott (1986) provides an excellent review of deltas with specific reference to their sedimentological characteristics, while Morgan (1970a) provides a brief but useful introduction and Morgan (1970b) a more detailed treatment of the topic. The global variation in delta morphology in relation to environmental controls is examined by Coleman (1981), Coleman and Wright (1975), Wright and Coleman (1973) and Galloway (1975), while Wright and Coleman (1974) look in detail at the development of the intensively studied Mississippi delta system.

Pye (1983) reviews the literature on coastal dunes, while Pye and Bowman (1984) consider the impact of Holocene sea-level rise on their development. Guilcher (1988) provides a comprehensive coverage of the geomorphology of coral reefs and Hopley (1982) presents a detailed study of the Great Barrier Reef of Australia. Further references on coral reef development are given at the end of Chapter 17.

References

Audley-Charles, M. G., Curray, J. R. and Evans, G. (1977) Location of major deltas. *Geology* 5, 341–4.

Bird, E. C. F. (1984) *Coasts: An Introduction to Coastal Geomorphology*. Blackwell, Oxford.

Bird, E. C. F. (1985) *Coastline Changes: A Global Review*. Wiley, Chichester and New York.

Bird, E. C. F. and Schwartz, M. L. (eds) (1985) *The World's Coastline*. Van Nostrand Reinhold, New York.

Bradley, W. C. and Griggs, G. B. (1976) Form, genesis, and deformation of central California wave-cut platforms. *Geological Society of America Bulletin* 87, 433–49.

Brenninkmeyer, B. (1976) Sand fountains in the surf zone. In: R. A. Davis and R. L. Ethington (eds) *Beach and Nearshore Sedimentation*. Society of Economic Paleontologists and Mineralogists Special Publication 24, 69–81.

Cangzi, L. and Min, C. (1987) The chenier plains of China. In: V. Gardiner *et al.* (eds) *International Geomorphology 1986: Proceedings of the First International Conference on Geomorphology* Part I. Wiley, Chichester and New York, 1269–79.

Carter, R. W. G. and Orford, J. D. (1984) Coarse clastic barrier beaches: a discussion of their distinctive dynamic and morphosedimentary characteristics. *Marine Geology* 60, 377–84.

Clark, M. W. (ed.) (1984) *Coastal Research: UK Perspectives*. Geo-Books, Norwich.

Coleman, J. M. (1981) *Deltas: Processes of Deposition and Models for Exploration*. Burgess, Minneapolis.

Coleman, J. M. and Wright, L. D. (1975) Modern river deltas: variability of processes and sand bodies. In: M. L. Broussard (ed.) *Deltas, Models for Exploration*. Houston Geological Society, Houston, 99–149.

Cooke, C. W. (1968) Barrier island formation: discussion. *Geological Society of America Bulletin* 79, 945.

Dalrymple, R. A. and Lanan, G. A. (1976) Beach cusps formed by intersecting waves. *Geological Society of America Bulletin* 87, 57–60.

Davies, J. L. (1980) *Geographical Variation in Coastal Development* (2nd edn). Longman, London and New York.

Davis, R. A. (ed.) (1985) *Coastal Sedimentary Environments*. Springer-Verlag, New York.

DuBois, R. N. (1972) Inverse relation between foreshore slope and mean grain size as a function of the heavy mineral content. *Geological Society of America Bulletin* 79, 1421–6.

Eddison, J. (1983) The evolution of the barrier beaches between Fairlight and Hythe. *Geographical Journal* 149, 39–53.

Elliott, T. (1986) Deltas. In: H. G. Reading (ed.) *Sedimentary Environments and Facies* (2nd edn). Blackwell Scientific Publications, Oxford, 113–54.

Emery, K. O. and Kuhn, G. G. (1982) Sea cliffs: their processes, profiles and classification. *Geological Society of America Bulletin* 93, 644–54.

Fisher, J. J. (1968) Barrier island formation. Discussion. *Geological Society of America Bulletin* 79, 1421–6.

Fitzgerald, D. M. and Rosen, P. S. (eds) (1987) *Glaciated Coasts*. Academic Press, San Diego.

Galloway, W. E. (1975) Process framework for describing the morphologic and stratigraphic evolution of the deltaic depositional systems. In: M. L. Broussard (ed.) *Deltas, Models for Exploration*. Houston Geological Society, Houston, 87–98.

Galvin, C. J. (1968) Breaker-type classification on three laboratory beaches. *Journal of Geophysical Research* 73, 3651–9.

Ginsburg, R. N. (ed.) (1975) *Tidal Deposits*. Springer-Verlag, Berlin.

Guilcher, A. (1988) *Coral Reef Geomorphology*. Wiley, Chichester and New York.

Guza, R. T. and Inman, D. L. (1975) Edge waves and beach cusps. *Journal of Geophysical Research* 80, 2997–3012.

Hopley, D. (1982) *The Geomorphology of the Great Barrier Reef: Quaternary Development of Coral Reefs*. Wiley, New York.

Hoyt, J. H. (1967) Barrier island formation. *Geological Society of America Bulletin* 78, 1125–36.

Hoyt, J. H. (1968) Barrier island formation: reply. *Geological Society of America Bulletin* 79, 947 and 1427–32.

Inman, D. L. and Nordstrom, C. E. (1971) On the tectonic and morphologic classification of coasts. *Journal of Geology* 79, 1–21.

King, C. A. M. (1972) *Beaches and Coasts*. Edward Arnold, London.

Komar, P. D. (1976) *Beach Processes and Sedimentation*. Prentice-Hall, Englewood Cliffs.

Komar, P. D. (1977) Selective longshore transport rates of different grain-size fractions within a beach. *Journal of Sedimentary Petrology* 47, 1444–53.

Komar, P. D. (ed.) (1983a) *CRC Handbook of Coastal Processes and Erosion*. CRC Press, Boca Raton.

Komar, P. D. (1983b) Rhythmic shoreline features and their origins. In: R. A. M. Gardner and H. Scoging (eds.) *Mega-Geomorphology*. Clarendon Press, Oxford, 92–112.

Komar, P. D. and Holman R. A. (1986) Coastal processes and the development of shoreline erosion. *Annual Review of Earth and Planetary Sciences* 14, 237–65.

Leatherman, S. P. (ed.) (1979) *Barrier Islands: From the Gulf of St. Lawrence to the Gulf of Mexico*. Academic Press, New York.

McLean, R. F. and Kirk, R. M. (1969) Relationship between grain-size, size-sorting and foreshore slope on mixed sand–shingle beaches. *New Zealand Journal of Geology and Geophysics* 12, 138–55.

Morgan, J. P. (1970a) Deltas – a resumé. *Journal of Geological Education* **18**, 107–17.

Morgan, J. P. (ed.) (1970b) *Deltaic Sedimentation; Modern and Ancient.* Society of Economic Paleontologists and Mineralogists Special Publication **15**.

Otvos, E. G. Jr and Price, W. A. (1979) Problems of chenier genesis and terminology – an overview. *Marine Geology* **31**, 251–63.

Pethick, J. S. (1984) *An Introduction to Coastal Geomorphology.* Edward Arnold, London and Baltimore.

Pye, K. (1983) Coastal dunes. *Progress in Physical Geography* **7**, 531–57.

Pye, K. and Bowman, G. M. (1984) The Holocene marine transgression as a forcing function in episodic dune activity on the eastern Australian coast. In: B. G. Thom (ed.) *Australian Coastal Geomorphology.* Academic Press, Sydney, 179–96.

Robinson, D. A. and Jerwood, L. C. (1987) Frost and salt weathering of chalk shore platforms near Brighton, Sussex, U.K. *Transactions, Institute of British Geographers* NS **12**, 217–26.

Scoffin, T. P. and Stoddart, D. R. (1983) Beachrock and intertidal cements. In: A. S. Goudie and K. Pye (eds) *Chemical Sediments and Geomorphology: Precipitates and Residua in the Near-Surface Environment.* Academic Press, London and New York, 401–25.

Spencer, T. (1988) Limestone coastal morphology: the biological contribution. *Progress in Physical Geography* **12**, 66–101.

Stoddart, D. R. (1969) Ecology and morphology of Recent coral reefs. *Biological Reviews of the Cambridge Philosophical Society* **44**, 433–98.

Suhayda, J. N. and Pettigrew, N. R. (1977) Observations of wave height and wave celerity in the surf zone. *Journal of Geophysical Research* **82**, 1419–24.

Thom, B. G. (1967) Mangrove ecology and deltaic geomorphology: Tabasco, Mexico. *Journal of Ecology* **55**, 301–43.

Thom, B. G. (1968) Coastal erosion in eastern Australia. *Australian Geographical Studies* **6**, 171–3.

Thom, B. G., Wright, L. D. and Coleman, J. M. (1975) Mangrove ecology and deltaic–estuarine geomorphology: Cambridge–Ord River region, Western Australia. *Journal of Ecology* **63**, 203–32.

Trenhaille, A. S. (1972) The shore platforms of the Vale of Glamorgan, Wales. *Transactions of the Institute of British Geographers* **56**, 127–44.

Trenhaille, A. S. (1978) The shore platforms of Gaspé, Quebec. *Annals of the Association of American Geographers* **68**, 95–114.

Trenhaille, A. S. (1987) *The Geomorphology of Rock Coasts.* Clarendon Press, Oxford.

Wentworth, C. K. (1938) Marine bench forming processes: Part 1, Water-level weathering. *Journal of Geomorphology* **1**, 6–32.

Wright, L. D. and Coleman, J. M. (1973) Variations in morphology of major river deltas as functions of ocean wave and river discharge regimes. *American Association of Petroleum Geologists Bulletin* **57**, 370–98.

Wright, L. D. and Coleman, J. M. (1974) Mississippi River mouth processes: effluent dynamics and morphologic development. *Journal of Geology* **82**, 751–78.

14

Climate, climatic change and landform development

14.1 Climate and landform development

The relationship of landforms to climate is a topic which has simultaneously been regarded as a focus for research in geomorphology and an approach which has fruitlessly occupied the time of geomophologists over the past several decades. The former view has characterized research by many French and German geomorphologists, whereas the latter has prevailed among Anglo-American researchers. The processes and landforms associated with fluvial, aeolian, glacial and periglacial environments reviewed in earlier chapters clearly demonstrate that *major* differences in climate have a profound effect on landscape development.

Disagreement comes with the finer distinctions between degrees of humidity and temperature evident in attempts to identify morphoclimatic zones (see Section 1.5.3), although recently there have been some rather more sophisticated efforts to cast climatic controls in terms of the magnitude and frequency of meteorological events. Although variations in both temperature and, more especially, precipitation can, in some instances, be sensitively reflected in processes such as those involved in weathering and fluvial activity, when landscapes as a whole are considered other factors related to tectonics and lithology often predominate. For instance, if we examine a range of humid tropical environments around the world, diversity of landscape is more apparent than uniformity (Table 14.1). A further problem in trying to

Table 14.1 Comparison of landscape characteristics in a range of humid tropical environments

FEATURE	NORTHERN AUSTRALIA	SRI LANKA	MALAYSIA	BORNEO	NEW GUINEA	HAWAII
Dominant age of landscape	Old	Old	Moderately old	Young	Very young	Very young
Proportion of labile rocks	Low	Low	Moderate	Moderate	High	Moderate
Recent tectonism	Practically none	Practically none	Negligible	Some	Intense	Infrequent
Cenozoic volcanism	Practically absent	Practically absent	Practically absent	Some	Abundant	Highly abundant
Tropical karst	Practically absent	Virtually absent	Residuals only	Abundant	Abundant	None
Relief	Low to moderate	Moderate	Moderate	Moderate (save for Kinabulu)	Very high	High
Rate of current regolith development	Slow	Slow	Slow to moderate	Moderate	Rapid	Rapid
Rate of current erosion and deposition	Low to moderate	Low to moderate	Moderate	High	Very high	Very high
Depth of colluvium	Generally shallow	Generally shallow	Variable	Variable	Variable	Variable but usually high
Ferricrete	Widespread	Widespread	Fragments	Possible	Absent	Absent
Pallid zone of deep weathering	Widespread	Widespread	Few	Unknown	Rare	Rare

Source: After I. Douglas (1978) in: C. Embleton, D. Brunsden and D. K. C. Jones (eds) *Geomorphology: Present Problems and Future Prospects.* Oxford University Press, Oxford, Table IV, p. 175.

correlate relatively small differences in climate with particular landforms or landscape types is the diversity of conditions under which apparently similar forms can develop.

Perhaps the most intractable problem in climatic geomorphology is the unambiguous linking of a particular landform with a specific set of climatic conditions. No landform better illustrates this dilemma than the pediment (see Section 7.6.2). Pediments are gently inclined slopes with gradients of between 2 and 10° extending from slopes cut in bedrock down to an area of sediment deposition at a lower elevation. The backing slope usually has a gradient of more than 20° and joins the pediment at a marked change in slope known as the **piedmont angle**. Some pediments have a concave-up long profile, and some have a partial veneer of sediment. Pediments can terminate upslope at a mountain front, or at a residual hill, known as an inselberg if large, and a tor if small.

Part of the confusion about the origin of pediments and their climatic affinities arises from the term being applied to two morphologically similar, but genetically distinct, forms. One type is associated with a resistant cap rock (either a sedimentary unit or a duricrust) overlying more readily erodible material (Figs 14.1 (A) and 14.2). The pediment forms through slope retreat and the rate at which this occurs is determined by the rate at which the cap rock

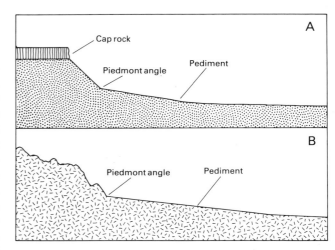

Fig. 14.1 *Two types of pediment: (A) associated with a sedimentary caprock or duricrust; (B) rock pediment.*

is undermined and collapses. Although such forms are most evident in semi-arid environments there is evidence from many regions that slope retreat and pediment growth is most active in more humid climatic interludes. This seems to be especially the case where the cap rock is permeable

Fig. 14.2 *Pediments extending from silcrete-capped surfaces, near Stuart Creek, South Australia.*

Fig. 14.3   *Granite bornhardt (domed inselberg) standing above a gravel-mantled rock pediment in the hyper-arid Namib Desert east of Walvis Bay, Namibia.*

since under humid conditions ground water emerges at the cap rock base and promotes undermining.

Much more controversy is associated with the other type of pediment, the **rock pediment**. In this form the pediment and the backing mountain range or inselberg are composed of the same rock type, usually a coarse-grained intrusive rock such as granite (Figs 14.1 (B) and 14.3). Rock pediments are most evident in, though not confined to, arid environments. One model suggests that they form through parallel slope retreat under present climatic conditions, and that the piedmont angle is attributable to the way granitic rocks break down into sand-sized material, or **grus**, without intermediate pebble and cobble-sized debris. Such small calibre material can be readily removed by sheet flow and therefore does not accumulate at the slope base. There are, however, two difficulties with this model as a general explanation of pediment formation. One is that many rock pediments are associated with pockets of deeply weathered regolith or are even cut in saprolite, and this clearly implies a more humid environment at some stage during their development. A second difficulty is that the dating of some rock pediment surfaces has demonstrated that they are

ancient forms which are essentially 'frozen' in their development in the prevailing arid climate.

Evidence of saprolite formation in association with rock pediment development comes from localities as diverse as the central Sahara, Australia and southern Arabia. In the Mojave Desert of southern California, USA, the presence of rock varnish on inselberg slopes indicates that they are inactive at present although surrounded by pediments several kilometres in length. Deep weathering under a humid climate is supported by the presence of tors and core stones, regolith stripping probably occurring when the climate became somewhat drier (but not arid). The piedmont angle is attributed to enhanced physical and chemical weathering related to concentrated water flow at the base of the back slope during and after regolith stripping. This interpretation is supported by palaeoclimatic evidence which indicates that desiccation of the Mojave region at the end of the Miocene was linked to the creation of a rain-shadow caused by the uplift of the Sierra Nevada and Transverse Ranges to the west. Prior to about 6 Ma ago the vegetation cover ranged from woodland to semi-desert in type, but since that time the climate has been distinctly more arid and pediment modification has been extremely slow, apparently involving an average of less than 100 m extension over the past 5 Ma. In the eastern Mojave Desert 60 basalt flows up to 4 Ma old partly bury pediment surfaces, and comparison of the relative positions of pediments covered by dated basalts and the modern pediment surfaces indicates the degree of modification that has occurred over this time interval. The conclusion is the amount of modification has been minimal, with downwearing predominating over backwearing.

Although the exhumation model of pediment formation, along with its requirement for a previously humid climate, seems to have widespread applicability there appear to be situations where it cannot apply because there is compelling evidence of prolonged aridity. One such case is provided by the pediments flanking the domed inselbergs, or **bornhardts**, of the Namib Desert in southern Africa (Fig. 14.3). Here the cold Benguela Current offshore in conjunction with the dominant sub-tropical high pressure system has created a hyper-arid coastal desert climate which has probably prevailed since at least the mid-Miocene. Although the present rate of slope retreat seems to be extremely slow there are no remnants of an earlier phase of significant rock weathering to support the idea of exhumation. Such pediment and inselberg forms represent one of the major enigmas of geomorphology.

14.2   The significance of climatic change

The presence of landforms, such as certain kinds of pediments, in climatic environments in which they could not have developed obviously raises the question of climatic change in landform interpretation. The state of the atmosphere varies from day to day and even hour to hour, and it is average weather conditions together with their variability that characterize the climate of a region. But climate also fluctuates at all time scales from decades to tens of millions of years. Whether a climatic change is of geomorphic significance depends on its magnitude and duration, and on the properties of the landform concerned. The larger or more resistant a landform, the longer, in general, it takes to adjust to a change in climate. The 'sensitivity' of different landscapes to climatic fluctuations is an important but complex question which affects the way we conceive long-term landscape development and we examine it in detail in Chapter 18 (see Section 18.2.3).

An important distinction is between 'active' landforms which seem to be in 'equilibrium' with the prevailing climatic environment, and inactive or relict landforms which could not have been formed under the current climatic regime. As will have been evident from our discussion of the pediment problem, this distinction is not always easy to make. Although we saw in Section 1.5.3 that it is possible to relate broad associations of geomorphic processes to major climatic zones, the overlap between categories is significant since most processes operate in most environments at least to some degree. In many cases we do not yet know enough about the development of specific landforms to be able to say with any confidence which are being actively formed in a particular climatic environment. The problem of distinguishing between active and relict forms can be particularly acute in those environments where even 'active' geomorphic processes operate sporadically. This is the case in many arid and semi-arid environments where the irregular patterns of precipitation mean that fluvial activity is very limited in duration, but, none the less, often very important in shaping the landscape. The question here is how long does a landform have to be inactive for it to be described as relict?

The past two decades have seen a quantum leap in our understanding of climatic change largely as a result of the exploration of the ocean floor, which contains a relatively complete and undisturbed climatic record, and of the development of new dating procedures and techniques of palaeo-environmental reconstruction. Our knowledge of climatic fluctuations on the continents, nevertheless, is still severely constrained by a lack of palaeoclimatic indicators capable of yielding datable material. In fact landforms themselves have been extensively used to reconstruct changes in temperature, precipitation and wind intensity and direction on the continents. The use of this kind of evidence actually raises a problem for geomorphologists concerned with interpreting the landscape in terms of climatic change because in many cases the climate changes themselves have been interpreted largely from landform evidence. Clearly there is a danger of circular arguments here and, if possible, we must use independent information on climatic change.

The consequences of climatic change for landform genesis can, very broadly, be divided into those effects related *primarily* to temperature changes, and those related *primarily* to changes in precipitation. The growth and decay of ice sheets is clearly influenced significantly by long-term changes in temperature, whereas surface runoff and fluvial processes, along with the level of aeolian activity, are highly responsive to precipitation changes. We must be careful, however, in making broad generalizations. For instance, although frequently resulting from a temperature fall, glacier growth can also occur as a result of an increase in precipitation when temperatures are stable, or even increase. Moreover, an increase in surface runoff can arise from a decrease in temperature (and hence rates of evapotranspiration) as well as an increase in precipitation. Nevertheless, it is convenient to separate the discussion of the expansion and contraction of glacial and periglacial morphoclimatic regimes, which are largely a response to temperature fluctuations, from the predominant impact of precipitation changes on fluvial and aeolian systems. Before we do this, however, it is necessary to review briefly our present understanding of the history of global climatic change.

14.3 The record of climatic change

A wide range of techniques are used to reconstruct past climates and they are employed on diverse types of evidence. For long-term climatic change information is acquired from the changing distributions of plants and animals determined from fossils, and from rock types which are attributable to particular climatic regimes. Evaporites and aeolian sandstones, for instance, are indicative of arid environments, whereas tillites demonstrate glacial conditions. Approaches to reconstructing more recent climatic change include the analysis of pollen (**palynology**) and landforms themselves.

One technique, however, has done more than any other over the past two decades or so to revolutionize our understanding of climatic change. In the late 1940s it was pointed out that the ratio of the two stable isotopes of oxygen, ^{16}O and ^{18}O, precipitated in carbonate would vary with water temperature. This idea was applied in 1955 by C. Emiliani to the analysis of calcium carbonate ($CaCO_3$) shells secreted by various sea creatures. The technique of **oxygen isotope analysis**, as it was termed, was most often applied to foraminifera, single-celled organisms mostly between 0.1 and 0.3 mm across. Emiliani found that the $^{18}O/^{16}O$ ratio varied in foraminifera shells recovered from ocean cores recording sedimentation over the past few hundred thousand years. Both planktonic (surface and near-surface) and benthic (bottom-dwelling) forms of foraminifera occur so it was possible to estimate temperature changes on both the sea surface and in the ocean depths. The changes in oxygen isotope ratios were dated using the palaeomagnetic time scale in combination with an assumed constant rate of sedimentation.

It was subsequently appreciated that the $^{18}O/^{16}O$ ratio of the sea water itself would also vary as the quantity of land ice changed with fluctuations in global temperatures. Such changes would be reflected in the foraminifera and would arise because the lower atomic mass of ^{16}O means that it is preferentially evaporated from the oceans. During glacials a proportion of this ^{16}O-enriched water would be locked up in ice sheets thus leaving the oceans relatively enriched in ^{18}O. It is difficult to separate these two effects – one indicating changes in ocean temperatures, the other fluctuations in the volume of ice sheets – and this has led to divergent views of climatic history, especially for the past 40 Ma.

14.3.1 The Cretaceous to Neogene record

There are probably few areas in the world where landforms or weathering deposits can be regarded as having survived more or less intact for more than the past 100 Ma, although it has been suggested that even more ancient forms exist in northern Australia. It is appropriate, therefore, to begin our brief survey of climatic change with the Cretaceous.

The climate of the Late Cretaceous – 100 to 65 Ma BP – is generally thought to have been warm and equable with a very much less marked temperature gradient with latitude compared with the present. Polar regions probably experienced a climate rather similar to mid-latitude and subtropical climates of the present day, although the amount of solar radiation received and the seasonality of climate would have been similar to equivalent latitudes today. However this widely accepted ice-free Cretaceous interpretation has been challenged by researchers who have interpreted large exotic blocks found within Early Cretaceous mudstones in Australia as **dropstones**, that is, debris dropped from floating ice. This evidence indicates that ice was present at sea level in high latitudes during the Cretaceous. Similar interpretations have been applied to exotic blocks in strata of a variety of ages and it has been argued that these indicate, contrary to conventional wisdom, that the Earth may only rarely have been ice free in the past. These interpretations have yet to be confirmed by other evidence, and meanwhile it is perhaps best to adopt the generally accepted ice-free Cretaceous model. Towards the end of the Cretaceous there is good evidence for a slight cooling, but this trend was reversed during the Paleocene and Eocene when a warm phase saw tropical soil formation extending to 45° latitude in both hemispheres and tropical species living in southern England and western Greenland.

The major point of dispute in the climatic record for the past 50 Ma is when the Antarctic ice sheet became established. One interpretation of the oxygen isotope data suggests that it first appeared about 15 Ma ago, but if $^{18}O/^{16}O$ ratio variations are caused predominantly by ice volume changes then a date of around 35 Ma is indicated (Fig. 14.4). This earlier date is supported by geomorphic and sedimentological evidence, including glacially abraded grains in offshore sediments, which indicate the first appearance of sea-level glaciation in East Antarctica at this time. By about 26 Ma ago the East Antarctic ice sheet was apparently more extensive than it is at the present day. An ice sheet in the Early Cenozoic also helps to account for the record of sea-level change during this period which is difficult to explain without the changes in ocean water volume generated by periodic fluctuations in ice volume (see Section 17.5.4).

The causes of the major climatic changes involved in the establishment of the Antarctic ice sheet are beyond the scope of this book, but continental drift and the drastic modifications of ocean currents that it generated were certainly important factors. Recall that the supercontinents of Laurasia and Gondwana broke up progressively throughout the Cretaceous (see Figure 2.16). In the northern hemisphere a temperate climate persisted until the mid-Miocene, but significant cooling can be traced back to at least 10 Ma ago and the Arctic Ocean became ice-covered by 5 Ma BP. Although some uncertainty remains, it appears that ice sheets became established in the northern hemisphere around 3.2 Ma BP. This date is indicated by the appearance of ice-rafted debris in the North Atlantic and North Pacific in sediments of this age.

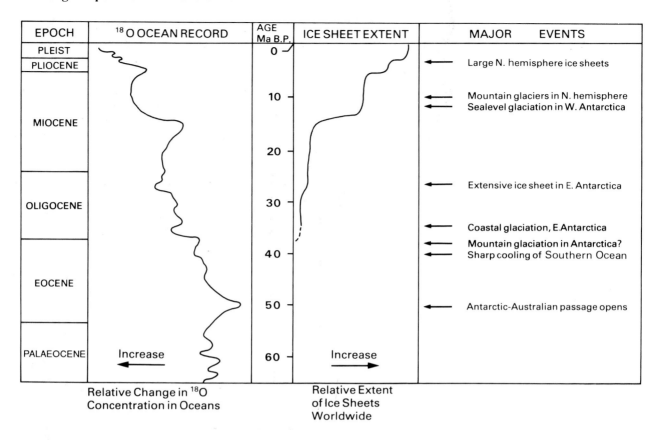

EPOCH	¹⁸O OCEAN RECORD	AGE Ma B.P.	ICE SHEET EXTENT	MAJOR EVENTS
PLEIST		0		
PLIOCENE				← Large N. hemisphere ice sheets
		10		← Mountain glaciers in N. hemisphere ← Sealevel glaciation in W. Antarctica
MIOCENE				
		20		
				← Extensive ice sheet in E. Antarctica
OLIGOCENE		30		
				← Coastal glaciation, E.Antarctica ← Mountain glaciation in Antarctica? ← Sharp cooling of Southern Ocean
		40		
EOCENE				
		50		← Antarctic-Australian passage opens
PALAEOCENE	Increase ←	60	Increase →	

Relative Change in ¹⁸O
Concentration in Oceans

Relative Extent
of Ice Sheets
Worldwide

Fig. 14.4 *History of the development of glaciation during the Cenozoic. Note that the variations in ice sheet extent are speculative due to lack of data, and that the dates at which glacial activity began at different localities is also subject to revision in the light of new information. (Based partly on D. E. Sugden (1987) in M. J. Clark, K. J. Gregory and A. M. Gurnell (eds)* Horizons in Physical Geography. *Macmillan, Basingstoke, Fig. 3.2.3, p. 217.)*

14.3.2   The Quaternary record

By 3 to 2.5 Ma BP large-scale oscillations in temperature reflecting glacial–interglacial cycles had become established, with a regular 100 ka cycle occurring over the past 700 ka. Many researchers attribute these regular climatic fluctuations largely to the effects of slight variations in the position and orientation of the Earth with respect to the Sun which affect the receipt of solar radiation; this is commonly known as the Milankovitch mechanism for glacial–interglacial cycles. Other research, however, has indicated that other factors, such as the organic productivity of the oceans and the concentration of CO_2 in the atmosphere, may also be important. The oxygen isotope data indicate that over the past 3 Ma the Earth has experienced a glacial climate for a significant majority of the time, with glacial episodes of around 80–100 ka duration alternating with 10 ka–20 ka interglacial interludes, at least over the past 700 ka. Significant changes in the $^{18}O/^{16}O$ ratio are now indicated by **oxygen isotope stage numbers** and these provide a good guide to the major changes in ice sheet volume and, in-

directly, global temperatures (Fig. 14.5). Key age markers for this ocean core oxygen isotope record have been provided by the palaeomagnetic time scale and there have also been attempts to link the ocean record with continental palaeoclimatic data. The most successful of these correlations has been achieved with loess sequences (see Section 10.3.7). The longest record is provided by the loess sequences of China which seem to give a continuous climatic record for the past 2.3 Ma. Enhanced rates of deflation and aeolian dust transport occurred during periods of semi-arid climate coinciding with glacial advances in the northern hemisphere. A correlation has now been made between the past 500 ka of a Chinese loess sequence with the ^{18}O record of a deep-sea core in the north-west Pacific and this has greatly improved the dating of the loess sequence (Fig. 14.6).

A major contribution to the understanding of the nature and magnitude of Quaternary climatic change has come from the CLIMAP (Climate/Long-Range Investigation Mapping and Prediction) project which attempted to document the global environment at the last glacial maximum 18 ka BP.

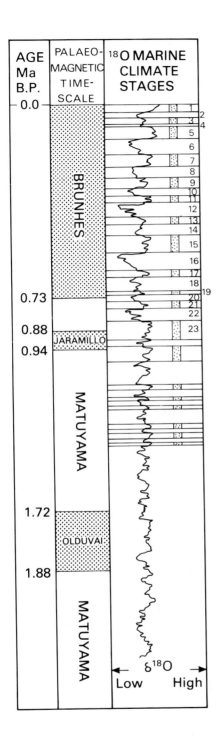

AGE
Ma
B.P.

PALAEO-
MAGNETIC
TIME-
SCALE

¹⁸O MARINE
CLIMATE
STAGES

0.0

BRUNHES

0.73

0.88

JARAMILLO

0.94

MATUYAMA

1.72

OLDUVAI

1.88

MATUYAMA

δ¹⁸O
Low   High

Fig. 14.5 *Oxygen isotope record for the about the past 2 Ma from the Pacific deep-sea core V28-239 analyzed by N. J. Shackleton and N. D. Opdyke correlated with the palaeomagnetic time scale of Berggren et al. The oxygen isotope stage numbers are shown, the odd-numbered shaded stages indicating relatively warm interludes. (Modified from W.B. Harland et al. (1982) A* Geologic Time Scale. *Cambridge University Press, Cambridge, Chart 2.17, p. 42.)*

Although concerned with just the most recent major glacial advance, the environmental changes identified were probably mirrored by earlier glacials of similar magnitude. Large ice sheets blanketed a significant proportion of the land area of the northern hemisphere and some 18 per cent of the Earth's surface was ice-covered (Fig. 14.7). In eastern North America the Laurentide ice sheet extended from the Rocky Mountains to the Atlantic Ocean and from the Arctic Ocean to the latitude of New York. In the ice free area to the south large lakes formed, such as those of the Great Basin of the western USA, in response to the supply of meltwater, and to lower evaporation as a result of lower temperatures.

Until the 1960s it was widely thought that the humid tropics had been little affected by the glacial–interglacial fluctuations of the higher latitudes. New data from a range of sources, including palynological studies of sediment cores and work on changing lake levels, have now shown that, during the latter part of the last glacial at least, much of the humid tropics were probably cooler and drier than at present (Fig. 14.8). Although the evidence is sparse in some regions, during the period around the last glacial maximum at about 18 000 a BP the tropical rain forests seem to have contracted to a smaller area than they presently cover. There was a compensating increase in the extent of savanna vegetation, indicating a significant reduction in precipitation. According to some interpretations of the sedimentological evidence up to half of the land area within 30° of the equator was covered by active ergs at this time, and lake levels, especially in the northern hemisphere, were generally very low in the period 14 000–21 000 a BP. The evidence for the magnitude of temperature depression is less certain with a greater temperature decrease being inferred from the lowering of the altitude of vegetation and glacier limits than is indicated by information from deep-sea cores.

It is in fact not a surprise to find evidence of tropical aridity coinciding with glacial episodes. We would expect a reduction in precipitation to occur as a result of reduced evaporation in response to lower ocean surface temperatures and a decrease in the area of the oceans as consequence of a global fall in sea level. Exceptions to this glacial aridity trend did occur though, with some regions such as North Africa being more humid (Fig. 14.8). This was due to the equatorward deflection of rain-bearing westerly weather systems by the large high pressure cells established over the extensive northern hemisphere ice sheets.

Climatic changes have continued since the end of the last glacial around 10 000 a BP but their magnitude has been much attenuated. Although mean global temperature does not appear to have varied by more than about 2 °C over the past 8000 a or so, evidence from fluctuating lake levels indicates that the tropics were generally much wetter in the period 12 000 to 5000 a BP than they were during the preceding 10 000 a, or have been over the past 5000 a. This humid episode was, however, punctuated by much drier

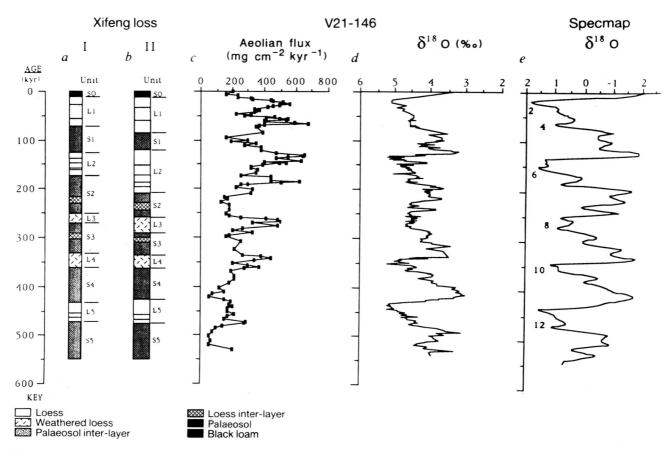

Fig. 14.6 *Correlation of the loess sequence at Xifeng, China, with the oxygen isotope record from west Pacific deep-sea core V21–146. Dating of the loess sequence in column (a) is based on a technique known as magnetic susceptibility and in (b) on correlation with the aeolian dust flux. Column (c) shows the aeolian flux record of the deep-sea core while column (d) illustrates the oxygen isotope record. Column (e) shows the global standard chronology for the oxygen isotope record (Specmap). (From S. A. Hovan et al. (1989) Nature 340, Fig. 3, p. 298.)*

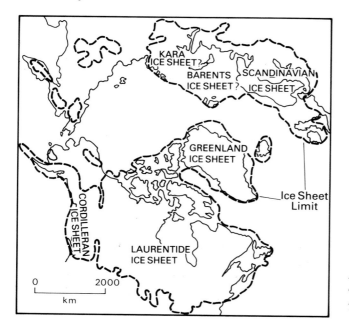

Fig. 14.7 *Probable extent of the northern hemisphere ice sheets at the last glacial maximum 18 000 a BP (Based largely on G. H. Denton and T. J. Hughes, (1981) The Last Great Ice Sheets. Wiley, New York, p.viii.)*

periods in particular regions. In Africa, for instance, there was a marked increase in aridity during the periods 11 000–10 000 a BP, 8500–6500 a BP and 6200–5800 a BP. It has been the subhumid and semi-arid tropical regions that have experienced the most significant post-glacial fluctuations in precipitation, so it is not surprising that these areas pose particular problems of landscape interpretation.

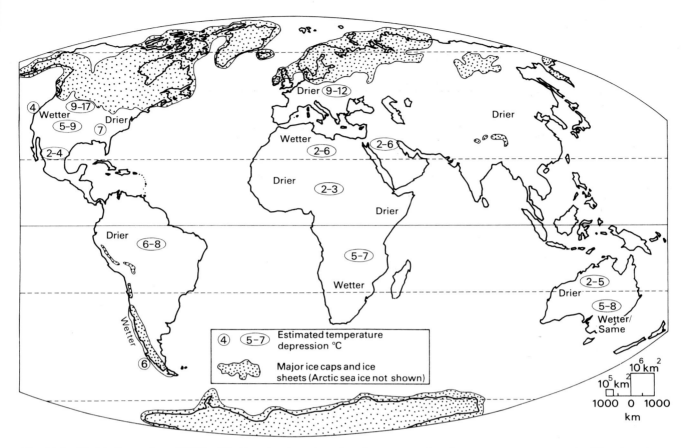

Fig. 14.8 *World climate about 18 000 a BP. Note that the palaeoclimatic reconstructions for some areas are uncertain. (Based partly on T. H. Van Andel (1985)* New Views on an Old Planet. *Cambridge University Press, Fig. 4.4, p. 61 and various other sources.)*

14.4 Effects arising predominantly from temperature changes

14.4.1 Landscapes of deglaciation

Virtually all glacial landforms are in a sense relict since both erosional and depositional glacial landforms are only fully exposed once the covering ice has retreated. Similarly, the onset of a glacial morphoclimatic regime requires a sustained cooling across the temperature threshold at which snow can accumulate from one year to the next. In Chapter 11 we looked at the development of specific glacial landforms, but here we will confine our attention to the broad nature of landscape modification that resulted from the expansion of the Late Cenozoic ice sheets into the mid-latitudes (Fig. 14.7) and the extension of upland ice caps and mountain glaciers to lower elevations.

14.4.1.1 Relict landscapes of glacial erosion

Mountain, or alpine, glaciation is characterized by a deepening of valleys by glacial erosion and active physical weathering on valley sides. Glacial troughs develop generally along the course of pre-existing river valleys (Fig. 11.16);

bedrock resistance influences glacial trough form with deeper and narrower valleys occurring in resistant lithologies. In regions of modest elevation, landscape modification during glacials may be confined to the formation of cirques. In the mid-latitudes of the northern hemisphere cirques are preferentially developed on north-east slopes. Snow accumulates here because of the combination of shade from summer insolation with protection from the predominant westerly winds. In the southern hemisphere south-east slopes are favoured for similar reasons.

The nature of landscape modification by ice sheets is controlled by ice sheet behaviour, the bedrock geology and the pre-existing form of the landscape (Table 14.2 and Fig. 14.9). The crucial factor seems to be the state of the basal ice. Active erosion can occur below warm-based ice and extensive areal scouring is accomplished where the pressure melting point is attained, normally beneath the centre of ice sheets and around the margins in lower latitudes. This broad pattern can be modified by the subglacial topography since upland regions will be more likely to be covered by thinner, cold-based ice. Only very limited erosion is achieved beneath the cold-based polar flanks of continental ice sheets.

Table 14.2 Main types of glacially eroded landscapes in relation to controlling variables

LANDSCAPE TYPE	ICE CHARACTERISTICS	LITHOLOGY	TOPOGRAPHIC CHARACTERISTICS
Areal scouring	Warm-based	Crystalline, jointed rocks; impermeable	Landscape of low relief with joint and fault control of deeper depressions. Up-glacier sides of ridges smoothed; lee sides rugged
Selective linear erosion	Warm-based in troughs, cold-based over plateaus	Not relevant	Troughs usually along pre-existing river valleys divide unmodified upland slopes and plateaus which may retain a mantle of pre-glacial regolith
Little or no sign of glacial erosion	Cold-based, no basal debris	Permeable? – reduces water present at ice–rock interface	Minimal glacial modification of pre-glacial forms with regolith-covered slopes and interfluves
Composite ice sheet/ alpine	Alternating ice sheet/ glacier types	Hardrocks	Cirques and other alpine forms well preserved where ice-sheet erosion predominantly linear, but highly modified where upland erosion has occurred
Alpine landscapes	Valley and cirque glaciers	Affects glacial trough morphology	Dendritic network of deep troughs with cirques at higher elevations separated by horns and arêtes

(Based largely on R. J. Chorley *et al.* (1984) *Geomorphology*. Methuen, London, Table 19.1, p. 515 and discussion in D. E. Sugden and B. S. John (1976) *Glaciers and Landscape*. Edward Arnold, London, Chap. 10.)

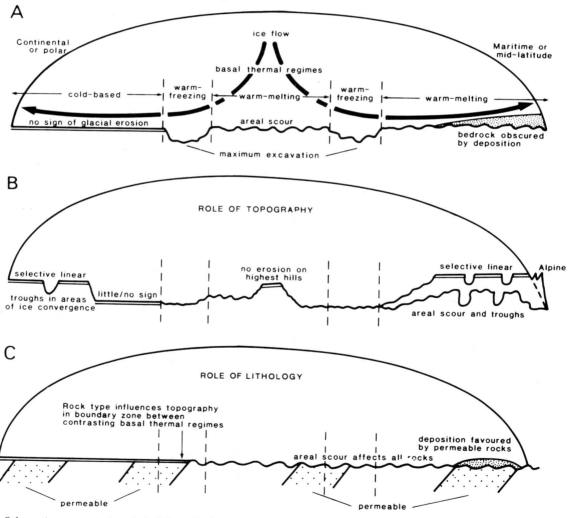

Fig. 14.9 *Schematic representation of glacial erosion by ice sheets. The left side represents polar or continental environments while the right side illustrates maritime or mid-latitude conditions. (A) The main basal regimes and the idealized erosional effects produced; (B) the likely effects of topography; (C) the effects of variations in bedrock permeability on erosion and deposition. (From R. J. Chorley et al. (1984) Geomorphology. Methuen, London, Fig. 19.8, p. 514.)*

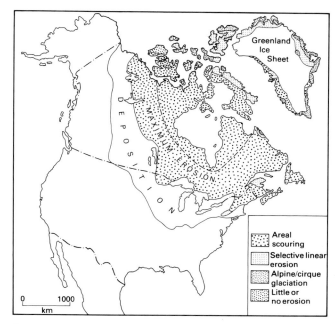

Fig. 14.10 *Pattern of glacial erosion associated with the Laurentide ice sheet of North America. (Based largely on B. J. Skinner and S. C. Porter (1987)* Physical Geology. *Wiley, New York, Fig. 19.27, p. 544 and D. E. Sugden (1982)* Arctic and Antarctic. *Blackwell, Oxford, Fig. 4.16, p. 84.)*

These anticipated patterns of erosion are apparent across the large area of North America affected by the Laurentide ice sheet (Fig. 14.10). Areal scouring is evident in those areas that lay beneath the centre of the ice sheet such as southern and western Baffin Island (Fig. 14.11). In some areas the local direction of ice movement is clearly evident from the orientation of lakes, fluted ground moraine and drumlin fields (Fig. 14.12). Selective linear erosion has occurred in higher regions such as eastern Baffin Island and north-east Greenland, while little erosion is evident in northern Greenland and over much of arctic Canada. Similar patterns occur at a smaller scale in Scotland which lay beneath the western margin of the Fenno-Scandian ice sheet. Areal scouring together with the erosion of deep troughs and cirques (largely a result of high relief) prevails in the west, whereas selective linear erosion has been predominant in the Cairngorm Mountains in eastern Scotland which retain pre-glacial weathering forms on plateau summits. Preservation of pre-glacial regolith is even more pronounced in the Buchan area on the north-east coast of Scotland.

14.4.1.2 Relict landscapes of glacial deposition

The development of landscapes of glacial deposition depends primarily on the duration of glaciation and the quantity of debris carried by a glacier or ice sheet. Two types of ice-margin landscape can be distinguished. Stagnant-ice landscapes are dominated by irregular hummocky moraine,

while active ice-retreat landscapes are characterized by sub-glacial landforms traversed by marginal ridges.

Retreating valley glaciers leave a series of terminal moraines which may be partially destroyed by meltwater activity. Stagnation of the glacier snout during retreat allows the development of kettle-holes and kames. Not surprisingly the patterns for ice sheets are more complex, but it is still possible to identify a sequence of depositional forms at the continental scale left by retreating ice sheets in regions such as North America (Fig. 14.13). This sequence can, however, be modified by topography. On a relatively featureless plain the succession of subglacial and ice-marginal depositional forms should be clearly developed, but where a retreating ice front encounters an irregular topography pockets of ice will eventually be isolated in depressions. This stagnated ice will give rise to a chaotic depositional topography characterized by moraine and modified by fluvioglacial deposits.

14.4.1.3 Glacio-isostasy

An important indirect effect of the fluctuations in ice sheet volume associated with glacial–interglacial oscillations is the changing load on the crust. As an ice sheet grows it loads the crust which subsides; when the ice melts the load is removed and the crust rebounds. This effect is referred to as **glacio-isostasy**.

Since the lithosphere possesses some flexural rigidity (see Section 2.2.4) it bends in response to a load, and this gives rise to an accompanying peripheral forebulge (Fig. 14.14). After the load is removed and the crust rebounds the forebulge subsides. These adjustments do not, of course, occur instantaneously in response to changes in load since the rate of crustal displacement is controlled by the rate at which the highly viscous mantle moves at depth to compensate for load changes. Regions such as Scandinavia and eastern Canada, which were covered by a great thickness of ice at the last glacial maximum, have so far only accomplished about half of the total vertical adjustment of between 500 and 800 m that will eventually occur in response to deglaciation which was completed by 8000–10 000 a BP.

The most obvious manifestation of such isostatic rebound is the displacement of shorelines (Fig. 14.15; see Chapter 17). Although the melting of ice sheets at the end of the last glacial led to a rapid rise in sea level this was outpaced by shorelines located in the vicinity of the major centres of ice accumulation. This retreat of shorelines is still continuing, and it is known that before the rebound of eastern Canada has been completed Hudson Bay will have become dry land.

14.4.2 Relict periglacial landscapes

Relict periglacial features are widely known from those regions marginal to the northern hemisphere ice sheets and

Fig. 14.11   *Aerial photo showing areal scouring by the Laurentide ice sheet in southern Baffin Island, northern Canada. Numerous small lakes are characteristic of areally scoured glaciated landscapes. The area covered is about 14 km across. (Photo courtesy The Photographic Survey Corporation Ltd, Toronto, Canada.)*

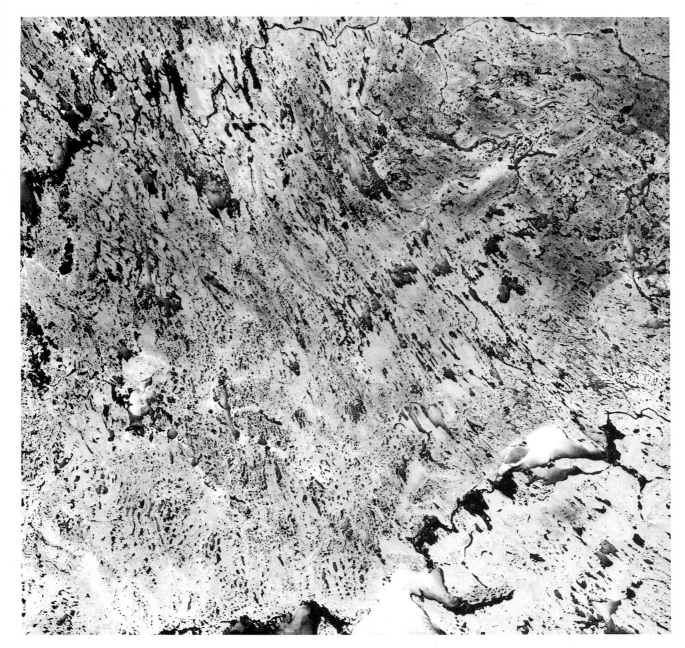

Fig. 14.12   *Orientated topography north of Aberdeen Lake, Keewatin District, Northwest Territories, Canada. The orientation is evident largely from numerous lakes and indicates a NNW–SSE ice flow direction. Although within a zone of areal scouring the region has a cover of till. The area covered is about 180 km across. (Landsat image courtesy R. S. Williams Jr.)*

from other areas which experienced a marked cooling during the Quaternary. Information for Europe is particularly detailed. Remnants of pingos have been identified in southern England and Wales where mean annual temperatures are around 9°C today. Ice-wedge casts are widely known and some are associated with patterned ground; polygonal patterns with meshes 8 m across have been identified near Evesham in southern England. Although the evidence is not so abundant in North America, a number of fossil peri-

glacial features have been recorded. The distribution of ice-wedge casts extends equatorwards about as far as the most southerly limit of the Laurentide ice sheet (Fig. 14.16).

Relict periglacial features have played an important part in the reconstruction of climatic conditions prevailing during the Quaternary. Such palaeoclimatic reconstruction depends on a knowledge of the environmental parameters within which contemporary periglacial features are forming, but this procedure must be undertaken with caution. It is important,

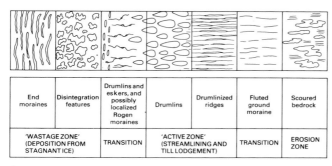

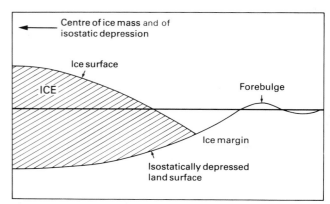

Fig. 14.13 *Schematic representation of the sequence of depositional landscapes to be expected in association with the mid-latitude periphery of Late Cenozoic continental ice sheets. (Modified from D. E. Sugden and B. S. John (1976)* Glaciers and Landscape. *Edward Arnold, London, Fig. 13.16, p. 275.)*

Fig. 14.14 *Loading of the crust by the growth of an ice sheet. Note the flexure of the lithosphere and the formation of a peripheral forebulge. (After J. J. Lowe and M. J. C. Walker (1984)* Reconstructing Quaternary Environments, *Longman, London, Fig. 2.30, p. 61.)*

for instance, to appreciate the variablity of periglacial conditions in different locations in the Quaternary. The southern margins of the ice sheet covering much of European Russia would have experienced an extremely cold and arid regime compared with the more maritime 'Icelandic'-type climate prevailing in the ice-free zone of southern Britain. A further

important point is the distinction that should be made between present-day high latitude periglacial regions with tundra or taiga vegetation, and those regions peripheral to the continental ice sheets of the Quaternary. Although some-

Fig. 14.15 *Raised beach at about 8 m above present mean sea level in the Isle of Arran, western Scotland. In this area of northern Britain uplift associated with isostatic rebound following deglaciation has outpaced the post-glacial rise in sea level.*

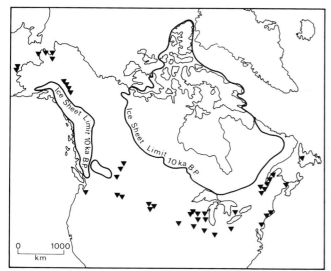

Fig. 14.16   Location of ice-wedge casts in North America. (After A.L. Washburn (1979) Geocryology. Edward Arnold, London, Fig. 14.20, p. 307 and R. J. E. Brown and T. L. Péwé (1973) in North American Contribution, Permafrost Second International Conference, National Academy of Sciences, Fig. 8, p. 90.)

times loosely regarded as analogous environments, they are in fact significantly different. The Quaternary ice sheets of the northern hemisphere extended well into the mid-latitudes (as far as 40 °N in eastern North America), and the climate of these areas would have been distinctly warmer, at least in summer, than present-day high latitude environments. The higher angle of the solar radiation would have promoted more rapid melting of snow and a greater depth of seasonal thaw. There is, indeed, no exact equivalent of such a climate today. A further problem is that some periglacial forms assumed to be active today may in fact be relict features and therefore incapable of indicating the critical temperature thresholds controlling their development. For instance, it has been argued that the present distribution of pingos does not necessarily indicate the temperature limit required for their development since most examples in North America and the USSR began to develop hundreds of thousands of years ago under temperatures greatly different from those of today.

In spite of these problems it is possible to indicate the approximate climatic parameters of certain landforms characteristic of periglacial environments (Table 14.3). The most useful features are: those confined to permafrost since these provide a clear indication of the maximum mean annual temperature; those that are not rapidly destroyed by processes active in a subsequent morphoclimatic regime or are preserved in some way; and those that cannot easily be confused with non-periglacial forms. Relict ice wedges and, in spite of reservations, pingos, most closely fulfil these requirements, being capable of giving minimum estimates of the temperature change to the present. They are consequently widely used in palaeoclimatic reconstructions. Ice-

wedge polygons are confined to the zone of continuous permafrost and can be preserved by infilling by sediment on thawing. Pingos similarly require permafrost for their development and can be identified as remnant collapsed features, although there may be some confusion with other thermokarst forms. Relict cryoplanation terraces and cryopediments have also been used, but these are less suitable because they are not necessarily associated with permafrost nor are they always readily distinguishable from structural benches and other non-periglacial features. Deposits and their associated structures may remain long after the morphological form with which they were originally associated has been degraded. Gelifluction deposits, involutions and loess blankets have all been employed as evidence of periglacial conditions although they do not provide very specific guides to temperature.

14.5   Effects arising predominantly from precipitation changes

14.5.1   Change in fluvial systems

Changes in climate during the Late Cenozoic have undoubtedly had a major impact on both global and regional patterns of runoff and sediment transport, but in most cases it is only the changes of the relatively recent past that can be recognized clearly in the morphology of abandoned or highly modified river channels. The link between changing climate and river channel behaviour is not a simple one since the fluvial system responds not just to a change in mean annual precipitation but to changes in a complex range of other factors which can affect the magnitude and temporal variability of runoff and sediment transport. These factors include alterations in the density and type of vegetation cover, changes in rates of evapotranspiration related both to temperature and vegetation characteristics, changes in the intensity of rainfall and its distribution throughout the year and changes in the areal extent of permafrost and the contribution of snow-melt to runoff. An additional factor is the lags that may occur between a change in climate and the response of the fluvial system (Fig. 14.17). Several responses are involved; first, the impact of climatic change on the vegetation cover; secondly, the effect of these external environmental factors on the sediment and discharge characteristics of streams; and finally, the morphological adjustments of the channel and valley system to these sediment and discharge variables.

Some idea of the drastic changes that can occur to fluvial systems as a consequence of climatic change can be gained from research on the eastern Sahara Desert. Parts of this region, especially the area known as the Selima Sand Sheet on the Egypt–Sudan border, are quite appropriately described as hyper-arid at the present day with no trace of surface water or fluvial activity. The Selima Sand Sheet itself is

Table 14.3 Environmental range of periglacial processes and landforms

Processes	LOWLANDS			HIGHLANDS			
	Polar	Subpolar	Mid-latitude	Polar	Subpolar	Mid-latitude	Low latitude
Frost action							
Frost wedging	***	****	**	***	****	***	**
Frost heaving and thrusting	****	***	**	***	****	**	*
Permafrost cracking	****	***	?	***	**	?	?
Other processes							
Frost creep	***	****	**	**	****	***	*
Gelifluction	****	***	**	***	****	**	*
Nivation	****	***	?	***	****	***	**
Wind action	****	***	**	****	***	**	**
Landforms							
Patterned ground							
Nonsorted circles	****	***	?	***	***	**	?
Sorted circles	****	**	?	***	****	?	?
Small nonsorted polygons	****	***	**	***	****	***	***
Large nonsorted polygons (ice wedge or sand wedge)	****	***	?	***	**	?	?
Small sorted polygons	****	***	*	***	****	**	***
Large sorted polygons	****	**	?	***	**	?	?
Small stripes	****	***	*	***	****	**	***
Large stripes	****	**	?	***	****	*	?
Other features							
Permafrost	****	***	?	****	***	**	*
Pingos	****	***	?	?	?	—	—
Palsas	***	****	?	?	***	**	?
Thaw slumps	****	****	?	**	**	?	?
Gelifluction lobes	****	***	*	***	****	**	*
Block fields and rock streams	***	***	?	****	****	**	*
Rock glaciers	?	?	?	***	****	****	*
Nivation benches and hollows	****	***	?	***	****	**	*
Cryoplanation terraces	***	***	?	****	****	?	?
Asymmetric valleys related to permafrost	****	***	?	****	***	?	?
Loess	***	****	**	*	*	*	*

Notes: Asterisks indicate relative abundance of individual landforms and relative significance of a particular process in a range of environments. **** Abundant, *** common, ** uncommon, * scarce, ? rare or absent.
Source: Modified from A. L. Washburn (1979) *Geocryology*. Edward Arnold, London, Table 13.1, pp. 277–8.

one of the most barren, featureless regions on Earth, consisting of an almost entirely flat plain with a cover of wind-blown sand and fine gravel. In 1981 an experimental radar system installed on board the space shuttle Columbia acquired several radar images of this area. On inspecting these images it became apparent that the radar pulses had penetrated the relatively thin and completely dry sand cover to reveal buried river valleys. Some of these are almost as broad as the present Nile Valley, and earlier work on the drainage history of this region indicates that they may be as much as 25 Ma old. Superimposed on these large valley systems are smaller incised channels which were probably periodically active during the wetter phases of the Quaternary.

This is obviously an extreme example of the affect that climatic change can have on fluvial systems, but far less dramatic climatic fluctuations can also lead to profound changes in channel morphology. In Chapter 9 we described the way in which channel morphololgy can change in response to changes in controlling variables, but it is important to emphasize that not all such adjustments are a consequence of natural processes. In examining changes that have occurred in the past few thousand years, and especially the last 200 a, it is important to keep in mind the dramatic impact that human activity can have on fluvial systems. This does not just involve the direct control of rivers through dam construction and the modification of channels, but also encompasses indirect effects, such as the increase in sediment supply to channels which often follows the replacement of natural vegetation by agricultural land (see Section 15.3.1.1).

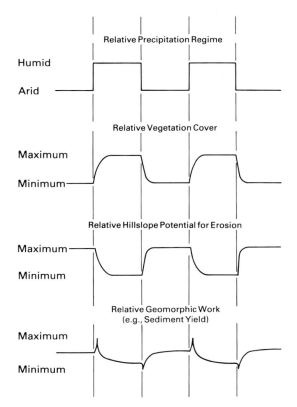

Fig. 14.17 *Schematic representation of the effect of an abrupt change in climatic regime on fluvial systems. Maximum erosion and sediment transport is associated with the change from arid to humid conditions when the increase in protective vegetation cover lags behind the increase in precipitation. Erosion rates subsequently decline as vegetation density increases and surface runoff is reduced. (After J. C. Knox (1972) Annals of the Association of American Geographers **62**, Fig. 6, p. 408.)*

14.5.1.1 Underfit streams

As a broad generalization we would expect valleys to be of a size proportionate to the dimensions of the river channels that they contain. In a number of instances, however, streams appear to be far too small with respect to their valleys. Such **underfit streams** or **misfit streams** could obviously arise where there has been a reduction in precipitation and runoff; the channel can adjust its size rather rapidly to a decrease in discharge but the response of the whole valley will be much slower. (It is, of course, possible to have an increase in discharge and therefore a misfit in the sense of a stream that is 'too large' for its valley, but in this case the valley will experience rapid erosion until it reaches an equilibrium size.) Mechanisms other than a decrease in precipitation can also produce underfit streams; discharge can be reduced by one river system diverting flow from another through river capture (see Section 16.2.3) and in permeable lithologies, such as limestone, flow can be diverted progressively underground thereby depleting surface discharge (see Section 9.6). Climatic mechanisms, however, are the most likely

where underfit streams are widespread in an area.

Of particular interest are misfit streams meandering within valleys which themselves exhibit a regular meandering pattern of much greater wavelength; such streams are described as **manifestly underfit**. If the relationship between discharge and meander wavelength established from meandering channels in alluvium is applied to these **valley meanders**, an estimate can be made of the discharge that would have been required to form them. Valley wavelengths are usually between 5 and 13 times the wavelengths of the sinuous underfit stream channels contained within them, and since discharge is proportional to the square of the wavelength the increased discharge required to form the valley meanders is between 25 and 169 times. Even the low end of this range is unrealistic if the change in discharge is brought about by a simple alteration in precipitation, so other factors need to be considered. In some cases large discharges could be produced by glacial meltwater, but it is also important to consider whether meanders at least partly cut in bedrock necessarily adjust in the same way to discharge as those formed in alluvium and whether the lateral migration of a meandering channel could, of itself, create a meandering valley pattern of longer wavelength.

In some cases relatively straight channels are seen to flow within wider meandering valleys, but more than the anticipated number of pools and riffles occur in the channel within each meander bend. Such streams are termed **Osage-type underfits** and appear to represent a situation where the channel is still adjusting to a decrease in discharge and has not yet decreased the amplitude of its meanders to match the reduction in flow.

14.5.1.2 Long-term river metamorphosis

Adjustments in river channel morphology which occur in response to changes in discharge and sediment load are sometimes referred to as river metamorphosis. One of the most frequently cited examples of river metamorphosis is the study of the Murrumbidgee River carried out by S. A. Schumm. The Murrumbidgee is a tributary of the Murray River and drains westwards across southern New South Wales on an alluvial plain with a mean slope of about 0.3 m km⁻¹. The presently active channel is around 60 m in width and it meanders across a floodplain which also contains many large oxbow lakes marking an earlier and much larger meandering channel, and the trace of a much less sinuous channel (Fig. 14.18). Schumm measured various morphological and sedimentological properties of these three distinct channels (Table 14.4) and interpreted them as reflecting adjustments to the hydrological regime and sediment load characteristics prevailing when each channel was active.

Evidence from landforms and soils from the alluvial plain indicates that the older low-sinuosity palaeochannel was apparently active in the Late Pleistocene when the cli-

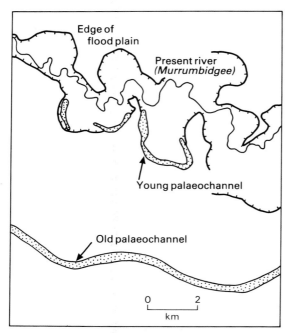

Fig. 14.18 *Active and palaeochannels of the Murrumbidgee River, New South Wales, Australia. Flow is to the left. (After S. A. Schumm (1969) in: R. J. Chorley (ed.) Water, Earth and Man. Methuen, London, Fig. 11.II.4, p. 531.)*

Table 14.4 Morphology of the active and palaeochannels of the Murrumbidgee River.

	MODERN MURRUMBIDGEE	YOUNGER MEANDERING PALAEOCHANNEL	OLDER LOW-SINUOSITY PALAEOCHANNEL
Channel width (m)	65	140	180
Channel depth (m)	6	11	3
Width/depth ratio	10	13	67
Sinuosity	2.0	1.7	1.1
Gradient (m km^{-1})	0.13	0.15	0.37
Meander wavelength (km)	0.85	2.14	5.49
Channel			
Silt-clay (%)	25	16	1.6
Bed load (%)	2.2	3.4	34
Bankfull discharge (m^3 s^{-1})	310	1 430	640
Estimated sand transport at bankfull discharge (t day^{-1})	2000	21 000	54 000

Source: After S. A. Schumm (1977) *The Fluvial System.* Wiley, New York, Table 5–5, p. 169.

mate in the region was drier than at present, whereas the younger meandering palaeochannel was functioning when conditions became wetter around 3000 a BP. Nevertheless, changes in channel morphology occurred not only in response to changes in discharge but also to changes in the nature of the sediment load which reflect indirect climatic influences through changes in the vegetation cover in the catchment. The active channel of the Murrumbidgee currently transports small quantities of silt, sand and clay and when conditions were wetter in the past a more dense vegetation cover would have reduced the quantity of sediment being transported in spite of the increase in runoff. In contrast, when conditions were drier the resulting reduction in vegetation cover would have allowed sediment supply to the channel to increase significantly and more sediment would have been transported in the low-sinuosity palaeochannel. Although the total runoff would have been lower, peak discharges would have been higher as a result of increased quick flow.

By assessing the morphological properties of the palaeochannels in the light of the relationships between channel properties, discharge and sediment transport evident from modern channels, Schumm showed how it is possible to estimate the discharge and the nature of the sediment load of the Murrumbidgee palaeochannels. Calculations show that the older low-sinuosity palaeochannel would have been capable of transporting large quantities of sand at bankfull discharge, and this channel is, as anticipated, full of sand. The large meandering palaeochannel would also have been competent to transport significant quantities of sand but it is filled largely with clay and silt. This suggests that it was only in the drier phase when the low-sinuosity channel was active that large quantities of sand were being supplied to the channel from the poorly vegetated catchment. The main response represented by the meandering palaeochannel to an increase in discharge is the increase in channel dimensions represented by a greater channel depth and width and longer meander wavelength.

An important aspect of this analysis of the Murrumbidgee River is the manner in which the increased transport of sand was accomplished during the dry episode, even though discharge and the depth of flow decreased. Clearly an increase in channel gradient would have been required, but this was not achieved by deposition and a steepening of the whole alluvial plain but by a decrease in channel length through a reduction in sinuosity. As shown in Table 14.4, the present meandering channel has a sinuosity almost twice as great as the low-sinuosity palaeochannel and consequently the gradient of the latter would have been nearly twice as high.

Other studies have broadly supported Schumm's conclusions about the response of river systems to climatic changes, at least in semi-arid environments. An investigation of the Colorado River in Texas, for instance, has shown a tendency during dry phases for an increase in channel width, meander wavelength and sediment size and a decrease in channel sinuosity (Fig. 14.19). But other studies from temperate environments suggest that these relationships are not necessarily universal. Indeed, complex sequences of change in river channel morphology may arise from different rates

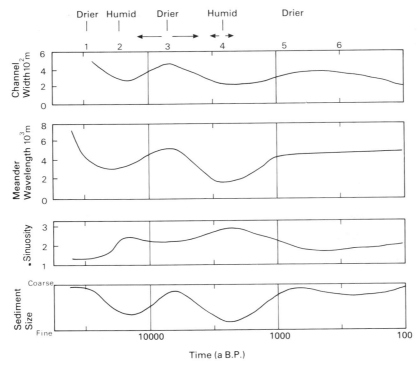

Fig. 14.19 *Changes in the Colorado River of central Texas in response to climatic fluctuations during the Late Quaternary. Six phases are identified: (1) relatively arid climate associated with infrequent, high-magnitude floods, high sediment yields, aggradation of coarse sediment and braiding, and low channel sinuosity; (2) cooler, wetter and more uniform precipitation associated with a decreasing magnitude of formative discharges, transport of finer material and progressive change to high-sinuosity channels, followed by more frequent but lower magnitude floods, vegetated slopes and less soil erosion, (3) warmer and more arid conditions with increasing seasonality of precipitation associated with incision followed by aggradation and braiding, and wide, low sinuosity channels; (4) more humid climate associated with lower formative discharges and more uniform flows, transport of finer material and narrow, sinuous channels with a reduced meander wavelength; (5) a relatively dry climate associated with streams carrying a coarse load, entrenchment followed by aggradation, and wider, less sinuous streams; (6) decrease in formative discharges, coarse load and progressive channel narrowing. (Modified from V. R. Baker and M. M. Penteado-Orellana (1977) Journal of Geology **85**, Fig. 11, p. 414.)*

of change in discharge and sediment transport and from changes in how frequently channel-forming floods occur. Clearly, more studies are needed from a wide variety of environments in order to assess the range of responses that can occur.

The necessity of expanding our understanding of fluvial systems which have experienced different climatic histories is underscored by research on the Amazon, the greatest river system in the world. At an estimated 230000 m³ s⁻¹ (that is, 15 per cent of the total flow of fresh water to the oceans) its mean annual discharge is over five times that of the next largest river, the Zaire (Congo). The gradient of the Amazon over a distance of more than 3000 km from the foothills of the eastern Andes in Peru to its mouth averages only 0.1 m km⁻¹; this is reduced to an almost imperceptible gradient of 0.03 m km⁻¹ over the lower 1440 km of its course below the city of Manaus.

In the Amazon system there is a marked contrast in the morphology of trunk rivers with high elevation sources in the Andes and rivers which drain the tropical lowlands. The latter, exemplified by the Juruá which rises at an elevation of only 600 m, display little variation in discharge throughout the year, transport a negligible bed load and a modest suspended load of extremely fine calibre, and have narrow floodplains and single meandering channels with variable but generally high sinuosity (Table 14.5). In comparison the rivers draining the Andes, such as the Solimões (the Amazon trunk river) (Fig. 14.20) and the Japurá carry significantly more bed load and a higher suspended load and have broad floodplains and anastomosing channels of low sinuosity. These contrasts have been interpreted as a consequence of the different abilities of these two types of river to transport the vast quantities of coarse alluvium that were mobilized during the much drier conditions which prevailed at the height of the last glacial. The relationships established for river metamorphosis in semi-arid environments do not account for the kinds of channel adjustments evident in the Amazon system where discharge characteristics and the active role of vegetation in stabilizing channel banks seem to be critical factors.

Table 14.5 Properties of rivers in the Amazon Basin

	JURUÁ (JCT. SOLIMÕES)	JAPURÁ (JCT. SOLIMÕES)	SOLIMÕES (NEAR FONTE BOA)
Environment	Tropical	Mixed	Mixed
Drainage area (10^3km^2)	217	289	1263
Discharge (10^{12}m^{3}a^{-1})	0.197	0.351	1.161
Solid load (% total load)	40	48	~30
Bed load (% solid load)	<1	2.3	~5
Sediment yield (10^3kg km^{-2}a^{-1})	49.4	120.2	Not available
Pattern	Moderate to tortuous sinuosity	Straight, moderately anastomosing	Moderately sinuous, highly anastomosing
Meander form	Compound asymmetric	Simple symmetric	Simple asymmetric
Sinuosity	1.8–3.0	1.1	1.5

Source: After V. R. Baker (1978) in: A. D. Miall (ed.) *Fluvial Sedimentology*. Canadian Society of Petroluem Geologists Memoir **5**, Table 2, p. 221.

14.5.2 Aeolian systems

As in the case of rivers, aeolian systems are sensitive to changes in precipitation, although changes in temperature can also influence the level of aeolian activity through the effect on evapotranspiration rates and thus soil moisture and the density of vegetation cover. Similarly, as with fluvial systems, the change in precipitation will have most impact if it occurs across a threshold range; in the case of aeolian systems this will be the precipitation level at which vegetation cover begins to restrict severely wind speeds at ground level and thereby curtail the entrainment of particles by the wind. The precise level of precipitation at which this occurs will vary depending on local conditions, but 200 mm seems to be a reasonable approximation in many areas. In addition to changes in surface characteristics brought about by changes in vegetation the variations in wind strength itself can also affect the level of aeolian activity, and alterations in the directional characteristics of the wind regime also promote adjustments in the alignment of depositional, and ultimately, erosional forms (see Chapter 10).

A wide range of evidence, but especially pollen records and dated fluctuations in lake levels in tropical latitudes, have demonstrated major changes in climate in the world's arid and semi-arid regions during the Quaternary, it now being well established that in general greater aridity at these latitudes coincided with glacials in the mid- and high latitudes. Indeed there may have been up to 25 arid–humid cycles in this period and no doubt many more smaller amplitude oscillations of the sort which have been evident in the past few thousand years. The major changes in the areal extent and intensity of the world's arid regions clearly must have had a profound effect on aeolian activity, and this is attested by the great expanse of relict ergs around the regions of currently active sand movement (see Chapter 10). It has been suggested that mean wind speeds were higher during glacials and that it is these strong winds that were responsible for creating the huge megadunes of the world's ergs. This provides an interesting alternative explanation to put aside the generally accepted notion of megadunes being the largest component of a hierarchy of aeolian bedforms (see Chapter 10), and is supported by some recent research on megadunes in a part of northern Saudi Arabia which interprets them as relict rather than active forms.

In the Kalahari Desert in southern Africa, three distinct arid phases have been recognized during the Late Quaternary on the basis of relict linear dune systems with differing morphologies and alignments (Fig. 14.21). The oldest (Group A) are located in north-western Botswana and adjacent areas of Namibia and Angola. They are cut by fluvial and lacustrine features associated with a wet period which affected the northern Kalahari from 30 to 20 ka BP and therefore must pre-date these landforms. A second phase of aridity during which dunes in north-eastern Botswana (Group B) were active followed this humid period and coincides with the dry episode recorded from many localities in the tropics which corresponds with the last glacial maximum around 18 ka BP. By about 12 ka BP the zone of aridity had shifted south to the southern Kalahari and the dunes in this region were probably active at this time and remained so into the early Holocene.

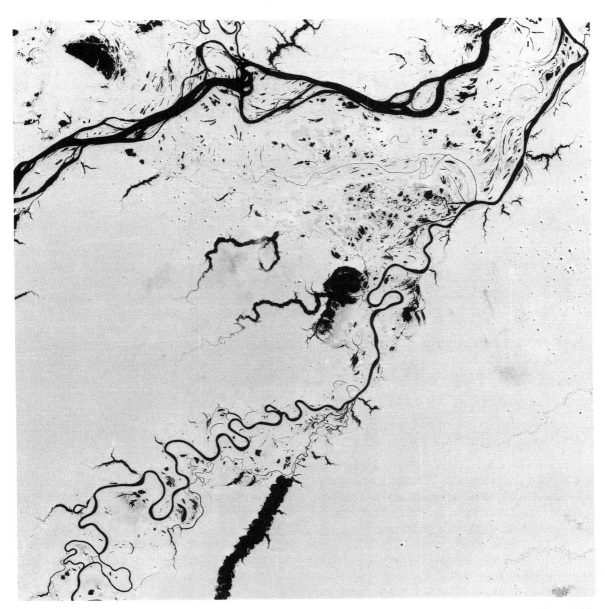

Fig. 14.20 *Landsat image of an area about 300 km south of Manaus in the Amazon Basin, Brazil, covering an area about 180 km across. The two major rivers visible are the Solimões (one of the two primary tributaries of the Amazon system – extreme top of image), and the smaller Purus. Both rivers are still transporting sediment mobilized during the last glacial and stored in the valleys of the Andes upstream. The enormous accumulation of sediments in the vast floodplain, or varzea, of the Solimões has diverted the course of the Purus. The lakes visible in the centre of the image are known as rias fluviales and are formed by the damming of streams flowing from interfluves by the floodplain sediments of the major channels. Post-glacial sea-level rise has also been important in the development of the Amazon system and this is discussed in Section 17.6.1. (Landsat image courtesy N. M. Short.)*

Further reading

The importance of climate in landform development is emphasized in the books by Büdel (1982) and Tricart and Cailleux (1972), but a critical appraisal of climatic geomorphology is provided by Stoddart (1969). A thoughtful analysis of how the magnitude and frequency characteristics of meteorological events might be used to identify meaningful morphoclimatic categories is presented by Ahnert (1987). The problem of pediment formation has generated an enormous literature but the brief reviews by Oberlander (1989) and Twidale (1981) are useful; the development of the intensively studied pediments of the Mojave Desert is discussed by Dohrenwend *et al.* (1987) and Oberlander (1972, 1974). There is no single comprehensive summary which covers

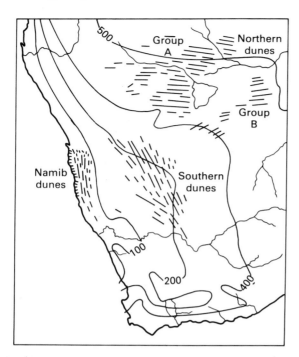

Fig. 14.21 *Distribution of major active and relict dune fields in southern Africa in relation to present-day mean annual precipitation (mm). (From N. Lancaster (1981)* Palaeogeography, Palaeoclimatology, Palaeoecology *33, Fig. 8, p. 338.)*

the impact of climatic change on landform development, but the volumes edited by Derbyshire (1976) and Douglas and Spencer (1985) cover a number of relevant topics.

On the long-term record of climatic change Frakes (1979) provides a detailed background and Frakes and Francis (1988) consider the evidence for Cretaceous glaciation. On the Quaternary record Lowe and Walker (1984) is a good introduction, while Imbrie and Imbrie (1986) cover the oxygen isotope record in detail. Denton and Hughes (1981) is an exhaustive treatment of the Quaternary ice sheets, while Fink and Kukla (1977), Kukla (1987) and Pye (1987) discuss the related record of climatic change represented by loess deposits. The correlation of the Chinese loess sequences with the deep-sea core ^{18}O record is considered by Hovan *et al.* (1989). An overview of Late Quaternary lake fluctuations is presented by Street-Perrott and Harrison (1984), while Sarnthein (1978) considers the expansion of ergs during glacial phases and Van der Hammen (1982) discusses the palaeoecological evidence for greater aridity in the humid tropics during the last glacial maximum.

Turning to the landforms themselves, a broad overview of the impact of glaciation with respect to both landscapes of glacial erosion and glacial deposition is provided by Sugden and John (1976), while Boulton *et al.* (1985) relate the dynamics of ice sheets to regional and continental-scale patterns of erosion and deposition. Bell and Laine (1985) discuss the nature of erosion associated with the Laurentide

ice sheet, the extent of glacial modification of landscapes in Scotland is considered by Hall and Sugden (1987) and Andrews (1987) assesses the effects of glacial isostasy. A comprehensive survey of relict periglacial landforms is provided by Washburn (1979), and Black (1976) specifically examines features indicative of permafrost. Reger and Péwé (1976) and Flemal (1976) provide, respectively, details on cryoplanation terraces and pingos as permafrost indicators, while MacKay (1988) questions the palaeoclimatic value of pingos. Boardman (1987) considers a number of relict periglacial forms in the British Isles.

Various aspects of the response of fluvial systems to climatic change are covered in the volume edited by Gregory (1983), and Knox (1972) provides a valuable analysis of the effect of fluctuations in precipitation. The remarkable buried channels of the eastern Sahara are described by McCauley *et al.* (1982). Dury (1977) presents a comprehensive review of the problem of underfit streams and Schumm (1969, 1977) considers various aspects of river channel metamorphosis. Schumm (1968) presents a detailed study of changes in channel morphology for the Murrumbidgee River in Australia and Baker and Penteado-Orellana (1977) present a similar investigation of the Colorado River of Texas. Starkel (1983), however, emphasises the complex nature of the adjustments that can occur, while Baker (1978) demonstrates the distinctive behaviour of some humid tropical rivers. Finally, the impact of climatic change on aeolian systems is considered in broad terms by Goudie (1983) and with respect to the Kalahari Desert by Lancaster (1981).

References

Ahnert, F. (1987) An approach to the identification of morphoclimates. In: V. Gardiner *et al.* (eds) *International Geomorphology 1986* Part II. Wiley, Chichester and New York, 159–88.

Andrews, J. T. (1987) Glaciation and sea level: a case study. In: R. J. N. Devoy (ed.) *Sea Surface Studies: A Global View.* Croom Helm, London and New York, 95–126.

Baker, V. R. (1978) Adjustment of fluvial systems to climate and source terrain in tropical and subtropical environments. In: A. D. Miall (ed.) *Fluvial Sedimentology.* Canadian Society of Petroleum Geologists Memoir **5**, 211–30.

Baker, V. R. and Penteado-Orellana, M. M. (1977) Adjustment to Quaternary climatic change by the Colorado River in central Texas. *Journal of Geology* **85**, 395–422.

Bell, M. and Laine, E. P. (1985) Erosion of the Laurentide region of North America by glacial and glaciofluvial processes. *Quaternary Research* **23**, 154–74.

Black, R. F. (1976) Features indicative of permafrost. *Annual Review of Earth and Planetary Sciences* **4**, 75–94.

Boardman, J. (ed.) (1987) *Periglacial Processes and Landforms in Britain and Ireland.* Cambridge University Press, Cambridge.

Boulton, G. S., Smith, G. D., Jones, A. S. and Newsome, J. (1985) Glacial geology and glaciology of the last mid-latitude ice sheets. *Journal of the Geological Society London* **142**, 447–74.

Büdel, J. (1982) *Climatic Geomorphology* (translated by L.

Fischer and D. Busche). Princeton University Press, Princeton, and Guildford.

Denton, G. H. and Hughes, T. J. (eds) (1981) *The Last Great Ice Sheets*. Wiley, New York.

Derbyshire, E. (ed.) (1976) *Geomorphology and Climate*. Wiley, London and New York.

Dohrenwend, J. C., Wells, S. G., McFadden, L. D. and Turrin, B. D. (1987) Pediment dome evolution in the eastern Mojave Desert, California. In: V. Gardiner *et al.* (eds) *International Geomorphology 1986* Part II. Wiley, Chichester and New York, 1047–62.

Douglas, I. and Spencer, T. (eds) (1985) *Environmental Change and Tropical Geomorphology*. Allen and Unwin, London and Boston.

Dury, G. H. (1977) Underfit streams: retrospect, perspect and prospect. In: K. J. Gregory (ed.) *River Channel Changes*. Wiley, Chichester and New York, 281–93.

Fink, J. and Kukla, G. J. (1977) Pleistocene climates of central Europe: at least seventeen interglacials after the Olduvai event. *Quaternary Research* **7**, 363–71.

Flemal, R. C. (1976) Pingos and pingo scars: their characteristics, distribution, and utility in reconstructing former permafrost environments. *Quaternary Research* **6**, 37–53.

Frakes, L. A. (1979) *Climate Throughout Geologic Time*. Elsevier, New York.

Frakes, L. A. and Francis, J. E. (1988) A guide to Phanerozoic cold polar climates from high-latitude ice-rafting in the Cretaceous. *Nature* **333**, 547–9.

Goudie, A. (1983) The arid Earth. In: R. Gardner and H. Scoging (eds) *Mega-Geomorphology*. Clarendon Press, Oxford, 152–71.

Gregory, K. J. (ed.) (1983) *Background to Palaeohydrology*. Wiley, Chichester and New York.

Hall, A. M. and Sugden, D. E. (1987) Limited modification of mid-latitude landscapes by ice sheets: the case of Northeast Scotland. *Earth Surface Processes and Landforms* **12**, 531–42.

Hovan, S. A., Rea, D. K., Pisias, N. G. and Shackleton, N. J. (1989) A direct link between the China loess and marine ^{18}O records: aeolian flux to the north Pacific. *Nature* **340**, 296–8.

Imbrie, J. and Imbrie, K. P. 1986. *Ice Ages: Solving the Mystery*. (Harvard University Press, Cambridge).

Knox, J. C. 1972. Valley alluviation in south-western Wisconsin. *Annals of the Association of American Geographers* **62**, 401–10.

Kukla, G. (1987) Loess stratigraphy in China. *Quaternary Science Reviews* **6**, 191–219.

Lancaster, N. (1981) Paleoenvironmental implications of fixed dune systems in southern Africa. *Palaeogeography, Palaeoclimatology, Palaeoecology* **33**, 327–46.

Lowe, J. J. and Walker, M. J. C. (1984) *Reconstructing Quaternary Environments*. Longman, London and New York.

MacKay, J. R. (1988) Pingo collapse and paleoclimatic reconstruction. *Canadian Journal of Earth Sciences* **25**, 495–511.

McCauley, J. F. *et al.* (1982) Subsurface valleys and geoarcheology of the eastern Sahara revealed by shuttle radar. *Science* **218**, 1004–20.

Oberlander, T. M. (1972) Morphogenesis of granitic boulder slopes in the Mojave Desert, California. *Journal of Geology*, **80**, 1–20.

Oberlander, T. M. (1974) Landscape inheritance and the pediment problem in the Mojave Desert of southern California. *American Journal of Science* **274**, 849–75.

Oberlander, T. M. (1989) Slope and pediment systems. In: D. S. G. Thomas (ed.) *Arid Zone Geomorphology*. Belhaven Press, London; Halsted Press, New York.

Pye, K. (1987) *Aeolian Dust and Dust Deposits*. Academic Press, London and Orlando.

Reger, R. D. and Péwé, T. L. (1976) Cryoplanation terraces: indicators of a permafrost environment. *Quaternary Research* **6**, 99–109.

Sarnthein, M. (1978) Sand deserts during glacial maximum and climatic optimum. *Nature* **272**, 43–6.

Schumm, S. A. (1968) River adjustment to altered hydrologic regimen – Murrumbidgee River and paleochannels, Australia. *United States Geological Survey Professional Paper* **598**.

Schumm, S. A. (1969) River metamorphosis. *American Society of Civil Engineers Journal of the Hydraulics Division* **95**, 255–73.

Schumm, S. A. (1977) *The Fluvial System*. Wiley, New York and London.

Starkel, L. (1983) The reflection of hydrologic changes in the fluvial environment of the temperate zone during the last 15,000 years. In: K. J. Gregory (ed.) *Background to Palaeohydrology*. Wiley, Chichester and New York, 213–35.

Stoddart, D. (1969) Climatic geomorphology: review and reassessment. *Progress in Geography* **1**, 159–222.

Street-Perrott, F. A. and Harrison, S. P. (1984) Temporal variations in lake levels since 30 000 BP: an index of the global hydrological cycle. In: J.E. Hansen and T. Takahasi (eds) *Climate Processes and Climate Sensitivity*. Geophysical Monograph **29**. American Geophysical Union, Washington DC.

Sugden, D. E. and John, B. S. (1976) *Glaciers and Landscape: A Geomorphological Approach*. Edward Arnold, London.

Tricart, J. and Cailleux, A. (1972) *Introduction to Climatic Geomorphology* (translated by C. J. Kiewiet de Jonge). Longman, London.

Twidale, C. R. (1981) Origins and environments of pediments. *Journal of the Geological Society of Australia* **28**, 423–34.

Van der Hammen, T. (1982) Paleoecology of tropical South America. In: G. T. Prance (ed.) *Biological Diversification in the Tropics*. Columbia University Press, New York.

Washburn, A. L. (1979) *Geocryology: A Survey of Periglacial Processes and Environments*. Edward Arnold, London.

Part IV

Endogenic – exogenic interactions

15

Rates of uplift and denudation

15.1 Tempo of geomorphic change

Having examined the operation of the various endogenic and exogenic geomorphic processes in Parts II and III we can now look at the way these two sets of processes interact in the formation of the landscape. It is appropriate to begin with the overall rates at which endogenic and exogenic processes operate since it is the balance between uplift and denudation at any particular point on the Earth's surface that determines whether that point becomes higher or lower with respect to sea level. Through our discussion of endogenic and exogenic progresses it will have become apparent that the rates at which they operate vary enormously; in fact a glance at Figure 15.1 shows that their rates range over some 16 orders of magnitude.

The rates of operation of geomorphic processes may be expressed in various ways. One approach is to view the transport of materials in terms of geomorphic work and to represent this in units of power. This can be done by relating the operation of a particular process to its duration. Although this approach is useful for comparing the relative importance of specific geomorphic processes, when considering the overall rates at which the landscape changes it is more usual to express this in terms of an average lowering or raising of the ground surface over a period of time. Since we are concerned in this chapter primarily with the net effect of various geomorphic processes in modifying the landscape this is the approach we will use here.

Changes in the average elevation of the ground surface are most commonly expressed in terms of mm ka^{-1}. For long periods of time of the scale of millions of years it is more appropriate to use the equivalent unit of m Ma^{-1} (note that 1 mm ka^{-1} equals 1 m Ma^{-1}). Very rapid processes may be better described in terms of mm a^{-1} or even m s^{-1}. It has become fashionable in recent years to express changes in ground elevation in 'Bubnoff units', often represented as B.

This is simply equivalent to a change in ground elevation expressed in mm ka^{-1} or m Ma^{-1} and seems to represent a rather needless proliferation of units.

Since rates of denudation are frequently estimated from the *mass* of sediment transported by rivers, glaciers or the wind, we need to be able to convert this mass to a *volume* in order to estimate the equivalent rate of ground lowering. This is done by relating the mass of material transported to its mean density, usually assumed to be the density of typical crustal rocks (approximately 2700 kg m^{-3}). Such a procedure, however, can raise difficulties when comparing the removal of solid and dissolved constituents by rivers. This problem is examined in Section 15.3.1.

15.2 Rates of uplift

Before we discuss how rates of uplift can be measured it is important to clarify what we actually mean when we use the term 'uplift'. This is necessary because it is used in two quite distinct senses. Here we will use the term **surface uplift** to refer to the upward movement of the landsurface with respect to a specific datum (normally sea level). What we will term **crustal uplift**, on the other hand, refers to the upward movement of the rock column with respect to a specific datum (also normally sea level).

As is apparent from Figure 15.2 the relationship between surface uplift and crustal uplift depends on the prevailing rate of denudation. Crustal uplift will only equal surface uplift if there is no denudation during the period of uplift. If, as is likely, denudation does occur then crustal uplift will be greater than surface uplift. But if crustal uplift is exactly matched by denudation then there will be no surface uplift. In fact crustal uplift is frequently exceeded by denudation, and in this situation surface elevation is obviously reduced.

In most cases surface uplift is associated with active tec-

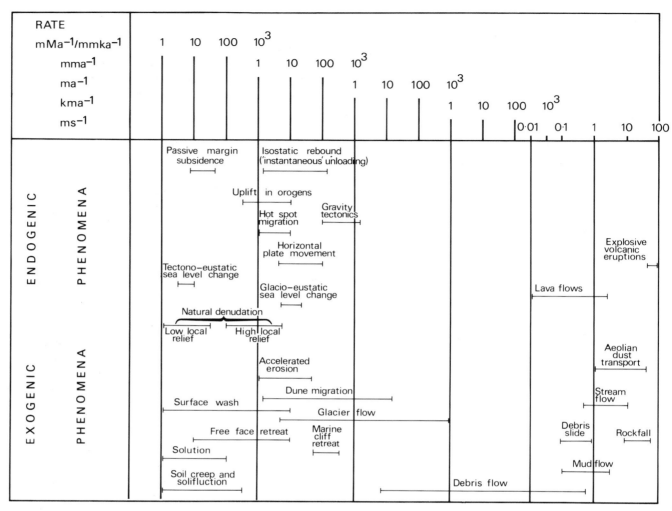

Fig. 15.1 *Comparison of rates of various endogenic and exogenic geomorphic processes. Note that rates of specific slope processes are discussed in Chapter 7 and are included here for comparative purposes.*

tonic processes, but crustal uplift can occur simply as the inevitable isostatic response to denudation; that is, the removal of material through denudation reduces the load on the crust and it moves upward to restore isostatic equilibrium. In general such crustal uplift resulting from isostatic rebound does not exceed the rate of denudation so there is no net surface uplift; nevertheless, where flexural effects are important (see Section 4.2.3), or where a landsurface is locally dissected by deep gorges, such isostatic recovery can lead to surface uplift of parts of a landscape.

15.2.1 Methods of measurement and estimation

Rates of uplift can be determined either directly or indirectly by a range of techniques appropriate to different time scales (Fig. 15.3). Some of these methods measure or estimate surface uplift, whereas others indicate rates of crustal uplift, and it is important to distinguish between them.

Geodesy is concerned with the precise determination of the Earth's surface form and **geodetic** methods can provide information on short-term rates of surface uplift. One way this can be done is through precise levelling surveys which are repeated after a number of years or, more often, several decades. Although systematic errors in levelling make it difficult in some cases to distinguish a real vertical movement from an artefact of the measuring procedure, such geodetic techniques provide an important basis for documenting surface uplift and subsidence over periods of a few years to a few decades. The technology is now available whereby satellites can determine with great accuracy the form of the Earth's surface (**satellite altimetry**) and changes in the geoid over time have already been recorded by this means.

Along coasts tide gauge records can provide high-quality information on vertical movements of the landsurface, as long as allowance is made for the slow, continuing, post-

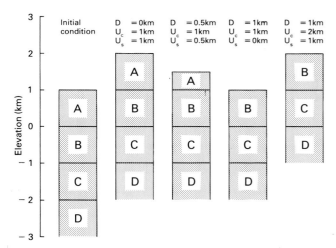

Fig. 15.2 *Schematic illustration of the distinction between surface uplift and crustal uplift. D designates the depth of denudation and U_c and U_s, respectively, the amount of crustal and surface uplift. Note that if local (Airy) isostasy is to be maintained then denudation of a given amount will be accompanied by crustal uplift of around 80 per cent of the depth of denudation (see Figure 4.7.)*

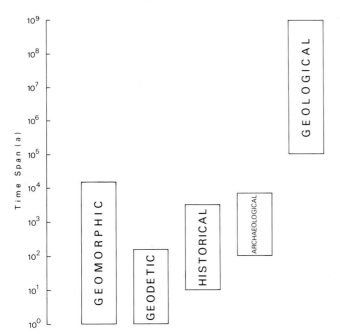

Fig. 15.3 *Appropriate time scales for different types of evidence used to determine rates of uplift.*

glacial sea level rise (currently about 1 mm a⁻¹). Tide gauge stations are relatively numerous in some areas of the world. In a few instances their records extend back into the last century, but the quality of some of these early records is too poor for their use in precise calculations of surface uplift or subsidence rates.

As the time period with which we are concerned becomes longer, so the means of determining rates of vertical movement become more indirect and less precise. Historical and archaeological evidence can be valuable in certain circumstances. Surface uplift in Iraq around AD 100–200 has been inferred from the breaching of an irrigation canal, and a similar instance has been reported of the disruption of an Inca canal in Peru. In the coastal regions of Greece, a tectonically active area, rapid vertical movements have been documented from the displacement of buildings and other constructions built in antiquity. Discrete phases of surface uplift and subsidence have been inferred from the displacement of the northern end of the ramp (*didkos*) used by the Greeks to haul ships across the Isthmus of Corinth.

Geomorphic evidence of vertical movements is represented by dated landform features which can be related to a previous absolute or relative elevation. Such evidence might include dated basalt flows displaced by faulting or elevated wave-cut platforms or beaches which can be dated from their associated deposits. One of the best-documented uplift histories in the world is to be found in the marine terraces of the Huon Peninsula, Papua New Guinea. Here a series of coral terraces has been cut by high sea levels during the Late Pleistocene as the coastline has risen rapidly. The rates of uplift can be established because the age of each coral terrace can be determined by radiometric dating (see Sections 15.2.2 and 17.2.1).

Geological evidence for uplift is usually relevant to periods of millions of years and is established from the elevation of rocks of known age formed at, or below, sea level, and from a variety of dating techniques which enable estimates to be made of the depth of burial of a rock at a known time in the past. The present-day height of marine deposits above present sea level can be used to calculate a long-term mean rate of surface uplift if the sea level at the time of the formation of a deposit is known. Such evidence, however, only provides an estimate of net surface uplift since there could have been intervening periods of subsidence.

Increasing use is now being made of radiometric dating to estimate long-term rates of rock uplift relative to the landsurface (Fig. 15.4(A)). The 'clocks' of radiometric dating procedures, such as K–Ar and Rb–Sr dating, are only set once the rock has cooled below a certain temperature, known as the **closure temperature**; this is the point at which the system becomes closed and the products of radioactive decay are retained (see Appendix B). If the geothermal gradient (typically 20–30 °C km⁻¹ in the upper crust) is known, the depth below the surface at which the radiometric 'clock' started can be calculated. The radiometric age of the rock then gives the time taken for denudation to remove the overlying strata and expose the dated rock at the surface. This information provides us with a mean denudation rate, and if we assume that there has been no change in surface elevation then this will be equal to the

A

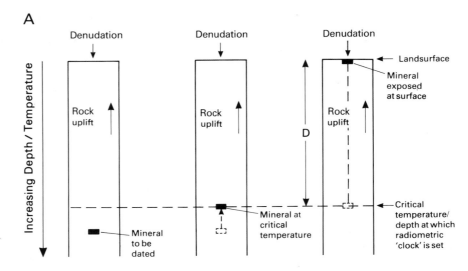

B

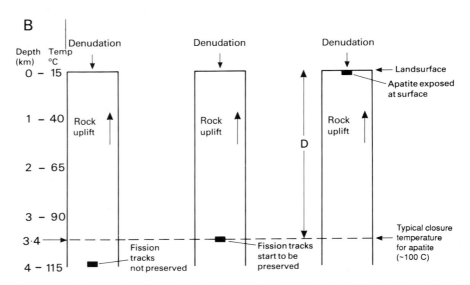

Fig. 15.4 Schematic illustration of the use of radiometric dating (A) and fission track dating (B) in the estimation of denudation and uplift rates relative to the landsurface. In (A) the age of a mineral at the surface is determined by the time elapsed since it passed through the critical isotherm at which its radiometric 'clock' was set. This isotherm, and therefore depth, varies for different minerals. The rate of denudation is calculated by dividing depth (D) by age. In (B) the estimation of denudation using fission track dating of apatite is illustrated. If we assume that the geothermal gradient is 25 °C km⁻¹ (as indicated) and the age of apatites exposed at the surface is 10 Ma then the mean denudation rate is 3400 (m)/10 = 340 m Ma⁻¹.

mean crustal uplift rate relative to sea level. Note that, used alone, this technique does not provide any direct information on the amount of absolute uplift that has occurred.

The estimation of crustal uplift rates using radiometric dating is most appropriately applied to orogenic belts since it is here that metamorphic events associated with significant heating of the crust are most likely to 'reset' the radiometric clock by overprinting the original radiometric dates established when the rocks were first formed. This enables a rock uplift rate relative to the landsurface to be estimated for a short period of a few million years since the resetting event. It is also only in active orogenic belts that

denudation and crustal uplift rates are likely to be approximately equal for a significant period of time.

An elaboration of this technique is to compare the radiometric ages of two different minerals which start their radiometric clocks at different temperatures. The radiometric clock for the mineral muscovite, for instance, starts about 200 °C higher than the radiometric clock for biotite. If the geothermal gradient is 20 °C km⁻¹ this is eqivalent to a difference in depth of 10 km, so the difference in age between the two minerals when exposed together at the surface represents the time taken for 10 km of denudation to occur. Again, assuming a rough equilibrium between crustal uplift and denudation rates, this provides an estimate of the former.

Another technique based on a similar principle is **fission track dating** (Fig. 15.4(B)). The fission of uranium-238, which is relatively abundant in minerals such as apatite found in granites and other basement rocks, involves the break up of the original nucleus into two fragments of roughly equal mass. These charged particles recoil from each other and move in opposite directions thereby creating microscopic dislocations or 'tracks' in the crystal lattice. The number of tracks formed is a function of the concentration of uranium present and the time elapsed since tracks began to be preserved. Determining the uranium concentration and counting the number of microscopic tracks formed enables an age to be estimated. The particular advantage of the technique for estimating rock uplift relative to the land-surface and denudation rates is that for the mineral apatite the fission tracks are destroyed, or **annealed**, at temperatures above about $100 \pm 20\,°C$. The fission track closure temperature for apatite is, therefore, located at a relatively shallow depth in the crust corresponding to this temperature. Depending on the geothermal gradient, this **fission track annealing zone** for apatite usually lies at a depth of between 3.5 and 5.5 km.

As with the radiometric methods already discussed, this technique can only provide estimates of crustal uplift rates indirectly through the calculation of denudation rates. Outside active orogenic belts where these two rates will rarely be similar, the main use of both radiometric and fission track techniques is in the estimation of denudation rates (see Section 15.4).

15.2.2   Spatial and temporal variations

From our discussion of the mechanisms of orogeny and epeirogeny in Chapters 3 and 4 we would expect considerable differences in their associated rates of uplift. Although this is clearly the case in the long term, evidence from relevelling surveys provides some equivocal data on rates of uplift in regions affected by epeirogenic movements.

15.2.2.1   Orogenic uplift
In active orogenic zones significant vertical movements of the surface up to several metres may occur in a single earthquake. In the major Alaskan earthquake of 1964, for example, an area of some $400\,000\,km^2$ was raised by an average of 2 m and a maximum of 12 m. Although there may be rapid variations in rates of vertical displacement in space and time along convergent and transform plate margins, high mean rates of uplift can be sustained for periods of hundreds of thousands of years (Table 15.1). An extreme example is the Ventura Anticline at the western margin of the Transverse Ranges in southern California which has experienced crustal uplift at an average rate of at least $16\,000\,m\,Ma^{-1}$ over the past 1 Ma. Rather more modest rates of surface uplift of up to $3300\,m\,Ma^{-1}$ have been recorded from the elevated coral terraces of the Huon Peninsula (Fig. 15.5).

Minimum crustal uplift rates in major orogenic belts, such as the Alps and the Himalayas, located at convergent plate margins range from 300 to about $800\,m\,Ma^{-1}$ averaged over periods of several million years (Table 15.1). But rates may be much higher than this; in southern Tibet, for

Table 15.1 Long-term mean uplift rates in orogenic zones determined by various methods

LOCATION	METHOD	RATE (m Ma^{-1})	PERIOD	SOURCE
Central Alps	Apatite fission track age	300–600*	6–10 Ma BP	Schaer et al. (1975) Tectonophysics **29**, 293–300
Central Alps	Rb and K–Ar apparent ages of biotite	400–1000*	10–35 Ma BP	Clark and Jäger (1969) American Journal of Science **267**, 1143–60
Kulu–Mandi Belt, Himalayas	Apparent Rb-Sr ages of coexisting biotites and muscovites	700*	25 Ma BP–Present	Mehta(1980) Tectonophysics **62**, 205–17
Southern Alps, New Zealand	Apparent K–Ar ages of schists	10 000*	1 Ma BP–Present	Adams (1981) Geol. Soc. Lond. Spec. Publ. **9**, 211–22
Southern Alps, New Zealand	Estimated ages of elevated marine terraces	5000–8000†	140 ka BP–Present	Bull and Cooper (1986) Science **234**, 1225–8
Huon Peninsula, Papua New Guinea	U-series and ^{14}C dating of elevated marine terraces	1000–3000†	120 ka BP–Present	Chappell (1974) J. Geophys Res. **79**, 456–64

* Rock uplift (relative to surface)
† Surface uplift.

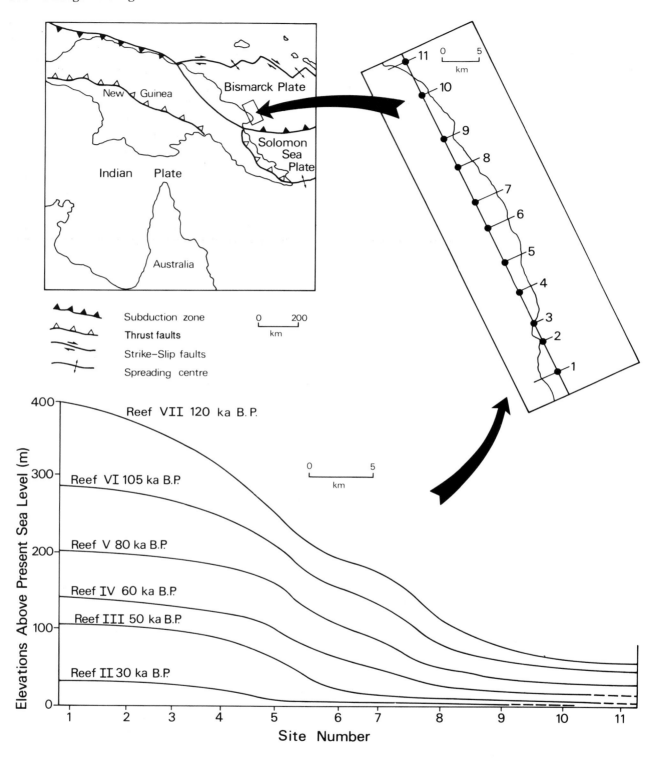

Fig. 15.5 *Tectonic setting and surface uplift history of the terraces of the Huon Peninsula, Papua New Guinea, determined from the* [14]C *and U-series dating of wave-cut notches formed during Late Pleistocene sea level highstands. Uplift rates can be seen to have varied along the coast with much higher rates towards the south-east. (Based on J. Chappell (1974)* Journal of Geophysical Research **79**, *Fig. 2, p. 392. Copyright by the American Geophysical Union.)*

instance, Late Pliocene – Early Pleistocene terrace deposits containing a fauna indicative of a lowland sub-tropical climate have since been elevated to a height of 4000–5000 m indicating surface uplift averaging more than 2000 m Ma^{-1}. Overall, crustal uplift rates in the Himalayas appear to be currently averaging around 5000 mm ka^{-1}. This is matched by long-term rates in some Andean ranges, such as the Cordillera Blanca in Peru which, on the basis of the exposure of a granite batholith emplaced at a depth of 8 km about 10 Ma BP, has experienced crustal uplift of around 4000–5000 m Ma^{-1} since that time.

Even these rates, though, are modest in comparison with those estimated for sections of the Southern Alps in New Zealand. This mountain range, located along the boundary of the Pacific and Indian Plates, has apparently sustained a rate of crustal uplift averaging up to 10 000 m Ma^{-1} over the past 1 Ma. This extraordinarily high uplift rate is, as in the case of the Transverse Ranges of southern California, related to the oblique convergence occurring along the associated plate boundaries (see Section 3.5.2).

For some mountain ranges the variations in local rates of surface uplift correspond well with their present-day topography; that is, the regions of highest elevation are rising at the greatest rate (Fig. 15.6). In other orogenic belts there is a mismatch between topography and present patterns of vertical movement. This is the case for the Alps where detailed relevelling surveys documenting vertical movements over the past 60 a or so show that the current maximum rates of surface uplift are to be found in parts of the alluvium filled valleys of the Rhône and the Rhine. This apparent

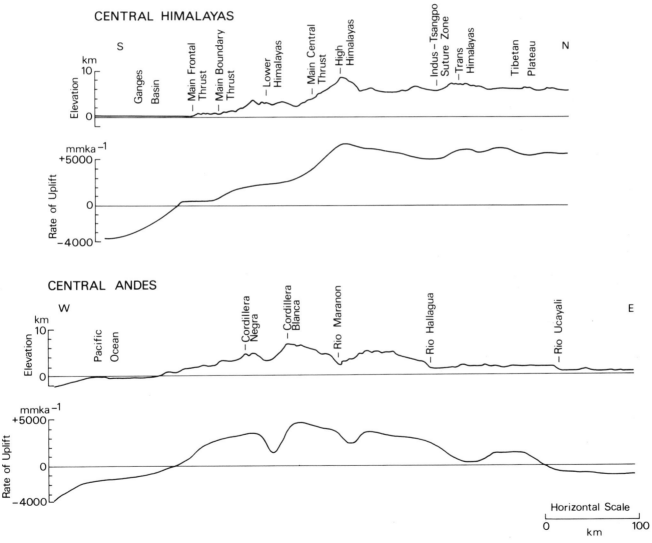

Fig. 15.6 *Relationship between topography and surface uplift rates along transects across the central Himalayas and central Andes. (Modified from A. Gansser (1983) in: K. J. Hsü (ed.)* Mountain Building Processes. *Academic Press, London, Fig. 2, p. 223.)*

anomaly may be at least partly explained by the continuing isostatic compensation for the removal of the ice loads occupying these valleys during the last glacial maximum.

15.2.2.2 *Epeirogenic uplift*

Perhaps because of the less obvious vertical movements involved, data on rates of epeirogenic uplift are rather sparse, at least for the long term. Rates for uplift averaged over millions of years appear to lie in the range of 10–200 m Ma^{-1}. The Colorado Plateau in the western USA, for instance, has risen about 500 m during the Late Cenozoic at a mean rate of around 100 m Ma^{-1}, while the Deccan Plateau of India has experienced uplift of 600 m or so extending over the past 40 Ma at a mean rate of about 15 m Ma^{-1}.

Although we would expect to see short-term rates that are similarly slow, this is apparently not the case. Detailed relevelling surveys carried out in the USA, for instance, demonstrate that over the past several decades there has been little to distinguish apparent rates of vertical crustal movement in the tectonically active western USA from the supposedly stable eastern part of the country (Fig. 15.7).

Such data suggest remarkably high rates of surface uplift of up to 20 mm a^{-1} at the present time in regions such as Florida. These rates exceed even those in the world's most active orogenic belts so we are led to ask whether they are credible or whether they are simply an artefact of systematic errors in the levelling measurements. Although in most cases the changes in elevation recorded exceed the random error limits expected in such surveys, systematic errors certainly cannot be ruled out. On the other hand, the patterns of apparent movement observed along the eastern margin of the USA correspond to known tectonic structures and various geomorphic features such as drainage divides and stream gradient anomalies.

If such movements are real the problem remains of finding a plausible mechanism to explain them. It is obvious that such high rates in 'passive' tectonic settings of generally subdued relief cannot be sustained for long periods of time, otherwise they would generate a quite unrealistic topography (a sustained uplift of 20 mm a^{-1} would give rise to a 2000 m high mountain range in just 100 000 a assuming no erosion). This implies that such epeirogenic movements

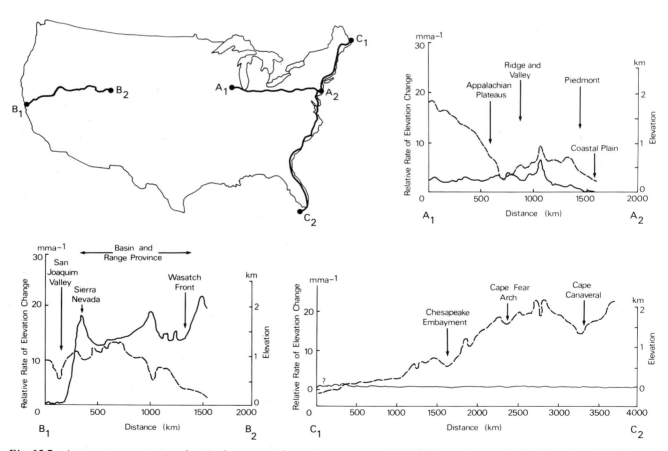

Fig. 15.7 *Apparent present rates of vertical movement for various transects in the USA based on precise levelling data. Solid lines indicate topography, dashed lines apparent vertical movement. Note that rates of movement are expressed in mm a^{-1}. (Based on L. D. Brown et al. (1980) in N.-A. Mörner (ed.)* Earth Rheology, Isostasy and Eustasy. *Wiley, Chichester, Fig. 1, p. 391 and Fig. 10, p. 399; L. D. Brown (1978)* Tectonophysics **44**, *Fig. 2, p. 211; and L. D. Brown and J. E. Oliver (1976)* Reviews of Geophysics and Space Physics **14**, *Fig. 11, p. 21.)*

are episodic or oscillatory, with periods of uplift alternating with periods of subsidence. Although episodic or oscillatory movements must be related in some way to epeirogenic mechanisms (along passive continental margins) or plate interactions (along active plate margins), the specific processes involved are unknown. Nevertheless, the considerable areal extent of the regions affected by these rapid epeirogenic movements (up to 1000 km across) implies that the cause cannot lie just in the crust but must be deep-seated and probably involves the entire thickness of the lithosphere and even possibly a part of the asthenosphere.

Although there is great uncertainty about the cause of short-term rapid uplift in many regions of the world, far more is known about the extremely high rates of uplift recorded from areas once covered by Pleistocene ice sheets. In fact the relationship between the temporal and spatial variations in rates of isostatic rebound and the thickness and location of the Pleistocene ice sheets has provided vital information about the nature of the sub-lithospheric mantle and the way in which it responds to the removal of a load. Detailed information on the decay of the great ice sheets of the last glacial and the resulting isostatic rebound of the crust is provided by raised shorelines containing datable deposits. As can be seen in Table 15.2, such evidence demonstrates that the highest rates of surface uplift are attained by crust lying below the thickest part of the ice sheet (where unloading generates the greatest degree of isostatic disequilibrium). It also shows that rates of isostatic rebound decline exponentially through time, being extremely rapid immediately after deglaciation, with rates of up to 100 000 mm ka^{-1}. Somewhat slower rates of surface uplift, at around 5000–10 000 mm ka^{-1}, continue for at least a further 15 000 a in the case of major continental ice sheets.

Table 15.2 Post-glacial rates of vertical movement (mm a^{-1}) in regions covered by Late Pleistocene ice sheets

	LAURENTIDE		FENNO-SCANDIAN CENTRE	SCOTTISH CENTRE
	CENTRE	EASTERN PERIPHERY		
10–8 ka BP	100–70	70	30	10
4–3 ka BP	30	3–5	10–15	4–5
Recent decades	>(5±2) (20±5?)	–(2–3)	9–10	3.8–5.8

Note: Post-glacial eustatic rise in sea level taken into account.
Source: Data from A. A. Nikonov (1980) in: N.–A. Mörner (ed.) *Earth Rheology, Isostasy and Eustasy*. Wiley, Chichester, Table 2, p. 347 compiled from various sources.

15.3 Present fluvial denudation rates

The analysis of contemporary rates of denudation raises major difficulties, but it provides the only means we have of determining which factors are most significant in controlling the rate at which landscapes are eroded. An assessment of fluvial denudation rates is important because the output of sediment and dissolved constituents from the usually well-defined boundaries of a drainage basin enables us to monitor the net effects of weathering, slope processes and fluvial transport across a specific area. Defining source regions for the removal of material by glaciers, and especially wind, is far more difficult. Perhaps the greatest limitation in using present-day denudation rates as a basis for understanding long-term landform development is that we cannot be sure how typical they are. This is why it is also vital to examine other evidence for long-term rates of denudation.

15.3.1 Methods of Measurement and Estimation

The basis for determining fluvial denudation rates is the estimation of the **solid** (sediment) and solute (dissolved) load carried by rivers (see Section 8.4.1). A very approximate estimate can be achieved simply by multiplying the mean sediment and solute concentration calculated from a small number of samples by mean discharge. More accurate estimates, however, must take into account the way in which the concentrations of sediment and dissolved constituents vary with discharge and, in particular, the way in which they vary with flood events. This is accomplished by using **sediment** and **solute rating curves** constructed from equations which describe the best fit relationships between sediment and solute concentrations and discharge. For greater accuracy separate rating curves can be used for rising or falling stage relationships and for seasonal flows. Solid and solute transport rates can then be calculated by relating the rating curve to either continuous stream-flow data, or flow-duration curves based on hourly, daily or even monthly data. Such a procedure is subject to errors, particularly where the frequency of sampling is low. In small basins transport rates can vary significantly from hour to hour, and even in large basins a high degree of variability can occur between individual years.

Since data on sediment and solute transport rates are not available for all drainage basins, estimates of denudation rates on a continent-wide or global basis must be founded on some form of extrapolation. This is usually based on empirical relationships observed between measured solid and solute transport rates and the factors thought to control these rates, especially those related to climate and relief. Where such relationships are found to be strong it is then possible to estimate the sediment and solute transport rates on the basis of climate and relief, and, in some cases, other variables.

This approach has been much favoured by some French geomorphologists. For instance, in an extensive survey published in 1960, F. Fournier estimated the global pattern of denudation by relating sediment yield data on 78 basins (ranging in area from 2460 to 1 060 000 km^2) in a variety of climatic zones to a seasonality of rainfall index. This was expressed as p^2/P, where p is the mean monthly maximum precipitation and P is the mean annual precipitation. The idea of such a seasonality of rainfall index is that it indicates the importance in sediment transport of peak discharges generated by intense seasonal rainfall. Fournier also incorporated the effect of relief by using separate regression equations for different relief types. Because the relationships between sediment and solute transport rates and the various climatic and relief variables used are not perfect, such an approach can only give an approximate picture of the variation in denudation rates.

Additional information on solid and solute loads for major river basins is constantly accumulating, and most estimates of contemporary global denudation are now increasingly based on these larger data sets rather than extrapolation from results for a limited number of basins. Nevertheless, there remain significant difficulties in converting solid and solute load data into meaningful estimates of denudation rates. A general problem is simply the unreliability of measurements for many of the world's major rivers. The data on solid and solute load are inadequate for basins such as the Amazon, Ganges and Brahmaputra which contribute large quantities of both sediment and dissolved constituents to the world's oceans. The data for Chinese rivers, which are also important sediment contributors, also used to be poor but are now much improved following the establishment of comprehensive monitoring programmes for several major basins. In some cases it is even difficult to find out details of the method of measurement used. A widely cited estimate of the sediment yield for the Irrawady River in Burma is in fact based on a few measurements taken by unknown methods in the 1870s. Even where modern techniques of measuring discharge and of sampling sediment and solute concentrations have been applied, there may still be significant errors in rating curves and insufficient information on the monthly and annual variability of discharge and the occurrence of rare peak discharges which may be particularly important in affecting the overall rate of sediment transport. Moreover, if there is no gauging station at the mouth of a river the total load transported to the ocean may be overestimated as there may be significant deposition or precipitation downstream of the sampling point. Other important problems relate specifically to the accurate estimation of solid and solute load, and we will now examine these in some detail.

15.3.1.1   Solid load

In most measurements of solid load transport only the suspended sediment is sampled and, because of the difficulties involved with large rivers, no attempt is made to measure bed load. In lowland tropical rivers this may not be a major problem; for example it has been estimated that bed load accounts for only 1 per cent of the total sediment load of

Table 15.3   Comparison of sediment yields (t km^{-2} a^{-1}) under natural and artificial conditions in various countries

LOCATION	NATURAL	CULTIVATED LAND	BARE SOIL
UK	10–50	10–300	1000–4500
USA	3–300	500–17 000	400–9000
China	<200	15 000–20 000	28 000–36 000
India	50–100	30–2000	1000–2000
Nigeria	50–100	10–3500	300–15 000
Ivory Coast	3–20	10–9000	1000–75 000

Source: Data from R. P. C. Morgan (1986) *Soil Erosion and Conservation*. Longman, London, Table 1.1, p. 5, based on various sources.

the Amazon River. In mountainous terrains, though, bedload transport can be very significant and in such environments the widely applied assumption that bed load is about 10 per cent of suspended load may be a drastic underestimate.

Another problem in estimating natural rates of denudation and in evaluating the factors that control them is that in many river basins sediment yields have been changed dramatically, at least at the local scale, through anthropogenic effects. Dam construction promotes a downstream reduction in sediment yield, but cultivation, surface mining and urban construction all lead to disturbance of the soil and consequently can give rise to sediment yields far in excess of those generated from naturally vegetated surfaces (Table 15.3). It has been estimated, for instance, that the conversion of forest to cropland in the states of the mid-Atlantic seaboard of the USA has led to a tenfold increase in sediment yield, but such increases within a basin are not necessarily reflected immediately by an increase in sediment load at the basin terminus. In the case just cited, for example, it has been calculated that over 90 per cent of the sediment stripped from the upland regions of the Piedmont region of the south-east USA remains stored as colluvium on hillslopes, and as alluvium in valley bottoms. In other words there is not a steady state between sediment supply and sediment transport. A similar situation exists in the upper Mississippi Basin, where a detailed sediment budget study has demonstrated that only 7 per cent of the sediment being removed from the upper parts of a tributary valley is currently reaching the Mississippi River (Fig. 15.8).

The figure of 7 per cent indicates the proportion of sediment being mobilized within the basin that actually reaches the basin outlet and represents the **sediment delivery ratio** for the basin. This is a central concept in understanding the dynamics of drainage basin erosion through time. In the long term the sediment delivery ratio must be approximately unity otherwise the river basin would become progressively choked with sediment. (In very large basins it may be less than unity, since sediment may progressively accumulate in the basin as the basin floor subsides under the load of overlying sediment.) In the short term, however, there

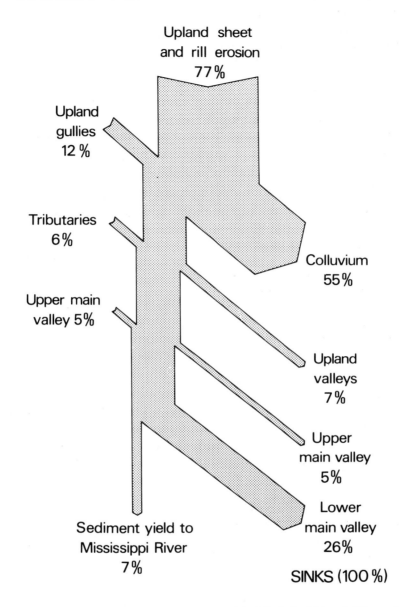

SOURCES (100 %)

Upland sheet
and rill erosion
77 %

Upland
gullies
12 %

Tributaries
6 %

Upper main
valley 5%

Sediment yield to
Mississippi River
7%

Colluvium
55 %

Upland
valleys
7 %

Upper
main valley
5 %

Lower
main valley
26 %

SINKS (100 %)

Fig. 15.8 *Sediment budget from 1938–75 for Coon Creek, a tributary of the Mississippi River in Wisconsin, USA. The figures show the percentage of sediment mobilized from different sources within the basin and the percentage deposited in different temporary storages (sinks). (Based on data from S. W. Trimble (1983)* American Journal of Science *283, 454–74.)*

may be transfers of material within a drainage basin but little export of material from it. Debris removed from slopes can be stored in fans, talus slopes and as river terraces and floodplain sediment.

The disparity between the amount of erosion occurring on hillslopes within a basin and the total amount of sediment being removed from it will tend to become more marked as drainage basin area increases, since larger catchments are likely to have a greater channel storage capacity in the form of extensive floodplains. An additional factor will be basin morphology with mountainous catchments generally having a smaller sediment storage capacity than lowland basins. Eventually this stored sediment will be mobilized and transported out of the basin, but this may take decades or centuries, or, in the case of very large basins, many thousands of years or more. Glacial and talus deposits generated in mountain valleys during the Late Pleistocene have still not been removed from many mid-latitude basins, while in the Amazon Basin only the major rivers are currently able to rework the coarser alluvium that was deposited during the more arid phases of the Quaternary.

15.3.1.2   Solute load

In view of the complexity of measuring separately each dissolved constituent in stream water, the easily measured electrical conductivity, or **specific conductance**, of the water is often used to provide an estimate of the solute concentration. Although there is a strong correlation between the concentration of ionic species in solution and electrical conductivity, the exact relationship varies depending on the concentrations present of particular dissolved constituents. Moreover, SiO_2, which is a significant component of many tropical lowland rivers, is not recorded by this technique. The solute load of rivers can also be enhanced by anthropogenic inputs. This is especially the case in basins containing industrial activities, and also in agricultural regions where the fertilizers applied to crops can find their way into stream waters in significant quantities. Human activities can also lead to an increase in rates of chemical weathering and thereby enhanced rates of solute input into stream waters.

The most important problem in estimating the contribution of solute transport to total denudation is the separation of the denudational and non-denudational components. Allowances must be made for dissolved constituents introduced into a basin through precipitation, since these represent a non-denudational component additional to the solutes released through bedrock weathering. An adjustment can either be made by using a knowledge of the chemical and mineralogical composition of the lithologies exposed in a basin to predict the solutes likely to be derived from bedrock weathering or, for greater accuracy, the chemical composition of the precipitation can be measured directly.

Table 15.4 gives the average composition of world river water and shows that two major non-denudational components introduced by precipitation are Na^+ and Cl^-, the two most abundant dissolved constituents in sea water. An even more important non-denudational component, however, is HCO_3^- which arises from the incorporation of atmospheric CO_2 during weathering reactions. Together these non-denudational components on average amount to about 40 per cent of the solute load of rivers but there are major departures from this figure. In regions of rapid rock weathering in highly reactive lithologies the non-denudational component is relatively minor, but in some tropical lowland rivers draining thoroughly leached and almost chemically inert weathering mantles the non-denudational component may be very significant. In the Rio Ucayali in the mountainous Andean region of the Amazon Basin it is estimated that only 4.8 per cent of Na^+, K^+, Ca^{2+} and Mg^{2+} comes from precipitation but in the Rio Tefé, a lowland tributary of the Amazon, 81 per cent of these four constituents is contributed by precipitation. Apart from their SiO_2 content the chemical composition of many of the lowland Amazon tributaries is in fact very similar to extremely dilute sea water, a reflection of the predominance of precipitation-derived solutes. An additional consideration relevant to many different environments and time scales is that the denudational component of the solute load can also be underestimated if solutes taken up in vegetation are removed from the basin as litter in streams.

15.3.1.3   Estimation of volumetric changes

A rather neglected problem concerning the estimation of denudation rates from solid and solute load data is the conversion of a measurement expressed as a mass per unit area per unit time (usually $t\ km^{-2}\ a^{-1}$) to a volumetric equivalent. Since we are concerned with the change in the *form* of the ground surface through time, it is the volumetric change which is of importance. At first sight this seems a trivial task; we simply divide the mass of material by the density of the bedrock (say $2700\ kg\ m^{-3}$ for rocks making up most of the upper continental crust) to give a volume ($m^3\ km^{-2}\ a^{-1}$). We then convert this to an *average* rate of lowering for the whole basin, acknowledging, of course, that the actual change in elevation will inevitably vary considerably from place to place. Conveniently, a volume change express-

Table 15.4   Average composition of world river water and estimates of denudational and non-denudational contributions for different constituents

	Ca^{2+}	Mg^{2+}	Na^+	K^+	Cl^-	SO_4^{2-}	HCO_3^-	SiO_2	Total
Average composition of world river water (concentration ($mg\ l^{-1}$))	13.5	3.6	7.4	1.35	9.6	8.7	52.0	10.4	106.6
Provenance of major solute components (%)									
Non-denudational:									
Precipitation (oceanic salts)	2.5	15	53	14	72	19	—	—	12
Atmospheric CO_2	—	—	—		—	—	57	—	28
Denudational:									
Chemical weathering	97.5	85	47	86	28	81	43	100	60

Source: Based largely on data in M. Meybeck (1983) in: *Dissolved Loads of Rivers and Surface Water Quantity/Quality Relationships* International Association of Hydrological Sciences Publication **141**, 173–192.

ed in $m^3 km^{-2} a^{-1}$ is equivalent to a mean rate of ground lowering expressed as $mm\ ka^{-1}$. The averaging over a period of 1000 a is largely conventional and has the advantage of giving an amount of lowering that can be readily appreciated. It does not imply that the rate of denudation has been measured over this length of time, and this is, of course, not the case when using stream load data.

Unfortunately, this simple procedure does not necessarily yield an accurate estimate of the equivalent average volume change actually occurring in the landscape. If soil or weathered material, which may have a bulk density between 1100 and 2000 $kg\ m^{-3}$, is being eroded, is it appropriate to use the density of the underlying bedrock in calculating the associated volume change? The answer is yes if the rate of rock weathering and removal of weathered material and soil are more or less in a steady state, but our understanding of the mode of landscape modification in many different geomorphic environments, and especially those of lowland tropical regions of subdued relief, suggests that periods characterized by low denudation rates, during which the weathering mantle increases in depth, are punctuated by phases of active erosion. In such cases rock weathering and weathering mantle erosion are not in equilibrium, and in this situation it is not clear that converting sediment removal into a rate of ground lowering on the basis of the original bedrock density is meaningful.

This problem is compounded when we consider chemical denudation. Conventionally, the specific solute load of streams measured in $t\ km^{-2} a^{-1}$ is converted into a rate of ground lowering in $mm\ ka^{-1}$ using an appropriate bedrock density.

Where solutes come from the direct solution of bedrock, as occurs to a large extent in limestone terrains, this is a valid procedure. But in many cases bedrock weathering takes place without any change in volume. Where such isovolumetric weathering occurs (see Section 6.2.4.1) solutes are lost to stream waters, but there are no associated volume changes as the transformation of fresh bedrock into saprolite composed largely of clay minerals is accompanied by a compensating decrease in bulk density. Even where weathering is not strictly isovolumetric there is almost invariably a significant compensation through a decrease in bulk density associated with the removal of dissolved constituents from the weathering mantle. Consequently, in lowland regions mantled by thick weathering profiles, it is misleading to view the removal of weathering-derived solutes *in the short term* as necessarily being directly related to chemical denudation in the sense of a reduction of the average elevation of a basin. In considering the rates of mechanical and chemical denudation discussed in the following section, it will be important to keep these points in mind.

15.3.2 Rates of mechanical and chemical denudation

Table 15.5 lists a range of estimates of total solid and solute load transport to the world's oceans and gives the equivalent global mean denudation rates (assuming an average source rock density of 2700 $kg\ m^{-3}$). It is important to point out here that these, and all other estimates of denudation rates, are subject to variable, and often large, errors. Although rates may be reported to a precision of one decimal place, this certainly does not imply that they are accurate to that degree.

The estimates in Table 15.5 are based on the total land area of $148 \times 10^6\ km^2$ and would be increased by about 40 per cent if only the global area of external drainage (about $105 \times 10^6\ km^2$) is considered. The various estimates in Table 15.5 are not directly comparable as the basis for their calculation differs, but it appears that the early estimate of Fournier, based largely on his suggested relationship between sediment yield and seasonality of precipitation, is too high. The more recent low estimate of Milliman and Meade, on the other hand, is for actual solid load transport to the oceans at the present day, and includes the effects of sediment entrapment by dams on rivers such as the Colorado, Nile and Zambezi.

What, then, is the best estimate we can make of current global denudation rates excluding the effects of human activities? Taking the Milliman and Meade estimate of 13 500 Mt a^{-1} for suspended sediment transport, we can add on 500 Mt a^{-1} as a reasonable estimate of the material trapped by dams and 1500 Mt a^{-1} to allow for unrecorded bed-load transport. In addition we perhaps should allow 500 Mt a^{-1} as a deduction to take account of the increase in erosion

Table 15.5 Estimates of total transport by rivers of solids and solutes to the oceans and equivalent estimated denudation rates

AUTHOR	MEAN LOAD		EQUIVALENT DENUDATION RATE[†]
	$(10^9\ t\ a^{-1})$	$(t\ km^{-2} a^{-1})$	$(mm\ ka^{-1})$
*Solid load**			*Mechanical*
Fournier (1960)	58.1	392.6	145.4
Jansen and Painter (1974)	26.7	180.4	66.8
Schumm (1963)	20.5	138.5	51.3
Holeman (1968)	18.3	123.6	45.8
Milliman and Meade (1983)	13.5	91.2	33.8
Lopatin (1952)	12.7	85.8	31.8
Solute load			*Chemical[‡]*
Goldberg (1976)	3.9	26.4	5.9
Livingstone (1963)	3.8	25.7	5.7
Meybeck (1979)	3.7	25.0	5.6
Meybeck (1976)	3.3	22.3	5.0
Alekin and Brazhnikova (1960)	3.2	21.6	4.8

* Suspended load only.
† Denudation rates based on a rock density of 2700 m^{-3}.
‡ Rates for chemical denudation assume that 40% of total solute load is from non-denudational sources.

rates as a consequence of human activities. These are admittedly rough approximations, but the total this gives of 15 000 Mt a^{-1} is probably a reasonable estimate of natural solid load transport to the oceans.

Estimates for global solute transport to the oceans are less variable, and we will take the recent figure established from a detailed global survey by M. Meybeck of 3700 Mt a^{-1} as being the most accurate available. From this we need to make a deduction to allow for the non-denudational component. As a global average 40 per cent is probably a reasonable figure and this gives an estimate of 2200 Mt a^{-1} for 'denudational' solute load transport. This gives a global mean annual transport of 17 200 Mt of material to the oceans which, averaged over the entire land area of the continents, gives a rate of 116 t km^{-2} a^{-1}. Assuming a mean source rock density of 2700 kg m^{-3}, this converts to a mean global denudation rate of 43 mm ka^{-1}. If we exclude areas of internal drainage where material is being transported but not removed from the continents, this figure rises to 61 mm ka^{-1}. Of this total about 85 per cent is accounted for by solid load transport and 15 per cent by the transport of solute load. It is reasonable to assume that, over large areas, denudation will lead to a compensatory isostatic rebound which reduces the 43 mm ka^{-1} mean rate of lowering with respect to sea level to only 8 mm ka^{-1}.

Such a global mean conceals great variations from area to area and basin to basin. Around 70 per cent of the total load transported to the oceans is provided from only 10 per cent of the land area and just three rivers, the Ganges, the Brahmaputra and the Huang He (Yellow) carry 20 per cent of the global fluvial sediment load. Figure 15.9 shows the results of a global survey of sediment yields based on measurements from more than 1500 sites and indicates that specific sediment yield rates exceed 1000 t km^{-2} a^{-1} (roughly equivalent to a mechnical denudation rate of 370 mm ka^{-1}) in the world's mountainous regions. By contrast, rates are less than 50 t km^{-2} a^{-1} (equivalent to approximately 19 mm ka^{-1}) in many lowland regions.

If we examine the world's largest drainage basins we find that present-day total denudation rates range from 3 mm ka^{-1} for the interior Chari Basin in Africa and 5 mm ka^{-1} for the Kolyma Basin in eastern Siberia, up to 529 mm ka^{-1} for the Huang He Basin of China and 677mm ka^{-1} for the Brahmaputra Basin draining the eastern Himalayas (Table 15.6, Fig. 15.10). These figures, however, by no means represent the extremes found in smaller basins. Minimum sediment yields lie well below 2 t km^{-2} a^{-1} (equivalent to <1 mm ka^{-1}) (1.7 t km^{-2} a^{-1}, for instance, for the Queanbeyan Basin (172 km^2) in the southern tablelands of the East Australian Highlands and less than 1 t km^{-2} a^{-1}

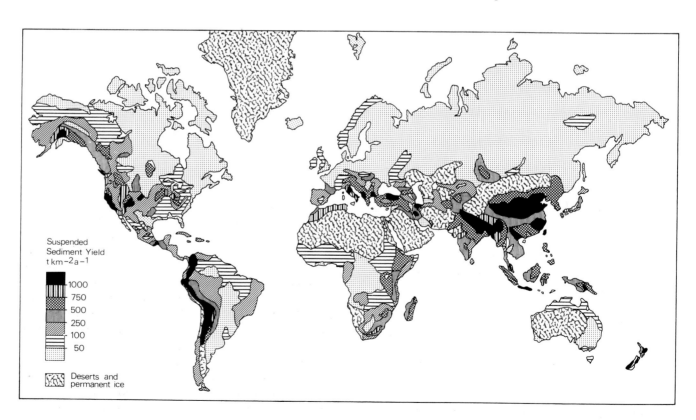

Fig. 15.9   *Global pattern of yields of suspended sediment. The values relate to intermediate-sized basins of 10^4–10^5 km^2. (After D. E. Walling and B. W. Webb (1983) in: K. J. Gregory (ed.)* Background to Palaeohydrology. *Wiley, Chichester, Fig. 4.2, p. 76.)*

Table 15.6   Estimated denudation rates for the world's thirty-five largest drainage basins based on solid and solute transport rates

	DRAINAGE AREA (10^6 km²)	RUNOFF (mm a⁻¹)	TOTAL DENUDATION (mm ka⁻¹)	MECHANICAL DENUDATION (mm ka⁻¹)	CHEMICAL* DENUDATION (mm ka⁻¹)	CHEMICAL DENUDATION AS % OF TOTAL
Amazon	6.15	1024	70	57	13	18
Zaire (Congo)	3.82	324	7	4	3	42
Mississippi	3.27	177	44	35	9	20
Nile	2.96	30	15	13	2	10
Paraná (La Plata)	2.83	166	19	14	5	28
Yenisei	2.58	217	9	2	7	80
Ob	2.50	154	7	2	5	70
Lena	2.43	206	11	2	9	81
Chiang Jiang (Yangtze)	1.94	464	133	96	37	28
Amur	1.85	175	13	10	3	22
Mackenzie	1.81	169	30	20	10	33
Volga	1.35	196	20	7	13	64
Niger	1.21	159	24	13	11	47
Zambezi	1.20	186	31	28	3	11
Nelson	1.15	96	—	—	—	—
Murray	1.06	21	13	11	2	18
St Lawrence	1.03	434	13	1	12	89
Orange	1.02	89	58	55	3	5
Orinoco	0.99	1111	91	78	13	14
Ganges	0.98	373	271	249	22	8
Indus	0.97	245	124	108	16	13
Tocantins	0.90	385	—	—	—	—
Chari	0.88	69	3	2	1	29
Yukon	0.84	232	37	27	10	28
Danube	0.81	254	47	31	16	35
Mekong	0.79	595	95	75	20	21
Huang He (Yellow)	0.77	63	529	518	11	2
Shatt-el-Arab	0.75	61	104	93	11	11
Rio Grande	0.67	5	9	6	3	38
Columbia	0.67	375	29	16	13	46
Kolyma	0.64	111	5	3	2	31
Colorado	0.64	31	84	78	6	7
São Francisco	0.60	151	—	—	—	—
Brahmaputra	0.58	1049	677	643	34	5
Dnepr	0.50	104	6	1	5	88

* Allowance made for non-denudational component of solute loads.
Source: Based primarily on data from M. Meybeck (1976) *Hydrological Sciences Bulletin* **21**, 265–89, and J. D. Milliman and R. H. Meade (1963) *Journal of Geology* **91**, 1–21.

for several rivers in Poland). Maximum sediment yields exceed 10 000 t km⁻² a⁻¹, with the Haast River draining a region of high precipitation and very rugged relief in the Southern Alps in New Zealand having a yield of 12 736 t km⁻² a⁻¹ (equivalent to 4717 mm ka⁻¹). But even this is exceeded by the 53 500 t km⁻² a⁻¹ of the Huangfu-chuan River, a tributary of the Huang He, which drains over 3000 km² of gullied loess-covered terrain in a region of sparse vegetation with a semi-arid climate characterized by periodic intense storms. This is equivalent to a denudation rate of 19 814 mm ka⁻¹, a figure which will only be sustained while supplies of readily erodible loess remain in the catchment.

Chemical denudation rates are in general less variable than those for mechanical denudation, but none the less still exhibit a wide range. Minimum solute load yields lie below 1 t km⁻² a⁻¹, whereas maximum yields of 6000 t km⁻² a⁻¹ occur in rare instances where rivers drain highly soluble deposits such as halite. More usual maxima lying below 1000 t km⁻² a⁻¹ (equivalent to 370 mm ka⁻¹) occur in limestone regions. High rates of chemical denudation are invariably observed in humid mountainous regions. In such environments rates of chemical weathering are high as the weathering front is at, or close to, the surface, but the accompanying high rates of mechanical denudation prevent the accumulation of thick weathering profiles (see Section 6.2.4.1). Conversely, minimum rates are recorded in semi-arid regions where runoff is very low (although solute concentrations may be very high), in lowland humid tropical regions where solute *concentrations* are generally extremely low and in high latitude lowland terrains where both runoff and solute concentrations are low.

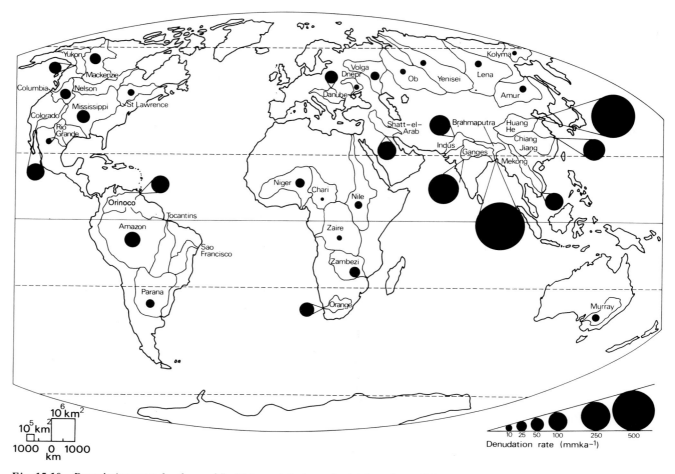

Fig. 15.10   *Denudation rates for the world's 35 largest drainage basins based on solid and solute load data. Allowance has been made for the non-denudational component of solute loads. Source rock density is assumed to be 2700 kg m⁻³ (Based on data in Table 15.6.)*

15.3.3   Relative importance of mechanical and chemical denudation

On the global scale we have already noted that the ratio of mechanical to chemical denudation is about 6 : 1, but what are the variations about this world mean? It is evident from Table 15.6 that for some basins, especially those in a predominantly humid lowland environment such as the great Siberian basins of the Ob, the Yenisei and the Lena, chemical denudation can actually greatly exceed mechanical denudation (although we must keep in mind the point made in Section 15.3.1.3). The other extreme is reached with those basins with extremely high sediment yields such as the Brahmaputra and the Huang He where chemical denudation represents 5 per cent or less of total denudation. Overall chemical denudation rates are less variable than those for mechanical denudation, ranging over two orders of magnitude rather than three or more. Rates of mechanical and chemical denudation overall show a fairly strong positive relationship, but as the ratio of solid to solute load per

unit area increases as total denudation increases, chemical denudation becomes *proportionally* less significant in drainage basins experiencing higher total denudation rates. This is apparent in Fig. 15.11 which illustrates the relative and absolute rates of solid and solute load transport for major basins. It is apparent that those rivers transporting the greatest solid load are in general those also transporting the highest solute load. Note also that the solute load tends to form a greater proportion of total load when the total load is small.

A fairly clear pattern emerges from Figure 15.11 which suggests the primary influences on relative rates of mechanical and chemical denudation. The highest rates for both are found in basins draining major orogenic belts such as the Brahmaputra and the Ganges. The Chiang Jiang (Yangtze) Basin, on the other hand, has a particularly high rate of chemical denudation presumably because it occupies large areas of limestone terrain in Szechwan Province in western China. Rivers draining largely semi-arid regions such as the Colorado, Orange and Shatt-el-Arab (Tigris and Euphrates)

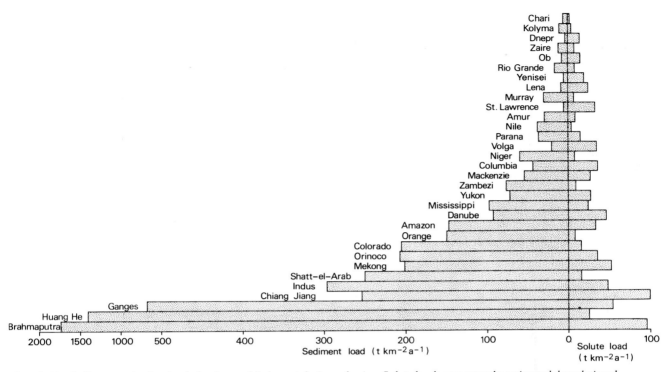

Fig. 15.11 *Sediment and solute loads for the world's largest drainage basins. Solute loads represent the estimated denudational component only. Data for the Nelson, Tocantins and São Francisco Basins are not available. (Based primarily on data in M. Meybeck (1976)* Hydrological Sciences Bulletin **21**, 265–89 *and J. D. Milliman and R. H. Meade (1983)* Journal of Geology **91**, 1–21.)

can be seen to have very low rates of solute transport in relation to total transport. This contrasts starkly with the basins of humid and subarctic regions such as the Lena, Yenisei, Ob and Dnepr which have high rates of solute transport in relation to total transport. The extremely high relative rate of solute transport for the St Lawrence Basin (some 87 per cent of total transport) is explained, at least in part, by the large proportion of sediment currently being trapped in the Great Lakes.

It is important to note that even with its predominantly semi-arid climate the Colorado Basin still has a higher rate of chemical denudation than either the humid tropical Zaire Basin or the largely subarctic Ob Basin. This is presumably because of the high rates of chemical denudation occurring in the high relief section of the Colorado Basin in the Rocky Mountains and exemplifies the generalization that, as for mechanical denudation, rates of chemical denudation appear to be much more strongly related to relief than to climate. This point is graphically illustrated in Figure 15.12 which compares the solid and solute loads of major tributaries of the Amazon. The Marañón and Ucayali Basins in the mountainous Andean part of the Amazon Basin show high rates of both solid and solute transport in comparison with the Negro Basin which drains the plateau country of the Guiana Highland, and the Xingu Basin which occupies a region of very low relief on the northern margin of the Mato Grosso. Some 85 per cent of the total solute load of

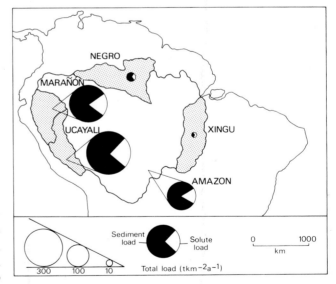

Fig. 15.12 *Rates of solid and solute load transport in sub-catchments of the Amazon Basin. Rates of solute load transport are for the estimated denudational component (Based on data in M. Meybeck (1976)* Hydrological Sciences Bulletin **21**, 265–9.)

the Amazon Basin originates from the Andean region of the catchment with the Marañón and Ucayali Basins alone jointly contributing 45 per cent.

Although we expect that there will frequently not be a

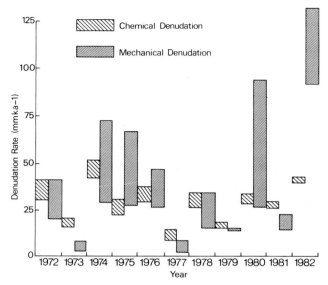

Fig. 15.13 *Annual range and variability in rates of mechanical and chemical denudation for four catchments in Idaho, USA. (Based on J. L. Clayton and W. F. Megahan (1986) Earth Surface Processes and Landforms 11, Fig. 3, p.394.)*

steady state between sediment supply and sediment transport rates, in general this is unlikely to be the case for the release of solutes and their transport out of the basin. This point needs to be emphasized since it must be taken into account in any comparison of rates of mechanical and chemical denudation. Another factor which has to be considered, especially when comparisons are based on short-term records, is the different ways in which sediment and solute transport rates vary through time. This is well illustrated by a detailed study of four small forested catchments on a coarse-grained granitic lithology in Idaho, USA (Fig. 15.13). Here the mean total denudation rate over an eleven year period was calculated to be 8.9 mm ka^{-1}. The 11 a mean mechanical denudation rate was found to exceed that for chemical denudation for three out of the four catchments. Nevertheless, because of the greater temporal variability of mechanical denudation it was more probable that in any one year chemical denudation would exceed mechanical denudation in three out of the four basins. In this case episodic high rates of mechanical denudation were considered to be related to high peak flows generated by snow-melt runoff after winters with heavy snowfalls.

15.4 Long-term fluvial denudation rates

15.4.1 Methods of estimation

The major problem with using sediment and solute load data as a basis for estimating long-term denudation rates is that we cannot be certain that we are not sampling an atypical period and therefore extrapolating from an unrepresentative

sample. In particular, we have already highlighted the problems of anthropogenic influences on present-day sediment and solute load data, as well as the complications arising from the lag between upstream sediment supply and sediment removal from basins. Fortunately, there are a number of other indirect means which we can use to estimate long-term rates of denudation.

15.4.1.1 Estimates from sediment volumes

The variability of sediment yields over periods of a few decades even in cool, humid temperate environments is evident from studies of sedimentation rates in small lakes and reservoirs. Denudation rates can be estimated on the basis of a known volume of sediment deposited over a known period of time and originating from a known source

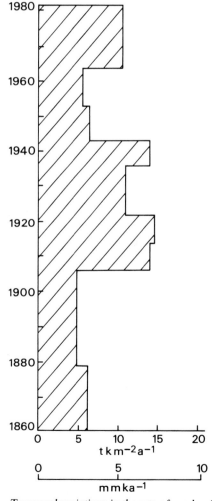

Fig. 15.14 *Temporal variations in the rate of mechanical denudation in the Merevale catchment, Warwickshire, UK, based on rates of lake sedimentation over nine time periods since 1861. The sedimentation rates have been adjusted to allow for non-catchment derived sediment. (Modified from I. D. L. Foster et al. (1985) Earth Surface Processes and Landforms 10, Fig. 9, p.59.)*

area. In a study of a small catchment in Warwickshire, UK, changes in the volumes of sediment deposited in a reservoir in nine periods since 1861 were calculated on the basis of 54 sediment cores dated radiometrically and correlated from magnetic measurements (Fig. 15.14). In this case the source area for the sediment was accurately known (being the present catchment area) and there was also good dating control.

In order to find out about rates of denudation over much longer periods of time we need to apply this approach at much greater temporal and spatial scales. We can, for example, attempt to equate denudation over large areas of continental drainage with the volume of marine sediment deposited offshore along the adjacent continental margin. In effect this is like treating offshore sedimentary basins as enormous, long-lived reservoirs.

The procedure involves four steps. First, the volume and age of sediments offshore must be determined from borehole data and seismic stratigraphy. The major errors here arise from an insufficient coverage of boreholes to give good dating control and the possibility of the addition and removal of sediment by ocean-bottom currents. Secondly, a deduction must be made to allow for the porosity of the sediments and their content of biogenic carbonate deposits (remains of marine organisms). Next, an estimate has to be made of the continental area from which the sediment was derived. This can be a particularly difficult task, especially if we are dealing with drainage areas in existence many tens of millions of years ago. Finally, the mean rate of mechanical denudation is calculated by relating the volume of land-derived sediments over a particular time period to the area of the assumed source region.

In one such investigation of the eastern seaboard of North America it has been estimated that during the Cenozoic about 1000 km³ of sediment has been produced for each kilometre of the 3000 km of coast from Georgia in the USA northwards to Newfoundland. Deducting 600 km³ km⁻¹ to allow for the contribution of porosity and biogenic carbonate gives 400 km³ km⁻¹ for the volume of land-derived sediment. Assuming that the present crest of the Appalachian Mountains represented the western limit of the sediment source this gives an estimate of 2 km of mechanical denudation over the past 65 Ma for the eastern seaboard of North America at a mean rate of just over 30 m Ma⁻¹. Such estimates are, of course, approximate and do not include chemical denudation for the simple reason that solutes are dispersed throughout the ocean (although they are eventually incorporated either into marine organisms (especially Ca^{2+}) or into deposits on the ocean floor).

15.4.1.2 Erosion of dated surfaces

A more direct and location-specific estimation of denudation rates can be made on the basis of the amount of lowering experienced by a surface of known age. This approach can be applied at a range of temporal scales. For instance, it is possible to estimate rates of soil loss from around the roots of trees dated by dendrochronology. Since some species of tree may have long life spans (the 2500 a or more recorded by some bristlecone pines, for example) this technique provides a useful time range. In one study carried out in the Piceance Basin, Colorado, USA, the date when the tree roots were first exposed by erosion was determined by a number of factors including the interpretation of the annual ring growth pattern and the earliest occurrence of reaction wood. Little variation in denudation rates was found between two sites, although a significant difference was found between north-facing slopes with a mean rate of 560 mm ka⁻¹ and south-facing slopes with a mean rate of 1180 mm ka⁻¹. The use of the exposure of datable materials can also be applied to human artefacts. Where archaeological remains of known age are present it is sometimes possible to estimate rates of ground lowering, or deposition, since their construction, or abandonment.

For longer time periods it is necessary to use dated surfaces. A classic study using this approach is the investigation by B. P. Ruxton and I. McDougall of the erosion of the Hydrographers Range, a dissected volcanic peak in north-east Papua New Guinea. On the basis of radiometric dating it is known that this composite andesitic volcano last erupted about 650 ka BP and, although the original form of the now eroded upper part of the cone is not known, its likely elevation was about 2000 m. Below about 1000 m the presence of young lavas on interfluves means that the original surface here can be reconstructed fairly accurately. By measuring the cross-sectional area of valleys cut into this surface at a range of altitudes, Ruxton and McDougall were able to estimate the volume of material removed over the past 650 ka. The denudation rate estimated in this case represented the combined effects of mechanical and chemical denudation and was found to increase with increasing local relief and slope gradient (Fig. 15.15).

15.4.1.3 Fission track and radiometric techniques

The final approach we need to consider is one which is likely to become increasingly important in studies of long-term denudation rates, namely the exposure of rocks known, on the basis of fission track or radiometric evidence, to have been at a specific depth below the surface at a particular time in the past. We have already outlined the principles of this approach in Section 15.2.1; as pointed out there, although these techniques have been primarily used to infer crustal uplift rates, the direct evidence they provide in fact relates to denudation rates.

It is interesting to see how estimates of denudation rates based on this approach compare with those determined from the volumes of sediment deposited offshore. In a study of the exposure of granite intrusions in northern New England on the northern seaboard of the eastern USA, differences in

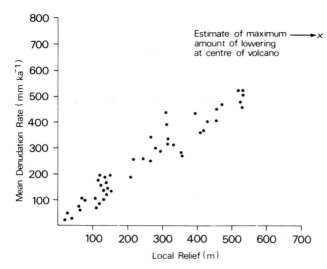

Fig. 15.15 *Mean rate of denudation as a function of local relief (difference in elevation between major ridge crests and adjacent valley bottoms) for the Hydrographers Range, Papua New Guinea based on the estimated dissection of dated surfaces. (Modified from R. P. Ruxton and I. McDougall (1967)* American Journal of Science *265, Fig. 5(A) p. 557.)*

fission track ages between apatite and zircon (the latter having a significantly higher annealing temperature) have been used to estimate their depth and temperature of emplacement. It was calculated that a presently exposed Jurassic intrusion dated at 180 Ma would have solidified at a depth of 5.3–7.6 km and a Cretaceous intrusion (115 Ma) at a depth of 3–3.6 km. These values convert to long-term rates of denudation of 42 to 29 m Ma[-1] since 180 Ma BP and 31 to 27 m Ma[-1] since 115 Ma BP. It gives one a sense of confidence that these estimates are broadly in line with the rate of 30 m Ma[-1] (for mechanical denudation only) for the past 65 Ma estimated on the basis of offshore sediment volumes for the eastern seaboard of North America.

Similarly, comparable results have been established for the Alps where differences in the radiometric ages of biotite and muscovite indicate a mean denudation rate of around 1000 m Ma[-1]. This contrasts with an estimated mean mechanical denudation rate of 100 m Ma[-1] for the Rhône Basin in the western Alps based on the sediment volume of the offshore submarine Rhône fan. These results become comparable if it is assumed, as seems likely, that the bulk of the sediment transported by the River Rhône originates from the mountainous Alpine section of its catchment. The highest long-term denudation rate yet recorded appears to be the 10 000 m Ma[-1] over the past 1 Ma determined from the K–Ar ages of exposed schists in the Southern Alps of New Zealand.

15.4.2 Variations in rates

We have just seen that, on the basis of fission track, radio-

metric and sediment volume estimates, the range of long-term denudation rates is broadly comparable to that observed for present-day rates of fluvial denudation. We now need to look in more detail at how long-term rates vary over time and space and assess the degree of equivalence between modern and past rates.

One basis for the estimation of long-term global trends in continental denudation is provided by the sediment cores recovered from the ocean basins during the various legs of the Deep Sea Drilling Project (DSDP). This evidence suggests that there have been globally synchronous fluctuations in sedimentation rates (and therefore presumably rates of continental erosion) (Fig. 15.16). High rates during the Middle Eocene (49 to 45 Ma BP) and the Middle Miocene to the present (from 14 BP) greatly exceed those for intervening periods during the Cenozoic. These fluctuations have been interpreted as the result of lower global preci-

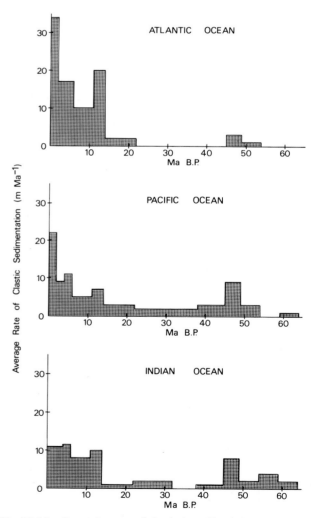

Fig. 15.16 *Cenozoic rates of clastic (non-biogenic) sedimentation in the world's ocean basins based on 110 Atlantic, 170 Pacific and 54 Indian Ocean DSDP cores. (Based on data in T. A. Davies* et al. *(1977)* Science *197, Fig. 1, p. 54.)*

pitation during these periods, although there is evidence that any increased continental denudation during arid phases is likely to be a result of enhanced aeolian erosion (see Section 15.6) rather than greater rates of fluvial sediment transport. However, it is also necessary to consider changes in rates of sediment deposition along continental margins since much of the coarser material removed from the continents is deposited there and fails to reach the deep ocean.

An overriding climatic control has also been invoked by another study examining variations in the rates of deposition of Al_2O_3 recorded in DSDP cores; Al_2O_3 is a good indicator of continental erosion because it is the major element most characteristic of non-biogenic sedimentation. A sixfold increase in the rate of Al_2O_3 accumulation in the northern and central Atlantic and Pacific Oceans over the past 15 Ma indicates a greatly increased rate of denudation during this period, although the increase in the southern Atlantic and Pacific is a more modest 100 per cent. Again, this may reflect primarily increased rates of aeolian denudation since Al_2O_3 could be derived from clay particles in soils deflated from regions experiencing phases of aridity. In a more recent analysis of DSDP cores it has been estimated that the total of denudationally derived sediment and dissolved load delivered to the world's oceans over their present lifetimes of 100–200 Ma is 1.65 km^3 a^{-1}. This suggests a rather low mean rate of continental denudation of 11 m Ma^{-1} (calculated on the basis of the present land area of 148×10^6 km^2).

Several estimates of long-term denudation rates in southeastern Australia based on a variety of methods indicate low rates of landscape modification since the Late Mesozoic (Fig. 15.17). River incision in the upper Lachlan valley of the Murray Basin into radiometrically dated basaltic lavas filling the valley indicates minimum rates of downcutting of 8 m Ma^{-1} for the past 20 Ma and 3–4 m Ma^{-1} for the previous 40 Ma. The rate for the past 20 Ma is broadly confirmed by the rate of sedimentation in the fan at the confluence of the Lachlan with the Murrumbidgee which indicates a rate of 4 m Ma^{-1}. A mean rate of 1–3 m Ma^{-1} is indicated for the whole of the Murray Basin during the Cenozoic, while fission track ages from apatites suggest a denudation rate for the southern part of the East Australian Highlands of 9 m Ma^{-1} over the past 230 Ma or so. Similarly modest rates of denudation of 15–30 m Ma^{-1} are indicated for the past 80–100 Ma following the uplift of the east Australian margin. This evidence is of interest both because of the low rates of denudation that it suggests for the tablelands of south-eastern Australia and for the lack of any evidence of a significant increase in erosion in the Late Cenozoic in parallel with the global trends indicated by the DSDP data (with the possible exception of the Quaternary). This is in spite of evidence indicating increasing aridity and climatic seasonality during this period. The Australian evidence clearly brings into question the global validity of the

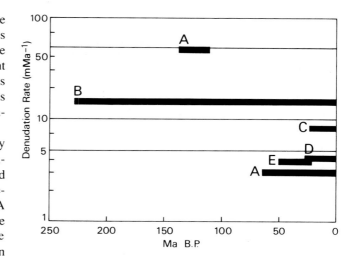

Fig. 15.17 *Rates of incision and denudation from the Murray Basin, south-east Australia, based on various methods: (A) basin sedimentation; (B) fission track ages of exposed granitic rocks; (C) river incision into dated basalts, (D) fan sedimentation; (E) river incision between basalts of two ages. (Modified from P. Bishop (1985)* Geology *13, Fig. 3, p. 481.)*

trends apparently present in DSDP records and especially the climatic interpretation placed upon them.

What, then, can we say about the present-day relationship between long-term and present-day denudation rates? Although we are living in a rather unusual period of geological time (the Quaternary) characterized by rapid changes in eustatic sea level and major oscillations in global climate, contemporary fluvial denudation rates, perhaps rather surprisingly, do not appear to be significantly different from the long-term average. The estimated present-day global mean of 43 m Ma^{-1} is somewhat higher than the long-term estimate of 11 m Ma^{-1} based on DSDP data, but the latter is very uncertain and is probably a significant underestimate of total continental denudation because it does not adequately incorporate continental margin sedimentation. Certainly, fission track estimates of mean denudation rates of around 30 m Ma^{-1} during the Cenozoic in regions of subdued relief are not incompatible with a global average in excess of 40 m Ma^{-1}. Moreover, the range of present-day denudation rates appears to be very similar to the range evident for the long term, with minima of around 1 m Ma^{-1} and maxima in excess of 5000 m Ma^{-1}. There is evidence that rates in the Late Cenozoic may have increased significantly in some regions as a result of active tectonism or climatic oscillations, and changes in areas of internal and external drainage must also have affected sediment input on to the continental margins.

Although at a global scale there is a broad correspondence between present-day and long-term rates of denudation, anthropogenic effects have clearly become overwhelming in some areas. In Natal in South Africa, for instance, a comparison of modern mechanical denudation rates with long-term rates based on the volume of offshore sediments suggests that the former are between 12 and 22 times the latter. Very high rates of mechanical denudation have arisen over the past few decades as a result of rapid soil erosion arising from poor land management practices associated with excessive population pressures in an environment characterized by steep slopes and periodic intense rainfall.

15.5   Factors controlling fluvial denudation rates

Some of the factors that influence rates of fluvial denudation have already been mentioned in passing and we now consider these factors in more detail. Variations in denudation rates can ultimately be explained by the way erosivity and erodibility interact. As we noted in Section 7.4.4, erosivity is the energy available at the surface to detach and transport regolith; it represents the potential of denudational systems to remove material from drainage basins. As well as the effectiveness of the export of material by streams it also includes the efficacy of other processes involved in sediment transport including rainsplash, sheet flow and rill erosion, and the various mechanisms of mass movement and other denudational processes. In terms of fluvial denudation the important factors affecting erosivity include runoff and the gravitational energy available for transporting material downslope, the latter being related to slope and stream gradients within the basin.

Erodibility is the susceptibility of materials at the surface to transport by denudational processes. This is related to a complex set of factors including mechanical strength, hardness, cohesion and particle size. In addition the rate of solute transport is influenced by the susceptibility of rocks and sediments at, or near, the surface to mechanisms of chemical weathering. The highest rates of denudation will be observed where both erosivity and erodibility are high and, conversely, minimal rates of denudation will occur where both are low. We now look at these factors further.

The strong relationship between relief and denudation rates is evident from the global pattern of denudation (Figs 15.9 and 15.10). It is important here to distinguish between elevation and local relief. The former itself has no direct influence on denudation rates, but the latter is closely related to local slope and thereby to erosivity. As noted in Section 9.2.2, local relief is the difference between the maximum and minimum elevation in an area of specific but limited size. A very high correlation has been established between denudation rate and local relief measured within 20×20 km (400 km^2) areas across a range of mid-latitude drainage basins (Fig. 15.18). Although it is true that regions at high elevations also tend to have high local relief, especially in orogenic belts, this is far from always the case. Take, for instance, the high plateau of southern Africa, where rather subdued local relief (with equally modest rates of denudation) is found up to 2000 m above sea level.

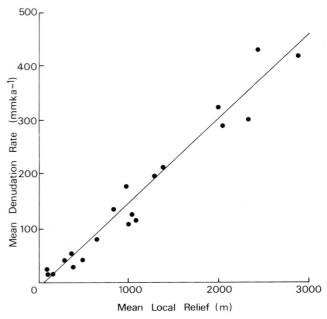

Fig. 15.18   *Relationship between mean local relief and mean denudation rate for 20 mid-latitude drainage basins. (Modified from F. Ahnert (1970)* American Journal of Science **268**, *Fig. 3, p.251.)*

Extremely low rates of denudation are recorded for many of the basins of tropical Africa, northern Eurasia and subarctic North America, in spite of substantial runoff, largely because the local relief in these areas is generally very low. In the Zaire Basin, for instance, gradients are so minimal in the lower part of the catchment that natural lakes have formed.

As we noted in Section 15.3.3 local relief also affects absolute rates of chemical denudation. This is presumably because in regions of high local relief, with steep slopes covered by thin soils or with bare rock surfaces, the potential for chemical weathering is greatest, assuming there is adequate precipitation. This contrasts with areas of low local relief in most climatic environments where the presence of a thick weathering mantle inhibits the movement of water at the weathering front where most of the weathering reactions occur.

Climate has been widely held to have a predominant influence on rates of denudation, although in the light of more recent data it is doubtful whether this view can now be sustained. Several attempts have been made to analyze the relationship between denudation rate and mean annual precipitation (Fig. 15.19). Although there is little agreement in detail between these postulated relationships certain common elements emerge – an initial peak of erosion in semi-arid environments and a progressive increase in denudation above a mean annual precipitation of around 1000 mm. The curve by Ohmori is based on the largest data set and indicates an initial maximum at 350–400 mm mean annual

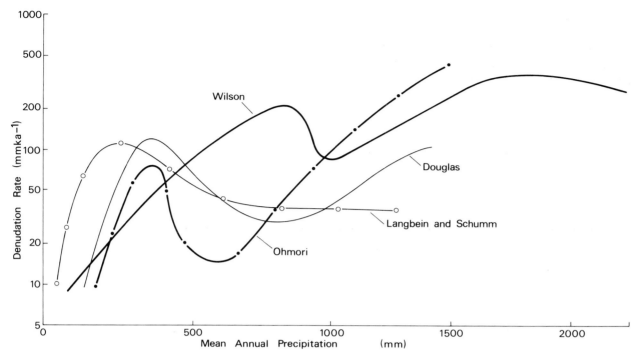

Fig. 15.19 *Various estimates of the relationship between (mechanical) denudation rate and mean annual precipitation. (Based on H. Ohmori (1983)* Bulletin, Department of Geography, University of Tokyo *No. **15**, Fig. 2, p. 84.)*

precipitation, followed by a minimum at 600 mm and then a progressive increase exceeding the initial maximum at around 1000 mm. This two-maxima curve can be explained by the way the amount of precipitation influences both erosivity and, through the mediation of vegetation, erodibility. The **biomass**, or weight of plant material per unit area, increases with mean annual precipitation up to about 2500 mm a^{-1} in the transition from desert through desert scrub to grassland and forest (Fig. 15.20). The presence of vegetation greatly reduces the erodibility of surface materials, and the abrupt retardation of mechanical denudation rates above a mean annual precipitation of around 400 mm is probably attributable to a marked increase in vegetation cover, as it begins to more than outweigh the effects of increasing runoff. Eventually, however, the maximum protective effect of vegetation is reached and beyond this point increasing precipitation, and therefore runoff, tends to lead to increasing rates of mechanical denudation. The importance of the protective role of vegetation in retarding erosion is dramatically illustrated by the effect of different land uses on denudation rates (Table 15.7).

The relationship between runoff and chemical denudation is rather different as there is a consistent positive correlation between solute load and annual runoff (Fig. 15.21). This is apparently because as precipitation increases there is more water available for chemical reactions in the regolith and solute release, and also greater runoff to transport these solutes. The relationship between chemical denu-

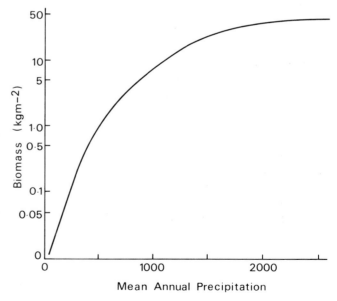

Fig. 15.20 *Generalized relationship between mean annual precipitation and biomass based on various studies. (Based on H. Ohmori (1983)* Bulletin, Department of Geography, University of Tokyo *No. **15**, Fig. 3, p. 86.)*

dation and temperature is very weak, however, apparently because the effect of temperature on weathering rates and solute release is overwhelmed by other variables, especially precipitation and local relief.

Table 15.7   Typical rates of erosion for various types of land use in the USA

LAND USE TYPE	SEDIMENT YIELD (t km^{-2} a^{-1})	DENUDATION RATE (mm ka^{-1})
Forest	8.5	4.2
Grassland	85.0	42.5
Cropland	1 700	850
Felled forest	4 250	2 125
Active open-cast mines	17 000	8 500
Construction sites	17 000	8 500

Source: Based on data in Environmental Protection Agency (1973) *Methods for Identifying and Evaluating the Nature and Extent of Nonpoint sources of Pollutants*, Washington, DC.

It has long been recognized that the seasonality of precipitation has a crucial effect on rates of mechanical denudation. This results from both the dramatic increase in sediment-carrying capacity in peak flows and the reduction in the protective role of vegetation during the dry season. The high sediment yields in semi-arid environments may also be influenced by the frequency of high intensity storms, which characterize such regions, as well as the lack of a continuous vegetation cover. In a detailed study in the upper Colorado Basin in the western USA, where the mean annual precipitation ranges from 150 to 1500 mm, rates of

mechanical denudation have been found to be negatively related to mean annual runoff but positively related to runoff variability (that is, the frequency of high magnitude flows; Fig. 15.22). The relationship between runoff variability and chemical denudation is, however, negative because as total runoff and, therefore, chemical denudation increases, river discharges become less variable.

A further factor which indirectly influences fluvial denudation rates is drainage basin size. Some studies have indicated that mechanical denudation rates are greater for smaller catchments than larger basins, at least up to a basin size of 2000 km^2 or so. There seem to be a number of reasons for this relationship. Small catchments are commonly in the upper parts of larger basins and they typically have steeper valley-side slopes and channel gradients. Such catchments also have smaller floodplains and less potential for sediment storage. Finally, it is possible for high-intensity storms generating major floods to cover a small basin entirely, whereas such storms will only affect part of a larger basin. Contrasting findings to this inverse relationship between basin area and mechanical denudation rate have come, however, from a study of sediment and solute yields in western Canada. Here the lowest denudation rates occur in the smallest basins (<100 km^2) and the highest in medium-sized catchments (1000–100 000 km^2). The largest

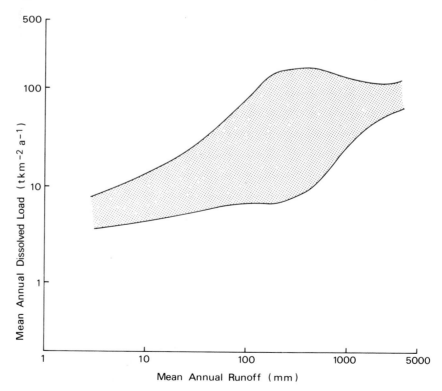

Fig. 15.21   *Generalized relationship between mean annual runoff and mean annual solute load for a sample of 496 rivers. The scatter in the relationship is probably due largely to the effects of varying lithology. (Based on D. E. Walling and B. W. Webb (1986) in: S. T. Trudgill (ed.)* Solute Processes, *Wiley, Chichester, Fig. 7.3, p. 260.)*

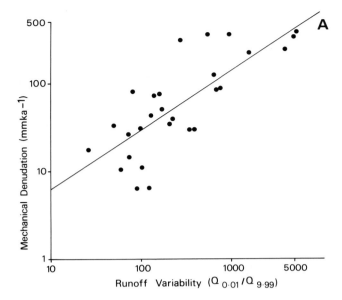

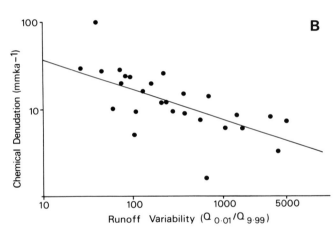

Fig. 15.22 *Relationship between runoff variability and mechanical (A) and chemical (B) denudation rates in the upper Colorado Basin, western USA. Runoff variability is represented as the ratio between the discharge that is equalled or exceeded 0.01 per cent of the time and the discharge that is equalled or exceeded 99.9 per cent of the time ($Q_{0.01}/Q_{99.9}$). (Modified from K-H Schmidt (1985)* Earth Surface Processes and Landforms **10**, *Fig. 6, pp. 504–5.)*

basins ($>100\,000\,km^2$) were found to have intermediate rates.

There are considerable uncertainties as to the dominant factors controlling rates of fluvial denudation, and much research is still needed to test some widely accepted, but not necessarily well supported, generalizations. None the less, it appears that at the global scale, relief and, to a lesser extent, climate are the main determinants of denudation rates (Table 15.8). At the regional and local scale, lithology and the specific factors of erodibility, which determine the supply of sediment and solutes, become more significant.

15.6 Rates of aeolian denudation

Few surveys of global denudation have adequately evaluated the relative importance of aeolian denudation, but it is clear from our discussion of aeolian processes and dust transportation (Chapter 10) that it is potentially significant, at least in arid regions. One reason for this relative neglect is that rates of aeolian denudation are extremely difficult to quantify. Present rates of dust transport from the continents to the oceans can be estimated from samples collected from ships, while individual dust storms can now be monitored by remote sensing techniques either from satellites or by photography from orbiting vehicles such as the space shuttle. Atmospheric transport can also be inferred from the presence of dust derived from remote land areas in the soils of oceanic islands. Such dust may be transported great distances – up to 6500 km from the Sahara to Barbados, 8000 km from the Sahara to Miami, 10 000 km from central Asia to Alaska and 11 000 km from central Asia to the north Pacific islands of Hawaii and Eniwetok (Fig. 15.23).

Estimates of rates of atmospheric dust transport vary enormously from 100 up to 5000 Mt a^{-1}. Assuming all this dust is deposited in the oceans (and therefore lost from the continents) and that the source material density is 2700 kg m^{-3}, this indicates a global mean aeolian denudation rate of between 0.25 and 13 mm ka^{-1}. Any deposition on land would lower these rates, but we must also remember that deflation of dust is confined to a rather small proportion of the total land area, so in these regions the aeolian denudation rate would be much higher than these global average figures. Recent data suggest that the high estimates of several thousand million tonnes per year are, in fact, overestimates, but present-day aeolian denudation rates may still be impressive at the regional scale. It is estimated, for instance, that the rate of dust transport from the Sahara to the Atlantic averages 146 Mt a^{-1}. This implies a sediment yield averaged over the area of the Sahara as a whole of around 16 t km^{-2} a^{-1} which converts to a denudation rate of 6 mm ka^{-1}. However, it is clear that the source of this dust is largely confined to particular regions within the Sahara, especially the Bodele Depression, the alluvial plains of Niger and Chad, southern Mauritania, northern Mali and central southern Algeria, southern Morocco and western Algeria, the southern fringes of the Mediterranean Sea in Libya and Egypt and northern Sudan. If the dust source is limited to these areas then they are probably experiencing a rate of aeolian denudation in excess of 20 mm ka^{-1}, just under half the mean global fluvial denudation rate. Similar estimates of aeolian denudation appear to be valid for east-central Asia, the source of around 20 Mt a^{-1} of fine soil material carried to the north-west Pacific.

Sediment in ocean cores enables past rates of dust deposition to be monitored and both in the north-west Pacific

Table 15.8   Solid and denudational solute load of major rivers and total denudation in relation to climate and relief

CLIMATE AND RELIEF ZONE	SOLID LOAD (t km^{-2} a^{-1})	DENUDATIONAL SOLUTE LOAD (t km^{-2} a^{-1})	TOTAL LOAD (t km^{-2} a^{-1})	TOTAL DENUDATION (mm ka^{-1})	TYPICAL SOLUTE LOAD AS % OF TOTAL
Mountainous, high precipitation	200–1500	70–350	250–2000	95–740	10
Mountainous, low precipitation	100–1000	10–60	120–1000	45–370	10
Moderate relief, temperate or tropical climate	40–200	25–60	80–300	30–110	35
Low relief, dry climate	10–100	3–10	15–100	5–35	10
Low relief, temperate climate	20–50	12–50	40–80	15–30	65
Low relief, subarctic climate	1.5–15	5–35	5–40	5–15	80
Low relief, tropical climate	1–10	2–15	4–30	1.5–10	50

Source: Based partly on data in M. Meybeck (1976) *Hydrological Sciences Bulletin* **21**, Table 2, p. 279.

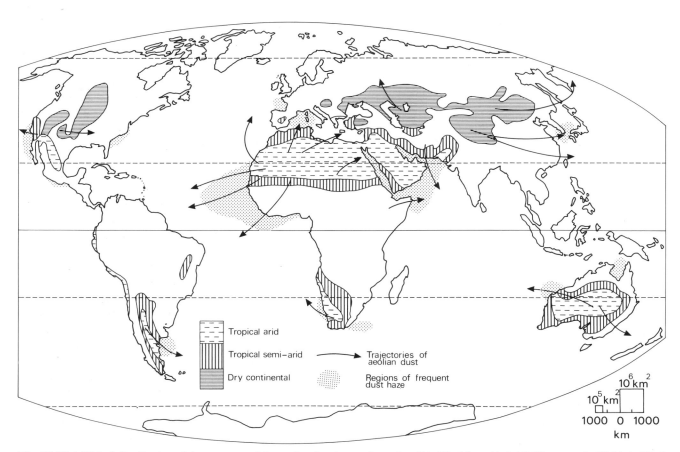

Fig. 15.23   *Global distribution of dust storm activity and major dust trajectories. (Modified from N. J. Middleton* et al. *(1986) in W. G. Nickling (ed.)* Aeolian Geomorphology. *Allen and Unwin, Boston, Fig. 4, p. 247.)*

and Atlantic Oceans this shows that current rates are representative of the past few thousand years. Taking the ocean core record further back, however, we find some interesting fluctuations in rates of deposition. Around 18 000 a BP rates of accumulation off the coast of West Africa were about twice those of the present, and similar increases at this time have been recorded off coasts adjacent to the Australian, Arabian and Thar Deserts. It is now possible to identify changing rates of aeolian denudation on the basis of atmospheric dust deposition in ocean cores as far back as the Cretaceous. Rates of deposition were apparently low during the Early Cenozoic, but increased after about 25 Ma BP and then accelerated markedly from 7 to 3 Ma BP The most dramatic increase occurred, however, around 2.5 Ma BP accompanying the onset of major glaciation in the northern hemisphere. For much of the Cenozoic the Sahara seems to have remained a major dust source.

Variations in rates of dust deposition in the geological past highlight some of the factors which control rates of aeolian denudation. The dramatic increase in rates at the beginning of the Pleistocene appears to be linked to the establishment of arid – humid climatic cycles at low latitudes at this time roughly in phase with the glacial – interglacial cycles at higher latitudes. This accords with evidence from current dust storm activity which suggests that the most abundant sources of dust are in areas that are presently changing to more arid climatic regimes. There was, for instance, a threefold increase in rates of dust transport out of North Africa during the droughts of the early 1970s and early 1980s. The observation that dust-storm frequency reaches a maximum where mean annual precipitation is around 100–200 mm (Fig. 15.24) is consistent with evidence that water is essential for the weathering processes capable of producing large quantities of the fine sedimentary particles susceptible to deflation. Many hyper-arid

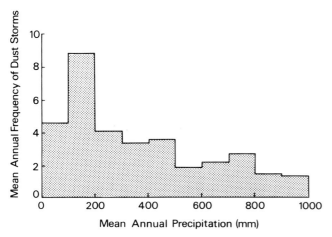

Fig. 15.24 *Global mean annual frequency of dust storms in relation to mean annual precipitation. (From A. S. Goudie (1983) Progress in Physical Geography 7, Fig. 11, p. 615.)*

areas, by contrast, are relatively 'blown out' with a surface cover of rocky plains or dune fields which yield little of the fine particles necessary for long distance aeolian transport.

15.7   Rates of glacial denudation

As with aeolian denudation, rates of glacial denudation have received relatively little attention and are similarly difficult to estimate. The methods used include the measurement of the sediment and solute load of glacial meltwater streams, reconstructions of pre-glacial or interglacial land surfaces, the relating of the volume of glacial drift in a specific region to an assumed glacial source area and the estimation of the contribution of glacial debris to marine sediments. There are comparatively few estimates of rates of mechanical denudation for active valley glaciers and these show an extremely broad range from 100 to 5000 mm ka^{-1}. The limited evidence available suggests that rates of solute transport in glacial meltwater streams are relatively high, with a chemical denudation rate of about 200 mm ka^{-1} being calculated for the South Cascade Glacier in the Cascade Range of the north-west USA. This is presumably related to the high surface area for chemical reactions provided by the fine particle size of rock flour, and may help to explain the high rates of chemical denudation estimated for partially glaciated mountain belts.

There is considerable uncertainty as to the rates of denudation associated with continental ice sheets. This is illustrated by the debate over the depth of erosion accomplished by the Pleistocene Laurentide ice sheet in eastern North America. One view is that the glacial scouring of this region has been largely confined to the exhumation of a weathering front formed under pre-glacial conditions and has been limited to an average of a few tens of metres at most. In support of this idea is the absence of the quantities of glacial till to be expected if deep erosion had occurred. On the other hand, it has been pointed out that much of the glacial debris produced could have been transported offshore by meltwater streams. Recent estimates of marine deposition rates along the eastern margin of North America do indeed suggest a significant input of glacially derived material during the Pleistocene. Although the precise source region for this sediment is impossible to define, it has been estimated that the volume deposited offshore implies a minimum mean depth of erosion of 120 m for the Laurentide region with a possible maximum of 200 m. This gives a mean denudation rate of between 48 and 80 m Ma^{-1} throughout the Pleistocene, and this is clearly rather higher than would be expected of fluvial denudation for such an area of subdued relief.

Rates of sediment production from 'permanent' ice masses, such as the Antarctic ice sheet, are very low and it appears that significant erosion is only accomplished by periodic ice sheets such as those of the northern hemisphere in the

Pleistocene. Whatever the general picture it is clear that erosion under even periodic ice sheets can be highly selective. This is illustrated in the Buchan area of north-east Scotland where, in spite of successive ice sheet advances across the region, are still preserved deep weathering mantles of pre-Pleistocene age.

15.8   Comparison of rates of uplift and denudation

Few topics in geomorphology have received a more confused analysis than the relationship between rates of uplift and denudation and the implication this has for the way landscapes evolve. We will leave the question of landscape evolution until Chapter 18, but we will conclude this chapter by considering relative rates of uplift and denudation.

In an influential paper published in 1963, S. A. Schumm asserted that modern rates of orogenic uplift at around 7500 mm ka[-1] are some eight times greater than average maximum denudation rates. This conclusion has subsequently been widely accepted and has been interpreted as meaning that the assumption that episodic phases of rapid uplift punctuate the continuous but more leisurely progress of erosion is broadly correct. Such an assumption has important implications for the understanding of landscape evolution so we must ask, is it really true? Schumm was working with a much smaller database than that now available and his estimates of maximum rates of denudation are unduly modest. We have seen, in fact, that maximum rates of both orogenic uplift and denudation lie between 5000 and 10 000 mm ka[-1]. Rates of glacio-isostatic rebound can, of course, be even higher than this but such vertical displacements can only amount to a total of a few hundreds of metres.

The evidence now available on rates of crustal uplift and denudation in orogenic belts indicate that the interaction between uplift and denudation has an inherent tendency towards a steady state where both are roughly equal in rate. During the uplift of a mountain range the denudation rate along its crest is initially slower than the rate of crustal uplift. This is because as we have seen, the denudation rate is largely a function of local relief, and this in turn is related to the degree of fluvial dissection. As this begins at the margins of an uplifted mountain range and progresses towards its crest by headward erosion, the summit region is relatively unaffected by denudation in these early stages of uplift. Consequently the elevation of the crest of the range increases, but as fluvial dissection works towards the crest the rate of denudation increases until eventually rates of uplift and denudation are roughly equal. In spite of further crustal uplift, in part promoted by isostatic rebound as a consequence of continuing denudational unloading, the mountain range cannot become any higher as long as the long-term rate of crustal uplift remains roughly constant and there are no significant changes in exogenic conditions. This equivalence of rates of crustal uplift and denudation in young mountain ranges is not fortuitous but simply reflects the achievement of a steady state, and the time taken to attain the steady state elevation is largely a function of the rate of uplift.

But how high can a mountain range become before this steady state is achieved? This seems to depend largely on the width of the mountain range, since the broader the range the longer it takes for fluvial dissection to reach the crestal regions, and therefore the longer the time during which the crustal uplift rate on the divide exceeds the rate of denudation. Consequently, it is no surprise to observe that the

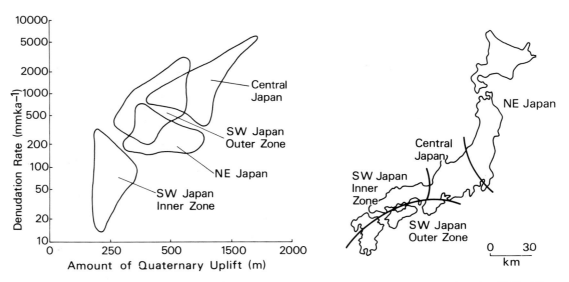

Fig. 15.25   *Relationship between rates of Quaternary uplift and rates of present denudation in Japan. (Based on T. Yoshikawa (1985) in A. Pitty (ed.) Themes in Geomorphology. Croom Helm, London, Fig. 12.3, p. 203, and Fig. 12.4, p. 204.)*

crest of the relatively narrow Southern Alps of New Zealand, which are only some 80 km across, has an average height of about 3000 m whereas the Himalayas, which have a width of around 350 km, have an average crestal elevation of around 7500 m.

The relationships between rates of crustal uplift and denudation in orogenic belts can be instructively examined by looking at New Zealand and Japan, both regions for which relatively good data are available. In Japan the central zone with crustal uplift rates of about 2200 mm ka^{-1} has attained a steady state and the outer zone nearly so, whereas in the north-east and south-west zone crustal uplift still exceeds denudation (Fig. 15.25). It appears that in the central and south-west zones the rapid rates of crustal uplift have prompted a rapid response in the rate of denudation, whereas in the other two zones the slower rates of crustal uplift have not yet led to a fully compensating increase in denudation rates. A similar picture is revealed in South Island, New Zealand, where rates of crustal uplift increase rapidly towards the west as the zone of oblique convergence marked by the Southern Alps is approached (see Section 3.5.2; Fig. 15.26). The crest of the Southern Alps seems to be in a steady state with very high rates of both crustal uplift and denudation in excess of 5000 mm ka^{-1}. On the eastern flanks of the Southern Alps, however, in areas such as eastern Otago, rates of crustal uplift are considerably lower (100–300 mm ka^{-1}), but still well in excess of rates of denudation at around 70 mm ka^{-1}. Here, then, there has apparently been insufficient time for a steady state between crustal uplift and denudation to be established.

So far we have limited our discussion of the relationship between crustal uplift and denudation rates to the situation occuring in orogenic belts. What of those regions experiencing epeirogenic uplift? Excluding areas of glacio-isostatic rebound, long-term rates of epeirogenic crustal uplift appear to be of the order of 100 mm ka^{-1} (in spite of possible short-term rates an order of magnitude higher than this). However, the low local relief of many areas experiencing

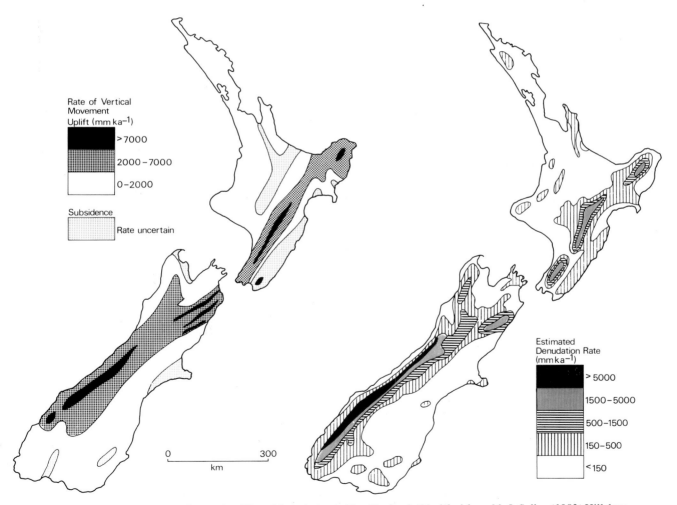

Fig. 15.26 *Comparison of rates of crustal uplift and denudation in New Zealand. (Modified from M. J. Selby, (1982)* Hillslope Materials and Processes. *Oxford University Press, Oxford, Fig. 11.11, p. 237.)*

epeirogenic uplift, such as the regions of broad continental warping in Africa and Australia, mean that denudation rates in such areas are often even lower. Consequently, there is little prospect of a steady state being achieved in these cases, a point which has interesting implications for models of landscape development as we will see in Chapter 18.

Further reading

There is an enormous range of literature on rates of uplift and denudation, but fortunately much of this is summarized in reviews. Most of the material on uplift rates is not set specifically in a geomorphic context, but the importance of such information for understanding landform development cannot be overemphasized. It is, however, important to be aware of the confusion between surface uplift and crustal uplift evident in much of the literature on uplift rates.

The most useful general assessment of crustal movements and the techniques used to monitor them, especially over the short to medium term, is the book by Vita-Finzi (1986). The ambiguities inherent in the measurement of present-day deformation using relevelling surveys are discussed by Brown et al. (1980), while Adams (1984) looks at a range of evidence for recent crustal movements along the northern Pacific coast of the USA. Schaer et al. (1975) compare modern and past rates of uplift in the Swiss Alps, while various aspects of the literature on rates of tectonic uplift are briefly reviewed in my short discussion of neotectonics and landform genesis (Summerfield, 1987). Rates of glacio-isostatic uplift are considered by Nikonov (1980). Clark and Jäger (1969) discuss the application of radiometric dating to estimating crustal uplift rates, while Gleadow and Fitzgerald (1987), Moore et al. (1986) and Parrish (1983) apply fission track dating techniques. Both present-day and long-term rates of uplift in orogenic belts are considered by Gansser (1983).

There have been several recent assessments of estimates of rates of mechanical and chemical denudation based on sediment and solute load data for rivers. Sediment yields are dealt with by Milliman and Meade (1983) and Walling and Webb (1983a), while the solute loads of rivers are considered at a predominantly local and regional scale by Walling and Webb (1986) and at a largely global scale by Walling and Webb (1983b). These wide-ranging reviews also consider the problems of estimating sediment and solute loads and of inferring patterns of denudation from them. On this theme Meybeck (1983) and Cryer (1986) deal in detail with the question of atmospheric inputs in river solutes, while Dunne (1978) provides an excellent investigation of chemical denudation rates carefully excluding non-denudational components. Turning to sediment loads, Douglas (1967) assesses the importance of anthropogenic factors, while Trimble (1977, 1983) and Meade (1982) address the

important effect that the disequilibrium between sediment supply and transport can have on estimates of denudation rates. Saunders and Young (1983) and Young and Saunders (1986) provide useful overviews of the relative importance of different denudational processes under various climates, and Meybeck (1976) and Clayton and Megaham (1986) provide, respectively, a general and specific assessment of the relative importance of mechanical and chemical denudation.

At the longer time scale, Carrara and Carroll (1979) provide a case study estimating erosion rates from tree root exposure, while Foster et al. (1985) demonstrate the variability in sediment transport evident in lake sediment records over periods of a few decades. Assessments of long-term global rates of sediment supply to the oceans include those by Davies et al. (1977) (updated by Worsley and Davies, 1979), Donnelly (1982) and Howell and Murray (1986). Matthews (1975) uses the offshore sedimentary record to examine Cenozoic rates of erosion in eastern North America, while Ruxton and McDougall (1967) document the rate of fluvial dissection of the Hydrographers Range in Papua New Guinea. The use of fission track and radiometric ages to determine long-term average denudation rates is illustrated in the papers already mentioned on the estimation of crustal uplift rates, and by Doherty and Lyons (1980). Bishop (1985) convincingly demonstrates the predominance of low rates of denudation throughout the Cenozoic in south-eastern Australia, while Young (1983) briefly examines the evidence for, and implications of, low denudation rates on a broader basis. At a more local scale Martin (1987) compares long-term sedimentation rates off the Natal coast with contemporary sediment yields. The factors determining denudation rates are examined briefly in the reviews of solid and solute load data mentioned above, but more detailed evaluations are provided by Wilson (1973) on climatic controls, Ahnert (1970) on the local relief factor, Schmidt (1985) on the importance of runoff variability, Schumm (1963) and Slaymaker (1987) on drainage basin area and Ohmori (1983) and Trimble (1988) on the role of vegetation. The influential paper by Langbein and Schumm (1958) is still worth a look as long as its limited applicability is appreciated.

The role of dust transportation is considered by Goudie (1983) and Middleton et al. (1986), while Lever and McCave (1983) use evidence from deep sea cores to trace the variations in aeolian deposition in the Atlantic since the Cretaceous. Rates of sediment transport in glacial streams are examined by Gurnell (1987), and the high solute concentrations typical of glacial meltwater are considered by Collins (1983). The debate concerning rates of ice sheet erosion is illustrated through the contributions of White (1972), Sugden (1976) and Bell and Laine (1985), while Hall and Sugden (1987) provide specific evidence of the inefficiency, at least locally, of ice sheets as denudational agents.

On the question of the relative rates of uplift and denudation it is appropriate to begin with Schumm's influential assessment (Schumm, 1963) and then to look at the excellent discussions by Adams (1985) on the Southern Alps and Yoshikawa (1985) on Japan. The factors determining the elevation of a mountain range are considered by Ahnert (1984).

References

Adams, J. (1984) Active deformation of the Pacific northwest continental margin. *Tectonics* **3**, 449–72.

Adams, J. (1985) Large-scale tectonic geomorphology of the Southern Alps, New Zealand. In: M. Morisawa and J. T. Hack (eds) *Tectonic Geomorphology*. Allen and Unwin, Boston and London, 105–28.

Ahnert, F. (1970) Functional relationships between denudation, relief and uplift in large mid-latitude drainage basins. *American Journal of Science* **268**, 243–63.

Ahnert, F. (1984) Local relief and the height limits of mountain ranges. *American Journal of Science* **284**, 1035–55.

Bell, M. and Laine, E. P. (1985) Erosion of the Laurentide region of North America by glacial and fluvioglacial processes. *Quaternary Research* **23**, 154–74.

Bishop, P. (1985) Southeast Australian late Mesozoic and Cenozoic denudation rates: a test for late Tertiary increases in continental denudation. *Geology* **13**, 479–82.

Brown, L. D., Reilinger, R. E. and Citron, G. P. (1980) Recent vertical crustal movements in the US: evidence from precise levelling. In: N.-A. Mörner (ed.) *Earth Rheology, Isostasy and Eustasy*. Wiley, Chichester and New York, 389–405.

Carrara, P. E. and Carroll, T. R. (1979) The determination of erosion rates from exposed tree roots in the Piceance Basin, Colorado. *Earth Surface Processes* **4**, 307–17.

Clark, S. P. and Jäger, E. (1969) Denudation rate in the Alps from geochronologic and heat flow data. *American Journal of Science* **267**, 1143–60.

Clayton, J. L. and Megahan, W. F. (1986) Erosional and chemical denudation rates in the southwestern Idaho batholith. *Earth Surface Processes and Landforms* **11**, 389–400.

Collins, D. N. (1983) Solute yield from a glacierized high mountain basin. In: B. W. Webb (ed.) *Dissolved Loads of Rivers and Surface Water Quantity/Quality Relationships*. International Association of Hydrological Sciences Publication **141**, 41–9.

Cryer, R. (1986) Atmospheric solute inputs. In: S. T. Trudgill (ed.) *Solute Processes*. Wiley, Chichester and New York, 15–84.

Davies, T. A., Hay, W. W., Southam, J. R. and Worsley, T. R. (1977) Estimates of Cenozoic oceanic sedimentation rates. *Science* **197**, 53–5.

Doherty, J. T. and Lyons, J. B. (1980) Mesozoic erosion rates in northern New England. *Geological Society of America Bulletin* **91**, 16–20.

Donnelly, T. W. (1982) Worldwide continental denudation and climatic deterioration during the late Tertiary: evidence from deep-sea sediments. *Geology* **10**, 451–4.

Douglas, I. (1967) Man, vegetation, and the sediment yield of rivers. *Nature* **215**, 925–8.

Dunne, T. (1978) Rates of chemical denudation of silicate rocks in tropical catchments. *Nature* **274**, 244–6.

Foster, I. D. L., Dearing, J. A., Simpson, A. and Carter, A. D. (1985) Lake catchment based studies of erosion and denudation in the Merevale catchment, Warwickshire, U.K. *Earth Surface Processes and Landforms* **10**, 45–68.

Gansser, A. (1983) The morphogenic phase of mountain building. In: K. J. Hsü (ed.) *Mountain Building Processes*. Academic Press, London and New York, 221–8.

Gleadow, A. J. W. and Fitzgerald, P. G. (1987) Uplift history and structure of the Transantarctic Mountains: new evidence from fission track dating of basement apatites in the Dry Valleys area, southern Victoria Land. *Earth and Planetary Science Letters* **82**, 1–14.

Goudie, A. S. (1983) Dust storms in space and time. *Progress in Physical Geography* **7**, 502–30.

Gurnell, A. M. (1987) Suspended sediment. In: A. M. Gurnell and M. J. Clark (eds) *Glacio-Fluvial Sediment Transfer*, Wiley, Chichester and New York, 305–54.

Hall, A. M. and Sugden, D. E. (1987) Limited modification of mid-latitude landscapes by ice sheets: the case of northeast Scotland. *Earth Surface Processes and Landforms* **12**, 531–42.

Howell, D. G. and Murray, R. W. (1986) A budget for continental growth and denudation. *Science* **233**, 446–9.

Langbein, W. B. and Schumm, S. A. (1958) Yield of sediment in relation to mean annual precipitation. *American Geophysical Union Transactions* **39**, 1076–84.

Lever, A. and McCave, I. N. (1983) Eolian components in Cretaceous and Tertiary North Atlantic sediments. *Journal of Sedimentary Petrology* **53**, 811–32.

Martin, A. K. (1987) A comparison of sedimentation rates in the Natal Valley, SW Indian Ocean with modern sediment yields in east coast rivers, southern Africa. *South African Journal of Science* **83**, 716–24.

Matthews, W. H. (1975) Cenozoic erosion and erosion surfaces of eastern North America. *American Journal of Science* **275**, 818–24.

Meade, R. H. (1982) Sources, sinks, and storage of river sediment in the Atlantic drainage of the United States. *Journal of Geology* **90**, 235–52.

Meybeck, M. (1976) Total annual dissolved transport by world major rivers. *Hydrological Sciences Bulletin* **21**, 265–89.

Meybeck, M. (1983) Atmospheric inputs and river transport of dissolved substances. In: *Dissolved Loads of Rivers and Surface Water Quantity/Quality Relationships*. International Association of Hydrological Sciences Publication **141**, 173–92.

Middleton, N. J., Goudie, A. S. and Wells, G. L. (1986) The frequency and source areas of dust storms. In: W.G. Nickling (ed.) *Aeolian Geomorphology*. Allen and Unwin, Boston and London, 237–59.

Milliman, J. D. and Meade, R. H. (1983) World-wide delivery of river sediment to the oceans. *Journal of Geology* **91**, 1–21.

Moore, M. E., Gleadow, A. J. W. and Lovering, J. F. (1986) Thermal evolution of rifted continental margins: new evidence from fission tracks in basement apatites from southeastern Australia. *Earth and Planetary Science Letters* **78**, 255–70.

Nikonov, A. A. (1980) Manifestations of glacio-isostatic processes in northern countries during the Holocene and at present. In: N.–A. Mörner (ed.) *Earth Rheology, Isostasy and Eustasy*. Wiley, Chichester and New York, 341–54.

Ohmori, H. (1983) Erosion rates and their relation to vegetation from the viewpoint of world-wide distribution. *Bulletin of the Department of Geography University of Tokyo* **15**, 77–91.

Parrish, R. R. (1983) Cenozoic thermal evolution and tectonics of the Coast Mountains of British Columbia 1, fission-track dating, apparent uplift rates, and patterns of uplift. *Tectonics* **2**, 601–32.

Ruxton, B. P. and McDougall, I. (1967) Denudation rates in northeast Papua from potassium–argon dating of lavas. *American Journal of Science* **265**, 545–61.

Saunders, I. and Young, A. (1983) Rates of surface processes on slopes, slope retreat and denudation. *Earth Surface Processes and Landforms* **8**, 473–501.

Schaer, J. P., Reimer, G. M. and Wagner, G. A. (1975) Actual and ancient uplift rate in the Gotthard region, Swiss Alps: a comparison between precise leveling and fission-track apatite age. *Tectonophysics* **29**, 293–300.

Schmidt, K.–H. (1985) Regional variation of mechanical and chemical denudation, upper Colorado River Basin, U.S.A. *Earth Surface Processes and Landforms* **10**, 497–508.

Schumm, S. A. (1963) The disparity between present rates of denudation and orogeny. *United States Geological Survey Professional Paper* **454-H**.

Slaymaker, O. (1987) Sediment and solute yields in British Columbia and Yukon: their geomorphic significance reexamined. In: V. Gardiner *et al.* (eds) *International Geomorphology 1986* Part I. Wiley, Chichester, 925–45.

Souchez, R. A. and Lemmens, M. M. (1987) Solutes. In: A. M. Gurnell and M. J. Clark (eds) *Glacio-Fluvial Sediment Transfer.* Wiley, Chichester and New York, 285–303.

Sugden, D. E. (1976) A case against deep erosion of shields by ice sheets. *Geology* **4**, 580–2.

Summerfield, M. A. (1987) Neotectonics and landform genesis. *Progress in Physical Geography* **11**, 384–97.

Trimble, S. W. (1977) The fallacy of stream equilibrium in contemporary denudation studies. *American Journal of Science* **277**, 876–87.

Trimble, S. W. (1983) A sediment budget for Coon Creek basin in the Driftless Area, Wisconsin, 1853–1977. *American Journal of Science* **283**, 454–74.

Trimble, S. W. (1988) The impact of organisms on overall erosion rates within catchments in temperate regions. In: H. A. Viles (ed.) *Biogeomorphology*. Blackwell, Oxford, 83–142.

Vita-Finzi, C. (1986) *Recent Earth Movements: An Introduction to Neotectonics.* Academic Press, London and New York.

Walling, D. E. and Webb, B. W. (1983a) Patterns of sediment yield. In: K. J. Gregory (eds) *Background to Palaeohydrology.* Wiley, Chichester and New York, 69–100.

Walling, D. E. and Webb, B. W. (1983b) The dissolved loads of rivers: a global overview. In: *Dissolved Loads of Rivers and Surface Water Quantity/Quality Relationships.* International Association of Hydrological Sciences Publication **141**, 3–20.

Walling, D. E. and Webb, B. W. (1986) Solutes in river systems. In: S. T. Trudgill (ed.) *Solute Processes.* Wiley, Chichester and New York, 251–327.

White, W. A. (1972) Deep erosion by continental ice sheets. *Geological Society of America Bulletin* **83**, 1037–56.

Wilson, L. (1973) Variations in mean annual sediment yield as a function of mean annual precipitation. *American Journal of Science* **273**, 335–49.

Worsley, T. R. and Davies, T. A. (1979) Sea-level fluctuations and deep-sea sedimentation rates. *Science* **203**, 455–6.

Yoshikawa, T. (1985) Landform development by tectonics and denudation. In: A. Pitty (ed.) *Themes in Geomorphology.* Croom Helm, London, 194–210.

Young, A. and Saunders, I. (1986) Rates of surface processes and denudation. In: A. D. Abrahams (ed.) *Hillslope Processes.* Allen and Unwin, Boston, and London, 3–27.

Young, R. W. (1983) The tempo of geomorphological change: evidence from southeastern Australia. *Journal of Geology* **91**, 221–30.

16

Tectonics and drainage development

16.1 Active and passive tectonic controls on drainage

Tectonic controls can be exerted on the development of drainage systems in two ways (Table 16.1). **Active tectonic controls** involve the response of fluvial systems to ongoing tectonic activity. Such controls are most evident in tectonically active environments, but even in apparently tectonically quiescent regions drainage systems can be significantly affected by active tectonic controls. Such controls include faulting as well as the tilting and deformation of the ground surface.

Passive tectonic controls (also sometimes referred to as structural controls) operate through the influence exerted by previous tectonic activity on subsequent drainage development. These controls are usually more subtle than those arising from active tectonism and are more difficult to unravel because later erosion may have removed much of the evidence of pre-existing structures. Passive controls most commonly operate through the effect that previous tectonic activity has on the structural organization (disposition and arrangement) of lithologies of varying degrees of resistance. Some of the world's most spectacular regional scale landforms are fold terranes which have been partially eroded by drainage systems (Fig. 3.17).

Although the distinction between active and passive controls is a useful one, it is important to emphasize that both kinds frequently operate together in influencing the development of drainage systems. A drainage network in a sedimentary terrane which is being actively folded will both respond directly to the ongoing ground deformation and to the arrangement of alternating weak and resistant beds exposed as the folds are progressively eroded.

16.2 Passive tectonic controls

16.2.1 Drainage patterns

Drainage patterns are influenced by many factors, including climatic and lithological variables, but geological structure is overall the most important factor. Indeed, drainage patterns are often so dramatically affected by passive tectonic controls that the patterns themselves are frequently used as a basis for unravelling geological structures from data sources such as aerial photographs and satellite imagery.

The major types of drainage pattern associated with tectonic controls are illustrated in Figure 16.1 and are listed, with their primary characteristics, in Table 16.2. The patterns illustrated are those that occur most commonly at the meso-

Table 16.1 Mesoscale tectonic controls of drainage systems

CONTROLS	EFFECTS
Active controls	
Active faulting	Linear, hanging and wineglass valleys; channel offsets, ponding and diversion; terraces and knickpoints
Active folding and tilting	Antecedent and dip drainage; water gaps; channel incision, aggradation and lateral shifting
Passive (structural) controls	
Fault traces	Linear, hanging and wineglass valleys; channel offsets, ponding and diversion; terraces and knickpoints
Tilting	Trellis drainage pattern; parallel, long dip and short anti-dip streams
Domes	Radial and annular drainage patterns; superimposed rivers
Anticlines and synclines	Trellis drainage pattern; superimposed rivers; water gaps
Lineation	Asymmetric valleys; linear channels
Joints	Rectangular drainage patterns

Source: Modified from M. Morisawa (1985) *Rivers*. Longman, London, Table 10.1, p. 157.

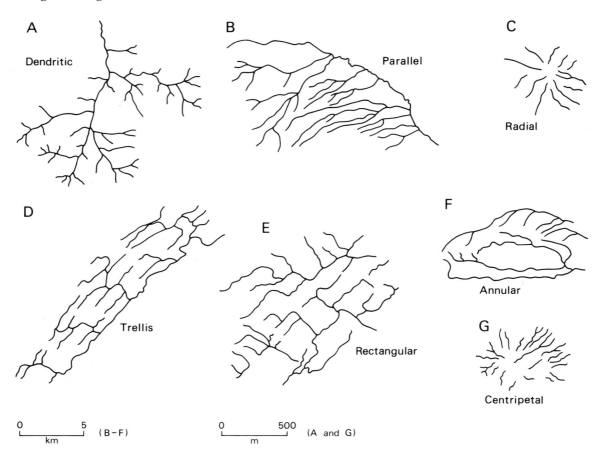

Fig. 16.1 *Major types of drainage pattern related to structural controls. These are actual examples taken from various localities in the USA: (A) dendritic – Big Flat, Arkansas; (B) parallel – East Wind River Range, Sage Creek; (C) radial – Mount Ellsworth, Utah; (D) trellis – Tuscarora Creek, Pennsylvania; (E) rectangular – Schroon and Bouquet Rivers, New York; (F) annular – Maverick Spring Dome, Wyoming; (G) centripetal – Menan Buttes, Idaho. (Modified from M. Morisawa (1985)* Rivers. *Longman, London, Fig. 10.3, p. 161.)*

Table 16.2 Classification of drainage patterns related to structural controls

	TYPE	DESCRIPTION OF PATTERN	STRUCTURAL CONTROL
A.	**Dendritic**	Spreading tree-like arrangement. No evident orientation of channels	Horizontal sediments or homogeneous crystalline rocks. Lack of structural control on rocks of uniform resistance.
B.	**Parallel**	Main channels regularly spaced and parallel, or subparallel, to each other. Tributaries join at very acute angles	Closely spaced faults, monoclines or isoclinal folds.
C.	**Radial**	Streams flow outward from centre	Volcanic cones, domes
D.	**Trellis**	A dominant drainage direction with a secondary direction perpendicular to it. Primary tributaries join main stream at right angles, secondary tributaries are parallel to main stem	Tilted or folded alternately resistant/weak sedimentary units
E.	**Rectangular**	Drainage forms a perpendicular net with the two directions equally developed.	Joints or faults
F.	**Annular**	Main rivers have circular pattern with subsidiary channels at right angles	Eroded dome in alternate resistant/weak sediments
G.	**Centripetal**	Streams flow inward to centre	Calderas, craters, tectonic basins

Source: Modified from M. Morisawa (1985) *Rivers*. Longman, London, Table 10.4, p. 160.

scale and macroscale. Continental drainage systems, however, are usually more complex and may include a variety of specific drainage pattern types.

The basis for comparison for structurally related drainage patterns is the **dendritic** type. This consists of a system of irregular branching tributaries forming a tree-like pattern with junctions at a variety of angles but usually well below 90°. Dendritic patterns develop in areas where there are no marked lithological or structural controls affecting drainage. They are frequently associated with horizontal, or nearly horizontal, sedimentary strata, or with massive igneous rocks. They may also be found on complex folded strata, or on contorted metamorphosed rocks where lithological variations (in terms of resistance to weathering and erosion) are insufficient to modify the dendritic pattern.

Where sedimentary rocks are tilted there may be a succession of relatively weak and relatively resistant lithologies exposed. River channel incision will tend to be more active on the less resistant lithology leading to the development of a **strike valley** flanked on the up-dip side by a **dip slope** and on the down-dip side by an escarpment (Fig. 16.2). This creates a **trellis** drainage pattern, with roughly parallel strike streams being joined at high angles by short **dip streams** and **anti-dip streams** (Fig. 16.3). As the fluvial system incises downwards it will also migrate laterally, a process known as **homoclinal shifting**. This arises from the following of a weak downward-dipping lithology by the strike stream and enhances the asymmetry of the strike valley,

as is evident from the anti-dip tributaries being shorter than the dip streams.

The angle of dip of alternating weak and resistant beds can obviously change, and the associated topographic forms vary as a consequence (Fig. 16.4). Fluvial erosion of resistant beds in horizontal strata leads to the development of steep-sided, flat-topped plateaus called **mesas**, and to similar but smaller **buttes** (Fig. 16.5). With modestly dipping beds an asymmetric **cuesta** is produced, while steeply dipping strata at an angle of around 45° lead to the formation of an approximately symmetric ridge known as a **hogback** (Fig. 16.3).

At its simplest, folded strata consist of pairs of limbs of roughly equal inclination. Fluvial dissection of a series of folds will initially produce a symmetric trellis drainage pattern with trunk streams following the synclinal axis. But major valleys are frequently seen to be developed along anticlinal axes with the synclinal axes forming ridges. A possible explanation for this inversion of topography is that the extension of strata across anticlines promotes joint widening, and that series of joints along anticlinal axes then become favoured sites for weathering and erosion. In many cases, however, it can be demonstrated that several kilometres of rock have been removed by fluvial erosion prior to the exhumation of a particular anticlinal structure and that many phases of topographic inversion and 'normal' relief development are likely to have occurred during the prolonged phase of landscape lowering. Whether an anti-

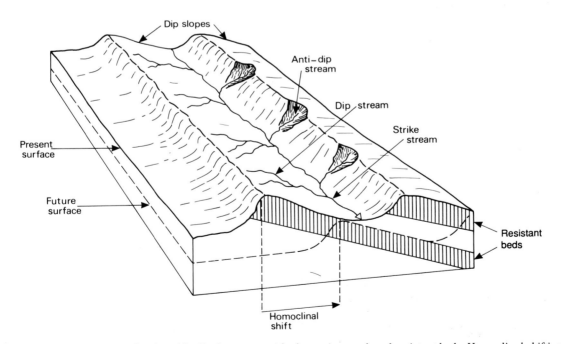

Fig. 16.2 *Trellis drainage pattern developed in dipping strata with alternating weak and resistant beds. Homoclinal shifting of the strike valley system down-dip is also shown. (Modified from A. L. Bloom (1978)* Geomorphology: A Systematic Analysis of Late Cenozoic Landforms. *Prentice-Hall, Englewood Cliffs, Fig. 11–10, p. 263.)*

Fig. 16.3   *Trellis drainage pattern developed on steeply dipping Precambrian strata, South Australia.*

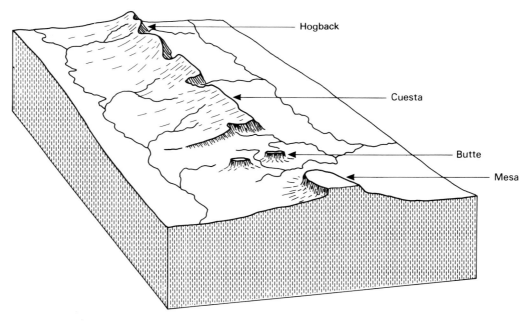

Fig. 16.4   *Relationship between topographic forms and angle of dip of strata with alternating weak and resistant beds. (Modified from A. L. Bloom (1978)* Geomorphology: A Systematic Analysis of Late Cenozoic Landforms. Prentice-Hall, Englewood Cliffs, Fig. 11–9, p. 262.)*

Fig. 16.5 *Mesas and buttes in Monument Valley, Utah, USA, in a fluvially eroded landscape developed in massive, flat-lying sandstone formations.*

cline is associated with a positive or negative topographic form probably reflects the relative resistance of the lithology exposed along its axis at a given time.

Obviously, folded strata are not infinitely continuous along strike. In some cases folds plunge along strike over short distances in relation to the length of their axes. In these circumstances a simple trellis pattern is modified by the bending of trunk streams around the limbs of folds. Except where folding is relatively recent, fluvial erosion of folded strata will give rise to a landscape of successive cuestas and hogbacks facing inward towards the fold axes. Where folds have very large dimensions and low amplitudes extensive cuesta and strike valley landscapes are developed; the Paris Basin of northern France and the Weald of southern England are well-studied examples. The Paris Basin is a large synclinal structure consisting of a succession of outward facing escarpments. The Weald, on the other hand, is a complex eroded anticlinal structure with inward-facing escarpments.

16.2.2 Fracture and joint control

Even on homogeneous lithologies, jointing may be suffi-

ciently well developed to produce linear zones which are more susceptible to weathering and erosion. In such cases downcutting rivers will tend to exploit these lines of weakness and the drainage pattern will be modified as a consequence. Such joint control is often evident on granitic rocks where an orthogonal jointing system is reflected in a **rectangular** drainage pattern characterized by numerous channel junctions with angles around 90°.

Joint and fracture orientations may correspond to either ancient stress fields or to recent patterns of stress in the lithosphere (the **neotectonic stress field**) which, at the broadest scale, is related to plate tectonic processes. There has long been controversy, for instance, over whether the trends of valleys in Switzerland are related to active tectonic controls or whether they simply represent the remnant of a drainage system created initially in the Miocene on a surface sloping to the north towards an ancestral River Rhine. Statistical analysis of valley orientations indicates a non-random distribution corresponding broadly to major joint directions and thus a strong tectonic control over drainage patterns (Fig. 16.6). Studies in many other parts of the world have demonstrated a similar correspondence between drainage orientations and stress fields, and such a

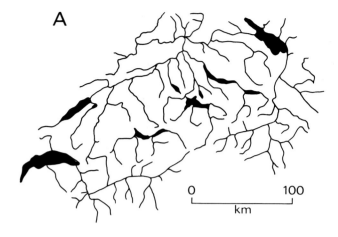

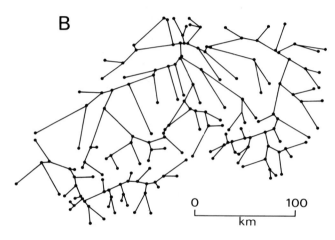

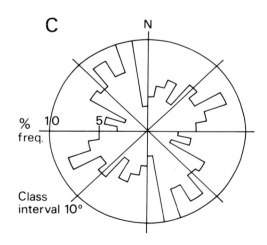

Fig. 16.6 *Correspondence between joint direction and orientation of river channels in Switzerland: (A) drainage network; (B) rectified drainage network constructed by joining tributary termini and junctions with straight lines from which river channel orientations are measured; (C) polar histogram of river channel orientations. (Modified from A. E. Scheidegger (1979)* Geographica Helvetica *34, Figs. 2, 3 and 4, p. 10.)*

tectonic control also appears to extend in some cases to glacial valleys.

A further interesting example of the relationship between fracture and joint systems and drainage is to be found in north-west Europe. A north-west-trending system of large-scale fractures in the Late Cretaceous and Paleogene strata of southern England and northern France appears to have been created by the stress field associated with the Late Neogene and Quaternary phase of lithospheric compression in the Alps. In northern France the system of fractures and joints is coincident with a rectangular drainage pattern. In southern England, however, no such correspondence is evident, probably because channel orientations there are more influenced by additional local fracture systems related to flexures which trend obliquely to the major north-west orientated fracture system.

Relationships between neotectonic stress fields and drainage patterns are most evident at the macroscale and megascale since at smaller scales local controls over joint patterns become predominant. The important point is that any significant deviation from a random (dendritic) drainage pattern at the megascale or macroscale should lead us at least to consider the possibility of tectonic control (although depositional landforms such as drumlins and dune ridges can also exert an important non-random influence over drainage patterns).

16.2.3   River capture

An important process in the development of drainage subject to structural controls is **river capture**, or **stream piracy** (although it is important to emphasize that the process can also operate in the absence of such controls). River capture occurs when one stream erodes more aggressively than an adjacent stream and captures its discharge by intersecting its channel. The faster rate of erosion of the 'capturing' stream may be due to it having a steeper gradient or discharge, or to the fact that it is eroding less resistant strata or to a combination of these factors (Fig. 16.7). At the macroscale an additional factor could be contrasts in precipitation either side of a mountain chain, with high discharge streams on its more humid side cutting back to capture rivers on its less humid side.

The point at which the capture occurs is frequently indicated by a sharp change in channel direction, often of the order of 90°, known as an **elbow of capture**. The removal of discharge from the captured stream (known as **beheading**) means that its source will move down the original valley; the old river course upstream will be marked by fluvial deposits and the abandoned river valley will form a **wind gap**.

River capture is only possible through the agency of **headward erosion** whereby valleys are extended towards and beyond their drainage basin divide. This process is well

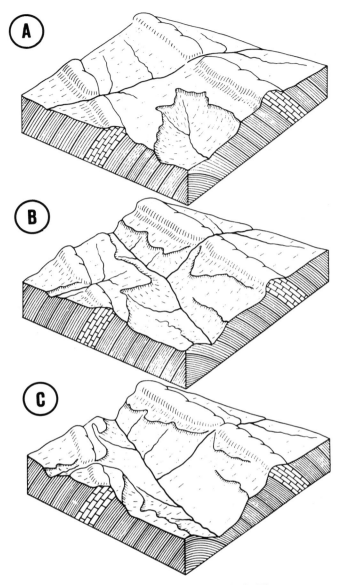

Fig. 16.7 *Three stages of stream capture (A–C). The stream at the lower elevation cuts back through less resistant strata to capture the stream flowing at a higher level on more resistant formations. (From S. A. Schumm (1977) The Fluvial System (Wiley, New York and London) Fig. 6.36, p.235 after O. P. Jenkins (1964) California Division of Mines Bulletin 135, Fig. 72, p. 196.)*

documented for microscale gully systems but its operation at larger scales is less clear. The key problem is how runoff sufficient to promote the channel erosion and valley downcutting necessary for headward erosion can be generated from the small area between existing valley heads and the drainage divide.

16.2.4 Transverse drainage

One of the most perplexing problems of drainage develop-

ment to unravel is that provided by **transverse drainage**. Such a drainage pattern, which is also known as **discordant drainage**, occurs when river channels cut across geological structures (Fig. 16.8). Such drainage, which is common in fold mountain belts, is often regarded as anomalous, although this is not really an appropriate description. Much of the early and influential research on transverse drainage focused on the Appalachian Mountains of the eastern USA. This was unfortunate in that these are ancient (Palaeozoic) mountains which have experienced prolonged erosion and no significant folding since the Permian. Consequently, it is difficult to determine their early history in detail. This is perhaps why several competing hypotheses were presented to explain their drainage characteristics and why they remained current for so long – the evidence to reject them was just not available.

The simplest explanation for transverse drainage is that streams cut back by headward erosion along transverse faults. This is a variant of the river capture mechanism since in cutting across a mountain belt such streams would intersect the accordant **strike drainage**. This hypothesis is testable with reference to existing field evidence. It has specifically been rejected for the central Appalachians, but may play an important role in drainage development in evolving orogens (see Section 16.6).

Another explanation for transverse drainage is **antecedence**. The idea here is that a river which exists prior to the uplift of a mountain belt and maintains sufficient erosive power to keep pace with uplift will eventually be transverse. Antecedence has been favoured as an explanation for transverse drainage in many of the world's Cenozoic orogens, for example the Alps, the Himalayas and the Cascade Mountains of the north-west USA. Evidence favouring this mechanism includes the general direction of drainage away from the oldest structures in an expanding zone of deformation. A variant of this mechanism is **drainage inheritance**. This occurs where a drainage system developed on an erosion surface lacking a sedimentary cover is rejuvenated and erodes down on to underlying structures which are discordant with respect to the inherited drainage orientation.

A third mechanism is **superimposition (superposition)**. Drainage developed on a sedimentary cover could be let down on to an underlying structure, and such superimposed drainage would, initially at least, be discordant with respect to this structure. Similarly, there might be marked structural differences between vertically adjacent crustal units, especially where major thrusting and sliding has occurred. Although a plausible mechanism, superimposition is extremely difficult to verify except in the case of very young orogens where vestiges of the original sedimentary cover remain. In ancient mountain belts, denudation will have removed all the evidence of any pre-existing sedimentary cover.

Fig. 16.8 *Glen Helen Gorge, in the western Macdonnell Ranges of central Australia, cut by a transverse southward-flowing headwater tributary of the Finke River through a resistant east–west-orientated ridge. This transverse drainage appears to have evolved through a combination of drainage inheritance and limited superimposition.*

16.3   Tectonics and drainage adjustment

The traditional approach to the analysis of drainage history assumed that drainage systems are originally formed on an initial slope before gradually becoming adjusted to structure. It is now widely acknowledged, however, that this kind of model of drainage development is unrealistic. Except in the rare circumstances where drainage is initiated almost instantaneously on a virgin surface (such as that provided by a newly formed volcanic island) there is likely to be an existing topography which will influence any subsequent drainage development. This will invariably be the case with drainage developed on a landsurface newly exposed by a fall in sea level. In these circumstances there is exposure of a pre-existing submarine topography which may have been partly formed during an earlier phase of sub-aerial exposure. There will also be an existing sub-aerial drainage system on the adjacent littoral zone which will extend over the newly emerged coastal margin. Rather than considering hypothetical starting points for drainage development, it is more fruitful to consider the progressive changes and modifications that can take place. There is a constant interplay

between tectonic deformation, erosion and deposition. The point is well summarized by the dictum that drainage systems have a heritage rather than an origin.

16.3.1   Effects of warping

The interplay between tectonics and drainage is clearly shown in the effects of warping. Drainage disruption will occur if the rate of uplift along the upwarp axis exceeds the rate of downcutting. The effects can be quite complex (Fig. 16.9) but the specific drainage patterns generated are, in some cases, sufficiently diagnostic to enable important elements of a region's tectonic history to be reconstructed from drainage pattern evidence. In areas of minimal relief, with very low stream gradients, quite modest tectonic movements can have profound effects on drainage patterns. This is partly because only a very slight tilt of the landsurface, perhaps by only a fraction of a degree, is necessary to cause **drainage reversal**. Another factor is that a river with a very low gradient, especially when it also has a small discharge, will be incapable of a rate of downcutting sufficient to maintain its original course.

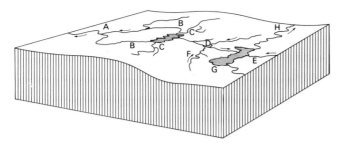

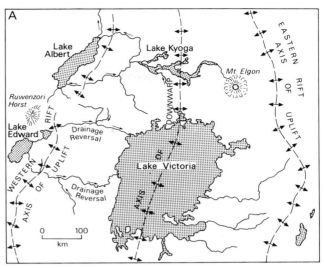

Fig. 16.9 *Effects of warping on drainage systems. Downstream of the upwarp the channel gradient will be increased along part of the channel (A), but the discharge from upstream of the upwarp axis will be cut off (beheading) (B). Along the axis of the upwarp channel gradients will be reduced and an area of disorganized drainage, possibly including shallow lakes, may become established (C). If upwarps are of low amplitude and long wavelength then regions of poorly integrated, low gradient drainage may be very extensive. On the upstream flank of the upwarp the drainage will be reversed (D) as far as the point where the initial downstream slope is re-established (E). Tributary streams may have sufficient erosive power to maintain their original direction of flow although their gradients will be reduced (F). This will produce a barbed drainage pattern (usually a diagnostic feature of drainage reversal). Lakes may also form where reversed drainage joins the upstream drainage which has maintained its original flow direction (G), but the development of strike drainage parallel to the upwarp axis (H) may prevent lakes from forming. (Based on C. Ollier (1981) Tectonics and Landforms. Longman, London, Fig. 12.7, p.168.)*

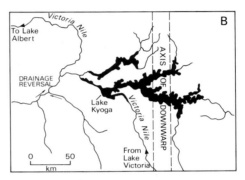

Fig. 16.10 *Effects of warping on drainage development in the East African Rift System (A). Detail of Lake Kyoga in Uganda formed by drainage reversal caused by the Late Pleistocene back-tilting of the plateau east of the Western Rift is also shown (B).*

The effect on drainage systems of recent, and quite rapid, warping are to be seen in the East African Rift System (Fig. 16.10(A)). A high rate of surface uplift occurred along a NNE-trending axis to the east of the Western Rift and deformed an Early Pleistocene landsurface. Rivers originally draining into the rift were **defeated** and drainage reversal occurred, giving rise to a **barbed** drainage pattern with junction angles greater than 90°. Lakes Victoria and Kyoga developed due to ponding of this reversed drainage until breaching by the present Victoria Nile re-established a connection with the Western Rift into Lake Albert. The peculiar shape of Lake Kyoga (Fig. 16.10(B)) is a result of the drowning of a dendritic drainage system as a consequence of the warping of Early Pleistocene surface. This particular example not only illustrates the effect of crustal warping on drainage but also indicates how the development of drainage patterns through time can be used to elucidate the sequence and location of tectonic activity.

16.3.2 Effects of faulting

Although in comparison with warping its effects are normally more localized, faulting can also cause significant drainage adjustments. A relative upward movement on its downstream side forming a fault scarp will, if of sufficient magnitude, behead a stream and create a **fault-line lake**. If

the relative movement is downward on the downstream side then a waterfall will be formed.

Strike-slip faulting also tends to involve some vertical movement and so may give rise to similar adjustments to normal and reverse faulting. The main effect of strike-slip faulting, however, is the creation of **offset drainage** along the line of the fault (Fig. 3.25). This is displayed in classic fashion along sections of the San Andreas Fault System where a series of crustal movements has led to repeated offsetting of streams traversing the fault. Various drainage patterns have been produced in consequence, including right-angle offsets, 'Z' patterns and trellis forms (Fig. 16.11). Such drainage adjustments can provide valuable evidence on the magnitude of strike-slip movement, but care must be taken in their interpretation since modest differential vertical movements of only a few centimetres across a fault can be sufficient to deflect small streams and produce **false offsets**, while vertical displacements of just a few metres are capable of offsetting even large river channels.

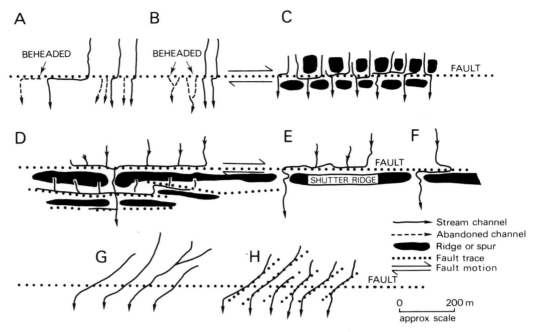

Fig. 16.11   *Diagrammatic representation of patterns of fault-related stream channels occurring along the San Andreas Fault in the Carrizo Plain area, California. Offset channels are shown in (A) and (B): (A) the misalignment of single channels directly related to the amount of fault displacement and the age of the channel – there is no ridge on the downslope side of the fault and beheading is common; (B) misaligned paired stream channels. Patterns resulting from a combination of offset and deflection are represented in (C)–(F): (C) compound offsets of ridge spurs and both the left and right offsetting and deflection of channels; (D) trellis drainage produced by multiple fault strands and shutter ridges; (E) indicates how a combination of offset and deflection by shutter ridges can give rise to exaggerated or reversed apparent offset; (F) shows how capture by an adjacent channel followed by right-lateral slip may produce a 'Z' pattern. Two types of false offset are represented by (G) and (H): (G) shows how differential uplift may deflect streams to produce false offsets, and (H) illustrates* en echelon *fractures over a fault zone followed by streams producing false offsets. (After R. E. Wallace (1975) in J. C. Crowell (ed.)* San Andreas Fault in Southern California. *California Division of Mines and Geology, Special Report **118**, Fig. 2, p. 243.)*

16.3.3   Drainage evolution in rift valleys

Traditional views of drainage evolution in rift valleys were based on the classical symmetric model of rift structure. These emphasized the importance of axial drainage along the rift and the role of marginal streams crossing the opposing fault scarps in constructing alluvial fans on either side of the rift. As we saw in Chapter 4 (see Section 4.3 and Figure 4.9) the symmetric model of rift development has been challenged on the basis of seismic data which indicate that some rift valleys, at least, have an asymmetric half-graben structure.

An important consequence of the half-graben model of rift structure is the back-tilting of drainage behind the footwall scarp (Fig. 16.12(A)). This occurs through a combination of the regional doming that frequently accompanies rifting and displacements associated with movements along the major listric boundary fault. If movement occurs along more than one major boundary fault, then small back-tilted blocks are formed which prevent direct drainage to the rift and divert rivers along strike. These rivers may eventually gain access to the rift at the site of transfer faults or other fault systems which accommodate the movement between half-graben units of opposite polarity (Fig. 16.12(B)). Large deltas, such as the Turkwel delta of the Turkana Rift in Kenya, can form where these diverted rivers enter a rift valley.

The drainage of Lake Tanganyika in East Africa provides an extreme case where listric boundary faults have acted as a barrier to drainage (Fig. 16.13). In other parts of the East African Rift System both diverted rivers and axial drainage are more common. Nevertheless, irrespective of their detailed structure, most of the sediment supplied to rifts with a half-graben structure generally comes not from streams draining the steep footwall scarp but from axial drainage or the much larger catchments of the opposing roll-over section of the rift. Consequently, the alluvial fans located at the footwall scarp are normally small in comparison with the extensive alluvial cones which develop on the roll-over zone.

16.3.4   History of the Colorado river

The Colorado River of the western USA provides an excel-

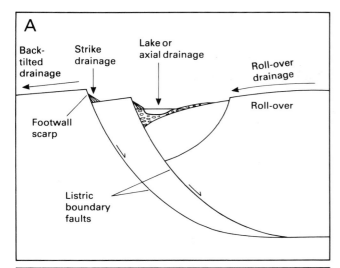

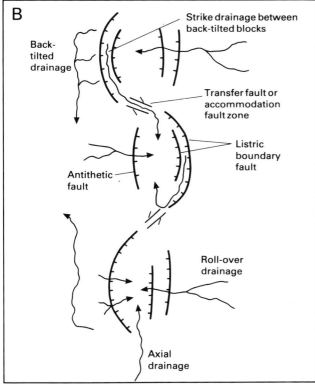

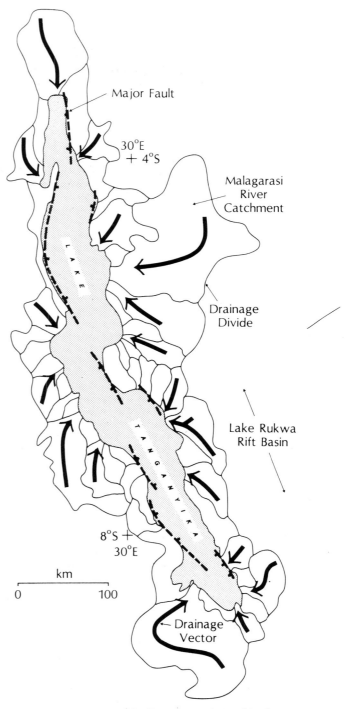

Fig. 16.12 *Schematic representation of the relationship between drainage patterns and half-graben rift structure showing cross-sectional (A) and plan view (B). (Modified from L. E. Frostick and I. Reid (1987) in L. Frostick and I. Reid (eds) Desert Sediments: Ancient and Modern. Geological Society Special Publication 35, Fig. 3, p. 56. Reproduced by permission of the Geological Society.)*

lent illustration of the way in which large-scale drainage evolution is intimately linked to interpretations of tectonic history (Fig. 16.14). The composition and structure of Tertiary gravel deposits in ancient stream channels along the southern margin of the Colorado Plateau indicate a source

Fig. 16.13 *Lake Tanganyika illustrating relationships between major listric boundary faults (dashed lines with ticks indicating downthrow side), drainage basin size and primary direction of drainage systems (drainage vectors). Notice how the largest catchments occur predominantly on the roll-over side of each half-graben unit. (From L. E. Frostick and I. Reid (1987) in: L. Frostick and I. Reid (eds) Desert Sediments: Ancient and Modern. Geological Society Special Publication 35, Fig. 5, p. 57. Reproduced by the permission of the Geological Society.)*

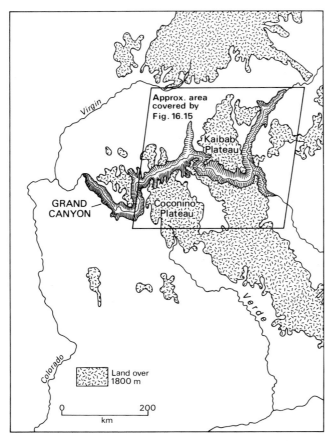

Fig. 16.14 *Course of the present-day Colorado River in relation to the uplifted Colorado Plateau.*

to the south and south-west. At this time central Arizona was more elevated than the region now occupied by the Grand Canyon. Radiometric dating of basalt pebbles in these deposits show that this situation lasted until at least 10 Ma BP. The cutting of the Grand Canyon by the Colorado River was promoted by the uplift of the Colorado Plateau (Fig. 16.15). This apparently occurred between 10 and 5 Ma BP since the age of valley-fill basalts indicates that the south-flowing drainage of the Verde River away from the uplift had developed by 5 Ma BP.

Setting these changes in a broader perspective, the Colorado Plateau was low relative to the elevated Basin and Range Province to the west until at least 18 Ma BP. Between 18 and 10 Ma BP faulting to the west of the Colorado Plateau led to the differentiation of the Basin and Range Province and the development of interior drainage within it. The ancestral Colorado only began to drain to the west between 10 and about 3 Ma BP, as the Colorado Plateau experienced uplift relative to the 'collapsing' Basin and Range terrain. By about 3 Ma BP a well-developed Grand Canyon had been formed and the Colorado River had almost attained its present depth at its downstream end.

16.4 Active tectonics and channel adjustment

As we saw in Chapter 9, alluvial channels are very sensitive to variations in discharge and sediment load characteristics, and many of the changes in channel morphology that occur over time can be attributed to these variables. It is important, none the less, to consider the possible role of active tectonics in affecting channel properties. As we have already noted, active faulting can have a dramatic impact on fluvial systems (see Section 16.3.2); yet slower, vertical deformations can lead to more subtle, but still significant, adjustments in channel morphology. Anomalies in stream channels which may be indicative of active tectonics include the local development of braiding or meanders, the creation of ponds or marshes, the accumulation of alluvial fills and the widening or narrowing of channel reaches. The deformation of fluvial channels can take many forms, but the primary effect is usually either a local reduction or increase in channel gradient, or lateral tilting of the channel leading to channel aggradation or incision.

16.4.1 Field examples

The Mississippi Basin in the USA provides several well-documented examples of the effects of slow, ongoing deformation on river channel characteristics. On the borders of the states of Louisiana, Arkansas and Mississippi an area known as the Monroe uplift has been active since the Tertiary, and repeated levelling surveys during the period 1934–1966 have shown that surface uplift is still continuing, at least on the southern flank of the uplift. Quaternary uplift is demonstrated by convexities in the longitudinal profiles of river terraces of this age, and continuing uplift is indicated by convex-up sections in the present-day valley floor profiles (Fig. 16.16). The response of an alluvial channel to this kind of slow crustal deformation can be illustrated by Big Colewa Creek, one of the streams shown in Figure 16.16. Several effects are apparent, including an abrupt threefold downstream increase in channel gradient at the axis of uplift which has led to channel incision and a corresponding increase in average bank height from 2 to 6 m. Channel sinuosity increases from 1.2 upstream of the uplift axis to 1.7 in the reach immediately downstream of the zone of maximum uplift, before gradually falling to 1.5.

Active tectonics may also have important indirect effects on river channel morphology. A notable example is provided by the New Madrid earthquake which occurred in the centre of the Mississippi Basin in 1812. It appears that this major earthquake promoted widespread collapse of unstable banks along the lower Mississippi which led to a large injection of sediment into the river. This in turn caused major downstream adjustments in channel morphology, including a reduction in meandering and the formation of

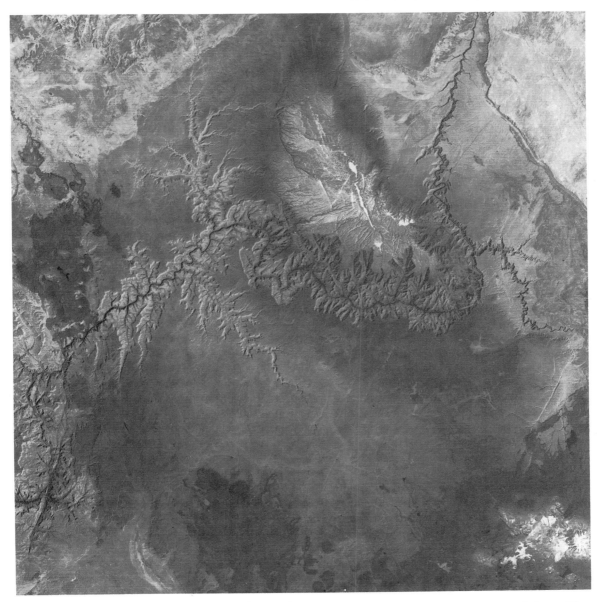

Fig. 16.15 *Landsat image showing the Grand Canyon carved into the uplifted Colorado Plateau by the Colorado River. The area covered by the image is shown on Figure 16.14. (Image courtesy N. M. Short.)*

islands and point bars. These changes were so significant that they persisted for several decades after the earthquake.

16.4.2 Experimental studies

In the field it is often very difficult to separate the effects on channel morphology of active tectonics from variations in discharge and sediment characteristics, since there is often a complex interaction between these groups of variables. Experimental studies in which the hydrological and sedimentological variables can be controlled thus provide a

useful means of isolating these effects, and thereby revealing changes in channel characteristics which are influenced by tectonic deformation of the channel bed.

One of the most sophisticated of such studies carried out to date has been that by S. Ouchi working at the Colorado State University. Using a large laboratory flume to investigate the effects of vertical deformation across alluvial channels, he found that changes produced depended not only on the rate and amount of displacement but also on the morphology and sediment properties of the channel (Table 16.3). Braided channels were found to be unable to change

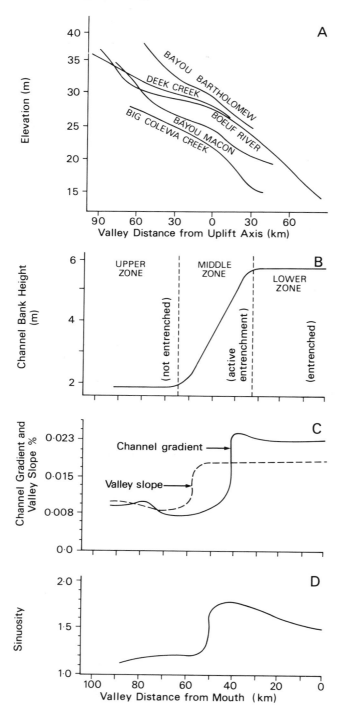

Fig. 16.16 *Affects of uplift on channel characteristics for streams crossing the Monroe uplift in the lower Mississippi Basin: (A) the longitudinal profiles of various stream channels; (B), (C) and (D) changes in channel bank height, channel gradient and channel sinuosity for Big Colewa Creek. (After A. W. Burnett and S. A. Schumm (1983) Science 222, Figs 3(b) and 4, p. 50.)*

their sinuosity in response to anticlinal uplift but rather adjusted by channel incision which in turn led to the forma-

tion of terraces. Meandering channels, by contrast, adjusted by an increase in sinuosity (resulting in greater channel length), bank erosion and point bar growth in the steepened reach downstream, and by flooding over channel bars in the flattened reach upstream of the uplift axis. Subsidence was found to lead to deposition in the reach astride the synclinal axis and the downstream development of transverse bars in braided channels but in meandering channels the flooding of bars downstream of the zone of maximum subsidence and an increase in sinuosity upstream was observed.

These experimental studies show that although active deformation commonly occurs at rates which are slow in comparison with changes in the other variables which affect channel morphology, if such deformation persists for only a few decades the displacements achieved will be sufficient to generate significant changes in channel morphology. River channel morphology, therefore, appears to provide a very sensitive means of detecting ongoing crustal deformation, assuming non-tectonic factors can be clearly isolated. In identifying the presence of active tectonics a wide range of responses of fluvial systems can be considered (Table 16.4).

16.5   Global tectonics and continental drainage

The geomorphic significance of the world's major rivers is illustrated by the fact that, excluding the largely ice-covered land masses of Greenland and Antarctica, some 47 per cent of the area of the continents is drained by only 50 river basins. Indeed the five largest basins – the Amazon, Zaire (Congo), Mississippi, Nile and Yenisei – alone account for 10 per cent, with the Amazon itself contributing 5 per cent. On examining drainage basins of this size it becomes immediately apparent that tectonics exerts a strong influence over their large-scale morphology and development through time.

At the global scale it is possible to identify five major types of drainage system in relation to tectonic setting (Table 16.5). These are illustrated in Figure 16.17 where the relationship between tectonic setting and drainage system characteristics is readily apparent. A **passive margin basin** drains to a passive continental margin from a cratonic interior which lacks a Mesozoic or Cenozoic mountain belt. This type of drainage system is particularly common on the African continent and for this reason is sometimes referred to as **Afrotype drainage**. **Passive margin – distal orogen basins** are so called because, whilst draining to a passive margin, a significant proportion of their tributaries have their source in a mountain belt along the distal margin of the basin. The classic example of this kind of drainage basin is the Amazon, but such drainage systems exist along most of the eastern margin of the American continent as a consequence of the the American Cordillera which marks the

Table 16.3 Response of experimental fluvial channels to uplift and subsidence

		UPSTREAM FROM ZONE OF DEFORMATION	UPSTREAM FLANK OF DEFORMATION	AXIS OF DEFORMATION	DOWNSTREAM FLANK OF DEFORMATION	DOWNSTREAM FROM ZONE OF DEFORMATION
Braided channel	Uplift	Aggradation Thalweg shift Submerged bars		Degradation Terrace formation Single bars		Aggradation Braided
	Subsidence	Degradation Single thalweg		Aggradation Braided	Flooding	Degradation Single thalweg
Meandering channel	Uplift	Aggradation Flooding Multiple channels (anastomosing)		Degradation Sinuosity increase Bank erosion		Aggradation
	Subsidence	Degradation Sinuosity increase Bank erosion		Aggradation Flooding, cutoffs Multiple channels		Local scour

Source: Modified from D. I. Gregory and S. A. Schumm (1987) *in: K. Richards (ed.) River Channels: Environment and Process*, (Blackwell, Oxford) Fig. 3.3, p. 46, after S. Ouchi (1985) *Geological Society of America Bulletin*, **96**, Table 1, p. 508.

Table 16.4 Types of adjustments of river channel characteristics relevant to the identification of active tectonics

Deformation of valley-floor longitudinal profile

Deformation of channel projected profile

Change of channel and valley gradient

Change of channel width and depth

Conversion of floodplain to low terrace

Active floodplain development

Variations in channel stability including reaches of active channel incision, deposition, channel shifting and formation of knickpoints

Indirect effects both upstream and downstream of the zone of deformation including deposition, channel incision, bank erosion and flooding

Table 16.5 Classification of major types of drainage system in relation to tectonic setting

TYPE	CHARACTERISTICS	EXAMPLE
Passive margin	Drain to passive continental margin from cratonic hinterland lacking Mesozoic or Cenozoic mountain belts	Zaire
Passive margin – distal orogen	Drain from Mesozoic or Cenozoic mountain belt to a passive margin	Amazon
Intra-orogen	Drain along strike within mountain belt	Irrawaddy
Extra-orogen	Drain along strike marginal to mountain belt	Ganges
Trans-orogen	Drain across strike of a mountain belt	Columbia

complex zone of plate interaction along its western periphery; such drainage basins are indeed sometimes referred to as **Amerotype**. The presence of a high mountain belt in the distal part of the basin is probably the reason why the drainage pattern of this type of basin is relatively unidirectional in comparison with the rather more centripetal drainage which characterizes Afrotype basins.

The often large drainage area of passive margin basins of both types is one of the reasons why deltas are preferentially located along passive margin and marginal sea coasts (Fig. 13.2). Both types of passive margin basin may also be modified by the presence of a degraded ancient orogen along the continental margin (formed during an episode of continental collision prior to the subsequent continental rifting). These may present a topographic barrier sufficient to prevent major continental drainage outlets developing along extensive sections of passive margins. An example of this effect is the Appalachian Mountains of the eastern USA which appear to 'deflect' the drainage of the Mississippi

Basin southwards towards the Gulf of Mexico.

The three other types of drainage basin are associated with mountain belts. An **intra-orogen basin** is located primarily within a mountain belt and the drainage network runs parallel to its strike. Drainage parallel to strike also characterizes an **extra-orogen basin**, but in this case the basin is located along the margins of a mountain belt. **Trans-orogen basins** differ from the other two in that the drainage traverses the strike (that is, runs across the grain) of the mountain belt.

While useful, this classification scheme presents an oversimplified view of the relationship between global tectonics and continental drainage. Some basins may contain elements of more than one type – the upper part of the Brahmaputra River, for instance, runs along the strike of the Himalayan orogen before finally turning at right angles across the eastern Himalayas to join the Ganges at the head of the Bay of Bengal. Moreover, other tectonic settings need to be considered. The cratonic interior basins of Africa, such as

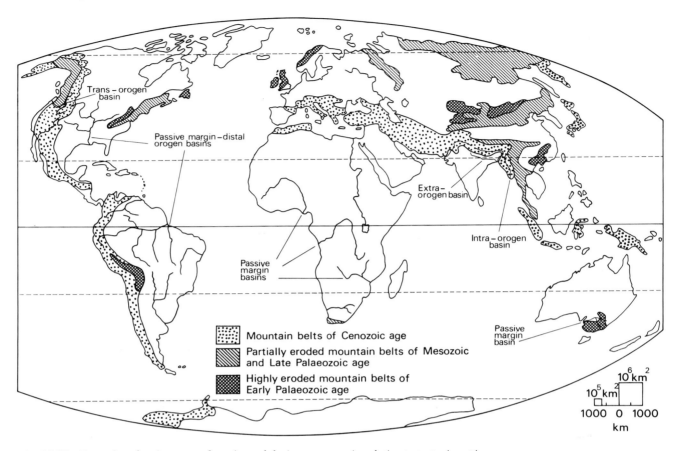

Fig. 16.17   *Examples of major types of continental drainage system in relation to tectonic setting.*

the Chad and Okavango Basins, the river systems of block-faulted terrains, such as those of the south-west USA, and the river systems of western China dominated by the effects of active strike-slip faulting all reflect specific tectonic controls.

16.6   Drainage development in orogens

Three components of structural control are typically evident in the development of drainage in orogens. One is the distinct separation of the erosional zone of the mountain mass from the adjacent aggradational basin by a zone of reverse faulting. Secondly, longitudinal fault systems within the orogen predispose the drainage to be similarly orientated along strike; moreover, drainage in the aggradational basin adjacent to the orogen also tends to be longitudinal except for comparatively short, but erosionally vigorous, streams draining directly from the mountain front. A third key structural element is the existence of salients and re-entrants along the boundary between the mountain mass and the adjacent basin.

Along the Himalayas such irregularities probably date from the collision of the Indian and Eurasian Plates. Re-entrants are structurally low regions and are significant

because they provide the foci for extensive drainage basins entering the alluvial plain from the adjacent mountain mass. They supply much more sediment than the smaller streams draining the intervening salients. This is exemplified by the drainage of the Himalayas where the Punjab re-entrant is occupied by the Indus, and the upper Assam re-entrant is occupied by the Brahmaputra (Fig. 16.18). Between these two major river systems only comparatively small rivers drain to the Ganges aligned longitudinally along the Indo-Gangetic plain.

16.6.1   Drainage of the Zagros Mountains

Modern orogens, which are actively growing and where primary structures have not yet been removed by erosion, provide the optimal conditions for studying the relationships between structure and tectonics in mountain belts. The Zagros Mountains of Iran, which represent an intercontinental collision orogen, provide an excellent example. Parts of this range have experienced major folding and uplift as recently as the Pliocene. Rivers rising near the north-eastern margin of the highland flow completely through the orogen on their way to the Mesopotamian–Persian Gulf trough (Fig. 16.19(A)). A detailed analysis of the drainage systems

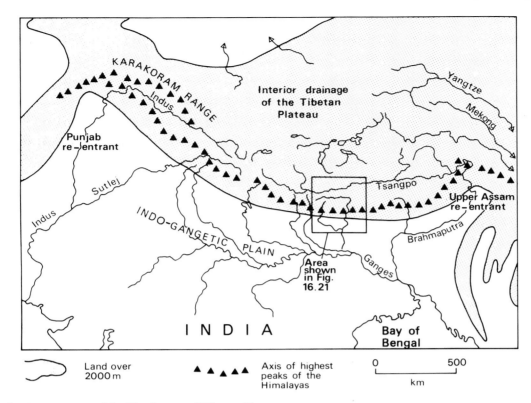

Fig. 16.18 *Drainage systems of the Himalayas and Tibetan Plateau.*

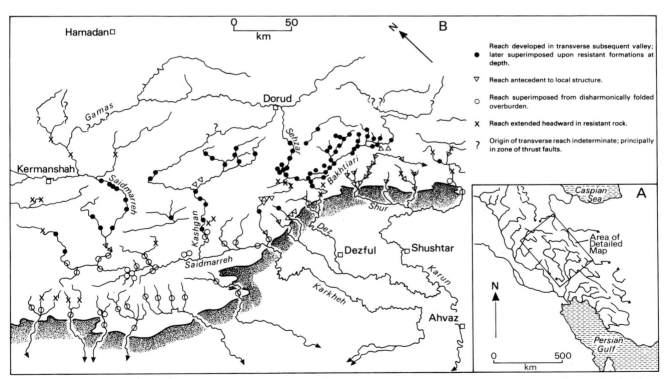

Fig. 16.19 *Drainage of the Zagros Mountains, Iran: (A) general tectonic setting of the major drainage systems; (B) most probable origins of stream reaches transverse to anticlinal structures of the central Zagros. (Modified from T. M. Oberlander (1965) The Zagros Streams Syracuse University Geographical Series 1, Fig. 1, p. 2 (A) and T. M. Oberlander (1985) in: M. Morisawa and J. T. Hack (eds) Tectonic Geomorphology. Allen and Unwin, Boston Fig. 3, p. 165 (B).)*

of the Zagros Mountains by T. Oberlander has demonstrated that no single hypothesis of transverse drainage can explain all the observed drainage patterns. None the less, every drainage anomaly in the region can be classified as some variant of one or other of the general models of transverse drainage development (Fig. 16.19(B)).

Superimposed drainage was found to be confined to specific structural associations and no evidence was found of a regional cover mass or erosion surface from which extensive superimposition or drainage inheritance could have occurred. Similarly, antecedence was diagnosed as being localized rather than regional. The most widespread mechanism leading to the development of transverse drainage has apparently been headward stream extension into zones

of relatively less resistant strata which were successively exposed as the folds were eroded. Far from being anomalous, the polygenetic development of transverse drainage in the Zagros Mountains producing spectacularly discordant drainage patterns is viewed by Oberlander as an inevitable consequence of *normal* stream development in the fold structures typical of intercontinental collision orogens (Fig. 16.20).

16.6.2 Drainage of the Himalayas

The traditional view of the transverse rivers of the Himalayas, which include the Indus and the Brahmaputra, is that they are antecedent, having originated on the high Tibetan

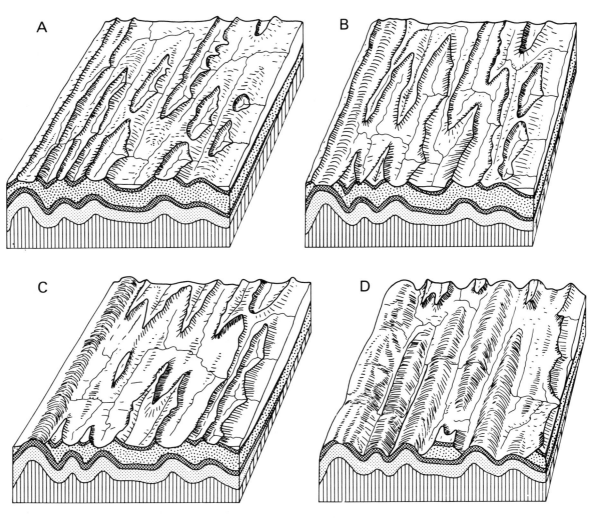

Fig. 16.20   *Model of drainage evolution in open folds, such as those which characterize much of the Zagros Mountains, in which the depth of readily erodible beds between resistant formations exceeds fold amplitudes: (A) the initial erosion of anticlines exposes erodible formations and leads to the expansion of axial basins; (B) the expanding axial basins begin to merge and so create outcrops of erodible strata which extend across fold axes; (C) through-flowing transverse streams occupy continuous outcrops of erodible beds and are superimposed on to lower resistant beds emerging in anticlinal cores; (D) anticlinal mountains which are resurrected on the lower resistant beds are breached by the superimposed transverse streams (After T. M. Oberlander (1985) in M. Morisawa and J. T. Hack (eds)* Tectonic Geomorphology. *Allen and Unwin, Boston, Fig. 8, p. 172.)*

Plateau prior to the major phase of uplift of the Himalayan ranges. The upper courses of both the Indus and the Brahmaputra (Tsangpo) closely follow the Indus–Tsangpo suture zone marking the initial line of collision between the Indian and Eurasian continents. These rivers traverse the axis of the Himalayas through spectacular gorges. Where the Indus leaves Kashmir it is only 1000 m above sea level, but is overlooked by the world's fourth highest mountain, Nanga Parbat, towering to over 8000 m. A number of other large rivers draining south from the Himalayas presently have source tributaries on the Tibetan Plateau and on the northern slopes of the Higher Himalayas. The antecedence model of Himalayan drainage requires that the river systems are ancient since they must have originated before the faulting, thrusting and folding that created the Himalayas.

A closer look at the relationships between river courses and geological structures in the Himalayas, however, indicates that there is not complete discordance, as would be anticipated from a simple antecedence model. The Indus and Brahmaputra, for example, cross the axis of the Himalayas exactly at the points where the trend of the range changes (Fig. 16.18). As for the other transverse

rivers, many of these are seen to coincide closely with transverse anticlines, that is, anticlinal structures which are transverse to the main axis of the Himalayas. The River Arun, which has its headwaters just a few kilometres from the Tsangpo and which has widely been regarded as antecedent, follows one of these structures (Fig. 16.21). The Arun transverse anticline, located east of Mount Everest, has an amplitude of 10 000 m and the River Arun has incised itself to a depth of 15 000 m into this structure as isostatic rebound has caused significant crustal uplift in response to denudational unloading. The transverse warps appear to have developed late in the history of the Himalayas, so if they have been instrumental in controlling the location of transverse drainage systems then these must also be recent.

Rather than being antecedent it seems likely, then, that these transverse Himalayan fluvial systems represent a structurally controlled form of superimposed drainage. Given a sufficient thickness of readily erodible strata within the transverse anticlines of the Himalayas the unroofing of these structures could have enabled stream systems to locate preferentially within them. Suitable weak lithologies exist within the Himalayas, but extensive erosion means that it is not certain that they were present in the hypothesized localities. None the less, the correspondence between transverse structures and river courses argues strongly against any simple model of antecedence.

16.7   Drainage development associated with passive margins

We have already noted how the major drainage basins of Africa drain to passive continental margins, but the large-scale morphology of these basins as well as the location of their outlet to the ocean also appears to be subject to tectonic control. The overall shape and size of the major drainage basins of Africa appear to be related quite closely to the basin and swell topography characteristic of the continent. Breaches of swells (crustal upwarps) by rivers are rare except where major river channels reach the coast. Here drainage outlets appear to be located either at structurally low points along the margin between domal upwarps (as in the case of the Orange and Zaire Rivers) or at aulacogens. The classic instance of the latter situation is provided by the Niger River which is located in the Benue Trough, an aulacogen created during the formation of the southern Atlantic Ocean. Another, perhaps less obvious, example is the Zambezi whose mouth is located within a complex rift structure associated with the splitting of Madagascar away from the African mainland.

Somewhat similar structural controls have also been suggested for the Amazon. Its mouth is also located in an aulacogen which apparently continues inland as an elongated structural low as far west as the Andes. Such continental-scale structures originated during the break-up

Axes of major ranges

Major gorges

Fig. 16.21   Drainage in the vicinity of Mount Everest in the eastern Himalayas. Note the discordant courses of the south-flowing rivers which traverse the axis of the High Himalayas. (Modified from A. Holmes (1978) Principles of Physical Geology *(3rd edn, revised D. L. Holmes). Chapman and Hall, London, Fig. 19.36, p. 387.)*

of Gondwana and may have also influenced the courses of other South American rivers (such as the Paraná and São Francisco) which drain towards the Atlantic.

The relationship between tectonics and drainage patterns in Africa and South America are illustrated in Figure 16.22. The two continents have been replaced into their approximate relative positions prior to rifting around 140 Ma BP. This reconstruction serves to emphasize the similarities between drainage patterns and the tectonic settings of the passive margins on either side of the present-day South Atlantic (although the pre-rift drainage would, of course, have been very different from that of the present). But how do drainage systems adjust to the major disruption caused by continental break-up? This is an important question because many of the world's major drainage basins drain to passive margins formed by the break-up of Pangaea which began about 200 Ma BP.

16.7.1   Drainage history of the African continent

Africa provides perhaps the best starting point for an investigation of the effects of continental break-up on drainage development. This is both because it has no significant relict orogenic belts blocking drainage outlets across its margins, and because, unlike South America, it is almost entirely rimmed by passive margins.

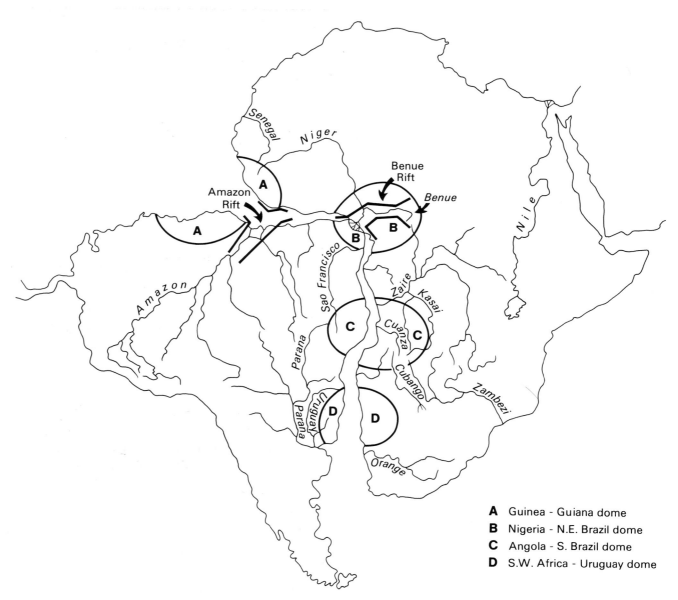

Fig. 16.22   *Relationship between present-day drainage and domal rift structures on the Atlantic margins of Africa and South America.*

A   Guinea - Guiana dome
B   Nigeria - N.E. Brazil dome
C   Angola - S. Brazil dome
D   S.W. Africa - Uruguay dome

Even a cursory examination of a topographic map of Africa shows that the drainage of the African continent has a number of apparently anomalous features. For instance, the drainage pattern of several of the major basins is to a large extent centripetal. This is particularly evident in the Zaire Basin where most of the major tributaries are orientated towards a point in the centre of the basin rather than towards its outlet. Other major rivers, such as the Niger, have long sections which flow directly *away* from a closely adjacent coastline. Africa, along with Australia, is also remarkable for the proportion of its total area occupied by internal drainage systems.

Undoubtedly climatic factors, especially oscillations between arid and humid climatic regimes, have played an important role in the development of these drainage characteristics. But tectonic controls also seem to have been crucial; not only has the tectonically induced basin and swell topography of Africa provided an overall constraint to drainage basin development, but the presence of a crustal upwarp extending around most of the margins of the continent has probably been of great importance to the post-Gondwana drainage evolution of the continent.

The effect of this marginal upwarp becomes very evident when we examine the hypsometric curves for Africa's major drainage basins (Fig. 16.23). Most obvious is the very small proportion of the area of each basin at low elevations. The form of these hypsometric curves is in stark contrast to the form we might expect for mature drainage basins which would have a significant proportion of their total basin area at low elevations. Remember that the drainage systems of Africa have had more than 100 Ma to evolve since the final emergence of Africa as a discrete continent after the break-up of Gondwana.

The long profiles of the major river channels also raise interesting questions. The gradient of the Zaire River along 2000 km of its middle course averages only 0.05 m km⁻¹, whereas from Kinshasa to Matadi near its mouth the mean gradient increases by over 15 times to 0.78 m km⁻¹ as the channel plunges down the knickpoint formed by the rapids at Stanley Pool. Other African rivers also have significant knickpoints such as the Victoria Falls on the Zambezi and the Augrabies Falls on the Orange (Fig. 16.24).

Fig. 16.24 The Augrabies Falls, a major knickpoint on the Orange River, southern Africa (top photo), which separates its low-gradient braided middle course (middle photo) from its deeply incised lower course (bottom photo).

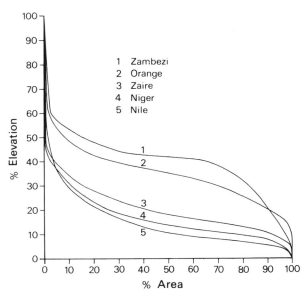

Fig. 16.23 Percentage hypsometric curves for the five largest drainage basins in Africa.

One possible explanation for these anomalous hypsometric curves and longitudinal profiles is that the margins of Africa have experienced significant uplift in the relatively recent past, perhaps in the Late Pliocene and Pleistocene. This view has certainly been put forward by several researchers in one form or another, but it is not supported unequivocally by the geological and geomorphic evidence and, moreover, it is difficult to reconcile with what we know about the mechanisms producing uplift along passive margins. As we saw in Chapter 4, significant passive margin uplift is likely to occur either immediately before continental break-up (in the active rifting model where domal uplift precedes rifting) or soon after (in the passive rifting model where uplift arises as a consequence of the rifting mechanism). The possibility remains, therefore, that the marginal upwarps that we see today influencing drainage were present in some form at the time of passive margin creation, or within a few million years thereafter. In most cases, any subsequent surface uplift has probably been largely confined to flexural effects associated with passive isostatic compensation as the coastal periphery of these passive margins has been gradually eroded.

This model of passive margin tectonics immediately suggests a radical alternative model for the drainage evolution of much of Africa. This model proposes that the upwarps around the margins of Africa significantly delayed the incorporation of much of the African continent into a fully developed external drainage system. The present-day pattern of drainage can best be explained by the large-scale capture of internal basins by the active headward erosion of streams draining the oceanward flanks of the marginal upwarps. This idea is not entirely new since various interpretations along these lines have been suggested previously, including the view that there was internal drainage over much of Africa until the Miocene (Fig. 16.25). Other versions of this drainage capture model have advocated rather earlier capture events. Certainly, the relatively recent development of the Zaire Basin into an external drainage system is suggested by the major knickpoint represented by the rapids below Stanley Pool and the extremely low gradients found in the central part of the basin. Further south there is some evidence of major shifts in the location of the mouth of the ancestral Orange River, and of dramatic changes in the organization of fluvial systems draining to the eastern margin of southern Africa (Fig. 16.26).

16.7.2   A model of passive margin drainage development

As yet the case for large-scale drainage capture is not proven, but it does fit in well with current ideas about the tectonic evolution of passive margins. Much more research is needed on this topic, but even now it is possible to put forward a plausible model of how megascale drainage capture might work.

Figure 16.27 provides a schematic representation of a possible sequence of drainage adjustments during continental rifting and subsequent passive margin subsidence. The model assumes that uplift precedes rifting and does not take into account the possible development of an aulacogen (see Section 4.5 for details of the tectonic evolution of passive margins). The model is also specifically confined to the evolution of rifted rather than sheared margins.

An additional factor which must be considered in the drainage evolution of passive margins is the role of aulacogens since they can be important foci for continental drainage. Of particular importance is the fact that they can be established relatively soon after rifting (the Niger delta, for instance, dates back at least 60 Ma) and thereby provide an early outlet for continental drainage to a newly developing ocean. The likely effects on drainage development of aulacogens and other components of structural variability along a passive margin are illustrated in Figure 16.28 which is based on the 'plan view' of the passive margin development discussed in Chapter 4 (see Figure 4.16).

16.7.3   Variation in passive margin drainage evolution

Although it is possible to predict the broad outline of drainage history from the likely tectonic evolution of passive

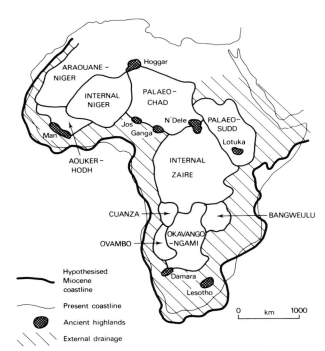

Fig. 16.25 *Speculative map of African drainage as it might have appeared immediately before the major uplift and rifting of the East African Rift System. (Modified from J. de Heinzelin (1964) in F. Clark Howell and F. Boulière (eds)* African Ecology and Human Evolution. *Methuen, London, Fig. 7, p. 650.)*

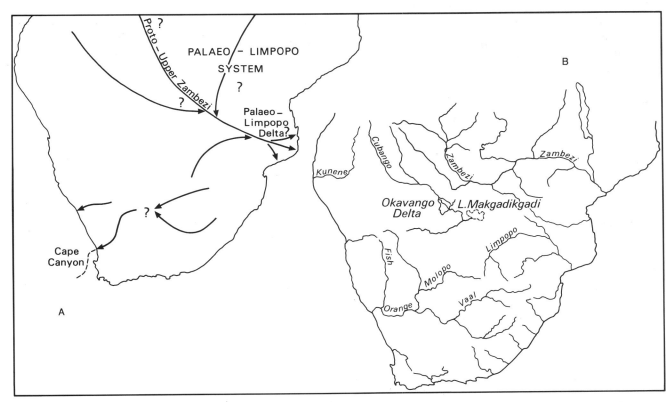

Fig. 16.26 *Highly conjectural reconstruction of the pre-Miocene drainage of southern Africa (A) compared with the modern drainage system (B). It has been suggested (C. J. H. Hartnady per.comm.) that the protrusion on the east coast in southern Mozambique may represent an enormous ancient delta which was fed by a very large drainage system possibly extending into central Africa. (Pre-Miocene drainage reconstruction based on various sources.)*

margins, careful examination of actual passive margin drainage reveals considerable variation. While major river systems penetrate the marginal upwarps along the African passive margin, on the subcontinent of India the drainage of the steep western flank of the Western Ghats is completely separated from river systems flowing eastwards. The drainage divide between the two systems coincides with a major escarpment and lies within 100 km of the west coast along much of the margin. Since the east coast of India is also a passive margin the dramatic asymmetry of the drainage of the subcontinent may be due to uplift having occurred earlier in the west.

The eastern margin of Australia provides a further example of passive margin drainage evolution. Here the drainage divide – separating eastward-flowing streams draining to the Tasman Sea and westward-flowing streams draining towards the Australian interior – does not coincide with the generally well-developed great escarpment but lies at a variable distance to the west of it. Moreover, there is evidence to suggest that the continental drainage divide has not shifted laterally to any significant extent since at least the Late Oligocene – Early Miocene. Clearly, there is considerable variation in the history of drainage evolution along passive margins, and it will be a significant research

challenge to reconcile this variability with tectonic models of passive margin development.

Further reading

The topics discussed in this chapter are rather patchily covered in the literature. Of the more general texts Ollier (1981) discusses aspects of both active and passive (structural) tectonic controls on drainage and Morisawa (1985), (Chapter 10) emphasizes the passive controls over drainage development. Structural controls are covered in detail in the books by Tricart (1974) and Twidale (1971), but these treatments do not include adequate discussions of recent ideas on the development of transverse drainage in orogens or of the evolution of passive margin basins.

Drainage patterns are examined in depth by Howard (1967) who illustrates how they can be used in the interpretation of geological structures. Joint control of drainage patterns in homogeneous lithologies is discussed by Schick (1965), and Scheidegger and Ai (1986) provide a useful introduction to the relationship between valley orientations and the joint and fracture patterns generated by local neotectonic stress fields. Useful regional studies of these relationships are provided by Bevan and Hancock (1986) on

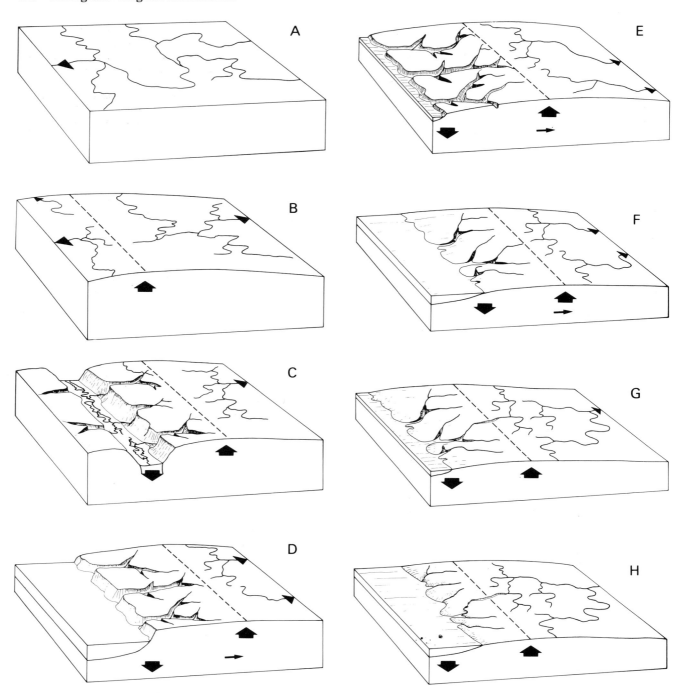

Fig. 16.27 *A speculative model of drainage development in passive margin basins. Note that this model assumes that uplift precedes rifting and does not incorporate the effects of aulacogens striking into the margin. An initial drainage across a low relief landscape is assumed (A). Initial epeirogenic warping disrupts the drainage system and causes drainage reversal on the 'upstream' (left) side of the upwarp (B). Such drainage disruption could be achieved by quite slow rates of uplift where channel gradients are low and the erosive power of rivers is minimal. Where stream power is greater (either due to higher channel gradients or higher (peak) discharges related to climatic conditions in the catchment) rivers might initially be able to maintain their courses across the upwarp. The development of significant down-faulting associated with rift development will, however, eventually bring about a complete disruption of pre-rift drainage (C). During this phase of rifting three drainage elements are created: (1) low gradient networks draining the flanks of the rift upwarp, most probably to newly formed internal basins; (2) steep gradient streams which cut deep canyons into the rift walls and achieve rapid headward erosion towards the drainage divide formed along the line of maximum uplift on the rift shoulder; (3) longitudinal drainage on the rift floor which redistributes sediments supplied from the rift walls. Continuing extension across the rift and further subsidence of the rift floor leads to the formation of a nascent ocean (D). The rift wall streams now form the new coastal*

drainage which is initially confined to a narrow belt along the newly created rift margin. Thermal subsidence and sedimentary loading will together induce progressive drowning of the margin but, perhaps more significantly, uplift will also be promoted inland by rotation and flexure of the margin (D–F). The zone of maximum uplift along the margin is also likely to migrate inland as the lithosphere cools and becomes more rigid. Uplift of the marginal upwarp will also be enhanced by denudational unloading as the steep gradient coastal streams continue to erode inland following the migrating zone of maximum uplift. Isostatic uplift could result in a net increase in the height of interfluves along the upwarp crest because erosion will probably be localized in canyons. Eventually the most aggressively headward-eroding coastal streams will begin to capture the inland drainage by breaching the crest of the marginal upwarp (G). Capture would initially be confined to headwaters, but eventually entire internal fluvial systems would be incorporated into coastal systems and a major knickpoint will be formed at the junction of the high gradient coastal channels and the much lower gradient trunk channels of the captured internal drainage network (H).

southern England and northern France, and by Scheidegger (1980) on Ontario, Canada. The development of structurally controlled drainage through time is discussed at length in the books by Tricart and Twidale mentioned above. An excellent case study of transverse drainage is that by Mabbutt (1966) who provides a detailed analysis of the drainage history of central Australia and illustrates the complex relationships that can exist between passive tectonic controls and the long-term development of fluvial systems.

The effects of warping and faulting on drainage have been inadequately investigated, but the book by Ollier (1981) (Chapter 12) provides some useful material on this topic. Frostick and Reid (1987) consider drainage development associated with rifts and specifically examine the impact of half-graben structures on drainage routeing. The history of the Colorado River provides an interesting case

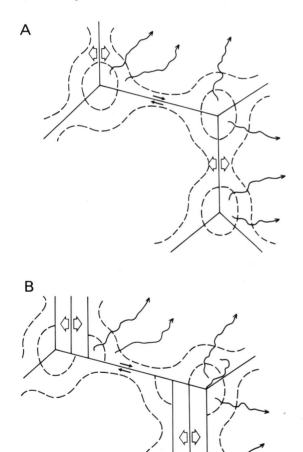

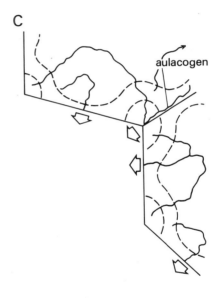

Fig. 16.28 *Schematic model of the effects on drainage development of structural and morphological variability along passive margins. A–C represent stages of development. Note the role of aulacogens as likely foci for external drainage early in the life of the margin. The dashed lines schematically indicate contours.*

study of the effects of warping and faulting on drainage development, and this example can be followed up in the papers by Lucchitta (1972, 1979) and McKee and McKee (1972). Until the 1980s little research had been undertaken on the adjustment of river channels to ongoing crustal deformation; now there are a number of both field and experimental studies on this topic. Gregory and Schumm (1987) and Schumm (1986) provide excellent introductions to the existing literature, while Burnett and Schumm (1983), Ouchi (1985) and Russ (1982) report specific field investigations and experimental studies.

To date the relationships between tectonic setting and drainage systems at the macro- and megascale have been assessed largely in terms of the sedimentary characteristics of the associated depositional basins. Miall (1981, 1984) presents an excellent overview of this approach which contains much of geomorphic interest; Audley-Charles *et al.* (1977) and Potter (1978), on the other hand, give greater emphasis to the fluvial systems themselves. The best analysis of drainage development in active orogens is that of Oberlander (1985) who also produced an earlier classic study of the drainage of the Zagros Mountains (Oberlander, 1965). Aspects of the drainage history of Africa are included in the papers by De Swardt and Bennet (1974), Dingle and Hendey (1984), Summerfield (1985) and Thomas and Shaw (1988). Kale and Rajaguru (1988) look at drainage development in western India while Bishop (1982, 1986) and Bishop *et al.* (1985) provide a detailed overview of the drainage history of the south-eastern margin of Australia.

References

Audley-Charles, M. G., Curray, J. R. and Evans, G. (1977) Location of major deltas. *Geology* 5, 341–4.

Bevan, T. G. and Hancock, P. L. (1986) A late Cenozoic regional mesofracture system in southern England and northern France. *Journal of the Geological Society, London* 143, 355–62.

Bishop, P. (1982) Stability or change: a review of ideas on ancient drainage in eastern New South Wales. *Australian Geographer* 15, 219–30.

Bishop, P. (1986) Horizontal stability of the Australian continental drainage divide in south central New South Wales during the Cainozoic. *Australian Journal of Earth Sciences* 33, 295–307.

Bishop, P., Young, R. W. and McDougall, I. (1985) Stream profile change and longterm landscape evolution – Early Miocene and modern rivers of the east Australian highland crest, New South Wales, Australia. *Journal of Geology* 93, 455–74.

Burnett, A. W. and Schumm, S. A. (1983) Active tectonics and river response in Louisiana and Mississippi. *Science* 222, 49–50.

De Swardt, A. M. J. and Bennet, G. (1974) Structural and physiographic development of Natal since the late Jurassic. *Transactions of the Geological Society of South Africa* 77, 309–22.

Dingle, R. V. and Hendey, Q. B. (1984) Late Mesozoic and Tertiary sediment supply to the eastern Cape Basin (SE Atlantic) and palaeo-drainage systems in southwestern Africa. *Marine Geology* 56, 13–26.

Frostick, L. E. and Reid, I. (1987) Tectonic control of desert sediments in rift basins ancient and modern. In: L. Frostick and I. Reid (eds) *Desert Sediments: Ancient and Modern.* Geological Society Special Publication 35, 53–68.

Gregory, D. I. and Schumm, S. A. (1987) The effect of active tectonics on alluvial river morphology. In: Richards, K. (ed.) *River Channels: Environment and Process.* Institute of British Geographers Special Publication 18. Blackwell, Oxford and New York, 41–68.

Howard, A. D. (1967) Drainage analysis in geologic interpretation: a summation. *American Association of Petroleum Geologists Bulletin* 51, 2246–59.

Kale, V. S. and Rajaguru, S. N. (1988) Morphology and denudation chronology of the Coastal and Upland river basins of western Deccan Trappean landscape (India): a collation. *Zeitschrift für Geomorphologie* 32, 311–27.

Lucchitta, I. (1972) Early history of the Colorado River in the Basin and Range Province. *Geological Society of America Bulletin* 83, 1933–48.

Lucchitta, I. (1979) Late Cenozoic uplift of the southwestern Colorado Plateau and adjacent lower Colorado River region. *Tectonophysics* 61, 63–95.

Mabbutt, J. A. (1966) Landforms of the western Macdonnell Ranges: A study of inheritance and periodicity in the geomorphology of arid central Australia. In: G. H. Dury (ed.) *Essays in Geomorphology.* Heinemann, London, 83–119.

McKee, E. D. and McKee, E. H. (1972) Pliocene uplift of the Grand Canyon Region – time of drainage adjustment. *Geological Society of America Bulletin* 83, 1923–32.

Miall, A. D. (1981) Alluvial sedimentary basins: tectonic setting and basin architecture. In: A. D. Miall (ed.) *Sedimentation and Tectonics in Alluvial Basins.* Geological Association of Canada Special Paper 23, 1–33.

Miall, A. D. (1984) *Principles of Sedimentary Basin Analysis.* Springer-Verlag, New York.

Morisawa, M. (1985) *Rivers.* Longman, London and New York.

Oberlander, T. M. (1965) *The Zagros Streams.* Syracuse University Geographical Series No. 1.

Oberlander, T. M. (1985) Origin of drainage transverse to structures in orogens. In: M. Morisawa and J. T. Hack (eds) *Tectonic Geomorphology.* Allen and Unwin, Boston and London, 155–82.

Ollier, C. (1981) *Tectonics and Landforms.* Longman, London and New York.

Ouchi, S. (1985) Response of alluvial rivers to slow active tectonic movement. *Geological Society of America Bulletin* 96, 504–15.

Potter, P. E. (1978) Significance and origin of big rivers. *Journal of Geology* 86, 13–33.

Russ, D. P. (1982) Style and significance of surface deformation in the vicinity of New Madrid, Missouri. *United States Geological Survey Professional Paper* 1236, 45–114.

Scheidegger, A. E. (1980) The orientation of valley trends in Ontario. *Zeitschrift für Geomorphologie* 24, 19–30.

Scheidegger, A. E. and Ai, N. S. (1986) Tectonic processes and geomorphological design. *Tectonophysics* 126, 285–300.

Schick, A. P. (1965) The effects of lineative factors on stream courses in homogenous bedrock. *Hydrological Sciences Bulletin* 10, 5–11.

Schumm, S. A. (1986) Alluvial response to active tectonics. In: *Active Tectonics.* National Academy Press, Washington DC, 80–94.

Summerfield, M. A. (1985) Plate tectonics and landscape development on the African continent. In: M. Morisawa and J. T. Hack (eds) *Tectonic Geomorphology*. Allen and Unwin, Boston, and London, 27–51.

Thomas, D. S. G. and Shaw, P. A. (1988) Late Cainozoic drainage evolution in the Zambezi Basin: geomorphological evidence from the Kalahari rim. *Journal of African Earth Sciences* **7**, 611–18.

Tricart, J. (1974) *Structural Geomorphology* (translated S. H. Beaver and E. Derbyshire). Longman, London and New York.

Twidale, C.R. (1971) *Structural Landforms*. MIT Press, Cambridge and London.

17

Sea-Level Change

17.1 Global and regional sea-level change

Sea-level change is a central concern of geomorphologists. Not only does the sea surface determine the base level for erosion, but relative vertical movements of land and sea can greatly alter the area of land exposed to exogenic sub-aerial geomorphic processes. Indeed, over the past 100 Ma the relative proportion of land and sea has changed dramatically. The study of sea-level change is also important because, as we have seen in previous chapters, it can provide key evidence of climatic change and can also give us a benchmark for estimating rates of tectonic uplift.

Satellite technology now enables us to measure the form of the sea surface (the ocean geoid – see Section 2.1.1) to a precision of a few millimetres. These accurate altimetric measurements can be used to monitor changes in sea surface elevation due to tides and currents which occur over short periods (hours, days or months). Sea-level change in a geological sense, on the other hand, occurs on a much longer time scale of thousands and millions of years. These changes in sea level through time can only be determined in terms of movements of the sea surface with respect to adjacent land areas. Consequently, it is shorelines, marking the interface between land and ocean, which are the focus of sea-level change studies. In many cases sea-level changes, which of course involve *vertical* movements of the sea surface, have to be inferred, often indirectly, from *horizontal* shoreline movements. **Transgressions** occur when the shoreline advances landward, whereas **regressions** represent the retreat of shorelines from the land.

A relative change in sea level is a rise or fall in the mean level of the sea surface with respect to the landsurface (or the sea bed). Either the sea surface, or the landsurface, or the two in combination, may rise or fall during a relative sea-level change. A relative rise of sea level is an apparent rise in the mean level of the sea surface with respect to a landsurface and can result from:

1. Sea level rising while the landsurface subsides, remains stationary or rises at a slower rate;
2. Sea level remaining stationary while the landsurface subsides; or
3. Sea level falling while the land surface subsides at a more rapid rate.

A relative fall of sea level is an apparent fall in the mean level of the sea surface with respect to a landsurface and can result from:

1. Sea level falling while the landsurface rises, remains stationary or subsides at a slower rate;
2. Sea level remaining stationary while the landsurface rises;
3. Sea level rising while the landsurface rises at a more rapid rate.

Relative sea-level changes can occur on a regional or global scale, and one of the primary tasks of research into sea-level fluctuations is to separate the regional components from global changes. Global sea-level changes result from mechanisms which affect the level of the sea surface world-wide. Such global changes in sea level are often described as **eustatic**. By contrast, regional changes are (with one important exception – see Section 17.5.3) the result of uplift or subsidence of the land. The importance of eustatic sea-level change was first highlighted in the late nineteenth century by the Austrian geologist Eduard Suess who remarked on the synchronous episodes of deposition of marine sediments around the world evident in the geological record. He interpreted these as being a result of world-wide changes in sea level, inferring that outside the world's orogenic belts the continents had exhibited long-term stability. This eustatic model soon attracted criticism from those who saw abundant evidence for widespread local and regional deformation of the continents, but it served to emphasize the distinction between global and regional causes of relative sea-level change.

17.2   Evidence for sea-level change

In order to unravel the complex history of sea-level changes we have to make use of a wide range of techniques; these involve both fixing the location of a past shoreline and dating it as accurately as possible. Different methods are appropriate to various time scales and, clearly, as we go further back in time the task of reconstructing the record of sea-level change, either locally or globally, becomes increasingly difficult

17.2.1   Quaternary sea level

It is only over the past 100 a or so that sea-level changes have been determined from accurate measurements of the height of the sea surface at tide gauge stations. Averaged across the world these data indicate a current sea-level rise of around 1 mm a^{-1}. In contrast to the direct information from tide gauge records, sea-level changes over longer periods of time must be based on indirect information. Relative falls in sea level are demonstrated by elevated shoreline features such as **raised beaches** which provide deposits which may yield material suitable for dating by radiometric techniques. Vertical sequences of erosional notches can often provide sufficient information to provide a relative chronology of sea-level fall, even if absolute dating is not possible. A relative rise in sea level is indicated by submerged coastal topography and by drowned river valleys or glacial troughs which, in some instances, can be traced seawards as submarine canyons and channel networks. Dating of such submarine features, however, must be based on associated deposits; these may include submerged weathering profiles which may be datable if they contain organic material. In regions affected by Quaternary glaciation it may be possible to identify glacial sediments which, although now below sea level, can be shown to have been deposited sub-aerially.

It has long been appreciated that the fluctuations in ice-sheet volume during the Quaternary have been the primary control of global sea level for at least the past 2 Ma or so. As we saw in Chapter 14, changes in global sea level during the Quaternary have been reflected in the oxygen isotope record preserved in the calcareous shells of microfossils which have accumulated in the sediments deposited on the ocean floor. The recovery of cores drilled at various locations across the world's oceans have thus made it possible to monitor global sea-level fluctuations over the last several hundred thousand years. The oxygen isotope variations provide a first order approximation of global ice volumes and hence global sea-level changes (a 0.01 per cent increase in the ^{18}O concentration indicating an approximately 10 m decrease in global sea level).

The oxygen isotope record can be more accurately calibrated by correlating it with detailed local studies of sea-level change based on accurately dated sedimentological or morphological features. There is abundant evidence of Quaternary sea-level change along the well-studied coast-lines in the mid-latitudes of the northern hemisphere but much of this region has been dominated by local tectonic effects as land areas have subsided and risen isostatically in response to the growth and decay of ice sheets.

Shorelines remote from the direct effects of glacial isostasy are, however, to be found in the tropics and over much of the southern hemisphere and if they occur in regions of active tectonic uplift they may provide a detailed record of global sea-level fluctuations. One notable example of such a tectonically active coast is the Huon Peninsula of Papua New Guinea (see Section 15.2.1 and Figure 15.5). Here there is the ideal combination of an actively rising shoreline (rates of 700–3300 mm ka^{-1}) and datable deposits (coral reef terraces) which range in age up to 120 ka old. As uplift has occurred along the coast each high sea-level stand has cut a new terrace on the shoreline. Since the age of each terrace can be determined radiometrically, the difference in height between present global sea level and these earlier eustatic highstands can be estimated if it is assumed that the rate of uplift has been constant at any given point along the shoreline (Box 17.1).

17.2.2   Pre-Quaternary sea levels

As we go back in time beyond the beginning of the Quaternary, morphological evidence becomes progressively more difficult to relate to particular changes in sea level, since landforms formed at such times have usually been severely modified by erosion; consequently, we have to rely largely on the record preserved by marine sediments.

17.2.2.1   Continental evidence

It is often possible to estimate approximately the depth at which a body of marine sediment was deposited by determining its **facies** – that is, those lithological and organic characteristics imparted by its environment of deposition. Where marine sediments are now exposed sub-aerially the change in relative sea level since deposition can be estimated by utilizing fossil and sedimentological evidence. Marine sediments containing rapidly evolving faunal assemblages can be dated to a resolution of around 1 Ma or less, so a fairly detailed record of sea-level change can be reconstructed by this means. As such, this technique only provides information on regional sea-level changes, but where marine sediments overlie a cratonic terrane which is thought to have been tectonically stable, then any sea-level changes inferred would be of global extent. Our confidence in world-wide sea-level changes is greatly strengthened if it is found that sea-level changes identified in different continents are found to have been synchronous.

An alternative approach to inferring the depth of deposition from the facies of marine sediments is simply to plot

Box 17.1   Estimating global sea-level change from Quaternary marine terrace sequences

Estimating global sea-level change from the record provided by the marine terrace sequence of the Huon Peninsula first requires that we separate the effect of surface uplift along the coast. This is done by using the 120 ka terrace formed during the highstand of the last inter-glacial as a reference point (labelled terrace VII on Figure B17.1(A)). The height of this terrace is determined at various points along the coast and the difference between this height and global sea level at 120 ka BP gives the total amount of uplift at each point during this period. It is known from studies of stable shorelines around Australia that sea level at 120 ka BP was between 2 and 8 m higher than at present. We can therefore calculate the mean rate of uplift for each point along the coast (U) from:

$$U = \frac{(H_T - H_S)}{120}$$

where H_T is the elevation (m) of the 120 ka terrace above present sea level and H_S the height (m) of the 120 ka BP global sea level above the present.

This gives the mean uplift rate in m ka^{-1} which varies from 3.3 to 0.7 along the coast (see Figure 15.5). If we assume that the rate of uplift has been constant over the past 120 ka, then we can calculate the fluctuations in global sea level by relating the actual elevation of each dated terrace to the mean uplift rate of the terrace sequence (Fig. B17.1(B)). Thus the relative sea level (S) indicated by each terrace of known age and present elevation is given by

$$S = h - Ut$$

where h is the elevation above present sea level (m), t the age (ka) of the terrace, and U the mean uplift rate (m ka^{-1}) at that point.

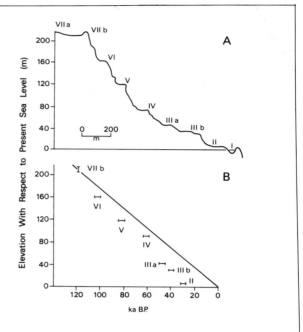

Fig. B17.1   *Typical cross-section of raised coral reef terrace on the Huon Peninsula, Papua New Guinea, with the same reef codes as used in Figure 15.5(A); illustration of the uniform uplift assumption used as the basis for the estimation of the pattern of sea-level change (B). (After J. Chappell (1987) in: M. J. Tooley and I. Shennan (eds) Sea-Level Changes.* Institute of British Geographers Special Publication **20**, Blackwell, Oxford, Fig. 10.11, p. 323.)

the area of the continents covered by marine strata for particular periods of geological time–in effect tracing changes in shorelines through time. If sea level rises relative to an adjacent land area there will be a transgression as the sea encroaches on to the land. Conversely, if sea level falls there will be a regression as the sea withdraws across the continental margin. Again, if such regressions and transgressions are synchronous across several continents we might feel confident in claiming that sea-level change had been of global extent.

One problem with this approach is that the hypsometry of the continents may change through time, and such changes in hypsometry can alter the areal extent of a regression or transgression arising from a given change in sea level. A second problem is that transgressions and regressions are not determined solely by the relative vertical movements of land and sea. Erosion can cause a coastline to recede and thus promote a transgression, while deposition of sediment, especially at the deltas of major rivers, can cause a regression. Such effects must be taken into account in assessing evidence for relative sea-level change and are discussed further in Section 17.5.4.

17.2.2.2   Seismic stratigraphy

Marine sediments deposited along continental margins contain a wealth of evidence which could help us determine the record of long-term sea-level changes, but how do we gain access to this information locked up in offshore sedimentary sequences? One approach is to drill boreholes. These give us a very good idea of the changes in age and characteristics of sediments with depth, but the information they provide is specific to the particular drill site and the cost of drilling offshore is very high. Although boreholes have been used extensively by oil companies in their search for oil along the world's continental margins, it is the application of seismic techniques that has revolutionized our knowledge of offshore stratigraphy by providing indirect information on submarine sedimentary sequences.

If sound waves are transmitted downwards from a ship and the time taken for them to bounce off the sea bed and return is measured it is possible to use the variations in 'echo' time to map the morphology of the sea floor. This is the well-known technique of echo-sounding. Using more powerful but lower frequency seismic waves, which can partially penetrate solid rock, it is possible to locate the boundaries, or discontinuities, between different sedimen-

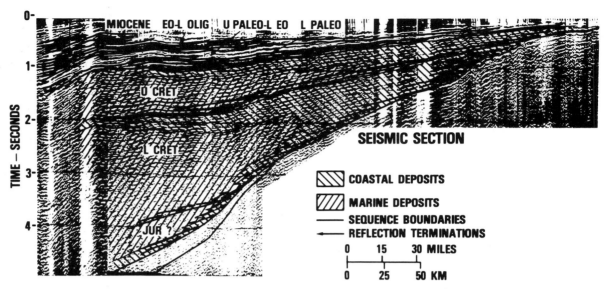

Fig. 17.1 *Offshore seismic section from the continental margin of north-west Africa. The section shows strata of probable Jurassic age laid down at an early stage of continental rifting overlain by thick marine and coastal sequences of Cretaceous sediments. The vertical scale is expressed in terms of the transmission time (in seconds) for the seismic waves to travel through the strata. The equivalent total depth of the section is around 12 km. (From P. R. Vail et al. (1977) in C. E. Payton (ed.)* Seismic Stratigraphy – Applications to Hydrocarbon Exploration. *American Association of Petroleum Geologists Memoir* **26** *Fig. 5, p. 69.)*

tary units from differences in the velocity with which the seismic waves travel through them. As a ship tracks across the ocean, seismic profiles are generated which, after computer processing, enable the underlying sedimentary structures and sequences to be identified, a technique known as seismic stratigraphy (Fig. 17.1; see Section 4.3.1).

The echo times from these discontinuities between sedimentary units, termed **seismic reflectors,** can be converted to a depth if we know (or can reasonably estimate) the velocity of seismic waves through the rock. Most seismic reflectors recorded in sedimentary sequences appear to be surfaces which, prior to being buried, formed the sea bed or an erosion surface on land. Such surfaces are therefore **unconformities,** representing depositional breaks caused by erosion or lack of deposition. Since most unconformities can be assumed to represent surfaces of uniform age they can be regarded as **chronostratigraphic markers.** The actual age of surfaces represented by seismic reflectors can only be determined by correlating them with borehole sequences which have been dated.

How, then, can seismic stratigraphy contribute to our knowledge of sea-level change? The most active researchers involved in the interpretation of data from seismic stratigraphy in terms of sea-level change have been based in the oil industry, and foremost among these has been the group at the Exxon Corporation originally led by P. R. Vail. This group has argued that processes of sub-aerial erosion and submarine deposition operate relative to a base level which is determined by the position of the sea surface. A relative change in sea level leads to a change in base level which is

reflected in the lateral and vertical displacement of sedimentary units (continental, littoral and marine) across the continental shelf and continental slope (Fig. 17.2). Since, as we have already noted, most seismic reflectors separating these different sedimentary units represent ancient sea floors, the highest point on each reflector indicates the sea

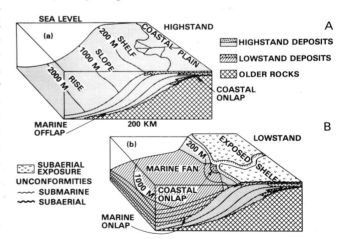

Fig. 17.2 *Typical depositional patterns (highly simplified) observed on seismic sections of passive continental margins: (A) the prograding sedimentary sequences associated with high relative sea level; (B) the onlap of marine deposits on to earlier sedimentary sequences during a period of low sea level when much sediment is carried beyond the edge of the continental shelf (From P. R. Vail et al. (1977) in C. E. Payton (ed.)* Seismic Stratigraphy – Applications to Hydrocarbon Exploration. *American Association of Petroleum Geologists Memoir* **26** *Fig. 9, p. 73.)*

level at the time it was formed. By correlating records of relative sea-level change recorded on different continental margins, the Exxon group have constructed what they believe to be a record of world-wide sea-level change throughout the Cenozoic and Mesozoic (see Section 17.4).

The detailed application of seismic stratigraphy is far more complex than has been indicated here because several factors – including variations in rates of sediment supply and removal and rates of relative sea-level change – can greatly complicate the relationship between the position of seismic reflectors and sea-level change. It also needs to be emphasized that many researchers have been extremely sceptical about certain aspects of the Exxon global sea-level curve. Apart from alternative interpretations of the data, the curve has some remarkable features which challenge the conventional wisdom about the mechanisms of sea-level change and some of the original data from which it has been constructed have been withheld for commercial reasons. None the less, the Exxon curve has been widely adopted as the best available representation of the record of global sea-level change, especially since some of the criticisms of its original presentation have been met by the publication in 1987 of a revised version in which the interpretation from seismic stratigraphy has been integrated with the record of sea-level change determined from marine sediments now exposed on land.

17.3   The record of Quaternary sea-level change

The *general* trend of Holocene sea-level change is well established and is clearly related to melting of the northern hemisphere ice sheets which began about 15 000 a BP. This led to an initially rapid rise in global sea level at rates of up to 12 000 mm ka^{-1}, which declined significantly from about 8000 a BP. The details of Holocene sea-level change have, however, been more difficult to resolve, especially for the past 8000 a. It has been variously argued that during this period global sea level has risen continuously but slowly, that it rose up to about 3600 a BP. but has subsequently been more or less stable, and, finally, that it rose to about 2 m above present sea level around 6000 a BP before declining. A significant increase in data on sea-level changes in the past few years, particularly in the tropics, has made it clear that there have been *real* differences in the Holocene 'eustatic' sea level record in different parts of the world and that these discrepancies cannot be entirely attributed to errors in dating or local tectonic effects. Moreover, as we shall see in Section 17.5.3, we now have an elegant explanation of how these variations are produced.

For the Quaternary as a whole the deep-sea ocean core record now provides the framework for studies of global sea-level change (Fig. 17.3; see Section 14.3.2). Major deglaciation events associated with sea-level highstands corre-

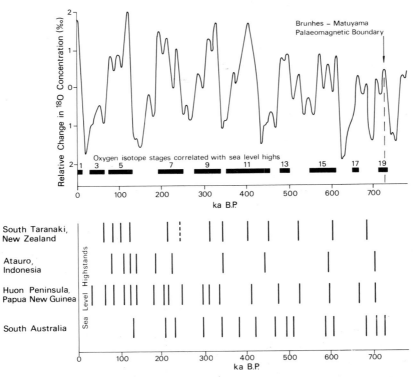

Fig. 17.3   *A composite record of variations in ^{18}O concentrations from ocean floor sediments for the past 700 ka compared with the chronology of highstands at various locations unaffected by glacio-isostatic rebound. (Based on J. Imbrie et al. (1984) in: A. Berger et al. (eds) Milankovitch and Climate. Reidel, Dordrecht, Vol. 1; and B. Pillans (1987) in: R. J. N. Devoy (ed.) Sea Surface Studies: A Global View. Croom Helm, London, Fig. 9.10, p. 281.)*

late with oxygen isotope stages 1, 5, 7, 9, 11, 13, 15, 17 and 19, while stage 3 represents a partial deglaciation accompanied by moderately high sea levels. The Brunhes–Matuyama palaeomagnetic boundary, which is identifiable in these cores, has been accurately located by K–Ar dating at 730 ka B P and provides an important reference point in the mid-Pleistocene. Since the major deglaciation events of the Pleistocene have a broadly similar isotopic signal, we can infer that the changes in ocean water volumes associated with earlier glacial–interglacial cycles were similar to the most recent transition from a glacial to interglacial climatic regime. Consequently, it is likely that the magnitude of sea-level change was also similar.

The correlation between the ocean-core record of global sea-level change and evidence of sea-level change from well-dated coral terrace sequences in the tropics is fairly good back to about 400 ka BP. In addition to the Huon Peninsula sequence, Barbados in the Caribbean provides an excellent location for reconstructing Late Pleistocene global sea-level change. It is tectonically active and has been rising rapidly at up to 500 mm ka^{-1} over the past several hundred thousand years. Fifteen separate coral terraces have been dated up to an elevation of 200 m above present sea level with a maximum age of 650 ka BP. The terrace sequence appears to indicate six or seven complex interglacial sea-level high-stands which correlate fairly well with oxygen isotope stages 5–17. Moreover, the 120 ka terrace from the Huon Peninsula has an equivalent of the same age in Barbados. Beyond about 400 ka BP. the correspondence between the ocean floor record and sea-level records from coastlines around the world is

more variable. Much of this discrepancy is probably due to errors in dating (especially in the ocean floor record). Dating errors for shoreline features can be as great as 50 ka at 370 ka BP, so correlations are fraught with difficulty once we go back beyond the past 200 ka or so.

17.4   The record of pre-Quaternary sea-level change

Most discussion of pre-Quaternary sea-level change is currently focused on the global record presented by the Exxon group. Figure 17.4 illustrates the most recent version of their global sea-level curve for the Cretaceous to the present and shows fluctuations at a variety of scales. The long-term rise evident for the Early Cretaceous was preceded by a prolonged period of generally low sea levels extending from the Late Palaeozoic some 320 Ma BP until the Late Jurassic about 150 Ma BP. Since the mid-Cretaceous there has been a fairly persistent trend of falling sea level, although estimates of the amount of this fall vary. The original analysis by the Exxon group used an estimate of a Late Cretaceous maximum some 350 m above present sea level, but their revised curve uses a more modest estimate of 250 m. Perhaps the most remarkable features of the Exxon curve are the dramatic sea-level falls evident throughout the Mesozoic and Cenozoic, most notably the very rapid Late Oligocene fall of around 150 m.

Most researchers concerned with global sea-level change

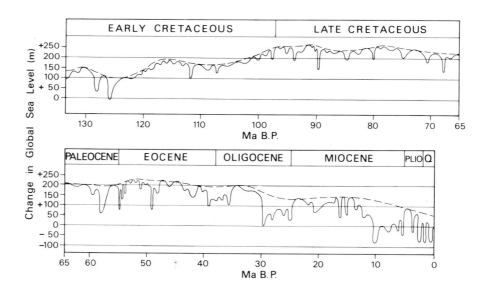

Fig. 17.4 *Global sea-level curves from the beginning of the Cretaceous produced by the Exxon group largely on the basis of seismic stratigraphy. The long-term trend is shown by a dashed line with shorter-term fluctuations being indicated by a solid line. (Modified from B. U. Haq et al. (1987)* Science *235, Figs 2 and 3, pp. 1159 and 1161.)*

have had little difficulty in accepting the general validity of the long-term trends depicted on the Exxon curve because they accord with other evidence of high global sea level in the Late Cretaceous (primarily the wide extent of marine strata of this age which outcrop across North America and Europe). It is the short-term fluctuations that have attracted criticism, it being argued, for instance, that the rate and magnitude of sea-level changes exhibited by the extremely rapid falls in the Exxon curve can only be explained by extensive glaciations. The major problem here is that there is little evidence for glacial episodes over the past 200 Ma until the mid-Cenozoic (but see Section 14.3.1). It has also been suggested that some of the short-term fluctuations may not be of worldwide extent, but may simply be an artefact of the bias arising from the large proportion of seismic data collected from passive margins with a similar history of subsidence. These criticisms are crucial to an understanding of long-term global sea-level change, and both can only be assessed with reference to possible mechanisms of sea-level change.

17.5 Mechanisms of global sea-level change

Our knowledge of global sea-level fluctuations over the past 100 Ma or so indicates that they have involved total movements in excess of 200 m and have occurred over a wide range of time scales. What, then, causes these global sea-level changes? There are several possible mechanisms, but they can be grouped into three basic types – those that alter the volume of water in the world's oceans, those that change the volume of the ocean basins and those that produce changes in the geoid (Fig. 17.5).

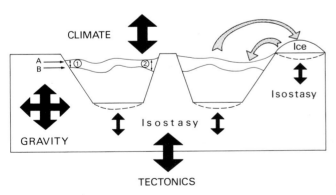

Fig. 17.5 *Schematic representation of the major types of controls of global sea-level change: (A) and (B) represent two sea levels which are associated with different amounts of change in different locations due to geoidal effects (see Section 17.5.3).*

17.5.1 Changes in ocean water volume

By far the most significant factor influencing the amount of water in the oceans is the waxing and waning of land-based ice sheets (ice shelves, pack ice and icebergs do not count as they displace their own mass of sea water). We have already referred to this mechanism, known as **glacio-eustasy**, in relation to Quaternary sea-level fluctuations. Using the total area of the oceans (3.6×10^8 km²), the density of sea water (1030 kg m⁻³) and the density of glacier ice (920 kg m⁻³), changes in global sea-level can be calculated if the volume of terrestrial ice that melts or accumulates is known. The volumes of present-day ice sheets are difficult to estimate accurately, and of course estimates for the extensive ice bodies of the Late Cenozoic are even less certain. Nevertheless, rough calculations can be made. A rise of 60–75 m would be produced by the melting of the present Antarctic ice sheet, and a further few metres would be added by the dissipation of the Greenland ice sheet, giving a total rise of bewteen 65 and 80 m. Melting of other land-based ice bodies would contribute only a trivial additional rise in sea level. This does not give the ultimate change in global sea level, however, because we have to take into account the effects of isostasy. The additional water added to the oceans will depress the crust of the ocean floor and the newly flooded continental margins, a process known as **hydro-isostasy**. This would reduce the overall rise in sea level to between 40 and 50 m for the melting of present-day ice sheets.

Various attempts have been made to calculate the equivalent changes for the growth and decay of the Pleistocene ice sheets. The limits of Pleistocene glaciations are fairly well known, and the form of ice sheets, and hence their approximate volumes, may be inferred from these limits. It is not certain, however, whether the major ice sheets reached their maximum size simultaneously. If we assume, for simplicity, that they did, then the maximum sea-level fall for the most extensive well-documented glacial (correlated with oxygen isotope stage 6 – about 120–190 ka BP) was probably of the order of 150 m, or about 100 m if an allowance is made for isostatic compensation. The rather less extensive last glacial probably produced a more limited maximum fall of around 115 m, or 75 m assuming isostatic compensation. These figures, it must be emphasized, are only estimates and some of the evidence suggests somewhat more extensive ice sheets and hence rather larger sea-level falls of perhaps as much as 175 m. None the less, an estimate of Quaternary glacio-eustatic changes of a maximum of 150 m (isostatically uncompensated) is probably not far from the mark.

The rates of Quaternary sea-level change are in accordance with likely rates of ice sheet growth and decay. The rapid sea-level rise between 15 000 and 6000 a BP attained a maximum rate of 12 000 mm ka⁻¹ and would have required a sustained melting of the predominantly northern hemisphere ice sheets at a rate of around 5000 km³ a⁻¹. Ice sheet growth at a similar rate is indicated by the rapid sea-level fall prior to 18 000 a BP. These rates of ice sheet growth and decay may seem unrealistic until we remember that some $1100 \, \text{km}^3 \, \text{a}^{-1}$ of water equivalent is presently added to the Antarctic ice sheet. This annual accumulation is currently more or less balanced by ablation but the potential for rapid volume changes with fluctuations in accumulation and ablation rates is clear. Consequently, we have no difficulty in accounting for very rapid global sea-level changes of a magnitude of up to 150 m during periods when the Earth is experiencing major glacial–interglacial cycles, but the problem remains of explaining the very rapid falls evident in the Exxon sea-level curve during the Mesozoic and Early Cenozoic.

Other factors influencing ocean-water volume involve changes of only 10 m or less. It is well known that the temperature of the oceans has changed frequently throughout the Phanerozoic, and such changes are fairly well documented for the Quaternary. Through its influence on the density of ocean water, and hence its volume, temperature fluctuations can be reflected in changes in sea level. Such changes, however, will be comparatively small; even an increase of 10 °C throughout the entire depth of the oceans would only produce a 10 m rise in global sea level.

17.5.2   Changes in ocean basin volume

Changes in the volume of the ocean basins were almost certainly the most significant factor in influencing the broad trend of global sea level throughout the Mesozoic and Early Cenozoic. Such changes can arise from a variety of causes, including orogenesis, the isolation and desiccation of small ocean basins, sedimentation and variations through time in the volume of the world-wide mid-oceanic ridge system. Estimates of the likely effects on global sea level of other mechanisms capable of altering the volume of the ocean basins indicate that these are probably of less significance. Extensive upwarps of the ocean floor related to concentrations of hot spots have probably had some effect on global sea level as they develop and subsequently decay. One suggestion is that massive mid-plate volcanism in the western Pacific between about 110 and 70 Ma BP may have led to a global sea-level rise of between 40 and 100 m. Changes in the hypsometry of the continents as a consequence of the break-up of Pangaea may also have caused major changes in global sea level with a possible rise of 130 m between about 140 and 50 Ma BP and a small fall of 10 m since then.

17.5.2.1   Variations in the volume of mid-oceanic ridges
As the volume of the world mid-oceanic ridge system is equivalent to about 12 per cent of the total volume of ocean water, changes in the overall ridge system volume through time are clearly a potentially important mechanism of global sea-level change. Variations in mid-oceanic ridge volume can arise both from alterations in the total length of the

ridge system world-wide as a result of the periodic reorganization of plate boundaries, and from changes in the rate of sea-floor spreading. Of these two mechanisms it is probably changes in sea-floor spreading rates that have been most significant, at least since the Late Cretaceous.

The effect of fluctuations in sea-floor spreading rates can be estimated on the basis of the age–depth relationship exhibited by oceanic lithosphere (Fig. 17.6). After it is formed new oceanic lithosphere cools, thickens, increases in density and gradually subsides. Where spreading rates are slow, cool lithosphere and therefore deep ocean occurs close to the ridge axis (Fig. 17.7); conversely, where spreading rates are high, the lithosphere is still relatively warm and buoyant some distance from the ridge axis and ocean depths here are consequently fairly shallow. If the spreading rate changes there will be a consequential change in ridge volume with an increase in spreading rate producing an increase in ridge volume and a rise in global sea level, and a decrease in spreading rate giving rise to a decrease in ridge volume and a sea-level fall. It is important to note, however, that the *rate* at which sea level can change as a result of variations in spreading rate is limited by how quickly oceanic lithosphere cools.

Evidence for this mechanism is provided by the correlation between periods of accelerated spreading rates, as indicated by the age of the ocean crust astride mid-oceanic ridges, and global sea-level rise. In particular, the major Late Cretaceous sea-level high coincides with a well-documented phase of rapid sea-floor spreading. Although there are uncertainties as to the exact magnitude and timing of changes in spreading rates, quantitative estimates can be made of the effect on global sea level. Initial calculations

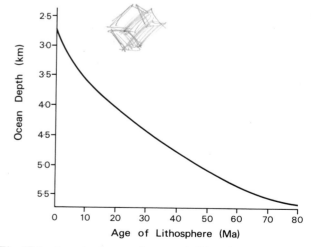

Fig. 17.6   *Age–depth curve for oceanic lithosphere. The process of subsidence with increasing lithospheric age and density is relatively simple, and for the first 70 Ma or so the increase in depth is proportional to the square root of the age of the lithosphere. (After W. C. Pitman, (1978) Geological Society of America Bulletin* **89**, *Fig. 1, p. 1390.)*

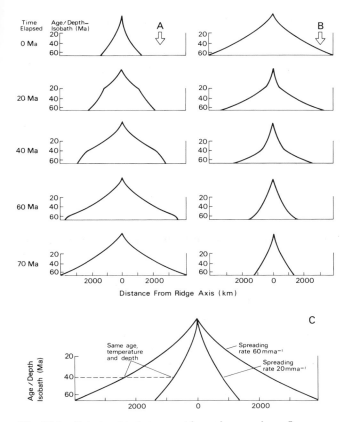

Fig. 17.7   *Relationship between ridge volume and sea-floor spreading rate. (A) The cross-section of a mid-oceanic ridge that has been spreading at 20 mm a⁻¹ for 70 Ma. At time 0 Ma the spreading rate is increased to 60 mm a⁻¹. The sequence of profiles shows the increasing cross-sectional area at 20, 40, 60 and 70 Ma after the increase in spreading rate. By 70 Ma the ridge has reached its new equilibrium profile which has three times the cross-sectional area of the ridge 70 Ma earlier. (B) The converse situation with a ridge that has been spreading at 60 mm a⁻¹ for 70 Ma. The spreading rate is then decreased at time 0 Ma to 20 mm a⁻¹. The sequence of profiles shows the decreasing cross-sectional area at 20, 40, 60 and 70 Ma after the decrease in spreading rate. By 70 Ma the new equilibrium profile of the ridge has attained a cross-sectional area only one third that of the ridge at time 0 Ma. (C) Cross-sections of ridges which have spread at 20 mm a⁻¹ and 60 mm a⁻¹ superimposed. (Adapted from W. C. Pitman (1978) Geological Society of America Bulletin 89, Fig. 2, p. 1391.)*

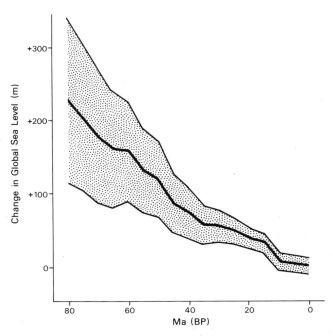

Fig. 17.8   *Change in global sea level since 80 Ma BP due to changes in mid-oceanic ridge volume and corrected for post-Miocene ice-volume change. The thicker line shows the best-estimate curve and the shaded area indicates the probable range of error in the estimates. (Modified from M. A. Kominz (1984) in: J. S. Schlee (ed.) Interregional Unconformities and Hydrocarbon Accumulation. American Association of Petroleum Geologists Memoir 36, Fig. 7, p. 121.)*

American craton (which appears to have been essentially stable since at least the mid-Cretaceous).

At certain times in the past, variations in the length of the world mid-oceanic ridge system may have been more important than changes in spreading rates in altering the volume of the ocean basins. Some of the rise through the Early Cretaceous may have been related to the extension of the world ridge system arising from the break-up of Gondwana. The modest sea-level rise in the Eocene, which occurred during a period of reduced rates of sea-floor spreading, may similarly have been a result of the development of ridge systems in the North Atlantic.

17.5.2.2   Sediment accumulation on the ocean floor
Sediment supplied to the oceans from the continents would, on its own, decrease ocean volume and therefore raise sea level. With current rates of sediment supply this would produce a rise of between 15 and 40 mm ka⁻¹, but we must take into account two compensating factors. These are the isostatic depression of the ocean floor as a result of the sediment load and the removal of sediments from the sea floor, either through uplift or subduction. Short-term changes in global sea level of a few tens of metres could possibly occur if these two factors do not exactly compensate for sedimentation.

by W. C. Pitman indicated a fall of global sea level from about 350 to 50 m above present sea level from 85 to 15 Ma BP and these data were used to calibrate the initial Exxon curve. Subsequently, more detailed calculations have highlighted the large errors inherent in estimating the effects of changing ridge volume and have indicated a rather smaller best-estimate for global sea-level fall since the Late Cretaceous of about 230 m (Fig. 17.8) (although the revised Exxon curve uses a slightly higher figure of about 250 m). These estimates are in fairly close agreement with the 270 m fall recently calculated independently from the elevation of sediments known to have been deposited near sea level on the North

Changes in global sea level over greater time scales could result from long-term variations in the amount of sediment supplied to the oceans from the continents as a consequence of climatic change or tectonic factors. Another factor affecting global sea level has been the apparently sharp increase in rates of carbonate accumulation in the ocean basins since the mid-Cretaceous (as a result of the more active growth of carbonate-secreting marine organisms). This may have resulted in a 300 m decrease in average ocean depth which would produce a global sea-level rise of about 55 m after isostatic compensation.

A further effect of sedimentation rates is related to sea-floor spreading. Since sea-floor spreading rates over the past few million years have been slow in comparison with the Cretaceous, the average age of the oceanic lithosphere is probably much greater now than around 100 Ma BP. Assuming constant rates of sedimentation this means that there would be proportionately less sediment in the ocean basins during the Cretaceous than today. It has been estimated that this extra sediment could produce a sea-level rise of 15 m and that the combination of all three factors gives a net rise of global sea level of 77 m since the mid-Cretaceous. As the volume of the mid-oceanic ridges has been decreasing during this period this extra sediment would have had the effect of reducing the expected fall in sea level by about 30 per cent.

17.5.2.3   Effects of orogeny

As orogeny is frequently associated with crustal shortening, which gives rise to both a thickening of the continental crust and a reduction in continental area, it can potentially have a role in global sea-level change. The collision of India and Eurasia leading to the formation of the Himalayas and the Tibetan Plateau represents the most significant orogenic event since the Palaeozoic. Assuming that the Tibetan Plateau is formed of continental crust of twice the average thickness, and that the reduction in continental area arising from the crustal shortening is replaced by ocean of average depth, then the resulting global sea-level fall would be about 26 m, although isostatic compensation would reduce this to 18 m. Other Mesozoic and Cenozoic orogenic events have probably had effects an order of magnitude smaller than this.

17.5.2.4   Desiccation of small ocean basins

The idea that the drying out of small ocean basins could occur and lead to a rapid, albeit rather modest, change in global sea level is a rather recent one. In the early 1970s K. J. Hsü cited the existence of thick evaporite deposits in the sediments of the Mediterranean, and the presence of deep submarine gorges extending from the mouths of major rivers, such as the Rhône and the Nile, as evidence for the evaporation of the entire Mediterranean Sea around 5 Ma BP. The resulting basin, looking something like a giant Death Valley, would have been about 2800 km in length, 850 km wide and up to 3 km deep. The water evaporated during such a desiccation event would have eventually been returned to the world's oceans and would have initially produced a global sea-level rise of about 15 m, which would have been reduced to about 10 m after the ocean floor had adjusted isostatically to the increased water load.

The initial isolation of the Mediterranean may have been caused by localized uplift around the Straits of Gibraltar where it is connected to the Atlantic Ocean, or it may have been due to a global sea-level fall, perhaps related to the accelerated growth of the Antarctic ice sheet in the Miocene. Although there are alternative possible explanations for the occurrence of evaporites on the floor of the Mediterranean, the additional presence of aeolian sediments seems to be conclusive evidence in favour of the desiccation hypothesis. Subsequent research has indicated that the Mediterranean may have dried out on several separate occasions between 5.5 and 5 Ma BP, with the Straits of Gibraltar periodically acting as a giant cataract as the Mediterranean Basin was refilled after each desiccation event. While the connection with the Atlantic Ocean was completely severed the likely rate of evaporation of 4000 km^3 a^{-1} could have reduced the Mediterranean to an arid basin in about 1000 a. Given these enormous evaporative losses it is likely that some 40 000 km^3 a^{-1} of water (about 1.25×10^6 m^3 s^{-1} or around seven times the mean annual discharge of the Amazon River) must have cascaded through the Straits of Gibraltar during the reflooding episodes.

Similar desiccation events may have occurred in the past, particularly during the early stages of continental break-up. It has been suggested that periodic drying out of the very large isolated ocean basins of the nascent South Atlantic in the Early Cretaceous, which is indicated by the presence of thick evaporite deposits, may have produced rapid global sea-level rises of up to 60 m.

17.5.3   Geoidal effects

We noted in Section 17.5.1 how changes in the volume of water in the oceans cause the crust to deform as it adjusts isostatically to the alterations in load. These crustal deformations, however, are not confined to simple vertical movements; they also involve a continuous redistribution of mass between and within the ocean basins. The record of relative sea-level change observed at any point on the Earth's surface consequently represents the net effects of a series of complex interactions between ocean water, ice masses and both lateral and vertical adjustments in the solid Earth. In the 1970s geophysicists working in conjunction with geomorphologists developed a mathematical model which predicted how these interactions would manifest themselves in different patterns of relative sea-level change around the

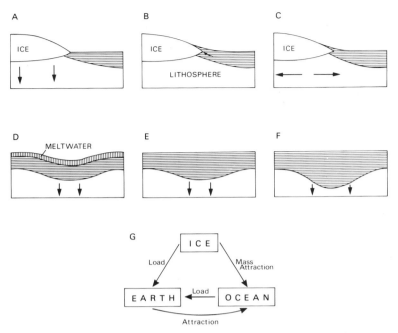

Fig. 17.9   *Schematic representation of the sequential adjustments to the geoid arising from ice sheet growth and decay and the associated changes in the volume of water in the world's oceans. The weight of ice depresses the crust (A) and the ice mass attracts the water (B); the resulting transfer of matter within the Earth further distorts the geoid (C). When the ice mass melts, the weight of meltwater depresses the Earth differentially (D) and more water flows into the depression, thereby increasing the water load (E); these changes lead to further deformation of the ocean floor and further transfers of matter within the Earth with additional adjustments to the geoid (F). The interactions between ice loads, water loads and deformation of the Earth are also illustrated (G). (Based on J. A. Clark et al. (1978) Quaternary Research **9**, Fig. 1, p. 266.)*

world during the Holocene (Fig. 17.9). This model predicts six zones across the world's oceans, each characterized by a different history of Holocene sea-level change as a result of both isostatic and geoidal effects (Fig. 17.10).

Prior to the development of this model it was widely thought that a *single* eustatic curve existed which described the changes in sea level brought about by the waxing and waning of ice sheets in all areas remote from the main ice sheets themselves (and consequently free from the effects of glacio-isostasy – see Section 14.4.1.3). The presence of raised beaches of Holecene age in regions remote from the sites of Pleistocene ice sheets were attributed to occasional intense storms, localized tectonic uplift or simply erroneous dating, rather than being viewed as evidence of real differences in sea-level history. It now appears that the geoidal model is able to explain such emergences and account for the slightly different patterns of Holocene sea-level change recorded around the world.

More extreme effects on sea level have also been attributed to changes in the geoid through time than those associated with the changes in ice and water loads occurring during glacial–interglacial cycles. The present total relief of the geoid is around 180 m (Fig. 2.1), and presumably the geoid must change, at least to some extent, through time, although the likely magnitude and rate of this change is unknown.

Geoidal changes may generate sea-level fluctuations on their own but may also superimpose themselves on sea-level movements produced by other mechanisms. If major changes in the geoid have occurred over geological time then this would greatly complicate the interpretation of fluctuations in global sea level. However, this mechanism is extremely difficult to verify with respect to the pre-Quaternary history of global sea-level change, and although it may be of significance its actual effects on sea-level change are unproven.

17.5.4   Explaining the long-term record of global sea-level change

Any proposed mechanism of global sea-level change must obviously accord with the rates and magnitudes of sea-level fluctuations suggested by the evidence. The only mechanisms which it is generally accepted can cause global sea-level changes in excess of 100 m are the growth and decay of major ice sheets and changes in the volume of the world mid-oceanic ridge system (Table 17.1). We might add long-term changes in the geoid or global hypsometry, but the importance of these mechanisms and the magnitude of sea-level changes that might be associated with them is very

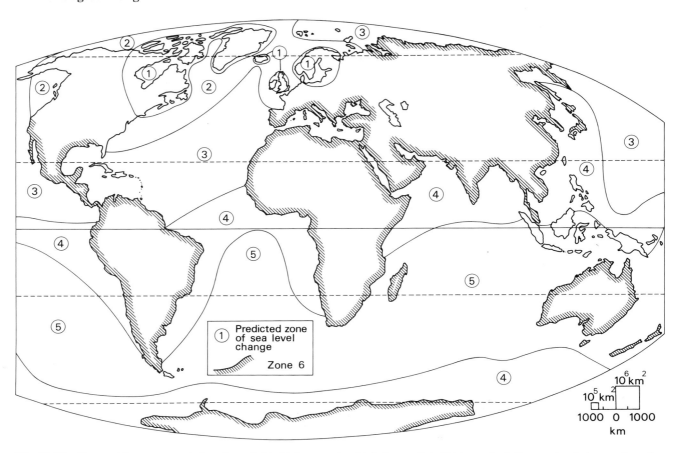

Fig. 17.10 *The world's oceans and shorelines divided into zones in which the predicted history of relative sea-level change since the beginning of the last deglaciation has a different form. Zone 1 comprises regions originally covered by ice in which isostatic uplift of the initially ice-covered land outpaces the global sea-level rise but at a declining rate. This produces a continuous fall in relative sea level which becomes progressively less rapid with time. Zone 2 is characterized by a continuous and rapid rise in relative sea level due to the progressive rise in global sea level and crustal downwarping as a result of the collapse of the ice marginal crustal upwarp (forebulge). In the transitional region between zones 1 and 2 there is an initial relative sea-level fall followed by a later rise. For zone 3 the predicted pattern is of an initial rise in relative sea level on deglaciation followed by a slight fall, giving maximum sea levels 0.5 m above the present. This pattern again results from the interaction of the gradually subsiding ice marginal forebulge and the redistribution of the water load on the Earth's surface following deglaciation. In this zone the model does not exactly fit the field evidence since many areas show a continuous sea-level rise. Zone 4 is characterized by a continuous sea-level rise while in zone 5 a small relative fall in sea level occurs immediately after ice sheet melting has ceased. Finally, in zone 6, which comprises all continental shorelines removed from ice sheet influences, a small relative sea-level fall is predicted as deglaciation is completed. This is a consequence of the depression of the ocean basins and submerged parts of the continental shelf, resulting from the increased load contributed by meltwater which causes a crustal tilt across the continental margins and a slight emergence of shorelines. (Based on W. R. Peltier (1987) in R. J. N. Devoy (ed.)* Sea Surface Studies: A Global View. *Croom Helm, London, Fig. 3.12, p. 77.)*

uncertain. Other mechanisms of fairly well known but more limited significance include orogenesis, the desiccation of small ocean basins, and the effects of sedimentation.

What, then, can we say about the rate of sea-level change? Glacio-eustasy operates so rapidly that it is effectively instantaneous on geological time scales, but of the other mechanisms of global sea-level change only ocean basin dessication generates changes at a comparable rate. All the other mechanisms cause sea-level changes which are slower by at least three orders of magnitude.

We seem to have no difficulty in explaining both the rates and magnitudes of changes in global sea level which have occurred during the past few million years of an ice-

age Earth. The problem arises when we try to account for the large, rapid sea-level changes which characterize the whole of the Exxon sea-level curve. Even in its more recent modified form it contains numerous dramatic sea-level falls throughout the apparently ice-free Mesozoic and Early Cenozoic. Perhaps the most remarkable of these is the Late Oligocene fall of around 150 m about 30 Ma BP. This seems to have occurred in less than 1 Ma at an average rate of 150 mm ka^{-1}. This is slow by the standards of glacio-eustasy, but far too rapid to be accounted for by changes in mid-oceanic ridge volume.

One possibility is that ice sheets have existed for a much greater proportion of geological time than previously thought

Table 17.1 Approximate magnitude and rate of global sea-level change associated with various eustatic mechanisms

Mechanism	Magnitude (m)	Rate (mm ka^{-1})	Confidence in estimate	Comments	Data sources for** magnitude estimates
Changes in ocean water volume					
Glacio-eustasy	<150*	<12 500	High		Donovan and Jones (1979)
Change in mean ocean temperature of 10 °C	10	Slow	High		Donovan and Jones (1979)
Changes in ocean basin volume					
Change in volume of mid-oceanic ridges	<230	<10	Moderate		Kominz (1984)
Imbalance between inorganic sediment deposition and removal	<7	<1.5	Moderate	Over past 5 Ma	Harrison *et al.* (1981)
Relationship between sea-floor spreading rates and mean sediment thickness	<15	<0.15	Moderate	Over past 100 Ma	Harrison *et al.* (1981)
Variations in rate of organic carbonate accumulation	55	0.55	Moderate	Over past 100 Ma	Harrison *et al.* (1981)
Orogenesis–formation of Himalayas and Tibetan Plateau	18	1.6	Moderate		Harrison *et al.* (1981)
Desiccation of Mediterranean Sea	10*	<15 000	High		Donovan and Jones (1979)
Hot-spot-related ocean-floor deformation	<100	Very slow	Low	110–70 Ma BP	Schlanger *et al.* (1981)
Changes in continental hypsometry	<130	Very slow	Low	140–50 Ma BP	Wyatt (1986)

* Isostatic compensation for change in water load included in estimate.
** Full source details can be found in the list of references at the end of the chapter.

and so can explain rapid global sea level changes before the Late Cenozoic. This may be true for the Early Cenozoic, but the palaeoclimatic evidence strongly indicates a rather uniformly warm ice-free Earth in the Mesozoic. The key problem is to determine when ice sheets first became established on Antarctica as we are fairly certain that major ice sheets did not become established in the northern hemisphere until the Late Pliocene. The beginning of glaciation on Antarctica is very uncertain, but the weight of evidence is currently in favour of it beginning between 45 and 20 Ma BP; the most recent evidence from the Ocean Drilling Program indicates that there was glacial activity at sea level in East Antarctica by 35 Ma BP (see Section 14.3.1).

A possible alternative explanation is that most of the rapid sea-level falls recorded in the Exxon curve are in fact regional regressions and do not indicate changes in global sea level. In order to understand how this might occur we have to look at the relationship between sedimentation, continental margin subsidence and relative sea-level change. As we noted in Section 17.2.2.1, whether a shoreline moves landward (transgression) or seaward (regression) depends not just on the relative movement of land and sea but also on the rate of deposition. Along a continental margin which is gradually subsiding and is receiving a more or less constant supply of sediment, the shoreline will be located where the rate of sea-level rise is balanced by the rate of deposition. If the rate of sea-level rise slows down then the shoreline will move seaward, causing a regression, since sedimentation will outpace the now slower rate of sea-level rise. Conversely, if the rate of sea-level rise increases the shoreline will move landward, producing a transgression as the rate of sea-level rise will exceed the rate of deposition.

We have been talking here about sea-level rise, although it is clear that the trend over the past 80 Ma or so has been for *global* sea level to fall. Nevertheless, if we look at many of the world's continental margins we find that a significant proportion of them are passive margins formed by the break-up of Gondwana and Laurasia. As they have cooled and become loaded by sediment offshore, these margins have been progressively subsiding (see Sections 4.4 and 4.5). Consequently most passive continental margins have experienced a *relative* sea-level rise over the past 100 Ma or so even though global sea level has been falling. Significantly, because of their potential for oil exploration, a large proportion of the data used to construct the Exxon sea level curve has come from these subsiding passive margins. It is possible, then, that all we need to explain the rapid regressions in the Exxon sea level curve are variations in the rate of global sea-level fall. This in turn would give rise to variations in the rate of relative sea-level rise along subsiding passive margins and consequently a succession of regressions and transgressions.

17.6   Sea-level change and landscape development

17.6.1   Drainage basin response

Relative changes in sea level have both direct and indirect effects on drainage basins. The direct effects are the adjustments of drainage systems to the changes in base level and the increase or decrease in drainage basin area brought about by a relative rise or fall in sea level. The indirect effects arise from the climatic changes associated with alterations

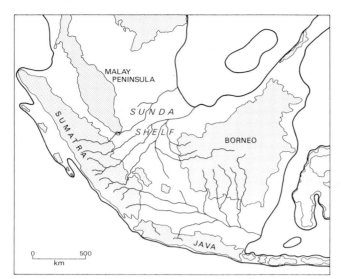

Fig. 17.11   *The geography of the Sunda Shelf in south-east Asia during the sea-level lowstands of the Pleistocene. The coastline is drawn in the position it would occupy if sea level were 100 m below that of the present. Much of the Sunda Shelf is dry land and is drained by rivers originating in present-day Sumatra and Borneo. (The reconstruction of the drainage is partly based on J. H. F. Umbgrove (1942) The Pulse of the Earth. Martinus Nijhoff, The Hague, Fig. 83, p. 131.)*

in the relative area of land and sea. A relative rise in sea level reduces the drainage basin area, whereas a relative fall can produce a marked downstream enlargement, especially where there is a shallow continental shelf offshore. In such cases the significant increase in land area is likely to lead to greater aridity in the continental hinterland and a consequential reduction in runoff. The sea-level fluctuations of the Late Cenozoic caused major shifts in drainage area and the periodic submergence of extensive drainage systems in the relatively brief interludes of high interglacial sea level such as we are experiencing at the present day (Fig. 17.11).

   Although the most widespread effects of sea-level fluctuations on drainage systems are associated with changes in base level, these effects are not always simple or self-evident. A relative fall in sea level will, of course, increase drainage basin relief and thus the potential energy available in the landscape (Fig. 1.5). But the response of fluvial systems to a relative change in sea level depends on the way in which the change is generated and on the morphology of the coastal zone across which the shoreline migrates. To understand these factors we need to examine two situations. The first is where the land is stationary, and the relative change in sea level is caused entirely by a change in the position of the sea surface; the second is where sea level is stationary, and the land rises or subsides. (In reality, of course, both effects may operate simultaneously.)

   Where the sea surface rises along a stable continental margin the effects are relatively easy to predict. The down-

stream part of the drainage basin will be submerged, and if the sediment supply is sufficient there will be deposition and a delta may be constructed. Further deposition may then occur above the delta as a new profile becomes established. This regrading of the river long profile will not

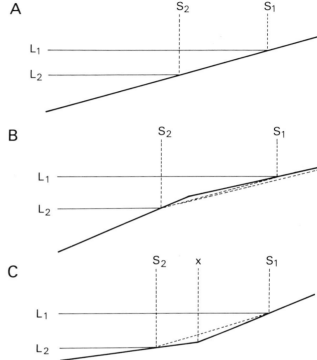

Fig. 17.12   *Schematic representation of drainage system response to a fall in sea level along continental margins with different configurations. In (A) there is a fall in sea level (L_1–L_2) at a margin with a constant gradient and the shoreline migrates from position S_1 to position S_2. Although deposition rates may consequently fall in the newly exposed zone between S_1 and S_2 we would not expect any significant incision here as the gradient of the margin is no greater than inland of the original shoreline. Certainly it is unlikely that channel incision is likely to be sufficient to cause the rejuvenation of river systems inland of the original shoreline (S_1). In (B) there is an increase in gradient between the initial (S_1) and subsequent (S_2) shoreline positions. In this case we would expect the greater erosional energy of rivers seaward of the point at which the gradient increases to lead to channel incision and a regrading of river long profiles (indicated by dashed line). Ultimately such regrading may extend inland beyond the original shoreline (S_1). In (C) the fall in sea level eventually exposes an area of lower gradient margin. In the upper section above the point at which the gradient decreases (x) we would expect a simple extension of river systems with little incision since there is no change in gradient above and below the initial shoreline (S_1). Below this slope break, however, the decrease in gradient would probably promote deposition. This could eventually lead to the regrading of the whole of the newly exposed coastal plain between S_1 and S_2 and deposition might eventually extend inland of the original shoreline. (From M. A. Summerfield, (1985) in: M. Morisawa and J. T. Hack (eds) Tectonic Geomorphology. Allen and Unwin, Boston, Fig. 3, p. 39.)*

necessarily extend far up the river system and consequently the effects of such a base level rise will not necessarily extend throughout an entire drainage basin.

A fall in sea level might be expected to have the opposite effect, that is, lead to erosion and the incision of drainage systems. Whether this occurs, however, depends on the gradient of the surfaces over which a newly extended drainage channel flows towards the now more distant shoreline. In particular, it depends on whether channel gradients across the newly exposed continental shelf are greater, less, or the same as those where the channel approached the original shoreline (Fig. 17.12). Only if gradients across the newly exposed shelf are greater than those immediately inland will significant river incision occur. This may in turn lead to rejuvenation of river systems inland of the original high sea level shoreline. Active incision of the continental shelf by rivers during periods of low sea level may cut deep valleys which, after a subsequent sea-level rise, become **submarine canyons**.

The possible effects of different channel gradients across a continental margin can be illustrated by looking at the consequences of a significant lowering of base level in southern Africa (Fig. 17.13). Imagine a relative sea-level fall of 500 m exposing the present submarine topography of the continental margin. The gradients of rivers extending over the newly exposed broad coastal plains of the western and southern margins would in general be less than the average gradients of the present-day channels at the coast. Only along a section of the eastern coast, where there is a very steep offshore slope, would there be a significant increase in seaward channel gradients and consequently active river incision into the exposed continental shelf. A situation similar to this (although with a rather less dramatic sea-level fall than we have assumed above for illustrative purposes) may have occurred during a possible major regression in the late Oligocene–Early Miocene. Interestingly, the numerous submarine canyons of the eastern margin of southern Africa appear to have been cut during this period.

The sea-level fluctuations of the Quaternary which have caused base levels to be lowered by up to 100 m for substantial periods of time during the past 1 Ma or so have promoted widespread valley incision along continental margins. Such valley downcutting is not normally evident today since the Holocene sea-level rise has led to valley aggradation and the burial of these incised valleys by sediment. Perhaps one of the best documented instances of valley

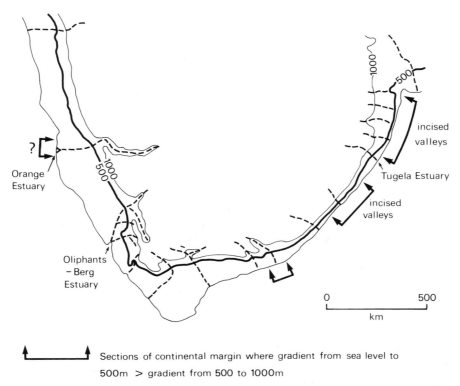

Fig. 17.13 *Hypothetical effect of a major sea-level fall on the drainage systems of southern Africa. The labelled contours indicate elevations above sea level in metres after a relative sea-level fall of 500 m. This means that the 500 m contour coincides with the present-day coastline. The dashed lines schematically represent the courses of major present day rivers across the exposed continental margin. Sections of the margin where a major relative fall in sea level would lead to a downstream increase in channel gradient, and hence the potential for channel incision, tend to coincide with actual areas characterized by incised valleys and submarine canyons. (Based on R. V. Dingle et al. (1983) Mesozoic and Tertiary Geology of Southern Africa. Balkema, Rotterdam, Fig. 190, p.313.)*

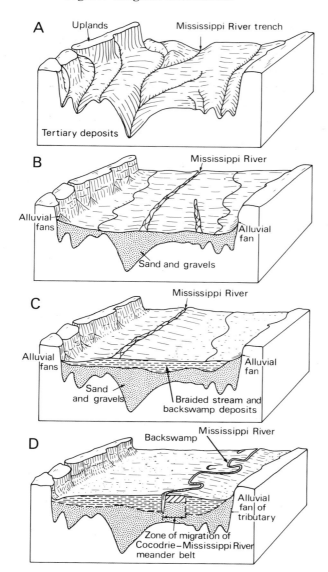

Fig. 17.14 *Late Quaternary history of the lower Mississippi Valley showing the transformation from an entrenched river developed during the low sea levels of the last glacial to the present-day meandering alluvial channel. (A) The situation around 15 000 a BP with the Mississippi entrenched by around 120 m in response to the lowering of base level and an increase in channel gradient to about 0.155 m km⁻¹. From 14 000 to 4000 a BP the post-glacial rise in sea level led to the deposition of mainly coarse sediments, supplied in prodigious quantities by meltwater from the decaying Laurentide ice sheet to the north, and the establishment of a braided channel with a gradient of around 0.140 m km⁻¹ (B). A progressive reduction in both the rate of sea-level rise and the size of sediment being carried by the river gradually led to the deposition of predominantly fine sediments and the reduction of the average channel gradient to about 0.127 m km⁻¹ (C). Finally, the stabilization of sea level and reduction in sediment load promoted the creation of a single meandering channel with an average gradient of approximately 0.112 m km⁻¹ (D). (After H. N. Fisk (1944) Geological Investigation of the Alluvial Valley of the Lower Mississippi River. Mississippi River Commission, Vicksburg.)*

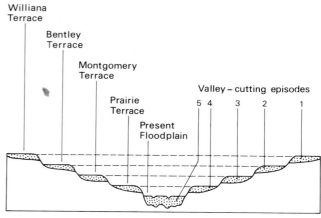

Fig. 17.15 *An interpretation of the Quaternary history of the lower Mississippi Valley based on an idealized representation of terrace relationships. Five successive stages of channel incision during low glacial sea levels and terrace formation during high interglacial sea-levels are shown. (After H. N. Fisk (1944) Geological Investigations of the Alluvial Valley of the Lower Mississippi River. Mississippi River Commission, Vicksburg, Fig. 77.)*

downcutting and subsequent aggradation is provided by the lower Mississippi Valley (Fig. 17.14). In this case there was an extremely high rate of deposition because of the enormous amounts of glacially-derived sediment being transported from the north by meltwater as the Laurentide ice sheet retreated. According to some interpretations the terrace deposits of the lower Mississippi Valley indicate several distinct phases of sedimentation interspersed with episodes of valley incision which can be correlated with alternating high and low Pleistocene sea levels (Fig. 17.15).

Another dramatic example of the effects of Quaternary sea-level change on drainage systems is provided by the Amazon system. The channel of the Amazon River itself has a remarkably gentle gradient of only 0.1 m km⁻¹ from the foothills of the Andes to the Atlantic Ocean. Downstream of Manaus, which is almost 1500 km from the ocean, the gradient is a mere 0.03 m km⁻¹. In this lower reach the Amazon has a very broad floodplain up to 100 km across. This extraordinary situation is largely an artefact of post-glacial sea-level rise. During the low sea levels of the last (and previous) glacials the Amazon was deeply entrenched, but the rise in post-glacial sea level initially transformed the lower Amazon into an enormous ria. Once global sea level began to stabilize in the Holocene the sediment-charged western Amazon rapidly infilled this embayment, thereby creating the wide, low-gradient floodplain of the present-day lower Amazon. The rate of sediment accumulation was, in fact, so rapid that floodplain deposits blocked the mouths of major lower Amazon tributaries, such as the Xingu and Tapajos.

If the relative change in sea level arises from vertical movements of the land rather than the sea, the effects on drainage systems will depend on the nature of the crustal

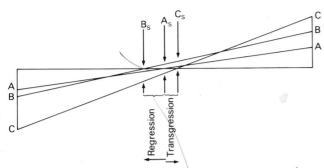

Fig. 17.16 *Effects of continental-margin flexure on coastal gradients and relative sea level. A–A shows the initial configuration of the margin with the shoreline positioned at A_s. B–B shows the situation after the flexure of the margin has increased and the axis of rotation has moved seaward. As a result the gradient across the margin increases and there is uplift along the coast and a regression as the shoreline moves seaward to B_s. C–C shows a further increase in flexure across the margin, but in this case the axis of rotation has moved inland. This also leads to a steepening of the gradient across the margin, but the landward migration of the axis of rotation means that there is a transgression as the coast subsides and the shoreline migrates landward to C_s. (Based on L. King (1982) The Natal Monocline. University of Natal Press, Pietermaritzburg, Fig. 12, p. 34.)*

deformation. Where the entire land area rises or subsides uniformly then the effects will be similar to a fall or rise in the sea surface. However, along passive margins, which are characterized by crustal flexure with uplift inland and subsidence offshore (see Section 4.5.2), the picture is more complex. An increase in flexure of the margin will always lead to a steepening of channel gradients in the coastal zone, but this can be associated with either a regression and relative sea-level rise, or a transgression and relative sea-level fall, depending on whether the axis of rotation of the margin moves seaward or landward (Fig. 17.16).

One further effect worthy of mention is hydro-isostasy. A fall in sea level will reduce the water load on a continental margin thus cause it to flex upwards. The vertical movement involved is probably insufficient in most cases to disrupt river systems draining across the margin, but in other instances the effects may be dramatic. It has been estimated, for example, that the desiccation of the Mediterranean Sea during the Late Miocene caused a flexural upwarp of its north-western and south-eastern margins of up to 450 m along the shelf edge. This would probably have been sufficient to have caused the landward reversal of rivers with small discharges and low coastal gradients. The Rhône and Nile, on the other hand, apparently had enough erosive energy to maintain their courses across the margin.

17.6.2   Coastal effects

Geomorphologists have long been aware of the importance of changes in sea level on coastal landforms, and this awareness led some to propose that all coasts could be usefully classified on the basis of whether they were emergent (associated with a relative fall in sea level) or submergent (associated with a relative rise in sea level). Unfortunately, such a classification is not as relevant as it might first appear for two main reasons. First, since most coastlines have experienced a relative rise in sea level during the Holocene, erosional landforms diagnostic of submergence, such as rias and fjords, are very widespread. Secondly, because many coastal landforms do not adjust quickly to changes in sea level, the rapidly fluctuating sea levels of the Quaternary have produced coasts which exhibit characteristics associated with both submergence and emergence. This is clearly the case with many erosional forms, but even applies to depositional forms which would be expected to respond more rapidly to changing sea levels. This has resulted in present depositional landforms related to high Holocene sea levels frequently located on sediments laid down during previous interglacial sea level highstands.

One feature of present-day coastlines around the world is the widespread erosion of beaches and other types of coastal depositional landforms. This, too, appears to be related to sea-level change. During the rise in sea level at the end of the last glacial, the transgression of shorelines across what had previously been coastal plains meant that new sources of sediment were constantly being encountered. The coarser fraction of this sediment was transported onshore to form beaches, or blown inland to create coastal dunes. As sea levels began to stabilize at roughly their present position around 4000 a BP new sources of sediment no longer became available and consequently many beaches have been experiencing a net loss of sediment since that time.

One of the most critical factors determining the response of coasts to a change in sea level is the morphology of the continental shelf and the coastal plain. Three distinct effects can be identified. The most obvious is the rate of shoreline migration. The lower the gradient of the continental shelf and coastal plain over which regressions and transgressions occur the greater will be the amount of horizontal movement of the shoreline for a given vertical change in sea level. Thus shoreline movements generated by global sea-level changes will be much more extensive and rapid along generally gently sloping rifted margins that along the steeper and narrower continental shelves of active margins. On continental shelves of very low gradient rates of shoreline migration during periods of interglacial sea-level rise have been prodigious. For instance, during the period 18 000 a BP to 6000 a BP the shoreline moved inland an average distance of over 300 km across the eastern margin of the USA at rates of up to 8 m a⁻¹ in New Jersey and 13 m a⁻¹ in Louisiana. It is along such gently sloping coasts, characterized by rapid rates of shoreline movement, that it is most likely that there will be insufficient time for littoral processes to construct equilibrium depositional landforms. This indeed appears to be the case along the eastern seaboard of the U S A where the continental shelf has a range of morphological zones which have not yet been fully modified by present-day shelf processes, and which appear to be related to previous shoreline positions.

Two further factors related to the interaction of sea-level change and the configuration of the continental shelf are wave energy budget and tidal range. The tidal range along a continental margin can be greatly amplified if a low– gradient shelf is submerged by a relative rise in sea level. Similarly, the rate of change in water depth as the shoreline is approached partly controls the rate at which wave energy is dissipated and hence affects the amount of geomorphic work that waves are able to accomplish (see Section 13.2). A change in sea level which alters the gradient of the shoreline, especially in the critical zone where water depth is less than half the wavelength of waves, can therefore indirectly affect the nature of wave activity along coasts.

17.6.3   Oceanic islands

Oceanic islands provide unparalleled 'laboratories' for exploring relative sea-level change, and, incidentally, for the investigation of other important geomorphic problems. Although they are distributed across the world's oceans they are far more numerous in the Pacific than elsewhere. In the mid-nineteenth century, Charles Darwin, who began his scientific career more as geologist than a biologist, recognized that oceanic islands were of just three types; volcanic islands, coral reefs and combinations of the two. (We now know this is not strictly correct as a small proportion of oceanic islands, such as the Seychelles in the Indian Ocean, rise from fragments of continental crust – see Figure 3.32 – but the classification is true for the oceanic islands we are concerned with here which rest on oceanic lithosphere.) Although coral reefs are highly variable in morphology, the ultimate form is the **atoll**, a ring of coral encircling a shallow lagoon.

The existence of coral atolls rising from the depths of the ocean provided a puzzle for nineteenth-century geologists. Charles Lyell, Darwin's geological mentor, regarded them as merely the craters of submarine volcanoes which had been overgrown by coral, but such an explanation did not account for many of the morphological characteristics of atolls. Although many hypotheses of atoll formation were proposed in the nineteenth and early twentieth centuries, it was Darwin who, in 1837, first put forward the theory linking volcanic islands and coral reefs that we know today to be essentially correct.

Darwin's coral reef theory was one of the most remarkable examples of scientific deduction in the history of the earth sciences in that all its essentials were worked out before he had ever seen an atoll. During his voyage around the world in the surveying ship H M S *Beagle* (1831–36) Darwin observed graphic evidence of recent significant uplift on the western coast of South America which he considered to be closely linked to the active volcanism of the Andes. While

on the voyage Darwin had been reading the first two volumes of Lyell's *Principles of Geology* which argued for a version of uniformitarianism incorporating the notion that the Earth's average morphology remained more or less the same over geological time - that is the Earth's surface as a whole existed in a 'steady state' in which uplift in one area would be compensated by subsidence in another. Darwin applied this idea to the oceans, postulating that regions of uplift would be matched by regions of subsidence, and then linked this proposition with the observation that areas of the ocean containing atolls tend to be quite separate from those with active volcanoes. Darwin then saw that the three types of oceanic islands could be explained as stages in a single sequence of development (Fig. 17.17).

The first stage is the growth of a volcano from the ocean floor and its emergence above sea level. Volcanic activity eventually ceases, and in warm tropical oceans coral can grow in the shallow water around the edge of the volcano and construct a fringing reef. As the volcano subsides the coral is able to build the reef upward sufficiently quickly to maintain its upper surface close to sea level. A barrier reef is consequently formed which is separated from the partially

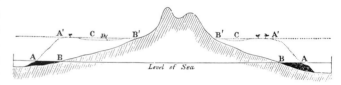

A A—Outer edge of the reef at the level of the sea.
B B—Shores of the island.
A'A'—Outer edge of the reef, after its upward growth during a period of subsidence.
C C—The lagoon-channel between the reef and the shores of the now encircled land.
B' B'—The shores of the encircled island.
   N.B.—In this, and the following woodcut, the subsidence of the land could only be represented by an apparent rise in the level of the sea.

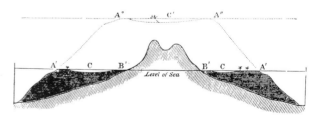

A'A'—Outer edges of the barrier-reef at the level of the sea. The coco -nut trees represent coral-islets formed on the reef.
C C—The lagoon-channel.
B' B'—The shores of the island, generally formed of low alluvial land and of coral detritus from the lagoon-channel.
A''A''—The outer edges of the reef, now forming an atoll.
C'—The lagoon of the newly-formed atoll. According to the scale the depth of the lagoon and of the lagoon-channel is exaggerated.

Fig. 17.17   *Diagrams from Darwin's book on coral reefs illustrating his theory of the development of atolls from volcanic islands. (From C. Darwin (1874)* The Structure and Distribution of Coral Reefs *(2nd edn) Smith Elder, London, Figs. 5 and 6, pp. 131 and 133.)*

submerged volcano by a lagoon. Eventually the volcano subsides completely below sea level to leave an atoll which continues to build upwards at the rate of subsidence. Darwin was well aware that his entire theory depended on the ability of coral to grow upwards at a rate at least as rapid as subsidence, and he was careful on his return to England to verify this assumption from empirical observations.

Darwin's coral reef theory is remarkable both for its simplicity and explanatory power, but it was only with the advent of plate tectonics that a full understanding of the process of volcano subsidence (which is such an important element of the theory) was achieved. Although drilling into atolls has demonstrated, as Darwin predicted, great thicknesses of carbonate of up to 1200 m resting on a volcanic base, there are a number of potential causes of subsidence. One is global sea-level rise which could cause apparent subsidence, but the vertical movements involved and the subsidence histories reconstructed for different atolls rule this out as the most common mechanism. Isostatic subsidence as a result of the load of the coral and volcanic edifice on the oceanic lithosphere is another possibility but again the vertical movement that this could produce is far smaller

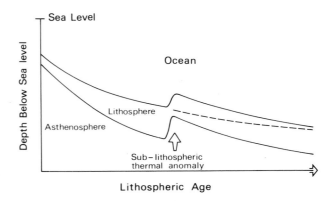

Fig. 17.18 *Changes in the elevation of the ocean floor as a result of heating associated with a sub-lithospheric thermal anomaly. Such a heating anomaly may give rise to a volcano which grows sufficiently to emerge above sea level. (After R. S. Detrick and S. T. Crough (1978)* Journal of Geophysical Research **83**, *Fig. 4(C), p. 1239. Copyright by the American Geophysical Union.)*

than the amount of subsidence experienced by many atolls. The only feasible mechanism is therefore the subsidence of the oceanic lithosphere as it ages and cools. As can be seen

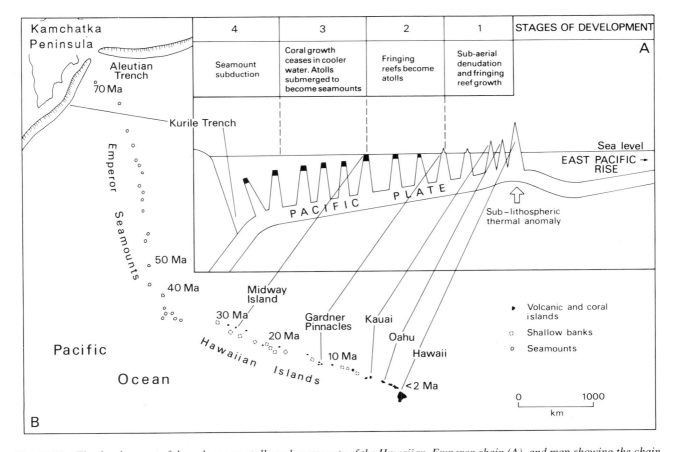

Fig. 17.19 *The development of the volcanoes, atolls and seamounts of the Hawaiian–Emperor chain (A), and map showing the chain extending from the island of Hawaii to the Kurile Trench off the Kamchatka Peninsula (B). The change in the orientation of the chain about halfway along is due to an alteration in the direction of movement of the Pacific Plate about 40 Ma BP. The dates along the chain indicate the age of volcanic activity. (Based partly on G. A. J. Scott and G. M. Rotondo (1983)* Coral Reefs **1**, *Fig. 5, p. 146 (A), and D. A. Clague et al. (1975)* Geological Society of America Bulletin **86**, *p. 991 (volcano ages in (B)).)*

from Figure 17.6, this can involve subsidence of nearly 3000 m over 80 Ma.

We can now construct a model of volcanic island and atoll development by considering the effects of a sub-lithospheric thermal anomaly (see Section 4.2.1) located beneath moving oceanic lithosphere (Fig. 17.18). Heating at the base of the lithosphere makes it less dense and causes it to rise, thereby interrupting its 'normal' cooling and subsidence behaviour. If the heating anomaly is sufficiently vigorous to create a hot spot, magma will penetrate the lithosphere and a volcano will form. The continuing motion of the lithosphere will, however, soon carry the volcano away from the heating anomaly and eruptive activity will cease. This volcano will then be carried gradually below sea level on the now recooling oceanic lithosphere.

Such a sequence is beautifully exemplified by the Hawaiian–Emperor seamount chain which can be traced as a line from the present centre of volcanic activity on the island of Hawaii north-westwards towards the Kamchatka Peninsula (Fig. 17.19). In tropical waters coral growth can keep pace with subsidence of this kind, but if atolls move into cooler waters (as in the case of the Hawaiian–Emperor seamount chain) corals cannot grow and atolls subside below sea level to form flat-topped seamounts, or guyots.

A survey of the world's coral reefs shows that there is a wide range of forms (Fig. 17.20). These include atolls that have apparently been raised by 50 m or more above sea level, so we must ask whether the simple model that we have just outlined provides a comprehensive explanation of coral reef evolution. The answer is no, but it does appear

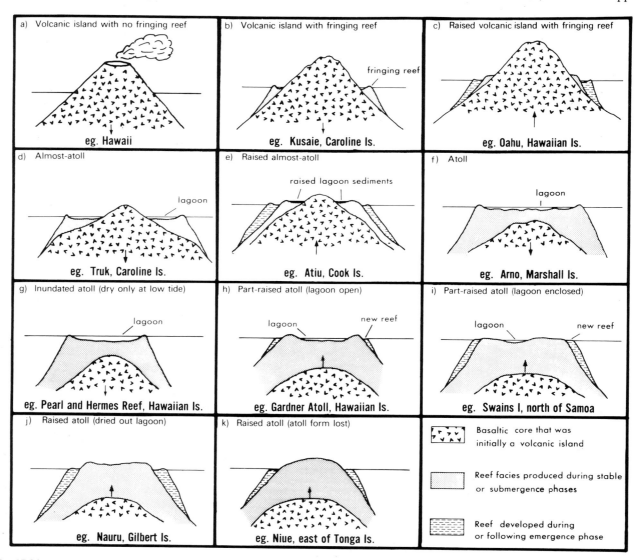

Fig. 17.20 *Types of oceanic island in the Pacific Ocean. Differences between the various types reflect the net effects of volcanism, oceanic lithosphere subsidence and flexure, sub-lithosperic heating and consequential uplift, eustatic sea-level change, coral growth and sub-aerial denudation. (From G. A. J. Scott and G. M. Rotondo (1983) Coral Reefs* **1***, Fig. 2, p. 141, based largely on H. J. Wiens (1962) Atoll Environment and Ecology. Yale University Press, New Haven and O. K. Leont' yev et al. (1975) USSR Oceanology* **14***, 840–846.)*

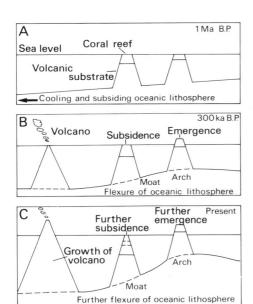

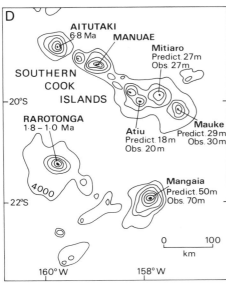

Fig. 17.21   Relative sea-level change associated with the flexure of the lithosphere as a result of loading by a volcano. (A) The initial situation with coral reefs growing on volcanic substrate subsiding on cooling oceanic lithosphere. (B) The initial effects of lithospheric flexure brought about by the development of a volcano near by with submergence in the 'moat' and emergence on the 'arch'. (C) Further submergence and emergence occurs as the mass of the volcano increases and the flexure of the lithosphere becomes more marked. Note the short time scale of such deformation. (D) The application of a flexural model to the southern Cook Islands in the southern Pacific. Loading has been caused by three volcanoes – Rarotonga, Aitutaki and Manuae – and the predicted uplift for the surrounding islands is compared with the observed uplift. The match is close except for Mangaia which may be within the zone of deformation of a fourth volcano to the south not included in the calculations. (Based on M. McNutt and H. W. Menard (1978) Journal of Geophysical Research 83, Fig. 2, p. 1207 and Fig. 4, p. 1210; and General Bathymetric Chart of the Oceans, 1984. Canadian Hydrographic Service (bathymetric contours).)

that most of the deviations from the model can be explained by known erosional, tectonic and eustatic effects. For instance, there are two ways in which atolls can be elevated. One is the possibility that an existing atoll may be carried over another sub-lithospheric thermal anomaly. It will then be elevated as the lithosphere on which it rests experiences uplift. A second possible mechanism is probably only relevant to rather modest amounts of uplift. This can arise if a new volcano appears amidst a group of atolls as they pass over a heating anomaly beneath the lithosphere (Fig. 17.21). The new volcano loads the lithosphere and causes it to flex, producing a moat immediately around it and a more distant peripheral arch – this is analogous to the crustal forebulge around an ice sheet. Islands within the moat will consequently subside whereas those on the arch will be elevated. In some cases several volcanoes can be seen to be simultaneously loading the lithosphere and this, not unexpectedly, produces very complex patterns of island uplift and subsidence (Fig. 17.21). We therefore have to be very careful when using oceanic islands as 'dipsticks' with which to monitor global sea-level change since closely adjacent islands may exhibit significantly different sea-level histories over time spans as short as the Holocene.

Further reading

Until the mid-1980s anyone with an interest in sea-level change had to search through a vast journal literature in order to get an up-to-date perspective on the subject. Fortunately, this is no longer the case, as there are now some excellent reviews of various aspects of the history and causes of sea-level change. The most recent and wide-ranging survey is that edited by Devoy (1987a), while the collections edited by Smith and Dawson (1983) and Tooley and Shennan (1987) provide more specialized treatments focusing on Quaternary sea-level change.

On the evidence for Quaternary sea-level change, Sutherland (1983, 1987) provides a comprehensive discussion of the problems of dating, while Chappell and Shackleton (1986) attempt to reconcile the coral terrace sequence of the Huon Peninsula first presented by Chappell (1974) with the oceanic oxygen isotope record. Turning to the pre-Quaternary record Bond (1979) and Harrison et al. (1981) demonstrate that sea-level reconstructions based on the amount of present-day land flooded at various times in the past must take into account changes in continental hypsometry, while Sahagian (1987) shows how the central cratonic region of North America can be used as a 'stable' frame of reference for the estimation of the position of global sea level in the Cretaceous. The procedures for calculating global sea-level fluctuations from seismic stratigraphy are described in detail in the seminal memoir edited by Payton (1977) (see especially Vail et al, 1977a, 1977b), and these are updated in Schlee (1984); Devoy (1987b) provides an accessible introduction to this topic. Detailed assessments of the problems of reconstructing global sea-level change from seismic stratigraphy are made by Burton *et al.* (1987), Hubbard (1988) and Miall (1984, 1986), while Haq *et al.* (1987) providès the most recent version of the 'Exxon curve' which incorporates evidence from continental outcrops of marine sediments.

The record of Quaternary sea-level change is covered in the general reviews by Mörner (1987a) for the northern hemisphere and Pillans (1987) for the southern hemisphere (although the former should be read with due regard for the author's enthusiasm for geoidal effects on global sea level). More detailed discussions of particular areas are to be found in Tooley and Shennan (1987) which includes an excellent evaluation by Chappell (1987) of the Late Quaternary record in the important Australian region. For the pre-Quaternary record the Exxon group summary by Haq *et al.* (1987) and the more wide-ranging review by Hallam (1984) are particularly useful.

Mechanisms of sea-level change are discussed in a brief but useful review by Donovan and Jones (1979), but for more detail it is necessary to refer to papers on specific mechanisms. The link between glacial–interglacial cycles and changes in ocean water volume as established by the oxygen isotope record is described by Shackleton and Opdyke (1973), while Matthews (1984) considers the relevance of glacio-eustasy to Early Cenozoic global sea-level change. Pitman (1978) provides an estimate of the effect on global sea level of changes in mid-oceanic ridge volume which is refined by Kominz (1984). Changes in sea level related to sediment accumulation in the oceans are examined by Harrison *et al.* (1981), while Schlanger *et al.* (1981) considers the effects of thermally induced uplift of the ocean floor. Wyatt (1986) speculates on the effects on global sea level of changes in continental hypsometry and the Mediterranean desiccation hypothesis is vividly presented by Hsü (1972). The geoidal model of sea-level fluctuations related to changes in ice and water masses around the globe is covered in an accessible way by Peltier (1987) who has also provided an overview of the complex mathematical analysis on which the model is based (Peltier, 1982). Assessments of the correspondence between the model and deglaciation sea-level histories for particular areas of the world include those by Andrews (1987) and Devoy (1987c), while the more speculative consequences of possible large-scale changes in the geoid for long-term sea-level change are discussed by Mörner (1981, 1987b). Finally, the problematic relationship between changing rates of passive margin subsidence and the transgressive and regressive sedimentary sequences which form the basis for sea-level curves derived from seismic stratigraphy is considered by Watts (1982) and Pitman and Golovchenko (1983).

The response of drainage basins to sea-level change has not yet received sufficient attention, but brief discussions of some of the relevant factors are to be found in Chappell (1983), Norman and Chase (1986) and Summerfield (1985). Coastal effects have been more fully considered, as is

evident in the detailed review by Orford (1987). Oceanic islands have recently attracted much attention, from geophysicists as well as geomorphologists. The beautifully illustrated book by Menard (1986) gives a superb introduction to the topic, while Guilcher (1988) provides a more conventional geomorphic treatment. On more specific topics Scott and Rotondo (1983) discuss the life history of oceanic islands as they subside on ageing lithosphere, and McNutt and Menard (1978) and Spencer *et al.* (1987) examine the uplift of atolls through the loading effects of new volcanic islands. The reading of Darwin's original work on coral reefs, which is available as a reprint (Darwin, 1984), is a rewarding experience for any geomorphologist not just for its detailed observations on atolls but also as an exemplar of the application of a scientific methodology appropriate to the study of long-term landform development. Finally, the potential of oceanic islands as a 'test-bed' for geomorphic theory is advocated by Nunn (1987) who points out the simplicity of controls on their landform development in comparison with continental areas.

References

Andrews, J. T. (1987) Glaciation and sea level: a case study. In: R. J. N. Devoy (ed.) *Sea Surface Studies: A Global View.* Croom Helm, London and New York, 95–126.

Bond, G. C. (1979) Evidence for some uplifts of large magnitude in continental platforms. *Tectonophysics* **61**, 285–305.

Burton, R., Kendall, C. G. St. C. and Lerche, I. (1987) Out of our depth: on the impossibility of fathoming eustasy from the stratigraphic record. *Earth-Science Reviews* **24**, 237–77.

Chappell, J. (1974) Geology of coral terraces, Huon Peninsula, New Guinea: a study of Quaternary tectonic movements and sea-level changes. *Geological Society of America Bulletin* **85**, 553–70.

Chappell, J. (1983) Aspects of sea levels, tectonics, and isostasy since the Cretaceous. In: R. A. M. Gardner and H. Scoging (eds) *Mega-Geomorphology*. Clarendon Press, Oxford and New York, 56–72.

Chappell, J. (1987) Late Quaternary sea-level changes in the Australian region. In: M. J. Tooley and I. Shennan (eds) *Sea-Level Changes*. Institute of British Geographers Special Publication **20**. Blackwell, Oxford and New York, 296–331.

Chappell, J. and Shackleton, N. J. (1986) Oxygen isotopes and sea level. *Nature* **324**, 137–40.

Darwin, C. (1984) *The Structure and Distribution of Coral Reefs.* University of Arizona Press, Tucson.

Devoy, R. J. N. (ed.) (1987a) *Sea Surface Studies: A Global View* Croom Helm, London and New York.

Devoy, R. J. N. (1987b) Hydrocarbon exploration and biostratigraphy: the application of sea-level studies. In: R. J. N. Devoy (ed.) *Sea Surface Studies: A Global View*. Croom Helm, London and New York, 531–68.

Devoy, R. J. N. (1987c). Sea-level changes during the Holocene: the North Atlantic and Arctic Ocean. In: R.J.N. Devoy (ed.) *Sea Surface Studies: A Global View*. Croom Helm, London and New York, 294–347.

Donovan, D. T. and Jones, E. J. W. (1979) Causes of world-wide changes in sea level. *Journal of the Geological Society London* **136**, 187–92.

Guilcher, A. (1988) *Coral Reef Geomorphology.* Wiley, Chichester and New York.

Hallam, A. (1984) Pre-Quaternary sea level changes. *Annual Review of Earth and Planetary Sciences* 12, 205–43.

Haq, B. U., Hardenbol, J. and Vail, P. R. (1987) Chronology of fluctuating sea-levels since the Triassic. *Science* 235, 1156–66.

Harrison, C. G. A., Brass, G. W., Saltzman, E., Sloan, J. II, Southam, J. and Whitman, J. M. (1981) Sea level variations, global sedimentation rates and the hypsographic curve. *Earth and Planetary Science Letters* 54, 1–16.

Hsü, K. J. (1972) When the Mediterranean dried up. *Scientific American* 277(6), 27–36.

Hubbard, R. J. (1988) Age and significance of sequence boundaries on Jurassic and Early Cretaceous rifted continental margins. *American Association of Petroleum Geologists Bulletin* 72, 49–72.

Kominz, M. A. (1984) Oceanic ridge volumes and sea-level change – an error analysis. In: J. S. Schlee (ed.) *Interregional Unconformities and Hydrocarbon Accumulation.* American Association of Petroleum Geologists Memoir 36. Tulsa, 109–27.

Matthews, R. K. (1984) Oxygen isotope record of ice-volume history: 100 million years of glacio-eustatic sea-level fluctuation. In: J. S. Schlee (ed.) *Interregional Unconformities and Hydrocarbon Accumulation.* American Association of Petroleum Geologists Memoir 36. Tulsa, 97–107.

McNutt, M. and Menard, H. W. (1978) Lithospheric flexure and uplifted atolls. *Journal of Geophysical Research* 83, 1206–12.

Menard, H. W. (1986) *Islands.* Scientific American Books, New York.

Miall, A. D. (1984) *Principles of Sedimentary Basin Analysis.* Springer-Verlag, New York.

Miall, A. D. (1986) Eustatic sea level changes interpreted from seismic stratigraphy: a critique of the methodology with particular reference to the North Sea Jurassic record. *American Association of Petroleum Geologists Bulletin* 70, 131–7.

Mörner, N.–A. (1981). Revolution in Cretaceous sea level analysis. *Geology* 9, 344–6.

Mörner, N.–A. (1987a) Quaternary sea-level changes: northern hemisphere data. In: R. J. N. Devoy (ed.) *Sea Surface Studies: A Global View.* Croom Helm, London and New York, 242–63.

Mörner, N.–A. (1987b) Models of global sea-level changes. In: M. J. Tooley and I. Shennan (eds) *Sea-Level Changes.* Institute of British Geographers Special Publication 20. Blackwell, Oxford and New York, 332–55.

Norman, S. E. and Chase, C. G. (1986) Uplift of the shores of the western Mediterranean due to Messinian desiccation and flexural isostasy. *Nature* 322, 450–1.

Nunn, P. D. (1987) Small islands and geomorphology: review and prospect in the context of historical geomorphology. *Institute of British Geographers Transactions,* N S 12, 227–39.

Orford, J. (1987) Coastal processes: the coastal response to sea-level variation. In: R. J. N. Devoy (ed.) *Sea Surface Studies: A Global View.* Croom Helm, London and New York, 415–63.

Payton, C. E. (ed.) (1977) *Seismic Stratigraphy - Applications to Hydrocarbon Exploration.* American Association of Petroleum Geologists Memoir 26. Tulsa.

Peltier, W. R. (1982) Dynamics of the ice age Earth. *Advances in Geophysics* 24, 1–146.

Peltier, W. R. (1987) Mechanisms of relative sea-level change and the geophysical responses to ice-water loading. In: R. J. N. Devoy (ed.) *Sea Surface Studies: A Global View,* Croom Helm, London and New York. 57–94.

Pillans, B. (1987) Quaternary sea-level changes: southern hemisphere data. In: R . J. N. Devoy (ed.) *Sea Surface Studies: A Global View.* Croom Helm, London and New York, 264–93.

Pitman, W.C. III, (1978) Relationship between eustacy and stratigraphic sequences of passive margins. *Geological Society of America Bulletin* 89, 1389–403.

Pitman,W.C. III, and Golovchenko, X. (1983) The effect of sea-level change on the shelfedge and slope of passive margins. *Society of Economic Paleontologists and Mineralogists Special Publication* 33, 41–58.

Sahagian, D. (1987) Epeirogeny and eustatic sea-level changes as inferred from Cretaceous shoreline deposits: applications to the central and western United States. *Journal of Geophysical Research* 92, 4895–904.

Schlanger, S. O., Jenkyns, H. C. and Premoli-Silva, I. (1981) Volcanism and vertical tectonics in the Pacific Basin related to global Cretaceous transgressions. *Earth and Planetary Science Letters* 52, 435–49.

Schlee, J. S. (ed.) (1984) *Interregional Unconformities and Hydrocarbon Accumulation.* American Association of Petroleum Geologists Memoir 36. Tulsa.

Scott, G. A. J. and Rotondo, G. M. (1983) A model to explain the differences between Pacific Plate island-atoll types. *Coral Reefs* 1, 139–50.

Shackleton, N. J Opdyke, N. D (1973) Oxygen isotope and paleomagnetic stratigraphy of Equatorial Pacific core V28–238: Oxygen isotope temperatures and ice volumes on a 10^5 to 10^6 year scale. *Quaternary Research* 3, 39–55.

Smith, D. E. and Dawson, A. G. (eds) (1983) *Shorelines and Isostasy.* Institute of British Geographers Special Publication 16. Academic Press, London and New York.

Spencer, T., Stoddart, D. R. and Woodroffe, C. D. (1987). Island uplift and lithospheric flexure: observations and cautions from the South Pacific. *Zeitschrift für Geomorphologie Supplementband* 63, 87–102.

Summerfield, M. A. (1985) Plate tectonics and landscape development on the African continent. In: M. Morisawa and J. T. Hack (eds) *Tectonic Geomorphology.* Allen and Unwin, Boston and London, 27–51.

Sutherland, D. G. (1983) The dating of former shorelines. In: D. E. Smith and A .G. Dawson (eds) *Shorelines and Isostasy.* Institute of British Geographers Special Publication 16. Academic Press, London and New York, 129–57.

Sutherland, D. G. (1987) Dating and associated methodological problems in the study of Quaternary sea-level changes. In: R. J. N. Devoy (ed.) *Sea Surface Studies: A Global View.* Croom Helm, London and New York, 165–97.

Tooley, M. J. and Shennan, I. (eds) (1987) *Sea-Level Changes.* Institute of British Geographers Special Publication 20. Blackwell, Oxford and New York.

Vail, P. R., Mitchum, R. M., Jr and Thompson, S. III (1977a) Seismic stratigraphy and global changes of sea-level, part 3: relative changes of sea level from coastal onlap. In: C. E. Payton (ed.) *Seismic Stratigraphy – Applications to Hydrocarbon Exploration.* American Association of Petroleum Geologists Memoir 26, Tulsa, 63–81.

Vail, P. R., Mitchum, R. M. Jr and Thompson, S. III (1977b) Seismic stratigraphy and global changes of sea-level, part 4: global cycles of relative changes of sea-level. In: C. E. Payton (ed.) *Seismic Stratigraphy – Applications to Hydrocarbon Exploration.* American Association of Petroleum Geologists Memoir 26. Tulsa, 83–97.

Watts, A. B. (1982) Tectonic subsidence, flexure and global changes of sea-level. *Nature* 297, 469–74.

Wyatt, A. R. (1986) Post-Triassic continental hypsometry and sea-level. *Journal of the Geological Society London* 143, 907–10.

18

Long-term landscape development

18.1 Models of landscape evolution

One of the most obvious questions we can ask about landscapes is how they came to be as they are. Indeed, the historical approach to landform analysis was the dominant perspective until the 1960s. Over the past two decades, however, the other obvious question – what are the processes operating in the landscape today and how do they relate to the landforms we see – has become pre-eminent to the extent that studies of landscape development through time have been rather neglected. The detailed work on surface processes over the past two decades has significantly increased our understanding of the relationships between process and form at the small scale and over short periods of time. But the gap between our understanding of landform genesis at this scale and our knowledge of how whole landscapes function at long time scales has been widely acknowledged. A much better appreciation of the role of tectonic and climatic controls over landscape development, coupled with the application of new dating techniques and major theoretical advances, is beginning, once again, to bring the problem of long-term landscape development centre stage. This chapter draws extensively on topics introduced earlier in the book and tries to show how new concepts and data are beginning to shed a fresh light on some long-standing problems in geomorphology.

Geomorphology has seen various attempts to systematize the development of landscapes through time by isolating the key factors which determine the way in which landforms evolve. These models of landscape evolution, the most influential of which have been those proposed by W. M. Davis, W. Penck, L. C. King and J. Büdel, have had a profound effect on the kinds of problems that geomorphologists have considered and the ways they have attempted to tackle them.

18.1.1 The Davisian cycle of erosion: peneplanation

The model of landscape evolution usually known as the cycle of erosion was developed by W. M. Davis between 1884 and 1899 and owed much to the evolutionary thinking that had permeated both the natural and social sciences in Britain and North America during the latter half of the nineteenth century. Davis considered that in a similar way to life forms, landforms could be effectively analyzed in terms of their evolution. He regarded landscapes as evolving through a progressive sequence of stages, each exhibiting characteristic landforms. In his view these sequential changes in form through time made it possible to infer the temporal stage of development of a landscape from its form alone.

A second key concept implicit in the cycle of erosion model (although not explicitly referred to by Davis) is that of thermodynamics. The development of the principles of thermodynamics had been a major achievement of nineteenth-century science with repercussions just as profound as those of evolution. This aspect of the cycle of erosion model has been highlighted only relatively recently and relates closely to systems analysis in geomorphology. The second law of thermodynamics states that in an **isolated system** (that is, one which cannot give off or receive either mass or energy) **entropy** can never increase. The concept of entropy has been applied in many different contexts, but a general definition is that it is a measure of the energy in a system which is unable to perform work. In a system in a state of low entropy there are large differences in the distribution of energy and the flows from areas of high to areas of low energy allow work to be performed. Conversely, in a high entropy system energy is much more evenly distributed and the flow of energy and the performance of work is correspondingly reduced. At the theoretical point of maximum entropy the distribution of energy is entirely uniform and no work can be done in the system.

In strict terms, landscapes are neither isolated nor even closed systems since they are constantly importing and exporting mass and energy. Nevertheless, it has been argued that as the potential energy created by the uplift of a landsurface is the major source of energy in the landscape system it is a justifiable simplification to regard landscapes *as if they are* isolated systems. Given this assumption, the cycle of erosion can be seen as representing a progressive decline in potential energy and increase in entropy as the landscape is eroded. Indeed Davis saw the slope angles and stream gradients at any particular point in the landscape as reflecting the distribution of potential energy expressed as local differences in elevation. The total potential energy of the landscape, and hence its stage of evolution at any given time, could be expressed in terms of its mean elevation above base level (Fig. 18.1(A)).

Although Davis acknowledged numerous complications that could affect the cycle of erosion, his detailed descrip-

tion of the anticipated sequence of forms (Fig. 18.2) was based on several important assumptions: that denudation occurs under a humid temperate climate (which Davis regarded as 'normal') on a uniform lithology, and that the cycle is initiated by the relatively brief and rapid uplift of a landsurface of minimal local relief which does not experience significant erosion during the uplift phase. Given these conditions, he described a series of stages in the cycle of erosion categorized by way of analogy to the stages of human life as **youth, maturity** and **old age**. Davis argued that there would be a progressive decline in slope angles (Fig. 7.25(A)) and stream gradients through time which would ultimately result in the production of a landsurface close to base level with very subdued relief. Such a surface he termed a **peneplain** and consequently Davis's model of landscape development characterized by declining surface gradients through time is often referred to as **peneplanation**. It is important to note, however, that the term peneplain is used by some writers much more broadly to refer to *any* low relief surface however formed. The alternative

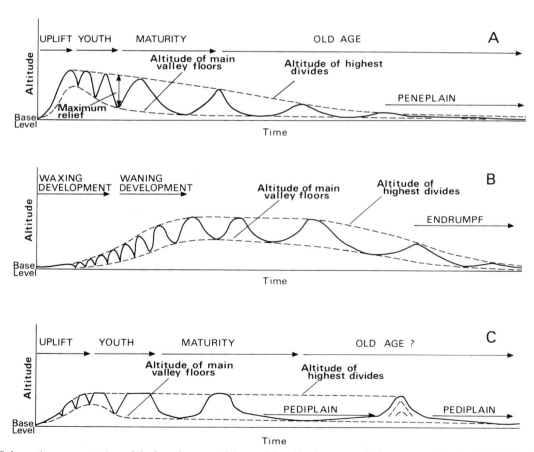

Fig. 18.1 *Schematic representation of the key elements of the models of landscape evolution proposed by Davis (A), Penck (B) and King (C). Note that for simplicity base level is assumed to be fixed through time and that the temporal scale is not necessarily comparable between diagrams. In the Davisian scheme the stage of old age should be regarded as many times longer than youth and maturity. (Modified from J. B. Thornes and D. Brunsden, (1977)* Geomorphology and Time. *Methuen, London, Fig. 6.2, p. 122.)*

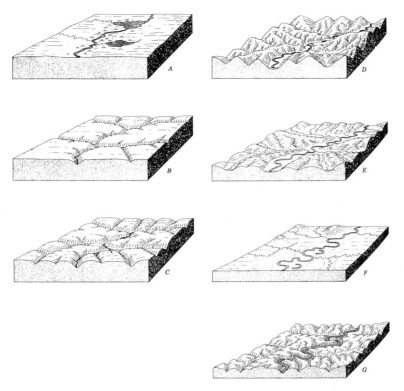

Fig. 18.2 *The Davisian cycle of erosion under a humid climate. The assumed starting-point is a landsurface with little local relief, either a peneplain developed during the previous cycle as shown in (A) or an emerged submarine surface. Uplift leads to rapid incision of the landsurface by rivers. In early youth (B) narrow river valleys separate broad areas of largely uneroded uplands and river gradients are irregular with waterfalls, rapids and lakes formed in response to lithological variations. These channel gradient irregularities have been eliminated by the end of middle youth, and by the end of late youth (C) major rivers are graded and lateral erosion enables the development of narrow floodplains in their lower courses. The flat uplands which have been steadily reduced in area during youth as the drainage network has grown are eliminated altogether by the beginning of maturity (D). This is the stage of maximum local relief and the drainage network becomes fully integrated and closely adjusted to structure. Hereafter local relief begins to decline as the graded river channels, which by this stage have spread far up tributary valleys, are lowered progressively less rapidly than interfluves. Associated with this change is the reduction in average slope angles, as the steep slopes of youth which are close to the stability angle of the partially weathered debris are transformed into lower gradient slopes as the active basal removal of debris ceases. Throughout maturity, floodplains become gradually wider and major rivers develop meandering channels. By late maturity (E) local relief has been significantly reduced and the landscape comprises gentle valley-side slopes and extensive floodplains. As old age is reached (F) the entire landscape is graded and floodplains are several times broader than the active meandering belts within them. The mean elevation of the landsurface, already close to base level, is lowered further only very gradually. Note, however, that in regions remote from the coastline to which rivers are flowing the developing peneplain will remain well above base level since river channels must have a certain minimum gradient in order to transmit water. Low rates of erosion allow the accumulation of thick weathering mantles which, in progressively masking the underlying bedrock, gradually free river channels from structural controls. None the less, particularly resistant lithologies may allow erosional residuals, known as* **monadnocks**, *to survive into late old age. Finally, renewed uplift will initiate a new cycle of erosion (G). (From A. N. Strahler (1969)* Physical Geography *(3rd edn.) Wiley, New York, Fig. 27.1, p. 466, drawn by E. Raisz.)*

spelling 'peneplane' which is used occasionally is certainly misleading as Davis in no sense envisaged the development of a planar surface as the ultimate product of the cycle of erosion.

As we have mentioned, Davis acknowledged the presence of factors that might complicate the stately progression of landscapes illustrated in Figure 18.2. The cycle might be interrupted by renewed uplift at any stage which would cause **rejuvenation** of the landscape through the develop-

ment of youthful forms which would coexist with older forms and thereby create a **polycyclic landscape**. The simplest assumption was that such uplift would only manifest itself as a fall in base level at the downstream extremity of drainage basins (normally at the coastline), and would lead to the gradual encroachment of steeper river gradients and slopes upstream through the drainage systems of the uplifted landscape (Fig. 18.2(G)).

A second complication Davis noted was climate. Davis

effectively represented landscape development under the humid temperate morphoclimatic regime of his home area of the north-east USA as 'normal'. But he accepted that the detailed nature of landform evolution under different prevailing climates would not be identical because of variations in the intensity of geomorphic processes under different morphoclimatic regimes. Consequently, he developed 'arid' and 'glacial' versions of the cycle of erosion while later disciples of his evolutionary approach added further variants.

A third complication was provided by lithology and structure which Davis saw as exerting specific controls on landscape evolution largely through their influence on drainage patterns. He maintained, however, that such controls would become progressively less significant as the cycle of erosion proceeded. In the case of limestone terrains, later workers found it necessary to develop a specific karst cycle of erosion. Nevertheless, in spite of these complications, Davis maintained the value of regarding landscapes primarily in terms of their evolutionary stage in a unidirectional temporal sequence.

Although his cyclic scheme never gained wide acceptance on the continent of Europe, it dominated Anglo-American geomorphology for several decades. Since the 1950s, however, both the theoretical utility and the empirical validity of the cycle of erosion have been increasingly challenged. What, then, are the major criticisms of the model? Although contemporary critics have tended to focus on the rather vague understanding of surface processes evident in Davis's formulation of landform development in general, and slope development in particular, perhaps the most fundamental problem with the cycle of erosion arises from the assumptions concerning the rates and occurrence of uplift.

Presumably due to the lack of quantitative data when he was writing, Davis was never very specific about actual rates of uplift and denudation. The estimates he did give, such as the 20–200 Ma for the peneplanation of the fault-block mountains of Utah, indicates that he envisaged extensive time scales. Our current knowledge of uplift rates (see Chapter 15) suggests that few areas of the world remain stable for periods of tens of millions of years or more, and therefore it seems that polycyclic landscapes are likely to be the norm rather than the exception. Furthermore, isostatic uplift is an inevitable consequence of denudational unloading as the cycle runs its course. As a result, continuous crustal uplift, albeit at a declining rate through time, will affect the entire duration of a cycle of erosion and greatly delay the attainment of full peneplanation. As we have seen in Chapters 3 and 4, inter-plate and intra-plate tectonic mechanisms give rise to quite different temporal and spatial patterns of uplift, and in neither case does the elevation of the landsurface take the form of geologically brief, discrete episodes of rapid surface uplift. Epeirogenic

movements characteristic of plate interiors usually involve slow, but prolonged surface uplift, while the high crustal uplift rates characteristic of convergent and oblique-slip plate margins persist for as long as the plate interactions giving rise to them are sustained.

Another major criticism of the Davisian model arises from its inability to accommodate the frequent and rapid climatic changes that have characterized the Quaternary. These have been of world-wide extent and, in conjunction with the frequent major changes in base level with which they have been associated through their effect on global sea level, they make it extremely unlikely that landscapes anywhere can be realistically viewed as representing a simple unidirectional sequence of forms.

18.1.2   The Penck model: uplift and denudation related

As has already been pointed out, the Davisian model never gained universal support, and geomorphologists on the continent of Europe found its assumptions – especially those concerning the nature of uplift – drastically over-simplified. In spite of these criticisms the only coherent alternative scheme of landform development to emerge prior to the Second World War was that of W. Penck. Penck's ideas have never been popular among English-speaking geomorphologists both because of his rather obscure writing style and terminology, and because the majority of geomorphologists unable to read German had to rely for several decades on misleading representations of his views by Davis and other writers.

Penck's ideas on uplift differed significantly from those of Davis (Fig. 18.1(B)). Whereas the latter assumed brief episodes of rapid uplift punctuating prolonged periods of stability, Penck argued that, in orogenic belts at least, active uplift could continue for a considerable time and in such situations Davis's notion of evolutionary stages of landscape development would be of dubious value. On the basis of the evidence from sedimentary sequences flanking the Alps, Penck considered that rates of active uplift initially increased slowly before reaching a maximum and then declining gradually.

In certain circumstances Penck thought that periods of increasing and decreasing rates of uplift might be reflected in slope forms. This link could arise from the effect changing rates of crustal uplift could have on rates of river incision. High rates of crustal uplift, Penck argued, would raise river channels further above base level and thus increase their gradients. This would lead to an acceleration in river downcutting until the rate of incision matched the rate of crustal uplift. The converse situation would apply during a decline in the rate of crustal uplift, with rates of river incision decreasing as downcutting reduced channel gradients. Penck considered that a uniform rate of river incision would give rise to straight slopes which would retreat at a constant angle. If the rate of downcutting were to increase, however, a phase of **waxing development** would ensue and slopes would steepen progressively from the base upwards to produce a convex profile. Conversely, a decrease in river downcutting could create a phase of **waning development** and slopes could become progressively less steep from the base upwards, creating concave profiles.

Penck's model of landscape evolution can thus be summarized as follows. An initial gradual increase in the rate of crustal uplift of a primary surface (**Primärrumpf**) leads to the widespread development of convex slopes. Further acceleration in the rate of uplift results in the formation of a series of benches (**Piedmottreppen**) around the margins of the primary uplifted surface. As the rate of uplift begins to decline there is a transition from waxing development, characterized by rapid downcutting, to waning development where the rate of stream incision is reduced and valley widening through the parallel retreat *of individual slope elements* gradually becomes dominant (Fig. 7.25(C)). As noted in Chapter 7 (see Section 7.6.2), this form of slope evolution is perhaps best described as slope replacement to distinguish it from the version of whole-slope parallel retreat advocated by King and misattributed by Davis to Penck (Fig. 7.25(B)). The steepest slope elements forming free faces retreat most rapidly leaving behind basal series of lower angle debris slope segments. The retreat of free faces eventually leads to the formation at drainage divides of large residual hills, or inselbergs, which are flanked by pediments. The eventual elimination of inselbergs leaves a landscape termed by Penck an **endrumpf** consisting entirely of slowly retreating, low angle concave slopes.

Although Penck's emphasis on the response of drainage systems to changing rates of uplift provides useful pointers as to how we might attempt to integrate tectonics into models of long-term landform development, his scheme as a whole is untenable as it pays insufficient attention to other factors affecting landform development. In particular it fails to acknowledge the importance of changes in river discharge which might arise as a result of climatic change, and it also underplays the role of lithology and the nature of weathering, both of which can significantly affect relationships between stream activity and slope form.

18.1.3 The King model: pediplanation

L. C. King's model of landscape evolution resembles Davis's in assuming that uplift is episodic and rapid in comparison with rates of denudation, and that the overall morphology of a landscape at any point in time is diagnostic of its evolutionary stage of development (Fig. 18.1(C)). The essential, and significant, difference in King's scheme lies in the mode of slope development he proposed. King initially developed his model to account for the landscapes of southern Africa. These are characterized by extensive, gently inclined surfaces dotted with inselbergs and separated by escarpments, and have developed under predominantly arid to tropical wet–dry morphoclimatic regimes. King's notion of slope development appears to owe much to Davis's misrepresentation of Penck's ideas (compare Figure 7.25(B), (D)). Rather than the sequential replacement of parallel retreating slope segments by lower angle elements, King envisaged the parallel retreat of a single free-face slope unit leaving a broad, concave pediment sloping at an angle of 6–7° or less at its base. Gradually over time, pediments coalesce to form **pediplains** and this mode of landscape development is therefore called **pediplanation.**

King considered that once pediment surfaces have been formed they persist with little change until the next phase of surface uplift promotes a new cycle of river incision and escarpment retreat which consumes existing pediplains and creates new ones. As in the Davisian model, the dating of such denudational episodes can be described in terms of the timing of the fall in base level initiating each new landscape cycle. None the less, the landsurface itself is diachronous because in King's model landscapes essentially develop through backwearing as escarpments experience parallel retreat; landsurfaces, therefore, are progressively older away from escarpments (Fig. 18.3). Consequently, it is possible to talk of the local age of a landsurface, and even to refer to a terminal age determined by the final removal of a pediplain remnant.

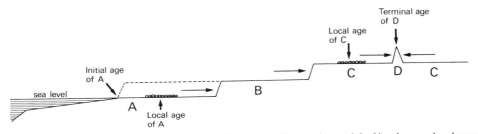

Fig. 18.3 *Different criteria for defining the ages of erosion surfaces according to the model of landscape development proposed by L. C. King. The surfaces labelled A–D were initiated during three episodes of base level fall. Each is diachronous and deposits on the surface may be capable of yielding a minimum local age at that point. The final elimination of the last remnant of a particular surface (D) gives its terminal age.*

King originally envisaged episodes of uplift occurring predominantly along continental margins as the result of a delayed isostatic response to denudational unloading. He considered that such isostatic uplift would only take effect once an escarpment had retreated over a critical threshold distance. This mechanism of discontinuous isostatic uplift is based on a misunderstanding by King (and other geomorphologists since) of how the crust responds to changes in load. The response is, of course, continuous on the geological time scale of denudational cycles (see Section 2.2.4), although flexural isostasy does provide a possible means of generating surface uplift along passive margins experiencing escarpment retreat (see Section 4.2.3). Subsequently, King advocated the somewhat ill-defined mechanism of **cymatogeny**, involving the upwarping and flexure of continental margins as a result of active subcrustal processes, as the means by which new cycles of pediplanation could be initiated. Although modern concepts of passive margin tectonics have now replaced King's ideas about uplift mechanisms, the notion of widespread and long-term escarpment retreat has gained new impetus from recent attempts to understand the evolving morphology of passive margins following continental break-up.

18.1.4 The Büdel model: etchplanation

Although having little impact on the development of Anglo-American geomorphology, the ideas of J. Büdel have exerted a considerable influence on workers in the continent of Europe, especially in Germany. Büdel's key notion concerning landscape development is that of a 'double surface of levelling'. In regions covered with thick weathering deposits, especially the relatively stable shield areas of the humid tropics, denudation of the landscape occurs simultaneously through the removal of material from the surface – largely, it is thought, by sheet wash – and by ongoing chemical decomposition at the weathering front. This combination of deep weathering and surface removal produces an **etchplain** (or an **etchsurface** where an uneven basal surface has been exposed) and the overall process is termed **etchplanation** (Fig. 18.4). Elements of this model of landscape evolution can be traced back to the British geologist, E. J. Wayland, who worked in Uganda in East Africa in the 1930s, but it is Büdel who has developed the concept of etchplanation most fully.

In the humid tropics an important element in landscape development is the spatial variability of the factors which determine weathering rates, especially lithology and drainage. As a result of variations in these controls the form of the weathering front is highly irregular, and the depth of the weathering mantle does not necessarily bear any relationship to the form of the ground surface. During periods of tectonic and climatic stability rates of weathering and denudation are roughly in balance and the depth of the weather-

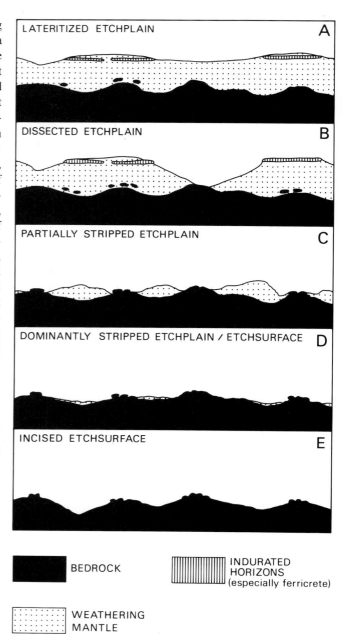

LATERITIZED ETCHPLAIN — A

DISSECTED ETCHPLAIN — B

PARTIALLY STRIPPED ETCHPLAIN — C

DOMINANTLY STRIPPED ETCHPLAIN / ETCHSURFACE — D

INCISED ETCHSURFACE — E

BEDROCK

INDURATED HORIZONS (especially ferricrete)

WEATHERING MANTLE

Fig. 18.4   The development of different types of etchplains and etchsurfaces. The diagrams do not necessarily represent an evolutionary sequence as repeated episodes of accelerated erosion may only succeed in partially removing the weathering mantle. The types of etchplains and etchsurfaces illustrated are: (A) lateritized etchplains comprising a surface of low local relief underlain by a thick weathering mantle, including indurated lateritic horizons (ferricretes), which has been subject to only limited stream incision; (B) dissected etchplains in which accelerated stream downcutting promoted by climatic change or uplift leads to the development of well-defined valleys, fringed in places by duricrust breakaways, and the very localized exposure of bedrock and the formation of tors; (C) partially stripped etchplains characterized by widespread stream dissection and the extensive stripping of the weathering mantle,

*including resistant duricrust layers, to reveal numerous rock
outcrops in the form of tors; (D) dominantly stripped etchplains
representing a very advanced stage of stripping in which the
weathering mantle is retained only in deep pockets along the
weathering front and where some of the exposed bedrock has also
been subject to erosion (forming an etchsurface where significant
relief is present); (E) incised etchsurfaces in which the basal
bedrock surface has been extensively modified by fluvial erosion,
almost certainly as a result of a significant change in base level
rather than climatic change. (Diagrams and descriptions of
etchplain and etchsurface types based on M. F. Thomas (1974)*
Tropical Geomorphology, *Macmillan, London, Fig. 41,
pp. 236–8.)*

ing mantle varies little. But a change in climate, or increase
in the rate of crustal uplift, can disrupt this steady state by
generating an increase in the rate of river incision or, in the
case of climatic change alone, through the disturbance of
the vegetation cover. During such a perturbation of the
geomorphic system the weathering mantle may be partially,
or even wholly stripped. As the landscape is lowered in
response to the more vigorous erosional activity, water
tables will fall. This will tend to increase rates of water
throughput at the weathering front and in turn lead to an
increase in the rate of weathering.

Etchplains can assume a range of forms depending on a
number of factors including the lithology and structure of
the local rocks (which influences the depth of the weathering mantle), the intensity and duration of erosional episodes
and the morphology of the basal weathering surface. Various
types of etchplain can be produced as a result (Fig. 18.4),
and this has led to some confusion over the application of
the term.

18.1.5   Classic models of landscape evolution: summary and assessment

At this point it is probably useful to summarize very briefly
the essential elements of what might be described as the
classic models of landscape evolution, before seeing in the
following sections how the problem of long-term landform
development is currently being tackled. The cycles of erosion
envisaged by Davis and King are similar in that they both
assume that surface uplift occurs as more or less discrete
pulses which punctuate the progressive erosional development of the landscape. Penck, on the other hand, explicitly
incorporated the idea of surface uplift occurring for much
longer periods of landscape history and playing an integral
role in how the landscape evolves rather than simply providing an initial input of potential energy. But in terms of
the changes in form that the landscape experiences through
time, King is much closer to Penck in proposing that backwearing generally predominates over downwearing (although,
as we have pointed out, Penck's and King's conceptions of
exactly how slopes retreat were different). With particular
respect to King's form of parallel retreat, it has been pointed

out that this could not occur in any strict sense over large
horizontal distances because of inevitable variations in rock
strength associated with what might be quite subtle changes
in lithology. While clearly accepting this point, it is useful
to retain the notion of parallel retreat in a broad sense in
order that it can be contrasted with the idea of a progressive
decline in slope gradients in the landscape through time.

The distinction between backwearing and downwearing
is important because of the different isostatic responses to
which we would expect them to give rise. Isostatic compensation of a landscape experiencing extensive downwearing
would not, in general, lead to any surface uplift, whereas
flexural effects along the kind of sharp topographic discontinuity formed by a major escarpment could lead to
localized surface uplift (see Section 4.2.3). It is, of course,
possible, and indeed likely, that both downwearing and
backwearing occur simultaneously, although the latter may
be slow with respect to the former. Slow downwearing is,
in fact, just what is implied by Büdel's notion of etchplanation, and it certainly seems that we cannot assume that
once pediments are created by escarpment retreat they
necessarily remain immune from the effects of weathering
and erosion. Indeed, if we accept the evidence for generally
warmer (and probably also wetter) climates in the Cretaceous
and Early Cenozoic (see Section 18.4.3) then very ancient
landsurfaces are unlikely to have remained untouched by
episodes of deep weathering even in regions which are now
predominantly arid.

Although the subject of intense debate during the first
half of this century, over the past two or three decades there
has been relatively little discussion among geomorphologists about the relative merits of the classic schemes of
landscape evolution discussed here. Many have regarded
them as being so deficient in their treatment of exogenic
geomorphic processes that they are barely worth serious
consideration. While accepting the value of the idea of progressive landscape change through time, others have rejected
the specific models of landform change proposed as over-
simplified and inadequate. Yet other geomorphologists have
pointed to the important effects that lithology or changing
morphoclimatic regimes have on the way landforms evolve
through time, and have argued that these factors render the
search for an all-embracing model of landscape evolution
futile. Finally, there has been the idea that in reality landscapes do not in fact evolve in any systematic manner but
simply oscillate around an equilibrium form.

Irrespective of the merits of these views, their effect has
been to direct attention away from problems of long-term
landscape development to the apparently more tractable
questions posed by the nature of shorter-term, and smaller
scale, geomorphic change. None the less, this situation is
now changing as a result of both conceptual and technical
developments since the mid-1970s. One has been the attempt
to integrate a modified version of the concept of dynamic

equilibrium into the notion of progressive landscape change embodied in the principle of evolution. Another has been the revolution in our knowledge of tectonic processes that has occurred over the past two decades and in particular the way that this has immeasurably improved our understanding of the nature and causes of uplift. Finally, there have been major advances in the dating of geomorphic events and the ability to estimate long-term rates of denudation (see Section 15.4). It is to these themes that we turn next.

18.2 Landscape stability and change

18.2.1 The Hack dynamic equilibrium model

One reaction to the evolutionary thinking embodied in Davis's notion of a cycle of erosion was the proposal by J. T. Hack that landscapes could be better understood in terms of 'dynamic equilibrium'. In rejecting the idea of progressive change in the form of the landscape through time, Hack resurrected the approach of G. K. Gilbert focusing on the continuous adjustment between force and resistance. He argued that in landscapes that have experienced a long period of denudation there will be a mutual adjustment between lithological controls and prevailing surface processes. In the ideal case where base level, surface processes and lithology remain constant through time, the form of the landsurface remains unchanged since the whole landscape is lowered at a constant rate. Relief, slope angles and stream gradients are adjusted in such a way that each unit area yields the same sediment load; regions of resistant rock have steep, rugged relief, whereas areas of less resistant lithologies have subdued relief and gentle slopes. (Note that in the sense we have already defined the terms (see Section 1.3.4) this model is essentially one of steady-state equilibrium.) The major shortcoming of this approach as a general landscape model is that, while this condition of uniform lowering might apply to particular areas of limited extent, it cannot apply to entire drainage basins in the long term. This is because the lowering of the surface of a drainage basin towards base level necessarily involves a reduction in the gradient of trunk streams, and this will eventually affect tributary basins. A further problem with Hack's dynamic equilibrium concept is that climatic change and tectonic activity are likely to lead to changes in the nature and rates of processes through time, while progressive surface lowering will expose different lithologies. The maintenance of a 'dynamic equilibrium' assumes a rapid adjustment to such changes but there is abundant evidence in some landscapes of the survival of relict landforms. It appears that the concept of 'dynamic equilibrium' is likely to be most applicable to parts of slowly eroding landscapes which have not experienced major climatic shifts and which are effectively isolated from base level changes.

18.2.2 The dynamic metastable equilibrium model

In Chapter 1 we discussed how the idea of equilibrium in a landscape was linked to the temporal and spatial scale being considered (see Section 1.3.4). However, we have yet to consider exactly how a landscape composed of individual components in a steady-state equilibrium can experience progressive lowering in the longer term. This problem has been addressed by S. A. Schumm who has proposed that these ideas of landscape stability and landscape change can be reconciled by incorporating the concept of episodic erosion (see Section 9.5.1) into a decay equilibrium model of landscape evolution. The key element of Schumm's model is that valley floors are lowered episodically rather than continuously. This could occur through the accumulation of sediment from the upper parts of a basin covering the bedrock of the valley floor. Periodically this sediment is removed and the bedrock of the valley floor is lowered. Such valley floor incision may be promoted by external factors, such as a major flood (see Section 9.5.2), but it might also arise as a result of the breaching of a geomorphic threshold due to sediment deposition, causing the channel gradient to reach a threshold of instability.

This situation can be described as one of **dynamic metastable equilibrium**, and differs from the concept of dynamic equilibrium in that the reduction in channel bed elevation takes place discontinuously not progressively. Under conditions of dynamic metastable equilibrium, phases of steady-state equilibrium are punctuated by adjustments involving the breaching of either extrinsic or geomorphic thresholds which shift the steady-state equilibrium to a new level (Fig. 18.5). Schumm argues, therefore, that it is possible to replace the Davisian notion of progressive, gradual change through time by a model of landscape evolution in which change occurs in a step-like manner (Fig. 18.6). Interesting-

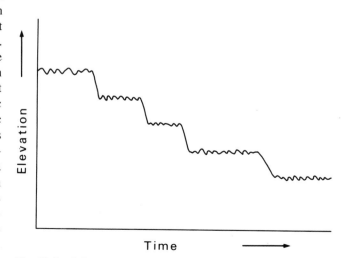

Fig. 18.5 *Schematic representation of dynamic metastable equilibrium. Compare with Figure 1.9(C).*

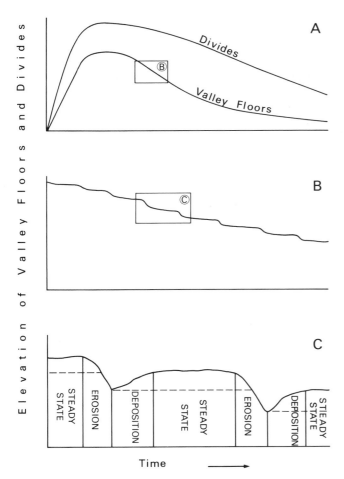

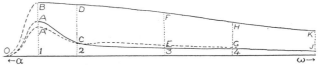

Fig. 18.7 *Davis's representation of the cycle of erosion incorporating the idea of valley aggradation (shown by the dashed line C–E–G) during drainage basin evolution (From W. M. Davis, (1899) Geographical Journal **14**, Fig. 1, p. 486.)*

Fig. 18.6 *Modified form of the cycle of erosion incorporating the idea of episodic erosion and dynamic metastable equilibrium: (A) cycle of erosion as envisaged by Davis; (B) discontinuous reduction in elevation of valley floors as a result of episodic erosion; (C) close-up of (B) showing periods of steady-state equilibrium interrupted by episodes of rapid erosion followed by deposition. (Based on S. A. Schumm (1975) in W. N. Melhorn and R. C. Flemal (eds) Theories of Landform Development. Allen and Unwin, London, Fig. 5, p. 77.)*

ly, Davis himself appears to have been aware that the reduction of elevation of valley floors was not necessarily a progressive and continuous process. In his 1899 presentation of the cycle of erosion he pointed out that at some stage after the initiation of a cycle trunk rivers might not be able to cope with the amount of sediment being generated by actively downcutting upper tributary streams and would adjust by aggrading. He even provided an illustration of this effect (Fig. 18.7) which in some respects anticipates the concept of episodic erosion.

18.2.3 Landscape sensitivity

Another necessary modification of our understanding of landscape development is that landscapes vary in both the

magnitude and frequency of external changes that they experience and in their ability to adjust to such changes. A core concept in geomorphology is that of characteristic form; distinct landform environments exist which are determined by a set of external variables including rock type and structure, climate, biological activity, and rates of tectonic activity and base level changes. These variables interact over time and space, giving rise to a characteristic magnitude and frequency of landforming events and producing a uniform and repetitive assemblage of landforms reflecting a strong interdependence between process and form which persists for as long as the controlling conditions remain constant.

These characteristic forms are, in reality, subject to significant changes in external conditions, either in the form of environmental change or more transient high magnitude – low frequency events, as well as inherent internal instabilities at a variety of time scales. In fact it has been argued that the conditions controlling landscape development rarely remain constant sufficiently long for characteristic forms to develop and that in such cases **transient forms** dominate the landscape. This is particularly likely to be true in environments where rare, large magnitude events can generate substantial long-lived changes in the landscape. Viewed in this way we can divide geomorphic time into three categories: **lag time**, which represents the time taken for the system to begin reacting to change; relaxation time, which designates the time taken to achieve a characteristic form; and **characteristic form time** which indicates the time over which characteristic forms persist (Fig. 18.8). The central, and as yet largely unanswered, question is what is the relative importance of characteristic and transient forms in the landscape?

Clearly, a crucial factor in deciding this question is the sensitivity of the landscape to change. **Landscape sensitivity** is defined as the propensity for the landscape to undergo a recognizable change in response to changes in the external variables controlling the geomorphic system. Whether a landscape experiences a change will depend on both its sensitivity (an inherent property of the landscape system) and the magnitude, frequency and duration of changes in the controlling variables. Traditionally, the resistance of a landscape to change has been seen largely in terms of either rock resistance (as reflected in rock strength, resistance to weathering and erodibility) or morphological resistance (in

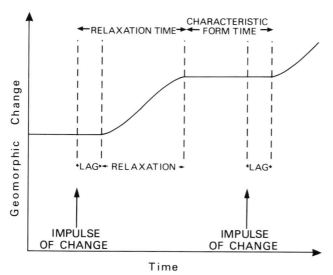

Fig. 18.8 *Schematic illustration of the concepts of lag time, relaxation time and characteristic form time. (After D. Brunsden, (1980)* Zeitschrift für Geomorphologie Supplementband *36, Fig. 1, p. 18.)*

Table 18.1 Characteristics associated with sensitive and insensitive landscapes

CHARACTERISTIC	SENSITIVE LANDSCAPES	INSENSITIVE LANDSCAPES
Nature of geomorphic system	Fast-responding	Slow-responding
Local relief	High	Low
Denudation rates	High	Low
Response to climatic change	Rapid	Slow
Relationship of landforms to contemporary morphoclimatic regime	Strong	Weak
Landform persistence	Short	Long
Magnitude of geomorphic event required to overcome barriers to change	Small	Large
Relative importance of high magnitude/low frequency events	Low	High
Abundance of characteristic forms	High	Low
Abundance of transient forms	Low	High

the sense that low gradient slopes and regions of low local relief will be less responsive to external changes).

Although these elements of landscape sensitivity are important it is now realized that other factors are also significant. These include channel gradient, drainage density and the strength of the linkages between the various components of the landscape system. Where channel gradients are steep and drainage density is high, changes (in, for instance, base level) can propagate rapidly throughout a drainage basin. In headwater areas of drainage basins there is a strong coupling between channel systems and slope systems, and consequently a high landscape sensitivity, whereas downstream floodplain development means that river channels are only weakly linked with valley-side slopes and interfluves, thus making these landscape components less sensitive to change. Landscape sensitivity can also be affected by the sequence of morphoclimatic regimes produced by climatic change. Characteristic forms generated under one morphoclimatic regime may be of too low a gradient to be significantly altered under the following regime.

We can now identify two end-members of landscape sensitivity: highly sensitive, fast-responding, mobile geomorphic systems which adjust rapidly to change, and insensitive, slow-responding systems which are highly resistant to change. The characteristics of these two extreme types of geomorphic system are outlined in Table 18.1. Within any landscape, sensitivity will inevitably vary since certain landscape elements, such as river channels and soil systems, will always be more responsive than others, such as plateaus and interfluves. Nevertheless, the gross contrasts *between* different landscapes are of critical importance in the interpretation of

their long-term development. Insensitive landscapes will be characterized by a persistence of landform patterns and of structural and lithological controls. The landscape will not reflect landform development under any one morphoclimatic regime; rather it will be a palimpsest of geomorphic systems consisting of a series of forms partially developed or modified under previous morphoclimatic conditions. Characteristic forms will be rare or absent; transient forms will predominate. Sensitive landscape systems, on the other hand, will display a much closer correspondence with a single morphoclimatic regime and lithological and structural controls will be less marked. Characteristic forms are much more likely to be present but transient forms may also be abundant, especially where changes in external controlling variables are frequent.

18.2.4 Time-dependent and time-independent landforms

The concept of landscape sensitivity can be linked to ideas about the role of temporal and spatial scale in understanding landform genesis to provide us with a very promising framework within which to address the problem of whether landscapes should be viewed as evolving or in a steady state. At one extreme the form of a landscape can be viewed solely in terms of its stage of development and thus regarded as **time-dependent**; at the other a landscape can be analyzed solely in terms of the interaction between the materials of which it is composed and the processes acting upon them – the **time-independent** approach. Some researchers regard Hack's dynamic equilibrium model to be time-independent, but this is not the case for landscapes as a whole. Hack suggested that landforms would retain a constant form over time, but since he also envisaged downwearing through time the landscape overall must gradually be reduced in elevation and is therefore ultimately time-dependent.

Whether a time-dependent or time-independent approach is an appropriate basis for landscape analysis will depend very much on the temporal and spatial scale of our study. But it also depends on the sensitivity of the landscape since, for a given magnitude and frequency of perturbations in external controls (such as climatic change and tectonic activity), a sensitive landscape will adjust much more rapidly than an insensitive landscape and will therefore spend a larger proportion of its time in equilibrium with prevailing environmental conditions.

We would expect the most extensive development of time-independent landscapes where local relief and denudation rates are high and where there is a strong linkage between landscape components, so that adjustments propagating through the drainage system can extend rapidly through slope systems and up to interfluves. Active orogenic belts characterized by extremely high rates of crustal uplift and denudation seem to be the only regions where these requirements are likely to be attained over a landscape as a whole.

One of the best examples is probably the Southern Alps in New Zealand. In Section 15.8 we noted the relative rates of uplift and denudation in these mountains and found them to be comparable. The topography of South Island, New Zealand, shows a transition from an incised erosion surface in the east to sharp-peaked mountains in the west, reflecting an increasing rate of crustal uplift and denudation as the plate boundary represented by the Alpine Fault is approached (Fig. 3.29). As the erosion surface has moved towards the plate boundary it has been uplifted, and the resulting increase in channel gradients has led to river incision and the widening of river valleys with a corresponding narrowing of plateau summits (Fig. 18.9). At an altitude of around 2000 m (at which point the crustal uplift rate is around 500 mm ka^{-1}) the original erosion surface has been completely removed and above this height interfluves consist of sharp ridges. Up to this point the morphology of the landscape is time-dependent and we can plot an evolutionary sequence of forms. Surface uplift is occurring because crustal uplift is increasing more rapidly than denudation as a result of the relatively poor linkage between river channels and interfluves; that is, the landscape is relatively insensitive to changes in external conditions (in this case the accelerating rate of crustal uplift).

Once valley deepening and widening has produced sharp-edged interfluves the morphology of the landscape changes comparatively little, although it continues to experience surface uplift up to the point where the rate of crustal uplift is balanced by the rate of denudation. Valley-side slopes are straight and at their threshold angle of stability, and therefore respond very rapidly to changes in conditions in

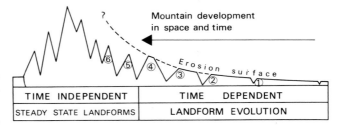

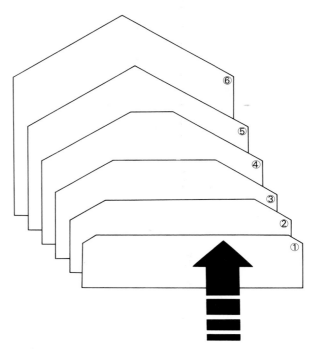

Fig. 18.9 *Schematic representation of the transition from time-dependent to time-independent landforms in the Southern Alps, New Zealand. The actual transition occurs between stages 4 and 5 when the flat-topped interfluves are finally eliminated as a result of valley deepening and widening. (Based partly on J. Adams, (1985) in: M. Morisawa and J. T. Hack (eds) Tectonic Geomorphology. Allen and Unwin, Boston, Fig. 4, p. 117.)*

the river channels. The highest parts of the Southern Alps can therefore be regarded as being in a true steady state where crustal uplift and denudation are in equilibrium and elevation oscillates around a constant value. As such it is a time-independent landscape which will retain a constant form for as long as external factors remain roughly constant.

We can extend this steady-state model applied to the Southern Alps to other active orogenic belts. The rugged mountains which form the backbone of the island of Taiwan appear to provide another example of an extensive steady-state landscape. Taiwan lies along the active boundary between the Philippine and Eurasian Plates and their oblique convergence has led to the southward propagation of a mountain belt, the Central Mountains, over the past 4 Ma.

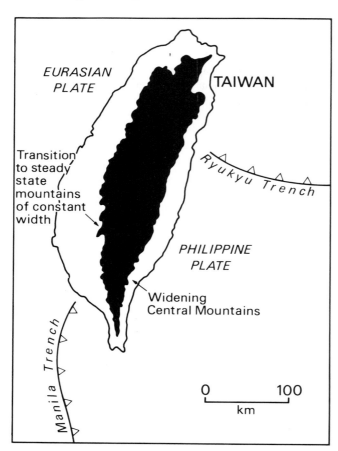

Fig. 18.10 *Topography of Taiwan in its plate tectonic setting showing the northward increase in the breadth of the Central Mountains to a constant width of 90 km.*

This is evident in the northward increase in width and elevation of the mountain range over a distance of around 120 km in the southern part of the island (Fig. 18.10). North of this point the mountain belt has a constant width of about 90 km and is in a steady state with crustal uplift and denudation both at rates of around 5000 mm ka^{-1}. The progressive northward widening of the Central Mountains provides an excellent case for the application of space–time substitution (see Section 1.4.2), since from the known rates of plate motion it can be demonstrated that the growth in width of the mountain belt over a distance of 120 km takes 1.3 Ma.

The elevation attained by steady-state topography will depend both on the rate of crustal uplift and lithological controls. If rates of crustal uplift are slow then an equilibrium with denudation is likely to be achieved before high mountains have been created. In weakly resistant rocks, on the other hand, even rapid rates of crustal uplift will not generate very high topography because higher rates of denudation can be generated on lower gradient slopes. Of great significance, however, is the width of the mountain belt (see Section 15.8) because changes in base level and

channel gradient brought about by uplift will take longer to propagate through large than through small drainage basins.

In intra-plate settings landscapes are very unlikely to contain extensive areas of truly steady-state, time-independent landforms, even though the very slow rates of landscape lowering may give the impression of elevation stability. The predominance of low channel gradients and poor linkages between fluvial and slope systems render such landscapes predominantly insensitive. This landscape insensitivity reaches an extreme form in landscapes in parts of Africa and Australia which retain a landform record of tectonic and climatic events extending back over tens of millions of years. More generally, the major escarpments which characterize many passive margins effectively decouple the coastal and interior geomorphic systems so that changes in base level are transmitted very inefficiently across these landscapes.

18.3   Denudation chronology

18.3.1   Principles of denudation chronology

Implicit in the Davisian model of landform evolution was the idea that a landscape should retain some evidence of earlier landscape cycles, or at least of separate phases of development associated with episodes of rejuvenation within a single cycle. It was thus possible, it was argued, to construct a landscape history, or **denudation chronology**, documenting the sequence of erosional events. Early attempts at such landscape histories were based largely on morphological evidence which could only yield a crude relative chronology.

In theory a denudation chronology would be expected to record the sequence of erosional and depositional events resulting from climatic as well as tectonic and sea-level controls. Although climatic oscillations clearly have a profound impact on landscape development, most denudation chronologies constructed within a Davisian framework were concerned with the reflection of major base-level changes in the formation of sequences of erosion surfaces over millions or tens of millions of years. Since the 1960s there has been a severe reaction to the production of such denudational schemes which have been widely regarded as going far beyond the limitations of the available data. Criticisms have been based on a number of grounds.

A major problem is the extrapolation of surfaces assumed to be of the same age over large distances between often small erosional remnants. Accordance of summit elevation as a basis for the correlation between, and relative dating of, erosion surfaces is complicated by the possibility that surfaces may be warped after they are formed. This may be obvious where extensive remnants remain, but can cause considerable difficulties in correlation where there are only small, isolated erosion surface residuals. Erosion surfaces, by their very nature, are often bereft of deposits which might provide a basis for dating. In some cases weathering deposits have been cited as an indicator of relative age, it being assumed that degree of weathering can be regarded as a measure of relative age. But it is now appreciated that local and regional factors of drainage and climate can have an overriding influence on the nature of weathering deposits.

Another problem is the isolating of structural controls from base-level controls in the formation of erosion surfaces. Although an erosion surface may locally truncate different lithologies, and thus appear to be related to purely erosional controls, this does not necessarily exclude the possibility of local structural controls on drainage channels which in turn influence landscape development upstream. Separating structural from drainage basin-wide base-level controls is a particular problem in regions, such as parts of eastern Australia (Fig. 18.11) and southern Africa (Fig. 18.12), which are dominated by flat-lying lithologies of varying resistance.

A further difficulty is that landsurfaces previously buried by sediments may be exhumed by subsequent erosion and significantly influence the course of later landscape evolution. In southern Africa and Western Australia, for instance, there is evidence of the exhumation of pre-Permian landsurfaces, and it may be difficult to differentiate such surfaces from those that have undergone continuous sub-aerial development. A final possible problem is that of the preservation of erosion surfaces over long periods of time. Given the recent evidence of how extraordinarily low rates of denudation can be on the plateau surfaces of areas such as south-eastern Australia, this is less of a difficulty than it once appeared. Moreover, it is evident that many low relief landsurfaces are armoured by duricrusts and are thereby very effectively protected from erosion. Clearly, the construction of denudation chronologies is going to be most appropriate where landscape sensitivity is low and where, therefore, there is going to be the greatest potential for the preservation of landforms related to earlier denudational events.

It now seems that the greatest obstacle to reconstructing landscape histories, that of attaching dates to denudational events, can be tackled, at least to a certain extent. One means of doing this is by applying geochronometric dating to materials on the landsurface (see Section 15.4.1.3). An example of this approach is the estimation of the ages of valleys and their subsequent lowering by the radiometric dating of basalt flows that they contain (see Section 18.3.3.4). Another valuable source of information applicable to the timing and depth of erosion over large continental areas is provided by the sedimentary deposits produced, particularly those that accumulate offshore.

18.3.2 Continental denudation and the marine stratigraphic record

One of the most exciting potential developments in the understanding of landform development through time over the past decade has been the vast improvement in our knowledge of the offshore sedimentary sequences of the continental margins. Such sequences have to analysed with care in view of the possibilities of sediment redistribution by submarine currents, and to allow for the effects of compaction and the contribution of biogenic material; none the less they contain a wealth of information which could provide insights into the variations in rates of denudation through time in the adjacent continental hinterland. Such sedimentary sequences have already been widely used to reconstruct sea-level changes (see Section 17.2.2.2) but their application to estimating long-term denudation rates is still in its infancy (see Section 15.4.1.1).

18.3.3 Regional case studies

By their very nature, studies attempting to reconstruct landscape histories involve detailed morphological analysis and the integration of numerous types of data. Such studies cannot, therefore, be adequately summarized in a brief form. The case studies outlined here merely indicate the approach taken and the major conclusions reached. They can be pursued, however, through the references provided at the end of the chapter.

18.3.3.1 South-east England

In 1939 S. W. Wooldridge and D. L. Linton published a detailed reconstruction of the denudation chronology of south-east England which was revised in 1955. Their landscape history covered a relatively small area consisting of Cenozoic sediments overlying chalk of Late Cretaceous age, and drew on a wide range of data, including morphological, lithological and structural evidence, drainage anomalies and datable deposits. They identified three distinct erosion sur-

Fig. 18.11 *Structural control of topography by a massive sandstone unit in the valley of the Shoalhaven River, East Australian Highlands, east of Canberra, Australia.*

Fig. 18.12   *Landsat image showing the structurally controlled Great Escarpment of western Cape Province, South Africa. The location of the photograph at the beginning of the chapter is arrowed. (Landsat image from Satellite Remote Sensing Centre, R.S.A.)*

faces: a warped, sub-Cenozoic unconformity trimmed by marine action ('Sub-Paleogene surface'); an unwarped Neogene erosion surface; and an unwarped Plio-Pleistocene marine platform. These provided the basis for a complex denudation chronology (Table 18.2) in which the Sub-Paleogene surface was uplifted and deformed prior to the formation of the Neogene erosion surface, the margins of which were in turn trimmed by a Late Pliocene–Early Pleistocene marine transgression (the Calabrian transgression). This incursion of the sea also laid down marine deposits from which drainage was superimposed. This led to the contrast between the structurally adjusted drainage of the

higher, sub-aerially eroded central part of the region (the Weald) and the discordant drainage of the periphery.

Wooldridge and Linton's interpretation of landscape history, involving distinct episodes of uplift interspersed with the formation of low relief erosion surfaces, was strongly cyclic in nature and followed clearly in the tradition of Davis. Subsequent criticisms have focused on the continuity of tectonic activity and the degree of landscape modification achieved by the Calabrian transgression (Table 18.2). It is now thought that tectonic instability throughout the Cenozoic was much more complex and prolonged, and the Calabrian transgression probably less extensive, and certainly much

Table 18.2 Comparison of two versions of the denudation chronology of south-east England

EPOCH/PERIOD	WOOLDRIDGE AND LINTON (1955)	JONES (1980)
Holocene	Impact of human activities. Post-glacial transgression leads to valley sedimentation and coastline development	
Pleistocene	Cyclic fall in sea level producing erosion surfaces and terraces. Periglacial phases.	Widespread incision in response to glacially lowered sea levels. Slow retreat of chalk escarpments and lowering of clay vales leads to increased local relief.
Pliocene	Calabrian transgression-development of discordant drainage pattern	Continued uparching of Weald but at a slower rate. Rivers cut down to within 100–150 m of sea level in the Weald by end of Pliocene. Widespread removal of Paleogene cover and Upper Cretaceous chalk leads to exposure of various Lower Cretaceous strata. Lithological control of drainage development becomes important. A Mio-Pliocene marine transgression leaves little impact on landscape
Miocene	Development of Mio-Pliocene peneplain	
Oligocene	Alpine folding phase. Drainage disruption	Low relief Paleogene surface with duricrusts. Further growth of Weald anticline. Progressive withdrawal of sea from structural basins promotes further sub-aerial erosion. Integration of drainage.
	Sub-aerial denudation. Development of sarsens (silcrete)	
Eocene	Re-establishment of down-dip drainage pattern	Continuing flexure leads to emergence of Weald and initiation of concordant radial drainage pattern on mantle of Paleocene deposits
Paleocene	Paleogene transgression–development of 'Sub-Paleogene' surface	Continuing tectonic deformation leading to predominantly eastward-flowing drainage. Transgression drowns structurally low areas. Up to 200 m of denudation.
	Sub-aerial denudation and initiation of down-dip drainage pattern	
Cretaceous	Uplift. Development of major structural pattern	300–450 m of uplift around 70 Ma BP along a NW–SE axis. Upwarping leads to sub-aerial exposure
	Aptian–Cenomanian transgression. Deposition of 300 m of chalk	

Source: Based largely on D. K. C. Jones (1980), in D. K. C. Jones (ed.) *The Shaping of Southern England*. Institute of British Geographers Special Publication **11**, 13–47.

less erosionally significant that Wooldridge and Linton envisaged.

18.3.3.2 The Appalachian region

The Appalachian region of the eastern USA has been a focus of historical landscape studies for over a century since the publication of Davis's classic *Rivers and valleys of Pennsylvania* in 1889. In spite of the pervasive relationship between morphology and lithology over much of the region, Davis and his disciples considered that there were sufficient departures from lithological controls to warrant a cyclic interpretation involving successive episodes of uplift and erosion surface formation. The major anomalies identified included the fact that although the tectonic structure of the Appalachians is broadly symmetric, the drainage is decidedly asymmetric with rivers crossing from west to east. This transverse drainage in turn suggested a history of antecedence, superimposition and headward erosion along less resistant lithological units. In addition, the apparent accordance of summit elevations in parts of the Appalachian region suggested the existence of erosion surfaces. The lack of any datable deposits between the Early Permian and the Pleistocene meant that there were effectively no critical tests available for the various schemes proposed. In some cases erosion surfaces were extrapolated entirely from unconformities in sedimentary sequences along the adjacent coastal plain, and extensive sedimentary covers were postulated (notably a marine Cretaceous cover) for areas in which no remaining outcrops existed (Fig. 18.13).

In stark contrast to the cyclic interpretations of Davis and his followers, Hack used the Appalachians as an example of the operation of 'dynamic equilibrium' in his initial presentation of the concept. Where the evolutionists saw cyclic erosion surfaces Hack saw accordant summits at different levels on different lithologies, and attributed them to the variable rates of erosion on rocks of contrasting resistance. More recently the long-standing problem of the geomorphic development of the Appalachian region has been looked at afresh with the benefit of new techniques and data able to yield estimates of actual rates of long-term denudation, and new tectonic concepts which place the landscape history in the context of early orogenic events and subsequent continental break-up and passive margin tectonics. Detailed research on deep subsurface structures has demonstrated that some correspond in a complex fashion to surface morphology, while new data on the tectonic and sedimentary history of the region have led to the abandonment by some researchers of the idea of an extensive cover of Cretaceous

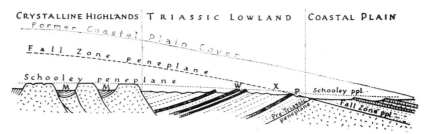

Fig. 18.13 *Scheme of landscape development proposed by D. W. Johnson for the eastern margin of the USA in the vicinity of New Jersey involving the formation of a pre-Cretaceous Fall Zone peneplain which was covered by a Cretaceous marine transgression before being upwarped and eroded to form the widespread Schooley peneplain. (From D. W. Johnson (1931)* Stream Sculpture on the Atlantic Slope. *Columbia University Press, New York, Fig. 15, p. 77.)*

strata over the region. Fission track dating inland in conjunction with data on sediment volumes offshore has shown that denudation over the region of eastward draining rivers has averaged around 2–2.5 km over the past 100 Ma, at rates ranging up to 60 m Ma^{-1}. Although rates of denudation have apparently varied over both time and space, there is no evidence for the succession of brief episodes of uplift and prolonged periods of erosion required by conventional cyclic models of landscape evolution.

The growing consensus is that, although there have been discontinuities in rates of crustal uplift, at least during the Late Cenozoic, landscape history since the opening of the North Atlantic and associated uplift of the new continental margin in the Late Cretaceous has probably been dominated by ongoing denudation and compensatory isostatic uplift. While a mean thickness of about 2.5 km of rock has been removed, isostatic rebound has reduced surface lowering to around 500 m. This situation has been described by some researchers as representing a steady state between weathering, erosion and crustal uplift, but this is not strictly true because local compensatory isostatic uplift alone leads to a progressive, albeit in many cases very slow, surface lowering. Flexural effects could allow a localized steady state to be maintained for a short period, but these have yet to be considered in relation to landscape development in this area.

18.3.3.3   Southern Africa

Perhaps the grandest schemes of denudation chronology have been those put forward by L. C. King. On the basis of detailed investigations of the morphology and sedimentary sequences of the coastal belt of Natal in South Africa, King proposed a chronological scheme which, he argued could be applied on a global basis (Fig. 18.14). Few would now accept the possibility of world-wide denudational episodes in the light of our present understanding of uplift patterns in plate interiors and the probable dismemberment of drainage by continental break-up (see Section 16.7.1). Nevertheless, the attempt by King to link onshore and offshore data anticipated what is likely to become an important approach in the documenting of landscape histories. Recently the chro-

Table 18.3 Denudation chronology of southern Africa according to T. C. Partridge and R. R. Maud (compare with Fig. 18.14)

EVENT	AGE
Post-African II cycle of erosion	Late Pliocene to Holocene
Major uplift of up to 900 m	Late Pliocene (~ 2.5 Ma BP)
Post-African I cycle of erosion	Early mid-Miocene to Late Pliocene (~ 2.5 Ma BP)
Moderate uplift of 150–300 m	End of Early Miocene (~ 18 Ma BP)
African cycle of erosion with minor tectonic interludes	Late Jurassic/Early Cretaceous to end of Early Miocene (~ 18 Ma BP)
Break up of Gondwana and initiation of Great Escarpment	Late Jurassic/Early Cretaceous

Source: Based on T. C. Partridge and R. R. Maud (1987) *South African Journal of Geology* **90**, Table 1, p. 187.

nology put forward by King has been subject to major revision (Table 18.3); it will be interesting to see to what extent this revised chronology can be related not only to detailed offshore depositional evidence but also to estimates of long-term denudation rates employing fission track dating and contemporary ideas on the tectonic history of passive margins.

Some idea of the likely value of offshore sedimentary sequences to reconstructing landscape histories can be gained from Figure 18.14 in which the denudation chronology for Africa proposed by King is compared with the record of offshore sedimentation. The most important point to note is the lack of any consistent correspondence between hiatuses (gaps) in marine deposition and either phases of uplift or erosion proposed by King. This lack of correspondence may be due to a number of factors. The denudation chronology outlined by King may simply be wrong; alternatively, the relationship between denudation inland and sedimentation offshore may be complex, in which case unconformities offshore cannot be simply equated with erosion surfaces inland. This is almost certainly the case for, as we saw in Section 17.6.1, river systems can respond in different ways

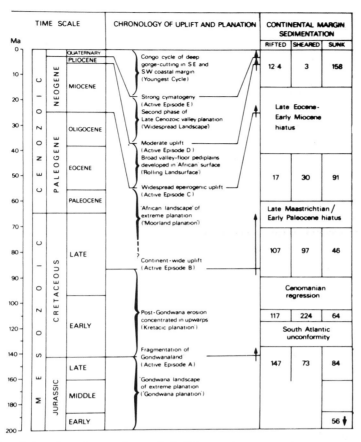

Fig. 18.14 *Comparison of denudation chronology and record of offshore sedimentation for Africa. The denudation chronology is according to L. C. King and refers primarily to central and southern Africa. King's later modified terminology for the various landsurfaces is shown in parentheses. Vertical arrows indicate approximate timing and duration of uplift according to King. Sedimentation rates are for the western, southern and eastern margins of Africa as estimated by Dingle (1982); figures refer to mean sedimentation rates in m Ma^{-1}. Sunk margins are those on the east coast which have experienced persistent subsidence since rifting. Hiatuses in deposition are also indicated. The time scale is based on Harland et al. (1982). (From M.A. Summerfield (1985) in M. Morisawa and J. T. Hack (1985) Tectonic Geomorphology. Allen and Unwin, Boston, Table 1, pp. 34–5. Primary sources: L.C. King (1967) The Morphology of the Earth (2nd edn.) Oliver and Boyd, Edinburgh; L. C. King (1976) Zeitschrift für Geomorphologie 20, 133–48; L. C. King (1983) Wandering Continents and Spreading Sea Floors on an Expanding Earth. Wiley, Chichester and New York; R. V. Dingle (1982) in R. A. Scrutton (ed.) Dynamics of Passive Margins. American Geophysical Union, Washington: Geological Society of America, Boulder 59–71; W. B. Harland et al. (1982) A Geologic Time Scale. Cambridge University Press, Cambridge.)*

to a fall in base level depending on the gradient across continental margins. Although seaward warping of a continental margin could increase channel gradients and promote faster rates of offshore sedimentation, a reduction in base level due to a eustatic sea-level fall might not generate a phase of rejuvenation in drainage basins along the margin if channel gradients decreased offshore.

18.3.3.4 *The East Australian Highlands*

The East Australian Highlands extend along almost the entire eastern periphery of the continent, although they attain their greatest elevation in the south where heights in excess of 2000 m occur. Over most of its length the highland belt has the form of a broad, asymmetric arch with a very

gradual descent to the west, but a generally sharp drop in elevation eastwards across a major escarpment to a coastal plain. The Highlands lie along a rifted margin, albeit one with a complex history.

Attempts to interpret the history of the East Australian Highlands in cyclic terms inferred a major Plio-Pleistocene surface uplift on the basis of the present elevation of a duricrusted erosion surface, assumed to have formed close to sea level in the Miocene. Deeply incised valleys cutting back from the coast, such as that of the Shoalhaven River (Fig. 18.11), added to the idea that uplift had been recent. The challenge to this cyclic model was based on both the possibility of significant lithological control and dating of Cenozoic lavas erupted on to the Highlands. Direct lithological control of the topography of the Highlands is locally provided by essentially horizontally bedded sediments (Fig. 18.11). Elsewhere, erosion surfaces bevelling steeply dipping strata may in some cases be graded to outcrops of resistant beds downstream. The presence of such lithological controls implies that the low relief surfaces of the Highlands need not have been formed close to sea level. Radiometric dating of Cenozoic lavas on the summit of the Highlands in New South Wales also demonstrate that the landsurfaces there are much older than the earlier cyclic models assumed. Recent research by both geologists and geophysicists concerning the tectonic history of the East Australian Highlands has not achieved a consensus, with at least three distinct models being suggested: (1) isostatic uplift of Mesozoic highlands in reponse to prolonged erosion; (2) active Late Mesozoic and/or Cenozoic uplift; and (3) passive margin subsidence of Mesozoic highlands punctuated by episodes of active uplift.

A particularly detailed study has been undertaken by P. Bishop in an area of the East Australian Highlands in central New South Wales. In this region, located in the headwaters of the westward-draining Lachlan River, basaltic lavas which erupted about 20 Ma BP and flowed into river valleys have been preserved as sinuous hill-top cappings. These basalt outcrops in places overlie fluvial sediments which show the flow direction of the pre-basalt streams to be similar to the present river channels. Together this evidence enables the longitudinal profiles of the pre-basalt and modern river valleys to be compared (Fig. 18.15). These show that there has been little change in longitudinal profiles over 20 Ma, during which time valley floors have been lowered by about 100 m. This extremely slow rate of incision is broadly in accord with other evidence of low denudation rates in the Murray Basin of which the Lachlan catchment forms a part (see Section 15.4.2). Such rates argue strongly against cyclic notions of landscape development and point to the predominantly slow rates of landscape modification in passive margin environments inland of actively eroding escarpment environments. Assuming local isostatic compensation, the mean denudation rates estimated for the Lachlan headwaters area indicate a surface lowering of no more than 2 m Ma^{-1} over the past 20 Ma.

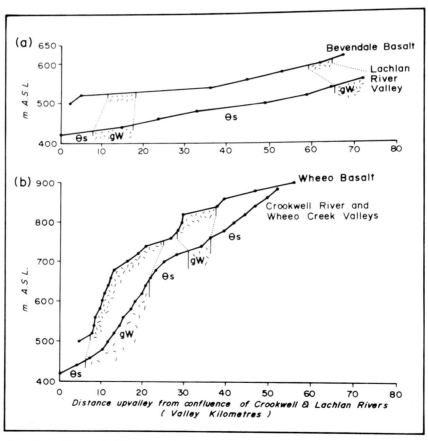

Fig. 18.15   *Longitudinal profiles of the Early Miocene and present Lachlan River valley (a), and tributary Crookwell River – Wheeo Creek valleys (b). Note that valley-floor incision from the upper (Early Miocene) profile has been relatively consistent downvalley and that lithological control of the form of the longitudinal profiles has been sustained. (From P. Bishop et al. (1985) Journal of Geology* **93**, *Fig. 7, p. 466.)*

18.4   Tectonics, climate and landform development

Various points made in this chapter underscore the importance of tectonics in controlling the way landscapes evolve; but the role of tectonics in long-term landform development is not confined to these essentially direct effects. Tectonic activity can also play an important part in influencing local, regional and global climates and thereby exert an additional indirect control over landscape evolution.

18.4.1   Continental drift and changing morphoclimatic regimes

One obvious way in which we might expect tectonics to influence climate is through continental drift. At the global scale, changes in the distribution of the continents over the past few hundred million years have certainly had a profound impact on climates world-wide, although the precise effects have yet to be ascertained. It seems clear that the movements of the southern continents played an important part in the transition from the apparently warm, equable climates of the Late Cretaceous to the distinctly cooler, and globally more diverse, climates of the Late Cenozoic (see Section 14.3.1).

An additional climatic consequence of continental drift could arise from changes in the palaeolatitudes of individual continents. In most cases, climatic changes associated with drift have been slow in comparison to changes related to major adjustments in global circulation patterns. But it has been suggested that this has not been the case for Australia, which has been moving rapidly equatorwards for about the past 40 Ma from a mean latitude of around 45°S to about 25°S at the present day. This has enabled it to maintain an essentially tropical or sub-tropical climate in spite of the marked global cooling of the Late Cenozoic. It is tempting to suggest that one consequence of this equatorward movement was Australia's entry into the latitudinal zone of aridity related to the sub-tropical high pressure cells during the Late Cenozoic. That this is not an adequate explanation of changing climatic conditions in Australia is indicated by evidence for a humid environment along the southern coast of Australia in the Miocene which extended deep into the interior. An alternative interpretation is that any climatic changes linked to drift have been subordinate to major changes in global circulation patterns, with Australia being overtaken by intensified sub-tropical high pressure cells migrating equatorwards which promoted aridity starting at about the Miocene–Pliocene boundary.

18.4.2   Orogenesis and changing morphoclimatic regimes

Another way in which tectonics can influence climate (and thereby indirectly landform development) is through oro-

genesis. Such influences can operate in various ways. Uplift occurring in an orogenic belt may cause the displacement of landforms created in a particular climatic environment into a new morphoclimatic regime. Such transitions can occur rather rapidly if the rate of surface uplift is high, and climatic landform relicts formed at lower elevations may be temporarily preserved as they pass into higher altitudes. In general, the high rates of denudation in most mountain belts would make either significant vertical displacement or prolonged survival of climatic relicts unlikely, but this is not necessarily so for montane plateaus such as the Tibetan Plateau and the Altiplano of the central Andes. In these cases the much more subdued local relief – and consequently more modest rates of denudation on the plateau summits – could permit the long-term preservation of relict landforms. Certainly, there is clear evidence from the reduction in the size of lakes for a progressively more arid climate on the Tibetan Plateau as it has apparently experienced rapid surface uplift during the Quaternary.

A gradual decrease in temperature with elevation is the most obvious effect of surface uplift. In extreme cases this can promote the growth of glaciers and the initiation of a glacial morphoclimatic regime. In most situations precipitation increases markedly with elevation, but in very high mountains there can be a gradual reduction above a height of 3000–3500 m where moisture-depleted air begins to be encountered. At the equator the mean annual air temperature is below 0 °C at an altitude of around 4500 m, but at higher latitudes this critical boundary approaches gradually closer to sea level. The Andes in Ecuador illustrate a full range of morphoclimatic regimes from lowland humid tropical to glacial. On the western slopes the upper, northern part below the snow line is characterized by intensely dissected topography created under prolonged humid conditions. This contrasts with the lower and more southerly section of the Ecuadorian Andes where landform development has occurred under a distinctly more arid climatic regime. As illustrated by the Andes, climatic transitions with altitude may not be uniform on either flank of a mountain range. Where an orogen is transverse to the prevailing rain-bearing winds there will tend to be a marked asymmetry in climate, with a drier altitudinal sequence of morphoclimatic environments on the leeward side.

18.4.3   Plate tectonics, climate and weathering

Recent research has begun to highlight a possible relationship between tectonics and climate that is even more pervasive than those relating to continental drift and orogenic uplift. This arises from variations in rates of sea-floor spreading which lead to changes in the amount of atmospheric CO_2; these, in turn, promote adjustments in global temperatures (Fig. 18.16). In the short term the concentration of atmospheric CO_2 is closely related to the life

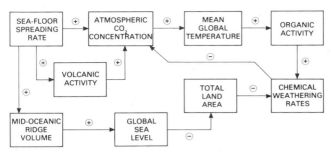

Fig. 18.16 *Schematic representation of the likely causal relationships between sea-floor spreading rates, atmospheric CO₂ concentration, climate and weathering rates.*

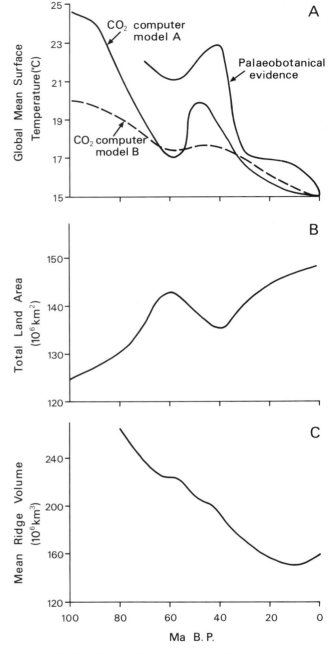

processes of organisms, but in the long term it appears that geological, or more specifically tectonic, controls are dominant. Over geological time the amount of CO_2 in the atmosphere represents a balance between its release by volcanic activity and rock metamorphism, and uptake in weathering reactions (see Section 6.2.3.3). The tectonic process which ultimately seems to control the release of CO_2 is sea-floor spreading. High rates of sea-floor spreading will be associated with both active volcanism along mid-oceanic ridges and, because of correspondingly high rates of subduction, considerable volcanic activity and metamorphism along steady-state convergent plate margins. In both cases increased rates of sea-floor spreading should be associated with increased rates of CO_2 outgassing.

There is fairly convincing evidence that rates of volcanic activity have varied significantly over periods of tens of millions of years, and there is certainly abundant data indicating that rates of sea-floor spreading have changed over the past 100 Ma or so (Fig. 18.17(C)). Although weathering reactions tend to absorb CO_2, the surface area of the continents available for these processes would be reduced during periods of high sea-floor spreading rates because of the corresponding rise in global sea levels as mid-oceanic ridge volumes grow (Fig. 18.17(B); see Section 17.5.2.1).

Atmospheric CO_2 is transparent to incoming short-wave radiation from the Sun, but efficiently absorbs outgoing long-wave radiation emitted from the Earth. This is the so-called 'greenhouse effect'. If the Cretaceous atmosphere had a significantly higher CO_2 content then the climate at that time should have been warmer world-wide, and a range of palaeontological, geochemical and palaeogeographical data certainly supports this interpretation. We would expect warmer temperatures to promote faster rates of chemical weathering, not primarily through an acceleration in weathering reaction rates, but rather as a result of enhanced rates of organic activity leading to a greater production of soil CO_2 and a consequential increase in soil acidity. Because the concentration of CO_2 in soil is generally so much higher than that in the atmosphere, changes in the overall concentration of CO_2 in the atmosphere probably have little *direct*

Fig. 18.17 *Trends in mid-oceanic ridge volume (indicative of sea-floor spreading rate), total land area and global mean surface temperature over the last 80–100 Ma. In (A) CO₂ computer model A is the predicted temperature fluctuation as a consequence of changes in CO₂ related to a variable rate of sea-floor spreading since 100 Ma BP whereas CO₂ computer model B is the temperature prediction based on a linear decrease in sea-floor spreading rate. (Based on R. A. Berner et al. (1983) American Journal of Science **283**, Fig. 12, p. 679 (A); E. J. Barron (1983) Earth-Science Reviews **19**, Fig. 4, p. 317 (B); and M. A. Kominz (1984) in J. S. Schlee (ed.) Interregional Unconformities and Hydrocarbon Accumulation. American Association of Petroleum Geologists Memoir **36** Fig. 5, p. 119 (C).)*

effect on weathering rates; rather the effect is indirect through the influence of temperature on organic activity and hence soil acidity.

Estimates of the concentration of atmospheric CO_2 during the Cretaceous vary enormously from three times to 100 times present values. Even assuming an increased concentration at the lower end of this range, the resulting likely higher global temperatures in the Cretaceous (Fig. 18.17(A)) – and the possibly significantly enhanced rates of chemical weathering – need to be borne in mind when we try and intepret landscapes whose evolution extends back to this period.

It is also worth noting that chemical weathering apparently provides a vital natural negative feedback mechanism limiting increases in the concentration of CO_2 in the atmosphere, and thereby helping to limit long-term fluctuations in global temperature. Since rates of chemical denudation, and therefore chemical weathering, are highest in regions of high local relief (see Section 15.5) we might also anticipate that long-term changes in continental morphology, in particular the rifting of continents and the growth and erosion of orogenic belts, might also influence the concentration of atmospheric CO_2. These possible relationships represent a fruitful field for future research linking climatic change to geomorphic processes and long-term landscape history.

18.4.4   Tectonics, climate and landscape development: a summary

It is useful to conclude this discussion of long-term landscape development by presenting an overview of the way in which tectonic processes are related to long-term fluctuations in global climate, which in turn generate changes in morphoclimatic regimes. A schematic representation of these relationships is provided in Figure 18.18; this indicates climatic changes arising from sea-floor spreading, continental drift, orogenesis and volcanicity. It is important to note that the relationships suggested in this diagram refer to periods of tens of millions of years. Within this temporal scale glacial–interglacial oscillations represent rapid fluctuations in morphoclimatic regimes within a framework of longer-term changes in climate. In the high and mid-latitudes the effects of these short-term climatic cycles dominate the landscape, but landform relicts are found from morphoclimatic regimes operative several tens of millions of years ago in the most insensitive landscapes of the tropics and sub-tropics.

Further reading

Models of long-term landscape evolution are discussed by Thornes and Brunsden (1977) and Thorn (1988), and in

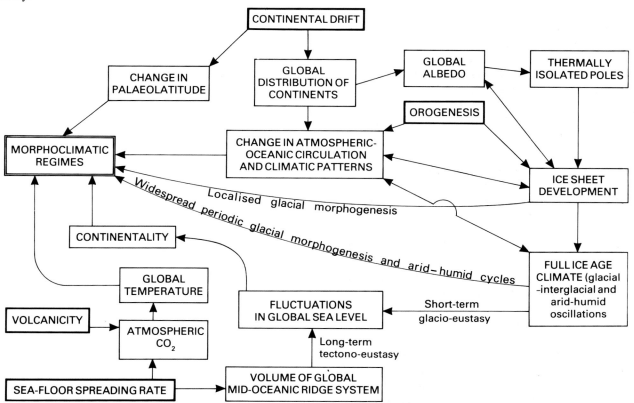

Fig. 18.18   *A schematic representation of some of the ways in which tectonic processes generate changes in morphoclimatic regimes over time. The relationships shown are to a certain extent speculative and are by no means comprehensive.*

various papers in the volume edited by Melhorn and Flemal (1975). Other contributions to the revival of interest in the theoretical problems posed by long-term landscape development include those by Chorley (1962), Howard (1965), Schumm and Lichty (1965), Thomas and Summerfield (1987) and Thornes (1983). As with all reading on this topic, great care has to be taken when using these sources in noting the sense in which the terms dynamic equilibrium, steady state and time-independent form are being applied since their precise usage varies.

The cyclic model of W. M. Davis is given exhaustive treatment in the comprehensive survey of his work by Chorley *et al.* (1973) while key writings by Davis himself include his general statement of the cycle (Davis, 1899) and his later collection of essays (Davis, 1954). The Davisian approach to landform analysis is assessed in the context of systems analysis by Chorley (1962). The ideas of Penck are less well recorded in the Anglo-American literature; the basic source is the translation of Penck's book originally published posthumously in German (Penck, 1953). It is also worth consulting the paper by Simons (1962) which tries to clear up the confusion which arose from Davis's misrepresentation of Penck's ideas concerning slope evolution.

King's model of landscape evolution is covered in a series of papers and books of which the most important are King (1953, 1957, 1967, 1976). Readers should be aware, however, of the various revisions that King made to his model, especially in terms of the timing and mechanisms of uplift, and should consult the papers by De Swardt and Bennet (1974) and Partridge and Maud (1987) which carefully document these changes. The concept of etchplanation is comprehensively covered by Büdel (1982) and Thomas (1974), while Kiewietdejonge (1984a, b) provides a wide-ranging review of Büdel's approach to geomorphology. Thomas (1989a, b) assesses the role of etching in landscape development and Thomas and Thorp (1985) apply the etchplanation model to the landscape of West Africa.

Hack's 'dynamic equilibrium' model is covered in Hack (1960, 1975), and the dynamic metastable equilibrium model and the concept of episodic erosion is developed by Schumm (1975, 1979). The concept of landscape sensitivity is discussed in detail by Brunsden and Thornes (1979), while Brunsden (1980) provides a briefer and more accessible treatment. The importance of relative rates of uplift and denudation in determining whether landscapes are essentially time-dependent or time-independent is emphasized by Adams (1985) and Ahnert (1984). Adams (1985), Whitehouse (1988) and Suppe (1987) provide case studies of steady-state landscapes along convergent plate boundaries in New Zealand and Taiwan.

Various issues relevant to denudation chronology are considered by Fairbridge and Finkl (1980), Mabbutt (1988), Ollier *et al.* (1988), Summerfield (1985), Twidale (1976) and Young (1983). Further details on the regions discussed

as case studies can be found in the following: Jones (1980), Small (1980) and Wooldridge and Linton (1955) on southeast England; Costa and Cleaves (1984), Davis (1889), Johnson (1931), Hack (1960, 1979, 1982) and Pavich (1985) on the Appalachians; De Swardt and Bennet (1974), Partridge and Maud (1987) and Summerfield (1985) on southern Africa; and Bishop (1988), Bishop *et al.* (1985), Ollier (1982), Wellman (1987, 1988) and Young (1983) on the East Australian Highlands.

Much of the rather limited literature on the relationships between tectonics, climate and landform development has not been written from a specifically geomorphic perspective, but none the less is still thought-provoking for the geomorphologist. The relationship of global climate to changing patterns in the distribution of the continents is discussed by Barron *et al.* (1984) and Cogley (1979), and the impact of orogenic uplift on montane climates is analyzed in terms of landform development by Garner (1974, 1983). The link between sea-floor spreading rates, atmospheric CO_2, global temperatures and rates of chemical weathering is examined by Berner and Lasaga (1989), while Barron (1983) reviews the evidence for a warm, equable Cretaceous climate. Finally, the idea of alternating geomorphic and sedimentological regimes arising from contrasting modes of tectonic activity through geological time is presented by Mackenzie and Pigott (1981).

References

Adams, J. (1985) Large-scale tectonic geomorphology of the Southern Alps, New Zealand. In: M. Morisawa and J. T. Hack (eds) *Tectonic Geomorphology*. Allen and Unwin, Boston, and London, 105–28.

Ahnert, F. (1984) Local relief and the height limits of mountain ranges. *American Journal of Science* **284**, 1035–55.

Barron, E. J. (1983) A warm, equable Cretaceous: the nature of the problem. *Earth-Science Reviews* **19**, 305–38.

Barron, E. J., Thompson, S. L. and Hay, W. W. (1984) Continental distribution as a forcing factor for global-scale temperature. *Nature* **310**, 574–5.

Berner, R. A. and Lasaga, A. C. (1989) Modeling the geochemical carbon cycle. *Scientific American* **260** (3), 54–61.

Bishop, P. (1988) The eastern highlands of Australia: the evolution of an intraplate highland belt. *Progress in Physical Geography* **12**, 159–82.

Bishop, P., Young, R. W. and McDougall, I. (1985) Stream profile change and longterm landscape evolution: Early Miocene and modern rivers of the East Australian Highland crest, central New South Wales, Australia. *Journal of Geology* **93**, 455–74.

Brunsden, D. (1980) Applicable models of long term landform evolution. *Zeitschrift für Geomorphologie Supplementband* **36**, 16–26.

Brunsden, D. and Thornes, J. B. (1979) Landscape sensitivity and change. *Transactions of the Institute of British Geographers* NS **4**, 463–84.

Büdel, J. (1982) *Climatic Geomorphology* (translated by L. Fischer and D. Busche). Princeton University Press, Princeton.

Chorley, R. J. (1962) Geomorphology and general systems theory. *United States Geological Survey Professional Paper* **500B**.

Chorley, R. J., Beckinsale, R. P. and Dunn, A. J. (1973) *The History of the Study of Landforms*. Volume Two: *The Life and Work of William Morris Davis*. Methuen, London.

Cogley, J. G. (1979) Albedo contrast and glaciation due to continental drift. *Nature* **279**, 712–13.

Costa, J. E. and Cleaves, E. T. (1984) The Piedmont landscape of Maryland: a new look at an old problem. *Earth Surface Processes and Landforms* **9**, 59–74.

Davis, W. M. (1889) The rivers and valleys of Pennsylvania. *National Geographic Magazine* **1**, 183–253.

Davis, W. M. (1899) The geographical cycle. *Geographical Journal* **14**, 481–504.

Davis, W. M. (1954) *Geographical Essays* (ed. by D. W. Johnson). Dover, New York.

De Swardt, A. M. J. and Bennet, G. (1974) Structural and physiographic development of Natal since the late Jurassic. *Transactions of the Geological Society of South Africa* **77**, 309–22.

Fairbridge, R. W. and Finkl, C. W. (1980) Cratonic erosional unconformities and peneplains. *Journal of Geology* **88**, 69–86.

Garner, H. F. (1974) *The Origin of Landscapes: A Synthesis of Geomorphology*. Oxford University Press, New York and London.

Garner, H. F. (1983) Large-scale tectonic denudation and climatic morphogenesis in the Andes mountains of Ecuador. In: R. Gardner and H. Scoging (eds) *Mega-Geomorphology*. Clarendon Press, Oxford and New York, 1–17.

Hack, J. T. (1960) Interpretation of erosional topography in humid temperate regions. *American Journal of Science* **258A**, 80–97.

Hack, J. T. (1975) Dynamic equilibrium and landscape evolution. In: W. N. Melhorn and R. C. Flemal (eds) *Theories of Landform Development*. Allen and Unwin, Boston and London, 87–102.

Hack, J. T. (1979) Rock control and tectonism – their importance in shaping the Appalachian Highlands. *United States Geological Survey Professional Paper* **1126-B**, B1–B17.

Hack, J. T. (1982) Physiographic divisions and differential uplift in the Piedmont and Blue Ridge. *United States Geological Survey Professional Paper* **1265**.

Howard, A. D. (1965) Geomorphological systems – equilibrium and dynamics. *American Journal of Science* **263**, 436–45.

Johnson, D. W. (1931) *Stream Sculpture on the Atlantic Slope*. Columbia University Press, New York.

Jones, D. K. C. (1980) The Tertiary evolution of south-east England with particular reference to the Weald. In: D. K. C. Jones (ed.) *The Shaping of Southern England*. Institute of British Geographers Special Publication **11**. Academic Press, London and New York, 13–47.

Kiewietdejonge, C. J. (1984a) Büdel's geomorphology I. *Progress in Physical Geography* **8**, 218–48.

Kiewietdejonge, C. J. (1984b) Büdel's geomorphology II. *Progress in Physical Geography* **8**, 365–97.

King, L. C. (1953) Canons of landscape evolution. *Bulletin of the Geological Society of America* **64**, 721–52.

King, L. C. (1957) The uniformitarian nature of hillslopes. *Transactions of the Edinburgh Geological Society* **17**, 81–102.

King, L. C. (1967) *The Morphology of the Earth* (2nd edn). Oliver and Boyd, Edinburgh.

King, L. C. (1976) Planation remnants upon high lands. *Zeitschrift für Geomorphologie* **20**, 133–48.

Mabbutt, J. A. (1988) Land-surface evolution at the continental time-scale: an example from interior Western Australia. *Earth-Science Reviews* **25**, 457–66.

Mackenzie, F. T. and Pigott, J. D. (1981) Tectonic controls of Phanerozoic sedimentary rock cycling. *Journal of the Geological Society London* **138**, 183–96.

Melhorn, W. N. and Flemal, R. C. (eds) (1975) *Theories of Landform Development*. Allen and Unwin, London and Boston.

Ollier, C. D. (1982) The Great Escarpment of eastern Australia: tectonic and geomorphic significance. *Journal of the Geological Society of Australia* **29**, 13–23.

Ollier, C. D., Gaunt, G. F. M. and Jurkowski, I. (1988) The Kimberley Plateau, Western Australia: a Precambrian erosion surface. *Zeitschrift für Geomorphologie* **32**, 239–46.

Partridge, T. C. and Maud, R. R. (1987) Geomorphic evolution of southern Africa since the Mesozoic. *South African Journal of Geology* **90**, 179–208.

Pavich, M. J. (1985) Appalachian piedmont morphogenesis: weathering, erosion, and Cenozoic uplift. In: M. Morisawa and J. T. Hack (eds) *Tectonic Geomorphology*. Allen and Unwin, Boston and London, 299–319.

Penck, W. (1953) *Morphological Analysis of Landforms* (translated by H. Czech and K. C. Boswell). Macmillan, London.

Schumm, S. A. (1975) Episodic erosion: a modification of the geomorphic cycle. In: W. N. Melhorn and R. C. Flemal (eds) *Theories of Landform Development*. Allen and Unwin, London and Boston, 69–85.

Schumm, S. A. (1979) Geomorphic thresholds: the concept and its applications. *Transactions of the Institute of British Geographers* NS **4**, 485–515.

Schumm, S. A. and Lichty, R. W. (1965) Time, space and causality in geomorphology. *American Journal of Science* **263**, 110–19.

Simons, M. (1962) The Morphological Analysis of Landforms: a new review of the work of Walther Penck (1888–1923). *Transactions of the Institute of British Geographers* **31**, 1–14.

Small, R. J. (1980) The Tertiary geomorphological evolution of south-east England: an alternative interpretation. In: D. K. C. Jones (ed.) *The Shaping of Southern England*. Institute of British Geographers **11**. Academic Press, London and New York, 49–70.

Summerfield, M. A. (1985) Plate tectonics and landscape development on the African continent. In: M. Morisawa and J. T. Hack (eds). *Tectonic Geomorphology* Allen and Unwin, Boston and London, 27–51.

Suppe, J. (1987) The active Taiwan mountain belt. In: J.-P. Schaer and J. Rodgers (eds) *The Anatomy of Mountain Ranges*. Princeton University Press, Princeton, 277–93.

Thomas, M. F. (1974) *Tropical Geomorphology*. Macmillan, London and Halstead Press, New York.

Thomas, M.F. (1989a) The role of etch processes in landform development I. Etching concepts and their applications. *Zeitschrift für Geomorphologie*, **33**, 129–42.

Thomas M. F. (1989b) The role of etch processes in landform development II. Etching and the formation of relief. *Zeitschrift für Geomorphologie* **33**, 257–74.

Thomas, M. F. and Summerfield, M. A. (1987) Long-term landform development: key themes and research problems. In: V. Gardiner (ed.) *International Geomorphology 1986: Proceedings of the First International Conference on Geomorphology* Part II. Wiley, Chichester and New York, 935–56.

Thomas, M. F. and Thorp, M. B. (1985) Environmental change and episodic etchplanation in the humid tropics of Sierra Leone: the Koidu etchplain. In: I. Douglas and T. Spencer (eds) *Environmental Change and Tropical Geomorphology*. Allen and Unwin, London and Boston, 239–67.

Thorn, C. E. (ed.) (1982) *Space and Time in Geomorphology*. Allen and Unwin, London and Boston.

Thorn, C. E. (1988) *Introduction to Theoretical Geomorphology*. Unwin Hyman, Boston and London.

Thornes, J. B. (1983) Evolutionary geomorphology. *Geography* **68**, 225–35.

Thornes, J. B. and Brunsden, D. (1977) *Geomorphology and Time*. Methuen, London.

Twidale, C. R. (1976) On the survival of paleoforms. *American Journal of Science* **276**, 77–95.

Wellman, P. (1987) Eastern Highlands of Australia; their uplift and erosion. *BMR Journal of Australian Geology and Geophysics* **10**, 277–86.

Wellman, P. (1988) Tectonic and denudational uplift of Australian and Antarctic highlands. *Zeitschrift für Geomorphologie* **32**, 17–29.

Whitehouse, I.E. (1988) Geomorphology of the central southern Alps, New Zealand: the interaction of plate collision and atmospheric circulation. *Zeitschrift für Geomorphologie Supplementband* **69**, 105–16.

Wooldridge, S. W. and Linton, D. L. (1955) *Structure, Surface and Drainage in South-East England* (2nd edn). George Philip, London.

Young, R. W. (1983) The tempo of geomorphological change: evidence from southeastern Australia. *Journal of Geology* **91**, 221–30.

Part V

Extraterrestrial landforms

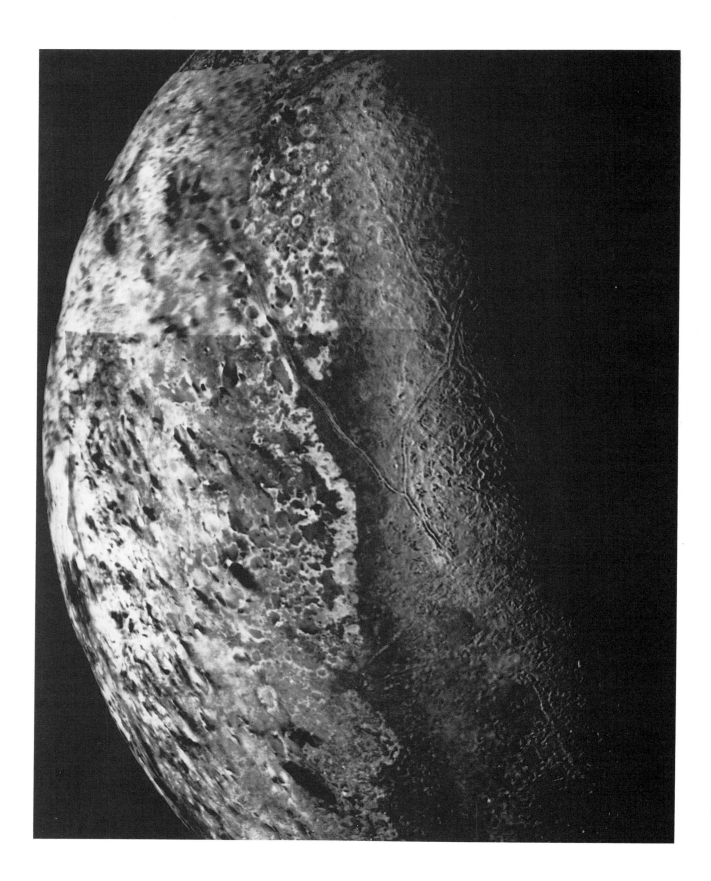

19

Planetary geomorphology

19.1 Approaches to planetary geomorphology

We conclude this survey of global geomorphology by expanding our horizons to the earth-like planetary bodies of the Solar System. By earth-like we mean those planets and large moons composed of a solid crust on which landforms can develop and be preserved. The term 'terrestrial' is reserved to designate those features actually occurring on the Earth. The interpretation of surface planetary features is carried out in part by analogy with terrestrial landforms, although some planetary landforms appear to have no terrestrial analogues. Many planetary bodies, such as the Earth's moon, retain landscape features formed very early in their history, and their study provides a context in which to interpret the long-term history of the Earth's surface. Moreover, comparative planetary geomorphology gives a basis for understanding how our Earth alone among the planets and moons of the Solar System came to develop its apparently uniquely complex surface form.

It is possible in this chapter to give only the very briefest outline of the exciting findings that have come from the various planetary orbiter and lander missions of the USA and the USSR. This is a fast-moving field of research, and although spectacular landforms have already been revealed by the initial examination of the mass of imagery returned to Earth, further more detailed analyses will no doubt provide a much fuller picture of the development of planetary surfaces.

Some of the global properties of the inner planets which directly or indirectly influence the operation of geomorphic processes are listed in Table 19.1. The mean distance of the Sun influences the amount of solar energy available at the

Table 19.1 Properties of the inner planets

	EARTH	MOON	MARS	MERCURY	VENUS
Mean distance from the Sun (km $\times$ 10^6)	149.6	0.3844*	227.9	57.9	108.2
Period of revolution (Earth days)	365.26	27.32*	687	88	224.7
Rotation period (Earth days)	0.9983	27.32	1.026	59	243 retrograde
Equatorial diameter (km)	12 756	3476	6787	4880	12 104
Mass (Earth = 1)	1	0.01226	0.108	0.055	0.815
Density (kg m^{-3})	5520	3340	3900	5400	5200
Atmosphere (main components)	Nitrogen, oxygen	None	Carbon dioxide	None	Carbon dioxide
Atmosphere (minor components)	Carbon dioxide, noble gases	None	Noble gases, nitrogen	None	Noble gases; hydrochloric hydrofluoric and sulphuric acids
Mean temperature at surface (°C)	15	107 (day) -153 (night)	-23	305 (day) -170 (night)	480
Atmospheric pressure at surface (millibars)	1000	0	6	<10^{-9}	90 000
Surface gravity (Earth = 1)	1	0.16	0.38	0.37	0.88

* Relative to the Earth.
Source: Modified from B. Murray *et al.* (1981) *Earthlike Planets*. W. H. Freeman, San Francisco, Table 2.1 p. 28.

surface, and together with the rotational period and the nature of the atmosphere this largely controls the average and extreme temperatures experienced on planetary surfaces. Atmospheric pressure and temperature are two crucial variables which determine whether water can exist in its liquid state on a planet's surface, and consequently they determine the nature of chemical and physical weathering and the presence or absence of fluvial activity. Surface gravity is a significant factor in influencing atmospheric pressure and indirectly in controlling the existence and nature of aeolian activity; moreover, it directly affects the operation of mass movement processes. There are in fact complex interactions between several of these factors, and we will consider these further as we discuss the geomorphology of individual planetary bodies.

19.2 The Moon

Knowledge of the broad outlines of the Moon's surface features extends back to the time when human beings first looked heavenwards. Contrasts between light and dark areas are clearly visible with the naked eye and, following the observations by telescope pioneered by Galileo and others

in the seventeenth century two major types of terrain were identified – the high albedo, heavily cratered uplands and the low albedo, smooth, lightly cratered lowlands, called **maria** (singular **mare**) because of their supposed resemblance to seas). The Ranger, Lunar Orbiter and Apollo programmes of the USA, together with a number of Soviet missions, have now provided an abundance of data on the Moon. Although many questions remain, we now have a fairly clear idea of the processes that have shaped the lunar surface and an approximate chronology for its development, based on the radiometric dating of rock samples returned to Earth.

Two major processes have fashioned the lunar landscape. One is **impact cratering**, which involves the transformation of the kinetic energy of an impacting object, or **bolide**, and the consequent formation of **impact craters** and associated features. Bolides can include comets and asteroids as well as meteorites. The second significant process is volcanic flooding. The maria are plains formed by extensive sheets of basaltic lava, erupted mainly between 3.9 and 3.1 Ga BP. Some 17 per cent of the surface is covered by lava plains, but by far the greater proportion of these occur on the near side of the Moon. The far side, which is always turned away

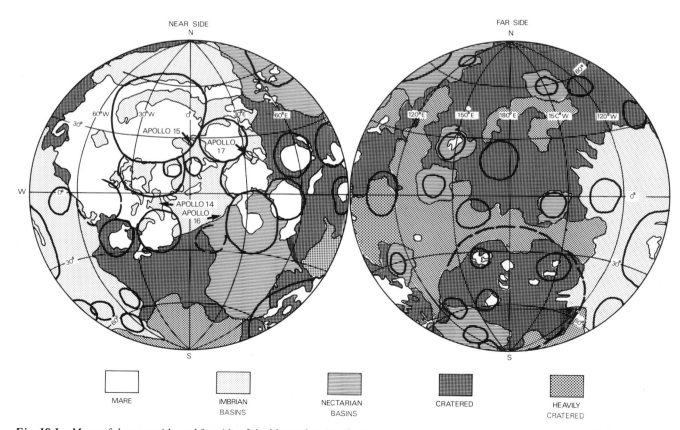

Fig. 19.1 *Maps of the near side and far side of the Moon showing the major geological and morphological provinces. The location of the landing sites for Apollo missions 14–17 are also indicated. (After K. A. Howard et al. (1974)* Reviews of Geophysics and Space Physics *12, Fig. 14 p. 322. Copyright by the American Geophysical Union.)*

from the Earth because of gravitational interactions arising from irregularities in the form of the two bodies, is almost completely covered by moderately to heavily cratered terrain (Fig. 19.1).

19.2.1 Impact cratering

The Moon has experienced a history of bombardment extending back to its formation 4.6 Ga BP. Craters formed by bolide impacts range from microscopic forms up to huge circular basins, such as the Mare Imbrium, which have subsequently been flooded by lava. A small number of craters of probable non-impact origin have also been identified on the Moon and are thought to be associated with volcanic activity.

Impacts occur at tremendous velocities of up to 150 000 km h^{-1}. The instantaneous conversion of the enormous quantities of kinetic energy released into heat as the object penetrates the surface, explodes and vaporizes means that small objects can produce surprisingly large craters. For example a 50 t meteorite about 3 m across can form a crater 150 m in diameter. As the kinetic energy of a projectile is proportional to the square of its velocity, bolide size is less important than speed in determining the size of the crater formed.

At the simplest level, lunar craters can be classified according to their size and characteristic morphology into four categories. Microcraters (< 10 mm in diameter) are found on rocks and boulders, and bombardment by very small meteorites is probably important in fracturing rocks and producing the fine-grained lunar regolith. Small craters (10 mm to 15 km across) have a simple form lacking major structural features, whereas large craters (15 to 300 km across) are complex features and typically show evidence of modification by molten rock, either related to the original impact or associated with subsequent volcanic activity. Some contain one or more central peaks probably formed through the rebound of the crater floor after impact (Fig. 19.2).

In craters more than about 120 km across, a series of rings formed by uplifted masses of rock tends to replace the central peak. In very large craters (> 300 km across), usually termed **basins**, up to five rings may be found. Most lunar mountain chains are formed from these large circular structures and most large fracture systems are related to them. The Montes Apennines, which reach some 5 km above the lunar surface, in fact represent part of the outermost ring of the vast Imbrium Basin. The origin of these multi-ringed basins is uncertain, but probably involved adjustments of the lunar crust after the impact of very large bolides during the early stages of the Moon's history.

19.2.2 Volcanism and tectonics

Although now quiescent, volcanic activity has been active

Fig. 19.2 *View of the relatively young, 27 km diameter lunar crater Euler located in the south-west of the Mare Imbrium. A central peak can be seen in the centre of the crater, probably formed as a result of rebound of the crater floor after impact. The effects of impact are seen to extend well beyond the primary crater since in the Moon's low gravity field ejecta may travel for hundreds of kilometres before returning to the surface to form large numbers of much smaller secondary craters. The bright bands that can be seen radiating away from the crater (particularly towards the top left) are known as **crater rays**. They are typical features of relatively young craters and are probably composed of fine material ejected on impact, including highly reflective fused silica (glass) beads formed by the shock waves generated by the primary impact explosion. With age these bright rays become darker and less conspicuous probably as a consequence of burial by ejecta from later impact events or changes in the crystal structure of the silica beads brought about by the intense ultra-violet radiation which reaches the unshielded lunar surface. (Apollo 17 image, World Data Center A for Rockets and Satellites.)*

on the Moon in the past. Most was associated with the formation of large impact basins, all of which were created early in lunar history. The Orientale Basin is the most recent and this was formed about 3.9 Ga BP. Puncturing of the early-formed crust by large bolides exposed the partially molten interior and led to the eruption of lava flows which proceeded to fill the basins. Eruption of basalts appears to have continued for a considerable period after basin formation.

During the Apollo 15 mission to Hadley Rille, a sinuous channel cutting across the Mare Imbrium, individual lava flows up to 60 m thick were observed. The Hadley Rille itself is thought to be a massive collapsed lava tunnel, smaller versions of which are known from lava fields on Earth (Fig. 1.1). Some idea of lava thicknesses can be gained by measuring the rim heights of craters partially buried by lava and comparing these with their likely original heights. Such analyses indicate that basin lava thicknesses of 300 m are typical. In addition to rilles, other landforms associated with volcanic activity include lobate flow fronts on lava flows, low elevation shield volcanoes up to 15 km or so across, and domed-shaped hills, which although of less certain volcanic origin, may have been formed from lavas more viscous than those which cover the mare basins.

Data from seismographs on the lunar surface show that the current level of seismic activity on the Moon is minimal, being only about one-millionth of that on the Earth. The Moon has in fact experienced little tectonic activity over the past 3 Ga apart from crustal dislocations associated with impact events. Nevertheless, some tectonic structures have been recorded; for instance, graben are locally abundant and some mare ridges may not be of purely volcanic origin but rather true tectonic features associated with vertical movements within the sub-basin crust.

19.2.3   Surface materials and processes

Apart from forming craters, bolide impacts also create the lunar regolith. This is made up of rock particles detached from bedrock, ejecta and molten rock from impact events (the latter consisting largely of fused silica in the form of glass beads) and minor amounts of bolide fragments. The most effective bolide size for generating regolith is the range which forms craters – about 10–1500 m in diameter. Larger bolides are too infrequent to be significant, while smaller objects are rarely able to penetrate the existing regolith cover and thus create any additional regolith material. None the less, small meteorites, down to those of microscopic dimensions, which can reach the lunar surface in the absence of an atmosphere, do play a role in pitting the rock surface and smoothing features at the small scale. The effectiveness of this mechanism, however, is minimal in comparison with processes of terrestrial denudation. This is apparent from the preservation of 10 m high features on lava flows over 3 Ga old observed by Apollo astronauts. The consequences of bolide impacts are not confined to fragmentation of the

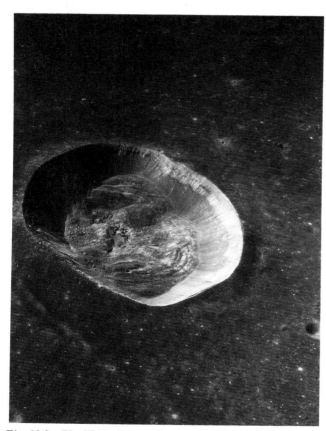

Fig. 19.3   *The 18 km diameter lunar crater Dawes showing clear evidence of mass movement in the form of extensive accumulations of debris on the crater floor. (Apollo 17 image, World Data Center A for Rockets and Satellites.)*

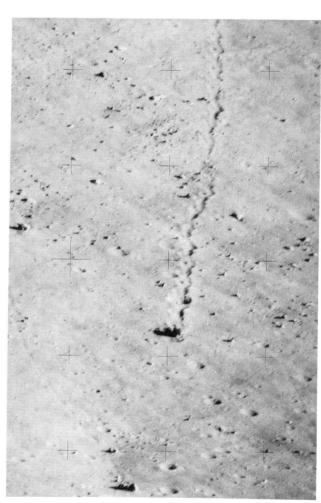

Fig. 19.4   *Boulder track on the slopes of the North Massif of the Moon near the Apollo 17 landing site. The track is nearly 1 km long and the largest boulders visible are about 5 m across. (Apollo 17 image, World Data Center A for Rockets and Satellites.)*

surface since the pressure and heat generated on impact is capable of fusing unconsolidated sediments together to form breccia.

In the absence of the transporting agents of water, wind and ice, movement of surface materials on the lunar surface (excluding the effects of bolide impacts and lava flows) is confined to mass movements. There is abundant evidence for such activity; slumping associated with impact events is indicated by the morphology of the terraced inner walls of large craters (Fig. 19.3) and tracks left by large boulders sliding down crater slopes have also been observed (Fig. 19.4). Landslides of unstable ejecta mantling crater wall slopes could be triggered by the seismic shocks generated by large bolide impacts, but downslope movement may also be precipitated by the subsequent impact of ejecta from such events. The scarcity of post-mare landslides, however, suggests that the occurrence of mass movement may have diminished after the early period of high cratering rates.

19.2.4   History of landscape development

The history of the lunar landscape has now been established in broad outline by calculating the density of cratering and the freshness of craters over different parts of the surface, and relating these variables to the radiometric dates provided from the Apollo lunar samples (Table 19.2). There are some uncertainties as to the accuracy of these dates (different radiometric techniques giving slightly different results) and there are also discrepancies between the various approaches taken to relate the density of cratering to the age of the surface. Nevertheless, the very low rate of modification of the lunar surface compared with the Earth means that we have a much clearer idea of its long-term geomorphic history.

After the formation of the Moon 4.6 Ga BP a process of differentiation of its constituents led to the formation of a crust which became sufficiently solidified by about 4.3 Ga BP to record impact events. Most of this very early Pre-Nectarian terrane (Table 19.2) is preserved as very subdued basins and craters on the far side of the Moon. Formation of the large lunar basins was concentrated between about 4.3 and 3.9 Ga BP and ended with the creation of the vast Imbrium Basin at the termination of the Nectarian System. Prior to the Apollo missions the lunar highlands were thought to be remnants of the earliest accretionary phase of the Moon's history when it was subject to continuous bombardment. Radiometrically dated Apollo samples, however, suggest a predominant age of around 4 Ga which indicates

Table 19.2   Chronology of landscape development on the Moon

TIME-STRATIGRAPHIC UNITS	AGE (Ga)	ROCK UNITS	EVENTS	NOTES
Copernican System		Few large craters	Tycho Aristarchus	Craters with bright rays and sharp features at all resolutions (e.g. Tycho, Aristarchus)
		Few large craters		Craters with bright rays and sharp features but now subdued at metre resolutions (e.g. Copernicus)
Eratosthenian System		Few large craters	Copernicus Eratosthenes	Craters with Copernican form but rays barely visible or absent
	3.2	Apollo 12 lavas		
	3.3	Apollo 15 lavas	Imbrium lavas	Few lavas with relatively fresh surfaces
	3.42	Luna 16 lavas		
Imbrian System		Mare lavas	Eruption of widespread lava sheets on near side: few eruptions on far side	Extensive piles of basaltic lava sheets with some intercalated impact crater ejects sheets
	3.6	Apollo 11 lavas		
	3.8	Apollo 17 lavas		
	3.9		Orientale Basin	
Nectarian System			Imbrium Basin Crisium Muscoviense Humorum	Numerous overlapping large impact craters and associated ejecta sheets together with large basin ejecta
	4.1		Nectaris } Basins Serenitatis Smythii Tranquillitatis Nubium	Any igneous activity at surface obscured by impact craters
Pre-Nectarian	4.6		Formation of Moon	Crystalline rocks formed by early igneous activity

Source: Modified from J. E. Guest and R. Greeley (1977) *Geology on the Moon*. Wykeham Publications, London, Fig. 1.5, p. 8.

that there was a period of frequent impacts by large bolides well after the initial accretionary phase.

This episode of intensive impact cratering is known as the **Late Heavy Bombardment**. The dating uncertainties already mentioned have led to a number of rather different interpretations of the details of this event. One suggestion is that the Late Heavy Bombardment represents the end of a 500 Ma period of continuous but declining impact activity. Another view is that there was a short-lived cataclysmic increase in the rate of bombardment around 4 Ga. Whatever the correct interpretation, evidence of the Late Heavy Bombardment on the Moon is of crucial importance as it provides a time scale for similar events which appear to have affected all the inner planets, and also possibly some of the more remote planetary bodies of the Solar System.

Infilling of the large lunar basins by lava flows occurred from about 3.8 to 3.2 Ga (Imbrian and Eratosthenian Systems). Geomorphic activity since then (Copernican System) has been confined to the formation of a few large craters and the continuing minor modification of the lunar surface through the impact of comparatively small bolides.

19.3   Mars

The first close view of the Martian surface was provided by the Mariner 4 fly-by of 1964. Two more fly-by missions in 1969 by Mariners 6 and 7 provided further information, but detailed views of the planet's fascinating landforms had to await the arrival of the Mariner 9 orbiter towards the end of 1971. At first the surface was tantalizingly obscured by a huge dust storm, but after several months the atmosphere cleared to reveal some of the most spectacular landscapes in the Solar System.

During a period of a single year the Mariner 9 orbiter imaged almost the entire surface of Mars at resolutions of 1–10 km, with selected areas being imaged at higher resolutions of up to 100 m. This mission provided the basis for the initial mapping of the planet's surface and was complemented by two Soviet orbiters in 1971 and 1973. The Viking 1 and Viking 2 combined orbiter and lander missions in 1976 provided more extensive high resolution imagery totalling nearly 60 000 individual frames. In addition to monitoring changes on the surface during the slowly

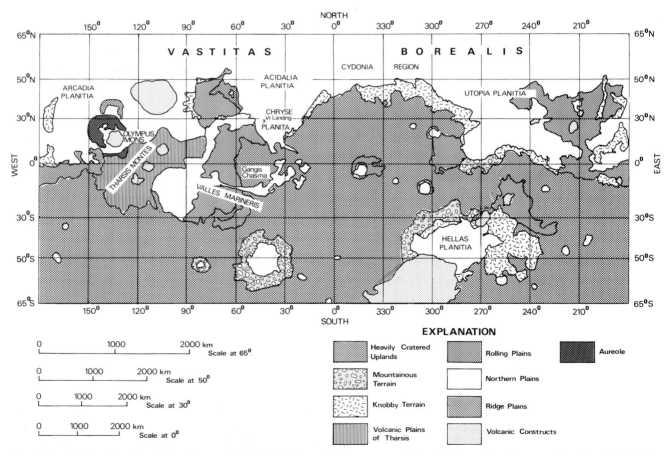

Fig. 19.5   *Geomorphic map of Mars between latitudes 65°N and 65°S constructed from Viking orbiter imagery. (Based on V. R. Baker, 1981,* Progress in Physical Geography *5, Fig. 1, p. 474, modified from the Geologic Map of Mars by D. H. Scott and M. H. Carr (1978)* United States Geological Survey, Miscellaneous Geologic Investigations, Map *I-1083.)*

changing Martian seasons the two Viking lander probes provided close-up views of landforms around the landing sites and the first analyses of the Martian regolith.

Maps constructed from Mariner and Viking orbiter imagery indicate a wide variety of terrain types on Mars and the operation of a range of geomorphic processes, including fluvial and aeolian erosion and deposition, weathering, mass movement and periglacial activity, as well as impact cratering. At the very broad scale the planet may be divided into two hemispheres separated by a great circle inclined at about 35° to the equator. The southern hemisphere is heavily cratered, whereas the northern hemisphere is on average lower and consists mostly of plains. The reason for this contrast is not known. At a more detailed scale nearly a dozen major morphological provinces have been identified (Fig. 19.5; Table 19.3).

19.3.1 Impact craters

Although impact craters are a very common element of most planetary landscapes, many of those on Mars have very special characteristics which make them unlike any-

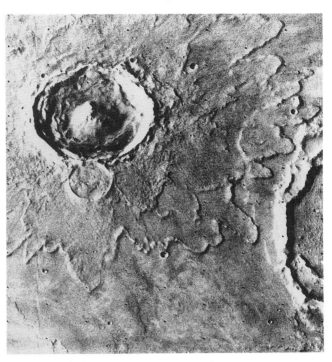

Fig. 19.6 *The 18 km diameter Martian crater Yuty located at latitude 22°N, longitude 34°. Several sheets of ejecta can be observed, each with well-developed lobes. These types of craters are common in a belt lying within 30–40° of the equator. Since the small crater adjacent to Yuty has experienced only partial burial the layers of ejecta must be relatively thin. (Viking 1 image, World Data Center A for Rockets and Satellites.)*

Table 19.3 Principal morphological provinces of Mars

MORPHOLOGICAL PROVINCE	MORPHOLOGICAL CHARACTERISTICS
Heavily cratered uplands	Widespread hilly and cratered terrains of the equatorial and southern highlands
Mountainous terrain	Ejecta and uplifted blocks of ancient terrain caused by large impacts
Knobby terrain	Fretted uplands and isolated flat-topped residuals formed as erosional remnants along the boundary between the heavily cratered uplands and northern plains
Volcanic plains of Tharsis	Relatively young lava flows originating from the Tharsis Montes volcanic province
Rolling plains	Intermediate age lava flows originating from the Elysium volcanoes or other sources
Northern plains	Complex lowland exhibiting evidence of ice-contact volcanism, permafrost features and aeolian modification
Ridged plains	Relatively old lava plains that show prominent ridges
Volcanic constructs	Individual large volcanoes including Olympus Mons, Alba Patera and Elysium Mons
Aureole	Complex terrain of elongate hills and ridges occurring around the margins of certain volcanoes
Chaotic terrain	Blocky, fractured terrain apparently developed through the collapse and subsidence of the heavily cratered terrain
Valleys and channels	Includes the highly modified landscape of Valles Marineris and various valley and channel features attributed to fluid flow erosion

Source: Based on terrain types identified in Fig. 19.5 and described in V. R. Baker (1981) *Progress in Physical Geography*, **5**, pp. 474–6.

thing observed elsewhere in the Solar System (Fig. 19.6). The form of the ejecta, which occurs as overlapping lobes, implies some form of fluid flow; indeed, these features have been termed 'splosh' craters. The most probable explanation for this unusual morphology is that the material disturbed on impact included subsurface ice (permafrost) which, either as a gas or liquid, became incorporated into the ejecta to form a debris flow. Secondary craters are also found which indicate that ballistic processes do operate to some extent. An intriguing finding is that the distribution of the debris lobe type craters appears to be related to latitude and altitude; this may be indirect evidence of a variation in the depth of permafrost over the planet, with ordinary ballistic impact type craters only forming where the permafrost is confined to great depths.

19.3.2 Volcanic and tectonic features

On the basis of our present knowledge Mars possesses the largest volcanoes in the Solar System. Olympus Mons reaches 26 km above the surrounding plain from which it is separated by a scarp up to 8 km in height. A number of collapse structures are found around the summit, and the caldera complex alone is some 80 km across (Fig. 19.7).

Fig. 19.7   *Oblique view of Olympus Mons, one of the largest volcanoes known in the Solar System. Based on the density of cratering it has been estimated that Olympus Mons is some 400 Ma old, although this may simply be the age of the youngest flows, the whole edifice probably having been built over billions of years. By comparison, volcanoes on Earth rarely survive more than a few million years before being destroyed by erosion. (Viking image, World Data Center A for Rockets and Satellites.)*

The morphology of Olympus Mons is broadly comparable to the shield volcanoes of Hawaii, but these two landforms differ dramatically in size with Hawaii (the largest shield volcano complex on Earth) rising only a mere 10 km from the ocean floor! (Fig. 19.8). The great height of some Martian volcanoes should not surprise us because the lower surface gravity on Mars (only 38 per cent of that of the Earth) means that large volcanic edifices there weigh less than their terrestrial counterparts and consequently do not load the underlying crust to the same extent.

The larger Martian volcanoes are all of the shield type and are associated with massive lava flows. Alba Petera is 1500 km across and dwarfs even Olympus Mons. It has a modest elevation of only 6 km so its flanks are inclined at a very shallow angle. It was thought initially to be an ancient degraded structure, but Viking imagery has revealed numerous juvenile features on its slopes including sheet flows and lobe fronts. These large shield volcanoes are concentrated in an area of probable 'mantle' upwelling known as the Tharsis Montes volcanic province (Fig. 19.5). This is a massive crustal bulge which formed about 3.9 Ga BP and its

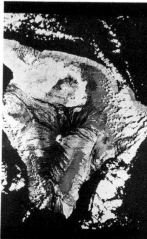

Fig. 19.8   *Olympus Mons (left) compared with the shield volcano of Hawaii (right). Olympus Mons is 650 km in diameter and is composed of thousands of individual lava flows. Its caldera complex alone is over half the size of the whole subaerial part of Hawaii shown here. (Olympus Mons – Viking 1 image, World Data Center A for Rockets and Satellites; Hawaii – Landsat mosaic courtesy N. M. Short.)*

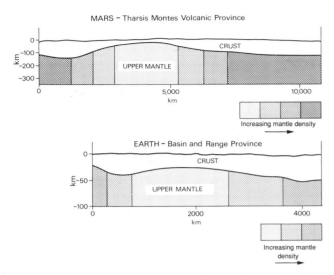

Fig. 19.9 Combined crustal density (Pratt) and crustal thickness (Airy) isostatic model for the Tharsis Montes volcanic province of Mars in comparison with a similar model for the Basin and Range Province in the western USA. (Modified from N. H. Sleep and R. J. Phillips (1979) Geophysical Research Letters 6, Fig. 2, p. 805. Copyright by the American Geophysical Union.)

origin may be similar to that of the Basin and Range Province of the western USA (Fig. 19.9).

In addition to extensive lava flows there is evidence in the form of blankets of tephra that many Martian volcanoes have experienced explosive episodes early in their history. This could be explained by the explosive consequences of magma penetrating a frozen, water-saturated regolith. Sub-sequently, the amount of near-surface water declined and later volcanism was confined to lava eruptions and the formation of dome and shield volcanoes.

A range of tectonic features have been identified on the Martian surface, many in association with volcanic structures. Far and away the most impressive of these is the system of canyons collectively called the Valles Marineris. This complex graben-like feature is some 5000 km long, averages around 200 km wide and is up to 7 km in depth. Some idea of the size of this landform can be gained by imagining a canyon four times as deep and about fifteen times as broad as the Grand Canyon, stretching coast to coast across the USA. The Valles Marineris has clearly formed through the effects of large-scale tensional stress in the Martian crust and its possible association with the Tharsis bulge suggests that it may be of a similar age.

19.3.3 Weathering

Although contrasting starkly with the Earth, the atmosphere and surface conditions on Mars are closer to our own planet than any of the other bodies of the Solar System. In spite of marked differences in atmospheric pressure, atmospheric composition and surface temperatures (Table 19.1) there are some apparent similarities in weathering processes. Analyses carried out by the Viking landers, although failing to confirm the presence of life, provided valuable data on the physical and chemical properties of surface materials. Some of the boulders observed in Viking lander imagery (Fig. 19.10) may have been formed by *in situ* weathering, but it

Fig. 19.10 Ground view across Chryse Planita from the Viking 1 landing site on Mars. Large numbers of boulders, up to 1 m or so across, can be seen in addition to drifts of fine-grained material. (Viking 1 mosaic, World Data Center A for Rockets and Satellites.)

is more likely that they are simply ejecta from impact events.

A possible explanation for the formation of the fine-grained material which mantles much of the surface at both the Viking landing sites is salt weathering since the Martian soil appears to be rich in sulphur, probably in the form of sulphate salts. The presence of a thin salt crust (sometimes misleadingly described as 'duricrust' by planetary geologists) is also indicated from the observation that surface material broke into small cohesive fragments where it was disturbed by the sampling arm attached to the Viking landers. Frost weathering may also be significant on Mars since both water and carbon dioxide ice have been identified on the surface, but as with other Martian geomorphic processes there is far more speculation than conclusive data. The closest terrestrial analogues are the cold, arid valleys of Antarctica. There is certainly a seasonal cycle of frost deposition on Mars and a thin, bright surface covering was seen to form during winter at the Viking 2 lander site.

19.3.4   Slopes and mass movement

The abundance of high escarpments on Mars provides considerable scope for the operation of mass movement mechanisms; certainly the scale of such processes exceeds anything occurring on Earth. In the Valles Marineris concave chutes separate the comparatively flat upland surface from the canyon itself (Fig. 19.11). These chutes lead down either into talus slopes or into a 'spur-and-gully topography' made up of alternating ridges and swales. The talus slopes typically have a mean angle of around 30° and in some cases have a vertical drop of several kilometres. In a study employing Viking imagery, 25 large landslides ranging in size from 40 to 7500 km² were identified in the Valles Marineris; one of these was seen to be recessed up to 30 km into the escarpment wall.

One explanation for these massive landslide features involves dry avalanching of unconsolidated material. This implies that, at least in some areas, the surface is covered by a thick, non-cohesive regolith possibly resulting from the shattering of bedrock through bolide impacts. Another, now more favoured, hypothesis is that the undermining of free faces by sapping, involving the evaporation of ground water or permafrost ice, is the primary landslide-triggering mechanism. This process may also have been instrumental in the formation of the **fretted terrain** located along the boundary between the ancient heavily cratered highlands and the sparsely cratered lowland plains. This transition zone consists of vast debris aprons with a characteristic convex lobate form and appears to represent an area of outward flow of debris from highlands to lowlands. The fairly uniform height of the eroding escarpments within the fretted terrain at between 1 and 2 km has been interpreted as indicating that scarp retreat may be controlled by the depth to which ice-rich permafrost extends.

Fig. 19.11   *Large landslides on the south wall of Gangis Chasma in the Valles Marineris region of Mars. The escarpment running left to right across the centre of the image is about 2 km high and separates a plateau region (top) from the floor of Gangis Chasma (bottom). Some of the landslides show backward rotation (arrowed), a feature characterizing rotational slumps on Earth. The area shown is about 60 km across. (Viking 1 mosaic, World Data Center A for Rockets and Satellites.)*

19.3.5   Aeolian processes and landforms

With a lack of surface water, the absence of any vegetation to stabilize the surface, and the presence of high winds and fine sediment it is not surprising that aeolian landforms are abundant on Mars (Fig. 19.12). There is a vast erg surrounding the north polar region, wind-laid dust and ice deposits at the poles themselves and numerous depositional and erosional forms in the equatorial zone. Although now recorded in detail by the imaging systems orbiting Mars, the great dust storms that periodically sweep the planet had long been observed from the Earth through high-powered telescopes. Such storms may shroud the whole planet, as occurred when Mariner 9 arrived in 1971, and can last for up to four months. Wind speeds can exceed 140 m s^{-1} (around 500 km h^{-1}) compared with a typical maximum of 30–40 m s^{-1} (about 120–160 km h^{-1}) on Earth; dust may rise up to 70 km above the Martian surface.

The much lower atmospheric pressure on Mars compared with Earth (Table 19.1) has important implications for the operation of aeolian activity. Threshold drag velocities, for instance, are much larger on Mars (Fig. 19.13); consequently saltating grains have about twenty times the velocity of similar-sized particles on Earth and the potential for aeolian

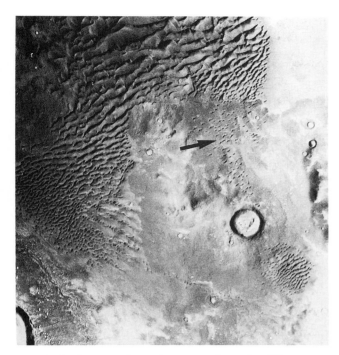

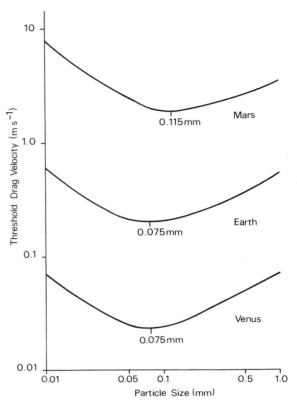

Fig. 19.12   *Dunes in the ancient cratered terrain of Mars located at latitude 47°S, longitude 340°. Most of the dunes are of the transverse ridge and barchanoid ridge type, but isolated barchans are also visible (arrowed). The area shown is about 60 km across. (Viking 2 image, World Data Center A for Rockets and Satellites.)*

Fig. 19.13   *Threshold drag velocity in air as a function of particle size on the Earth, Mars and Venus. Note that for 0.2 mm sized particles this velocity may be ten times greater on Mars than on Earth. Saltating grains on Mars have about ten times the momentum and one hundred times the kinetic energy of those on Earth. The reasons for the higher threshold drag velocity for particle sizes less than about 0.1 mm are not known for certain but are probably associated with interparticle cohesion and aerodynamic effects. (Modified from R. Greeley (1985)* Planetary Landscapes, *Allen and Unwin, London, Fig. 3.36, p. 66.)*

erosion is therefore considerable. Silt to clay-sized particles seem to predominate on the Martian surface, but sand-sized aggregates can apparently form through the electrostatic bonding of this finer material. The dunes observed on Mars are probably formed largely from these aggregates and are therefore in some respects analogous to clay dunes on Earth. Wind-tunnel experiments suggest, however, that such sand-sized aggregates may have a short life span since at slightly above Martian threshold drag velocities they rapidly break up into fine fragments around 20 μm in diameter. The image of the destruction of these aggregates as they smash into rock surfaces at high velocities has given rise to the vividly descriptive term **kamikaze grain**.

Inspection of Mariner and Viking orbiter imagery has revealed a wide variety of both erosional and depositional aeolian forms on Mars. Morphologically, many of these are very similar to landforms recorded from terrestrial deserts, but the great size of some features exceeds any analogous forms occurring on Earth. Various dune patterns have been identified; those of the Hellespontus area, for instance, have a crescentic ridge form reminiscent of terrestrial dunes in central Asia. Massive barchanoid and transverse forms have also been recorded as well as the coalescence of individual barchans, a feature well known from terrestrial deserts (Fig. 19.14).

Aeolian erosional landforms include modified crater rims, linear grooves, streamlined ridges and fluted cliffs. An aeolian origin is implied where parallel series of linear ridges and grooves extend over large distances, some of the forms seen on Mars being morphologically comparable to terrestrial yardangs. The distribution of aeolian erosional landforms is localized and this suggests that they may only form where there are friable surface materials. Estimates of rates of aeolian erosion on Mars range from a negligible 0.001 mm ka^{-1} for the lowland plains to a maximum of around 0.1 mm ka^{-1} in parts of the heavily cratered highlands. Although the apparent lower density of craters at high latitudes may be due in part to the blanketing effect of aeolian deposits, and thus imply the active deflation of material from elsewhere, measurements from the Viking landers have demonstrated that current rates of aeolian erosion and deposition are extremely low.

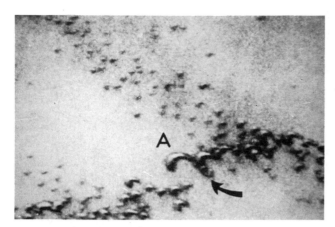

Fig. 19.14   The arm of a migrating barchan (A) on Mars coalescing with another barchan downwind (arrowed) The labelled barchan is about 500 m across and 200 m long and is situated near the southern edge of the erg located at latitude 73°N, longitude 39°50'. All the dunes visible are migrating towards the bottom of the image, that is, in an approximately easterly direction. (Viking 2 image, World Data Center A for Rockets and Satellites.)

19.3.6   Periglacial processes and landforms

It could be argued that 'periglacial' is an inappropriate term to apply to geomorphic processes on Mars since there is no firm evidence that glaciers have ever existed on the planet. None the less, in its broader sense, as applied in Chapter 12, the term is appropriate as it refers to environments characterized by low temperatures with fluctuations above and below freezing. Although water in its liquid state is known not to exist in significant quantities on the present-day Martian surface, locally it may occur at shallow depths in the regolith in the equatorial zone. But freeze–thaw activity and associated frost creep may have been widespread in the past when mean temperatures were probably rather higher and surface water may have been more abundant. If salt is an important constituent of the present-day surface, freeze–thaw may be currently active since liquid saline solutions might be able to exist during the day at low latitudes.

The possible extent of periglacial activity on Mars is suggested by the presence of patterned ground recorded in high resolution Viking imagery. Although the polygonal patterns observed are certainly similar to terrestrial patterned ground, some occur at a much larger scale than analogous forms on Earth. Ice-wedge polygons on Earth are typically 1–100 m across, but some of the Martian polygons are 5–10 km in diameter, the cracks themselves being hundreds of metres across. If they are indeed periglacial phenomena it is likely that they have been generated by temperature changes of considerable duration extending to great depths within the Martian permafrost, rather than by the seasonal temperature fluctuations largely responsible for ice-wedge

polygon formation on the Earth. Alternative explanations for these phenomena exist, however, including the contraction of cooling lava and extension associated with tectonic mechanisms.

The present water content of the Martian permafrost is not known but there is abundant morphological evidence for a high ice content in the past. Likely thermokarst features include alases (irregular depressions) and thermocirque topography (embayed escarpments; see Section 12.3.2.4). Such landforms, arising from the melting of ground ice, are best developed in terrestrial periglacial environments where the ice content is high. The **chaotic terrain** occurring in parts of the Martian equatorial zone has been widely interpreted as the result of large-scale ground collapse but the size of these features on Mars is much greater than terrestrial thermokarst forms (Fig. 19.15).

There is a clear latitudinal control over the operation of permafrost-related processes on Mars. Estimates of the likely heat flow from the planet's interior indicate that the maximum permafrost depth ranges from about 1 km at the equator to around 3 km at the poles. Under the present climatic regime permanent ice can probably exist to within a few centimetres of the Martian surface at latitudes above about 40°. Local melting of ice-rich permafrost to produce thermokarst and related landforms could result from a variety of mechanisms including volcanism and bolide impacts. More widespread melting might be precipitated by climatic changes, while scarp retreat could lead to the release of liquid water confined at depth within the permafrost layer.

19.3.7   Polar terrains

The frost caps of Mars expand to cover up to 30 per cent of the planet's surface area during the Martian winter but retreat to small residual 'ice caps' covering only 1 per cent of its area in the summer. The remnant north polar ice cap is largely composed of water ice, but the nature of the permanent southern cap is less certain and it may consist of both frozen water and carbon dioxide. The seasonal frost blanket is predominantly formed by the freezing of atmospheric carbon dioxide, although it probably also includes dust particles coated by water ice.

The perennial ice caps contain layered deposits which have been partially exposed by the formation of deep canyons. This is the **layered terrain** in which alternating bright and dark bands up to 30 m thick have been observed exposed along canyon walls. This banding may represent changes in the ratio of dust to ice deposited on the ice caps, and it has been suggested that such fluctuations ultimately have a climatic origin. Indeed, it has been speculated that Mars may have experienced cycles of climatic change similar to those on Earth during the Late Cenozoic Ice Age and that there may even be a common cause in the Milankovitch

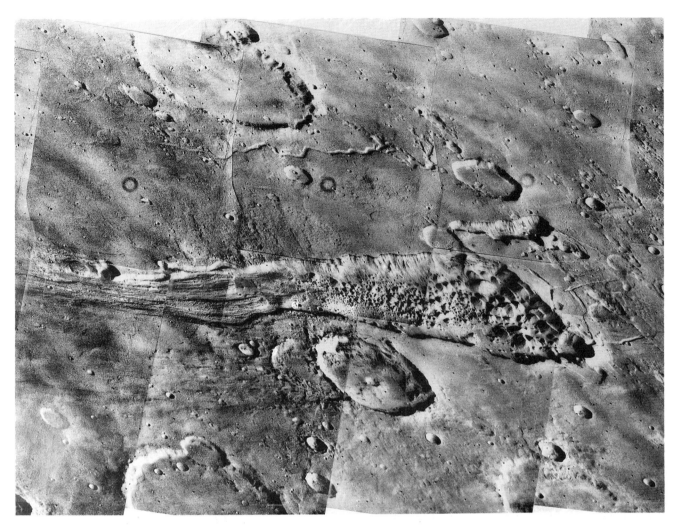

Fig. 19.15 *The head of a Martian outflow channel containing an extensive area of chaotic terrain (right centre of image). Detailed examination of this region has revealed slumped and collapsed blocks up to 10 km long lying at the base of steep escarpments rimmed by arcuate features. Vast quantities of debris have clearly been removed and this may have been accomplished by the catastrophic release of water from the melting of ice-rich permafrost. The area shown is about 300 km across. (Viking 1 mosaic, World Data Center A for Rockets and Satellites.)*

mechanism (see Section 14.3.2). The lack of fresh impact craters in the layered terrain certainly indicates that in Martian terms the surface features in this region are relatively young and that the layered deposits must be accumulating rapidly compared with the recent cratering rate.

As already mentioned, there is no unequivocal evidence for glacial erosion on Mars and it certainly seems that the existence of flowing ice is ruled out in the present Martian environment. Nevertheless, glacial processes have been proposed as an explanation for the so-called outflow channels on Mars (see Section 19.3.8) because their forms are, in some respects, similar to terrestrial glacially eroded valleys. An argument in favour of the glacial theory is the comparable scale of the Martian and terrestrial forms, but a glacial origin for outflow channels is difficult to reconcile

with their concentration within 30° of the Martian equator. As we shall now see, there is no shortage of alternative hypotheses to explain these and other channel forms on Mars.

19.3.8   Channels

No landforms on the Martian surface have generated more controversy than the range of channel features first seen on Mariner 9 imagery and subsequently investigated in detail using the higher-resolution Viking orbiter data. In nearly all cases these forms are most accurately described as valleys, but as the term 'channel' has gained wide currency among planetary geomorphologists we will retain it here. Three main types of channel have been recognized.

Fig. 19.16 *Nirgal Vallis, an 800 km long runoff channel on Mars located at latitude 28°S, longitude 40° and incised into old cratered terrain. See Figure 1.2 for detail from part of the channel network towards the left-hand edge of the area shown here. (Viking 1 mosaic, World Data Center A for Rockets and Satellites.)*

Fretted channels have wide, smooth floors and steep walls and are found in the fretted terrain marking the transition between the heavily cratered uplands and the lightly cratered lowland plains. The presence of large debris flows where these channels extend into the lowlands indicates that they have formed through escarpment retreat by mass movement.

Runoff channels, which are connected to form dendritic networks, are in many ways similar to terrestrial fluvial sys-

tems; they are found throughout the older cratered terrain (Fig. 19.16). When first observed they were simply attributed to surface runoff processes and were thus considered to be ancient features related to a past pluvial epoch on Mars. The junction angles of these networks are, however, much less acute than in most terrestrial fluvial systems. By analogy with morphologically similar forms which occur in certain environments on the Earth, such as on the Colorado Plateau in the south-west USA (Fig. 1.3), headward erosion by spring sapping is now thought to be a more likely mode of formation (Fig. 1.2). Meltwater from ice-rich permafrost is the likely source of the required water.

Outflow channels are much larger features tens of kilometres wide and hundreds of kilometres long (Fig. 19.15). They generally lack tributaries and most originate within chaotic terrain or large concave depressions. Of particular interest is the variety of bedforms seen on the floors of outflow channels including scour marks, longitudinal grooves and streamlined upland remnants (Fig. 19.17). Scour marks and other erosional and depositional forms extend into the plains on to which the mouths of outflow channels open. Although they are widely regarded as having been formed

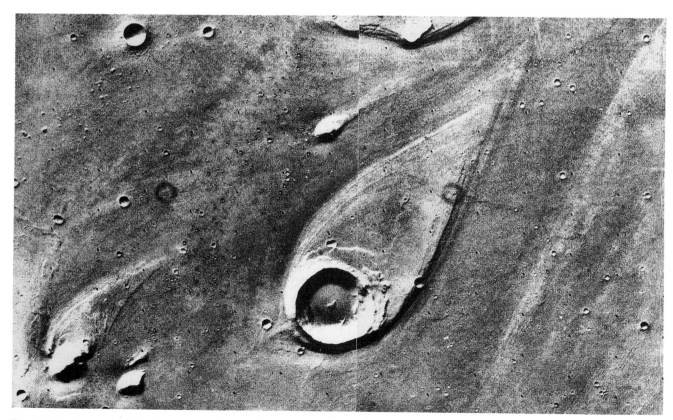

Fig. 19.17 *A streamlined 'island' about 40 km in length representing an eroded remnant of a former plateau located at the mouth of Ares Vallis, an outflow channel on Mars located at latitude 20°N, longitude 31°. The flow from the lower left to the upper right diverged around the crater rim forming a tapering tail downstream. (Viking 1 mosaic, World Data Center A for Rockets and Satellites.)*

by some kind of fluid flow there is little agreement as to the particular fluid agent involved. The erosive agents proposed include water, ice, wind, debris flows, liquefied crustal material and low-viscosity lava.

Detailed comparisons between the form of outflow channels and the morphology of the Channeled Scabland of eastern Washington and Oregon in the USA (see Section 11.4.1) have led to the suggestion that the Martian channels similarly owe their form to the action of catastrophic floods. Although the outflow channels are generally much larger than their terrestrial equivalents, this may be explained by the significantly different conditions of surface gravity and atmospheric pressure on Mars. The lower surface gravity, for instance, could allow the transportation of cobble-sized material in suspension, and it is likely that catastrophic floods could attain a sediment concentration of up to 60–70 per cent by volume. Formation of the outflow channels could, therefore, have been accomplished very quickly, perhaps in a matter of a few days or weeks. Even accepting the dominant role of some kind of water flow in sculpturing their overall form, outflow channels have clearly also been extensively modified by a range of other processes, including thermokarst development, impact cratering, aeolian erosion, spring sapping and a variety of mass movement processes.

If outflow channels are accepted as essentially water-eroded landforms, a crucial problem remains the source of such water which clearly must have been present in very large quantities. An initial proposal was that volcanic eruptions below glaciers led to rapid melting and the catastrophic release of water, but this hypothesis requires the existence of significant ice sheets outside the polar regions. A related idea is the melting of ice-rich permafrost associated with volcanic activity, although this would be a far less efficient process for rapidly generating large quantities of meltwater.

Another hypothesis involves the catastrophic release of artesian water under high pressure. This possibility arises from the postulated existence of an aquifer system extending from about 1 km to 5 or 10 km below the Martian surface and confined by a thick, impermeable, overlying permafrost layer. This mechanism has the advantage of simultaneously accounting for catastrophic water release and the formation of chaotic terrain at the head of outflow channels through the collapse of the surface on the removal of the underlying water.

19.3.9 History of landscape development

In the absence of dated surface materials from samples returned to Earth, the chronology of landscape development on Mars rests on estimates of cratering rates. A range of time scales have been proposed; one of these gives the age of the oldest crust as 4.2–4.3 Ga, volcanic activity in the Tharsis region at between 3 and 1 Ga BP, and the major phase

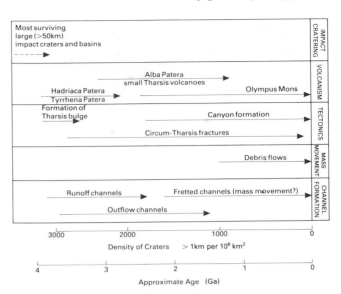

Fig. 19.18 *A possible chronology for the Martian landscape. (Modified from M. H. Carr (1981) The Surface of Mars. Yale University Press, New Haven, Fig. 16.1, p. 202.)*

of channel formation from 4 to 3.7 Ga. Another chronology suggests more recent activity with highland lava emplacement between 3.9 and 0.3 Ga BP and Tharsis region volcanism starting about 0.5 Ga BP. More detailed investigations of cratering rates in the outflow channels suggest that they formed from 0.5 to 2 Ga BP, while the runoff channels are considered to be rather older. The most modern features of the Martian landscape are some of the dune formations which, as repeated Viking orbiter observations show, are currently active. Figure 19.18 gives a very tentative chronology for the main landforming episodes on Mars which will no doubt be considerably revised when rock samples eventually become available for radiometric dating.

19.4 Mercury

Before the fly-by Mariner 10 mission in 1974 Mercury was the least known of the inner planets. Because of its comparatively small size and proximity to the Sun no surface details could be discerned from the Earth. The Mariner 10 fly-by provided adequate coverage of nearly half of the planet at resolutions from 4 km to 100 m. First impressions are that the surface is very like that of the Moon (Fig. 19.19). Impact craters are the predominant landform and most of the features associated with lunar craters are also found on Mercury. Minor differences can be attributed largely to the higher surface gravity of Mercury; ballistic ejecta, for instance, are found closer to the primary impact crater.

A major contrast with the morphology of the Moon, however, is provided by the occurrence of lobate scarps. These appear to be associated with thrust faults resulting from crustal compression of global extent. It has been sug-

Fig. 19.19 *The surface of Mercury as seen on approach during the Mariner 10 fly-by mission. (Mosaic from the* Atlas of Mercury, *Fig. 18, World Data Center A for Rockets and Satellites.)*

gested that an episode of global contraction may have resulted from the cooling and shrinking of Mercury's large core and the stresses imparted by the slowing of the planet's rate of rotation due to its gravitational interaction with the Sun.

Impact features on Mercury range in size from the smallest craters resolvable in Mariner 10 imagery up to the Caloris Basin, a complex feature more than 1300 km across encircled by mountains up to 2 km high. Antipodal to the Caloris Basin is an anomalous region of hilly and lineated terrain which appears to disrupt earlier landforms. The creation of this **weird terrain**, as it has been termed, may have occurred as a consequence of the focusing of seismic waves generated by the massive impact event which formed the Caloris Basin itself.

The smooth plains of Mercury, which cover about 15 per cent of its surface, are similar in appearance to the lunar maria. Although no unequivocal volcanic features have

been identified on Mariner 10 imagery, indirect evidence suggests that, like their lunar counterparts, these plains are at least primarily of volcanic origin. There are, for instance, many examples of material from the smooth plains filling craters clearly formed at an earlier stage.

In the absence of the kind of radiometrically based landscape chronology established for the Moon, any dating scheme for Mercury must rest on the assumption that it has experienced a lunar-type history of bombardment. Although there are similarities between the size frequencies and areal densities of impact craters on Mercury and the Moon, it has been questioned whether this necessarily indicates a temporal equivalence in bombardment history. It is perhaps safer, therefore, to talk in terms of a relative rather than an absolute landscape chronology for Mercury (Fig. 19.20).

19.5   Venus

Similarities in size, density and distance from the Sun make Venus in these respects the closest equivalent of Earth in the Solar System. There is a stark contrast, however, in the atmospheric properties of the two planets. Venus is blanketed by a thick atmosphere with a surface pressure some 90 times that of the Earth, while surface temperatures are around 480 °C. The dense atmosphere has prevented direct visual observation of the planet's surface from the Earth or by conventional orbiting satellite imaging systems. Instead our limited knowledge of the surface has so far come from the Soviet Venera landers and both Earth-based and orbiting satellite-based radar altimeters which are able to penetrate the atmosphere and provide data on variations in elevation over the planet.

The presently available radar data are of a resolution sufficient to pick out major morphological features, but inadequate to provide the detailed altitudinal information needed to determine unequivocally whether Venus has an

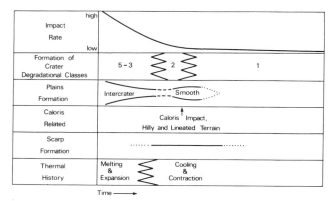

Fig. 19.20 *A relative chronology for the development of the surface features of Mercury. Note that other interpretations are possible. (After R. G. Strom (1979)* Space Science Reviews *24, Fig. 38, p. 65.)*

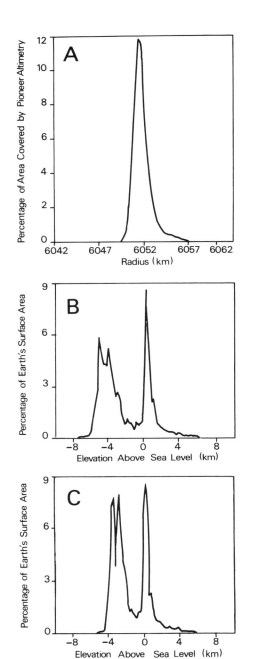

Fig. 19.21 *Hypsometric curves for Venus (A) and the Earth (B and C). Curve (B) represents the Earth's actual topography, whereas curve (C) represents the hypsometry with the loading effect of the oceans on the crust removed, the latter curve being the more appropriate for comparisons with Venus. (After J. W. Head* et al. *(1981)* American Scientist *69, Fig. 3, p. 617.)*

latter has arisen largely from a combination of density contrasts between continental and oceanic crust, and the effects of plate tectonic mechanisms in continually generating new dense oceanic crust and sweeping lighter material on to the continental margins. We can surmise, therefore, that if plate tectonics does occur on Venus then it must be of a rather different form from that active on the Earth.

Examination of the gross topography of Venus does in fact suggest a number of features with at least a superficial resemblance to tectonic structures on the Earth (Fig. 19.22). For instance, a plateau feature known as Beta Regio appears to be a large domal uplift traversed by a rift with associated volcanoes, and indeed it has been likened to the East African Rift System. There are also extensive arcuate and linear troughs concentrated in the upland massifs, but these seem to be more analogous to Martian canyons. The extensive elevated area named Terra Ishtar reaches up to 11 km above the median elevation of Venus, and around this massif are a number of depressions which appear to be of volcanic origin. Another significant feature is a huge cone about 600 km across with an apparent summit caldera and this has been compared with the Martian shield volcano Olympus Mons. Numerous crater-like forms have also been recorded in the lowland areas; some of these may be volcanic, others impact craters, but most are of uncertain origin.

Further information bearing on the possibility of plate tectonics on Venus has been provided by the Soviet Venera 13 and Venera 14 landers. Chemical analyses of surface materials by these probes indicate the existence of tholeiitic basalts at the Venera 14 landing site, rocks typical of mid-oceanic spreading ridges on the Earth, and the presence of high-alkali basalts characteristic of terrestrial continental hot-spot volcanism at the Venera 13 site some 1000 km away. Although the presently available topographic data do not provide unequivocal evidence of the existence of plate tectonic structures on Venus, estimates of the likely rate of heat flow to the Venusian surface imply that a very large number of volcanic hot-spot centres must be present if spreading ridge systems are not available to dissipate heat.

Images of the surface returned by the Venera probes show a variety of terrains. The Venera 14 site is unusual in lacking unconsolidated material, but the other sites display fine to coarse debris resting on bedrock. Evidence of erosion of rock surfaces and the presence of angular rock fragments in Venera lander images indicate that the Venusian surface continues to experience geomorphic activity. Data on the extraordinary atmosphere provided by the Venera probes certainly suggest that chemical weathering should be extremely active. Since the partial pressure of H_2O increases with elevation the most vigorous weathering probably occurs in the upland regions.

In the absence of liquid water on the surface, at least in recent geological time, wind is presumably the main denudational agent. Theoretical calculations, in conjunction with

Earth-like topography dominated by the effects of plate tectonics, a Mars-like surface characterized by impact craters and volcanic landforms or a combination of the two. Nevertheless, one major morphological distinction is clear; the hypsometric curve for Venus displays a single peak in contrast to the Earth's bimodal distribution (Fig. 19.21). The

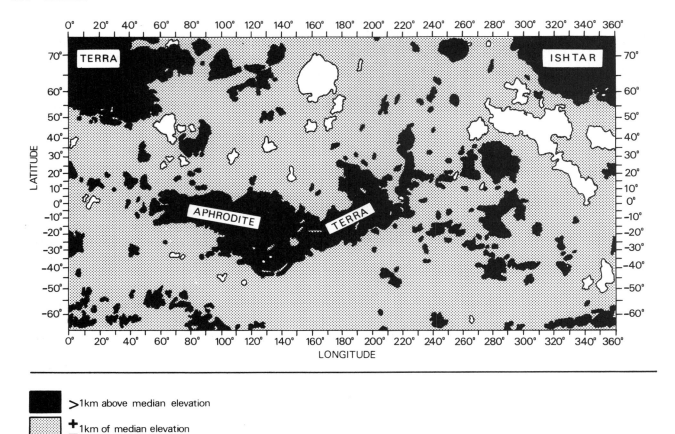

Fig. 19.22 *Major topographic features of Venus based on radar altimetry. Note that the Mercator projection used greatly exaggerates the area covered by features at high latitudes. (After G. E. McGill (1982) Nature, 296, Fig. 2, p. 14.)*

satellite data, indicate low wind speeds on the surface of Venus of around 1 m s⁻¹, but because the atmosphere is so dense particles up to 10 mm across can probably be transported (Fig. 19.13). Earlier in its history Venus may have been capable of retaining liquid water on its surface and so ancient fluvial features may yet be found. Of crucial importance to an understanding of the evolution of the Venusian surface is whether large impact basins remain from the Late Heavy Bombardment. This would indicate the extent to which endogenic and exogenic processes have subsequently modified the surface, but answers to these and other intriguing questions about the smaller-scale morphological features of the planet will have to await the return of high resolution radar imaging data from the Magellan probe launched from the space shuttle Atlantis in May 1989.

19.6 The moons of the outer planets

Beyond the inner Solar System lie the giant planets Jupiter, Saturn, Uranus and Neptune. Although these bodies lack a solid surface, they all possess orbiting satellites, some of which are of planet-sized dimensions. Until the Voyager programme of planetary exploration which began in the late 1970s little was known of the surface features of the rocky and icy satellites of the outer planets, and the first close-up images of the moons of Neptune were only received from Voyager 2 in 1989. The images that have so far been returned from the hugely successful Voyager probes have revealed landscapes with extraordinary characteristics, and which, in many cases, have no terrestrial analogues.

19.6.1 The Galilean moons of Jupiter

The four large (Galilean) satellites of Jupiter are comparable in size to Mercury and our own Moon and represent extremes of planetary processes and morphology (Table 19.4). Each has features preserved from different periods in its history, and each exhibits the effects of quite distinct geomorphic processes. Io, the innermost of the four, is one of the most bizarre planetary bodies in the Solar System (Fig. 19.23). It has a high albedo and a striking reddish-

Fig. 19.23 *View of Io taken at a range of 862 200 km. The lack of impact craters on the surface, together with evidence of continuing volcanism, show Io to be an extremely active body. (Voyager 1 image, World Data Center A for Rockets and Satellites.)*

Table 19.4 Properties of the major moons of Jupiter and Saturn

	DIAMETER (km)	DISTANCE (km)	DENSITY(kg m⁻³)
Jovian Moons		(from Jupiter)	
Io	3640	421 600	3530
Europa	3130	670 900	3030
Ganymede	5280	1 070 000	1930
Callisto	4840	1 880 000	1790
Saturnian moons		(from Saturn)	
Tethys	1060	294 700	1210
Dione	1120	377 400	1430
Rhea	1530	527 100	1330
Titan	5150	1 221 900	1880
Iapetus	1460	3 560 800	1160

certainly comes from radioactive decay within the planet's rocky interior, additional energy is also supplied from its strong tidal interaction with the giant mass of Jupiter close by. The surface is thus constantly being renewed by fresh eruptions of both silicate-rich and sulphur-rich lavas; any impact craters formed earlier in its history have consequently been obliterated.

Europa, the next satellite out, is, like Io, composed primarily of rock. Its very high albedo and slightly lower density in comparison with Io suggest a 50–100 per cent covering of frost and ice, probably a few kilometres thick, forming its remarkably smooth surface (Fig. 19.25). There are very few impact craters indicating that resurfacing, possibly also promoted by tidal energy derived from the moon's relative proximity to Jupiter, has occurred since the Late Heavy Bombardment (assuming that this event affected this part of the Solar System).

Callisto, the outermost of the Galilean moons, has a much more familiar looking surface dominated by impact craters and multi-ringed basins, presumably formed during the Late Heavy Bombardment and since preserved (Fig. 19.26). Its low density indicates a high ice content, but its low albedo suggests that any surface ice must be heavily contaminated with rock.

Ganymede, lying between the orbits of Callisto and Europa, has the most complex surface of the four Galilean satellites representing an amalgam of heavily cratered Callisto-like terrain and the linear fracture patterns reminiscent of Europa. Its low density indicates a high ice content and its high albedo suggests a 20–60 per cent surface covering of fresh ice. It is conjectured that during the Late Heavy Bombardment the crustal layer of Ganymede was very thin but it subsequently thickened, giving rise to a range of tectonic structures including strike-slip faults and rift zones into which fresh, clean ice was injected.

19.6.2 The moons of Saturn

Of the moons of Saturn, Titan is by far the largest and is exceptional in having a significant atmosphere. It is for this reason that the surface is obscured and consequently its surface morphology is unknown. In spite of their broadly similar density and composition there is no lack of variety in the geomorphology of the other Saturnian satellites imaged by the Voyager fly-bys.

Tethys is the closest to Saturn of the planet's five large moons (diameter greater than 1000 km) (Table 19.4). It has a heavily cratered surface and its most remarkable structure is an enormous impact scar over 400 km across, that is, some 40 per cent of the diameter of the satellite itself. As this feature now shows little vertical relief, it is hypothesized that the interior of Tethys was sufficiently warm and mobile early in its history to accommodate what was obviously an enormous impact without shattering. The burial of some

yellow colour, but quite the most remarkable observation made by the Voyager 1 probe on its transit through the Jovian system was of an active volcano hurling incandescent plumes hundreds of kilometres above its surface (Fig. 19.24). The lack of impact craters on its surface, together with the abundant morphological evidence of widespread volcanism, show Io to be the most active planetary body in the Solar System observed so far. Although some of the energy for this high level of endogenic activity

Fig. 19.24 *A volcanic eruption in progress on Io. The brightness of the volcanic plume that can be seen on the horizon rising 100 km from the surface has been greatly increased by computer enhancement. (Voyager 1 image, World Data Center A for Rockets and Satellites.)*

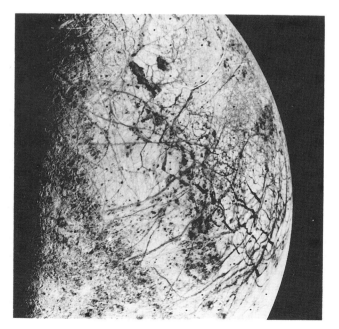

other large craters also indicates a degree of resurfacing on the satellite.

Almost identical in size to Tethys is Dione, orbiting some 80 000 km further out from Saturn. It is the densest moon after Titan and detailed evaluation of its heavily cratered surface suggests that it has been bombarded by two distinct populations of projectiles. The trailing hemisphere, that is, the side facing backward with respect to the direction of its orbital motion around Saturn, has impact craters similar to the lunar highlands and with all the hallmarks of the Late Heavy Bombardment. A second phase of impact cratering with a greater proportion of small projectiles is attributed to

Fig. 19.25 *View of Europa showing a region 600 × 800 km. Particularly noticeable is the network of lineaments on the surface representing low relief ridges and grooves, and contrasts in albedo. These patterns appear to have been caused by stress in the surface crust generated by various possible mechanisms including tidal deformation, global expansion due to dehydration of the interior and freezing of an early ocean. (Voyager 2 image, World Data Center A for Rockets and Satellites.)*

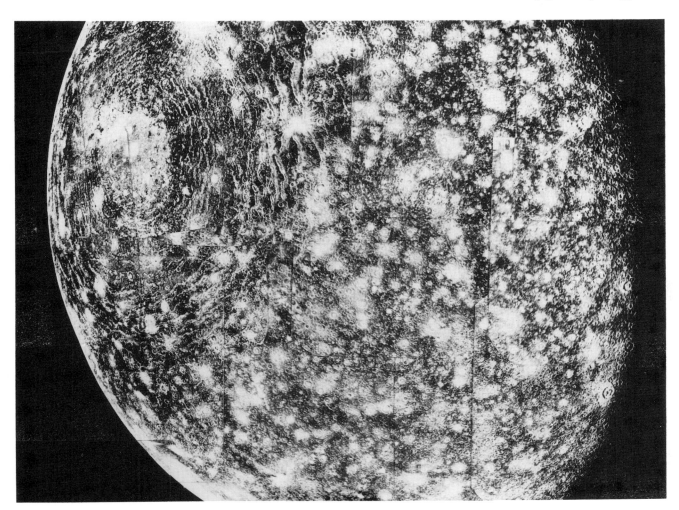

Fig. 19.26 *View of Callisto from a range of 202 000 km. The bright circular region to the upper left of the image is an impact basin about 300 km across. (Voyager 1 mosaic, World Data Center A for Rockets and Satellites.)*

debris generated through collisions within the Saturnian system of moons. This has given rise to a moderate cratering of the plains on Dione formed by resurfacing after the Late Heavy Bombardment. A similar history appears to have been experienced by Rhea which also shows evidence in its distribution of crater sizes for two populations of impacting projectiles. As on Dione, part of the trailing hemisphere shows evidence of resurfacing.

Closely equivalent in size to Rhea is Iapetus, the most distant of Saturn's large moons. It has an incredibly low density of about 1100 kg m^{-3}, and a startling contrast in albedo from around 50 per cent on its trailing hemisphere to an extraordinarily low 3–5 per cent on its leading face; this is possibly a result of dusting by particles originating from Phoebe, the most remote of the Saturnian satellites. An alternative explanation is that the dark material, which has been observed in circular features similar to craters filled by lava flows on the Moon and Mars, was extruded from the interior of the satellite.

19.6.3 The moons of Uranus

Although the images returned from the Voyager probes as they passed through the Jovian and Saturnian systems provided plenty of surprises, the moons of Uranus examined at close quarters by Voyager 2 in January 1986 yielded yet more bizarre surface forms. The 1500 km diameter moon Oberon was seen to have large impact craters with bright rays, and one mountain peak observed was estimated to be some 6 km high. Its comparatively low density of about 1600 kg m^{-3} indicates that it has an icy composition. The slightly smaller moon Ariel with a diameter of about 1200 km has a similarly low density. One of its most remarkable features are canyons which appear to have been flooded by eruptions of ice. These and similar ice-filled valley forms may owe their origin to the extrusion of ice on to the surface as a result of tectonic activity.

The strangest world in the Uranian system is Miranda. Just 480 km in diameter and with a density of only

$1200 \, \mathrm{kg \, m^{-3}}$, it is the nearest of the large satellites to Uranus. Its surface is composed of two contrasting types of terrain: one is bright and heavily cratered, whereas the other is darker and lower in elevation and is made up of a series of discrete and apparently unrelated regions of grooved terrain. At present the most widely accepted explanation of this topography is that Miranda suffered a cataclysmic impact early in its history which caused it to break up and then subsequently reaggregate. The now juxtaposed distinctive terrains are thought to have formed as a result of the late accretion of dense fragments which sank slowly through Miranda's outer icy 'mantle' towards its rocky core.

19.6.4   The moons of Neptune

After the startling images acquired of the surfaces of the moons of Jupiter, Saturn and Uranus, it was with great anticipation that researchers awaited the arrival of Voyager 2 at Neptune. In late August 1989 their patience was rewarded by views of a world stranger than anything that had been previously encountered in the epic 12-year journey of Voyager 2. Although initial attention was focused on Neptune itself, interest rapidly shifted to Triton, its largest moon with a diameter of 2700 km. Its fascinatingly complex landscape, which seems to be an assemblage of terrain types previously encountered on other planetary bodies, immediately precipitated a wide range of hypotheses attempting to explain its history.

Predictions prior to the arrival of Voyager 2 were that Triton should be a cold, dead world retaining the scars of the Late Heavy Bombardment. Yet the surface showed much evidence of resurfacing with far less than the expected density of impact craters (Fig. 19.27). Triton, it was thought, had neither the tidal energy of moons like Io, nor the radio-

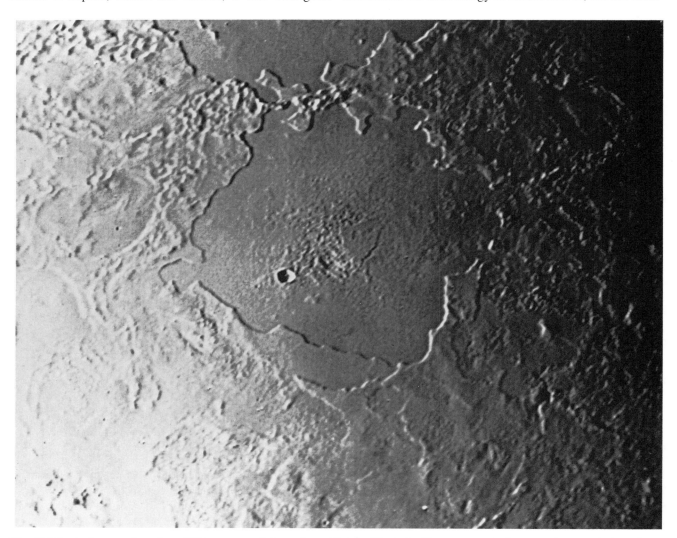

Fig. 19.27   *A close-up view about 500 km across of the surface of Triton. The lack of large craters indicates active resurfacing after the Late Heavy Bombardment, but the presence of numerous small craters suggests that generally there has been only a low level of activity over the past 2 Ga. (Voyager 2 mosaic courtesy JPL and NASA.)*

active energy of large, rocky planetary bodies to generate sufficient internal heat to have sustained 'volcanic' activity long after its formation. The key to this paradox may lie in Triton's peculiar orbit — it moves opposite to the direction of Neptune's rotation and at an inclination of 21° to the planet's equator. Such an orbit suggests that Triton was captured by Neptune and was pulled into a highly eccentric orbit which alternately took it close to the planet and far away. This type of orbit would have indeed created the kind of tidal energy that drives Io's intense volcanism at the present day. After a few hundred million years the orbit would have become less eccentric, and the reduction in gravitational energy would have caused Triton to cool rapidly. Consequently, the effects of the massive bombardment experienced by all planetary bodies early in the history of the Solar System have been erased from Triton's surface which now only retains evidence of less catastrophic impact events. More recent signs of resurfacing post-dating the Late Heavy Bombardment may be a result of the flowage of methane and nitrogen 'ice' precipitated out of Triton's thin atmosphere.

19.7   Comparative planetary geomorphology

The quantum leap in our knowledge of planetary bodies over the past two decades presents us for the first time with the possibility of making interplanetary comparisons of landforms and the processes that shape them. As we have already emphasized, much of the interpretation of the surface forms of the earth-like planets has been accomplished through analogies with terrestrial landforms. This has served to highlight certain common elements in landform genesis, but for the most part has revealed the considerable contrasts in both the nature and relative intensity of the operation of geomorphic processes throughout the Solar System. The comparative approach to the study of the evolution of planetary surfaces is in its infancy, but it is already possible to suggest some broad generalizations concerning the primary controls of the morphology of the earth-like planetary bodies (Table 19.5).

Until the exploration of the moons of the outer planets, a key variable in the geomorphic evolution of planetary bodies was considered to be size. The Earth has sufficient mass to

Table 19.5 Comparative geomorphology of some earth-like planetary bodies

PLANET/MOON	EXOGENIC PROCESSES	IMPACT CRATERS	VOLCANISM/RESURFACING	TECTONICS
Earth	Active chemical and physical weathering. Mass movement and fluvial, glacial and aeolian erosion and deposition	Very rare–largely obliterated by denudation, volcanism and tectonics	Locally active	Active plate tectonics
Moon	Regolith production by micrometeorite impact and mass movement	Abundant	Locally active early in history	Limited faulting early in history
Venus	Active chemical weathering and probably aeolian action and mass movement	Probably present	Probably active	Uncertain–possibly a form of plate tectonics
Mars	Physical and (limited) chemical weathering. Mass movement and aeolian activity. Phase of fluvial erosion probably early in history.	Fairly abundant	Locally abundant	Major structural lineaments
Mercury	Probably regolith production by micro-meteorite impact and mass movement	Abundant	Possibly active early in history	Thrust faulting associated with crustal compression early in history
Io	?	None? presumably obliterated by volcanic resurfacing	Very active	Related to volcanism
Europa	?	Uncommon–presumably obliterated by resurfacing	Resurfacing active early in history?	Linear patterns related to crustal stress
Ganymede	?	Relatively abundant	None	Lineaments with evidence of rifts and strike-slip faulting–form of plate tectonics?
Callisto	Mixing of rock with surface ice (associated with impact cratering?)	Abundant including multi-ringed basins	None	?
Tehtys	?	Abundant	Limited resurfacing	?
Dione	?	Abundant	Limited resurfacing	?
Rhea	?	Abundant	Limited resurfacing	?
Titan	Presence of atmosphere suggests some surface weathering	?	?	?
Iapetus	?	Abundant	Resurfacing by low albedo material	?

retain a significant atmosphere and thereby allow the operation of a wide range of denudational processes. The relatively low surface/volume ratio of the Earth also means that the rate of heat loss from its surface to space has been sufficiently slow to ensure continuing volcanic and tectonic activity promoted by the high temperatures of a molten interior. Smaller bodies, such as the Moon and Mercury, have a higher surface/volume ratio and consequently have experienced a much more rapid rate of cooling than the Earth. A rigid outer crust formed early in their history which was thus able to retain the scars of the Late Heavy Bombardment. Moreover, both the Moon and Mercury are too small, and therefore have insufficient surface gravity, to have prevented the escape of gas molecules and thus retained an atmosphere. Mars is roughly intermediate in size between the Earth and the Moon and Mercury and, as would be expected, it has both an intermediate atmospheric pressure and level of tectonic and volcanic activity. On this basis we would expect Venus to have a level of endogenic activity not very dissimilar to the Earth, since it is of comparative mass.

In broad terms, then, size (or more strictly mass) appears to be a significant factor in controlling the mode of geomorphic development of planetary bodies. This simple relationship, however, breaks down when we turn to the larger satellites of Jupiter and Saturn. In Figure 19.28 most of the larger earth-like planets and moons of the Solar System are plotted in a three-dimensional diagram, the axes of which

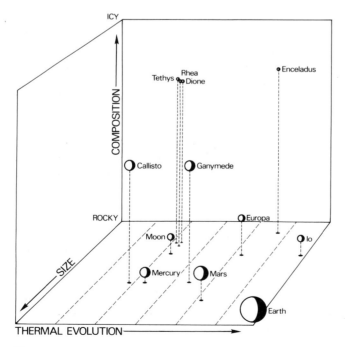

Fig. 19.28 *A comparison of major characteristics of some earth-like planets. (Adapted from L. A. Soderblom and T. V. Johnson, (1982)* Scientific American **246**(1), *p. 86. Copyright ©* (1982) *by Scientific American, Inc. All rights reserved.)*

represent size, composition and state of thermal evolution (that is, the nature and history of tectonic and volcanic activity). The inner planets all have a similar rocky composition, but several of the moons of Saturn are low-density bodies composed predominantly of ice. Even though they are very small in comparison with the Moon and Mercury, these satellites (Tethys, Dione, Rhea and Iapetus) show some evidence of resurfacing after the Late Heavy Bombardment. It seems likely that this activity, in spite of a lack of mass, is a consequence of their icy composition since ice has a far lower melting-point than rock and is therefore capable of producing tectonic activity at much lower temperatures. Hence on our diagram these satellites plot as about as tectonically evolved as the Moon, even though they are much smaller.

Turning to the Galilean moons of Jupiter we appear to have more exceptions to the size rule. These satellites are of the same order of size as the Moon, but they range in their degree of thermal evolution from the highly evolved Io to the poorly evolved Callisto. If we look more carefully, though, we see that the level of thermal evolution of each of the Galilean moons decreases with their increasing distance from Jupiter. This is apparently because Io, the innermost satellite, derives a significant proportion of its internal energy from its tidal interaction with its giant mother planet. This may also apply, albeit to much smaller extent, to Europa.

Perhaps the most remarkable satellite in the Saturnian and Jovian systems is Enceladus, a small moon orbiting close to Saturn. Although it has a diameter of only 500 km and has only 0.1 per cent of the mass of the Moon, it has a complex surface suggesting comparatively recent tectonic activity. Again this is probably associated with the generation of tidal energy due to the close proximity of the large mass of Saturn. Further examples of small planetary bodies with active resurfacing processes are provided by the moons of Uranus.

19.8   Global geomorphology

It is appropriate to conclude this chapter on planetary geomorphology by briefly considering what impact the study of planetary landscapes has had on geomorphology as a whole and, perhaps more importantly, what developments it is likely to precipitate in the future. One obvious contribution has been in the development, and more particularly the use, of remote sensing techniques in landform analysis. Planetary exploration has been founded, the manned Apollo missions notwithstanding, on data collection by unmanned satellites carrying remote sensing equipment. This has prompted the greater application of satellite imagery to the investigation of terrestrial landforms, as exemplified by the production of a landform atlas by NASA researchers and others based on satellite imagery, although there is much scope for further developments in this field. Global surveys of terrestrial dune systems have now been undertaken, based largely on Landsat imagery and prompted by the need for a comparative database for the interpretation of images of Martian dune systems. A further development has been the space shuttle imaging radar programme which, although so far attaining only a limited geographical coverage, has demonstrated the value of radar techniques in providing images of landforms in regions such as the humid tropics where persistent cloud cover limits the value of conventional Landsat imagery.

A second contribution provided by the exploration of the inner planets has been the stimulation of research on particular geomorphic processes. The realization of the important role played by aeolian processes on Mars, for instance, arising from the identification of vast dune systems and wind-eroded terrain on Pioneer and Viking imagery, has led to a significant growth in laboratory and field studies of terrestrial aeolian activity and landforms.

A third contribution of planetary geomorphology to the study of terrestrial landscapes has been the emphasis placed in the former on catastrophic events. Not only meteorite impact craters but also the formation of features such as the Martian outflow channels seem to be explicable only in terms of the operation of rapidly acting, highly energetic and devastating geomorphic processes. Such an emphasis has led to a re-evaluation of the role of catastrophic processes in shaping the Earth's relief.

Perhaps the most significant stimulus provided by the exploration of planetary landscapes, however, is one which has yet to be fully realized; it is an enlargement of the conceptions of the spatial and temporal scales with which geomorphologists should be concerned. Coupled with major advances in other fields within the earth sciences, such as those represented by plate tectonics and contemporary theories of long-term climatic change, the investigation of planetary landscapes provides us with an exciting new global perspective for geomorphology.

Further reading

Before the space missions of the 1960s and 1970s little was known of the surface features of the earth-like planets, but since 1970 there has been a flood of data from a variety of planetary orbiter and lander missions. There are now hundreds of articles on aspects of planetary geomorphology and only a very small selection is mentioned here. Virtually all of these have been published outside the normal outlets for geomorphic research, the publications *Icarus* and *Journal of Geophysical Research* being particularly useful sources. For a general introduction to planetary geomorphology there is no better starting point than the extensively illustrated book by Greeley (1986).

The results of the Apollo missions to the Moon have appeared in the proceedings of the various lunar science conferences published as supplements of *Geochimica et Cosmochimica Acta*. Burnett (1975) and El-Baz (1975) also provide assessments of the major results of the Apollo programme while Wilhelms (1987) has presented a superbly illustrated survey of lunar geomorphology. Contrasting estimates of cratering rates throughout the history of the Moon are to be found in Guiness and Arvidson (1977) and Neukum and König (1976), while crater morphology is considered by Hale and Grieve (1982), Head (1976a) and Pike (1977). Lunar volcanism is reviewed by Head (1976b), and the origin of lunar rilles and mare ridges is discussed by Lucchitta (1976) and Solomon and Head (1979).

The discussion of Martian geomorphology in this chapter draws extensively from the excellent review by Baker (1981a); other general surveys include the introductory article by Arvidson *et al.* (1978) and the detailed and comprehensively illustrated book by Carr (1981). A large number of papers on various aspects of the geomorphology of Mars are to be found in two supplements of the *Journal of Geophysical Research* (**84**, 7909–8519 (1979), and **87**, 9717–10 305 (1982)).

Impact craters on Mars and their associated layered debris are discussed by Carr *et al.* (1977a) and Mouginis-Mark (1979), while Carr (1976) provides an introduction to the volcanic features on the planet, a treatment which is supplemented by a more detailed discussion in Carr *et al.* (1977b) and a comparison with terrestrial volcanism in Malin (1977). The Valles Marineris structure is considered by Blasius *et al.* (1977), Sleep and Phillips (1979) suggest an isostatic model for the formation of the Tharsis Montes volcanic province, and the broader question of Martian tectonics is tackled by Wise *et al.* (1979). Clark and Van Hart (1981) and Berkley and Drake (1981) present ideas on weathering, and Sharp and Malin (1984) provide an overview of our present knowledge of deposits on the Martian surface based on Viking lander data. Mass movement on Mars is discussed in papers by Lucchitta (1979) on the spectacular landslides of the Valles Marineris and Kochel and Peake (1984) on the morphometry of the debris flows of the fretted terrain.

Aeolian features on Mars have been extensively investigated through analogies with terrestrial forms by Breed (1977), Breed *et al.* (1979) and Ward (1979). Ward *et al.* (1985) provide a comprehensive assessment of the occurrence of aeolian features on the planet, and the processes of wind abrasion are considered in the context of terrestrial aeolian action by Greeley *et al.* (1985). The best overall coverage of planetary aeolian processes, not only for Mars and the Earth but also for Venus, is the book by Greeley and Iversen (1985).

Permafrost on Mars is reviewed by Carr and Schaber (1977) and Lucchitta (1981) draws comparisons between periglacial features on Mars and Earth. The layered deposits of the poles and other problems of the polar terrain are discussed in Cutts *et al.* (1979), while Lucchitta *et al.* (1981) speculate on the role of glacial erosion on Mars and Clifford (1987) assesses the role of melting in generating chaotic terrain and outflow channels.

The enigmatic Martian channels have generated a large literature. A useful starting point is Baker (1981b), while a selection of views concerning the origin of outflow channels include those in Masursky *et al.* (1977), Carr (1979), Komar (1980), Nummedal and Prior (1981) and Cutts and Blasius (1981). The origin and modification of small Martian valleys are discussed by Baker and Partridge (1986) and Brakenridge *et al.* (1985). In view of the wide range of proposed mechanisms of channel formation it is fortunate that there is also a comprehensive overview available (Mars Channel Working Group, 1983). Baker (1985a) looks at the role of fluvial activity on Mars, while Carr (1987) takes a broad look at the role of water in the planet's history.

A useful assessment of the results of the Mariner 10 mission to Mercury is provided by Strom (1979), while the initial results of this project are contained in Murray (1975) and a special issue of *Icarus* (28, 429–609 (1976)). The possible gravitational effects of the Sun on the tectonics of Mercury is examined in Burns (1976), while McCauley *et al.* (1981) review the Caloris Basin and Dzurisin (1978) presents a general discussion of the tectonic features of the planet.

The morphology of the Venusian surface as revealed by radar altimetry is discussed by Barsukov *et al.* (1986), Basilevsky and Head (1988) and Head *et al.* (1985). An evaluation of the data collected by the various Soviet Venera landers is provided by Garvin *et al.* (1984), while White (1981) and Nozette and Lewis (1982) speculate on the nature of weathering and erosion. The intriguing question of whether there is a form of plate tectonics operative on Venus is addressed by Brass and Harrison (1982), Head *et al.* (1981) and Solomon and Head (1982), while Head and Wilson (1986) examine the possibilities of volcanic landforms.

Results from the Voyager missions to Jupiter and Saturn have rapidly generated a large literature. Useful surveys of the initial data on the larger satellites are to be found in Soderblom (1980) (Jovian moons), Soderblom and Johnson (1982) (Saturnian moons) and Johnson *et al.* (1987) (Uranian moons), but for a more detailed coverage the book by Greeley (1986) should be consulted.

The broader topic of comparative planetary geomorphology, as well as being addressed by Greeley (1986), is considered from a tectonic perspective by Head and Solomon (1981). Rossbacher and Rhodes (1987) provide an excellent example of the analogies that can be drawn between terrestrial and planetary landforms attributable to catastrophic flooding. It is to the essay by Baker (1985b), however, that readers should turn for a creative, optimistic and thorough-

ly convincing view of the potential value of this infant research field for geomorphology as a whole.

References

Arvidson, R. E., Binder, A. B. and Jones K. L. (1978) The surface of Mars. *Scientific American* **238**(3), 76–89.

Baker, V. R. (1981a) The geomorphology of Mars. *Progress in Physical Geography* **5**, 473–513.

Baker, V. R. (1981b) *The Channels of Mars.* University of Texas Press, Austin.

Baker, V. R. (1985a) Models of fluvial activity on Mars. In: M. J. Woldenberg (ed.) *Models in Geomorphology.* Allen and Unwin, Boston and London, 287–312.

Baker, V. R. (1985b) Relief forms on planets. In: A. Pitty (ed.) *Themes in Geomorphology*, Croom Helm, London, 245–59.

Baker, V. R. and Partridge, J. B. (1986) Small Martian valleys: Pristine and degraded morphology. *Journal of Geophysical Research* **91**, 3561–72.

Barsukov, V. L. *et al.* (1986) The geology and geomorphology of the Venus surface as revealed by the radar images obtained by Veneras 15 and 16. *Journal of Geophysical Research* **91**, D378–D398.

Basilevsky, A. T. and Head, J. W. III (1988) The geology of Venus. *Annual Review of Earth and Planetary Sciences* **16**, 295–317.

Berkley, J. L. and Drake, M. J. (1981) Weathering on Mars: Antarctic analog studies. *Icarus* **45**, 231–49.

Blasius, K. R., Cutts, J. A., Guest, J. E. and Masursky, H. (1977) Geology of the Valles Marineris, first analysis of imaging from the Viking 1 orbiter primary mission. *Journal of Geophysical Research* **82**, 4067–91.

Brakenridge, G. R., Newsom, H. E. and Baker, V. R. (1985) Ancient hot springs on Mars: origins and paleoenvironmental significance of small Martian valleys. *Geology* **13**, 859–62.

Brass, G. W. and Harrison, C. G. A. (1982) On the possibility of plate tectonics on Venus. *Icarus* **20**, 326–40.

Breed, C. S. (1977) Terrestrial analogs of the Hellespontus dunes, Mars. *Icarus* **20**, 326–40.

Breed, C. S., Grolier, M. J. and McCauley, J. F. (1979) Morphology and distribution of common 'sand' dunes on Mars: comparisons with Earth. *Journal of Geophysical Research* **84**, 8183–204.

Burnett, D. S. (1975) Lunar science: The Lunar legacy. *Reviews of Geophysics and Space Physics* **13**, 13–34.

Burns, J. A. (1976) Consequences of the tidal slowing of Mercury. *Icarus* **28**, 453–8.

Carr, M. H. (1976) The volcanoes of Mars. *Scientific American* **234**(4), 32–43.

Carr, M. H. (1979) Formation of Martian flood features by release of water from confined aquifers. *Journal of Geophysical Research* **84**, 2995–3007.

Carr, M. H. (1981) *The Surface of Mars.* Yale University Press, New Haven, and London.

Carr, M. H. (1987) Water on Mars. *Nature* **326**, 30–5.

Carr, M. H., Crumpler, L. S., Cutts, J. A., Greeley, R., Guest, J. E. and Masursky, H. (1977a) Martian impact craters and emplacement of ejecta by surface flow. *Journal of Geophysical Research* **82**, 4055–65.

Carr, M. H., Greeley, R., Blasius, K. R., Guest, J. E. and Murray, J. B. (1977b) Some Martian volcanic features as viewed from the Viking orbiters. *Journal of Geophysical Research* **82**, 3985–4015.

Carr, M. H. and Schaber, G. G. (1977) Martian permafrost features. *Journal of Geophysical Research* **82**, 4039–54.

Clark, B. C. and Van Hart, D. C. (1981) The salts of Mars. *Icarus* **45**, 370–8.

Clifford, S. M. (1987) Polar basal melting on Mars. *Journal of Geophysical Research* **92**, 9135–52.

Cutts, J.A and Blasius, K.R. (1981) Origin of Martian outflow channels: The eolian hypothesis. *Journal of Geophysical Research* **86**, 5075–102.

Cutts, J. A., Blasius, K. R. and Roberts, W. J. (1979) Evolution of Martian polar landscapes: interplay of long-term variations in perennial ice cover and dust storm intensity. *Journal of Geophysical Research* **84**, 2975–94.

Dzurisin, D. (1978) The tectonic and volcanic history of Mercury as inferred from studies of scarps, troughs and other lineaments. *Journal of Geophysical Research* **83**, 4883–906.

El-Baz, F. (1975) The Moon after Apollo. *Icarus* **25**, 495-537.

Garvin, J. B., Head, J. W., Zuber, M. T. and Helfenstein, P. (1984) Venus: The nature of the surface from Venera panoramas. *Journal of Geophysical Research* **89**, 3381–99.

Greeley, R. (1986) *Planetary Landscapes* (revised edn). Allen and Unwin, London and Boston.

Greeley, R. and Iversen, J. D. (1985) *Wind as a Geological Process*. Cambridge University Press, Cambridge and New York.

Greeley, R., Williams, S. H., White, B. R., Pollack, J. B. and Marshall, J. R. (1985) Wind abrasion on Earth and Mars. In: M. J. Woldenberg (ed.) *Models in Geomorphology*. Allen and Unwin, Boston and London, 373–422.

Guiness, E. A. and Arvidson, R. E. (1977) On the constancy of the lunar cratering flux over the past 3.3×10^9 yr. *Geochimica et Cosmochimica Acta Supplement* **8**, 3475–94.

Hale, W. S. and Grieve, R. A. F. (1982) Volumetric analysis of complex lunar craters: implications for basin ring formation. *Journal of Geophysical Research* **87**, A65–A76.

Head, J. W. III (1976a) The significance of substrate characteristics in determining the morphology and morphometry of lunar craters. *Geochimica et Cosmochimica Acta Supplement* **7**, 2913–30.

Head, J. W. III (1976b) Lunar volcanism in space and time. *Reviews of Geophysics and Space Physics* **14**, 265–300.

Head, J. W. III, Peterfreund, A. R., Garvin, J. B. and Zisk, S. H. (1985) Surface characteristics of Venus derived from Pioneer Venus altimetry, roughness and reflectivity measurements. *Journal of Geophysical Research* **90**, 6873–85.

Head, J. W. III and Solomon, S. C. (1981) Tectonic evolution of the terrestrial planets. *Science* **213**, 62–76.

Head, J. W. III and Wilson, L. (1986) Volcanic processes and landforms on Venus: Theory, predictions and observations. *Journal of Geophysical Research* **91**, 9407–46.

Head, J. W. III, Yuter, S. E. and Solomon, S. C. (1981) Topography of Venus and Earth: A test for the presence of plate tectonics. *American Scientist* **69**, 614–23.

Johnson, T. V., Brown, R. H. and Soderblom, L. A. (1987) The moons of Uranus. *Scientific American* **256**(4), 40–52.

Kochel, R. C. and Peake, R. T. (1984) Quantification of waste morphology in Martian fretted terrain. *Journal of Geophysical Research* **89**, C336–C350.

Komar, P. D. (1980) Modes of sediment transport in channelized water flows with ramifications to the erosion of the martian outflow channels. *Icarus* **42**, 317–29.

Lucchitta, B. K. (1976) Mare ridges and related highland scarps – result of vertical tectonics. *Geochimica et Cosmochimica Acta Supplement* **7**, 2761–82.

Lucchitta, B. K. (1979) Landslides in Valles Marineris, Mars. *Journal of Geophysical Research* **84**, 8097–113.

Lucchitta, B. K. (1981) Mars and Earth: Comparison of cold-climate features. *Icarus* **45**, 264–303.

Lucchitta, B. K., Anderson, D. M. and Shoji, H. (1981) Did ice streams carve Martian outflow channels? *Nature* **290**, 759–63.

Malin, M. C. (1977) Comparison of volcanic features of Elysium (Mars) and Tibesti (Earth) *Geological Society of America Bulletin* **88**, 908–19.

Mars Channel Working Group. (1983) Channels and valleys on Mars. *Geological Society of America Bulletin* **94**, 1035–54.

Masursky, H., Boyce, J. M., Dial, A. L., Schaber, G. G. and Strobell, M. E. (1977) Classification and time of formation of Martian channels based on Viking data. *Journal of Geophysical Research* **82**, 4016–38.

McCauley, J. F., Guest, J. E., Schaber, G. G., Trask, N. J. and Greeley, R. (1981) Stratigraphy of the Caloris basin, Mercury. *Icarus* **47**, 184–202.

Mouginis-Mark, P. J. (1979) Martian fluidized crater morphology: variations with crater size, latitude, altitude and target material. *Journal of Geophysical Research* **84**, 8011–22.

Murray, B. C. (1975) Mercury. *Scientific American* **233**(3), 58–68.

Neukum, G. and König, B. (1976) Dating of individual lunar craters. *Geochimica Cosmochimica Acta Supplement* **7**, 2867–81.

Nozette, S. and Lewis, J. S. (1982) Venus: Chemical weathering of igneous rocks and buffering of atmospheric composition. *Science* **216**, 181–3.

Nummedal, D. and Prior, D. B. (1981) Generation of martian chaos and channels by debris flow. *Icarus* **45**, 77–86.

Pike, R. J. (1977) Apparent depth/apparent diameter relation for lunar craters. *Geochimica et Cosmochimica Acta Supplement* **8**, 3427–36.

Rossbacher, L. A. and Rhodes, D. D. (1987) Planetary analogs for geomorphic features produced by catastrophic flooding. In: L. Mayer and D. Nash (eds) *Catastrophic Flooding*. Allen and Unwin, Boston and London, 289–304.

Sharp, R. P. and Malin, M. C. (1984) Surface geology from Viking landers on Mars: A second look. *Geological Society of America Bulletin* **95**, 1398–412.

Sleep, N. H. and Phillips, R. J. (1979) An isostatic model for the Tharsis province, Mars. *Geophysical Research Letters* **6**, 803–6.

Soderblom, L. A. (1980) The Galilean moons of Jupiter. *Scientific American* **242**(1), 68–83.

Soderblom, L. A. and Johnson, T. V. (1982) The moons of Saturn. *Scientific American* **246**(1), 73–86.

Solomon, S. C. and Head, J. W. III (1979) Vertical movements in mare basins: relation to mare emplacement, basin tectonics and lunar thermal history. *Journal of Geophysical Research* **84**, 1667–82.

Solomon, S. C. and Head, J. W. III (1982) Mechanisms for lithospheric heat transport on Venus: Implications for tectonic style and volcanism. *Journal of Geophysical Research* **87**, 9236–46.

Strom, R. G. (1979) Mercury: A Post-Mariner assessment. *Space Science Reviews* **24**, 3–70.

Ward, A. W. (1979) Yardangs on Mars: evidence of recent wind erosion. *Journal of Geophysical Research* **84**, 8147–66.

Ward, A. W., Doyle, K. B., Helm, P. J., Weisman, M. K. and Witbeck, N. E. (1985) Global map of eolian features on Mars. *Journal of Geophysical Research* **90**, 2038–56.

White, B. R. (1981) Venusian saltation. *Icarus* **46**, 226–32.

Wilhelms, D.E. (1987) The geologic history of the Moon. *United States Geological Survey Professional Paper* **1348**.

Wise, D. U., Golombek, M. P. and McGill, G. E. (1979) Tectonic evolution of Mars. *Journal of Geophysical Research* **84**, 7934–9.

Appendix A

Units of measurement

The units used in this book are based on the Système International d'Unités (SI) which has now become the standard for scientific literature world-wide. The only major deviation from SI units is the recording of temperature in °C (degrees celsius – sometimes expressed as centigrade) rather than K (degrees kelvin).

Multiples of SI units are expressed as follows:

10^{-6}	micro	μ
10^{-3}	milli	m
10^{-2}	centi	c*
10^{3}	kilo	k
10^{6}	mega	M
10^{9}	giga	G

Time

s	second
h	hour
a	year
ka	10^{3} years
Ma	10^{6} years
Ga	10^{9} years

Length

m	metre
cm	centimetre
mm	millimetre
μm	micrometre (micron)
km	kilometre

Velocity

m s^{-1}	metres per second
km h^{-1}	kilometres per hour (to convert to m s^{-1} multiply by 3.6)

* Not strictly an SI multiple.

Acceleration

m s^{-2}	metres per second per second

Mass

g	gram
kg	kilogram
t	tonne (= 10^{3} kg)

Density

kg m^{-3}	kilograms per cubic metre (to convert to g cm^{-3} multiply by 0.001)

Force

N	newton (the force required to give a mass of 1 kg an acceleration of 1 m^{-2}.

Pressure

Pa	pascal (= 1 Nm^{-2}) (to convert to mb (millibars) multiply by 100)

Work/energy

J	joule (the work done when the point of application of a force of 1 N is displaced through a distance of 1 m in the direction of the force; since changes in energy result in work being done, energy and work have the same units)

Power

W	watt (= 1 J s^{-1})

Dynamic viscosity

N s m^{-2}	newton second per metre squared

Temperature

°C	celsius (to convert to K (kelvin) add 273.15)

Appendix B

Dating techniques

A wide range of techniques is used in the dating of events in Earth history. Although some of these techniques are more directly relevant to landform analysis than others, it is important for geomorphologists to have a broad understanding of dating methods.

Relative dating

Relative dating methods indicate the order in which events have occurred, but not the time elapsed between them. The basis for the relative dating of rocks and sediments are the principles of stratigraphic succession. The **principle of original horizontality** states that sediments deposited from fluids under the influence of gravity are laid down in horizontal, or nearly horizontal, layers. The **principle of stratigraphic superposition** states that in a vertical profile through sedimentary units which have not been subsequently overturned as a result of tectonic or other mechanisms, their age increases from the top to the base. The recognition that fossil faunas in sedimentary strata succeed one another through time in a recognizable order enabled geologists in the early nineteenth century to extend relative dating in the horizontal dimension by tracing sedimentary units characterized by specific fossil assemblages across country. By the 1830s this technique was being extended to the correlation of sedimentary units originally defined on the basis of the fossil content of British strata to places as far afield as Russia and the USA.

To a limited extent the relative dating of geomorphic events can be achieved with reference to morphological evidence, supplemented, where possible, by evidence provided from associated weathering materials. Such relative dating was the primary basis for the reconstruction of landscape histories, known as denudation chronology, prior to the advent of modern geochronometric dating techniques (see Section 18.3).

Geochronometric dating

Radiometric dating

The establishment of specific ages for events in Earth history had to await the development of **radiometric dating** methods involving the measurement of radioactive isotopes (although various attempts to estimate the age of the Earth had been made by various indirect methods before radiometric techniques were developed earlier this century). Most atomic nuclei are stable, but several elements have inherently unstable, or radioactive, nuclei which undergo spontaneous transformation into other types of atomic nuclei. There are various ways in which this can occur, but the processes responsible for all such transformations are called **radioactive decay**. The decay of a radioactive isotope can involve the acquisition or emission of certain kinds of atomic particles, the emission of electromagnetic radiation, or both. The atom experiencing radioactive decay is termed the **parent isotope**, and the product is called the **daughter isotope**.

An example of radioactive decay is the transformation of ^{40}K (potassium-40) into ^{40}Ar (argon-40). Most of this transformation of ^{40}K occurs through the emission of single electrons (called β particles) to form ^{40}Ca (calcium-40), but 12 per cent of the ^{40}K atoms are transformed into ^{40}Ar by electron capture. Such processes of radioactive decay can be used as a basis for dating because the transformation of certain types of atomic nuclei into other types occurs at a predictable rate which is unaffected by any chemical changes, or any normal changes in temperature and pressure. The rate of decay of a given mass of a radioactive isotope declines exponentially; the time taken for the number of atoms of the parent isotope to be reduced by one half is known as the **half-life** of that particular isotope. In each successive half-life period the remaining number of atoms of the parent isotope is further reduced by half. The half-life of ^{40}K is 1.3 Ga, so after 2.6 Ga only one-quarter of the number of atoms of the parent isotope remain. Radiometric dating techniques are sometimes said to provide absolute ages, but this implies a greater degree of accuracy than is possible with present techniques. **Geochronometric dating** is a more appropriate term for such methods.

A number of factors have to be taken into consideration when radiometric dating is applied to geological materials. All radiometric dates have a margin of error. This can range from ±5 per cent for some methods to ±20 per cent for others. Another factor is the stage at which the radioactive decay system becomes closed; that is, the point where all the daughter isotopes being produced are trapped in the material and cannot escape. For example, at high temperatures argon diffuses out of minerals, so the amount of ^{40}Ar measured in a sample will only reflect that produced since the sample cooled below the critical closure temperature (this varies for different minerals). A radiometric age, therefore, indicates the length of time that the products of radioactive decay have been retained within the material being dated. K–Ar dating is often applied to basaltic lavas so the K–Ar age gives the time since the lava cooled below the critical temperature. If the lava is subsequently reheated then some of the argon may escape and the radiometric clock will be reset. Since rocks cool as erosion strips away the overburden and they are brought up to the surface, K–Ar dating provides a means of estimatimg long-term rates of denudation and rock uplift (see Sections 15.2.1 and 15.4.1.3).

Radiocarbon (^{14}C) dating, which is widely applied to organic remains, is a particularly useful dating method because of the very short half-life of ^{14}C (5370 a). During life, organisms contain the same proportion of ^{14}C as the atmosphere through the processes of photosynthesis, respiration and feeding. Once they die organisms acquire no more ^{14}C and the amount present at death declines through the radioactive decay of ^{14}C into ^{14}N. Although invaluable for dating organic remains up to 50 ka old (and with new, more sensitive techniques up to 100 ka) samples have to be carefully selected to overcome the problem of contamination.

The more important isotope transformations used in radiometric dating and the types of material to which they can be applied are listed in Table B1.

Fission track dating

Fission track dating is based on the accumulation of radiation damage from the spontaneous nuclear fission of ^{238}U in uranium-bearing minerals, such as apatite, zircon and sphene. These fission events lead to the emission of the two fragments of the ^{238}U nucleus into the surrounding crystal lattice which is thereby damaged. The damaged areas can be etched out by acid and the resulting fission tracks

Table B1 Some important isotopes used in radiometric dating

ISOTOPE	HALF-LIFE	DATING RANGE	DATABLE MATERIALS/MINERALS
Carbon-14	5730 a	<50 ka	Wood, charcol, peat, bone, animal tissue, shells, speleothems, ground water, ocean water, glacier ice
Thorium-230	75 ka	<200 ka	Organic carbonate
Uranium-234	250 ka	50–100 ka	Coral
Potassium-40	1.3 Ga	>100 ka	Muscovite, biotite, hornblende, whole volcanic rock
Uranium-238	4.5 Ga	>10 Ma	Zircon, uraninite
Rubidium-87	47 Ga	>10 Ma	Muscovite, biotite, potassium-feldspar, whole igneous or metamorphic rock

counted under a microscope. The number of tracks per unit area is determined by the amount of parent isotope and the time elapsed since tracks were first preserved. Above a certain temperature fission tracks are annealed, that is, they are destroyed, so the fission track dating clock only begins once a sample cools below a critical temperature. These critical temperatures vary for different minerals, but the mineral apatite has a low annealing temperature of around 100 °C which makes it especially valuable for estimating long-term rates of denudation and rock uplift (see Sections 15.2.1, and 15.4.1.3). Depending on the uranium content of the rock, fission track dating can be used to date rocks over a very broad time span. In certain types of volcanic rocks, dating down to a few thousand years is possible, but it can also be widely applied to rocks hundreds of millions of years old.

Palaeomagnetic dating

Palaeomagnetic dating is based on the fact that the properties of the Earth's magnetic field vary over time. Changes occur in its declination (the angle between magnetic and geographical north), its dip (inclination to the horizontal) and its intensity. The most dramatic changes are represented by complete reversals in the polarity of the magnetic field (see Section 2.3.2), but other more subtle changes also occur. These variations in the magnetic field can be used to date rocks or sediments where magnetically susceptible minerals or particles containing iron have become orientated with the magnetic field as they are formed or deposited. These rocks and minerals 'freeze in' the prevailing magnetic field, and since they can be dated by independent means (such as the radiometric dating of basalt flows) it is possible to construct a palaeomagnetic time scale which can then be used as a basis for future dating on palaeomagnetic evidence alone.

Varve dating

Varves are annually deposited layers of sediment. Each varve consists of a layer of coarse sediment which becomes finer upwards. The most common type used in dating are glacial varves. These are formed where glacial meltwater brings sediment into a body of still water. Each summer meltwater rivers carry a mixed load of sand and clay, and the sand is deposited first. Such annual layers can often be counted back over several thousand years and they clearly indicate changes in sediment load over time. Other types of varves may form as result of the production of organic matter over an annual cycle.

Dendrochronology

The dating technique of **dendrochronology** is based on the annual growth rings of trees. Although the seasonal growth of new wood in the spring and summer is affected by prevailing weather conditions, and in extreme cases of drought a growth ring may be eliminated, dendrochronology is a useful dating technique for historic time and an invaluable means of cross-checking ^{14}C dates within this time range.

Amino acid racemization dating

The recently developed technique of **amino-acid racemization dating** is based on certain time-dependent chemical changes (termed racemization) experienced by proteins preserved in the remains of organisms. The main problem with the method is that the rate of chemical change is also temperature dependent, so environments with a uniform temperature, such as deep caves, are required. Used alone it can yield only a relative age, but if calibrated by ^{14}C it can provide useful geochronometric ages. Dating as far back as 200 ka is possible.

Appendix C Geological time scale

Era	Sub-era/period/sub-period/Epoch			Age (A) (Ma BP)	Age (B) (Ma BP)
CENOZOIC	Quaternary		Holocene		
				0.01	0.01
			Pleistocene		
				1.64	1.6
	Tertiary	Neogene	Pliocene		
				5.2	5.3
			Miocene — Late		
				14.2	15.8
			Miocene — Early		
				23.3	23.7
		Paleogene	Oligocene		
				35.4	36.6
			Eocene		
				56.5	57.8
			Paleocene		
				65.0	66.4
MESOZOIC	Cretaceous		Late		
				97.0	97.5
			Early		
				145.6	144
	Jurassic		Late		
				157.1	163
			Middle		
				178.0	187
			Early		
				208.0	208
	Triassic				
				245.0	245
PALAEOZOIC	Permian				
				290.0	286
	Carboniferous				
				362.5	360
	Devonian				
				408.5	408
	Silurian				
				439.0	438
	Ordovician				
				510.0	505
	Cambrian				
				570.0	570
PRECAMBRIAN					

Source: Ages according to W. B. Harland *et al.* (1990) *A Geologic Time Scale 1989*, Cambridge University Press, Cambridge (A); and Geological Society of America (1983) *Decade of North American Geology Geologic Time Scale*. Geological Society of America, Boulder (B).

Appendix D

Particle size

Table D1 indicates the main size classes used in the description of sediments. More detailed information can be obtained from any textbook on sedimentology. The phi (φ) scale is widely used in the description of sediments. It is an inverse logarithmic scale where a particle size of 1 mm is given a φ value of zero. The convenience of the φ scale arises from its small numerical range in comparison with the metric equivalent. Grain sizes have traditionally been determined using square-meshed sieves or, for fine particles, pipette and hydrometer techniques. Automated sediment size analyzers employing laser technology are, however, now becoming available.

Table D1 Major class intervals used in description of sediment sizes

mm	φ	SIZE CLASSES	
		Boulder	
256	-8		
64	-6	Cobble	Gravel
32	-5		
16	-4	Pebble	
4	-2		
2.83	-1.5	Granule	
2.00	-1.0		
1.41	-0.5	Very coarse sand	
1.00	0.0		
0.71	-0.5	Coarse sand	
0.50	1.0		
0.35	1.5	Medium sand	Sand
0.25	2.0		
0.177	2.5	Fine sand	
0.125	3.0		
0.088	3.5	Very fine sand	
0.0625	4.0		
0.031	5.0	Coarse silt	
0.0156	6.0		Silt
0.0078	7.0	Fine silt	
0.0039	8.0		
		Clay	Clay

Source: Based on F. J. Pettijohn *et al.* (1972) *Sand and Sandstone*. Springer-Verlag, New York, Table 3–2, p. 71.

Appendix E

Behaviour of materials

Stress is defined as force per unit area and is measured in pascals (Pa) (1 Pa = 1 Nm^{-2}). **Strain** is a measure of the deformation experienced by a material as the result of an applied stress and is expressed in terms of the relative change in shape or volume. Three fundamental types of stress can be distinguished depending on the relative direction of the force being applied (Fig. E1). **Tensile stress** is an extensional force which tends to stretch or pull material apart. Such a stress is capable of changing the shape of the material. It may also increase its volume if the stretching leads to an opening of voids and hence a reduction of bulk density. **Compressive stress** is a force which tends to compress material and thereby change its shape. There can also be a reduction in volume if pores and fractures are closed up, or if mineral grains are reorganized to pack more closely. Material at depth is subject to compressive stress proportional to the weight of the overburden of rock. Finally, **shear stress** is a force which acts so as to tend to deform a mass of material by one part sliding over another along one or more failure planes. Although shear stress causes a change in the shape of the material, it generally has no significant effect on volume.

Materials respond in different ways to an applied stress depending on their inherent physical properties and the magnitude and duration of the stress (Fig. E2). A material exhibits **elastic behaviour** if it entirely recovers its original form when the deforming stress is removed. Over a large range of applied stresses, strain is proportional to the applied stress, a relationship known as **Hooke's law**. Beyond a certain point, known as the **elastic limit**, the deformation is no longer recovered when the stress is removed and the material exhibits **plastic behaviour**. In brittle rocks plastic deformation is very limited and failure occurs just beyond the elastic limit.

In a perfectly plastic material no deformation occurs until a threshold level of stress (known as the **yield stress**) is attained; thereafter strain occurs at a constant rate as long

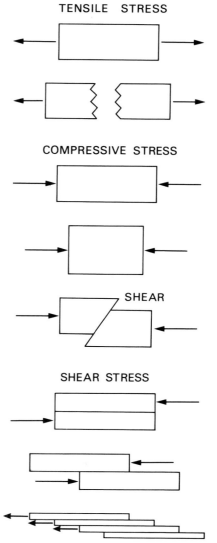

Fig. E1 *Schematic illustration of tensile, compressive and shear stress.*

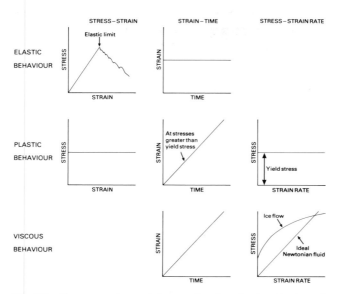

Fig. E2 *Response of materials experiencing elastic, plastic and viscous behaviour. (Modified from I. Statham (1977)* Earth Surface Sediment Transport. *Clarendon Press, Oxford, Fig. 2.13, p. 53.)*

as the stress is constant. In purely plastic behaviour a material completely retains the form assumed during deformation when the applied stress is relaxed. **Viscous behaviour** occurs when the rate of deformation, or **strain rate**, is proportional to the applied stress. This occurs in an ideal **Newtonian fluid** where deformation occurs in response to extremely small stresses. We normally associate viscous behaviour with liquids, but a form of non-Newtonian viscous behaviour can also apply to solids if the stress is applied over a sufficient period of time; note, for instance, that glaciers composed of solid ice can flow.

In reality materials making up lithosphere do not exhibit a single type of behaviour, but rather respond to stresses in a complex manner. Nevertheless, the fundamental distinctions between elastic, plastic and viscous behaviour are important in analyzing the response of lithospheric materials to stress.

Appendix F

Geological structures

Faults

Stress in the crust can be accommodated either by the flow and deformation of rocks over long periods of time or by fracture, the latter being common in the brittle rocks of the upper crust. One manifestation of fracture is the development of **joints**; these are structures of small dimensions lacking any significant movement. The other involves the creation of **faults**, representing fractures along which major movements have occurred. These range from a few metres in total, affecting only a shallow depth of rock or sediment, to large movements of tens or hundreds of kilometres along major faults which extend to the base of the lithosphere. The surface trace of the fault is known as the **fault line** and its horizontal orientation is described as its **strike** (Fig. F1(A)).

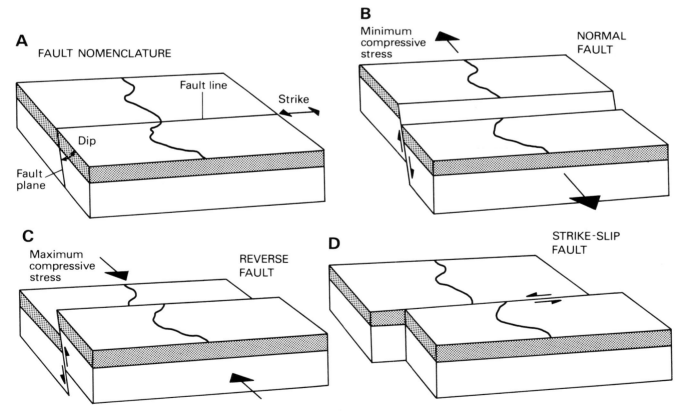

Fig. F1 *Schematic representation of the major types of fault: (A) illustration of dip, strike, fault line and fault plane; (B) normal fault; (C) reverse fault; (D) strike-slip fault.*

Normal faults develop under a pattern of predominantly tensional stress. Movement along the fault involves extension of the crust as the **dip** of the **fault plane** (Fig. F1(A)) is less than 90° (Fig. F1(B)). A down-faulted block between a pair of more or less parallel normal faults is known as a **graben**, and an up-faulted block is termed a **horst**. **Reverse faults** are normally associated with zones of compression, and some crustal shortening occurs as one block rides over the other along the fault (Fig. F1(C)). Where the angle of dip is low the term **thrust fault** (Fig. F2) is often applied, and this kind of faulting is frequently associated with intense crustal folding in orogenic belts. Where the mean compressive stress is vertical, **strike-slip faults** are formed (equivalent terms are **wrench fault** and **transcurrent fault**) (Fig. F1(D)). The fault plane is essentially vertical, and motion along the fault is horizontal. **Sinistral** strike-slip faults exhibit left-lateral movement across the fault (as illustrated in Figure F1(D)), while motion across **dextral** strike-slip faults is right-lateral.

While these represent the basic types of fault, other varieties exist. Where both horizontal and vertical movements are significant the term **oblique-slip fault** is applied; these can be either normal or reverse. Seismic studies of some block-faulted terrains and sedimentary basins have indicated that the dips of their 'normal' faults often decrease with depth until they become more or less horizontal. These are described as **listric faults** and their identification has provided important evidence bearing on the problem of crustal extension during continental rifting and break-up. **Normal listric faults** are associated with crustal thinning, but a change from a tensional to a compressive regional stress pattern can lead to a reversal of movement and the development of **reverse listric faults**.

Folding

While fracture is a common response to stress at the surface, sustained stress under high confining pressures can cause bedrock deformation and folding. Most folds originate at some depth, the simplest forms produced being the **monocline, anticline** and **syncline** (Fig. F2). In a **recumbent fold** the strata are overturned and both limbs of the fold are nearly horizontal. The horizontal compression that creates recumbent folds may eventually lead to the shearing of the upper part of the fold along a thrust fault. The strata moved forward over the thrust fault is called a **nappe**.

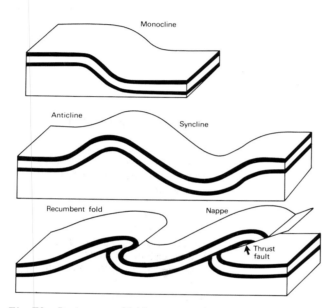

Fig. F2 *Basic types of fold structures.*

Index

Note that a page number in bold type indicates where the main definition of the indexed term is to be found.